Complexity and Control

Towards a Rigorous Behavioral Theory
of Complex Dynamical Systems

SERIES ON STABILITY, VIBRATION AND CONTROL OF SYSTEMS

About the Series

Rapid developments in system dynamics and control, areas related to many other topics in applied mathematics, call for comprehensive presentations of current topics. This series contains graduate level textbooks, monographs, and a collection of thematically organized research or pedagogical articles addressing key topics in applied dynamics.

The material is ideal for a general scientific and engineering readership, and is also mathematically precise enough to be a useful reference for research specialists in mechanics and control, nonlinear dynamics, and in applied mathematics and physics.

Reporting on academic/industrial research from institutions around the world, the SVCS series reflects technological advances in mechanics and control. Particular emphasis is laid on emerging areas such as modeling of complex systems, bioengineering, mechatronics, structronics, fluidics, optoelectronic sensors, micromachining techniques, and intelligent system design.

Selected Volumes in Series A

Vol. 14 Stability of Stationary Sets in Control Systems with Discontinuous Nonlinearities
Authors: V. A. Yakubovich, G. A. Leonov and A. Kh. Gelig

Vol. 15 Process Modelling and Simulation with Finite Element Methods
Author: W. B. J. Zimmerman

Vol. 16 Design of Nonlinear Control Systems with the Highest Derivative in Feedback
Author: V. D. Yurkevich

Vol. 17 The Quantum World of Nuclear Physics
Author: Yu. A. Berezhnoy

Vol. 18 Multiphysics Modeling with Finite Element Methods
Author: W. B. J. Zimmerman

Vol. 19 New Trends in Control Theory
Authors: V. G. Ivancevic and T. T. Ivancevic

Vol. 20 Complexity and Control: Towards a Rigorous Behavioral Theory of Complex Dynamical Systems
Authors: Vladimir G. Ivancevic and Darryn J. Reid

Selected Volume in Series B

Vol. 14 Impact and Friction of Solids, Structures and Intelligent Machines
Editor: A. Guran

Vol. 15 Advances in Mechanics of Solids: In Memory of Prof. E. M. Haseganu
Editors: A. Guran, A. L. Smirnov, D. J. Steigmann and R. Vaillancourt

SERIES ON STABILITY, VIBRATION AND CONTROL OF SYSTEMS

Series A Volume 20

Founder & Editor: Ardéshir Guran

Co-Editor: M. Cloud & W. B. Zimmerman

Complexity and Control

Towards a Rigorous Behavioral Theory of Complex Dynamical Systems

Vladimir G Ivancevic
Darryn J Reid

Joint & Operations Analysis Division
Defence Science & Technology Organisation, Australia

World Scientific

NEW JERSEY • LONDON • SINGAPORE • BEIJING • SHANGHAI • HONG KONG • TAIPEI • CHENNAI

Published by

World Scientific Publishing Co. Pte. Ltd.
5 Toh Tuck Link, Singapore 596224
USA office: 27 Warren Street, Suite 401-402, Hackensack, NJ 07601
UK office: 57 Shelton Street, Covent Garden, London WC2H 9HE

British Library Cataloguing-in-Publication Data
A catalogue record for this book is available from the British Library.

Series on Stability, Vibration and Control of Systems, Series A — Vol. 20
COMPLEXITY AND CONTROL
Towards a Rigorous Behavioral Theory of Complex Dynamical Systems

ISBN 978-981-4635-86-8

In-house Editor: Rhaimie Wahap

Typeset by Stallion Press
Email: enquiries@stallionpress.com

Printed in Singapore

Preface

Complexity and Control is a graduate-level monographic textbook, intended to be a novel and rigorous contribution to modern Complexity Theory.

The book contains 11 Chapters. Most Chapters have their own Appendices (of graduate-level textbook nature). Chapter 1. is an Introduction, setting up the stage for rigorous and controllable complexity theory. Chapter 2. introduces a general geometrical machinery for complexity and control, in the form of dynamics on Kähler manifolds; it includes Lagrangian dynamics on Riemannian manifolds, Hamiltonian dynamics on symplectic manifolds, and quantum dynamics in complex Hilbert spaces. Chapter 3. introduces global categorical framework for complexity and control, from mathematical, computer-scientific and (bio)physical perspectives. Chapter 4. introduces the basic crowd dynamics and control in the complex plane and from it develops a unique framework for simulation, optimization, control, learning and logic of generic complex systems. Chapter 5. is devoted to hierarchical self-similarity in group and crowd behaviors. Chapter 6. presents hybrid topological adaptation in evolving energy landscapes. Chapter 7. explores complexity and control in solitary pulse conduction of various nature. Chapter 8. develops quantum-computational architecture for perceptual control theory. Chapter 9. explores complexity and control in entropic and stochastic self-organization. Chapter 10. presents the latest research in human biodynamics and crash simulation. Chapter 11. gives the additional code samples (in several programming languages) used in complexity research.

The book *Complexity and Control* is designed as a one-semester course for engineers, applied and pure mathematicians, theoretical and experimental physicists, computer and economic scientists, theoretical chemists and biologists, as well as all mathematically educated scientists and students, both in industry and academia, interested in predicting and controlling complex dynamical systems of arbitrary nature.

Adelaide,
May 2014

Dr. Vladimir Ivancevic and Dr. Darryn Reid
Command and Control Branch
Joint and Operations Analysis Division
Defence Science & Technology Organisation
Australia
Vladimir.Ivancevic@dsto.defence.gov.au
Darryn.Reid@dsto.defence.gov.au

Acknowledgments

The authors are very grateful to Joint & Operations Analysis Division, Defence Science & Technology Organisation, Australia, and in particular to Dr. Jason Scholz, Research Leader Command and Control, for their valuable support.

We also express our gratitude to *World Scientific Publishing Company*, and especially to Professor Ardeshir Guran, Editor of the Series on Stability, Vibration and Control of Systems.

Our deepest thanks goes to our wives and families for their love and endurance.

Contents

Glossary of Frequently Used Symbols

General

– 'iff' means 'if and only if';
– 'r.h.s' means 'right hand side'; 'l.h.s' means 'l.h.s.';
– nD means n-dimensional [e.g., nD manifold $\equiv$ n-manifold];
– ODE means ordinary differential equation, PDE means partial differential equation, SDE means stochastic differential equation;
– imaginary unit: $\mathrm{i} = \sqrt{-1}$; bar means the complex-conjugate [e.g., $\bar{z} = x - \mathrm{i}y$ is the complex-conjugate of a complex number $z = x + \mathrm{i}y$];
– overdot means time derivative [e.g., $\dot{x}(t) = dx(t)/dt$];
– indicial notation with respect to (t, x, y, z) means the partial derivative [e.g., $u_x = \partial u/\partial x$];
– indicial notation with respect to $(i, j, k, l, ..., \alpha, beta, \mu, \nu, ...)$ means the covariant tensorial subscripts [e.g., T_{ij} or $S_{\mu\nu}$];
– *Einstein's summation convention over repeated indices* (not necessarily one up and one down) *is assumed in the whole text,* unless explicitly stated otherwise.

Sets

$\mathbb{N}$ – natural numbers;
$\mathbb{Z}$ – integers;
$\mathbb{R}$ – real numbers;
$\mathbb{C}$ – complex numbers;
$\mathbb{H}$ – quaternions;
$\mathbb{K}$ – number field of real numbers, complex numbers, or quaternions.

Maps

$f : A \to B$ – a function, (or map) between sets $A \equiv \mathrm{Dom}\, f$ and $B \equiv \mathrm{Cod}\, f$;

$$\begin{aligned} \mathrm{Ker}\, f &= f^{-1}(e_B) - \text{a kernel of } f; \\ \mathrm{Im}\, f &= f(A) - \text{an image of } f; \\ \mathrm{Coker}\, f &= \mathrm{Cod}\, f / \mathrm{Im}\, f - \text{a cokernel of } f; \\ \mathrm{Coim}\, f &= \mathrm{Dom}\, f / \mathrm{Ker}\, f - \text{a coimage of } f; \end{aligned}$$

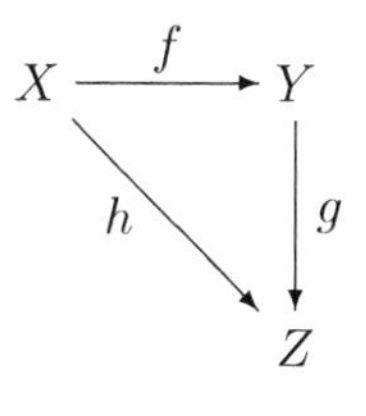

– a commutative diagram, requiring $h = g \circ f$.

Derivatives

$C^k(A, B)$ – set of k–times differentiable functions between sets A to B;
$C^k(A, B)$ – set of *smooth* functions between sets A to B;
$C^0(A, B)$ – set of *continuous* functions between sets A to B;
$f'(x) = \frac{df(x)}{dx}$ – derivative of f with respect to x;
$\dot{x}$ – total time derivative of x;
$\partial_t \equiv \frac{\partial}{\partial t}$ – partial time derivative;
$\partial_{x^i} \equiv \partial_i \equiv \frac{\partial}{\partial x^i}$ – partial coordinate derivative;
$\dot{f} = \partial_t f + \partial_{x^i} f \, \dot{x}^i$ – total time derivative of the scalar field $f = f(t, x^i)$;
$u_t \equiv \partial_t u, \ u_x \equiv \partial_x u, \ u_{xx} \equiv \partial_{x^2} u$ – only in PDEs;
$L_{x^i} \equiv \partial_{x^i} L, \ L_{\dot{x}^i} \equiv \partial_{\dot{x}^i} L$ – coordinate and velocity partials of Lagrangian;
$\mathcal{L}_X T$ – Lie derivative of a tensor–field T in the direction of a vector-field X;
$[X, Y]$ – Lie bracket (commutator) of two vector-fields X and Y;
$[F, G]$, or $\{F, G\}$ – Poisson (or, Lie–Poisson) bracket of two $C^1(\mathbb{R})$ functions F and G;
d – exterior derivative;
$dH = i_{X_H}\omega$ – Hamiltonian 1-form with symplectic 2-form ω, Hamiltonian vector-field X_H and inner product (contraction) $i_{X_H}\omega \equiv \langle X_H, \omega \rangle$;
$\mathcal{L}_{X_H}\omega_H \equiv d i_{X_H}\omega_H + \mathrm{i}_{X_H} d\omega_H = 0$ – Lie derivative;
$dz^k = dx^k + \mathrm{i}dy^k, \quad d\bar{z}^k = dx^k - idy^k$ – complex differentials;
$\partial_{\bar{z}} f$, [for $f \in C^1(\mathbb{C})$] – Wirtinger derivative, equivalent to the Cauchy-Riemann equations: $\partial_x \varphi = \partial_y \phi, \quad \partial_y \varphi = -\partial_x \phi$, [for $\phi, \varphi \in C^1(\mathbb{R})$]
$\partial \equiv \partial_j$ and $\bar{\partial} \equiv \partial_{\bar{j}}$ – Dolbeault operators;
∂_n – boundary operator;
d^n – coboundary operator;
$\nabla = \nabla(g)$ – affine Levi–Civita connection on a smooth manifold M with Riemannian metric tensor $g = g_{ij}$;
Γ^i_{jk} – Christoffel symbols of the affine connection ∇;
$\nabla_X T$ – covariant derivative of a tensor–field T with respect to the vector-field X, defined by means of Γ^i_{jk};
$T_{;x^i} \equiv T_{|x^i}$ – covariant derivative of the tensor–field T with respect to the coordinate basis $\{x^i\}$;
$\dot{\bar{T}} \equiv \frac{DT}{dt} \equiv \frac{\nabla T}{dt}$ – absolute (intrinsic, or Bianchi) derivative of the tensor–field T upon the parameter t; e.g., acceleration vector is the *absolute time derivative* of the velocity vector, $a^i = \dot{\bar{v}}^i \equiv \frac{Dv^i}{dt}$; note that in general, $a^i \neq \dot{v}^i$ – this is crucial for covariant definition of the Newtonian force.

Smooth Manifolds and Fibre/Vector Bundles

Unless otherwise specified, all *manifolds* $M, N, ...$ are assumed C^∞–smooth, real, finite–dimensional, Hausdorff, paracompact, connected and without boundary,[1] while all maps are assumed smooth (C^∞). We use the symbols $\otimes$, $\vee$, $\wedge$ and $\oplus$ for the tensor, symmetrized and exterior products, as well as the Whitney sum[2], respectively, while $\rfloor$ denotes the interior product (contraction) of (multi)vectors and p–forms, and $\hookrightarrow$ denotes a manifold imbedding (i.e., both a submanifold and a topological subspace of the codomain manifold). The symbols ∂_B^A denote partial derivatives with respect to coordinates possessing multi–indices ${}_A^B$ (e.g., $\partial_\alpha = \partial/\partial x^\alpha$);

TM – tangent bundle of the manifold M;
T^*M – cotangent bundle of the manifold M;
$\mathcal{K} = (M, \omega) = (M, g)$ – Kähler manifold: $\mathcal{K} = TM + \mathrm{i}T^*M$;
$\omega = \mathrm{i}g_{i\bar{j}}\, dz^i \wedge dz^{\bar{j}}$ – Kähler form;
$g = \mathrm{i}g_{i\bar{j}}\, dz^i \otimes dz^{\bar{j}}$ – Kähler metric;
$\pi_M : TM \to M$ – natural tangent-bundle projection;
$\pi : E \to M$ – projection from the fibre/vector bundle bundle E to the base manifold M;
(E, π, M) – fibre/vector bundle with the total space E, base space M and projection π;

We use the following kinds of *manifold maps*: immersion, imbedding, submersion, and projection. A map $f : M \to M'$ is called the *immersion* if the tangent map Tf at every point $x \in M$ is an injection (i.e., '1–1' map). When f is both an immersion and an injection, its image is said to be a submanifold of M'. A submanifold which also is a topological subspace is called imbedded submanifold. A map $f : M \to M'$ is called *submersion* if the tangent map Tf at every point $x \in M$ is a surjection (i.e., 'onto' map). If f is both a submersion and a surjection, it is called *projection* or *fibre bundle.*

[1] The only 1D manifolds obeying these conditions are the real line $\mathbb{R}$ and the circle S^1.

[2] *Whitney sum* $\oplus$ is an analog of the direct (Cartesian) product for vector bundles. Given two vector bundles Y and Y' over the same base X, their Cartesian product is a vector bundle over $X \times X$. The *diagonal map* induces a vector bundle over X called the Whitney sum of these vector bundles and denoted by $Y \oplus Y'$.

Lie Groups and (Co)Homology Groups

G – usually a general Lie group;
$GL(n)$ – general linear group with real coefficients in dimension n;
$SO(n)$ – group of rotations in dimension n;
T^n – toral (Abelian) group in dimension n;
$Sp(n)$ – symplectic group in dimension n;
$T(n)$ – group of translations in dimension n;
$SE(n)$ – Euclidean group in dimension n;
$H_n(M) = \operatorname{Ker}\partial_n / \operatorname{Im}\partial_{n-1}$ – nth homology group of the manifold M;
$H^n(M) = \operatorname{Ker} d^n / \operatorname{Im} d^{n+1}$ – nth cohomology group of the manifold M.
In particular,

$$H_d^2(M, \mathbb{R}) = \frac{\{d\text{-closed real (1,1)-forms}\}}{\{d\text{-exact real (1,1)-forms}\}}$$

is the 2nd-order de Rham cohomology group of the complex manifold M, which is equivalent to the Dolbeault cohomology group of M:

$$H_{\bar{\partial}}^{1,1}(M, \mathbb{R}) = \frac{\{\bar{\partial}\text{-closed real (1,1)-forms}\}}{\{\bar{\partial}\text{-exact real (1,1)-forms}\}}.$$

Other Spaces and Operators

$\mathbb{C}^n$ – nD complex space;
$\mathcal{H}(\mathbb{C})$ – complex Hilbert space;
$C^k(M)$ – space of k–differentiable functions on the manifold M;
$\Omega^k(M)$ – space of k–forms on the manifold M;
$\mathfrak{g}$ – Lie algebra of a Lie group G, i.e., the tangent space of G at its identity;
$\triangleright$ – semidirect (noncommutative) group product; e.g., $SE(3) = SO(3) \triangleright \mathbb{R}^3$;
$Ad(g)$ – adjoint endomorphism; recall that *adjoint representation* of a Lie group G is the linearized version of the action of G on itself by conjugation, i.e., for each $g \in G$, the inner automorphism $x \mapsto gxg^{-1}$ gives a linear transformation $Ad(g) : \mathfrak{g} \to \mathfrak{g}$, from the Lie algebra $\mathfrak{g}$ of G to itself.

$\sum\!\!\!\!\!\!\int$ – Feynman's path-integral symbol, denoting integration over continuous spectrum of smooth paths and summation over discrete spectrum of Markov chains; e.g., $\sum\!\!\!\!\!\!\int \mathcal{D}[x]\, \mathrm{e}^{\mathrm{i}S[x]}$ denotes the path integral (i.e., sum–over–histories) over all possible paths $x^i = x^i(t)$ defined by the Hamilton's action: $S[x] = \frac{1}{2}\int_{t_0}^{t_1} g_{ij}\, \dot{x}^i \dot{x}^j\, dt$.

More generally, Euclidean and Lorentzian versions of the path integral over all possible fields $\Phi^i = \Phi^i(x)$, are respectively given, using some classical field action (e.g., Maxwell) $A[\Phi]$, by:

$$Z_{\mathrm{Euc}}(\Phi) = \sum\!\!\!\!\!\!\int \mathcal{D}[\Phi]\, \mathrm{e}^{-A[\Phi]}, \qquad Z_{\mathrm{Lor}}(\Phi) = \sum\!\!\!\!\!\!\int \mathcal{D}[\Phi]\, \mathrm{e}^{\mathrm{i}A[\Phi]}.$$

Categories

$\mathcal{S}$ – all sets as objects and all functions between them as morphisms;
$\mathcal{PS}$ – all pointed sets as objects and all functions between them preserving base point as morphisms;
$\mathcal{V}$ – all vector spaces as objects and all linear maps between them as morphisms;
$\mathcal{B}$ – Banach spaces over $\mathbb{R}$ as objects and bounded linear maps between them as morphisms;
$\mathcal{G}$ – all groups as objects, all homomorphisms between them as morphisms;
$\mathcal{A}$ – Abelian groups as objects, homomorphisms between them as morphisms;
$\mathcal{AL}$ – all algebras (over a given number field $\mathbb{K}$) as objects, all their homomorphisms between them as morphisms;
$\mathcal{T}$ – all topological spaces as objects, all continuous functions between them as morphisms;
$\mathcal{PT}$ – pointed topological spaces as objects, continuous functions between them preserving base point as morphisms;
$\mathcal{TG}$ – all topological groups as objects, their continuous homomorphisms as morphisms;
$\mathcal{M}$ – all smooth manifolds as objects, all smooth maps between them as morphisms;
$\mathcal{M}_n$ – nD manifolds as objects, their local diffeomorphisms as morphisms;
$\mathcal{LG}$ – all Lie groups as objects, all smooth homomorphisms between them as morphisms;
$\mathcal{LAL}$ – all Lie algebras (over a given field $\mathbb{K}$) as objects, all smooth homomorphisms between them as morphisms;
$\mathcal{TB}$ – all tangent bundles as objects, all smooth tangent maps between them as morphisms;
$\mathcal{T}^*\mathcal{B}$ – all cotangent bundles as objects, all smooth cotangent maps between them as morphisms;
$\mathcal{VB}$ – all smooth vector bundles as objects, all smooth homomorphisms between them as morphisms;
$\mathcal{FB}$ – all smooth fibre bundles as objects, all smooth homomorphisms between them as morphisms;
Symplec – all symplectic manifolds (i.e., physical phase–spaces), all symplectic maps (i.e., canonical transformations) between them as morphisms;
Hilbert – all Hilbert spaces and all unitary operators as morphisms.

1

Introduction

We introduce our motive for writing this book on *complexity and control* with a popular "*complexity myth*," which seems to be quite wide spread among chaos and complexity theory fashionistas:

> Low-dimensional systems usually exhibit complex behaviours (which we know from May's studies of the Logistic map), while high-dimensional systems usually exhibit simple behaviours (which we know from synchronisation studies of the Kuramoto model).

We admit that this *naive* view on complex (e.g., human) systems versus simple (e.g., physical) systems might seem compelling to various technocratic managers and politicians; indeed, the idea makes for appealing sound-bites. However, it is enough to see both in the equations and computer simulations of pendula of various degree [1] – (i) a single pendulum, (ii) a double pendulum, and (iii) a triple pendulum – that this popular myth is *plain nonsense*. The only thing that we can learn from it is what every tyrant already knows: by using force as a strong means of control, it is possible to effectively synchronise even hundreds of millions of people, at least for a while.

---

[1] If the reader needs a bit stronger argument to dismiss the validity of this popular complexity myth, they might recall that the famous *Liouville theorem* describes the basic characteristic of any n-dimensional Hamiltonian system described by the phase-space coordinates $q^i \in M$ and the corresponding momenta $p_i \in M$: the time-evolution of this system preserves the *phase-space volume* (or, *Liouville measure*) defined by the following 'wedge' products on the phase-space manifold M:

$$\mathrm{vol} = dq^1 \wedge \cdots \wedge dq^n \wedge dp_1 \wedge \cdots \wedge dp_n.$$

This preservation of volume vol *causes structural instability* of the system, i.e., the *phase-space spreading effect*, by which small phase regions $R_t(\in M)$ will tend to get distorted from the initial one $R_o(\in M)$ (during the system evolution). The problem is much more serious in higher dimensions than in lower dimensions, since there are so many 'directions' in which the region $R_t(\in M)$ can locally spread (see, e.g. [Pen89] for the popular explanation).

More generally, the *complexity theory*, in its current stage of development, is still fascinated by *emergence* of wonderful behavioural qualities of high-dimensional self-organised systems. For example, the 'science of fractal images' is surely very nice, bridging the gap between science and visual arts, and creating illustrations that serve to inspire and fascinate. Yet we are left still wanting for something we contend is fundamentally important, namely a necessary *capability for prediction and control*[2]. This requirement was introduced into the world of physical sciences, at the end of the seventeenth century, by Sir Isaac Newton, and is the reason for the enduring usefulness of his theory, its empirical refutation notwithstanding. So, we can regard even the world of visually-appealing cellular automata, which are supposed to be able to efficiently model any kind of a complex system, as actually belonging to pre-Newtonian scientific era, because they are not capable of predicting and controlling individual agents' behaviours.

We are arguing neither for nor against cellular automata and derived ostensibly soft engineering technologies[3], which we acknowledge have produced many useful contributions to modern science and technology. In this book, we are proposing an alternative approach to behavioural complexity theory, intended to be especially rigorous, and designed for prediction and control of both crowds – or, indeed, large sets of agents of any kind – and of the individual agents comprising such groupings. We hope that this approach will also find its way to fields such as economics, where, in our view, the need for more sophisticated models such as this is especially pressing. In simple words, our goal is develop a rigorous framework for modelling very complex systems, potentially with thousands of individual agents, while not ignoring the psycho-physical individuality of each included agent. This novel, 'hard maths' approach to behavioural complexity theory can be called *high-dimensional Kähler manifold complexity.* It is a multi-fold mathematical approach, combining manifold

[2] In most approaches, a successful *predictability* is a necessary condition for a successful *controllability*. In other words, a comprehensive model, reflecting an aspiration of a 'natural law', should be capable of making successful definitive predictions; which is the same as saying that it should be refutable. Having such a model, it is just a matter of adding a controller function to make the system both controllable and observable. However, things can be more complicated than this in practice: among the most successful controllers are model-free fuzzy-logic controllers, which use linguistic variables to control the system at hand - without even knowing the system's model (e.g., to balance a double inverted pendulum using a fuzzy-logic controller without defining the Lagrangian equations of motion, it is necessary only to specify and tune-up a set of fuzzy If-Then rules, such as "If the two angles are medium left and small right, respectively, while the two angular speeds are high and medium, respectively, then push the cart with medium force to the left"). In this book we will focus on the modern and efficient *modeling theory*, rather than on the *model-free technology*, including fuzzy logic, as well as neural, Bayesian, Markovian and small-world networks.

[3] including genetic algorithms, evolutionary computation, artificial life, etc.

geometry, loop topology, complex analysis and gauge theory, and has already been successfully tested in modern physics. From the signals-and-systems perspective, it is a nonlinear generalization of Kalman's linear state-space control and filter theory; from dynamical perspective it is both Lagrangian and Hamiltonian; from geometrical perspective it is both Riemannian and symplectic; from physical perspective it is both classical and quantum; from ODE/PDE perspective it is both deterministic and stochastic. It is a uniquely promising internal framework for a wide variety of nonlinear systems of ODEs and PDEs, both real- and complex-valued. We submit that it is ready to become available as a new foundation of a *rigorous behavioral complexity theory.*

The other, external side, a 'wrapper', of this new approach to complexity is the *commutative flow framework* based on functional composition. Briefly, composing maps (or, functions, processes, signals, systems, transformations) is like following directed paths from one object to another (e.g., from set to set). In general, a diagram is commutative *iff* any two paths along arrows that start at the same point and finish at the same point yield the same 'homomorphism' via compositions along successive arrows. Commutativity of the whole diagram follows from commutativity of its triangular components (depicting a *commutative flow,* see Figure 1).

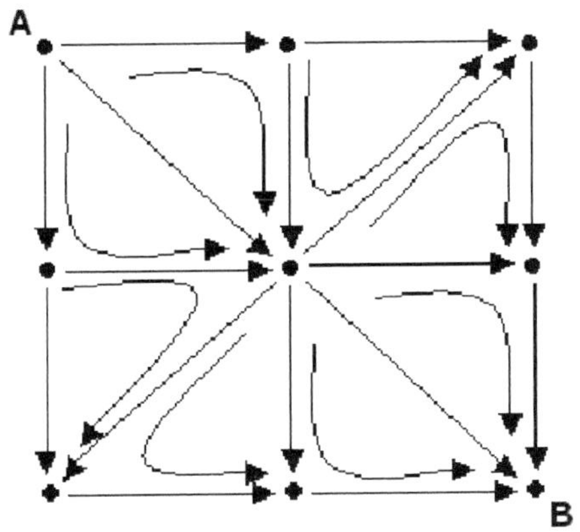

Fig. 1.1. *A commutative flow (including all curved arrows) on a triangulated digraph, defining the resulting path* ***AB****. Note that the commutativity of the whole diagram, and therefore the definition of the path* ***AB****, follows from the commutativity of its triangular components.*

The result, we hope, is a rigorous theory of complex behaviour and control that can address some of the shortcomings of contemporary ideas in complexity. Our approach is founded on the principle of making definitive predictions, both by which we may act in the world and by which we may subject our models to the harshest responsible logical and empirical scrutiny we can muster. Indeed, it is our enduring hope that our approach will eventually be supplanted by one or more improved theories of greater explanatory and predictive power; such is the nature of the growth of knowledge.

2

Local Geometrical Machinery for Complexity and Control

In this Chapter, we present local geometrical machinery for studying complexity and control, consisting of dynamics on Kähler manifolds, which combine three geometrical structures – Riemannian, symplectic and complex (Hermitian) – in a mutually compatible way. In other words, every Kähler manifold is simultaneously Riemannian, symplectic and complex (Hermitian). It is well known that Riemannian manifolds represent the stage on which Lagrangian dynamics is set, symplectic manifolds represent the stage for Hamiltonian dynamics, and complex (Hermitian) varieties comprise the stage for quantum dynamics. Therefore, Kähler manifolds represent the richest dynamical stage available where Lagrangian, Hamiltonian, and quantum dynamics all dance together.

2.1 Introduction: Why Kähler Manifolds?

A set of attractor-following dynamics of n agents (for arbitrary n) in the complex plane $\mathbb{C}$ generate an n-dimensional (nD) *crowd configuration space* (CCS), which is locally homeomorphic (i.e., topologically equivalent) to the nD complex space $\mathbb{C}^n$. In some special cases, including point attractors as well as line-segment attractors, the CCS is also globally homeomorphic to $\mathbb{C}^n$, so we can say that the CCS is $\mathbb{C}^n$. Incidentally, this also happens to be the finite-dimensional Hilbert space $\mathcal{H}(\mathbb{C}) \simeq \mathbb{C}^n$, that we need for the finite-dimensional control, optimization and learning, as well as for quantum computation (e.g., the Hilbert space for a single qubit is $\mathcal{H}_2 \simeq \mathbb{C}^2$, for two qubits it is $\mathcal{H}_2 \otimes \mathcal{H}_2 \simeq \mathbb{C}^4$, for three qubits it is $\mathcal{H}_2 \otimes \mathcal{H}_2 \otimes \mathcal{H}_2 \simeq \mathbb{C}^6$, etc.). So, the linear space $\mathbb{C}^n$ is both algebraic and geometrical basis for complex-valued crowd dynamics, control and learning.

However, as soon as we have some closed-line attractors in the complex plane with either one loop (a circle), or two loops ('figure 8'), or three or more loops, then the CCS (of agents following these attractors) becomes a more complicated, nonlinear geometrical object (a complex hyper-surface) which is characterized by the notion of a high-dimensional genus: it is said to be genus 0 in the linear case of an attractor without loops, genus 1 for a one-loop attractor, genus 2 for a two-loop attractor, genus 3 for a three-loop attractor, etc. Similarly, if we have a (quite common) case of several closed-loop attractors working/attracting simultaneously, then the total genus of the CCS will be equal to the sum of all attractor loops. So, if there exists one or several attractor-loops in the crowd dynamics, the CCS is no more globally equivalent to $\mathbb{C}^n$, but becomes a more general, nonlinear, complex n-manifold. In the simplest case of a single agent, the CCS represents the so-called *Riemann surface* (i.e., a donut with several holes), which is locally (at every point) homeomorphic to $\mathbb{C}^1$, that is the complex plane itself. In case of n agents, we clearly need a concept of a complex manifold of higher dimension that generalizes the Riemann surface to n dimensions. The rigorous mathematical framework for such nonlinear complex n-spaces (which includes all high-dimensional generalizations of Riemann surfaces while obeying necessary topological conditions, and possessing very rich geometry, topology, analysis and algebra), is the concept of a high-dimensional *Kähler manifold.*

In this chapter, we will develop a geometrical and topological theory of crowd's Kähler manifolds. Informally, a Kähler manifold $\mathcal{K}$ is such a configuration space of a crowd-dynamics system evolving in the complex plane, which is locally homeomorphic to $\mathbb{C}^n$, but also comprises three simultaneous and compatible geometrical and dynamical structures: Lagrangian dynamics (with its Riemannian geometry), Hamiltonian dynamics (with its symplectic-Poisson geometry) and quantum-like complex dynamics (with its geometric complex structure). If we have a CCS that is endowed with all three structures, then it is the Kähler n-manifold. This concept enables topological as well as algebraic, geometrical and analytic considerations.

2.1.1 Dynamical Prologue: Complex Hamiltonian Dynamics

This subsection has a motivational character only, to show that there is a complex structure hidden in the classical Hamiltonian formalism, which can serve as a bridge between the complex plane $\mathbb{C}$ and our proposed Kähler formalism. Both Lagrangian and Hamiltonian formalisms will be developed properly later (see Appendix), as the Kähler formalism contains them both (as its special cases).

Hamiltonian energy function $H(\mathbf{x},\mathbf{p}) \in \mathbb{R}$, describing planar Newtonian dynamics of a crowd, group, or team of n agents in the complex-plane $\mathbb{C} \cong \mathbb{R}\times\mathbb{R}$, is given in vector form by:

$$H(\mathbf{x},\mathbf{p}) = \frac{1}{2m}\left\|\mathbf{p}\right\|^2 + V(\mathbf{x}),$$

where $\mathbf{x} = \mathbf{x}(t) \in \mathbb{C}^n \cong \mathbb{R}^n\times\mathbb{R}^n$ is an n-vector of individual agents' generalized coordinates $x_i(t)$, $\mathbf{p}(t) = m\,\dot{\mathbf{x}}(t) \in \mathbb{C}^n$ is the corresponding n-vector of the generalized momenta $p_i(t)$,[1] while $V(\mathbf{x}) \in \mathbb{R}$ is crowd's (scalar) potential function. Then crowd's $\mathbb{C}^n$-dynamics, classically governed by the Newton's second law:

$$m\,\ddot{\mathbf{x}} = -\partial_{\mathbf{x}}V, \qquad \left(\partial_{\mathbf{x}}V \equiv \frac{\partial V}{\partial \mathbf{x}}\right)$$

obtains the form of classical Hamilton's equations of motion:[2]

$$\dot{\mathbf{x}} = \partial_{\mathbf{p}}H, \qquad \dot{\mathbf{p}} = -\partial_{\mathbf{x}}H. \tag{2.1}$$

If we introduce the symplectic matrix:[3]

$$J = \begin{bmatrix} 0 & I_n \\ -I_n & 0 \end{bmatrix}, \qquad \left(\text{with } \det J = 1 \text{ and } J^{-1} = J^T = -J\right)$$

(where I_n is the $n \times n$ identity matrix), then Hamilton's equations (2.72) can be rewritten in symplectic form as:

$$\dot{\boldsymbol{\xi}} = J\,\frac{\partial H}{\partial \boldsymbol{\xi}}, \qquad \text{with} \quad \boldsymbol{\xi} = (\mathbf{x}, \mathbf{p}) \in \mathbb{C}^n. \tag{2.2}$$

Alternatively, in complex notation (by setting: $\mathbf{z} = \mathbf{x} + \mathrm{i}\,\mathbf{p}$, $\mathrm{i} = \sqrt{-1}$), the symplectic Hamilton's equations (2.2) obtain the complex Hamiltonian form in $\mathbb{C}^n$ (see [AM78, CD82, AMR88, MR99]):

$$\dot{\mathbf{z}} = -2\mathrm{i}\frac{\partial H}{\partial \bar{\mathbf{z}}}, \qquad \text{with} \quad \frac{\partial}{\partial \bar{\mathbf{z}}} := \frac{1}{2}\left(\frac{\partial}{\partial \mathbf{x}} - \mathrm{i}\frac{\partial}{\partial \mathbf{p}}\right), \tag{2.3}$$

where bar denotes complex-conjugation.

The complex Hamiltonian crowd equations (2.3) represent dynamical starting point for the construction of the *crowd Kähler manifold* $\mathcal{K}$, which is a symplectic manifold with an integrable complex structure compatible with the symplectic structure J (alternatively, crowd Kähler manifold $\mathcal{K}$ can be defined as a Hermitian manifold with a closed Hermitian form called the Kähler metric).

---

[1] Note that in Newtonian and Lagrangian mechanics, velocities are derived quantities, defined as time derivatives of the coordinates, $\mathbf{v}(t) = \dot{\mathbf{x}}(t)$. In Hamiltonian mechanics, however, the momenta $\mathbf{p}(t)$ corresponding to the (generalized) coordinates are usually postulated as another set of coordinates on the equal footing as the coordinates themselves, but can also be derived from the Lagrangian function, or simply as $\mathbf{p}(t) = m\,\dot{\mathbf{x}}(t)$.

[2] We remark that this same dynamical picture the basis of quantum mechanics as well, where all dynamical observables (coordinates, momenta and Hamiltonian energy function itself) are replaced by the Hermitian operators in the associated Hilbert space.

[3] Note that, in some resources, the symplectic matrix J is alternatively denoted by Ω, although in geometrical references, Ω usually represents the space of all de Rham's exterior differential forms, the most important being the closed symplectic 2-form ω.

2.1.2 Unified Behavioral Picture: Kähler Geometrodynamics

By experiences from modern physics, both classical and quantum, we have learned that any kind of a multi-dimensional dynamical system can be formulated as a generalized Lagrangian or Hamiltonian system (that is not restrictive to conservative dynamics only, but also includes dissipative and forced dynamics; not only finite-dimensional, but infinite-dimensional as well). Now, instead of arguing which formulation would be more efficient (from theoretical and computational perspectives) for our general behavioral-complexity dynamics, would it be a generalized Lagrangian or a generalized Hamiltonian formalism, we propose here a unique formulation that comprises them both.

More precisely, our *Kähler behavioral geometrodynamics* (*BGD*) is given by the following $4n$D complex structure:

$$\begin{bmatrix} K\ddot{a}hler \\ \mathcal{K}\text{-}BGD \end{bmatrix} \simeq \begin{bmatrix} Riemannian\ Geo \\ TM \quad \text{with} \\ Lagrangian\ Dyn \end{bmatrix} + \mathrm{i} \begin{bmatrix} Symplectic\ Geo \\ T^*M \quad \text{with} \\ Hamiltonian\ Dyn \end{bmatrix},$$

such that the real part, $\mathrm{Re}(\mathcal{K})$, consists of a real Riemannian $2n$-manifold TM with Lagrangian nD crowd-complexity dynamics defined on it, while its imaginary part, $\mathrm{Im}(\mathcal{K})$, consists of a real symplectic $2n$-manifold T^*M with Hamiltonian $2n$D crowd-complexity dynamics defined on it. In addition, both component-manifolds, TM and T^*M, are themselves derived geometrical structures (a tangent bundle TM and a cotangent bundle T^*M), derived from the basic crowd configuration space M, which is an n-manifold with attractor dynamics of n agents defined in the complex plane $\mathbb{C}$.

In other words, standard approach of classical physics and quantum mechanics is given by the commutative diagram:[4]

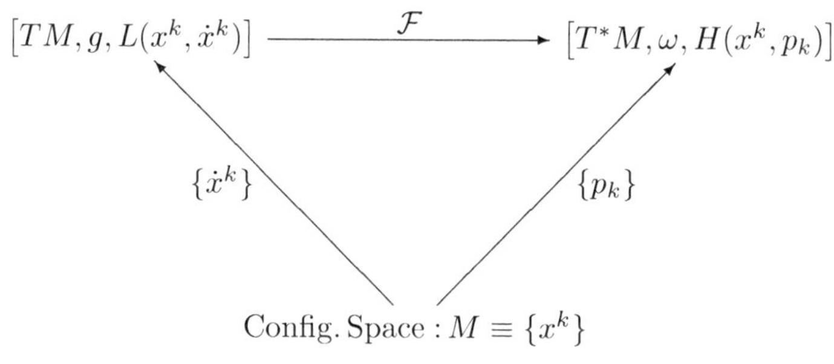

where the contravariant left-hand map: $M \to TM$, given by $\{x^k\} \mapsto \{x^k, \dot{x}^k\}$, defines the Riemannian geometry with the metric tensor $g = g_{jk}$ and Lagrangian function $L(\mathbf{x}, \mathbf{v}) = L(x^k, \dot{x}^k)$, while the covariant right-hand map $M \to T^*M$, given by $\{x^k\} \mapsto \{x^k, p_k\}$ defines the symplectic geometry with the symplectic closed 2-form $\omega = dx^k \wedge dp_k$ and Hamiltonian function

[4] See Chapter 3 for technical details on commutative diagrams.

$H(\mathbf{x}, \mathbf{p}) = H(x^k, p_k)$; the horizontal map $\mathcal{F}$ is called the fiber derivative, or Legendre transformation.[5]

Here are some more details on this geometrical-mechanics picture, which we will need for the future development. The left-hand branch of the above mechanical commutative diagram is usually called *Lagrangian dynamics*. To each nD smooth[6] configuration manifold M there is associated its $2n$D velocity phase-space manifold, denoted by TM and called the tangent bundle of M. The original configuration manifold M is called the base of TM. There is an *onto* map $\pi : TM \to M$, called the projection. Above each point $x \in M$ there is a tangent space $T_xM = \pi^{-1}(x)$ to M at x, which is called a *fibre*. The fibre $T_xM \subset TM$ is the subset of TM, such that the total tangent bundle, $TM = \bigsqcup_{x \in M} T_xM$, is a disjoint union of tangent spaces T_xM to M for all points $x \in M$. From dynamical perspective, the most important quantity in the tangent bundle concept is the smooth map $v : M \to TM$, which is an inverse to the projection π, i.e, $\pi \circ v = \mathrm{Id}_M$, $\pi(v(x)) = x$. It is the velocity vector-field $v^k = \dot{x}^k$. Its graph $(x, v(x))$ represents the cross-section of the tangent bundle TM; thus, velocity vector-fields are cross-sections of the tangent bundle TM. The Lagrangian (kinetic minus potential energy) function $L(x^k, \dot{x}^k)$ is a natural energy function on the tangent bundle TM, which is itself a smooth $2n$-manifold, having its own tangent bundle, TTM. Cross-sections of the second tangent bundle TTM are the acceleration vector-fields. The Lagrangian dynamics are governed by the set of n second-order equations of motion on TM:

$$\frac{d}{dt}\partial_{\dot{x}^k} L = \partial_{x^k} L, \qquad \text{with} \qquad L(x^k, \dot{x}^k) = \frac{1}{2} g_{jk}\, \dot{x}^j \dot{x}^k - V(x^k),$$

where $V(x^k)$ is the potential energy function and summation convention (over repeated indices) is in place.

The right-hand branch of the mechanical commutative diagram is usually called *Hamiltonian dynamics*. It takes place in the cotangent bundle T^*M, defined as follows. A dual notion to the tangent space T_xM to a smooth manifold M at a point x is its cotangent space T_x^*M at the same point x. Similarly to the tangent bundle TM, for any smooth configuration nD manifold M,

---

[5] Note that the Legendre transformation $\mathcal{F}$ is here mainly for historical reasons, which are also respected in virtually all mechanical textbooks: usually one first defines Lagrangian dynamics as a generalization of Newtonian dynamics, and then derives from it the corresponding Hamiltonian dynamics, via Legendre transformation. For our purpose, it is more important that it makes this diagram commutative. Also, it is proven in advanced texts (see [AM78]) that the map $\mathcal{F}$ is invertible, so it represents a two-way connection between the second-order Lagrangian formalism on TM with the first-order Hamiltonian formalism on T^*M. Later in the text, we will properly define both formalisms.

[6] Briefly, a smooth manifold is an nD topological space with a C^∞-smooth structure defined on it (for more technical details, see [II06b, II07]).

there is associated its $2n$D momentum phase-space manifold, denoted by T^*M and called the cotangent bundle. T^*M is the disjoint union of all its cotangent spaces T_x^*M at all points $x \in M$, i.e., $T^*M = \bigsqcup_{x\in M} T_x^*M$. Therefore, the cotangent bundle of an n-manifold M is the vector bundle $T^*M = (TM)^*$, which is the real dual of the tangent bundle TM. Momentum 1-forms (or, covector-fields) p_k are cross-sections of the cotangent bundle T^*M. The Hamiltonian (that is, kinetic plus potential energy) function $H(x^k, p_k)$ is a natural energy function on the cotangent bundle T^*M, which is itself a smooth $2n$-manifold, which has its own tangent bundle, TT^*M. Cross-sections of the mixed-second bundle TT^*M are the force 1-forms $F_k = \dot{p}_k$. The (forced and dissipative) Hamiltonian dynamics are governed by the set of $2n$ first-order equations of motion on T^*M:

$$\dot{x}^k = \partial_{p_k} H + \partial_{p_k} R, \qquad \dot{p}_k = F_k - \partial_{x^k} H + \partial_{x^k} R,$$

where F_k are external forces and $R = R(x^k, p_k)$ is the general dissipative function.

So, this is a picture from mechanics (for more technical details, see [II06b, II07]). On the other hand, a twice as powerful approach of modern gauge theory, which we will follow in the development of our behavioral complexity theory, unites both the Lagrangian manifold TM and the Hamiltonian manifold T^*M, in the form of a single Kähler manifold:

$$\mathcal{K} = TM + \mathrm{i}T^*M\,.$$

2.2 Complex Manifolds and Their Vector Bundles

2.2.1 Behavioral Dynamics on Complex Plane

In this section, we firstly, informally set-up the complex-plane stage for general behavioral dynamics, and secondly, formally define a minimum of complex analysis necessary for development of complex behavioral manifolds.

Behavioral dynamics in the complex plane $\mathbb{C}$

If we have a set of n agents' behaviors:

$$\psi(t) = \{\psi_1(t) + \psi_2(t) + ..., \psi_k(t);\; k \in n\}$$

(for arbitrary n), representing *behavioral dynamics* of various vessels, vehicles, people, robots, etc., naturally defined on the Euclidean plane $\mathbb{R}^2$, which can be conveniently identified with the complex plane $\mathbb{C} \cong \mathbb{R} \times \mathbb{R}$, is governed by the following set of standard rules:

$$\begin{aligned}
\boldsymbol{\psi}(t) &\equiv \mathbf{x}(t) + \mathrm{i}\mathbf{y}(t) = \boldsymbol{\rho}\left[\cos\boldsymbol{\theta}(t) + \mathrm{i}\sin\boldsymbol{\theta}(t)\right] \equiv \boldsymbol{\rho}\operatorname{cis}\boldsymbol{\theta} = \boldsymbol{\rho}\,\mathrm{e}^{\mathrm{i}\boldsymbol{\theta}}, \\
\text{where}\ :\ \boldsymbol{\theta}(t) &= \{\theta_1(t) + \theta_2(t) + \ldots, \theta_k(t);\ k \in n\} \qquad \text{are heading-angles}, \\
\boldsymbol{\rho}(t) &= \{\rho_k(t);\ k \in n\} = |\boldsymbol{\psi}(t)| = \sqrt{\mathbf{x}^2(t) + \mathbf{y}^2(t)} \qquad \text{are radius-vectors}, \\
\boldsymbol{\theta}(t) &\equiv \arg\left[\boldsymbol{\psi}(t)\right] = \arctan\frac{\mathbf{y}(t)}{\mathbf{x}(t)} = -\mathrm{i}\ln\frac{\boldsymbol{\psi}(t)}{\boldsymbol{\rho}}; \qquad \boldsymbol{\rho}^2 = |\boldsymbol{\psi}(t)|^2 = PDF.
\end{aligned}$$

Each individual complex number $\psi(t) \in \boldsymbol{\psi}(t) \in \mathbb{C}$ represents a "2D vector on steroids," for which not only addition/subtraction is defined (by the parallelogram law), but also multiplication, division, any powers and any roots. This means, e.g. that $\psi(t)$ can model an object's (e.g. ship, aircraft, etc.) course in time (while $\boldsymbol{\psi}(t)$ can model a fleet of vessels), and we can perform with it any kind of discrete- and/or continuous-time calculations. Besides, the complex-conjugation, $\overline{\psi(t)} = x(t) - \mathrm{i}y(t)$, means the flip around the real (horizontal, equatorial) axis X, while multiplication by imaginary unit, $\mathrm{i} = \sqrt{-1}$, means the rotation anti-clockwise by $\pi/2 = 90°$. So, e.g. if a ship motion/course is modeled by the complex number $\psi(t)$ then its course-correction for 45 degrees clockwise is modeled by $-0.5\,\mathrm{i}\,\psi(t)$, etc. Not to mention that the angles $\boldsymbol{\theta}(t)$, by definition, determine the instantaneous headings of all the agents, while the radius-vectors $\boldsymbol{\rho}(t)$, also by definition, determine their corresponding instantaneous distances from the origin $(0, 0)$ in the complex plane $\mathbb{C}$.

For these same reasons, the complex plane $\mathbb{C}$ has become the stage of modern physics (quantum gauge-, string-, and computation-theories). Conceptual and computational machinery from these theories (e.g. quantum Ising-spin chain used in quantum computation), applied to real-world problems, comprise "physical mathematics." Its *core* concept is R. Feynman's generalization of a complex vector (see Appendix):

$$\mathcal{I}_{\text{paths}} = \int \mathcal{D}[x]\,\mathrm{e}^{\mathrm{i}S[x]},$$

called the *path integral* (or transition amplitude, or sum-over all possible discrete + continuous paths $x^i = x^i(t)$ in the complex plane $\mathbb{C}$, both deterministic and random), determined by classical Hamilton's action $S[x]$ of the Lagrangian $L[x]$ (which is crowd's kinetic minus potential energy, with the metric tensor g_{ij}):

$$S[x] = \int L[x^i(t)]\,dt = \frac{1}{2}\int_{t_0}^{t_1} \left[g_{ij}\,\dot{x}^i\dot{x}^j - V(x)\right]\,dt.$$

More generally, the following symbol:

$$\mathcal{I}_{\text{fields}} = \int \mathcal{D}[\Phi]\,\mathrm{e}^{\mathrm{i}S[\Phi]}$$

denotes the sum-over-fields $\Phi^i = \Phi^i(x)$ in $\mathbb{C}$, which is both discrete + continuous and deterministic + random, defined by some field action $S[\Phi]$. In both cases, $\mathcal{D}[x]$ and $\mathcal{D}[\Phi]$ define probabilistic Lebesgue measures on the sets of all possible paths and fields in the complex plane $\mathbb{C}$.

In this way, we have the full dynamics of agents' paths $[x]$ + environmental fields $[\Phi]$:

$$\text{Behavioral Dynamics in } \mathbb{C} = \mathcal{I}_{\text{paths}} + \mathcal{I}_{\text{fields}}$$
$$= \int \mathcal{D}^{\text{agents}'}_{\text{paths}}[x]\, \mathrm{e}^{\mathrm{i}S[x]} + \int \mathcal{D}^{\text{environ}}_{\text{fields}}[\Phi]\, \mathrm{e}^{\mathrm{i}S[\Phi]},$$

in the sense of K. Lewin's 'topological' psycho-social force-fields and behaviors (see [Gol99] and Lewin's own references therein).

Holomorphic functions and their differential operators in $\mathbb{C}$

Recall (from undergraduate complex analysis) that *holomorphic functions* are complex-valued functions (of one or more complex variables), which are complex-differentiable in some open subset $U \subset \mathbb{C}$. The required *complex-differentiability* is established in the following way: if a complex function $f(z) \in U$ given by:

$$f(z) \equiv f(x + \mathrm{i}\, y) = \varphi(x, y) + \mathrm{i}\, \phi(x, y)$$

is holomorphic, then its real-valued components, φ and ϕ, both have continuous first partial derivatives, $\partial_x\varphi \equiv \partial\varphi/\partial x$ and $\partial_y\varphi \equiv \partial\varphi/\partial y$, which satisfy the *Cauchy-Riemann equations*:

$$\partial_x\varphi = \partial_y\phi, \qquad \partial_y\varphi = -\partial_x\phi, \qquad \Longrightarrow \qquad \partial_{\bar{z}} f = 0, \tag{2.4}$$

where $\partial_{\bar{z}} f$ is the *Wirtinger derivative*. The set of all holomorphic functions defined on $U \subset \mathbb{C}$ is denoted by $\mathcal{O}(U)$.

Next, complex-valued differential operators are defined in the following way. On the Euclidean plane $\mathbb{R}^2$ endowed with (x, y)-coordinates, the standard *de Rham differentials* ([Rha84]; see Appendix) are dx and dy. On the complex plane $\mathbb{C}$, identified with $\mathbb{R}^2$ by $z \mapsto (x, y)$ when $z = x + \mathrm{i}\, y$, the corresponding de Rham differentials are:

$$dz = dx + \mathrm{i}\, dy \qquad \text{and} \qquad d\bar{z} = dx - \mathrm{i}\, dy.$$

Let $U \subset \mathbb{C}$ be an open subset in $\mathbb{C}$, and $f \in C^\infty(U)$ be a smooth function defined on U, then the following facts hold:

1. the *differential operator* df is defined on U by:

$$df = \partial_z f\, dz + \partial_{\bar{z}} f\, d\bar{z},$$

where the *Wirtinger derivatives* $\partial_z f$ and $\partial_{\bar{z}} f$ (corresponding to the *Cauchy-Riemann equations* (2.4)) are defined by:

$$\partial_z f \equiv \frac{\partial f}{\partial z} = \frac{1}{2}\left(\frac{\partial f}{\partial x} - \mathrm{i}\frac{\partial f}{\partial y}\right), \qquad \partial_{\bar{z}} f \equiv \frac{\partial f}{\partial \bar{z}} = \frac{1}{2}\left(\frac{\partial f}{\partial x} + \mathrm{i}\frac{\partial f}{\partial y}\right). \tag{2.5}$$

2. $f \in \mathcal{O}(U)$ iff $\partial_{\bar{z}} f = 0$; that is, if $f \in \mathcal{O}(U)$ then $df = \partial_z f \, dz$.
3. if an open subset $U \subset \mathbb{C}$ is bounded, with a smooth boundary ∂U, $f \in \mathcal{O}(U)$, and a is an isolated interior point in U, then $(z-a)^{-1}$ is holomorphic and the *Cauchy integral formula* is valid:

$$f(a) = \frac{1}{2\pi \mathrm{i}} \int_{\partial U} \frac{f \, dz}{z - a}.$$

Residues of meromorphic functions in $\mathbb{C}$

A *meromorphic function* $f(z)$ defined on an open subset $U \subset \mathbb{C}$ is a function that is holomorphic on all U except at a finite set of isolated points [the so-called *poles* of $f(z)$], at each of which the function $f(z)$ must have a *Laurent-series expansion*.

Recall that a residue $\operatorname*{Res}_{z=a} f(z)$ of a meromorphic function $f(z)$ at the point $a \in \mathbb{C}$ is defined by the *Cauchy Residue Theorem*:

$$\operatorname*{Res}_{z=a} f(z) = \frac{1}{2\pi \mathrm{i}} \oint f(z) \, dz,$$

with integration along a closed curve around $z = a$ with the winding number[7] 1. In particular, if $f(z)$ is given by a Laurent expansion at $z = a$,

$$f(z) = \sum_{k=-\infty}^{\infty} a_k \, (z-a)^k \quad \text{then} \quad \operatorname*{Res}_{z=a} f(z) = a_{-1}.$$

However, if $a = \infty$, then the Laurent expansion

$$f(z) = \sum_{k=-\infty}^{\infty} a_k \, \frac{1}{z^k} \quad \text{gives} \quad \operatorname*{Res}_{z=\infty} f(z) = -a_{-1}.$$

Here we give several examples of residues, calculated using computer algebra systems *Reduce* and *Mathematica*:

[7] A winding number is an integer which counts how many times the curve winds around the point a.

$$\operatorname*{Res}_{z=\sqrt{2}}\left[\frac{z}{z^2-2}\right]=\frac{1}{2},\qquad \text{pole order}=1,$$

$$\operatorname*{Res}_{z=0}\left[\frac{\sin(z)}{z^2}\right]=1,\qquad \text{pole order}=1,$$

$$\operatorname*{Res}_{z=\sqrt{2}}\left[\frac{\sin(z)}{z^2-2}\right]=\frac{\sin[\sqrt{2}]}{2\sqrt{2}},\qquad \text{pole order}=1,$$

$$\operatorname*{Res}_{z=2}\left[\frac{1}{(z-1)^m\,(z-2)^2}\right]=-m,\qquad \text{pole order}=2,$$

$$\operatorname*{Res}_{z=\pi/2}\,[\tan(z)]=-1,\qquad \text{pole order}=1,$$

$$\operatorname*{Res}_{z=\pi/2}\left[\frac{\tan(z)}{\sec(z-\pi/2)}+\sec(z)\right]=-2,\qquad \text{pole order}=1.$$

Residue-based techniques have many applications in engineering and science. For example, they are heavily used for calculating the inverse Laplace transforms.[8]

2.2.2 3D Rotations: Spinors

In principle, any kind of rotations in our natural 3D space can be represented by *spinors*, complex-valued 2D vectors of a general form:

$$\{c_i, c_j\} \in \mathbb{C}^2, \quad \text{such that} \quad (c_i, c_j \in \mathbb{C}).$$

They are usually defined with respect to some reference basis in $\mathbb{C}^2$. In particular, the basis spinors, $\{\{1,0\},\{0,1\}\}$, are interpreted as 'spin-up' and 'spin-down', with respect to the z-axis of a Cartesian basis in $\mathbb{R}^3$. Another standard basis is: $\{\{\sqrt{2}^{-1},\sqrt{2}^{-1}\},\{-\sqrt{2}^{-1},\sqrt{2}^{-1}\}\}$, with the spinor components: $\{\sqrt{2}^{-1},-\sqrt{2}^{-1}\}$.

---

[8] Recall that the *forward Laplace transform* is a *linear integral operator* $\mathcal{L}:\mathbb{R}\to\mathbb{C}$, defined by the *improper integral*:

$$F(s)=\mathcal{L}\{f(t)\}=\int_0^\infty \mathrm{e}^{-st} f(t)\,dt,$$

while the corresponding *inverse Laplace transform* $\mathcal{L}^{-1}:\mathbb{C}\to\mathbb{R}$ is defined by the *complex Bromwich integral* (see, e.g. [II12]):

$$f(t)=\mathcal{L}^{-1}\{F(s)\}=\frac{1}{2\pi\mathrm{i}}\int_{\gamma-\upsilon\infty}^{\gamma+\mathrm{i}\infty} \mathrm{e}^{st}F(s)ds,$$

which is directly solved as:

$$f(t)=\sum \operatorname{Res}\left[\mathrm{e}^{st}F(s)\right] \text{ at poles of } F(s).$$

Like all vectors, the spinors have their norms (e.g. $Norm[\{2, -\mathrm{i}\}] = \sqrt{5}$) and also they can be normalized to a spinor with a unit norm:

$Normalize[\{2, -\mathrm{i}\}] = \{2/\sqrt{5}, -\mathrm{i}/\sqrt{5}\}$, such that $Norm[\{2/\sqrt{5}, -\mathrm{i}/\sqrt{5}\}] = 1$.

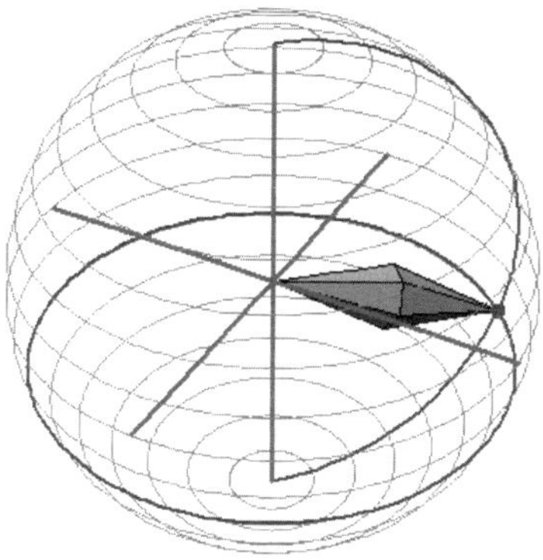

Fig. 2.1. *A sample spinor shown as a vector on the surface of the Bloch-sphere.*

A spinor can be shown as a vector on the surface of the *Bloch-sphere* (see Figure 2.1).

Related to spinors are *Pauli's sigma matrices*, Hermitian 2 by 2 unitary matrices of the form:

$$\sigma_x = \begin{pmatrix} 0 & 1 \\ 1 & 0 \end{pmatrix}, \quad \sigma_y = \begin{pmatrix} 0 & -\mathrm{i} \\ \mathrm{i} & 0 \end{pmatrix}, \quad \sigma_z = \begin{pmatrix} 1 & 0 \\ 0 & -1 \end{pmatrix}.$$

which represent quantum-mechanical *observables* for the spin measured with respect to the x-, y-, and z-axis of a Cartesian basis in $\mathbb{R}^3$. They obey both the commutation and anticommutation relations, given respectively by:

$$[\sigma_a, \sigma_b] = 2\mathrm{i}\varepsilon_{abc}\,\sigma_c\,, \qquad \{\sigma_a, \sigma_b\} = 2\delta_{ab}\,I,$$

where ε_{abc} is the Levi-Civita tensor, δ_{ab} is the Kronecker tensor, while I is the 2 by 2 identity matrix.

Spinors are also related to $SU(2)$, the rotation Lie group. For example,

$$SU(2)\mathrm{rot}[\{1,1,1\}] = \begin{pmatrix} \cos\left(\frac{\sqrt{3}}{2}\right) - \frac{\mathrm{i}\sin\left(\frac{\sqrt{3}}{2}\right)}{\sqrt{3}} & -\frac{(1+\mathrm{i})\sin\left(\frac{\sqrt{3}}{2}\right)}{\sqrt{3}} \\ \frac{(1-\mathrm{i})\sin\left(\frac{\sqrt{3}}{2}\right)}{\sqrt{3}} & \cos\left(\frac{\sqrt{3}}{2}\right) + \frac{\mathrm{i}\sin\left(\frac{\sqrt{3}}{2}\right)}{\sqrt{3}} \end{pmatrix}.$$

Here is a textbook example of the $SU(2)$-rotation:

$$SU(2)\mathrm{rot}[\phi,\{n_1,n_2,n_3\}] = \begin{pmatrix} \cos\left(\frac{\phi}{2}\right) - \mathrm{i}n_3\sin\left(\frac{\phi}{2}\right) & -\mathrm{i}(n_1 - \mathrm{i}n_2)\sin\left(\frac{\phi}{2}\right) \\ (n_2 - \mathrm{i}n_1)\sin\left(\frac{\phi}{2}\right) & \cos\left(\frac{\phi}{2}\right) + \mathrm{i}n_3\sin\left(\frac{\phi}{2}\right) \end{pmatrix}.$$

2.2.3 Attractor Dynamics on Riemann Surfaces

Why Riemann surfaces?

Quick answer: because of some 'strange' topological reasons outlined informally below.

As this might look as a nontrivial thing to explain, let us first consider a slightly easier example of a planar double pendulum. The motion of this 2-component dynamical system (which is chaotic, yet simple enough to visualize) is defined by 2 time-dependent angles $\theta_1(t)$ and $\theta_2(t)$. Topologically speaking (topology is popularly called 'rubber sheet geometry'), each of these two angles defines a circle S^1 in a plane of pendulum's oscillation, that is to say, a configuration space is a circle S^1 for each of the two segments. What about the whole double pendulum: its configuration space is the Cartesian product of the two circles: $CS = S^1 \times S^1 = T^2$, which is an ordinary torus. By induction, a configuration space (or more precisely, a configuration manifold) of an n-tuple pendulum (with n segments) is an n-torus $T^n = S^1 \times S^1 \times ... \times S^1$.

Well, this already looks similar to some very simple crowd dynamics. As stated in the previous subsection, the behavior of each individual agent in the crowd (or group, or team), is defined by a time-dependent complex number, $\rho e^{\mathrm{i}\theta(t)}$. So, in a trivial case, $\rho = \rho\left[x(t), y(t)\right] = 1$, and thus a configuration space for each agent is again a circle S^1 (in either Euclidean or complex plane). This means, for example, that a pair of agents is homeomorphic (i.e., topologically equivalent, see below) to a 'flexible' double pendulum, with a 'flexible torus' T^2_{flex} as a configuration manifold. By induction, a configuration manifold for a group of n trivial agents (with $\rho_k(x_k, y_k) = 1$) - is a 'flexible' n-torus T^n_{flex}. Provided a complex structure is defined on it, T^n_{flex} represents a simple (Abelian group) example of a Kähler n-manifold.

Now, let us have a more serious look at a single agent's behavior. From mechanical perspective, each agent is a rigid body constrained to a Euclidean (or complex) plane, i.e., its proper configuration space is $SE(2)$, a (special) Euclidean group of plane rigid motions. $SE(2)$ is a 3-parameter group, including two translational coordinates $[x(t), y(t)]$ and a heading angle $\theta(t)$. Topologically, $SE(2)$ is a Cartesian product of an (x, y)-plane $\mathbb{R}^2$ and a circle θ, which is a cylinder Cyl.[9] So, a natural (uncontrolled) dynamics of n agents

[9] More formally, a Lie group $SE(2) \equiv SO(2) \times \mathbb{R}$ is a set of all $3 \times 3-$ matrices of the form:

$$\begin{bmatrix} \cos\theta & \sin\theta & x \\ -\sin\theta & \cos\theta & y \\ 0 & 0 & 1 \end{bmatrix},$$

is topologically represented by an nD cylinder: $\mathrm{Cyl}^n = \prod_{k=1}^{n} \mathrm{Cyl}_k$, which is a Cartesian product of n cylinders Cyl_k.[10]

Finally, instead of a natural dynamics, we really want an attractor-driven crowd (team, or group) dynamics, that is, a controlled behavioral complexity. Let us start slowly, with a single agent following an *a priori* given attractor. The above picture of a topological cylinder is still valid for all open-line attractors. However, as soon as we have a closed loop in the attractor, it is again a circle S^1, so the configuration space of an attractor-driven agent is a Cartesian product of a cylinder and a circle, which can be again approximated by a 'flexible torus' T^2_{flex}. Thus, a one-hole attractor followed by a single agent forms T^2_{flex} as its configuration manifold.

What if the attractor has more than one hole in it? In that case, we have a similar but more complicated topological surface which is 'flexible torus with several holes'. This is a Riemann surface, which accidently has a complex structure already defined on it. So, a configuration manifold for a loop-attractor dynamics of a single agent (that is just a simple rigid body moving in the Euclidean or complex plane) is a Riemann surface. Therefore, without an attractor (or with a non-loop attractor), dynamics of a single plane-agent is constrained to a cylinder. With a loop-attractor, it is a Riemann surface, which is another simple example of a Kähler manifold.

At this stage we can only imagine how complex can be dynamics of n agents following a set of closed-loop attractors, each with several S^1-holes in the complex plane $\mathbb{C}$. However, no matter how complicated this crowd dynamics can be, its configuration manifold is still a Kähler n-manifold.

2.2.4 Complex Manifolds and Vector Bundles

The first step in the development of the Kähler manifold concept is its 'parent concept' of a complex manifold. As mentioned in the Introduction, the configuration space of a single agent's attractor dynamics in the complex plane $\mathbb{C}$ is a *Riemann surface* (that is, a donut with several holes), which is locally equivalent to $\mathbb{C}$. Its nD generalization is called a *complex manifold* (with a

---

including both rigid translations (i.e., Cartesian $x, y-$coordinates) and rotation matrix $\begin{bmatrix} \cos\theta & \sin\theta \\ -\sin\theta & \cos\theta \end{bmatrix}$ in Euclidean plane $\mathbb{R}^2$ (see [II06b, II07]).

[10] More formally, a crowd configuration n-manifold M is defined as a Cartesian product of n $SE(2)$-groups for all individual agents [IR10a]:

$$M = \prod_{k=1}^{n} SE(2)^k \equiv \prod_{k=1}^{n} SO(2)^k \times \mathbb{R}^k \simeq \mathrm{Cyl}^n,$$

coordinated by $\mathbf{x}^k = \{x^k, y^k, \theta^k\}$, (for $k = 1, 2, ..., n$).

certain number of holes, possibly countable infinity of them), which is a *topological space*[11] M endowed with a special kind of a C^∞-smooth geometrical *atlas* (see Appendix) in which the transition functions $\phi_{\alpha\beta}$ in the local open charts $U_{\alpha\beta} \subset M$ satisfy the Wirtinger derivatives (2.5).

More precisely, two charts (U_1, ϕ_1) and (U_2, ϕ_2) in a topological manifold M, such that $U_1 \cap U_2 \neq \varnothing$ are called *compatible* if $\phi_1(U_1 \cap U_2)$ and $\phi_2(U_2 \cap U_1)$ are open subsets of $\mathbb{C}^n$. A family $(U_\alpha, \phi_\alpha)_{\alpha\in A}$ of compatible charts on M such that the U_α form a covering of M is called an *atlas*. The maps $\phi_{\alpha\beta} = \phi_\beta \circ \phi_\alpha^{-1} : \phi_\alpha(U_{\alpha\beta}) \to \phi_\beta(U_{\alpha\beta})$ are called the transition functions, for the atlas $(U_\alpha, \phi_\alpha)_{\alpha\in A}$, where $U_{\alpha\beta} = U_\alpha \cap U_\beta$, so that the following triangle is commutative:

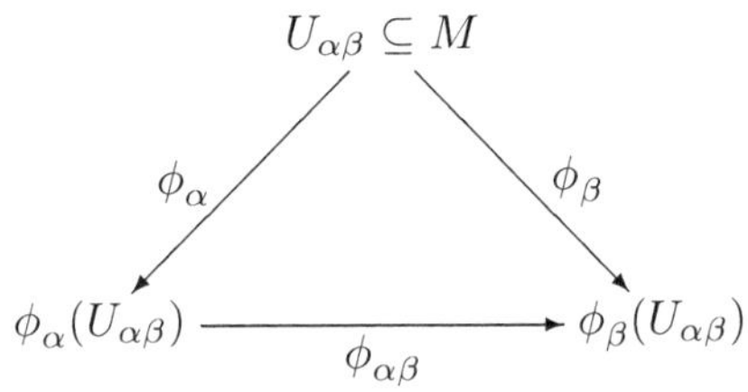

11 Topological spaces are structures that allow one to formalize concepts such as *convergence, connectedness* and *continuity*. They appear in virtually every branch of modern mathematics and are a central unifying notion. Technically, a *topological space* is an ND set X together with a collection T of subsets of X satisfying the following three axioms:
(i) the empty set and X are in T;
(ii) the union of any collection of sets in T is also in T; and
(iii) the intersection of any pair of sets in T is also in T.
The collection T is called the *topology* on X. The sets in T are the *open* sets, and their complements in X are the closed sets. The elements of X are called *points*. By induction, the intersection of any finite collection of open sets is open.

A function between topological spaces is said to be C^0-*continuous* iff the inverse image of every open set is open. This is an attempt to capture the intuition that there are no 'breaks' or 'separations' in the function. A *homeomorphism* is a bijection (that is, an invertible $1-1$ and *onto* function), which is continuous both ways. Two spaces are said to be homeomorphic if there exists a homeomorphism between them, and they are (from the standpoint of topology) essentially identical.

Now, our complex manifold M is a separable ND topological space, which is locally equivalent to $\mathbb{C}^n$, with the following additional properties:

(i) every point $x \in M$ has a neighborhood $U \subset M$ that is diffeomorphic (i.e., $1-1$, *onto* and C^∞-*smooth*) to an open subset $V \subset \mathbb{C}^n$;

(ii) M is a *Hausdorff space*: for every pair of points $x_1, x_2 \in M$, there are disjoint open subsets $U_1, U_2 \subset M$ such that $x_1 \in U_1$ and $x_2 \in U_2$;

(iii) M is a *second-countable space*: there exists a countable basis for the topology of M; and

(iv) a map $\phi : \mathbb{C}^n \supset V \to U \subset M$ is called a (local) coordinate map on M, and its component functions $(x^1, ..., x^n)$ defined by $\phi(m) = (x^1(m), ..., x^n(m))$ are called *local coordinates* on $U \subset M$.

An atlas $(U_\alpha, \phi_\alpha)_{\alpha\in A}$ on a manifold M is holomorphic, iff all transition functions $\phi_{\alpha\beta} : \phi_\alpha(U_{\alpha\beta}) \to \phi_\beta(U_{\alpha\beta})$ are holomorphic. Two holomorphic atlases are equivalent, iff their union is again a holomorphic atlas for M. An equivalence class of holomorphic atlases is called a *complex structure* on M.

This implies that holomorphic (or, analytic) functions $f : X \to \mathbb{C}$ can be defined on M, by using the coordinate representatives $f \circ \phi_i^{-1}$. On any even-dimensional *smooth manifold* M,[12] complex coordinates of the form: $z^j = x^j + \mathrm{i}\, y^j$ $(j = 1, ..., d)$ can always be introduced in local open charts $U_i \subset M$, with holomorphic transition functions between the charts, thus forming a holomorphic atlas for a complex manifold M of complex dimension $d = n/2$. This atlas represents the *complex structure* on M (see [Gre96, II06b, II07] and references therein).

Now, the so-called *biholomorphism* defines an equivalence between two complex manifolds, in much the same way as a *diffeomorphism* (that is, a C^∞-smooth homeomorphism) defines an equivalence between two C^∞-smooth manifolds. In other words, if M and N are two complex manifolds, they are equivalent if there is a map $\phi : M \to N$ which is both a diffeomorphism and a holomorphic map.

More precisely, a map $\varphi : M \to N$ between two complex manifolds M and N, with $M \ni m \mapsto \varphi(m) \in N$, is called a *holomorphic map*, if we have the following charting:

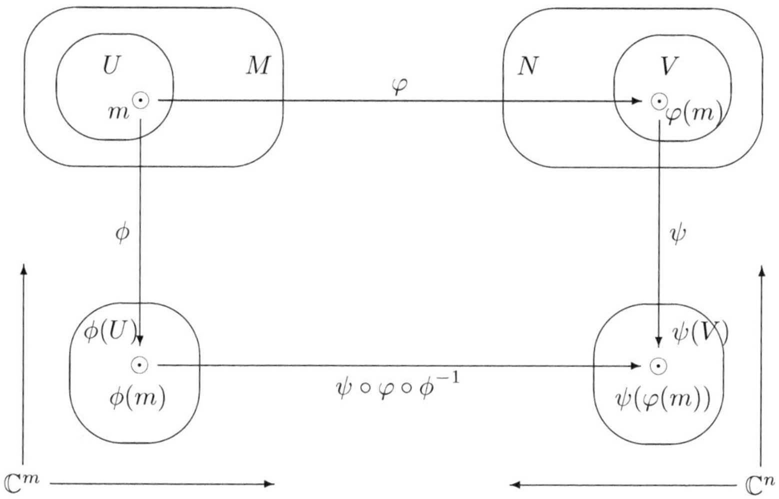

This means that for each $m \in U \subset M$ and each chart (V, ψ) on N, together with $\varphi(m) \in V \subset N$, there is a chart (U, ϕ) on M with $\varphi(U) \subset \mathbb{C}^m$ and also the map $\Phi = \psi \circ \varphi \circ \phi^{-1}$ is holomorphic, i.e., the following diagram commutes:

[12] A smooth manifold is a topological manifold M with a C^∞-smooth atlas $(U_\alpha, \phi_\alpha)_{\alpha\in A} \subset M$ (see Appendix).

$$\begin{array}{ccc} M \supseteq U & \xrightarrow{\ \varphi\ } & V \subseteq N \\ \downarrow{\scriptstyle \phi} & & \downarrow{\scriptstyle \psi} \\ \mathbb{C}^m \supset \varphi(U) & \xrightarrow[\ \Phi\ ]{} & \psi(V) \subset \mathbb{C}^n \end{array}$$

2.2.5 (Co)Tangent Spaces to the Complex Manifold

We now proceed with the development of the complex manifold concept. Firstly, a *tangent space* T_mM to any smooth or complex manifold M at a point $m \in M$ (that is, the closest flat approximation to M at its point m, see Appendix) has a real vector basis consisting of n linearly-independent partial derivatives:

$$T_mM : \{\partial_{x^1}|_m, ..., \partial_{x^n}|_m\}, \tag{2.6}$$

such that any vector $v \in T_mM$ can be expressed as: $v = v^\alpha \partial_{x^\alpha}|_m$.

Its dual, *cotangent space* T_m^*M at a point $m \in M$, has a real vector basis consisting of de Rham's exterior one-forms [Rha84] (see Appendix):

$$T_m^*M : \{dx^1|_m, ..., dx^n|_m\}, \tag{2.7}$$

where, by definition, $dx^i : T_mM \to \mathbb{R}$ is a linear map with $dx^i_m(\partial_{x^j}|_m) = \delta^i_j$.

In the case of a complex manifold M (of complex dimension $d = n/2$), the so-called *complexified tangent space* $T_mM^{\mathbb{C}}$ is the same real tangent space T_mM, but now extended in such a way that complex coefficients can be used in the vector-space manipulations, which is most easily done by defining $T_mM^{\mathbb{C}}$ as the tensor product of the real tangent space and the complex plane: $T_mM^{\mathbb{C}} = T_mM \otimes \mathbb{C}$. So, we can still use the real vector basis (2.6) with an arbitrary vector $v \in T_mM^{\mathbb{C}}$ expressed as $v = v^\alpha \partial_{x^\alpha}|_m$, but its components v^α can now be complex numbers. So, we can conveniently rearrange the basis vectors in (2.6) as [Gre96, II06b, II07]:

$$\begin{aligned} T_mM^{\mathbb{C}} : \{&(\partial_{x^1} + \mathrm{i}\,\partial_{x^{d+1}})|_m, ..., (\partial_{x^d} + \mathrm{i}\,\partial_{x^{2d}})|_m, \\ &(\partial_{x^1} - \mathrm{i}\,\partial_{x^{d+1}})|_m, ..., (\partial_{x^d} - \mathrm{i}\,\partial_{x^{2d}})|_m\}. \end{aligned}$$

In terms of complex coordinates: $z^j = x^j + \mathrm{i}\,y^j$, we can write this basis as:

$$T_mM^{\mathbb{C}} : \{\partial_{z^1}|_m, ..., \partial_{z^d}|_m, \partial_{\bar{z}^1}|_m, ..., \partial_{\bar{z}^d}|_m\},$$

where the real vector spaces $\partial_{x^j}|_m$ and $\mathrm{i}\,\partial_{x^j}|_m$ are considered linearly-independent, so that $T_mM^{\mathbb{C}}$ has the real dimension $2d = n$.

Also, analogous to the dual real basis (2.7) is the dual complex vector basis: $T_m^*M^{\mathbb{C}} = T_m^*M \otimes \mathbb{C}$, consisting of the complex exterior one-forms:

$$T_m^*M^{\mathbb{C}} : \{dz^1|_m, ..., dz^d|_m, d\bar{z}^1|_m, ..., d\bar{z}^d|_m\}.$$

2.2.6 Complex Vector Bundles

Now, as we proceed in developing more general and more abstract structures, we still want to keep a connection with a real-world dynamics. Firstly, we can recall that in classical Lagrangian mechanics, to each nD configuration manifold M there is associated its $2n$D *velocity phase-space* TM, called the *tangent bundle* of M, while the original configuration manifold M is called the *base* of TM. The core concept in the tangent bundle construction is the *onto* map $\pi : TM \to M$ called the *projection*. Above every point $m \in M$ there is a tangent space $T_mM = \pi^{-1}(m)$ to M at m, which is called a *fibre*. The fibre $T_mM \subset TM$ is the subset of TM, such that the total tangent bundle, $TM = \bigsqcup_{m\in M} T_mM$, is a *disjoint union* of tangent spaces T_mM to M for all points $m \in M$. From mechanical perspective, the most important quantity in the tangent bundle concept is the smooth map $v : M \to TM$, which is an inverse to the projection π, i.e, $\pi \circ v = \mathrm{Id}_M$, $\pi(v(m)) = m$. This map is called the *velocity vector-field*; its *graph* $(m, v(m))$ represents the *cross-section* of the tangent bundle TM. The tangent bundle TM is the simplest example of a real *vector bundle* concept, endowed with *Riemannian metric* g (see e.g. [II06b, II07]).

The second example of a real vector bundle, endowed with the symplectic two-form ω, is the dual concept to the tangent bundle TM of a smooth configuration manifold M, that is the *cotangent bundle* T^*M, called the *momentum phase-space* of a $2n$D Hamiltonian dynamical system. T^*M is defined as the disjoint union of all its cotangent spaces T^*_mM at all points $m \in M$, i.e., $T^*M = \bigsqcup_{m\in M} T^*_mM$. Therefore, the cotangent bundle of an n-manifold M is the vector bundle $T^*M = (TM)^*$, the (real) dual of the tangent bundle TM (see e.g. [II06b, II07]).

More generally, a *complex vector bundle* $\mathcal{K}$ is a vector bundle $\pi : \mathcal{K} \to M$ whose fiber bundle $\pi^{-1}(m)$ is a complex vector space. If a complex vector bundle has the structure of a holomorphic complex manifold, then it is called a holomorphic vector bundle. A *Hermitian metric* h on a holomorphic vector bundle $\mathcal{K}$ assigns a *Hermitian inner product* to every fiber bundle $\pi^{-1}(m)$.[13] By a partition of unity, any complex vector bundle has a Hermitian metric.

In particular, the *complexified tangent bundle* $\mathcal{K} \equiv TM \otimes \mathbb{C}$ is the the disjoint union of the complexified tangent spaces: $TM \otimes \mathbb{C} = \bigsqcup_{m\in M} T_mM^{\mathbb{C}}$. The holomorphic vector bundle $\mathcal{K} \equiv TM \otimes \mathbb{C}$ may have a Hermitian metric h, such that its real part is a Riemannian metric g and its imaginary part is a nondegenerate de Rham's exterior p-form ω which may be closed ($d\omega = 0$), therefore it is a symplectic form. In other words, the real part of $\mathcal{K} \equiv TM \otimes \mathbb{C}$

---

[13] The simplest example of a holomorphic vector bundle is the trivial bundle $\pi : U \times \mathbb{C}^2 \to U$, where U is an open set in $\mathbb{R}^n$. Here, a positive-definite Hermitian matrix H defines a Hermitian metric by the following inner product: $\langle v, w\rangle = v^T H\bar{w}$.

is the tangent bundle TM (with the Riemannian geometry and Lagrangian dynamics acting on it) and its imaginary part is the cotangent bundle T^*M (with the symplectic geometry and Hamiltonian dynamics acting on it), that is:

$$\mathcal{K} \equiv TM \otimes \mathbb{C} = TM + \mathrm{i}\,T^*M. \tag{2.8}$$

If this is the case, the symplectic two-form ω becomes the so-called *Kähler form* and the vector bundle $\mathcal{K}$ becomes the *Kähler manifold*.[14]

Note that the initial Kähler geometric relation (2.8) is beginning to realize our big picture of the *Kähler behavioral geometrodynamics*:

$$\begin{bmatrix} \textit{Kähler} \\ \textit{BGD} \end{bmatrix} \simeq \begin{bmatrix} \textit{Riemannian Geo} \\ TM \quad \text{with} \\ \textit{Lagrangian Dyn} \end{bmatrix} + \mathrm{i} \begin{bmatrix} \textit{Symplectic Geo} \\ T^*M \quad \text{with} \\ \textit{Hamiltonian Dyn} \end{bmatrix}.$$

In the next section, we will give a proper definition of a Kähler n-manifold.

2.3 From Kalman Systems to Riemann Manifolds

In this section we will show how the (left) Riemannian component of the Kähler dynamics naturally rises from our everyday work performed say with Matlab/Simulink. In the following section, we will do a similar modeling exercise with the (right) symplectic component of the general Kähler dynamics.

So, to start with, consider the standard Kalman's decomposable linear state-equation (see Appendix, section 2.8.2) for an arbitrary MIMO-system (e.g., related to some cognitive-control or perceptual-control problem):

$$\begin{aligned} \dot{\mathbf{x}} &= \mathbf{A}\,\mathbf{x} + \mathbf{B}\,\mathbf{u}, \\ \mathbf{y} &= \mathbf{C}\,\mathbf{x} + \mathbf{D}\,\mathbf{u}, \end{aligned} \tag{2.9}$$

with the state n-vector $\mathbf{x} = \mathbf{x}(t) \in \mathbb{X} \subset \mathbb{R}^n$, input $m-$vector $\mathbf{u} = \mathbf{u}(t) \in \mathbb{U} \subset \mathbb{R}^m$, output $k-$vector $\mathbf{y} = \mathbf{y}(t) \in \mathbb{Y} \subset \mathbb{R}^k$, state $n \times n$ matrix $\mathbf{A} = \mathbf{A}(t) : \mathbb{X} \to \mathbb{X}$, input $n \times m$ matrix $\mathbf{B} = \mathbf{B}(t) : \mathbb{U} \to \mathbb{X}$, output $k \times n$ matrix $\mathbf{C} = \mathbf{C}(t) : \mathbb{X} \to \mathbb{Y}$ and input-output $k \times m$ matrix $\mathbf{D} = \mathbf{D}(t) : \mathbb{U} \to \mathbb{Y}$.

Notwithstanding both the inherent 'beauty' and the practical usefulness of the state equation (2.54), which is the basis of $Matlab^{TM}$ Control and Signal toolboxes (and also used by half-a-dozen of other *Matlab* toolboxes), we might realize that, in real life nothing is linear, so the linear decomposable state-equation (2.54) can only be the first approximation to some more realistic nonlinear MIMO-system. Technically speaking, we can generalize (or, 'lift-up') the linear model (2.54) that lives in nD linear Euclidean space $\mathbb{R}^n$, into

[14] The simplest example of a Kähler manifold is already mentioned Riemann surface, that is a complex manifold of dimension 1, in which the imaginary part of any Hermitian metric must be a closed form since all 2−forms are closed on a real 2D manifold (see, e.g. [Gre96]).

the tensor (or, 'covariant') equation that lives on nD nonlinear Riemannian manifold M with the metric tensor g_{ij} and the metric form:

$$ds^2 = g_{ij}\, dx^i dx^j, \qquad (i, j = 1, ..., n),$$

where the summation convention (summing over repeated indices) is in place.

Using this 'nonlinear lift', from the linear state equation (2.54) we obtain the following covariant state equation:

$$\begin{aligned} \dot{\bar{x}}^i &= a^i_j\, x^j + b^i_s\, u^s, \quad (s = 1, ..., m) \\ y^t &= c^t_j\, x^j + d^t_s\, u^s, \quad (t = 1, ..., k) \end{aligned} \tag{2.10}$$

where $\dot{\bar{x}}^i = \dot{\bar{x}}^i(t)$ is the absolute (covariant) time-derivative, which includes Christoffel's symbols Γ^i_{jk} (i.e., the Levi-Civita connection associated with the metric form ds^2) of the Riemannian manifold M:

$$\dot{\bar{x}}^i := \dot{x}^i + \Gamma^i_{jk} x^j x^k.$$

The covariant state equation (2.55), which can be simulated in *Mathematica*^{TM} (or, *Maple*^{TM}), in case of the flat connection ($\Gamma^i_{jk} = 0$), reduces to the Kalman equation (2.54). Also, some higher-nonlinearity connection (e.g., Cartan) can be used instead of the Levi-Civita connection Γ^i_{jk}.

2.4 Basic Kähler Geometry

2.4.1 Essential Kähler Tensors and Cohomology Groups

Kähler differential geometry (see, e.g. [Yau06, Don99, CL08, Mor07, Son12] and references therein) represents a complexified synthesis of Riemannian geometry and symplectic geometry. It includes a variety of Kähler tensor-fields and complex exterior differential forms, which are constructed using the tensor product $\otimes$ and the wedge product $\wedge$, respectively (see Appendix). These tensor-fields represent complex generalizations of the standard tensor-fields (and their corresponding multilinear forms) from these two geometries. We remind the reader that the summation convention (over repeated indices) is always in place.

As a start-off, let M be a compact (i.e., closed and bounded[15]) complex n-manifold (which is our candidate for the Kähler manifold $\mathcal{K}$), with *local holomorphic coordinates* $\{z_1, \cdots, z_n\}$, defined in an open chart $U \subset M$, and the *Hermitian metric tensor-field* $g_{i\bar{j}} = g_{i\bar{j}}(z^i, z^{\bar{j}})$, where $(g_{i\bar{j}})$ is the *Hermitian matrix* function.

[15] Closed set contains all its limit points; its complement is an open set. An example of a bounded set is a set S of real numbers that is bounded (from above) if there is a real number k such that $k \geq s$ $(\forall s \in S)$.

The (main) *Kähler form* ω is defined on M as a closed ($d\omega = 0$) and positive ($\omega > 0$) exterior (1,1)-form,[16] given (in local holomorphic coordinates $z_1, \cdots, z_n$[17] of an open chart $U \subset M$) by:

$$\omega = \mathrm{i} g_{i\bar{j}}\, dz^i \wedge dz^{\bar{j}},$$

such that the corresponding *Kähler metric* g is a positive and symmetric (1,1)-form:

$$g = \mathrm{i} g_{i\bar{j}}\, dz^i \otimes dz^{\bar{j}}.$$

The pair $\mathcal{K} = (M, \omega) = (M, g)$ is called the *Kähler manifold*.

For example, in case of $\mathcal{K} = \mathbb{C}^N$, the Kähler form ω can be written as:

$$\begin{aligned}\omega &= \mathrm{i}\left(dz_1 \wedge \overline{dz_1} + dz_2 \wedge \overline{dz_2} + ... + dz_N \wedge \overline{dz_N}\right) \\ &= -2\left(dx_1 \wedge dy_1 + dx_2 \wedge dy_2 + ... + dx_N \wedge dy_N\right) \\ &= \sum_{k=1}^{N} dx_k \wedge dy_k.\end{aligned}$$

while the corresponding Kähler metric g is given by:

$$\begin{aligned}g &= \mathrm{i}\left(dz_1 \otimes \overline{dz_1} + dz_2 \otimes \overline{dz_2} + ... + dz_N \otimes \overline{dz_N}\right) \\ &= -2\left(dx_1 \otimes dy_1 + dx_2 \otimes dy_2 + ... + dx_N \otimes dy_N\right) \\ &= \sum_{k=1}^{N} dx_k \otimes dy_k = \delta^{ik} dx_k dy_k\,, \qquad (\delta^{ik} = \text{Kronecker delta}).\end{aligned}$$

[16] We emphasize that the so-called *Kähler condition*, which is independent of the choice of a holomorphic coordinate system on M, requires ω to be both closed and positive (1,1)-form, that is:

$$\frac{\partial g_{i\bar{k}}}{\partial z^j} = \frac{\partial g_{j\bar{k}}}{\partial z^i} \qquad \text{and} \qquad \frac{\partial g_{k\bar{i}}}{\partial z^{\bar{j}}} = \frac{\partial g_{k\bar{j}}}{\partial z^{\bar{i}}} \qquad (\text{for all } i, j, k = 1, 2, \cdots, n).$$

[17] To define *complex differential forms* in local holomorphic coordinates $z^1, \cdots, z^n$ on a complex manifold M, we use the following three-step procedure:

1. Decompose the complex coordinates into their real and imaginary parts: $z^k = x^k + \mathrm{i}y^k$ for each $j = 1, ..., n$;
2. Complex differentials are then defined as:

$$dz^k = dx^k + \mathrm{i}dy^k, \qquad d\bar{z}^k = dx^k - \mathrm{i}dy^k;$$

   and
3. Any complex exterior differential form α can now be defined (using two arbitrary real and smooth vector-functions f_j and g_j) by the following sum:

$$\alpha = f_j dz^k + g_j d\bar{z}^k.$$

To proceed with Kähler differential geometry, we need some basic cohomology groups, both real (*de Rham cohomology group*) and complex (*Dolbeault cohomology group*). Firstly, the *real Kähler class* of ω is the cohomology class $[\omega] \in H^2_d(M,\mathbb{R})$, where

$$H^2_d(M,\mathbb{R}) = \frac{\{d\text{-closed real (1,1)-forms}\}}{\{d\text{-exact real (1,1)-forms}\}}$$

is the 2nd-order *de Rham cohomology group* of $\mathcal{K} = (M,\omega)$. By the *Hodge theorem* (see Appendix, as well as, e.g. [II06b, II07]), any other positive Kähler form $\omega_\varphi \in [\omega]$ is given by:

$$\omega_\varphi = \omega + \mathrm{i}\frac{\partial^2\varphi}{\partial z^i \partial z^{\bar{j}}} \equiv \omega + \mathrm{i}\partial_i\partial_{\bar{j}}\varphi \equiv \omega + \mathrm{i}\partial\bar{\partial}\varphi,$$

where $\partial \equiv \partial_j$ and $\bar{\partial} \equiv \partial_{\bar{j}}$ are the so-called *Dolbeault differential operators*,[18] while φ is a real smooth function called the *Kähler potential*. The *functional space of Kähler potentials* is the set:

$$\mathcal{P}(M,\omega) = \{\varphi \mid \omega_\varphi = \omega + \mathrm{i}\partial\bar{\partial}\varphi > 0\}.$$

Alternatively, the Kähler class $[\omega]$ can be defined in terms of Dolbeault's operators as follows. Any p-form α defined on the Kähler manifold $\mathcal{K} = (M,\omega)$ is called *$\bar{\partial}$-closed* iff $\bar{\partial}\alpha = 0$ and *$\bar{\partial}$-exact* iff $\alpha = \bar{\partial}\eta$ for some $(p-1)$-form η on $\mathcal{K}$. The *Dolbeault cohomology group* $H^{1,1}_{\bar{\partial}}(M,\mathbb{R})$ is a complexification of the de Rham cohomology group $H^2_d(M,\mathbb{R})$, defined on $\mathcal{K}$ as a quotient:

$$H^{1,1}_{\bar{\partial}}(M,\mathbb{R}) = \frac{\{\bar{\partial}\text{-closed real (1,1)-forms}\}}{\{\bar{\partial}\text{-exact real (1,1)-forms}\}}.$$

A Kähler form ω on $\mathcal{K}$ defines a nonzero element $[\omega] \in H^{1,1}_{\bar{\partial}}(M,\mathbb{R})$. If a cohomology class $\alpha \in H^{1,1}_{\bar{\partial}}(M,\mathbb{R})$ can be written as $\alpha = [\omega]$ for some Kähler form ω on $\mathcal{K}$ then we say that α is a *complex Kähler class* (and write $\alpha > 0$).

---

[18] *Dolbeault's differential operators* can be defined using the *exterior derivative* (see Appendix) and the *canonical projection* of the vector bundle $\Omega^{p,q}$ of complex exterior (p,q)-forms, given respectively by:

$$d : \Omega^{p,q} \to \Omega^{p+1,q} + \Omega^{p,q+1}, \qquad \pi^{p,q} : \bigoplus_{p+q} \Omega^{p,q} \to \Omega^{p,q}, \qquad \text{as:}$$

$$\partial = \pi^{p+1,q} \circ d : \Omega^{p,q} \to \Omega^{p+1,q}, \qquad \bar{\partial} = \pi^{p,q+1} \circ d : \Omega^{p,q} \to \Omega^{p,q+1}.$$

They obey the following properties:

$$d = \partial + \bar{\partial}, \qquad \partial^2 = \bar{\partial}^2 = \partial\bar{\partial} + \bar{\partial}\partial = 0.$$

Next, the *volume form* ω^n (corresponding to the Kähler form ω) is given on $\mathcal{K} = (M, \omega) = (M, g)$ by:

$$\omega^n = \frac{1}{n!} \mathrm{i}^n \det\left(g_{i\bar{j}}\right) dz^1 \wedge dz^{\bar{1}} \wedge \cdots \wedge dz^n \wedge dz^{\bar{n}}. \tag{2.11}$$

The complex affine connections, i.e., *Christoffel symbols* (see Appendix) corresponding to the Kähler metric g are given [for all $i, j, k = 1, 2, \cdots, n$ on (M, g)] by:

$$\Gamma^k_{i\,j} = g^{k\bar{l}} \frac{\partial g_{i\bar{l}}}{\partial z^j} \quad \text{and} \quad \Gamma^{\bar{k}}_{\bar{i}\,\bar{j}} = g^{\bar{k}l} \frac{\partial g_{l\bar{i}}}{\partial z^{\bar{j}}}. \tag{2.12}$$

Given the third-order connection symbols (2.12), we can define the fourth-order *curvature tensor* on $\mathcal{K}$ in the following two steps (see Appendix):

(1) Firstly, we define the mixed curvature (1,3)-tensor on $\mathcal{K}$, by:

$$R_{i\ \ k\bar{\ell}}^{\ m} = -\partial_{\bar{\ell}} \Gamma^m_{ik}; \qquad \text{and}$$

(ii) Secondly, we define the covariant curvature (0,4)-tensor on $\mathcal{K}$, by:

$$R_{i\bar{j}k\bar{\ell}} = g_{m\bar{j}} R_{i\ \ k\bar{\ell}}^{\ m},$$

which is locally [in a open chart $U \subset \mathcal{K}$ with coordinates $z_1, \cdots, z_n$] defined by:[19]

$$R_{i\bar{j}k\bar{l}} = -\frac{\partial g_{i\bar{j}}}{\partial z^k \partial z^{\bar{l}}} + g^{p\bar{q}} \frac{\partial g_{i\bar{q}}}{\partial z^k} \frac{\partial g_{p\bar{j}}}{\partial z^{\bar{l}}} = -\partial_i \partial_{\bar{j}} g_{k\bar{\ell}} + g^{\bar{q}p} (\partial_i g_{k\bar{q}})(\partial_{\bar{j}} g_{p\bar{\ell}}),$$

(for all $i, j, k, l = 1, 2, \cdots, n$). It has the following three symmetries:

1. $\overline{R_{i\bar{j}k\bar{\ell}}} = R_{j\bar{i}\ell\bar{k}}$ (conjugate symmetry);
2. $R_{i\bar{j}k\bar{\ell}} = R_{k\bar{j}i\bar{\ell}} = R_{i\bar{\ell}k\bar{j}}$ (I Bianchi identity);
3. $\nabla_m R_{i\bar{j}k\bar{\ell}} = \nabla_i R_{m\bar{j}k\bar{\ell}}$ (II Bianchi identity);

Now, the *Ricci curvature* (or, contracted Riemannian curvature, see Appendix) of the Kähler metric g is a covariant (0,2)-tensor defined as:

$$R_{i\bar{j}} = g^{\bar{\ell}k} R_{i\bar{j}k\bar{\ell}} = g^{\bar{\ell}k} R_{k\bar{\ell}i\bar{j}} = R_{k\ \ i\bar{j}}^{\ k},$$

which has the trace in the form of the *scalar curvature* $R = g^{\bar{j}i} R_{i\bar{j}}$. The covariant Ricci tensor is locally [in an open chart $U \subset (M, g)$ with holomorphic coordinates $z_1, \cdots, z_n$] given by:

[19] The Kähler metric g possess a nonnegative *bisectional curvature tensor* $R_{i\bar{j}k\bar{l}}$ if (for all non-zero vectors v and w in the holomorphic tangent bundle TM of M) there is the following quadric form:

$$R_{i\bar{j}k\bar{l}} v^i v^{\bar{j}} w^k w^{\bar{l}} \geq 0.$$

$$R_{i\bar{j}} = -\frac{\partial^2 \log\left[\det(g_{k\bar{l}})\right]}{\partial z_i \partial \bar{z}_j} = -\partial_i \partial_{\bar{j}} \log\left[\det(g_{k\bar{l}})\right].$$

Associated to the tensor $R_{i\bar{j}}$ is the following closed Ricci (1,1)-form on $\mathcal{K}$:

$$\mathrm{Ric}(\omega) = \mathrm{i} R_{i\bar{j}}(\omega)\, dz^i \wedge dz^{\bar{j}} = -\mathrm{i}\partial\bar{\partial} \log\left(\det g\right).$$

The main purpose of the Ricci curvature is to define the *Kähler-Ricci flow*.

2.4.2 Global Kähler Geometry

Fano manifolds

The richest and most useful structure in Kähler geometry is the *Fano manifold*, that is a compact (i.e., closed and bounded) Kähler manifold with positive *first Chern class*. Recall that the *Chern classes*[20] [Che46] are characteristic classes associated to *complex vector bundles*. In the most important case of a *complex line bundle* (see [Gro58]), the only nontrivial Chern class is the first Chern class, which is an element of the second cohomology group of the underlaying manifold.[21] Formally, given a complex Hermitian vector bundle E (of complex rank n) over a smooth manifold M, a representative of each Chern class $c_k(E)$[22] of E are given as the coefficients of the *characteristic polynomial*:

$$\det\left(\frac{\mathrm{i}s\,\Omega}{2\pi} + I\right) = c_k(E)\, s^k,$$

(with the $n \times n$ identity matrix I and the scalar s used here only as an indeterminate to generate the sum from the determinant), of the *Cartan curvature 2-form*:

$$\Omega = d\omega + \tfrac{1}{2}[\omega, \omega],$$

where ω is the *connection form* for the gauge group of E. For example [MS74, Wik13d], consider a holomorphic local coordinate z on the *Riemann sphere*, that is a 1D complex projective space CP^1, with the complex line tangent bundle $E = TCP^1$. We use the Kähler metric:

$$h = \frac{dz d\bar{z}}{(1 + |z|^2)^2},$$

such that the Cartan curvature 2-form is:

[20] The Chern classes are usually used, in conjunction with the *Riemann-Roch theorem* and the *Atiyah-Singer index theorem*, in algebraic topology, differential geometry and algebraic geometry, for counting how many linearly-independent sections are there in some vector bundle E.

[21] As it is the top Chern class, it equals the *Euler class* of the bundle E, that is an element of the integral cohomology group: $H^r(E; \mathbf{Z})$.

[22] $c_k(E)$ is usually called the *Chern form* of a complex *Hermitian vector bundle* E.

$$\Omega = \frac{2dz \wedge d\bar{z}}{(1+|z|^2)^2},$$

while the first Chern class is:

$$c_1 = \left[\frac{\mathrm{i}}{2\pi}\mathrm{tr}\ \Omega\right].$$

To show that $E = TCP^1$ is a non-trivial vector bundle, we need to prove that the cohomology class c_1 is non-zero; this is shown by calculating its integral over the CP^1:

$$\int c_1 dz \wedge d\bar{z} = \frac{\mathrm{i}}{\pi}\int \frac{dz \wedge d\bar{z}}{(1+|z|^2)^2} = 2.$$

By the Stokes theorem, an exact form would integrate to 0, so the cohomology class c_1 is nonzero (thiss proves the so-called *hairy ball theorem* for the Riemann sphere CP^1).

It has been one of main problems in Kähler geometry to study if a Fano manifold admits a Kähler-Einstein metrics since the *Aubin-Yau theorem* on Kähler-Einstein metrics with negative scalar curvature and the *Calabi-Yau theorem* on Ricci-flat Kähler metrics in 70's (see [Tia97, Tia12, Ti13]).

Kähler structures

According to [Don08a], within a fixed Kähler class on a compact complex manifold M, there exist the following four different kinds of special Kähler metrics:

1. *Extremal Kähler metrics*, due to Calabi [Cal83], which are critical points (i.e., local minima) of the Calabi functional:

$$\int_M |\mathrm{Riem}(\omega)|^2 d\mu_\omega,$$

   where ω varies over the Kähler metrics in a fixed Kähler class and Riem is the Riemann curvature tensor. The Euler-Lagrange equation is

$$\bar{\partial}(\nabla S_\omega) = 0,$$

   where ∇ is the gradient operator defined by ω and $S(\omega)$ is the scalar curvature. In other words, the vector-field ∇S_ω should be a holomorphic vector-field. On the face of it, this is a sixth order partial differential equation for the Kähler potential ψ.
2. *Constant scalar curvature Kähler metrics* are the extremal metrics with constant S_ω.[23]

[23] If M has only trivial holomorphic vector-fields then there exist a constant scalar curvature.

3. *Kähler-Einstein metrics* are, by definition, those where the Ricci tensor Ric is a multiple of $\lambda\omega$.[24]
4. *Kähler-Ricci solitons* are metrics for which[25]

$$\mathrm{Ric} - \omega = L_v \omega,$$

where L_v is the Lie derivative along a holomorphic vector-field v.

Two Kähler structures and two alternative points of view

Consider a complex vector bundle E over a complex manifold M, as well as the following two structures:

- A Hermitian metric h on E;
- A holomorphic structure on E defined by the $\overline{\partial}$-operator

$$\overline{\partial} : \Omega^0(E) \to \Omega^{0,1}(E).$$

These two structures together define a unique compatible *unitary connection*, in the sense that the $\overline{\partial}$-operator is the $(0,1)$-component of the covariant derivative (see [Don08a]).

From the traditional point of view of complex algebraic geometry [GH78], a holomorphic structure is firstly fixed and then various Hermitian metrics are considered on it.[26] Alternatively, from the point of view of general Yang-Mills theory [AB82], the Hermitian metric is firstly fixed, and then various various $\overline{\partial}$-operators are considered.

Consider two Kähler structures on an underlying compact manifold M: a complex structure and a symplectic form, which are required to be algebraically compatible in the sense that the symplectic form is the imaginary part of a Hermitian metric. Again, from the conventional point of view, the complex structure is firstly fixed and then the Kähler form is varyed. If a reference form ω_0 is chosen and varyed in the fixed cohomology class then any other form can be represented by the following *Kähler potential*:

$$\omega_\psi = \omega_0 + \mathrm{i}\partial\overline{\partial}\psi.$$

For the alternative, Yang-Mills, point of view, the symplectic form ω is firstly fixed and then the space $\mathcal{J}$ of algebraically-compatible almost-complex structures on M is considered. Then the group SDiff of symplecto-morphisms of (M, ω) acts on $\mathcal{J}$, and this is the analogue of the unitary gauge group $U(E)$ in the previous case[27] [Don08a].

---

[24] These metrics occur only in the *Fano manifold* case.

[25] Ibid.

[26] Here, we have the following formula for the curvature tensor in a local holomorphic trivialisation:

$$F_h = \overline{\partial}(h^{-1}\partial h),$$

where the metric is defined by a matrix-valued function h.

[27] We consider the subset $\mathcal{J}_{\mathrm{int}}$ of integrable almost complex structures, which is preserved by SDiff. This is partitioned into equivalence classes under the relation

2.4.3 Local Kähler Geometry

Complex coordinates

Consider the complex coordinates in an open chart on a Kähler manifold M [Don08a]:

$$\tau_a = \frac{1}{2}(t_a + \mathrm{i}\theta_a).$$

Locally (in the neighborhood of a point in the free orbit $M_0 \in M$), the isometry group acts by translations in the θ_a directions, and the Kähler metric is given by $\mathrm{i}\partial\overline{\partial}\phi$ for a function ϕ of the complex variables τ_a. If we write $\phi = \phi(t_a)$ then the tensor $\mathrm{i}\partial\overline{\partial}\phi$ is (using summation convention, as always):

$$\mathrm{i}\partial\overline{\partial}\phi = \partial_{t_a t_b}\phi \, d\tau_a d\overline{\tau}_b, \quad \text{where} \quad \partial_{t_a t_b}\phi = \frac{\partial^2\phi}{\partial t_a \partial t_b},$$

and this defines a positive Hermitian form iff the Hessian matrix of ϕ is positive definite (in other words ϕ is a *convex* function of the real variables τ_a).[28] The Hessian of ϕ can be rewritten in index notation as $\nabla^2\phi = (\phi^{ab})$, while (ϕ_{ab}) is the inverse matrix. Then, the symplectic form ω is given by:

$$\omega = \frac{1}{2}\phi^{ab} dt_a \wedge d\theta_b,$$

while the Riemannian metric g is:

$$g = \frac{1}{2}\left(\phi^{ab} dt^a dt^b + \phi^{ab} d\theta^a d\theta^b\right).$$

Then the curvature tensor R (regarded as an element of $\Lambda^2 \otimes \Lambda^2$) is

$$R = R^{abcd} d\tau_a d\overline{\tau}_b \otimes d\tau_c d\overline{\tau}_d\,, \quad \text{where} \quad R^{abcd} = \phi^{abcd} - \phi^{ac\lambda}\phi^{bd\mu}\phi_{\lambda\mu}\,.$$

Symplectic coordinates

A different point of view was outlined by [Don94, Abr98] as follows. Consider an open set in the *symplectic manifold* $Q \times V^*$ (where $Q \subset V = \mathbb{R}^n$ is convex) with linear coordinates $\{x^a, \theta_a\}$, and the standard *symplectic form* (that is preserved by the translations in the θ_a):

$$\Omega = \frac{1}{2} dx^a d\theta_a.$$

$J_1 \sim J_2$ if $(M, J_1), (M, J_2)$ are isomorphic as complex manifolds. Although the group SDiff does not have a true complexification one can argue that the equivalence classes in $\mathcal{J}_{\text{int}}$ are formally the orbits of such a (mythical) complexified group, in the sense that they behave that way at the level of tangent spaces and Lie algebras [Don97].

[28] Thus the theory of convex functions on Euclidean spaces is embedded, as this translationally invariant case, in the theory of Kähler geometry.

Formally, we have a Hamiltonian action of the group $G = V^*$ on $Q \times V^*$, such that the moment map [MR99] is the projection to Q, with components the coordinates x^a. Consider G-invariant almost-complex structures on $Q \times V^*$, algebraically compatible with Ω. At each point of $Q \times V^*$ such a structure is specified by a subspace of the complexified cotangent bundle $T^*(Q \times V^*)$ which has a unique basis of the form (see [Don08a]):

$$\epsilon_a = d\theta_a + Z_{ab}dx^b,$$

[where (Z_{ab}) is a symmetric complex matrix with positive definite imaginary part] and the almost-complex structure is represented by a matrix-valued function (Z_{ab}), while its G-invariance specifies that Z is a function of the variables x^a. The following 2-forms

$$d\epsilon_a = \partial_{x^c} Z_{ab}\, dx^c dx^b$$

can be expressed as $\alpha_{ab} \wedge \epsilon_b$,[29] thus giving the integrability condition:

$$\partial_{x^c} Z_{ab} = \partial_{x^b} Z_{ac}. \tag{2.13}$$

Next, by the elementary *criterion for an exact differential*, the condition (2.13) implies that there are complex-valued functions $\mathrm{i}t_a$ such that:

$$Z_{ab} = \mathrm{i}\partial_{x^b} t_a.$$

Furthermore, as Z_{ab} is symmetric, then (by the same criterion) there is a single complex valued function F such that $t_a = \partial_{x^a} F$,[30] i.e.:

$$Z_{ab} = \partial_{x^a x^b} F.$$

Write u for the imaginary part of the function F above, so

$$Y_{ab} = \partial_{x^a x^b} u = u_{ab}.$$

Some linear algebra shows that the metric defined by the almost complex structure and the fixed form Ω is [Don08a]:

$$\frac{1}{2} u_{ij} dx^i dx^j + u^{ij} d\theta_i d\theta_j,$$

where (u^{ij}) is the matrix inverse of the Hessian (u_{ij}).

---

29 This happens only when all the $d\epsilon_a$ are zero, since $d\epsilon_a$ does not contain any terms involving $d\theta_i$.

30 If we let f be minus the real part of F then the action of $f \in \mathcal{G}$ takes the structure (Z_{ab}) to a new structure with zero real part. So, taking account of this diffeomorphism group, we can reduce to considering $Z = \mathrm{i}Y$, with Y real and positive-definite. Now the functions t_a are real and $\epsilon_a = d(t_a + \mathrm{i}\theta_a)$ so $t_a + \mathrm{i}\theta_a$ are local complex coordinates.

Thus we have two natural coordinate systems to use for this local Kähler geometry. In the symplectic picture, we can set

$$F_{ijkl} = u_{ia}u_{jb}\,\partial_{x^k x^l}u^{ab},$$

from which we can calculate the Riemann curvature tensor as:

$$F = F_{ijkl}\eta^i \wedge \eta^k \otimes \eta j \wedge \eta^l, \quad \text{where} \quad \eta^a = dx^a + \mathrm{i}u^{ab}d\theta_b.$$

For example, the norm of the Riemann curvature tensor is the same as the natural norm of F:

$$|F|^2 = F_{ijkl}F_{abcd}\,u^{ia}u^{jb}u^{kc}u^{ld}.$$

Similarly, the Ricci tensor is equivalent to the tensor:

$$G_{ij} = F_{ijkl}u^{kl}, \qquad \text{that is,} \quad G_{ij} = \partial_{x^i x^j}L, \quad \text{where} \quad L = \log\det(u_{ij}).$$

Finally, the scalar curvature is given by another contraction, yielding the Abreu formula [Abr98, Don02]:

$$S = G_{ij}u^{ij} = \partial_{x^i x^j}u^{ij}.$$

2.4.4 Invariant Hamiltonian Dynamics on Kähler Manifolds

In this subsection we consider symplectic Lie group actions on Kähler manifolds that are Lie groups (the so-called G-spaces).

Recall that the Kähler structure on a $2n$D symplectic manifold M with a symplectic form ω is given by a *compatible triple* (ω, J, g) on M, provided g is a Riemannian metric and J is an *almost complex structure*.[31]

In addition, let a Lie group G (with its Lie algebra $\mathfrak{g}$) act on (M, ω) by *symplectomorphisms*.[32] Then, any element $\xi \in \mathfrak{g}$ generates a vector-field ξ_M on M, called the *infinitesimal generator*, given by:

$$\xi_M(x) := \left.\frac{d}{dt}\right|_{t=0} \exp(t\xi)\cdot x, \qquad (x \in M),$$

whereexp: $\mathfrak{g} \to G$ is the *exponential map*. The G-action on M is said to be *Hamiltonian action* if there exists a smooth *equivariant map* $\mu\colon M \to \mathfrak{g}^*$, called the *momentum map*, such that for all $\xi \in \mathfrak{g}$ we have the *inner product*:

$$i_{\xi_M}\omega = \omega(\xi_M, \cdot) = d\langle \mu, \xi\rangle,$$

[31] An almost complex structure J on a Kähler manifold M is a tangent bundle automorphism $J\colon TM \to TM$ satisfying $J^2 = -I$, such that $g(\cdot,\cdot) = \omega(\cdot, J\cdot)$ and I is an identity matrix.

[32] Symplectomorphisms are *diffeomorphisms* which preserve the symplectic form ω.

where $\langle \cdot, \cdot \rangle : \mathfrak{g}^* \times \mathfrak{g} \to \mathbb{R}$ is the *duality pairing* of the Lie algebra $\mathfrak{g}$ with its dual Lie algebra $\mathfrak{g}^*$. In this case, the compatible triple (ω, J, g) on M is called G-invariant.

For example, if $G = SO(3)$ is the *rotation group* of a rigid body, parameterized by three Euler angles (roll, pitch and yaw), then its tangent Lie algebra $g = so(3)$ contains the corresponding three angular velocities (which are infinitesimal generators of the Euler angles), while its dual, cotangent Lie algebra $g^* = so(3)^*$ contains the three corresponding angular momenta. The momentum map is:

$$\mu \colon SO(3) \to \mathfrak{so}(3)^*.$$

For example, if $G = T^n$, is an n-torus, the existence of a momentum map μ is equivalent to the exactness of the one-forms $i_{\xi_M}\omega$ for all $\xi \in \mathfrak{g}$. In this case the obstruction of the action to being Hamiltonian lies in the first de Rham cohomology group of M.[33] If the complex structure J preserves harmonic one-forms and the G-action has fixed points on every connected component, then the action is Hamiltonian [PR10].

If the manifold is Kähler, the associated complex structure automatically preserves the space of harmonic one-forms. For example, the S^1-action on T^2 given by:

$$\mathrm{e}^{2\mathrm{i}\varphi} \cdot (\mathrm{e}^{2\mathrm{i}\theta_1}, \mathrm{e}^{2i\theta_2}) = (\mathrm{e}^{2\mathrm{i}\theta_1}, \mathrm{e}^{2\mathrm{i}(\theta_2+\varphi)})$$

is a symplectic action on a Kähler manifold which is free and hence has no fixed points.

Since for compact manifolds one always has the *Hodge decomposition* for the measure ω^n, the following statement is valid [PR10]: Let (M, ω) be a compact symplectic G-space. If M is a Kähler manifold, then the space of harmonic one-forms is invariant under the complex structure J. If the G-action has fixed points on every connected component of M then it is Hamiltonian. For the proofs and more technical details, see [PR10] and references therein.

2.5 Dynamics on Quaternion-Kähler Manifolds

2.5.1 Quaternion-Kähler Manifolds

Recall that famous Irish mathematical physicist, Sir William Rowan Hamilton (1805–1865), reformulated Newtonian and Lagrangian mechanics into a powerful *Hamiltonian formalism* that united classical mechanics and geometrical optics, thus setting-up the stage for the advent of quantum mechanics. Besides, in 1843 Hamilton invented the *quaternions*, a number system that generalizes

[33] The simplest example of a S^1-Hamiltonian action is rotation of the sphere S^2 about the polar axis. The flow lines of the infinitesimal generator defining this action are the latitude circles.

(or, extends) the complex numbers and also represents the four-parameter rotations.[34] He carved his fundamental formula for quaternion multiplication:[35]

$$i^2 = j^2 = k^2 = ijk = -1$$

into the stone of Brougham Bridge.[36] While Hamilton treated quaternions geometrically, modern mathematics prefers to treat the field of quaternions $\mathbb{H}$ algebraically, formally as the *quaternion group*, a non-Abelian group of order eight, defined by the following set:

$$\mathrm{Q} = \left\{-1, i, j, k : (-1)^2 = 1,\ i^2 = j^2 = k^2 = ijk = -1\right\},$$

with 1 as the identity element, while -1 commutes with the other elements of the group (for technical details, see [Qua13] and references and links therein).

From modern geometrical perspective, the so-called (almost) *quaternion manifold*, denoted by (M, V), is such an n-manifold M, which has a 3D vector bundle V, associated to the tangent bundle TM of M, consisting of tensors of type $(1,1)$ and in any coordinate neighborhood U of M, there exists a local canonical basis $\{F, G, H\}$ of the bundle V in $U \subset M$, such that [Tek09a]:

$$F^2 = G^2 = H^2 = FGH = -I,$$

[where I denotes the identity tensor of type $(1,1)$]. In such a way defined 3D vector bundle V represents an almost *quaternion structure* on M, and (M, V) is an almost quaternion manifold of dimension $n = 4m$ (for $m \geqslant 1$).

Note that there is also a dual quaternion structure on M, given by a dual local canonical basis $\{F^*, G^*, H^*\}$ of the 3D vector co-bundle V^*, associated to the cotangent bundle T^*M of M, which satisfies the condition:

$$F^{*2} = G^{*2} = H^{*2} = F^*G^*H^* = -I,$$

(it will be used later for explicit quaternion Kähler relations in the coframes of T^*M).

---

[34] Recall that three spatial rotations are usually described by three *Euler angles*: roll, pitch and yaw. However, Euler angles have the so-called *gimbal lock* (singularities at 180 degrees), while quaternions do not have such a disadvantage, and because of this, quaternions are used in various fields, including computer graphics, orbital mechanics, computer vision, attitude control and robotics.

[35] Note that the *extended Kalman filter* (see 2.8.2 in the Appendix) is frequently used, in the form of the so-called *quaternion filter*, for *sensor fusion*, including tracking of positions and quaternion-defined orientations of aircrafts, spacecrafts and submarines, as well as in medical applications requiring the *hand-eye calibration* (see, e.g. [Lei04] and references/links therein).

[36] Brougham Bridge crosses the Royal Canal in Cabra, Dublin, Ireland. Hamilton crossed this bridge and carved his fundamental quaternion formula on it when he was on his way to the Royal Irish Academy to present his discovery.

In any almost quaternion manifold (M, V), there is an almost quaternion metric structure, which is (for any cross-section ϕ of the bundle V and any vector-fields $X, Y \in M$) given by the Riemannian metric g such that:

$$g(\phi X, Y) + g(X, \phi Y) = 0.$$

In that case, (M, V) becomes an almost quaternion metric manifold (M, g, V). In addition, each of F, G, H in the local canonical basis $\{F, G, H\}$ of V is *almost Hermitian* with respect to the metric tensor g. Thus (for any vector-fields $X, Y \in M$), by setting:

$$\Phi(X, Y) = g(FX, Y), \quad \Psi(X, Y) = g(GX, Y), \quad \Theta(X, Y) = g(HX, Y),$$

we see that Φ, Ψ, Θ are local 2-forms on M.

Finally, to reach our target, the *quaternion Kähler manifold*, we have to assume that the Riemannian (Levi-Civita) connection ∇ of the quaternion metric manifold (M, g, V) satisfies the following condition [YK84]: For any vector-field $X \in M$, if ϕ is a cross-section of the bundle V, then $V_X \phi$ is also a cross-section of V. In the local canonical basis $\{F, G, H\}$ of V, this condition is equivalent to the following three relations:

$$\nabla_X F = r(X)G - q(X)H,$$
$$\nabla_X G = -r(X)F + p(X)H, \quad \nabla_X H = q(X)F - p(X)G,$$

where p, q, r are local 1-forms on M. If this condition is satisfied, the 3D vector bundle V becomes a *quaternion Kähler structure* of M, while (M, g, V) becomes the quaternion Kähler manifold.

More explicitely, following [Bur08], let $\{x_i, x_{n+i}, x_{2n+i}, x_{3n+i}\}$, for $i = 1, ..., n$ be a real coordinate system on a neighborhood $U \subset M$, such that: $\{\partial_{x_i}, \partial_{x_{n+i}}, \partial_{x_{2n+i}}, \partial_{x_{3n+i}}\}$ is the corresponding frame on the tangent bundle TM, while $\{dx_i, dx_{n+i}, dx_{2n+i}, dx_{3n+i}\}$ is the corresponding coframe on the cotangent bundle T^*M of M. Then the following frame-relations hold in the local canonical basis $\{F, G, H\}$ of V:

$$\begin{aligned}
F(\partial_{x_i}) &= \partial_{x_{n+i}}, \quad F(\partial_{x_{n+i}}) = -\partial_{x_i}, \\
F(\partial_{x_{2n+i}}) &= \partial_{x_{3n+i}}, \quad F(\partial_{x_{3n+i}}) = -\partial_{x_{2n+i}}, \\
G(\partial_{x_i}) &= \partial_{x_{2n+i}}, \quad G(\partial_{x_{n+i}}) = -\partial_{x_{3n+i}}, \\
G(\partial_{x_{2n+i}}) &= -\partial_{x_i}, \quad G(\partial_{x_{3n+i}}) = \partial_{x_{n+i}}, \\
H(\partial_{x_i}) &= \partial_{x_{3n+i}}, \quad H(\partial_{x_{n+i}}) = \partial_{x_{2n+i}}, \\
H(\partial_{x_{2n+i}}) &= -\partial_{x_{n+i}}, \quad H(\partial_{x_{3n+i}}) = -\partial_{x_i}.
\end{aligned}$$

In a dual local canonical basis $\{F^*, G^*, H^*\}$ of the 3D vector co-bundle V^* (associated to T^*M) the following coframe-relations hold:

$$
\begin{aligned}
F^*(dx_i) &= dx_{n+i},\ F^*(dx_{n+i}) = -dx_i,\\
F^*(dx_{2n+i}) &= dx_{3n+i},\ F^*(dx_{3n+i}) = -dx_{2n+i},\\
G^*(dx_i) &= dx_{2n+i},\ G^*(dx_{n+i}) = -dx_{3n+i},\\
G^*(dx_{2n+i}) &= -dx_i,\ G^*(dx_{3n+i}) = dx_{n+i},\\
H^*(dx_i) &= dx_{3n+i},\ H^*(dx_{n+i}) = dx_{2n+i},\\
H^*(dx_{2n+i}) &= -dx_{n+i},\ H^*(dx_{3n+i}) = -dx_i.
\end{aligned}
$$

2.5.2 Hamiltonian Dynamics on Quaternion-Kähler Manifolds

To start with, recall (see, e.g. [AM78]) that if M is an nD configuration manifold of a complex physical system and $H : T^*M \to \mathbb{R}$ is its (regular) Hamiltonian energy function defined on the cotangent bundle T^*M with the symplectic form $\omega \in T^*M$, called the momentum phase-space of M, then there is a unique Hamiltonian vector-field X on T^*M such that Hamilton's dynamical equations are given by the inner product:

$$i_X\omega = dH, \tag{2.14}$$

where ω is the symplectic form on T^*M. The triple (T^*M, ω, X) is called *Hamiltonian dynamical system* on the cotangent bundle T^*M.

Now, following [Tek09a], we will develop Hamiltonian formalism on the quaternion Kähler manifold $M = (M, g, V)$ as follows. Let F^* be a component of an almost quaternion structure V^*, λ_{F^*} be a *Liouville form* on M, and α_{F^*} be a 1-form on M. Then

$$
\begin{aligned}
\alpha_{F^*} &= \frac{1}{2}(x_i dx_i + x_{n+i}dx_{n+i} + x_{2n+i}dx_{2n+i} + x_{3n+i}dx_{3n+i}) \qquad \text{and}\\
\lambda_{F^*} &= F^*(\alpha_{F^*}) = \frac{1}{2}(x_i dx_{n+i} - x_{n+i}dx_i + x_{2n+i}dx_{3n+i} - x_{3n+i}dx_{2n+i}).
\end{aligned}
$$

Therefore, if ω_{F^*} is a closed *Kähler form* on a quaternion Kähler manifold M, then ω_{F^*} is also a *symplectic structure* on M.

On a quaternion Kähler manifold $M = (M, g, V)$, the following Hamiltonian structures can be defined:

1. The Hamiltonian vector-field is given by:

$$X = X^i\partial_{x_i} + X^{n+i}\partial_{x_{n+i}} + X^{2n+i}\partial_{x_{2n+i}} + X^{3n+i}\partial_{x_{3n+i}},$$

2. The differential of Hamiltonian energy H is:

$$dH = \partial_{x_i}H\, dx_i + \partial_{x_{n+i}}H\, dx_{n+i} + \partial_{x_{2n+i}}H\, dx_{2n+i} + \partial_{x_{3n+i}}H\, dx_{3n+i},$$

3. The symplectic 2-form on M is given by:

$$\omega_{F^*} = -d\lambda_{F^*} = dx_{n+i}\wedge dx_i + dx_{3n+i}\wedge dx_{2n+i},$$

4. The inner product (2.14) is given by:

$$i_X\omega_{F^*} = \omega_{F^*}(X) = X^{n+i}dx_i - X^i dx_{n+i} + X^{3n+i}dx_{2n+i} - X^{2n+i}dx_{3n+i}.$$

Combining these relations, the Hamiltonian vector-field X can be rewritten as:

$$X = -\partial_{x_{n+i}}H\,\partial_{x_i} + \partial_{x_i}H\,\partial_{x_{n+i}} - \partial_{x_{3n+i}}H\,\partial_{x_{2n+i}} + \partial_{x_{2n+i}}H\,\partial_{x_{3n+i}}.$$

Now, if a smooth curve $\alpha : I \subset \mathbb{R} \to M$ is an *integral curve* of the Hamiltonian vector-field X, i.e., $X(\alpha(t) = \dot{\alpha}$ (for $t \in I$), then in local coordinates: $\{x_i, x_{n+i}, x_{2n+i}, x_{3n+i}\} \subset M$ we have:

$$\alpha(t) = (x_i, x_{n+i}, x_{2n+i}, x_{3n+i}) \qquad \text{and}$$
$$\dot{\alpha}(t) = \dot{x}_i\,\partial_{x_i} + \dot{x}_{n+i}\,\partial_{x_{n+i}} + \dot{x}_{2n+i}\,\partial_{x_{2n+i}} + \dot{x}_{3n+i}\,\partial_{x_{3n+i}},$$

from which the following Hamilton's equations follow [Tek09a]:

$$\dot{x}_i = -\partial_{x_{n+i}}H, \;\; \dot{x}_{n+i} = \partial_{x_i}H,$$
$$\dot{x}_{2n+i} = -\partial_{x_{3n+i}}H, \;\; \dot{x}_{3n+i} = \partial_{x_{2n+i}}H.$$

These are dynamical equations defined with respect to a component F^* of an almost quaternion structure V^* on a quaternion Kähler manifold $M = (M, g, V)$. This means that the triple (M, ω_{F^*}, X) represents a *Hamiltonian dynamical system* on a quaternion Kähler manifold M. For more technical details, see [Tek09a] and references therein.

2.5.3 Lagrangian Dynamics on Quaternion-Kähler Manifolds

To start with, recall (see [AM78, LR89, Tek05]) that if M is an nD configuration manifold of a complex physical system and $L : TM \to \mathbb{R}$ is its (regular) Lagrangian energy function defined on the tangent bundle TM, called the velocity phase-space of M, then there is a unique vector-field ξ on TM such that dynamical equations are given by the inner product:

$$i_\xi \Phi_L = dE_L, \tag{2.15}$$

where Φ_L is the symplectic form on TM and E_L is the total energy function. The triple (TM, Φ_L, ξ) is called *Lagrangian dynamical system* on the tangent bundle TM, while (2.15) represents the 2nd-order ODE system.

Now, following [Tek09b], we will develop Lagrangian formalism on the quaternion Kähler manifold $M = (M, g, V)$ as follows. Let a component F of an almost quaternion structure V has coordinate functions: $\{x_i, x_{n+i}, x_{2n+i}, x_{3n+i}\} \subset M$. The Lagrangian vector-field ξ defines a *semispray* (see, e.g. [AM78]) given by:

$$\xi = X^i\partial_{x_i} + X^{n+i}\partial_{x_{n+i}} + X^{2n+i}\partial_{x_{2n+i}} + X^{3n+i}\partial_{x_{3n+i}}, \qquad \text{where}$$
$$X^i = \dot{x}_i, \quad X^{n+i} = \dot{x}_{n+i}, \quad X^{2n+i} = \dot{x}_{2n+i}, \quad X^{3n+i} = \dot{x}_{3n+i}.$$

The associated vector-field:

$$V_F = F(\xi) = X^i \partial_{x_{n+i}} - X^{n+i} \partial_{x_i} + X^{2n+i} \partial_{x_{3n+i}} - X^{3n+i} \partial_{x_{2n+i}}$$

is called the *Liouville vector-field* on the quaternion Kähler manifold $M = (M, g, V)$. The kinetic energy on M is the map $E_K : M \to \mathbb{R}$, given by:

$$E_K = \frac{1}{2} m_i (\dot{x}_i^2 + \dot{x}_{n+i}^2 + \dot{x}_{2n+i}^2 + \dot{x}_{3n+i}^2),$$

while the potential energy on M is the map $E_P : M \to \mathbb{R}$, given by: $E_P = m_i g h$, where m_i, g and h represent masses of a dynamical system, the gravity acceleration and distance to the origin of a dynamical system on the quaternion Kähler manifold M, respectively. Then the map $L : M \to \mathbb{R}$, given by: $L = E_K - E_P$ is the *Lagrangian function* on M; The associated function given by $E_L^F = V_F(L) - L$, is the total energy function.

Furthermore, the so-called *vertical derivative* d_F is defined as the Lie bracket (i.e., commutator) of the inner product i_F and the exterior derivative d,

$$d_F = [i_F, d] = i_F d - d i_F, \quad \text{in components given by:}$$
$$d_F = \partial_{x_{n+i}} dx_i - \partial_{x_i} dx_{n+i} + \partial_{x_{3n+i}} dx_{2n+i} - \partial_{x_{2n+i}} d_{3n+i}.$$

This implies that for each F, the corresponding Kähler form Φ_L^F is the closed 2-form on M given by:

$$\Phi_L^F = -dd_{_F} L.$$

By means of Φ_L^F, after heavy calculations performed in [Tek09b], the *Euler-Lagrangian equations* have been derived on a quaternion Kähler manifold $M = (M, g, V)$ in the form:

$$\partial_t (\partial_{x_i} L) + \partial_{x_{n+i}} L = 0, \qquad \partial_t (\partial_{x_{n+i}} L) - \partial_{x_i} L = 0,$$
$$\partial_t (\partial_{x_{2n+i}} L) + \partial_{x_{3n+i}} L = 0, \qquad \partial_t (\partial_{x_{3n+i}} L) - \partial_{x_{2n+i}} L = 0.$$

This means that the triple (M, Φ_L^F, ξ) represents a *Lagrangian dynamical system* on a quaternion Kähler manifold M. For more technical details, see [Tek09b] and references therein.

2.6 Kähler-Ricci-Flow Framework

2.6.1 Motivation for the Kähler-Ricci Flow

Recall (from Appendix) that the Riemann-Ricci flow (introduced by R. Hamilton [Ham82, Ham86, Ham88]), is a nonlinear, heat-like, geometric evolution equation on a Riemannian n-manifold M:

$$\partial_t g_{ij} = -2R_{ij}, \tag{2.16}$$

for a time-dependent Riemannian metric $g = g_{ij}(t)$ with the Ricci curvature tensor R_{ij}. At any time t, we can choose local harmonic coordinates so that the Ricci flow takes the general form:

$$\partial_t g_{ij} = \Delta_M g_{ij} + Q_{ij}(g, \partial g), \tag{2.17}$$

where $Q = Q_{ij}(g, \partial g)$ is a lower-order term quadratic in g and (its first order partial derivatives ∂g), while Δ_M is the Laplace-Beltrami differential operator, defined on C^2–functions on an M (with respect to the Riemannian metric g_{ij}) by:

$$\Delta_M \equiv \frac{1}{\sqrt{\det(g)}} \frac{\partial}{\partial x^i} \left(\sqrt{\det(g)} g^{ij} \frac{\partial}{\partial x^j} \right). \tag{2.18}$$

The quadratic Ricci-flow equation (2.17) has been proposed in [II11b] as a unique geometric framework for all real-valued nonlinear reaction-diffusion systems of the general form:

$$\begin{array}{ccccc} \partial_t \mathbf{u} & = & \mathbf{D}\Delta\mathbf{u} & + & \mathbf{R}(\mathbf{u}) \\ \updownarrow & & \updownarrow & & \updownarrow \\ \partial_t g_{ij} & = & \Delta_M g_{ij} & + & Q_{ij}(g, \partial g) \end{array}$$

where $\mathbf{u}(\mathbf{x}, t)$ is the the concentration state vector, $\mathbf{D}$ is a (symmetric) positive-definite matrix of diffusion coefficients and $\mathbf{R}(\mathbf{u})$ includes local reactions.

In this section, we will generalize this Riemann-Ricci flow framework for real-valued nonlinear reaction-diffusion systems to the complex-valued nonlinear reaction-diffusion systems, using a generalized form of the Ricci flow, called the Kähler-Ricci flow. This will be the unique framework for various nonlinear systems of complex PDEs, including: nonlinear Schrödinger equations, Gross-Pitaevskii equations, complex Ginzburg-Landau equations, open quantum Liouville equations, etc.

2.6.2 Ricci Flow on a Kähler Manifold

Recall that the Ricci flow, introduced by R. Hamilton in [Ham82], provides an indispensable tool for deforming Riemannian metrics towards canonical metrics (such as Einstein ones), by which (it is hoped that) one can further understand geometric and topological structures of underlying manifolds (see [Ham93b] for more technical detail).

In the richest particular case when the underlying manifold is a Kähler manifold, the normalized Ricci flow in a canonical *Kähler class*[37] preserves the Kähler class, which allows the Ricci flow to be reduced to a fully nonlinear parabolic equation on almost pluri-subharmonic functions, usually called the *Kähler-Ricci flow* [CT00]:

---

[37] A Kähler class is canonical if the first Chern class is proportional to this Kähler class.

$$\partial_t \varphi = \log\left(\det\left[\frac{(\omega + \partial\bar{\partial}\varphi)^n}{\omega^n}\right]\right) + \varphi - h_\omega, \tag{2.19}$$

where φ is the evolved Kähler potential and ω is the fixed *Kähler metric* in the canonical Kähler class. The corresponding *Ricci form*, $\mathrm{Ric}(\omega) \equiv \mathrm{Ric}(g)$, satisfies:

$$\mathrm{Ric}(\omega) - \omega = \partial\bar{\partial}h_\omega \quad \text{such that} \quad \int_M (e^{h_\omega} - 1)\omega^n = 0.$$

The normalized Kähler-Ricci flow (2.19) is usually written as:

$$\partial_t g = g - \mathrm{Ric}(g). \tag{2.20}$$

For a recent review, see [TZ13].

It was proved by [Cao85] that the Kähler-Ricci flow (2.20) has a global solution $g(t)$ in the case that $g(0) = g_0$ has canonical Kähler class, i.e., $2\pi c_1(M)$ as its Kähler class. In other words, Cao reproved in [Cao85] the famous *Calabi-Yau theorem* [Yau78]. The main problem here is to understand the limit of $g(t)$ as t tends to ∞. A desirable picture for the limit is given in the following *Tian folklore conjecture* [Tia97]: $(M, g(t))$ converges (at least along a subsequence) to a shrinking Kähler-Ricci soliton with mild singularities. This conjecture implies, in the case of Fano manifolds, the *Yau-Tian-Donaldson conjecture*, which states that a Fano manifold M admits a Kähler-Einstein metric iff it is K-stable (see [Tia97, Tia12, Ti13], as well as [CDS12a, CDS12b, CDS12c]).

2.6.3 Definition of the Kähler-Ricci Flow

We start with the minimalist definition, necessary for the further development (see, e.g. [CL08]). The general Kähler-Ricci flow is a complexified and generalized Ricci flow (2.67)-(2.17) that is lifted to a Kähler manifold $\mathcal{K} = (M, g_0)$ of complex dimension n with the complexified Laplace-Beltrami differential operator Δ_M lifted from equation (2.98) in the Appendix. In a local open chart $U \subset (M, g)$, starting from some smooth initial Kähler metric $g_0 = g_{i\bar{j}}(0)$, the Kähler-Ricci flow (2.22) is given (for $i, j = 1, \cdots, n$) by:

$$\partial_t g_{i\bar{j}}(t) = g_{i\bar{j}}(t) - R_{i\bar{j}}(t). \tag{2.21}$$

The Kähler-Ricci flow (2.21) preserves the Kähler class $[\omega] \in H^2(M, \mathbb{R})$, so it can be expressed in terms of the Kähler potentials $\varphi = \varphi(t)$, starting from some smooth initial potential value $\varphi_0 = \varphi(0)$, as:

$$\partial_t \varphi = \log \frac{{\omega_\varphi}^n}{\omega^n} + \varphi - h_\omega,$$

where both ω^n and ${\omega_\varphi}^n$ are given by (2.11), while the scalar function h_ω is defined by [CL08]:

$$\mathrm{i}\partial\bar{\partial}\,h_\omega = \mathrm{Ric}(\omega) - \omega, \qquad \int_M (\mathrm{e}^{h_\omega} - 1)\,\omega^n = 0.$$

From the flow (2.21), it follows that the evolution equation for the bisectional curvature $R_{i\bar{j}k\bar{l}} = R_{i\bar{j}k\bar{l}}(t)$ becomes:

$$\partial_t R_{i\bar{j}k\bar{l}} = \Delta_M R_{i\bar{j}k\bar{l}} - R_{i\bar{j}k\bar{l}} + R_{i\bar{j}m\bar{n}} R_{n\bar{m}k\bar{l}} - R_{i\bar{m}k\bar{n}} R_{m\bar{j}n\bar{l}} + R_{i\bar{l}m\bar{n}} R_{n\bar{m}k\bar{l}}\,,$$

starting from some smooth initial curvature $R_{i\bar{j}k\bar{l}}(0)$ and using the *evolved frames* [CL08].

The corresponding evolutions of the Ricci curvature $R_{i\bar{j}} = R_{i\bar{j}}(t)$ and the scalar curvature $R = R(t)$ are governed by:

$$\partial_t R_{i\bar{j}} = \Delta_M R_{i\bar{j}} + R_{i\bar{j}p\bar{q}} R_{q\bar{p}} - R_{i\bar{p}} R_{p\bar{j}}\,,$$
$$\partial_t R = \Delta_M R + R_{i\bar{j}} R_{j\bar{i}} - R, \quad \text{respectively,}$$

starting from some smooth initial Ricci and scalar curvatures, $R_{i\bar{j}}(0)$ and $R(0)$, respectively.

It was proved by [Cao85, CC99] (following the pioneering work of [Yau06]) that the Kähler-Ricci flow (2.21) exists globally for any smooth initial Kähler metric $g_0 = g_{i\bar{j}}(0)$.

2.6.4 Evolution of the Kähler-Perelman Entropy

Recall that that the richest and most useful structure in Kähler geometry is the Fano manifold, that is a compact (i.e., closed and bounded) Kähler n-manifold (M, g) with positive first Chern class $c_1(M)$ (for a recent review, see [TZ13] and references therein). Consider the normalized Kähler-Ricci flow on a Fano n-manifold (M, g):

$$\partial_t g(t) \;=\; g(t) \;-\; \mathrm{Ric}\,[g(t)]). \tag{2.22}$$

In a local open chart $U \subset (M, g)$, starting from some smooth initial Kähler metric $g_0 = g_{i\bar{j}}(0)$, the Ricci flow (2.22) is given by:

$$\partial_t g_{i\bar{j}}(t) = g_{i\bar{j}}(t) - R_{i\bar{j}}(t), \qquad (\text{for } i, j = 1, \cdots, n).$$

It was proved in [Cao85] that (2.22) has a global solution $g(t)$ in the case that g_0 has canonical Kähler class, i.e., $2\pi c_1(M)$ as its Kähler class. In particular, by the $\partial\bar{\partial}$-lemma, there exists a family of real-valued functions $u(t)$, called *Ricci potentials* (see, e.g. [MT07]) of $g(t)$, which are determined by:

$$g_{i\bar{j}} - R_{i\bar{j}} \;=\; \partial_i \partial_{\bar{j}} u, \qquad \frac{1}{V}\int \mathrm{e}^{-u(t)} dv_{g(t)} \;=\; 1,$$

where $V = \int dv_g$ denotes the volume of the Kähler-Ricci flow.

A Riemannian manifold (M, g) represents a *shrinking Ricci soliton* iff (see, e.g. [MT07]):

$$\mathrm{Ric}(g) + \mathrm{Hess}(u) = \lambda g, \qquad (\lambda > 0).$$

In particular, if (M, g) is a Fano manifold with $g \in 2\pi c_1(M)$, it is a *shrinking Kähler-Ricci soliton* iff $\lambda = 1$ and $u = u(t)$ is the Ricci potential; that is, iff (see [TZ13] and references therein):

$$\nabla\nabla u = 0,$$

or, applying the Bianchi identity, iff the following *Shur-identity* holds:

$$\triangle u - |\nabla u|^2 + u = a.$$

For any Kähler metric $g \in 2\pi c_1(M)$ with scalar curvature s and any smooth real time-function $u = u(t)$, define the *Kähler-Perelman entropy* $\mu(g)$ defined by the following infimum of sets of entropy functionals (compare with Perelman's original definition [Per02]):

$$\mu(g) = \inf\left\{\mathcal{W}(g,u) : \int_M \mathrm{e}^{-u} dv = V\right\}, \qquad \text{where}$$
$$\mathcal{W}(g,u) = \frac{1}{V}\int_M (s + |\nabla u|^2 + u - n)\,\mathrm{e}^{-u} dv.$$

A smooth minimizer of the entropy μ always exists (though not necessarily unique, see [Rot81]), while μ admits a (natural) upper bound:

$$\mu(g) \leq \frac{1}{V}\int_M u\,\mathrm{e}^{-u} dv = a \leq 0.$$

This implies the Kähler–Perelman monotonicity condition on the geometric entropy [TZ13]:

$$(\forall t \geq 0), \qquad \mathcal{E}(g_0) \leq \mathcal{E}\,[g(t)] \leq 0.$$

2.7 Summary of Kähler-Ricci Geometrical Dynamics

A Kähler manifold, $M \equiv (M, g) \equiv (M, \omega)$, is a Hermitian n-manifold[38] [that admits three mutually compatible dynamical structures: (i) Riemannian/Lagrangian, (ii) symplectic/Hamiltonian, and (iii) complex/quantum], given by the following data:

[38] A complex-valued function $f : \mathbb{C}^n \to \mathbb{C}$ is called holomorphic if $f = f_1 + \mathrm{i} f_2$ satisfies the Cauchy–Riemann relations:

$$\partial_{x^j} f_1 = \partial_{y^j} f_2, \qquad \partial_{x^j} f_2 = -\partial_{y^j} f_1, \qquad (\partial_{x^j} \equiv \partial/\partial x^j)$$

for each holomorphic coordinate, $z^j = x^j + \mathrm{i}y^j$ (with $\mathrm{i} = \sqrt{-1}$). A complex manifold M is a manifold with an atlas consisting of charts $t_{ij} : U_i \cap U_j \longrightarrow \mathbb{C}^n$, such that the transition functions $t_{ij}(z)$ are holomorphic and satisfy the cocycle condition: $t_{ik}(z) = t_{ij}(z) t_{jk}(z)$ on triple overlaps $U_i \cap U_j \cap U_k$. An almost complex structure J is defined on a complex manifold M as:

$$J\partial_{z^j} = \mathrm{i}\partial_{z^j}, \qquad J\partial_{\bar{z}^j} = -\mathrm{i}\partial_{\bar{z}^j}, \qquad J^2 = -1.$$

1. As an introductory example of a Kähler manifold $M_X = (M, g)$ consider a complexified tangent bundle, that is, a disjoint union of complexified tangent spaces at all points $x \in X$,

$$M_X \equiv TX \otimes \mathbb{C} = \sqcup_{x\in X} T_x X^{\mathbb{C}},$$

of some real smooth manifold X, with the standard tangent bundle, $TX = \sqcup_{x\in X} T_x X$, and its dual, cotangent bundle, $T^*X = \sqcup_{x\in X} T_x^* X$. Then the Kähler manifold $(M, g) = TX \otimes \mathbb{C}$ admits a Hermitian metric form g,[39] such that its real part is a Riemannian metric form $g_R \in TX$ and its imaginary part is a symplectic form $\omega_S \in T^*X$:

$$(M, g) = TX + \mathrm{i}T^*X \qquad \text{with} \qquad g = g_R + \mathrm{i}\omega_S.$$

2. An open chart $U \subset M$ defines a set of n holomorphic coordinates which locally identify $\mathbb{C}^n$ with $\mathbb{R}^{2n}$, as: $\{z^j = x^j + \mathrm{i}y^j \mid k = 1, \cdots, n\}$, with the corresponding holomorphic differentials (or, exterior 1-forms):

$$dz^j = dx^j + \mathrm{i}dy^j \qquad \text{and} \qquad d\bar{z}^j = dx^j - \mathrm{i}dy^j.$$

3. A Hermitian metric tensor, $g_{i\bar{j}} = g_{i\bar{j}}(z^i, z^{\bar{j}}) \in M$, obeys the following Kähler condition [independent of the choice of local holomorphic coordinates $z^j \in U$]:

$$\partial_j g_{i\bar{k}} = \partial_i g_{j\bar{k}} \qquad \text{and} \qquad \partial_{\bar{j}} g_{k\bar{i}} = \partial_{\bar{i}} g_{k\bar{j}}\,, \qquad (\partial_j \equiv \partial/\partial z^j, \;\; \partial_{\bar{j}} \equiv \partial/\partial z^{\bar{j}})$$

where $\partial_j \equiv \partial$ and $\partial_{\bar{j}} \equiv \bar{\partial}$ are Dolbeault's differential operators, which are the components of the standard exterior derivative on M: $d = \partial + \bar{\partial}$.[40]

A Hermitian manifold (M, g) is a complex manifold M with a Hermitian metric tensor $g_{i\bar{j}} = g_{i\bar{j}}(z^i, z^{\bar{j}})$, such that $\{g_{i\bar{j}}\}$ is a positive-definite Hermitian matrix function, $\{g_{i\bar{j}}\} = \{g_{i\bar{j}}\}^* > 0$. A Hermitian manifold (M, g) is a Kähler manifold iff the almost complex structure J satisfies: $\nabla_k J = 0$, where ∇_k is the Levi-Civita connection on (M, g).

39 All three structures defined on $(M, g) = TX \otimes \mathbb{C}$, that is, the Hermitian metric g, the Riemannian metric g_R, and the symplectic form ω_S, preserve the almost complex structure J. In other words, for any two complex tangent vectors $(\psi, \chi) \in M$ we have:

$$g(J\psi, J\chi) = g(\psi, \chi), \qquad g_R(J\psi, J\chi) = g_R(\psi, \chi), \qquad \omega_S(J\psi, J\chi) = \omega_S(\psi, \chi).$$

40 Being components of the exterior derivative d on M, Dolbeault's differential operators are the maps on the space $\Omega^{p,q}(M)$ of exterior forms on M, $\partial : \Omega^{p,q}(M) \to \Omega^{p+1,q}(M)$ and $\bar{\partial} : \Omega^{p,q}(M) \to \Omega^{p,q+1}(M)$.

In a local z^k-coordinate chart $U \subset M$, for any holomorphic function $f \in U$, ∂ and $\bar{\partial}$ operators are given by:

$$\partial f = \left(\partial_{x^k} f - \mathrm{i}\partial_{y^k} f\right) dz^k, \qquad \bar{\partial} f = \left(\partial_{x^k} f - \mathrm{i}\partial_{y^k} f\right) d\bar{z}^k.$$

4. The Kähler metric form g, corresponding to the Hermitian metric tensor $g_{i\bar{j}}$ is a positive-definite, symmetric (1,1)-form on M defined as:[41]

$$g = \mathrm{i}g_{i\bar{j}}\,dz^i \otimes dz^{\bar{j}} > 0, \quad (i,j = 1,\cdots,n), \quad \text{while}$$
$$g_R = g_{i\bar{j}}\,dz^i \otimes dz^{\bar{j}} = \text{Riemannian metric.}$$

5. The associated (symplectic) Kähler form ω, is the closed ($d\omega = 0$, follows directly from the Kähler condition) and positive-definite, exterior (1,1)-form[42] on M:

$$\omega = \mathrm{i}g_{i\bar{j}}\,dz^i \wedge dz^{\bar{j}} > 0, \qquad \text{while}$$
$$\omega_S = g_{i\bar{j}}\,dz^i \wedge dz^{\bar{j}} = \text{symplectic form.}$$

6. Locally, in an open chart $U \subset M$, the Kähler form ω is given by:

$$\omega = \mathrm{i}\partial_i\partial_{\bar{j}}\varphi = \mathrm{i}\partial\overline{\partial}\varphi > 0\,,$$

   where $\varphi \in U$ is a real-valued and smooth function called the Kähler potential.
7. Any p-form α defined on M is called $\overline{\partial}$-closed iff $\overline{\partial}\alpha = 0$ and $\overline{\partial}$-exact iff $\alpha = \overline{\partial}\eta$ for some (p-1)-form η on M. The associated Dolbeault cohomology group $H^{1,1}_{\overline{\partial}}(M,\mathbb{R})$ is a complexification of the standard de Rham cohomology group $H^2_d(M,\mathbb{R})$, defined on M as the quotient:

$$H^{1,1}_{\overline{\partial}}(M,\mathbb{R}) = \frac{\{\overline{\partial}\text{-closed real (1,1)-forms}\}}{\{\overline{\partial}\text{-exact real (1,1)-forms}\}}.$$

8. A Kähler form ω on M defines a nonzero element $[\omega] \in H^{1,1}_{\overline{\partial}}(M,\mathbb{R})$. If a cohomology class $\alpha \in H^{1,1}_{\overline{\partial}}(M,\mathbb{R})$ can be written as $\alpha = [\omega]$ for some Kähler form ω on M then we say that α is a Kähler class (and write $\alpha > 0$). Therefore, the Kähler class of ω is its cohomology class $[\omega] \in H^{1,1}_{\overline{\partial}}(M,\mathbb{R})$. Alternatively, in terms of $H^2_d(M,\mathbb{R})$, the Kähler class of ω is its cohomology class $[\omega] \in H^2_d(M,\mathbb{R})$. Usually, all this is simply written: the Kähler class of ω is the cohomology class $[\omega] \in H^2(M,\mathbb{R})$ represented by ω.
9. The $\partial\overline{\partial}$-Lemma [the holomorphic version of the Poincaré Lemma, also follows from the Hodge theory]: Let M be a compact Kähler manifold and suppose that $0 = [\alpha] \in H^{1,1}_{\overline{\partial}}(M,\mathbb{R})$ for a real smooth $\overline{\partial}$-closed $(1,1)$-form α. Then there exists a real smooth Kähler potential φ (uniquely determined up to the addition of a constant) with $\alpha = \mathrm{i}\partial\overline{\partial}\varphi$. In other

[41] In these notes, we are consistently using summation convention over repeated tensor indices.

[42] The Kähler form ω is also harmonic ($\delta\omega \equiv *d*\omega = 0$), a result from the Kähler-Hodge theory.

words, a real $(1,1)$-form α is $\overline{\partial}$-exact iff it is $\partial\overline{\partial}$-exact. It is an immediate consequence of the $\partial\overline{\partial}$-Lemma that if ω and ω_φ are Kähler forms in the same Kähler class on M, then $\omega_\varphi = \omega + \mathrm{i}\partial\overline{\partial}\varphi$ for some smooth Kähler potential function φ.

10. The nth power ω^n of the Kähler form ω is the volume form on M, given by:

$$\omega^n = \frac{1}{n!}\mathrm{i}^n \det\left(g_{i\bar{j}}\right) dz^1 \wedge dz^{\bar{1}} \wedge \cdots \wedge dz^n \wedge dz^{\bar{n}},$$
$$V = \int_M \omega^n \quad \text{is the standard volume.}$$

11. The space $\mathcal{K}_{[\omega]}$ of Kähler forms ω with the same Kähler class $[\omega]$ is given by:

$$\mathcal{K}_{[\omega]} = \{[\omega] \in H^2(M,\mathbb{R}) \mid V = 0, \quad \omega + \mathrm{i}\partial\overline{\partial}\varphi > 0\},$$

e.i., the functional space $\mathcal{P}(\omega)$ of Kähler potentials φ on (M,ω) is given by:

$$\mathcal{P}(\omega) = \{\varphi \in C^\infty(M,\mathbb{R}) \mid \omega_\varphi = \omega + \mathrm{i}\partial\overline{\partial}\varphi > 0\}.$$

12. The complex Christoffel symbols corresponding to the Kähler metric g on M are:

$$\Gamma^k_{ij} = g^{k\bar{l}}\partial_j g_{i\bar{l}} \qquad \text{and} \qquad \Gamma^{\bar{k}}_{\bar{i}\bar{j}} = g^{\bar{k}l}\partial_{\bar{j}} g_{l\bar{i}}, \qquad (i,j,k = 1,\cdots,n).$$

13. The corresponding mixed and covariant Riemannian curvature tensors on M are:

$$R^m_{ik\bar{l}} = -\partial_{\bar{l}}\Gamma^m_{ik} \qquad \text{and} \qquad R_{i\bar{j}k\bar{l}} = g_{m\bar{j}}R^m_{ik\bar{l}}, \qquad (i,j,k,l = 1,\cdots,n).$$

14. Locally, in an open chart $U \subset M$, the covariant Riemann tensor reads:

$$R_{i\bar{j}k\bar{l}} = -\partial_i\partial_{\bar{j}}g_{k\bar{l}} + g^{\bar{q}p}(\partial_i g_{k\bar{q}})(\partial_{\bar{j}}g_{p\bar{l}}),$$

and has the following three symmetries:
   a) $\overline{R_{i\bar{j}k\bar{l}}} = R_{j\bar{i}l\bar{k}}$ (complex-conjugate);
   b) $R_{i\bar{j}k\bar{l}} = R_{k\bar{j}i\bar{l}} = R_{i\bar{l}k\bar{j}}$ (I Bianchi identity); and
   c) $\nabla_m R_{i\bar{j}k\bar{l}} = \nabla_i R_{m\bar{j}k\bar{l}}$ (II Bianchi identity), where $\nabla_i \equiv \partial_i + g_{k\bar{q}}g^{\bar{q}j}\Gamma^k_{ij}$ is the complex covariant derivative on (M,g).

15. The Ricci curvature of the Kähler metric g is defined as:

$$R_{i\bar{j}} = g^{\bar{l}k}R_{i\bar{j}k\bar{l}} = g^{\bar{l}k}R_{k\bar{l}i\bar{j}} = R^k_{k\,i\bar{j}}\,.$$

Its trace is the scalar curvature: $R = g^{\bar{j}i}R_{i\bar{j}}$.

16. Locally, in an open chart $U \subset M$, Ricci tensor is given by:

$$R_{i\bar{j}}(g) = -\partial\overline{\partial}\log[\det(g_{i\bar{j}})].$$

17. The associated Ricci form $\mathrm{Ric}(\omega)$ is the closed (1,1)-form on M given by:

$$\mathrm{Ric}(\omega) \equiv \mathrm{Ric}(g) = \mathrm{i}R_{i\bar{j}}(g)\, dz^i \wedge dz^{\bar{j}} = -\mathrm{i}\partial\bar{\partial}\log[\det(g_{i\bar{j}})].$$

18. The first Chern class, denoted by $c_1(M)$, of a Kähler manifold (M, g), is defined as the cohomology class $[\mathrm{Ric}(g)] \in H^{1,1}_{\bar{\partial}}(M, \mathbb{R})$.
19. Metric g on M is called the Kähler–Einstein metric iff :

$$\mathrm{Ric}(\omega) = \lambda\omega, \qquad \text{for a real constant } \lambda = \frac{2\pi}{V}\int_M c_1(M) \wedge \omega^{n-1},$$
$$\text{if} \quad \mathrm{Ric}(g) = 0 \quad \text{then} \quad g \quad \text{is a Ricci–flat metric,}$$

where $c_1(M)$ is the first Chern class of M. If M admits a Ricci–flat metric $[\mathrm{Ric}(g) = 0]$, then its first Chern class must vanish $[c_1(M) = 0]$. This is Calabi conjecture, proven by S.-T. Yau.
20. A compact (i.e., closed and bounded) Kähler manifold (M, g) with positive first Chern class, $c_1(M) > 0$, is called the Fano manifold, in which case, $[\omega] = \pi c_1(M)$. A compact Kähler manifold with vanishing first Chern class, $c_1(M) = 0$, is called the Calabi–Yau manifold. A compact Kähler manifold with negative first Chern class, $c_1(M) < 0$, admits the Kähler–Einstein metric g, defined by:

$$g = -\mathrm{Ric}(g).$$

21. A Fano n-manifold (M, g) admits the (normalized) Kähler–Ricci flow:

$$\partial_t g(t) = g(t) - \mathrm{Ric}\,[g(t)]\,, \tag{2.23}$$

which is locally, in an open chart $U \subset M$, and starting from some smooth initial Kähler metric $g_0 = g_{i\bar{j}}(0)$, given by:

$$\partial_t g_{i\bar{j}}(t) = g_{i\bar{j}}(t) - R_{i\bar{j}}(t), \qquad (i, j = 1, \cdots, n).$$

22. The Kähler–Ricci flow (2.23) preserves the Kähler class $[\omega]$. It has a global solution $g(t) \equiv \omega(t)$ when $g_0 = g_{i\bar{j}}(0)$ has $[\omega] = 2\pi c_1(M)$ as its Kähler class [which is written as $g_0 \in 2\pi c_1(M)$].
23. In particular, by the $\partial\bar{\partial}$-lemma, there exists a family of real-valued functions $u(t)$, called Ricci potentials of the metric $g(t)$, which are special Kähler potentials. They are determined by:

$$g_{i\bar{j}} - R_{i\bar{j}} = \partial_i\partial_{\bar{j}}u, \qquad \frac{1}{V_R}\int_M \mathrm{e}^{-u(t)} dv_g = 1,$$

where $V_R = \int dv_g$ is the volume of the Kähler–Ricci flow (2.23).
24. At the level of Kähler potentials φ, the Kähler–Ricci flow (2.23) becomes:

$$\partial_t\varphi = \varphi + \log\frac{\omega^n_\varphi}{\omega^n} - h_\omega, \tag{2.24}$$

where the Hermitian metric form h_ω is defined by:

$$\mathrm{i}\partial\overline{\partial}h_\omega = \mathrm{Ric}(\omega) - \omega \quad \text{and} \quad \int_M (\mathrm{e}^{h_\omega} - 1)\omega^n = 0.$$

25. The corresponding evolutions of the Ricci curvature $R_{i\bar{j}} = R_{i\bar{j}}(t)$ and the scalar curvature $R = R(t)$ on (M, g) are respectively governed by:

$$\partial_t R_{i\bar{j}} = \Delta R_{i\bar{j}} + R_{i\bar{j}p\bar{q}}R_{q\bar{p}} - R_{i\bar{p}}R_{p\bar{j}}, \qquad \partial_t R = \Delta R + R_{i\bar{j}}R_{j\bar{i}} - R,$$

starting from some smooth initial Ricci and scalar curvatures, $R_{i\bar{j}}(0)$ and $R(0)$.
26. The existence of the Kähler–Ricci flow in a time interval $t \in [0, t_1)$ can be established as follows: If $\omega(t)$ is a solution of the Kähler–Ricci flow:

$$\partial_t \omega = -\mathrm{Ric}(\omega), \qquad \omega(0) = \omega_0, \tag{2.25}$$

then the corresponding cohomology class $[\omega(t)]$ evolves as an ODE:

$$\partial_t[\omega(t)] = -c_1(M), \qquad [\omega(0)] = [\omega_0],$$
$$\text{with the solution}: \ [\omega(t)] = [\omega_0] - t\,c_1(M),$$

so the Kähler–Ricci flow (2.25) exists for $t \in [0, t_1)$ iff $[\omega_0] - t\,c_1(M) > 0$ for $t \in [0, t_1)$.
27. The Kähler-Ricci flow (2.25) can be rewritten as a parabolic complex Monge–Ampère equation:

$$\partial_t \varphi = \log \frac{(\omega_\varphi + \mathrm{i}\partial\overline{\partial}\varphi)^n}{\omega^n}, \qquad [\text{with } \omega_\varphi + \mathrm{i}\partial\overline{\partial}\varphi > 0,\ \varphi(0) = 0].$$

28. The normalized Kähler–Ricci flow (2.23), or:

$$\partial_t \omega = -[\omega + \mathrm{Ric}(\omega)], \qquad \omega(0) = \omega_0,$$

can be rewritten as a normalized parabolic complex Monge–Ampère equation:

$$\partial_t \varphi = \log \frac{(\omega_0 + \mathrm{i}\partial\overline{\partial}\varphi)^n}{\omega^n} - \varphi, \qquad [\text{with } \omega_0 + \mathrm{i}\partial\overline{\partial}\varphi > 0,\ \varphi(0) = 0].$$

29. A Riemannian manifold (M, g) generates a shrinking Ricci soliton iff [TZ13]:

$$\mathrm{Ric}(g) + \mathrm{Hess}(u) = \lambda g, \qquad (\lambda > 0).$$

Similarly, if (M, g) is a Fano manifold with $g \in 2\pi c_1(M)$, it is a shrinking Kähler–Ricci soliton iff $\lambda = 1$ and for any $u \equiv u(t) \in C^\infty(M, \mathbb{R})$ called the Ricci potential; that is, iff: $\nabla\nabla u = 0$, or, applying the Bianchi identity, iff the following Shur identity holds:

$$\Delta u - |\nabla u|^2 + u = a.$$

30. For any Kähler metric $g \in 2\pi c_1(M)$ with scalar curvature R and for any $u \equiv u(t) \in C^\infty(M,\mathbb{R})$, the Kähler–Perelman entropy $\mathcal{E} \equiv \mathcal{E}(g)$ is defined by:

$$\mathcal{E}(g) = \inf\left\{\mathcal{W}(g,u) : \int_M \mathrm{e}^{-u} dv = V\right\}, \qquad \text{where}$$

$$\mathcal{W}(g,u) = \frac{1}{V}\int_M (R + |\nabla u|^2 + u - n)\,\mathrm{e}^{-u} dv, \quad V = \int_M dv.$$

31. A smooth minimizer of the entropy $\mathcal{E}$ always exists (but is not necessarily unique), so that $\mathcal{E}$ admits a natural upper bound:

$$\mathcal{E}(g) \le \frac{1}{V}\int_M u\,\mathrm{e}^{-u} dv = a \le 0.$$

This implies the Kähler–Perelman monotonicity condition on the geometric entropy [TZ13]:

$$(\forall t \ge 0), \qquad \mathcal{E}(g_0) \le \mathcal{E}\left[g(t)\right] \le 0.$$

32. A general positive energy functional is defined (within $V = \int_M \omega^n$) by [CT00b]:

$$J_\omega(\varphi) = \frac{1}{V}\sum_{i=0}^{n-1}\int_M \mathrm{i}\,\partial\varphi \wedge \bar{\partial}\varphi \wedge \omega^i \wedge \omega_\varphi^{n-1-i},$$

which in the simplest case $n = 1$, reduces to the Dirichlet energy:

$$J_\omega(\varphi) = \frac{1}{2V}\int_M \mathrm{i}\partial\varphi \wedge \bar{\partial}\varphi = \frac{1}{2V}\int_M |\partial\varphi|^2\omega.$$

33. The derivative of J_ω along a path $\varphi(t) \in \mathcal{P}(M,\omega)$,

$$\partial_t\, J_\omega(\varphi) = -\frac{1}{V}\int_M \partial_t\varphi\left(\omega_\varphi^n - \omega^n\right),$$

shows that J_ω does not satisfy the cocycle condition: $\omega_\varphi^n = \omega^n$. However, the extended energy functional F_ω, defined by:

$$F_\omega(\varphi) = J_\omega(\varphi) - \frac{1}{V}\int_M \varphi\,\omega^n - \log\left[\frac{1}{V}\int_M \mathrm{e}^{(h_\omega - \varphi)}\right]$$

satisfies the cocycle condition and its critical points are Kähler–Einstein metrics [CT00b]. In the simplest case $n = 1$, we have $M = S^2$ so F_ω becomes:

$$F_\omega(\varphi) = \frac{1}{2V}\int_M |\partial\varphi|^2\omega - \frac{1}{V}\int_{S^2}\varphi\,\omega - \log\left[\frac{1}{V}\int_{S^2} \mathrm{e}^{(h_\omega - \varphi)}\right].$$

34. For further technical details on the Kähler geometry see, e.g. [Yau78, Yau06, Don99, CL08, Mor07, Son12, Tia97, Tia12, Ti13, Nak03] and references therein. For further technical details on the Riemann-Ricci flow, see [Ham82, Ham86, Ham88, Ham93a, Ham93b, Ham99, CC99] and on the Kähler–Ricci flow see [Cao85, CT00, CT00b, Ye07, TZ13] and references therein.

2.8 Appendix

2.8.1 Real Banach and Complex Hilbert Spaces

Here, we outline the concept of Banach and Hilbert spaces,[43] which are vector spaces equipped with a norm that defines a complete metric space.

Recall that a *norm* on a vector space V is defined as a map:

$$\|\cdot\| : V \to \mathbb{R}, \quad \mathbf{v} \mapsto \|\mathbf{v}\|, \tag{2.26}$$

such that the following axioms hold (see, e.g. [CD82, AMR88]):

$$\begin{array}{ll} \textbf{N1}.\ \text{Positive definiteness}: & (\forall\, \mathbf{v} \in V) \quad \|\mathbf{v}\| \geq 0, \quad \|\mathbf{v}\| = 0 \Rightarrow \mathbf{v} = 0 \\ \textbf{N2}.\ \text{Homogeneity}: & (\forall\, \mathbf{v} \in V,\ \lambda \in \mathbb{R}) \quad \|\lambda\mathbf{v}\| = |\lambda|\, \|\mathbf{v}\| \\ \textbf{N3}.\ \text{Triangle inequality}: & (\forall\, \mathbf{v}_1, \mathbf{v}_2 \in V) \quad \|\mathbf{v}_1 + \mathbf{v}_2\| \leq \|\mathbf{v}_1\| + \|\mathbf{v}_2\| \end{array} .$$

In particular, the triangle inequality axiom **N3** has the following important consequence:

$$(\forall\, \mathbf{v}_1, \mathbf{v}_2 \in V) \quad |\, \|\mathbf{v}_1 - \mathbf{v}_2\|\, | \leq \|\mathbf{v}_1 - \mathbf{v}_2\|, \qquad \text{where} \quad |\, \|\mathbf{v}\|\, | \equiv \text{abs}(\|\mathbf{v}\|).$$

A *normed space*, in which there is a length-measure for vectors, is formally defined as a pair $(V, \|\cdot\|)$. A very common example is the Euclidean space $\mathbb{R}^n$ with the standard *Euclidean* L^2 *norm:*[44]

$$\|\mathbf{x}\| = \sqrt{\sum_{i=1}^{n} x_i^2}\,, \quad \text{where } \mathbf{x} = (x_1, ..., x_n), \tag{2.27}$$

in Fortran 2008 ≡ norm2(x) = sqrt(sum(x(1:n)**2)) .

Closely related to norms are metrics (or, distances). Recall that a *metric* on a vector space V is a map:

$$\mathbf{d} : V \times V \to \mathbb{R}, \quad (\mathbf{v}_1, \mathbf{v}_2) \mapsto \mathbf{d},$$

such that the following axioms hold (see, e.g. [CD82, AMR88]):

$$\begin{array}{ll} \textbf{M1}.\ \text{Definiteness}: & (\forall\, \mathbf{v}_1, \mathbf{v}_2 \in V) \quad \mathbf{d}(\mathbf{v}_1, \mathbf{v}_2) = 0 \quad \text{iff} \quad \mathbf{v}_1 = \mathbf{v}_2 \\ \textbf{M2}.\ \text{Symmetry}: & (\forall\, \mathbf{v}_1, \mathbf{v}_2 \in V) \quad \mathbf{d}(\mathbf{v}_1, \mathbf{v}_2) = \mathbf{d}(\mathbf{v}_2, \mathbf{v}_1) \\ \textbf{M3}.\ \text{Triangle inequality}: & (\forall\, \mathbf{v}_1, \mathbf{v}_2, \mathbf{v}_3 \in V) \quad \mathbf{d}(\mathbf{v}_1, \mathbf{v}_3) \leq \mathbf{d}(\mathbf{v}_1, \mathbf{v}_2) + \mathbf{d}(\mathbf{v}_2, \mathbf{v}_3) \end{array} .$$

A *metric space*, in which distances can be measured, is formally defined as a pair $(V, \mathbf{d})$. A very common example is again the Euclidean space $\mathbb{R}^n$ with the standard *Euclidean metric:*

---

[43] We are assuming that the reader has the basic familiarity with the concept of vector spaces.

[44] If the axiom **N1** is relaxed to: $(\forall\, \mathbf{v} \in V) \ \|\mathbf{v}\| \geq 0,$ the map (2.26) is called a *seminorm* and the corresponding space $(V, \|\cdot\|)$ is called a *seminormed space*. An example of a seminorm is the function $f : \mathbb{R}^2 \to \mathbb{R}, \ \|(x, y)\| = |x|$.

$$\mathbf{d}(\mathbf{x},\mathbf{y}) = \sqrt{\sum_{i=1}^{n} (x_i - y_i)^2} \,. \tag{2.28}$$

Let $(V, \mathbf{d})$ be a metric space and $\{u_n\}$ a sequence in V. Then $\{u_n\}$ is called a *Cauchy sequence* if for all real $\varepsilon > 0$, there is an integer N such that: $(n, m \geq N) \Rightarrow \mathbf{d}(un, um) < \varepsilon$. The metric space $(V, \mathbf{d})$ is called *complete* (or, a *Cauchy space*) if every Cauchy sequence in V converges in V. Norm and metric are 'close cousins'. Let $(V, \|\cdot\|)$ be a normed space and define

$$\mathbf{d}(\mathbf{v}_1, \mathbf{v}_2) = \|\mathbf{v}_1 - \mathbf{v}_2\| \,.$$

Then $(V, \mathbf{d})$ represents a metric space corresponding to $(V, \|\cdot\|)$ (for the proof, see [CD82, AMR88]).

Closely related to norms are also inner products. Recall that an *inner product* on a vector space V is a map:

$$\langle \cdot, \cdot \rangle : V \times V \to \mathbb{R}, \quad (\mathbf{v}_1, \mathbf{v}_2) \mapsto \langle \mathbf{v}_1, \mathbf{v}_2 \rangle \,, \tag{2.29}$$

such that the following axioms hold (see, e.g. [CD82, AMR88]):

I1. $(\forall \mathbf{v}, \mathbf{v}_1, \mathbf{v}_2 \in V) \quad \langle \mathbf{v}, \mathbf{v}_1 + \mathbf{v}_2 \rangle = \langle \mathbf{v}, \mathbf{v}_1 \rangle + \langle \mathbf{v}, \mathbf{v}_2 \rangle$
I2. $(\forall \mathbf{v} \in V, \lambda \in \mathbb{R}) \quad \langle \mathbf{v}, \lambda \mathbf{v}_1 \rangle = \lambda \langle \mathbf{v}, \mathbf{v}_1 \rangle$
I3. $(\forall \mathbf{v}_1, \mathbf{v}_2 \in V) \quad \langle \mathbf{v}_1, \mathbf{v}_2 \rangle = \langle \mathbf{v}_2, \mathbf{v}_1 \rangle$
I4. $(\forall \mathbf{v} \in V) \quad \langle \mathbf{v}, \mathbf{v} \rangle \geq 0, \quad \mathbf{v} = 0 \Rightarrow \langle \mathbf{v}, \mathbf{v} \rangle = 0$

An *inner-product space*, in which angles between vectors can be measured (as well as their lengths), is formally defined as a pair $(V, \langle \cdot, \cdot \rangle)$. For example, in $\mathbb{R}^n$, the standard *Euclidean inner product* is:

$$\langle \mathbf{x}, \mathbf{y} \rangle = \sum_{i=1}^{n} x_i y_i \equiv \text{dot\_product(x,y) in F95.} \tag{2.30}$$

In an inner product space $(V, \langle \cdot, \cdot \rangle)$, two vectors $\mathbf{v}_1, \mathbf{v}_2 \in V$ are called *orthogonal* iff $\langle \mathbf{v}_1, \mathbf{v}_2 \rangle = 0$.The most important theorem for inner products is the *Cauchy-Schwartz Inequality*:

$$|\mathbf{v}_1, \mathbf{v}_2| \leq \sqrt{\langle \mathbf{v}_1, \mathbf{v}_2 \rangle} \sqrt{\langle \mathbf{v}_1, \mathbf{v}_2 \rangle} \,,$$

which holds iff the vectors $\mathbf{v}_1, \mathbf{v}_2$ are linearly dependent (for the proof, see [CD82, AMR88]).

Norm and inner product are another 'close cousins'. Let $(V, \langle \cdot, \cdot \rangle)$ be an inner product space and

$$(\forall \mathbf{v} \in V) \quad \|\mathbf{v}\| = \sqrt{\langle \mathbf{v}, \mathbf{v} \rangle}.$$

Then $(V, \|\cdot\|)$ is a normed space corresponding to $(V, \langle \cdot, \cdot \rangle)$ (for the proof, see [CD82, AMR88]).

If $(V, \langle\cdot,\cdot\rangle)$ is an inner-product space whose corresponding metric is complete, then $(V, \langle\cdot,\cdot\rangle)$ is called a *Hilbert space*. The most common example is $\mathbb{R}^n$, which is a Hilbert space $\mathcal{H}$ with the standard inner product (2.30) (see, e.g. [CD82, AMR88]).

For example, in common `LM-optimization` problems in $\mathbb{R}^n$ (including general data fitting, output tracking control and function approximation in multilayer neural networks), either L^2-norm (2.27) or metric **d** (2.28) is effectively minimized.

In particular, in quantum mechanics, the essential property of any complex Hilbert space $\mathcal{H}(\mathbb{C})$ is the *Hermitian inner* dot-product $\langle\psi,\phi\rangle$. In physics, it is usually denoted by Dirac's celebrated *bra-ket* $\langle\psi|\phi\rangle$, or Feynman's *probability amplitude*$\langle\mathrm{Out}|\mathrm{In}\rangle$, or statistical-mechanical *partition function* Z. Thus, all these description are more-or-less equivalent: $\langle\psi,\phi\rangle \equiv \langle\psi|\phi\rangle \equiv \langle\mathrm{Out}|\mathrm{In}\rangle \equiv Z$. This inner product can be applied to any pair of $\mathcal{H}$-vectors, $|\psi\rangle$ and $|\phi\rangle$, to produce a single complex number $\langle\psi,\phi\rangle \equiv \langle\psi|\phi\rangle \in \mathbb{C}$, which satisfies a number of simple algebraic properties (compare to the axioms **I1-I4** above):

$$\begin{aligned}
\langle\psi+\phi,\varphi\rangle &= \langle\psi,\varphi\rangle + \langle\phi,\varphi\rangle \\
\langle\psi,\phi+\varphi\rangle &= \langle\psi,\phi\rangle + \langle\psi,\varphi\rangle \\
\langle a\,\psi,\phi\rangle &= a\,\langle\psi,\phi\rangle\,, \qquad \langle\psi,a\,\phi\rangle = \bar{a}\,\langle\psi,\phi\rangle \\
\langle\psi,\phi\rangle &= \overline{\langle\phi,\psi\rangle} \qquad \text{(where bar denotes complex-conjugate)} \\
\langle\psi,\psi\rangle &\geq 0, \quad \langle\psi,\psi\rangle = 0 \text{ only if } \psi = 0 \text{ (positive definite)} \\
\langle\psi,\phi\rangle &= \sum_i \psi_i\bar{\phi}_i \in \mathbb{C}^n, \qquad (\text{for } i = 1,...,n).
\end{aligned}$$

Probability of finding a quantum system at a certain ket-state $|\psi\rangle$ is given by its *probability density function* (PDF), which is defined as the absolute square of the wave ψ-function:

$$PDF := |\psi|^2 \equiv \left||\psi\rangle\right|^2 \equiv \langle\psi|\psi\rangle \equiv \langle\psi,\psi\rangle,$$

which is simply the squared length of the ket-vector $|\psi\rangle$ in $\mathcal{H}(\mathbb{C})$ and a normalized state is given by a $\mathcal{H}$-vector $|1\rangle$ (whose squared length is unity). More generally, the PDF-description of any microscopic process, or even any (quantum-influenced) macroscopic process, is the In$\rightarrow$Out map determined by the absolute square of the probability amplitude:

$$PDF : \mathrm{In} \mapsto \mathrm{Out} := \left|\langle\mathrm{Out}|\mathrm{In}\rangle\right|^2 \equiv |\psi|^2.$$

2.8.2 From Linear to Nonlinear MIMO Control Theory

Kalman's Linear State-Space Formalism

It is a well-known fact (implemented in $Matlab^{TM}$) that linear multiple-input multiple-output (MIMO) control systems can always be put into Kalman's canonical (modular) state-space form of order n, with m inputs and k outputs (see [KFA69]). In case of continuous-time systems we have the state and output equation of the form

$$
\begin{aligned}
d\mathbf{x}/dt &= \mathbf{A}(t)\,\mathbf{x}(t) + \mathbf{B}(t)\,\mathbf{u}(t), \\
\mathbf{y}(t) &= \mathbf{C}(t)\,\mathbf{x}(t) + \mathbf{D}(t)\,\mathbf{u}(t),
\end{aligned}
\tag{2.31}
$$

while in case of discrete time systems we have the state and output equation of the form

$$
\begin{aligned}
\mathbf{x}(n+1) &= \mathbf{A}(n)\,\mathbf{x}(n) + \mathbf{B}(n)\,\mathbf{u}(n), \\
\mathbf{y}(n) &= \mathbf{C}(n)\,\mathbf{x}(n) + \mathbf{D}(n)\,\mathbf{u}(n).
\end{aligned}
\tag{2.32}
$$

Both in (2.31) and in (2.32) the variables have the following meaning:

$\mathbf{x}(t) \in \mathbb{X}$ is an $n-$vector of state variables belonging to the *vector state-space* $\mathbb{X} \subset \mathbb{R}^n$;

$\mathbf{u}(t) \in \mathbb{U}$ is an $m-$vector of inputs belonging to the *vector input space* $\mathbb{U} \subset \mathbb{R}^m$;

$\mathbf{y}(t) \in \mathbb{Y}$ is a $k-$vector of outputs belonging to the *vector output space* $\mathbb{Y} \subset \mathbb{R}^k$;

$\mathbf{A}(t) : \mathbb{X} \to \mathbb{X}$ is an $n \times n$ matrix of state dynamics;

$\mathbf{B}(t) : \mathbb{U} \to \mathbb{X}$ is an $n \times m$ matrix of input map;

$\mathbf{C}(t) : \mathbb{X} \to \mathbb{Y}$ is a $k \times n$ matrix of output map;

$\mathbf{D}(t) : \mathbb{U} \to \mathbb{Y}$ is a $k \times m$ matrix of input-output transform.

Input $\mathbf{u}(t) \in \mathbb{U}$ can be empirically determined by trial and error; it is properly defined by quadratic optimization process called *Kalman regulator*, or more generally (in the presence of noise), by (extended) *Kalman filter* [Kal60].

Basics of Kalman Filtering

For a nice introduction to Kalman filtering, see [WB95]. Here, we give only a brief overview.

Recall that the *Kalman linear–quadratic regulator*, widely used in state–space control theory, represents a *linear state feedback control law*

$$u = -Kx,$$

for the linear MIMO system:

$$\dot{x} = Ax + Bu,$$

which minimizes a *quadratic cost function*

$$J = \int_0^\infty \left(x(t)^T Q\, x(t) + u(t)^T R\, u(t)\right) dt.$$

The control law is called *optimal with respect to the cost function* J.

Now, one might ask whether there is an optimal design technique for a *state estimator*. That is, is there an approach to *observer design* which is equivalent, in some sense, to the linear quadratic regulator?

Given the *observable system*

$$\dot{x} = Ax + Bu, \qquad y = Cx,$$

one may define the *dual system*

$$\dot{\theta} = A^T\theta + C^T\gamma,$$

and design an LQR controller to minimize the quadratic cost function

$$J = \int_0^\infty \left(\theta(t)^T Q\,\theta(t) + \gamma(t)^T R\,\gamma(t)\right) dt.$$

However, it is unclear how one should 'penalize' θ and γ in the cost function. Instead, consider the *extended observable system*

$$\dot{x} = Ax + Bu + w, \qquad y = Cx + v,$$

in which the dynamics are subject to random disturbances w and the measurements are subject to random noise v. In parallel with the development of the linear quadratic regulator, Rudolph Kalman examined the following *optimal estimator problem*: Construct a full state observer which minimizes the combined effect of the disturbances and the noise, thus providing a 'most likely' estimate of the system state. Solving this problem requires some information about the random processes. If the processes are zero–mean, Gaussian white noise processes, then the optimal estimator design problem becomes perfectly analogous to the LQR control design problem. In 1960, Kalman published his famous paper describing a recursive solution to the discrete–data linear filtering problem [Kal60]. Since that time, due in large part to advances in digital computing, the Kalman filter has been the subject of extensive particularly in the area of autonomous or assisted navigation (see, e.g., [Hay01, GWA01]).

The Kalman filter is a *discrete–time, two–step process*, the steps of which are usually called *predictor* and *corrector*, thus resembling a popular *Adams–Bashforth–Moulton integrator* for ODEs. The *predictor*, or *time update*, projects the current system's state estimate ahead in time. The *corrector*, or *measurement update*, adjusts the projected state estimate by an actual system's measurement at that time. In this way, the correction step makes corrections to an estimate, based on new information obtained from sensor measurements . The continuous–time version is usually referred to as *Kalman–Bucy filter* or *smoother* [SL03].

Consider a generic linear, discrete–time dynamical system. The concept of *discrete state* is fundamental to this description. The *state vector*, denoted by x_k, is defined as the minimal set of data that is sufficient to uniquely describe the unforced dynamical behavior of the system; the subscript k denotes discrete time. In other words, the state is the least amount of data on the past behavior of the system that is needed to predict its future behavior. Typically, the state x_k is unknown. To estimate it, we use a set of observed data, denoted by the *observable* vector z_k.

The state–space model of a generic linear, discrete–time dynamical system includes the *process equation* (2.33) and the *measurement equation* (2.34)

$$x_{k+1} = F_{k+1,k}x_k + w_k, \tag{2.33}$$
$$z_k = H_k x_k + v_k, \tag{2.34}$$

where $F_{k+1,k}$ is the *transition matrix* taking the state x_k from time k to time $k+1$, H_k is the *measurement sensitivity matrix*, while w_k and v_k are independent, additive, zero–mean, white *Gaussian noise* processes, defined below.

The covariance matrix of the *process noise* w_k is defined by:

$$E[w_n, w_k^T] = \begin{cases} Q_k, & \text{for } n = k, \\ 0, & \text{for } n = k. \end{cases}$$

Similarly, the covariance matrix of the *measurement noise* v_k is defined by

$$E[v_n, v_k^T] = \begin{cases} R_k, & \text{for } n = k, \\ 0, & \text{for } n = k. \end{cases}$$

The *Kalman filtering problem*, namely, the problem of jointly solving the process and measurement equations for the unknown state in an optimum manner may now be formally stated as follows: Use the entire observed data, consisting of the vectors $z_1, z_2, ..., z_k$, to find for each $k \geq 1$ the minimum mean–square error estimate of the state x_i. The problem is called *filtering* if $i = k$, *prediction* if $i > k$, and *smoothing* if $1 \leq i < k$.

The derivation of the Kalman filter is based on the following two theorems (see [Kal60, Hay01]):

- *Conditional mean estimator.* If the stochastic processes $\{x_k\}$ and $\{z_k\}$ are jointly Gaussian, then the optimum estimate $\hat{x}_k$ that minimizes the mean–square error J_k is the conditional mean estimator:

$$\hat{x}_k = E[x_k | z_1, z_2, ..., z_k].$$

- *Principle of orthogonality.* Let the stochastic processes $\{x_k\}$ and $\{z_k\}$ be of zero means; that is,

$$E[x_k] = E[z_k] = 0, \qquad \text{for all } k.$$

Then:
(i) the stochastic process $\{x_k\}$ and $\{z_k\}$ are jointly Gaussian; or

(ii) if the optimal estimate $\hat{x}_k$ is restricted to be a linear function of the observables and the cost function is the mean–square error,
(iii) then the optimum estimate $\hat{x}_k$, given the observables $z_1, z_2, ..., z_k$, is the orthogonal projection of x_k on the space spanned by these observables.

The *Kalman filter design algorithm* consists of (see [Kal60, Hay01]):

1. *Initialization*: For $k = 0$, set

$$\hat{x}_0 = E[x_0], \qquad P_0 = E[(x_0 - E[x_0])(x_0 - E[x_0])^T].$$

and

2. *Computation*: For $k = 1, 2, ...,$ compute:
(i) *State estimate propagation*

$$\hat{x}_{\bar{k}} = F_{k,k-1}\, \hat{x}_{\bar{k}-1};$$

(ii) *Error covariance propagation*

$$P_{\bar{k}} = F_{k,k-1}\, P_{k-1} F_{k,k-1}^T + Q_{k-1};$$

(iii) *Kalman gain matrix*

$$K_k = P_{\bar{k}} H_k^T [H_k P_{\bar{k}} H_k^T + R_k]^{-1};$$

(iv) *State estimate update*

$$\hat{x}_k = \hat{x}_{\bar{k}} + K_k(z_k - H_k \hat{x}_{\bar{k}});$$

(v) *Error covariance update*

$$P_k = (I - K_k H_k) P_{\bar{k}}.$$

Therefore, the basic Kalman filter is a linear, discrete–time, finite–dimensional system, which is endowed with a recursive structure that makes a digital computer well suited for its implementation. A key property of the Kalman filter is that it is the minimum mean–square (variance) estimator of the state of a linear dynamical system. The model is stochastic owing to the additive presence of process noise and measurement noise, which are assumed to be Gaussian with zero mean and known covariance matrices.

Extended (Nonlinear) Kalman Filter

The Kalman filtering problem considered so far has addressed the estimation of a state vector in a linear model of a dynamical system. If, however, the model is nonlinear, we may extend the use of Kalman filtering through a linearization procedure. The resulting filter is referred to as the *extended Kalman filter* (EKF) (see, e.g., [Hay01]). Such an extension is feasible by virtue of the fact that the Kalman filter is described in terms of difference equations in the case of discrete–time systems. While the ordinary (i.e., linear) Kalman filter is defined in terms of the measurement sensitivity matrix H_k, the extended Kalman filter can be defined in terms of a suitably differentiable vector–valued measurement sensitivity function $h(k, x_k)$.

To set the stage for a development of the extended Kalman filter, consider a nonlinear dynamical system described by the state–space model

$$x_{k+1} = f(k, x_k) + w_k, \qquad z_k = h(k, x_k) + v_k, \tag{2.35}$$

where, as before, w_k and v_k are independent zero–mean white Gaussian noise processes with covariance matrices R_k and Q_k, respectively. Here, however, the functional $f(k, x_k)$ denotes a nonlinear transition matrix function that is possibly time–variant. Likewise, the functional $h(k, x_k)$ denotes a vector–valued measurement sensitivity function, i.e., a nonlinear measurement matrix that may be time–variant, too [Hay01].

The basic idea of the extended Kalman filter is to linearize the state–space model (2.35) at each time instant around the most recent state estimate, which is taken to be either $\hat{x}_k$ or $\hat{x}_{\bar{k}}$, depending on which particular functional is being considered. Once a linear model is obtained, the standard Kalman filter equations are applied.

The *EKF design algorithm* consists of [Hay01]:

1. The *discrete state–space model* (2.35).
2. *Definitions*

$$F_{k,k} = \left.\frac{\partial f(k, x)}{\partial x}\right|_{x=x_k}, \qquad H_k = \left.\frac{\partial h(k, x)}{\partial x}\right|_{x=x_{\bar{k}}}.$$

3. *Initialization*: For $k = 0$, set

$$\hat{x}_0 = E[x_0], \qquad P_0 = E[(x_0 - E[x_0])(x_0 - E[x_0])^T].$$

4. *Computation*: For $k = 1, 2, ...,$ compute:
   (i) *State estimate propagation*

$$\hat{x}_{\bar{k}} = F_{k,k-1}\, \hat{x}_{\bar{k}-1};$$

   (ii) *Error covariance propagation*

$$P_{\bar{k}} = F_{k,k-1}\, P_{k-1} F^T_{k,k-1} + Q_{k-1};$$

   (iii) *Kalman gain matrix*

$$K_k = P_{\bar{k}} H_k^T [H_k P_{\bar{k}} H_k^T + R_k]^{-1};$$

   (iv) *State estimate update*

$$\hat{x}_k = \hat{x}_{\bar{k}} + K_k (z_k - H_k \hat{x}_{\bar{k}});$$

   (v) *Error covariance update*

$$P_k = (I - K_k H_k) P_{\bar{k}}.$$

Ensemble Kalman Filter and Nonlinear Estimation

The so-called *Ensemble Kalman Filter* (EnKF, firstly introduced by [Eve94] and later discussed in detail by [EL96, HM98]) is a Monte-Carlo (MC) implementation of the following *Bayesian update problem* [Man09]: Given a PDF of the state of the modeled system (the *prior*; in geophysics, it is usually called the *forecast*) and the data likelihood, the *Bayes theorem* is used to to obtain PDF after the data likelihood has been taken into account (the *posterior*). In other words, the EnKF gives such an MC-approach to *Kalman filtering* that estimates the covariances between observed variables and the model state variables through an ensemble of predictive model forecasts. EnKF is a recursive filter suitable for problems with a large number of variables, such as discretizations of PDEs (e.g. in geophysical models).

Here, following [Eve03, GRF12], we briefly outline how the EnKF can be used for *parameter estimation* of a nonlinear dynamical system. Let $\mathbf{p} \in \mathbb{R}^{\ell}$ be a vector holding the different model parameters, and $\mathbf{x}^f \in \mathbb{R}^n$ be the model state forecast. Let $\left(\mathbf{p}_i, \mathbf{x}_i^f\right)$ for $i = 1 \ldots N$ be an ensemble of model parameters and state forecasts, and $\mathbf{y}^o \in \mathbb{R}^m$ a vector of m observations, then the estimated parameter values $\mathbf{p}_i^a$ given by the EnKF equations are

$$\mathbf{p}_i^a = \mathbf{p}_i + \tilde{\mathbf{K}}\left(\mathbf{y}_i^o - \mathbf{H}\mathbf{x}_i^f\right), \qquad (i = 1, \ldots, N) \tag{2.36}$$

$$\tilde{\mathbf{K}} = \mathbf{C}^T\mathbf{H}^T\left(\mathbf{H}\mathbf{P}^f\mathbf{H}^T + \mathbf{R}\right)^{-1}, \tag{2.37}$$

where the matrix $\tilde{\mathbf{K}} \in \mathbb{R}^{\ell \times m}$ is a modified *Kalman gain matrix*, $\mathbf{P}^f \in \mathbb{R}^{n \times n}$ is the model forecast covariance matrix, $\mathbf{C} \in \mathbb{R}^{n \times \ell}$ is the cross-correlation matrix between the model forecast and parameters, $\mathbf{R} \in \mathbb{R}^{m \times m}$ is the observations covariance matrix, and $\mathbf{H} \in \mathbb{R}^{m \times n}$ is an observation operator matrix that maps state variables onto observations. In the EnKF, the vector $\mathbf{y}_i^o$ is a *perturbed observation* vector defined as

$$\mathbf{y}_i^o = \mathbf{y}^o + \varepsilon_i, \tag{2.38}$$

where $\varepsilon_i \in \mathbb{R}^m$ is a random vector sampled from a normal distribution with zero mean and a specified standard deviation σ. Usually σ is taken as the variance or error in the observations.

One of the main advantages of the EnKF is that the model forecast covariance matrix is approximated using the ensemble of model forecasts,

$$\mathbf{P}^f \approx \frac{1}{N-1}\sum_{i=1}^{N}\left(\mathbf{x}_i^f - \bar{\mathbf{x}}^f\right)\left(\mathbf{x}_i^f - \bar{\mathbf{x}}^f\right)^T, \tag{2.39}$$

where $\bar{\mathbf{x}}^f \in \mathbb{R}^n$ is the model forecast ensemble average. The use of an ensemble of model forecast to approximate $\mathbf{P}^f$ enables the evolution of this matrix for large nonlinear models at a reasonable computational cost. Additionally, the cross-correlation matrix $\mathbf{C}$ is defined as

$$\mathbf{C} = \frac{1}{N-1} \sum_{i=1}^{N} \left(\mathbf{x}_i^f - \bar{\mathbf{x}}^f \right) \left(\mathbf{p}_i - \bar{\mathbf{p}} \right)^T, \tag{2.40}$$

where $\bar{\mathbf{p}} \in \mathbb{R}^\ell$ is the parameter ensemble average.

The procedure used to estimate the parameters is the following: Let $t_1, \ldots, t_k$ be the time instances where observations are available. For each time instance t_j, $j = 1, \ldots, k$, the EnKF data assimilation provides parameter estimates for the ensemble, $\mathbf{p}_i^a(t_j)$, $i = 1 \ldots N$. A final parameter estimate is then computed by first taking the ensemble average and then the time average of the parameters:

$$\mathbf{p}^a = \frac{1}{k} \sum_{j=1}^{k} \left[\frac{1}{N} \sum_{i=1}^{N} \mathbf{p}_i^a(t_j) \right] \tag{2.41}$$

This approach avoids the problems of *parameter collapse* and *filter divergence*, since the data assimilation is used to estimate the parameters at each time instance independently. Additionally, since the state is not being updated in the assimilation, and only the parameters are being estimated, localization is not required for the EnKF. For more technical details, see [Eve03, GRF12].

Affine Nonlinear Control System

Now me move to modern geometric nonlinear control theory. A nonlinear extension of Kalman's linear state-space control theory is a nonlinear control theory, comprising a set of geometrical (manifold) techniques developed under the name of *affine nonlinear MIMO-systems*, which are of the general form (see, e.g. [Isi89, NS90, Lew95, LM97]):

$$\dot{x}(t) = f_0(x(t)) + u^i(t) f_i(x(t)), \qquad (i = 1, ..., m) \tag{2.42}$$

where $t \mapsto x(t)$ is a curve in a system's *state manifold* M. The vector-field f_0 is called the *drift vector-field*, describing the dynamics of the system in the absence of controls, and the vector-fields $f_1, ..., f_m$ are the *input vector-fields* or *control vector-fields*, indicating how we are able to actuate the system.

Lie Derivative and Lie Bracket in Control Theory

Given a scalar function $h(x)$ and a vector-field $f(x)$ on some configuration nD-manifold M (see below), we can define a new scalar function, $\mathcal{L}_f h = \nabla h f$, which is the *Lie derivative* of h w.r.t. f, i.e., the directional derivative of h along the direction of the vector f. Repeated Lie derivatives can be defined recursively:

$$\mathcal{L}_f^0 h = h, \qquad \mathcal{L}_f^i h = \mathcal{L}_f \left(\mathcal{L}_f^{i-1} h \right) = \nabla \left(\mathcal{L}_f^{i-1} h \right) f, \qquad (\text{for } i = 1, 2, ...)$$

Or, given another vector-field, g, then $\mathcal{L}_g\mathcal{L}_f h(x)$ is defined as

$$\mathcal{L}_g\mathcal{L}_f h = \nabla(\mathcal{L}_f h)\, g.$$

For example, if we have a control system

$$\dot{x} = f(x), \qquad y = h(x),$$

with the state $x = x(t)$ and the output y, then the derivatives of the output are:

$$\dot{y} = \frac{\partial h}{\partial x}\dot{x} = \mathcal{L}_f h, \qquad \text{and} \qquad \ddot{y} = \frac{\partial L_f h}{\partial x}\dot{x} = \mathcal{L}_f^2 h.$$

In particular, the Lie derivative of one vector-field with respect to another vector-field is called the Lie bracket. Given two vector-fields, $f(x)$ and $g(x)$, their Lie bracket is defined by:

$$[f, g] = Ad_f g = \nabla g f - \nabla f g = \frac{\partial g}{\partial x} f - \frac{\partial f}{\partial x} g,$$

where $\nabla f = \partial f / \partial x$ is the Jacobian matrix. We can define Lie brackets recursively,

$$Ad_f^0 g = g, \qquad Ad_f^i g = [f, Ad_f^{i-1} g], \qquad (\text{for } \; i = 1, 2, ...)$$

Lie brackets have the properties of bilinearity, skew-commutativity and Jacobi identity, so they form a Lie algebra (from a given space of vector-fields on a manifold M).

Example: Car-Parking Using Lie Brackets

In this popular example, the driver has two different transformations at his disposal. He/she can turn the steering wheel, or he/she can drive the car forward or back. Here, we specify the state of a car by four coordinates: the (x, y) coordinates of the center of the rear axle, the direction θ of the car, and the angle ϕ between the front wheels and the direction of the car. L is the constant length of the car. Therefore, the configuration manifold of the car is 4D, $M = (x, y, \theta, \phi)$.

Using (2.42), the driftless car kinematics can be defined as:

$$\dot{x} = g_1(x)\, u_1 + g_2(x)\, u_2, \tag{2.43}$$

with two vector-fields $g_1, g_2 \in \mathcal{X}^k(M)$.

The infinitesimal transformations will be the vector-fields

$$g_1(x) \equiv \text{DRIVE} = \cos\theta\frac{\partial}{\partial x} + \sin\theta\frac{\partial}{\partial y} + \frac{\tan\phi}{L}\frac{\partial}{\partial\theta} \equiv \begin{pmatrix} \cos\theta \\ \sin\theta \\ \frac{1}{L}\tan\phi \\ 0 \end{pmatrix},$$

$$\text{and} \qquad g_2(x) \equiv \text{STEER} = \frac{\partial}{\partial\phi} \equiv \begin{pmatrix} 0 \\ 0 \\ 0 \\ 1 \end{pmatrix}.$$

Now, STEER and DRIVE do not commute; otherwise we could do all your steering at home before driving of on a trip. Therefore, we have a Lie bracket

$$[g_2, g_1] \equiv [\text{STEER}, \text{DRIVE}] = \frac{1}{L\cos^2\phi}\frac{\partial}{\partial\theta} \equiv \text{ROTATE}.$$

The operation $[g_2, g_1] \equiv$ ROTATE $\equiv$ [STEER,DRIVE] is the infinitesimal version of the sequence of transformations: steer, drive, steer back, and drive back, i.e.,

$$\{\text{STEER}, \text{DRIVE}, \text{STEER}^{-1}, \text{DRIVE}^{-1}\}.$$

Now, ROTATE can get us out of some parking spaces, but not tight ones: we may not have enough room to ROTATE out. The usual tight parking space restricts the DRIVE transformation, but not STEER. A truly tight parking space restricts STEER as well by putting your front wheels against the curb.

Fortunately, there is still another commutator available:

$$[g_1, [g_2, g_1]] \equiv [\text{DRIVE}, [\text{STEER}, \text{DRIVE}]] = [[g_1, g_2], g_1] \equiv$$
$$[\text{DRIVE}, \text{ROTATE}] = \frac{1}{L\cos^2\phi}\left(\sin\theta\frac{\partial}{\partial x} - \cos\theta\frac{\partial}{\partial y}\right) \equiv \text{SLIDE}.$$

The operation $[[g_1, g_2], g_1] \equiv$ SLIDE $\equiv$ [DRIVE,ROTATE] is a displacement at right angles to the car, and can get us out of any parking place. We just need to remember to steer, drive, steer back, drive some more, steer, drive back, steer back, and drive back:

$$\{\text{STEER}, \text{DRIVE}, \text{STEER}^{-1}, \text{DRIVE}, \text{STEER}, \text{DRIVE}^{-1}, \text{STEER}^{-1}, \text{DRIVE}^{-1}\}.$$

We have to reverse steer in the middle of the parking place. This is not intuitive, and no doubt is part of the problem with parallel parking.

Thus from only two controls u_1 and u_2 we can form the vector-fields DRIVE $\equiv g_1$, STEER $\equiv g_2$, ROTATE $\equiv [g_2, g_1]$, and SLIDE $\equiv [[g_1, g_2], g_1]$, allowing us to move anywhere in the configuration manifold M. The car kinematics $\dot{x} = g_1 u_1 + g_2 u_2$ is thus expanded as:

$$\begin{pmatrix} \dot{x} \\ \dot{y} \\ \dot{\theta} \\ \dot{\phi} \end{pmatrix} = \text{DRIVE} \cdot u_1 + \text{STEER} \cdot u_2 \equiv \begin{pmatrix} \cos\theta \\ \sin\theta \\ \frac{1}{L}\tan\phi \\ 0 \end{pmatrix} \cdot u_1 + \begin{pmatrix} 0 \\ 0 \\ 0 \\ 1 \end{pmatrix} \cdot u_2 .$$

The *parking theorem* says: One can get out from any parking lot that is larger than the car. For more technical details, see [Isi89, NS90, Lew95, LM97].

2.8.3 Basic Complexity Geometrodynamics

Basic Tensor Machinery

To introduce tensors, consider a standard linear nD matrix system, $\mathbf{A}\mathbf{x} = \mathbf{b}$. It can be rewritten in the so-called *covariant form* (using the summation convention upon repeated indices) as

$$a_{ij}x^j = b_i\,, \qquad (i,j=1,...,n). \tag{2.44}$$

Here, i is a *free index* and j is a *dummy index* to be summed upon, so the expansion of (2.44) gives

$$\begin{aligned} a_{11}x^1 + a_{12}x^2 + ... + a_{1n}x^n &= b_1\,, \\ a_{21}x^1 + a_{22}x^2 + ... + a_{2n}x^n &= b_2\,, \\ &... \\ a_{n1}x^1 + a_{n2}x^2 + ... + a_{nn}x^n &= b_n\,, \end{aligned}$$

as expected from the original matrix form $\mathbf{Ax} = \mathbf{b}$. This indicial notation can be more useful than the matrix one, like e.g., in computer science, where indices would represent loop variables. However, the full potential of tensor analysis is to deal with nonlinear multivariate systems, which are untractable by linear matrix algebra and analysis. The core of this *nonlinear multivariate analysis* is *general functional transformation.*

Transformation of Coordinates

Suppose that we have two sets of curvilinear coordinates that are single-valued, continuous and smooth functions of time, $x^j = x^j(t)$, $(j=1,...,m)$ and $\bar{x}^i = \bar{x}^i(t)$, $(i=1,...,n)$, respectively, representing trajectories of motion of some mechanical system. Then a general $(m \times n)$D transformation (i.e., a nonlinear map) $x^j \mapsto \bar{x}^i$ is defined by the set of transformation equations

$$\bar{x}^i = \bar{x}^i(x^j), \qquad (i=1,...,n;\ j=1,...,m). \tag{2.45}$$

In case of the square transformation, $m = n$, we can freely exchange the indices, like e.g., in general relativity theory. On the other hand, in the general case of rectangular transformation, $m \neq n$, like e.g., in robotics, and we need to take care of these 'free' indices.

Now, if the *Jacobian determinant* of this *coordinate transformation* is different from zero,

$$\left|\frac{\partial \bar{x}^i}{\partial x^j}\right| \neq 0,$$

then the transformation (2.45) is reversible and the inverse transformation,

$$x^j = x^j(\bar{x}^i),$$

exists as well. Finding the inverse transformation is the problem of matrix inverse: in case of the square matrix it is well defined, although the inverse might not exist if the matrix is singular. However, in case of the square matrix, its proper inverse does not exist, and the only tool that we are left with is the so-called *Moore-Penrose pseudoinverse*, which gives an optimal solution (in the least-squares sense) of an overdetermined system of equations.

Every (overdetermined) rectangular coordinate transformation gives rise to a *redundant system.*

For example, in Euclidean 3D space $\mathbb{R}^3$, transformation from Cartesian coordinates $y^k = \{x, y, z\}$ into spherical coordinates $x^i = \{\rho, \theta, \varphi\}$ is given by

$$y^1 = x^1 \cos x^2 \cos x^3, \qquad y^2 = x^1 \sin x^2 \cos x^3, \qquad y^3 = x^1 \sin x^3, \tag{2.46}$$

with the Jacobian matrix given by

$$\left(\frac{\partial y^k}{\partial x^i}\right) = \begin{pmatrix} \cos x^2 \cos x^3 & -x^1 \sin x^2 \cos x^3 & -x^1 \cos x^2 \sin x^3 \\ \sin x^2 \cos x^3 & x^1 \cos x^2 \cos x^3 & -x^1 \sin x^2 \sin x^3 \\ \sin x^3 & 0 & x^1 \cos x^3 \end{pmatrix} \tag{2.47}$$

and the corresponding Jacobian determinant, $\left|\frac{\partial y^k}{\partial x^i}\right| = (x^1)^2 \cos x^3$.

An inverse transform is given by

$$x^1 = \sqrt{(y^1)^2 + (y^2)^2 + (y^3)^2}, \qquad x^2 = \arctan\left(\frac{y^2}{y^1}\right),$$

$$x^3 = \arctan\left(\frac{y^3}{\sqrt{(y^1)^2 + (y^2)^2}}\right), \qquad \text{with} \qquad \left|\frac{\partial x^i}{\partial y^k}\right| = \frac{1}{(x^1)^2 \cos x^3}.$$

As a main mechanical example, we have a rectangular transformation from 6 DOF external, end-effector (e.g., hand) coordinates, into n DOF internal, joint-angle coordinates. In most cases this is a redundant manipulator system, with infinite number of possible joint trajectories.

Scalars, Vectors and Covectors

A *scalar invariant* (or, a zeroth order tensor) with respect to the transformation (2.45) is the quantity $\varphi = \varphi(t)$ defined as

$$\varphi(x^i) = \bar{\varphi}(\bar{x}^i),$$

which does not change at all under the coordinate transformation. In other words, φ is *invariant* under (2.45). Biodynamic examples of scalar invariants include various energies (kinetic, potential, biochemical, mental) with the corresponding kinds of work, as well as related thermodynamic quantities (free energy, temperature, entropy, etc.).

Any geometric object $v^i = v^i(t)$ that under the coordinate transformation (2.45) transforms as

$$\bar{v}^i = v^j \frac{\partial \bar{x}^i}{\partial x^j}, \qquad \text{(remember, summing upon } j\text{–index)},$$

represents a *vector*, traditionally called a *contravariant vector*, or, a first-order contravariant tensor. Standard mechanical examples include both translational and rotational velocities and accelerations.

On the other hand, any geometric object $v_i = v_i(t)$ that under the coordinate transformation (2.45) transforms as

$$\bar{v}_i = v_j \frac{\partial x^j}{\partial \bar{x}^i},$$

represents a *one-form* or *covector*, traditionally called a *covariant vector*, or, a first order covariant tensor. Standard mechanical examples include both translational and rotational momenta, forces and torques.

Second-Order Tensors

Any geometric object $t^{ik} = t^{ik}(t)$ that under the coordinate transformation (2.45) transforms as

$$\bar{t}^{ik} = t^{jl} \frac{\partial \bar{x}^i}{\partial x^j} \frac{\partial \bar{x}^k}{\partial x^l}, \qquad (i, k = 1, ..., n; \; j, l = 1, ..., m),$$

represents a *second-order contravariant tensor*. It can be obtained as an *outer product* of two contravariant vectors, $t^{ik} = u^i v^k$.

Any geometric object $t_{ik} = t_{ik}(t)$ that under the coordinate transformation (2.45) transforms as

$$\bar{t}_{ik} = t_{jl} \frac{\partial x^j}{\partial \bar{x}^i} \frac{\partial x^l}{\partial \bar{x}^k},$$

represents a *second-order covariant tensor*. It can be obtained as an outer product of two covariant vectors, $t_{ik} = u_i v_k$.

Any geometric object $t^i_k = t^i_k(t)$ that under the coordinate transformation (2.45) transforms as

$$\bar{t}^i_k = t^j_l \frac{\partial \bar{x}^i}{\partial x^j} \frac{\partial x^l}{\partial \bar{x}^k},$$

represents a *second-order mixed tensor*. It can be obtained as an outer product of a covariant vector and a contravariant vector, $t^i_k = u^i v_k$.

Standard mechanical examples include:

1. The fundamental (material) covariant metric tensor $\mathbf{g} \equiv g_{ik}$, i.e., inertia matrix, given usually by the transformation from Cartesian coordinates y^j to curvilinear coordinates x^i,

$$g_{ik} = \frac{\partial y^j}{\partial x^i} \frac{\partial y^j}{\partial x^k}, \qquad \text{(summing over } j).$$

   It is used in the quadratic metric form ds^2 of the space in consideration (e.g., a certain mechanical configuration space)

$$ds^2 \equiv dy^j dy^j = g_{ik} dx^i dx^k,$$

   where the first term on the r.h.s denotes the *Euclidean metrics*, while the second term is the *Riemannian metric* of the space, respectively.

2. Its inverse $\mathbf{g}^{-1} \equiv g^{ik}$, given by

$$g^{ik} = (g_{ik})^{-1} = \frac{G_{ik}}{|g_{ik}|}, \qquad G_{ik} \text{ is the cofactor of the matrix } (g_{ik});$$

3. The *Kronecker-delta* symbol δ^i_k, given by

$$\delta^i_k = \begin{cases} 1 \text{ if } i = k \\ 0 \text{ if } i \neq k \end{cases},$$

used to denote the metric tensor in Cartesian orthogonal coordinates. δ^i_k is a discrete version of the *Dirac* δ*−function.* The *generalized Kronecker-delta symbol* δ^{ijk}_{lmn} (in 3D) is the product of *Ricci antisymmetric tensors* ε^{ijk} and ε_{lmn},

$$\delta^{ijk}_{lmn} = \varepsilon^{ijk}\varepsilon_{lmn} = \begin{cases} 0 & \text{if at least two indices are equal} \\ +1 & \text{if both } ijk \text{ and } lmn \text{ are either even or odd} \\ -1 & \text{if one of } ijk,\ lmn \text{ is even and the other is odd} \end{cases}.$$

For example, to derive components of the metric tensor $\mathbf{g} \equiv g_{ij}$ in standard spherical coordinates, we use the relations (2.46-2.47) between the spherical coordinates $x^i = \{\rho, \theta, \varphi\}$ and the Cartesian coordinates $y^k = \{x, y, z\}$, and the definition, $g_{ij} = \frac{\partial y^k}{\partial x^i}\frac{\partial y^k}{\partial x^j}$, to get the metric tensor (in matrix form)

$$(g_{ij}) = \begin{pmatrix} 1 & 0 & 0 \\ 0 & (x^1)^2\cos^2 x^3 & 0 \\ 0 & 0 & (x^1)^2 \end{pmatrix} = \begin{pmatrix} 1 & 0 & 0 \\ 0 & \rho^2\cos^2\varphi & 0 \\ 0 & 0 & \rho^2 \end{pmatrix}, \tag{2.48}$$

and the inverse metric tensor

$$(g^{ij}) = \begin{pmatrix} 1 & 0 & 0 \\ 0 & \frac{1}{(x^1)^2\cos^2 x^3} & 0 \\ 0 & 0 & \frac{1}{(x^1)^2} \end{pmatrix} = \begin{pmatrix} 1 & 0 & 0 \\ 0 & \frac{1}{\rho^2\cos^2\varphi} & 0 \\ 0 & 0 & \frac{1}{\rho^2} \end{pmatrix}. \tag{2.49}$$

Given a tensor, we can derive other tensors by raising and lowering its indices, by their multiplication with covariant and contravariant metric tensors. In this way, the so-called *associated tensors* to the given tensor are be formed. For example, v^i and v_i are associated tensors, related by

$$v_i = g_{ik}v^k \qquad \text{and} \qquad v^i = g^{ik}v_k.$$

Given two vectors, $\mathbf{u} \equiv u^i$ and $\mathbf{v} \equiv v^i$, their inner (dot, or scalar) product is given by

$$\mathbf{u} \cdot \mathbf{v} \equiv g_{ij}u^i v^j,$$

while their vector (cross) product (in 3D) is given by

$$\mathbf{u} \times \mathbf{v} \equiv \varepsilon_{ijk}u^j v^k.$$

Higher-Order Tensors

As a generalization of above tensors, consider a geometric object $R^i_{kps} = R^i_{kps}(t)$ that under the coordinate transformation (2.45) transforms as

$$\bar{R}^i_{kps} = R^j_{lqt} \frac{\partial \bar{x}^i}{\partial x^j} \frac{\partial x^l}{\partial \bar{x}^k} \frac{\partial x^q}{\partial \bar{x}^p} \frac{\partial x^t}{\partial \bar{x}^s}, \qquad \text{(all indices } = 1, ..., n). \tag{2.50}$$

Clearly, $R^i_{kjl} = R^i_{kjl}(x,t)$ is a fourth order tensor, once contravariant and three times covariant, representing the central tensor in Riemannian geometry, called the *Riemann curvature tensor*. As all mechanical configuration spaces are Riemannian manifolds, they are all characterized by curvature tensors. In case $R^i_{kjl} = 0$, the corresponding Riemannian manifold reduces to the Euclidean space of the same dimension, in which $g_{ik} = \delta^i_k$.

If one contravariant and one covariant index of a tensor a set equal, the resulting sum is a tensor of rank two less than that of the original tensor. This process is called *tensor contraction*.

If to each point of a region in an nD space there corresponds a definite tensor, we say that a *tensor-field* has been defined. In particular, this is a *vector-field* or a *scalar-field* according as the tensor is of rank one or zero. It should be noted that a tensor or tensor field is not just the set of its components in one special coordinate system, but all the possible sets of components under any transformation of coordinates.

Tensor Symmetry

A tensor is called *symmetric* with respect to two indices of the same variance if its components remain unaltered upon interchange of the indices; e.g., $a_{ij} = a_{ji}$, or $a^{ij} = a^{ji}$. A tensor is called *skew-symmetric* (or, *antisymmetric*) with respect to two indices of the same variance if its components change sign upon interchange of the indices; e.g., $a_{ij} = -a_{ji}$, or $a^{ij} = -a^{ji}$. Regarding tensor symmetry, in the following we will prove several useful propositions.

(i) *Every second-order tensor can be expressed as the sum of two tensors, one of which is symmetric and the other is skew-symmetric.* For example, a second order tensor a_{ij}, which is for $i,j = 1, ..., n$ given by the $n \times n-$matrix

$$a_{ij} = \begin{pmatrix} a_{11} & a_{12} & ... & a_{1n} \\ a_{21} & a_{22} & ... & a_{n2} \\ ... & ... & ... & ... \\ a_{n1} & a_{n2} & ... & a_{nn} \end{pmatrix},$$

can be rewritten as

$$
\begin{aligned}
a_{ij} &= \frac{1}{2}a_{ij} + \frac{1}{2}a_{ij} + \frac{1}{2}a_{ji} - \frac{1}{2}a_{ji}\,, \quad \text{that can be rearranged as} \\
&= \frac{1}{2}a_{ij} + \frac{1}{2}a_{ji} + \frac{1}{2}a_{ij} - \frac{1}{2}a_{ji}\,, \quad \text{which can be regrouped as} \\
&= \frac{1}{2}(a_{ij} + a_{ji}) + \frac{1}{2}(a_{ij} - a_{ji}), \quad \text{which can be written as} \\
&= a_{(ij)} + a_{[ij]}\,,
\end{aligned}
$$

where $a_{(ij)}$ denotes its symmetric part, while $a_{[ij]}$ denotes its skew-symmetric part, as required.

(ii) *Every quadratic form can be made symmetric.* For example, a quadratic form $a_{ij}x^i x^j$, that (for $i, j = 1, ..., n$) expands as

$$
\begin{aligned}
a_{ij}x^i x^j &= a_{11}x^1 x^1 + a_{12}x^1 x^2 + ... + a_{1n}x^1 x^n + \\
&+ a_{21}x^2 x^1 + a_{22}x^2 x^2 + ... + a_{2n}x^2 x^n + \\
&\cdots \\
&+ a_{n1}x^n x^1 + a_{n2}x^n x^2 + ... + a_{nn}x^n x^n,
\end{aligned}
$$

with a non-symmetric second order tensor a_{ij}, can be made symmetric in the following way.

$$
\begin{aligned}
a_{ij}x^i x^j &= \frac{1}{2}a_{ij}x^i x^j + \frac{1}{2}a_{ij}x^i x^j. \\
&\text{If we swap indices in the second term, we get} \\
&= \frac{1}{2}a_{ij}x^i x^j + \frac{1}{2}a_{ji}x^j x^i\,, \quad \text{which is equal to} \\
&= \frac{1}{2}(a_{ij} + a_{ji})\,x^i x^j. \\
&\text{If we now use a substitution,} \\
\frac{1}{2}(a_{ij} + a_{ji}) &\equiv b_{ij} = b_{ji}, \quad \text{we get} \\
a_{ij}x^i x^j &= b_{ij}x^i x^j,
\end{aligned}
$$

where a_{ij} is non-symmetric and b_{ij} is symmetric, as required.

(iii) *Every second order tensor that is the sum* $a^{ij} = u^i v^j + u^j v^i$, or, $a_{ij} = u_i v_j + u_j v_i$ *is symmetric.* In both cases, if we swap the indices i and j, we get $a^{ji} = u^j v^i + u^i v^j$, (resp. $a_{ji} = u_j v_i + u_i v_j$), which implies that the tensor a^{ij} (resp. a_{ij}) is symmetric.

(iv) *Every second order tensor that is the difference* $b^{ij} = u^i v^j - u^j v^i$, or, $b_{ij} = u_i v_j - u_j v_i$ is *skew-symmetric.* In both cases, if we swap the indices i and j, we get $b^{ji} = -(u^j v^i - u^i v^j)$, (resp. $b_{ji} = -(u_j v_i - u_i v_j)$), which implies that the tensor b^{ij} (resp. b_{ij}) is skew-symmetric.

Tensor Derivatives on Riemannian Manifolds

Consider now some nD Riemannian manifold M (see below) with the metric form (i.e., line element) $ds^2 = g_{ik}dx^i dx^k$, as a configuration space for a certain physical system.

Christoffel's Symbols

Partial derivatives of the metric tensor g_{ik} form themselves special symbols that do not transform as tensors (with respect to the coordinate transformation (2.45)), but nevertheless represent important quantities in tensor analysis. They are called *Christoffel symbols of the first kind*, defined by

$$\Gamma_{ijk} = \frac{1}{2}(\partial_{x^i} g_{jk} + \partial_{x^j} g_{ki} + \partial_{x^k} g_{ij}), \qquad \left(\text{remember, } \partial_{x^i} \equiv \frac{\partial}{\partial x^i}\right)$$

and *Christoffel symbols of the second kind*, defined by

$$\Gamma^k_{ij} = g^{kl}\Gamma_{ijl}.$$

The Riemann curvature tensor R^l_{ijk} (2.50) of the manifold M, can be expressed in terms of the later as

$$R^l_{ijk} = \partial_{x^j}\Gamma^l_{ik} - \partial_{x^k}\Gamma^l_{ij} + \Gamma^l_{rj}\Gamma^r_{ik} - \Gamma^l_{rk}\Gamma^r_{ij}.$$

For example, in 3D spherical coordinates, $x^i = \{\rho, \theta, \varphi\}$, with the metric tensor and its inverse given by (2.48, 2.49), it can be shown that the only nonzero Christoffel's symbols are:

$$\begin{aligned} &\Gamma^2_{12} = \Gamma^2_{21} = \Gamma^3_{13} = \Gamma^3_{31} = \frac{1}{\rho}, \qquad \Gamma^3_{23} = \Gamma^2_{32} = -\tan\theta, \\ &\Gamma^1_{22} = -\rho, \qquad \Gamma^1_{33} = -\rho\cos^2\theta, \qquad \Gamma^2_{33} = \sin\theta\cos\theta. \end{aligned} \tag{2.51}$$

Geodesics

From the Riemannian metric form $ds^2 = g_{ik}dx^i dx^k$ it follows that the distance between two points t_1 and t_2 on a curve $x^i = x^i(t)$ in M is given by

$$s = \int_{t_1}^{t_2} \sqrt{g_{ik}\dot{x}^i\dot{x}^k}\,dt.$$

That curve $x^i = x^i(t)$ in M which makes the distance s a minimum is called a *geodesic* of the space M (e.g., in a sphere, the geodesics are arcs of great circles). Using the calculus of variations, the geodesics are found from the differential *geodesic equation*,

$$\ddot{x}^i + \Gamma^i_{jk}\dot{x}^j\dot{x}^k = 0, \tag{2.52}$$

where overdot means derivative upon the line parameter s.

For example, in 3D spherical coordinates $x^i = \{\rho, \theta, \varphi\}$, using (2.51), geodesic equation (2.52) becomes a system of three scalar ODEs,

$$\ddot{\rho} - \rho\dot{\theta}^2 - \rho\cos^2\theta\dot{\varphi}^2 = 0, \qquad \ddot{\theta} + \frac{2}{\rho}\dot{\rho}\dot{\varphi} + \sin\theta\cos\theta\dot{\varphi}^2 = 0,$$
$$\ddot{\varphi} + \frac{2}{\rho}\dot{\rho}\dot{\varphi} - 2\tan\theta\dot{\theta}\dot{\varphi} = 0. \tag{2.53}$$

The Covariant Derivative

Ordinary total and partial derivatives of vectors (covectors) *do not transform as vectors* (covectors) with respect to the coordinate transformation (2.45). For example, let y^k be Cartesian coordinates and x^i be general curvilinear coordinates of a dynamical system (with $i, k = 1, ..., n$). We have: $x^i(t) = x^i[y^k(t)]$, which implies that

$$\frac{dx^i}{dt} = \frac{\partial x^i}{\partial y^k}\frac{dy^k}{dt}, \qquad \text{or equivalently,} \qquad \dot{x}^i = \frac{\partial x^i}{\partial y^k}\dot{y}^k,$$

that is a transformation law for the contravariant vector, which means that the velocity $v^i \equiv \dot{x}^i \equiv \frac{dx^i}{dt}$ *is* a proper contravariant vector. However, if we perform another time differentiation, we get

$$\frac{d^2x^i}{dt^2} = \frac{\partial x^i}{\partial y^k}\frac{d^2y^k}{dt^2} + \frac{\partial^2 x^i}{\partial y^k \partial y^m}\frac{dy^k}{dt}\frac{dy^m}{dt},$$

which means that $\frac{d^2x^i}{dt^2}$ *is not* a proper vector.

$\frac{d^2x^i}{dt^2}$ is an acceleration vector only in a special case when x^i are another Cartesian coordinates; then $\frac{\partial^2 x^i}{\partial y^k \partial y^m} = 0$, and therefore the original coordinate transformation is linear, $x^i = a^i_k y^k + b^i$ (where a^i_k and b^i are constant).

Therefore, $\frac{d^2x^i}{dt^2}$ represents an acceleration vector only in terms of Newtonian mechanics in a Euclidean space $\mathbb{R}^n$, while it is not a proper acceleration vector in terms of Lagrangian or Hamiltonian mechanics in general curvilinear coordinates on a smooth manifold M^n. And we know that Newtonian mechanics in $\mathbb{R}^n$ is sufficient only for fairly simple mechanical systems.

The above is true for any tensors. So we need to find another derivative operator to be able to preserve their tensor character. The solution to this problem is called the *covariant derivative*.

The covariant derivative $v^i_{;k}$ of a contravariant vector v^i is defined as

$$v^i_{;k} = \partial_{x^k} v^i + \Gamma^i_{jk} v^j.$$

Similarly, the covariant derivative $v_{i;k}$ of a covariant vector v_i is defined as

$$v_{i;k} = \partial_{x^k} v_i - \Gamma^j_{ik} v_j.$$

Generalization for the higher order tensors is straightforward; e.g., the covariant derivative $t^j_{kl;q}$ of the third order tensor t^j_{kl} is given by

$$t^j_{kl;q} = \partial_{x^q} t^j_{kl} + \Gamma^j_{qs} t^s_{kl} - \Gamma^s_{kq} t^j_{sl} - \Gamma^s_{lq} t^j_{ks}.$$

The covariant derivative is the most important tensor operator in general relativity (its zero defines *parallel transport*) as well as the basis for defining other differential operators in mechanics and physics.

Covariant Form of Gradient, Divergence, Curl and Laplacian

Gradient. If $\varphi = \varphi(x^i, t)$ is a scalar field, the gradient one-form $\mathrm{grad}(\varphi)$ is defined by

$$\mathrm{grad}(\varphi) = \nabla\varphi = \varphi_{;i} = \partial_{x^i}\varphi.$$

Divergence. The divergence $\mathrm{div}(v^i)$ of a vector-field $v^i = v^i(x^i, t)$ is defined by contraction of its covariant derivative with respect to the coordinates $x^i = x^i(t)$, i.e., the contraction of $v^i_{;k}$, namely

$$\mathrm{div}(v^i) = v^i_{;i} = \frac{1}{\sqrt{g}}\partial_{x^i}(\sqrt{g}v^i).$$

Curl. The curl $\mathrm{curl}(\theta_i)$ of a one-form $\theta_i = \theta_i(x^i, t)$ is a second order covariant tensor defined as

$$\mathrm{curl}(\theta_i) = \theta_{i;k} - \theta_{k;i} = \partial_{x^k}\theta_i - \partial_{x^i}\theta_k.$$

Laplacian. The Laplacian $\Delta\varphi$ of a scalar invariant $\varphi = \varphi(x^i, t)$ is the divergence of $\mathrm{grad}(\varphi)$, or

$$\Delta\varphi = \nabla^2\varphi = \mathrm{div}(\mathrm{grad}(\varphi)) = \mathrm{div}(\varphi_{;i}) = \frac{1}{\sqrt{g}}\partial_{x^i}(\sqrt{g}g^{ik}\partial_{x^k}\varphi).$$

The Absolute Derivative and 3D Curve Geometry

The *absolute derivative* (or *intrinsic*, or *Bianchi's derivative*) of a contravariant vector v^i along a curve $x^k = x^k(t)$ is denoted by $\dot{\bar{v}}^i \equiv Dv^i/dt$ and defined as the inner product of the covariant derivative of v^i and $\dot{x}^k \equiv dx^k/dt$, i.e., $v^i_{;k}\dot{x}^k$, and is given by

$$\dot{\bar{v}}^i = \dot{v}^i + \Gamma^i_{jk} v^j \dot{x}^k.$$

Similarly, the absolute derivative $\dot{\bar{v}}_i$ of a covariant vector v_i is defined as

$$\dot{\bar{v}}_i = \dot{v}_i - \Gamma^j_{ik} v_j \dot{x}^k.$$

Generalization for the higher order tensors is straightforward; e.g., the absolute derivative $\dot{\bar{t}}^j_{kl}$ of the third order tensor t^j_{kl} is given by

$$\dot{\bar{t}}^{j}_{kl} = \dot{t}^{j}_{kl} + \Gamma^{j}_{qs} t^{s}_{kl} \dot{x}^{q} - \Gamma^{s}_{kq} t^{j}_{sl} \dot{x}^{q} - \Gamma^{s}_{lq} t^{j}_{ks} \dot{x}^{q}.$$

The absolute derivative is the most important operator in mechanics, as it is the basis for the *covariant form* of both Lagrangian and Hamiltonian equations of motion of many mechanical systems.

Now that we have defined the absolute derivative, given three unit vectors: *tangent* τ^i, *principal normal* β^i, and *binormal* ν^i, as well as two scalar invariants: curvature K and torsion T, of a curve $\gamma(s) = \gamma[x^i(s)]$, the so-called *Frenet-Serret formulae* are valid[45]

$$\begin{aligned}
\dot{\bar{\tau}}^i &\equiv \dot{\tau}^i + \Gamma^i_{jk} \tau^j \dot{x}^k = K\beta^i, \\
\dot{\bar{\beta}}^i &\equiv \dot{\beta}^i + \Gamma^i_{jk} \beta^j \dot{x}^k = -(K\tau^i + T\nu^i), \\
\dot{\bar{\nu}}^i &\equiv \dot{\nu}^i + \Gamma^i_{jk} \nu^j \dot{x}^k = T\beta^i.
\end{aligned}$$

Covariant state equation

Consider again Kalman's linear state-equation (for an arbitrary MIMO-system):

$$\begin{aligned}
\dot{\mathbf{x}} &= \mathbf{A}\,\mathbf{x} + \mathbf{B}\,\mathbf{u}, \\
\mathbf{y} &= \mathbf{C}\,\mathbf{x} + \mathbf{D}\,\mathbf{u},
\end{aligned} \tag{2.54}$$

with the state n-vector $\mathbf{x} = \mathbf{x}(t) \in \mathbb{X} \subset \mathbb{R}^n$, input $m-$vector $\mathbf{u} = \mathbf{u}(t) \in \mathbb{U} \subset \mathbb{R}^m$, output $k-$vector $\mathbf{y} = \mathbf{y}(t) \in \mathbb{Y} \subset \mathbb{R}^k$, state $n \times n$ matrix $\mathbf{A} = \mathbf{A}(t) : \mathbb{X} \to \mathbb{X}$, input $n \times m$ matrix $\mathbf{B} = \mathbf{B}(t) : \mathbb{U} \to \mathbb{X}$, output $k \times n$ matrix $\mathbf{C} = \mathbf{C}(t) : \mathbb{X} \to \mathbb{Y}$ and input-output $k \times m$ matrix $\mathbf{D} = \mathbf{D}(t) : \mathbb{U} \to \mathbb{Y}$.

Notwithstanding both the inherent 'beauty' and the practical usefulness of the linear state equation (2.54) (which is the basis of *Matlab*^{TM} Control and Signal toolboxes), we might still realize that in real life nothing is linear, so the linear state-equation (2.54) can only be the first approximation to some more realistic nonlinear MIMO-system. Technically speaking, we can generalize (or, 'lift-up') the linear model (2.54) that lives in nD linear Euclidean space $\mathbb{R}^n$, into the tensor (or, 'covariant') equation that lives on nD nonlinear Riemannian manifold M with the metric tensor g_{ij} and the metric form:

$$ds^2 = g_{ij}\, dx^i dx^j, \qquad (i, j = 1, ..., n),$$

where the summation convention (summing over repeated indices) is in place.

Using this 'nonlinear lift', from the linear state equation (2.54) we obtain the following covariant state equation:

$$\begin{aligned}
\dot{\bar{x}}^i &= a^i_j\, x^j + b^i_s\, u^s, \quad (s = 1, ..., m) \\
y^t &= c^t_j\, x^j + d^t_s\, u^s, \quad (t = 1, ..., k)
\end{aligned} \tag{2.55}$$

---

[45] Here overdot denotes the total derivative with respect to the line parameter s (instead of time t).

where $\dot{\bar{x}}^i = \dot{\bar{x}}^i(t)$ is the absolute (covariant) time-derivative, which includes Christoffel's symbols Γ^i_{jk} (i.e., the Levi-Civita connection associated to the metric form ds^2) of the
Riemannian manifold M:

$$\dot{\bar{x}}^i := \dot{x}^i + \Gamma^i_{jk} x^j x^k.$$

The covariant state equation (2.55), which can be simulated in Mathematica® (or, *Maple*®), in case of the flat connection ($\Gamma^i_{jk} = 0$), reduces to the Kalman equation (2.54). Also, some more-nonlinear connection can be used instead of the Levi-Civita connection Γ^i_{jk}.

Basic Exterior Machinery

From Green's to Stokes' theorem

Now recall (from advanced calculus) that *Green's theorem* in the region C in the $(x, y)-$plane $\mathbb{R}^2$ connects a line integral $\oint_{\partial C}$ (over the boundary ∂C of a closed line C) with a double integral $\iint_C$ (over the line C; see e.g., [MT03])

$$\oint_{\partial C} Pdx + Qdy = \iint_C \left(\frac{\partial Q}{\partial x} - \frac{\partial P}{\partial y} \right) dxdy.$$

In other words, if we define two differential forms (integrands of $\oint_{\partial C}$ and $\iint_C$) as

$$\begin{aligned} 1-\text{form} : \mathbf{A} &= Pdx + Qdy, \qquad \text{and} \\ 2-\text{form} : \mathbf{dA} &= \left(\frac{\partial Q}{\partial x} - \frac{\partial P}{\partial y} \right) dxdy, \end{aligned}$$

(where $\mathbf{d}$ denotes the *exterior derivative* that makes a $(p+1)-$form out of a $p-$form, see next subsection), then we can rewrite Green's theorem as *Stokes' theorem*:

$$\int_{\partial C} \mathbf{A} = \int_C \mathbf{dA}.$$

The integration domain C is in topology called a *chain*, and ∂C is a one-dimensional (1D) boundary of a 2D chain C. In general, we have the celebrated BBZ law: the *boundary of a boundary is zero* (see [MTW73, CW95]), that is, $\partial(\partial C) = 0$, or formally $\partial^2 = 0$.

Exterior derivative

The exterior derivative $\mathbf{d}$ is a generalization of ordinary vector differential operators (*grad*, *div* and *curl* see [Rha84, Fla63]) that transforms $p-$forms $\boldsymbol{\omega}$ into $(p+1)-$forms $\mathbf{d}\boldsymbol{\omega}$ (see next subsection), with the main property:
$\mathbf{dd} = \mathbf{d}^2 = 0$, so that in $\mathbb{R}^3$ we have:

- any scalar function $f = f(x, y, z)$ is a 0–form;
- the gradient $\mathbf{d}f = \boldsymbol{\omega}$ of any smooth function f is a 1–form

$$\boldsymbol{\omega} = \mathbf{d}f = \frac{\partial f}{\partial x}dx + \frac{\partial f}{\partial y}dy + \frac{\partial f}{\partial z}dz;$$

- the curl $\boldsymbol{\alpha} = \mathbf{d}\boldsymbol{\omega}$ of any smooth 1–form $\boldsymbol{\omega}$ is a 2–form

$$\boldsymbol{\alpha} = \mathbf{d}\boldsymbol{\omega} = \left(\frac{\partial R}{\partial y} - \frac{\partial Q}{\partial z}\right) dydz + \left(\frac{\partial P}{\partial z} - \frac{\partial R}{\partial x}\right) dzdx + \left(\frac{\partial Q}{\partial x} - \frac{\partial P}{\partial y}\right) dxdy;$$

if $\boldsymbol{\omega} = \mathbf{d}f \Rightarrow \boldsymbol{\alpha} = \mathbf{dd}f = 0$.

- the divergence $\boldsymbol{\beta} = \mathbf{d}\boldsymbol{\alpha}$ of any smooth 2–form $\boldsymbol{\alpha}$ is a 3–form

$$\boldsymbol{\beta} = \mathbf{d}\boldsymbol{\alpha} = \left(\frac{\partial A}{\partial x} + \frac{\partial B}{\partial y} + \frac{\partial C}{\partial z}\right) dxdydz; \qquad \text{if } \boldsymbol{\alpha} = \mathbf{d}\boldsymbol{\omega} \Rightarrow \boldsymbol{\beta} = \mathbf{dd}\boldsymbol{\omega} = 0.$$

In general, for any two smooth functions $f = f(x, y, z)$ and $g = g(x, y, z)$, the exterior derivative $\mathbf{d}$ obeys the *Leibniz rule* [II06b, II07]:

$$\mathbf{d}(fg) = g\,\mathbf{d}f + f\,\mathbf{d}g,$$

and the *chain rule*:

$$\mathbf{d}\,(g(f)) = g'(f)\,\mathbf{d}f.$$

Exterior forms

In general, given a so–called 4D *coframe*, that is a set of coordinate differentials $\{dx^i\} \in \mathbb{R}^4$, we can define the space of all p–forms, denoted $\Omega^p(\mathbb{R}^4)$, using the exterior derivative $\mathbf{d} : \Omega^p(\mathbb{R}^4) \to \Omega^{p+1}(\mathbb{R}^4)$ and Einstein's summation convention over repeated indices (e.g., $A_i\,dx^i = \sum_{i=0}^{3} A_i\,dx^i$), we have:

1-form – a generalization of the Green's 1–form $Pdx + Qdy$,

$$\mathbf{A} = A_i\,dx^i \in \Omega^1(\mathbb{R}^4).$$

For example, in 4D electrodynamics, $\mathbf{A}$ represents electromagnetic (co) vector potential.

2-form – generalizing the Green's 2–form $(\partial_x Q - \partial_y P)\,dxdy$ (with $\partial_j = \partial/\partial x^j$),

$$\begin{aligned} \mathbf{B} &= \mathbf{dA} \in \Omega^2(\mathbb{R}^4), && \text{with components} \\ \mathbf{B} &= \frac{1}{2}B_{ij}\,dx^i \wedge dx^j, && \text{or} \\ \mathbf{B} &= \partial_j A_i\,dx^j \wedge dx^i, && \text{so that} \\ B_{ij} &= -2\partial_j A_i = \partial_i A_j - \partial_j A_i = -B_{ji}. \end{aligned}$$

where $\wedge$ is the anticommutative exterior (or, 'wedge') product of two differential forms; given a p–form $\boldsymbol{\alpha} \in \Omega^p(\mathbb{R}^4)$ and a q–form $\boldsymbol{\beta} \in \Omega^q(\mathbb{R}^4)$, their exterior product is a $(p+q)-$form $\boldsymbol{\alpha}\wedge\boldsymbol{\beta}\in\Omega^{p+q}(\mathbb{R}^4)$; e.g., if we have two 1−forms $\mathbf{a}=a_i dx^i$, and $\mathbf{b}=b_i dx^i$, their wedge product $\mathbf{a}\wedge\mathbf{b}$ is a 2−form $\boldsymbol{\alpha}$ given by

$$\boldsymbol{\alpha}=\mathbf{a}\wedge\mathbf{b}=a_i b_j\, dx^i\wedge dx^j=-a_i b_j\, dx^j\wedge dx^i=-\mathbf{b}\wedge\mathbf{a}.$$

The exterior product $\wedge$ is related to the exterior derivative $\mathbf{d}=\partial_i dx^i$, by

$$\mathbf{d}(\boldsymbol{\alpha}\wedge\boldsymbol{\beta})=\mathbf{d}\boldsymbol{\alpha}\wedge\boldsymbol{\beta}+(-1)^p\boldsymbol{\alpha}\wedge\mathbf{d}\boldsymbol{\beta}.$$

3-form

$$\begin{aligned}
\mathbf{C}&=\mathbf{dB}\ (=\mathbf{ddA}\equiv 0)\in\Omega^3(\mathbb{R}^4), &&\text{with components}\\
\mathbf{C}&=\frac{1}{3!}C_{ijk}\, dx^i\wedge dx^j\wedge dx^k, &&\text{or}\\
\mathbf{C}&=\partial_k B_{[ij]}\, dx^k\wedge dx^i\wedge dx^j, &&\text{so that}\\
C_{ijk}&=-6\partial_k B_{[ij]}, &&\text{where } B_{[ij]} \text{ is the skew−symmetric part of } B_{ij}.
\end{aligned}$$

For example, in the 4D electrodynamics, $\mathbf{B}$ represents the field 2−form *Faraday*, or the Liénard−Wiechert 2−form (in the next section we will use the standard symbol $\mathbf{F}$ instead of $\mathbf{B}$) satisfying the sourceless magnetic Maxwell's equation,

$$\text{Bianchi identity}:\quad \mathbf{dB}=0,\quad \text{in components}\quad \partial_k B_{[ij]}=0.$$

4-form

$$\begin{aligned}
\mathbf{D}&=\mathbf{dC}\ (=\mathbf{ddB}\equiv 0)\in\Omega^4(\mathbb{R}^4)\ \ (\text{if } \mathbf{B}\neq\mathbf{dA}), &&\text{with components}\\
\mathbf{D}&=\partial_l C_{[ijk]}\, dx^l\wedge dx^i\wedge dx^j\wedge dx^k, &&\text{or}\\
\mathbf{D}&=\frac{1}{4!}D_{ijkl}\, dx^i\wedge dx^j\wedge dx^k\wedge dx^l, &&\text{so that}\\
D_{ijkl}&=-24\partial_l C_{[ijk]}.
\end{aligned}$$

Stokes theorem

Generalization of the Green's theorem in the plane (and all other integral theorems from vector calculus) is the Stokes theorem for the $p-$form $\boldsymbol{\omega}$, in an oriented nD domain C (which is a $p-$chain with a $(p-1)-$boundary ∂C, see next section)

$$\int_{\partial C}\boldsymbol{\omega}=\int_C \mathbf{d}\boldsymbol{\omega}.$$

For example, in the 4D Euclidean space $\mathbb{R}^4$ we have the following three particular cases of the Stokes theorem, related to the subspaces C of $\mathbb{R}^4$:
The 2D Stokes theorem:

$$\int_{\partial C^2}\mathbf{A}=\int_{C^2}\mathbf{B}.$$

The 3D Stokes theorem:

$$\int_{\partial C^3} \mathbf{B} = \int_{C^3} \mathbf{C}.$$

The 4D Stokes theorem:

$$\int_{\partial C^4} \mathbf{C} = \int_{C^4} \mathbf{D}.$$

Basic Manifold Machinery

Now, we drop boldface symbols, as we move into more advanced geometrical machinery. First, recall that a smooth n-manifold is a curved nD space which is locally equivalent to $\mathbb{R}^n$. To sketch its formal definition, consider a set M (see Figure 2.2) which is a *candidate* for a manifold. Any point $x \in M$ has its *Euclidean chart*, given by a 1-1 and *onto* map $\varphi_i : M \to \mathbb{R}^n$, with its *Euclidean image* $V_i = \varphi_i(U_i)$. Formally, a chart φ_i is defined by

$$\varphi_i : M \supset U_i \ni x \mapsto \varphi_i(x) \in V_i \subset \mathbb{R}^n,$$

where $U_i \subset M$ and $V_i \subset \mathbb{R}^n$ are open sets.

Any point $x \in M$ can have several different charts (see Figure 2.2). Consider a case of two charts, $\varphi_i, \varphi_j : M \to \mathbb{R}^n$, having in their images two open sets, $V_{ij} = \varphi_i(U_i \cap U_j)$ and $V_{ji} = \varphi_j(U_i \cap U_j)$. Then we have *transition functions* φ_{ij} between them,

$$\varphi_{ij} = \varphi_j \circ \varphi_i^{-1} : V_{ij} \to V_{ji}, \qquad \text{locally given by} \qquad \varphi_{ij}(x) = \varphi_j(\varphi_i^{-1}(x)).$$

If transition functions φ_{ij} exist, then we say that two charts, φ_i and φ_j are *compatible*. Transition functions represent a general (nonlinear) *transformations of coordinates*, which are the core of classical *tensor calculus*.

A set of compatible charts $\varphi_i : M \to \mathbb{R}^n$, such that each point $x \in M$ has its Euclidean image in at least one chart, is called an *atlas*. Two atlases are *equivalent* iff all their charts are compatible (i.e., transition functions exist between them), so their union is also an atlas. A *manifold structure* is a class of equivalent atlases.

Finally, as charts $\varphi_i : M \to \mathbb{R}^n$ were supposed to be 1-1 and onto maps, they can be either *homeomorphisms*, in which case we have a *topological* (C^0) manifold, or *diffeomorphisms*, in which case we have a *smooth* (C^k) manifold.

Recall that the velocity phase-space TM of some configuration nm-manifold M has the Riemannian geometry with the *local metric form*:

$$\langle g \rangle \equiv ds^2 = g_{ij} dx^i dx^j, \tag{2.56}$$

where $g_{ij}(x)$ is the material metric tensor defined by the configuraton system's *mass-inertia matrix* and dx^i are differentials of the local joint coordinates x^i on M. Besides giving the local distances between the points on the manifold M, the Riemannian metric form $\langle g \rangle$ defines the system's kinetic energy:

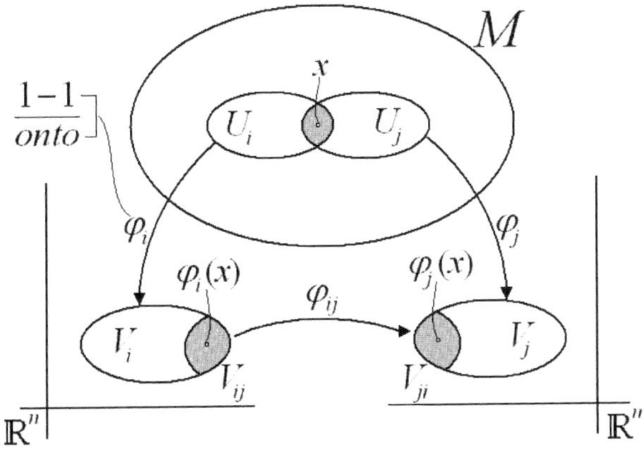

Fig. 2.2. *Depicting the manifold concept.*

$$T = \frac{1}{2} g_{ij} \dot{x}^i \dot{x}^j,$$

giving the *Lagrangian equations* of the conservative skeleton motion with kinetic-minus-potential energy Lagrangian $L = T - V$, with the corresponding *geodesic form*

$$\frac{d}{dt} L_{\dot{x}^i} - L_{x^i} = 0 \qquad \text{or} \qquad \ddot{x}^i + \Gamma^i_{jk} \dot{x}^j \dot{x}^k = 0, \tag{2.57}$$

where subscripts denote partial derivatives, while Γ^i_{jk} are the Christoffel symbols of the affine Levi-Civita connection of the configuraton manifold M.

The corresponding momentum phase-space $P = T^*M$ provides a natural *symplectic structure* that can be defined as follows. As the configuraton configuration space M is a smooth $n-$manifold, we can pick local coordinates $\{dx^1, ..., dx^n\} \in M$. Then $\{dx^1, ..., dx^n\}$ defines a basis of the cotangent space $T^*_x M$, and by writing $\theta \in T^*_x M$ as $\theta = p_i dx^i$, we get local coordinates $\{x^1, ..., x^n, p_1, ..., p_n\}$ on T^*M. We can now define the canonical symplectic form ω on $P = T^*M$ as:

$$\omega = dp_i \wedge dx^i,$$

where '$\wedge$' denotes the wedge or exterior product of exterior differential forms.

Tensor Fields and Bundles

A *tensor bundle* $\mathcal{T}$ associated to a smooth n-manifold M is defined as a *tensor product* of tangent and cotangent bundles:

$$\mathcal{T} = \bigotimes^q T^*M \otimes \bigotimes^p TM = \overbrace{TM \otimes ... \otimes TM}^{p \text{ times}} \otimes \overbrace{T^*M \otimes ... \otimes T^*M}^{q \text{ times}}.$$

Tensor bundles are special case of more general fibre bundles.

A *tensor-field* of type (p, q) (see Appendix) on a smooth n-manifold M is defined as a *smooth section* $\tau : M \to \mathcal{T}$ of the tensor bundle $\mathcal{T}$. The coefficients of the tensor-field τ are smooth (C^∞) functions with p indices up and q indices down. The classical position of indices can be explained in modern terms as follows. If (U, ϕ) is a chart at a point $m \in M$ with local coordinates $(x^1, ..., x^n)$, we have the *holonomous frame field*

$$\partial_{x^{i_1}} \otimes \partial_{x^{i_2}} \otimes ... \otimes \partial_{x^{i_p}} \otimes dx^{j_1} \otimes dx^{j_2} ... \otimes dx^{j_q},$$

for $i \in \{1, ..., n\}^p$, $j = \{1, ..., n\}^q$, over U of this tensor bundle, and for any $(p, q)-$tensor-field τ we have

$$\tau|U = \tau^{i_1 ... i_p}_{j_1 ... j_q} \partial_{x^{i_1}} \otimes \partial_{x^{i_2}} \otimes ... \otimes \partial_{x^{i_p}} \otimes dx^{j_1} \otimes dx^{j_2} ... \otimes dx^{j_q}.$$

For such tensor-fields the *Lie derivative* $\mathcal{L}_v$ along any vector-field v is defined, and it is a *derivation* (i.e., both linearity and Leibniz rules hold) with respect to the tensor product. Tensor bundle $\mathcal{T}$ admits many natural transformations. For example, a 'contraction' like the trace $T^*M \otimes TM = L(TM, TM) \to M \times \mathbb{R}$, but applied just to one specified factor of type T^*M and another one of type TM, is a natural transformation. And any 'permutation of the same kind of factors' is a natural transformation.

The tangent bundle $\pi_M : TM \to M$ of a manifold M (introduced above) is a special tensor bundle over M such that, given an atlas $\{(U_\alpha, \varphi_\alpha)\}$ of M, TM has the *holonomic atlas*

$$\Psi = \{(U_\alpha, \varphi_\alpha = T\varphi_\alpha)\}.$$

The associated linear bundle coordinates are the induced coordinates $(\dot{x}^\lambda)$ at a point $m \in M$ with respect to the *holonomic frames* $\{\partial_\lambda\}$ in tangent spaces $T_m M$. Their transition functions read

$$\dot{x}'^\lambda = \frac{\partial x'^\lambda}{\partial x^\mu} \dot{x}^\mu.$$

Technically, the tangent bundle TM is a tensor bundle with the structure Lie group $GL(\dim M, \mathbb{R})$.

Recall that the cotangent bundle of M is the dual T^*M of TM. It is equipped with the induced coordinates $(\dot{x}_\lambda)$ at a point $m \in M$ with respect to *holonomic coframes* $\{dx^\lambda\}$ dual of $\{\partial_\lambda\}$. Their transition functions read

$$\dot{x}'_\lambda = \frac{\partial x'^\mu}{\partial x^\lambda} \dot{x}_\mu.$$

Basic de Rham Machinery

Exact and closed forms and chains

In general, a $p-$form β is called *closed* if its exterior derivative $d = \partial_i dx^i$ is equal to zero,

$$d\beta = 0.$$

From this condition one can see that the closed form (the *kernel* of the exterior derivative operator d) is conserved quantity. Therefore, closed $p-$forms possess certain invariant properties, physically corresponding to the conservation laws (see e.g., [AMR88]).

Also, a $p-$form β that is an exterior derivative of some $(p-1)-$form α,

$$\beta = d\alpha,$$

is called *exact* (the *image* of the exterior derivative operator d). By Poincaré lemma, exact forms prove to be closed automatically,

$$d\beta = d(d\alpha) = 0.$$

Since $d^2 = 0$, *every exact form is closed.* The converse is only partially true, by Poincaré lemma: every closed form is *locally exact*.

Technically, this means that given a closed $p-$form $\alpha \in \Omega^p(U)$, defined on an open set U of a smooth manifold M (see Figure 2.2), any point $m \in U$ has a neighborhood on which there exists a $(p-1)-$form $\beta \in \Omega^{p-1}(U)$ such that $d\beta = \alpha|_U$.

In particular, there is a Poincaré lemma for contractible manifolds: Any closed form on a smoothly contractible manifold is exact.

The Poincaré lemma is a generalization and unification of two well–known facts in vector calculus:

1. If $\operatorname{curl} F = 0$, then locally $F = \operatorname{grad} f$; and
2. If $\operatorname{div} F = 0$, then locally $F = \operatorname{curl} G$.

A *cycle* is a $p-$chain, (or, an oriented $p-$domain) $C \in \mathcal{C}_p(M)$ such that $\partial C = 0$. A *boundary* is a chain C such that $C = \partial B$, for any other chain $B \in \mathcal{C}_p(M)$. Similarly, a *cocycle* (i.e., a *closed form*) is a cochain ω such that $d\omega = 0$. A *coboundary* (i.e., an *exact form*) is a cochain ω such that $\omega = d\theta$, for any other cochain θ. All exact forms are closed ($\omega = d\theta \Rightarrow d\omega = 0$) and all boundaries are cycles ($C = \partial B \Rightarrow \partial C = 0$). Converse is true only for smooth contractible manifolds, by Poincaré lemma.

De Rham duality of forms and chains

Integration on a smooth manifold M should be thought of as a nondegenerate bilinear pairing $\langle , \rangle$ between $p-$forms and $p-$chains (spanning a finite domain on M). Duality of $p-$forms and $p-$chains on M is based on the De Rham's 'period', defined as [Rha84, CD82]

$$\text{Period} := \int_C \omega := \langle C, \omega \rangle ,$$

where C is a cycle, ω is a cocycle, while $\langle C,\omega\rangle = \omega(C)$ is their inner product $\langle C,\omega\rangle : \Omega^p(M) \times \mathcal{C}_p(M) \to \mathbb{R}$. From the Poincaré lemma, a closed $p-$form ω is exact iff $\langle C,\omega\rangle = 0$.

The fundamental topological duality is based on the Stokes theorem,

$$\int_{\partial C} \omega = \int_C d\omega \qquad \text{or} \qquad \langle \partial C,\omega\rangle = \langle C, d\omega\rangle \,,$$

where ∂C is the boundary of the $p-$chain C oriented coherently with C on M. While the *boundary operator* ∂ is a global operator, the coboundary operator d is local, and thus more suitable for applications. The main property of the exterior differential,

$$d\circ d \equiv d^2 = 0 \quad \Longrightarrow \quad \partial\circ\partial \equiv \partial^2 = 0, \qquad \text{(and converse)},$$

can be easily proved using the Stokes' theorem (and the above 'period notation') as

$$0 = \left\langle \partial^2 C,\omega\right\rangle = \langle \partial C, d\omega\rangle = \left\langle C, d^2\omega\right\rangle = 0.$$

De Rham cochain and chain complexes

In the Euclidean 3D space $\mathbb{R}^3$ we have the following De Rham *cochain complex*

$$0 \to \Omega^0(\mathbb{R}^3) \xrightarrow[\text{grad}]{d} \Omega^1(\mathbb{R}^3) \xrightarrow[\text{curl}]{d} \Omega^2(\mathbb{R}^3) \xrightarrow[\text{div}]{d} \Omega^3(\mathbb{R}^3) \to 0.$$

Using the *closure property* for the exterior differential in $\mathbb{R}^3$, $d\circ d \equiv d^2 = 0$, we get the standard identities from vector calculus

$$\text{curl}\cdot\text{grad} = 0 \qquad \text{and} \qquad \text{div}\cdot\text{curl} = 0.$$

As a duality, in $\mathbb{R}^3$ we have the following *chain complex*

$$0 \leftarrow \mathcal{C}_0(\mathbb{R}^3) \xleftarrow{\partial} \mathcal{C}_1(\mathbb{R}^3) \xleftarrow{\partial} \mathcal{C}_2(\mathbb{R}^3) \xleftarrow{\partial} \mathcal{C}_3(\mathbb{R}^3) \leftarrow 0,$$

(with the closure property $\partial\circ\partial \equiv \partial^2 = 0$) which implies the following three boundaries:

$$C_1 \overset{\partial}{\mapsto} C_0 = \partial(C_1), \qquad C_2 \overset{\partial}{\mapsto} C_1 = \partial(C_2), \qquad C_3 \overset{\partial}{\mapsto} C_2 = \partial(C_3),$$

where $C_0 \in \mathcal{C}_0$ is a $0-$boundary (or, a point), $C_1 \in \mathcal{C}_1$ is a $1-$boundary (or, a line), $C_2 \in \mathcal{C}_2$ is a $2-$boundary (or, a surface), and $C_3 \in \mathcal{C}_3$ is a $3-$boundary (or, a hypersurface). Similarly, the de Rham complex implies the following three coboundaries:

$$\omega^0 \overset{d}{\mapsto} \omega^1 = d(\omega^0), \qquad \omega^1 \overset{d}{\mapsto} \omega^2 = d(\omega^1), \qquad \omega^2 \overset{d}{\mapsto} \omega^3 = d(\omega^2),$$

where $\omega^0 \in \Omega^0$ is 0-form (or, a function), $\omega^1 \in \Omega^1$ is a 1-form, $\omega^2 \in \Omega^2$ is a 2-form, and $\omega^3 \in \Omega^3$ is a 3-form.

In general, on a smooth nD manifold M we has the following de Rham cochain complex [Rha84]

$$0 \to \Omega^0(M) \xrightarrow{d} \Omega^1(M) \xrightarrow{d} \Omega^2(M) \xrightarrow{d} \Omega^3(M) \xrightarrow{d} \cdots \xrightarrow{d} \Omega^n(M) \to 0,$$

satisfying the closure property on M, $d\circ d \equiv d^2 = 0$.

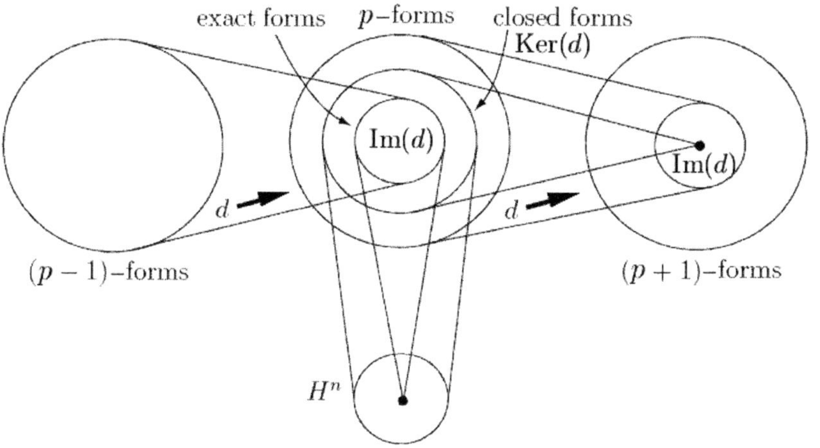

Fig. 2.3. *A small portion of the De Rham cochain complex, showing a homomorphism of cohomology groups.*

De Rham cohomology vs. chain homology

Briefly, the de Rham cohomology is the (functional) space of closed differential $p-$forms modulo exact ones on a smooth manifold.

More precisely, the subspace of all closed $p-$forms (cocycles) on a smooth manifold M is the kernel $\mathrm{Ker}(d)$ of the De Rham $d-$homomorphism (see Figure 2.3), denoted by $Z^p(M) \subset \Omega^p(M)$, and the sub-subspace of all exact $p-$forms (coboundaries) on M is the image $\mathrm{Im}(d)$ of the de Rham homomorphism denoted by $B^p(M) \subset Z^p(M)$. The *quotient space*

$$H^p_{DR}(M) := \frac{Z^p(M)}{B^p M} = \frac{\mathrm{Ker}\left(d : \Omega^p(M) \to \Omega^{p+1}(M)\right)}{\mathrm{Im}\left(d : \Omega^{p-1}(M) \to \Omega^p(M)\right)}, \tag{2.58}$$

is called the pth de Rham *cohomology group* of a manifold M. It is a topological invariant of a manifold. Two $p-$cocycles $\alpha,\beta \in \Omega^p(M)$ are *cohomologous*, or belong to the same *cohomology class* $[\alpha] \in H^p(M)$, if they differ by a $(p-1)-$coboundary $\alpha - \beta = d\theta \in \Omega^{p-1}(M)$. The dimension $b_p = \dim H^p(M)$ of the de Rham cohomology group $H^p_{DR}(M)$ of the manifold M is called the Betti number b_p.

Similarly, the subspace of all $p-$cycles on a smooth manifold M is the kernel $\mathrm{Ker}(\partial)$ of the $\partial-$homomorphism, denoted by $Z_p(M) \subset \mathcal{C}_p(M)$, and the sub-subspace of all $p-$boundaries on M is the image $\mathrm{Im}(\partial)$ of the $\partial-$homomorphism, denoted by $B_p(M) \subset \mathcal{C}_p(M)$. Two $p-$cycles $C_1,C_2 \in \mathcal{C}_p$ are *homologous*, if they differ by a $(p-1)-$boundary $C_1 - C_2 = \partial B \in \mathcal{C}_{p-1}(M)$. Then C_1 and C_2 belong to the same *homology class* $[C] \in H_p(M)$, where $H_p(M)$ is the homology group of the manifold M, defined as

$$H_p(M) := \frac{Z_p(M)}{B_p(M)} = \frac{\mathrm{Ker}(\partial : \mathcal{C}_p(M) \to \mathcal{C}_{p-1}(M))}{\mathrm{Im}(\partial : \mathcal{C}_{p+1}(M) \to \mathcal{C}_p(M))},$$

where Z_p is the vector space of cycles and $B_p \subset Z_p$ is the vector space of boundaries on M. The dimension $b_p = \dim H_p(M)$ of the homology group $H_p(M)$ is, by the de Rham theorem, the same Betti number b_p.

If we know the Betti numbers for all (co)homology groups of the manifold M, we can calculate the *Euler-Poincaré characteristic* of M as

$$\chi(M) = \sum_{p=1}^{n} (-1)^p b_p.$$

For example, consider a small portion of the De Rham cochain complex of Figure 2.3 spanning a space-time 4−manifold M,

$$\Omega^{p-1}(M) \xrightarrow{d_{p-1}} \Omega^p(M) \xrightarrow{d_p} \Omega^{p+1}(M)$$

As we have seen above, cohomology classifies topological spaces by comparing two subspaces of Ω^p: (i) the space of $p-$cocycles, $Z^p(M) = \operatorname{Ker} d_p$, and (ii) the space of $p-$coboundaries, $B^p(M) = \operatorname{Im} d_{p-1}$. Thus, for the cochain complex of any space-time 4−manifold we have,

$$d^2 = 0 \quad \Rightarrow \quad B^p(M) \subset Z^p(M),$$

that is, every $p-$coboundary is a $p-$cocycle. Whether the converse of this statement is true, according to Poincaré lemma, depends on the particular topology of a space-time 4−manifold. If every $p-$cocycle is a $p-$coboundary, so that B^p and Z^p are equal, then the cochain complex is exact at $\Omega^p(M)$. In topologically interesting regions of a space-time manifold M, exactness may fail [Wis06], and we measure the failure of exactness by taking the pth cohomology group

$$H^p(M) = Z^p(M)/B^p(M).$$

Exterior Lagrangian Systems

Let $\Omega^p(M) = \phi_I dx^I$ denote the space of differential p-forms on a n-manifold M, i.e., if a multi-index $I \subset \{1,\ldots,n\}$ is a subset of p elements then we have a p-form: $dx^I = dx_1^i \wedge dx_2^i \wedge \cdots \wedge dx_p^i$ on M. We define the exterior derivative d on M as:

$$d\phi = \partial_p \phi_I \, dx_p \wedge dx^I.$$

Consider an n-DOF dynamical system Ξ, evolving in time on its configuration n-manifold M (with local coordinates x^i, $i = 1,...,n$) as well as on its tangent bundle TM (with local coordinates $(x^i; \dot{x}^i)$, where overdot denotes time derivative). For the system Ξ we consider a *well-posed variational problem* $(I, \omega; \varphi)$ on the associated $(2n+1)$-dimensional *jet manifold* $X = J^1(\mathbb{R}, M) \cong \mathbb{R} \times TM$, with local canonical variables $(t; x^i; \dot{x}^i)$.

Here, (I, ω) is called the *Pfaffian exterior differential system* on X, given locally as:

$$\begin{cases} \theta^i = dx^i - \dot{x}^i \omega = 0, \\ \omega \equiv dt \neq 0, \end{cases}$$

with structure equations:

$$d\theta^i = -d\dot{x}^i \wedge \omega.$$

Integral manifolds $N \in J^1(\mathbb{R}, M)$ of the Pfaffian system (I, ω) are locally 1-jets: $t \to (t, x(t), \dot{x}(t))$ of curves $x = x(t) : \mathbb{R} \to M$.

φ in $(I, \omega; \varphi)$ is a 1-form: $\varphi = L\,\omega$, where $L = L(t, x, \dot{x})$ is the system's *Lagrangian* function defined on X, having both coordinate and velocity partial derivatives, respectively denoted by L_{x^i} and $L_{\dot{x}^i}$.

A variational problem $(I, \omega; \varphi)$ is said to be *well-posed* (or, strongly non-degenerate) if the determinant of the matrix of mixed velocity partials of the Lagrangian is positive definite: $\det \|L_{\dot{x}^i \dot{x}^j}\| > 0$.

The *extended Pfaffian system:*

$$\begin{cases} \theta^i = 0, \\ dL_{\dot{x}^i} - L_{x^i}\,\omega = 0, \\ \omega \neq 0. \end{cases}$$

generates classical *Euler-Lagrange equations*:

$$\partial_t L_{\dot{x}^i} = L_{x^i}\,, \qquad \text{(where } \partial_t = d/dt). \tag{2.59}$$

If an *integral manifold* N satisfies the Euler-Lagrange equations (2.59) of a well-posed variational problem $(I, \omega; \varphi)$ then the following holds:

$$\partial_t \left(\int_{N_t} \varphi \right)_{t=0} = 0,$$

for any *admissible variation* $N_t \in N$ that satisfies the *endpoint conditions*: $\omega = \theta^i = 0$.

Theorem (Griffiths): *Under the above conditions, both the Lagrangian dynamics with initial conditions:*

$$\begin{cases} \partial_t L_{\dot{x}^i} = L_{x^i}, \\ x(t_0) = x_0, \quad \dot{x}(t_0) = \dot{x}_0 \end{cases}$$

and the Lagrangian dynamics with endpoint conditions:

$$\begin{cases} \partial_t L_{\dot{x}^i} = L_{x^i}, \\ x(t_0) = x_0, \quad x(t_1) = x_1 \end{cases}$$

have unique solutions.

If M is a Riemannian manifold its metric $g = <.>$ is locally given by a positive definite quadratic form:

$$ds^2 = g_{ij}(x)\, dx^i dx^j\,, \tag{2.60}$$

where the metric tensor is a C^∞ symmetric matrix: $g(x) = \|g_{ij}(x)\|$.

Kinetic energy of the system Ξ is a function $T = T(x, \dot{x})$ on the tangent bundle TM, which induces a positive definite quadratic form in each fiber $T_xM \subset TM$. In local coordinates on T_xM, it is related to the Riemannian metric (2.60) by:

$$T\,\omega^2 = \frac{1}{2}\,ds^2.$$

If the potential energy of the system Ξ is a function $U = U(x)$ on M, then the autonomous Lagrangian is defined as kinetic minus potential energy: $L(x, \dot{x}) = T(x, \dot{x}) - U(x)$.

The condition of well-posedness is satisfied, as:

$$\det \|L_{\dot{x}^i \dot{x}^j}\| = \det \|g_{ij}(x)\| > 0.$$

The *covariant Euler-Lagrange equations* (2.59) expand as:

$$\partial_t \left(g_{ij}(x(t))\,\dot{x}^j(t)\right) = \frac{1}{2}\left(\partial_i g_{jk}(x(t)\,\dot{x}^j(t)\,\dot{x}^k(t)\right) - F_i(x(t)), \tag{2.61}$$

where $F_i(x(t)) = \partial_{\dot{x}^j} U(x)$ denotes the *gradient force 1-form.*

Letting $\|g^{ij}(x)\|$ be the inverse matrix to $\|g_{ij}(x)\|$ and introducing the *Christoffel symbols* (of the Levi-Civita connection Γ on M):

$$\Gamma^i_{jk} = g^{il}\frac{1}{2}\left(\partial_j g_{kl} + \partial_k g_{jl} - \partial_l g_{jk}\right)$$

the equations (2.61) resolve to the *contravariant form*

$$\ddot{x}^i(t) + \Gamma^i_{jk}(x(t))\,\dot{x}^j(t)\,\dot{x}^k(t) = -F^i(x(t)), \tag{2.62}$$

where: $F^i(x(t)) = g^{ij}(x)\,\partial_{\dot{x}^j} U(x)$ denotes the *gradient force vector-field.*

Griffiths Theorem implies that both the covariant dynamics with initial conditions:

$$\begin{cases} \ddot{x}^i(t) + \Gamma^i_{jk}(x(t))\,\dot{x}^j(t)\,\dot{x}^k(t) = -F^i(x(t)) \\ \qquad x(t_0) = x_0, \quad \dot{x}(t_0) = \dot{x}_0 \end{cases}$$

and the covariant dynamics with endpoint conditions:

$$\begin{cases} \ddot{x}^i(t) + \Gamma^i_{jk}(x(t))\,\dot{x}^j(t)\,\dot{x}^k(t) = -F^i(x(t)) \\ \qquad x(t_0) = x_0, \quad x(t_1) = x_1 \end{cases}$$

have unique solutions.

Basic Gauge Fields

Recall that a *gauge theory* is a theory that admits a symmetry with a local parameter. For example, in every quantum theory the global phase of the wave $\psi-$function is arbitrary and does not represent something physical.

Consequently, the theory is invariant under a global change of phases (adding a constant to the phase of all wave functions, everywhere); this is a global symmetry. In quantum electrodynamics, the theory is also invariant under a local change of phase, that is, one may shift the phase of all wave functions so that the shift may be different at every point in space-time. This is a local symmetry. However, in order for a well-defined derivative operator to exist, one must introduce a new field, the *gauge field*, which also transforms in order for the local change of variables (the phase in our example) not to affect the derivative. In quantum electrodynamics this gauge field is the electromagnetic potential 1-form A (or, a covector), in components within the nD coframe $\{dx^\mu\}$ on a smooth manifold M (dual to the frame, i.e., basis of tangent vectors $\{\partial_\mu = \partial/\partial x^\mu\}$, given by

$$A = A_\mu dx^\mu, \quad \text{such that} \quad A_{\text{new}} = A_{\text{old}} + df, \qquad (f \text{ is any scalar function})$$

- leaves the electromagnetic field 2-form $F = dA$ unchanged. This change df of local gauge of variable A is termed *gauge transformation.* In quantum field theory the excitations of fields represent particles. The particle associated with excitations of the gauge field is the *gauge boson.* All the fundamental interactions in nature are described by gauge theories. In particular, in quantum electrodynamics, whose gauge transformation is a local change of phase, the gauge group is the circle group $U(1)$ (consisting of all complex numbers with absolute value 1), and the gauge boson is the photon (see. e.g. [Fra86]).

The *gauge field* of classical electrodynamics, given in a local coframe $\{dx^\mu\}$ on M as an electromagnetic potential 1-form

$$A = A_\mu dx^\mu = A_\mu dx^\mu + df, \qquad (f = \text{ arbitrary scalar field}),$$

is globally a *connection* on a $U(1)-$bundle of M. The corresponding electromagnetic field, locally the 2-form on M,

$$F = dA, \qquad \text{in components given by}$$
$$F = \frac{1}{2} F_{\mu\nu}\, dx^\mu \wedge dx^\nu, \qquad \text{with} \quad F_{\mu\nu} = \partial_\nu A_\mu - \partial_\mu A_\nu$$

is globally the *curvature* of the connection A under the *gauge-covariant derivative*,

$$D_\mu = \partial_\mu - \mathrm{i}eA_\mu, \tag{2.63}$$

where e is the charge coupling constant.[46] In particular, in 4D space-time electrodynamics, the 1-form *electric current density* J has the components

[46] If a gauge transformation is given by

$$\psi \mapsto \mathrm{e}^{\mathrm{i}\Lambda}\psi$$

and for the gauge potential

$$A_\mu \mapsto A_\mu + \frac{1}{e}(\partial_\mu \Lambda),$$

$J_\mu = (\rho, \mathbf{j}) = (\rho, j_x, j_y, j_z)$ (where ρ is the charge density), the 2-form *Faraday* F is given in components of electric field $\mathbf{E}$ and magnetic field $\mathbf{B}$ by[47]

$$F_{\mu\nu} = \begin{pmatrix} 0 & E_x & E_y & E_z \\ -E_x & 0 & -B_z & B_y \\ -E_y & B_z & 0 & -B_x \\ -E_z & -B_y & B_x & 0 \end{pmatrix}, \qquad \text{with} \qquad F_{\nu\mu} = -F_{\mu\nu},$$

while its dual 2-form *Maxwell* $\star F$ has the following components

then the gauge-covariant derivative,

$$D_\mu = \partial_\mu - \mathrm{i}eA_\mu$$

transforms as

$$D_\mu \mapsto \partial_\mu - \mathrm{i}eA_\mu - \mathrm{i}(\partial_\mu \Lambda)$$

and $D_\mu \psi$ transforms as

$$D_\mu \mapsto \partial_\mu - \mathrm{i}eA_\mu - \mathrm{i}(\partial_\mu \Lambda).$$

47 Recall that in the 19th Century, Maxwell unified Faraday's electric and magnetic fields. Maxwell's theory led to Einstein's special relativity where this unification becomes a spin-off of the unification of space end time in the form of the *Faraday tensor* [MTW73]

$$F = E \wedge dt + B,$$

where F is electromagnetic 2−form on space-time, E is electric 1−form on space, and B is magnetic 2−form on space. Gauge theory considers F as secondary object to a connection-potential 1−form A. This makes half of the Maxwell equations into tautologies, i.e.,

$$F = dA \quad \Longrightarrow \quad dF = 0 \quad : \quad \text{Bianchi identity},$$

but does not imply the second half of Maxwell's equations,

$$\delta F = -4\pi J \quad : \quad \text{dual Bianchi identity}.$$

To understand the deeper meaning of the connection-potential 1−form A, we can integrate it along a path γ in space-time, $x \xrightarrow{\gamma} y$. Classically, the integral $\int_\gamma A$ represents an *action* for a charged point particle to move along the path γ. Quantum-mechanically, $\exp\left(\mathrm{i}\int_\gamma A\right)$ represents a *phase* (within the unitary Lie group $U(1)$) by which the particle's wave-function changes as it moves along the path γ, so A is a $U(1)$−connection.

In other words, Maxwell's equations can be formulated using complex line bundles, or principal bundles with fibre $U(1)$. The connection ∇ on the line bundle has a curvature $F = \nabla^2$ which is a 2-form that automatically satisfies $dF = 0$ and can be interpreted as a field-strength. If the line bundle is trivial with flat reference connection d, we can write $\nabla = d + A$ and $F = dA$ with A the 1-form composed of the electric potential and the magnetic vector potential (see [BM94]).

$$\star F_{\mu\nu} = \begin{pmatrix} 0 & -B_x & -B_y & -B_z \\ B_x & 0 & -E_z & E_y \\ B_y & E_z & 0 & -E_x \\ B_z & -E_y & B_x & 0 \end{pmatrix}, \qquad \text{with} \qquad \star F_{\nu\mu} = - \star F_{\mu\nu},$$

so that classical electrodynamics is governed by the *Maxwell equations*, which in modern exterior formulation read

$$\begin{aligned} dF &= 0, \qquad \delta F = -4\pi J, \qquad \text{or in components,} \\ F_{[\mu\nu,\eta]} &= 0, \qquad F_{\mu\nu},^{\mu} = -4\pi J_{\mu}, \end{aligned}$$

where $\star$ is the Hodge star operator and δ is the *Hodge codiferential* (see Appendix), comma denotes the partial derivative and the 1-form of electric current $J = J_{\mu} dx^{\mu}$ is conserved, by the electrical *continuity equation*,

$$\delta J = 0, \qquad \text{or in components,} \qquad J_{\mu},^{\mu} = 0.$$

The first, sourceless Maxwell equation, $dF = 0$, gives vector magnetostatics and magnetodynamics,

$$\begin{aligned} \text{Magnetic Gauss' law} &: \operatorname{div} \mathbf{B} = 0, \\ \text{Faraday's law} &: \partial_t \mathbf{B} + \operatorname{curl} \mathbf{E} = 0. \end{aligned}$$

The second Maxwell equation with source, $\delta F = J$, gives vector electrostatics and electrodynamics,

$$\begin{aligned} \text{Electric Gauss' law} &: \operatorname{div} \mathbf{E} = 4\pi\rho, \\ \text{Ampère's law} &: \partial_t \mathbf{E} - \operatorname{curl} \mathbf{B} = -4\pi \mathbf{j}. \end{aligned}$$

The standard *Lagrangian* for the free electromagnetic field, $F = dA$, is given by [II06b, II07]

$$\mathcal{L}(A) = \frac{1}{2}(F \wedge \star F),$$

with the corresponding *action functional*

$$S(A) = \frac{1}{2} \int F \wedge \star F.$$

Basic Lie Groups and Their Lie Algebras

In this subsection we give a minimalist definition of Lie groups and their Lie algebras (for more technical expose, see e.g. [II06b, II07] and references therein).

A *Lie group* G is a smooth n-manifold M that has at the same time a group $G-$structure consistent with its manifold $M-$structure in the sense that *group multiplication* $\mu : G \times G \to G$, $(g, h) \mapsto gh$ and the *group inversion* $\nu : G \to G, \quad g \mapsto g^{-1}$ are smooth maps. A point $e \in G$ is called the group identity element.

A Lie group can *act* on a smooth manifold M by moving the points of M, denoted by $G \times M \to M$.Group action on a manifold defines the *orbit* of a point m on a manifold M, which is the set of points on M to which m can be moved by the elements of a Lie group G. The orbit of a point m is denoted by $Gm = \{g \cdot m | g \in G\}$.

Let G be a real Lie group. Its *Lie algebra* $\mathfrak{g}$ is the tangent space TG_e to the group G at the identity e provided with the *Lie bracket* (*commutator*) operation $[X, Y]$, which is bilinear, skew-symmetric, and satisfies the *Jacobi identity* (for any three vector-fields $X, Y, Z \in \mathfrak{g}$):

$$[[X, Y], Z] = [X, [Y, Z]] - [X, [Y, Z]].$$

Note that in Hamiltonian mechanics, Jacobi identity is satisfied by Poisson brackets, while in quantum mechanics it is satisfied by operator commutators.

For example, $G = SO(3)$ is the group of rotations of 3D Euclidean space, i.e. the configuration space of a rigid body fixed at a point. A motion of the body is then described by a curve $g = g(t)$ in the group $SO(3)$. Its Lie algebra $\mathfrak{g} = \mathfrak{so}(3)$ is the 3D vector space of angular velocities of all possible rotations. The commutator in this algebra is the usual vector (cross) product.

A Lie group G acts on itself by left and right translations: every element $g \in G$ defines diffeomorphisms of the group onto itself (for every $h \in G$):

$$L_g : G \to G, \qquad L_g h = gh; \qquad R_g : G \to G, \qquad R_g h = hg.$$

The induced maps of the tangent spaces are denoted by:

$$L_{g*} : TG_h \to TG_{gh}, \qquad R_{g*} : TG_h \to TG_{hg}.$$

The diffeomorphism $R_{g^{-1}} L_g$ is an inner automorphism of the group G. It leaves the group identity e fixed. Its derivative at the identity e is a linear map from the Lie algebra $\mathfrak{g}$ to itself:

$$Ad_g : \mathfrak{g} \to \mathfrak{g}, \qquad Ad_g (R_{g^{-1}} L_g)_{*e}$$

is called the *adjoint representation* of the Lie group G.

Referring to the previous example, a rotation velocity $\dot{g}$ of the rigid body (fixed at a point) is a tangent vector to the Lie group $G = SO(3)$ at the point $g \in G$. To get the angular velocity, we must carry this vector to the tangent space TG_e of the group at the identity, i.e. to its Lie algebra $\mathfrak{g} = \mathfrak{so}(3)$. This can be done in two ways: by left and right translation, L_g and R_g. As a result, we obtain two different vector-fields in the Lie algebra $\mathfrak{so}(3)$:

$$\omega_c = L_{g^{-1}*} \dot{g} \in \mathfrak{so}(3) \qquad \text{and} \qquad \omega_x = R_{g^{-1}*} \dot{g} \in \mathfrak{so}(3),$$

which are called the 'angular velocity in the body' and the 'angular velocity in space,' respectively.

Now, left and right translations induce operators on the cotangent space T^*G_g dual to L_{g*} and R_{g*}, denoted by (for every $h \in G$):

$$L_g^* : T^*G_{gh} \to T^*G_h, \qquad R_g^* : T^*G_{hg} \to T^*G_h.$$

The transpose operators $Ad_g^* : \mathfrak{g} \to \mathfrak{g}$ satisfy the relations $Ad_{gh}^* = Ad_h^* Ad_g^*$ (for every $g, h \in G$) and constitute the *co-adjoint representation* of the Lie group G. The co-adjoint representation plays an important role in all questions related to (left) invariant metrics on the Lie group. According to A. Kirillov, the orbit of any vector-field X in a Lie algebra $\mathfrak{g}$ in a co-adjoint representation Ad_g^* is itself a symplectic manifold and therefore a phase space for a Hamiltonian mechanical system.

A Riemannian metric on a Lie group G is called left-invariant if it is preserved by all left translations L_g, i.e., if the derivative of left translation carries every vector to a vector of the same length. Similarly, a vector-field X on G is called left-invariant if (for every $g \in G$) $L_g^* X = X$.

Again referring to the previous example of the rigid body, the dual space $\mathfrak{g}^*$ to the Lie algebra $\mathfrak{g} = \mathfrak{so}(3)$ is the space of angular momenta $\boldsymbol{\pi}$. The kinetic energy T of a body is determined by the vector-field of angular velocity in the body and does not depend on the position of the body in space. Therefore, kinetic energy gives a left-invariant Riemannian metric on the rotation group $G = SO(3)$.

2.8.4 Advanced Complexity Geometrodynamics

Basic Riemannian-Ricci Machinery

Riemann and Ricci curvatures on a smooth manifold

Recall that proper differentiation of vector and tensor fields on a smooth Riemannian $n-$manifold is performed using the *Levi-Civita covariant derivative* (see, e.g. [II06b, II07]). Formally, let M be a Riemannian $n-$manifold with the tangent bundle TM and a local coordinate system $\{x^i\}_{i=1}^n$ defined in an open set $U \subset M$. The covariant derivative operator, $\nabla_X : C^\infty(TM) \to C^\infty(TM)$, is the unique linear map such that for any vector-fields X, Y, Z, constant c, and function f the following properties are valid:

$$\begin{aligned} \nabla_{X+cY} &= \nabla_X + c\nabla_Y, \\ \nabla_X(Y + fZ) &= \nabla_X Y + (Xf)Z + f\nabla_X Z, \qquad \text{with} \\ \nabla_X Y - \nabla_Y X &= [X, Y], \qquad \text{(torsion free property)} \end{aligned}$$

where $[X, Y]$ is the Lie bracket of X and Y (see, e.g. [Iva04]). In local coordinates, the metric g is defined for any orthonormal basis ($\partial_i = \partial_{x^i}$) in $U \subset M$ by

$$g_{ij} = g(\partial_i, \partial_j) = \delta_{ij}, \qquad \partial_k g_{ij} = 0.$$

Then the affine *Levi-Civita connection* is defined on M by

$$\nabla_{\partial_i}\partial_j = \Gamma^k_{ij}\partial_k, \qquad \text{where} \qquad \Gamma^k_{ij} = \frac{1}{2}g^{kl}\left(\partial_i g_{jl} + \partial_j g_{il} - \partial_l g_{ij}\right)$$

are the (second-order) *Christoffel symbols.*

Now, using the covariant derivative operator ∇_X we can define the *Riemann curvature* $(3,1)-$tensor $\mathfrak{Rm}$ by (see, e.g., [II06b, II07])

$$\mathfrak{Rm}(X,Y)Z = \nabla_X\nabla_Y Z - \nabla_Y\nabla_X Z - \nabla_{[X,Y]}Z.$$

$\mathfrak{Rm}$ measures the curvature of the manifold by expressing how noncommutative covariant differentiation is. The $(3,1)-$components R^l_{ijk} of $\mathfrak{Rm}$ are defined in $U \subset M$ by

$$\mathfrak{Rm}\left(\partial_i, \partial_j\right)\partial_k = R^l_{ijk}\partial_l, \qquad \text{which expands (see [MTW73]) as}$$
$$R^l_{ijk} = \partial_i\Gamma^l_{jk} - \partial_j\Gamma^l_{ik} + \Gamma^m_{jk}\Gamma^l_{im} - \Gamma^m_{ik}\Gamma^l_{jm}.$$

Also, the Riemann $(4,0)-$tensor $R_{ijkl} = g_{lm}R^m_{ijk}$ is defined as the $g-$based inner product on M,

$$R_{ijkl} = \left\langle \mathfrak{Rm}\left(\partial_i, \partial_j\right)\partial_k, \partial_l \right\rangle.$$

The first and second Bianchi identities for the Riemann $(4,0)-$tensor R_{ijkl} hold,

$$R_{ijkl} + R_{jkil} + R_{kijl} = 0, \qquad \nabla_i R_{jklm} + \nabla_j R_{kilm} + \nabla_k R_{ijlm} = 0,$$

while the twice contracted second Bianchi identity reads

$$2\nabla_j R_{ij} = \nabla_i R. \tag{2.64}$$

The $(0,2)$ *Ricci tensor* $\mathfrak{Rc}$ is the trace of the Riemann $(3,1)-$tensor $\mathfrak{Rm}$,

$$\mathfrak{Rc}(Y,Z) + \mathrm{tr}(X \to \mathfrak{Rm}(X,Y)Z),$$
$$\text{so that} \qquad \mathfrak{Rc}(X,Y) = g(\mathfrak{Rm}(\partial_i, X)\partial_i, Y),$$

Its components $R_{jk} = \mathfrak{Rc}\left(\partial_j, \partial_k\right)$ are given in $U \subset M$ by the contraction [MTW73]

$$R_{jk} = R^i_{ijk}, \qquad \text{or, in terms of Christoffel symbols,}$$
$$R_{jk} = \partial_i\Gamma^i_{jk} - \partial_k\Gamma^i_{ji} + \Gamma^i_{mi}\Gamma^m_{jk} - \Gamma^i_{mk}\Gamma^m_{ji}.$$

Being a symmetric second-order tensor, $\mathfrak{Rc}$ has $n+12$ independent components on an $n-$manifold M. In particular, on a 3-manifold, it has 6 components, and on a 2-surface it has only the following 3 components:

$$R_{11} = g^{22}R_{2112}, \qquad R_{12} = g^{12}R_{2121}, \qquad R_{22} = g^{11}R_{1221},$$

which are all proportional to the corresponding coordinates of the metric tensor,

$$\frac{R_{11}}{g_{11}} = \frac{R_{12}}{g_{12}} = \frac{R_{22}}{g_{22}} = -\frac{R_{1212}}{\det(g)}. \tag{2.65}$$

Finally, the scalar curvature R is the trace of the Ricci tensor $\mathfrak{Rc}$, given in $U \subset M$ by: $R = g^{ij}R_{ij}$.

Ricci flow on a Riemannian manifold

Parabolic *reaction-diffusion* systems are abundant in mathematical biology. They are mathematical models that describe how the concentration of one or more substances distributed in space changes under the influence of two processes: local chemical reactions in which the substances are converted into each other, and diffusion which causes the substances to spread out in space. More formally, they are expressed as semi-linear parabolic partial differential equations (PDEs, see e.g. [PBL05]). The evolution of the state vector $\mathbf{u}(\mathbf{x}, t)$ describing the concentration of the different reagents is determined by anisotropic diffusion as well as local reactions [II11b]:

$$\partial_t \mathbf{u} = \mathbf{D} \Delta \mathbf{u} + \mathbf{R}(\mathbf{u}), \qquad (\partial_t = \partial/\partial t), \tag{2.66}$$

where each component of the state vector $\mathbf{u}(\mathbf{x}, t)$ represents the concentration of one substance, Δ is the standard Laplacian operator, $\mathbf{D}$ is a symmetric positive-definite matrix of diffusion coefficients (which are proportional to the velocity of the diffusing particles) and $\mathbf{R}(\mathbf{u})$ accounts for all local reactions. The solutions of reaction-diffusion equations display a wide range of behaviors, including the formation of travelling waves and other self-organized patterns like *dissipative solitons* (DSs).

On the other hand, the *Ricci flow equation* (or, the parabolic Einstein equation), introduced by R. Hamilton in 1982 (see [Ham82, Ham86]), is the nonlinear heat-like evolution equation:

$$\partial_t g_{ij} = -2R_{ij}, \tag{2.67}$$

for a time-dependent Riemannian metric $g = g_{ij}(t)$ on a smooth real $n-$manifold M with the Ricci curvature tensor R_{ij}.[48] This equation roughly says that we can deform any metric on a 2-surface or $n-$manifold by the negative of its curvature; after *normalization*, the final state of such deformation will be a metric with constant curvature. However, this is not true in general since, in addition to the presence of singularities, the limits could be Ricci solitons (see below). The factor of 2 in (2.67) is more or less arbitrary, but the negative sign is essential to insure a kind of global *volume exponential decay*,[49] since

[48] This particular PDE (2.67) was chosen by Hamilton for much the same reason that A. Einstein introduced the Ricci tensor into his gravitation field equation,

$$R_{ij} - \frac{1}{2} g_{ij} R = 8\pi T_{ij},$$

where T_{ij} is the energy-momentum tensor. Einstein needed a symmetric 2-index tensor which arises naturally from the metric tensor g_{ij} and its first and second partial derivatives. The Ricci tensor R_{ij} is essentially the only possibility. In gravitation theory and cosmology, the Ricci tensor has the volume-decreasing effect (i.e., convergence of neighboring geodesics, see [H96P]).

[49] This complex geometric process is globally similar to a generic exponential decay ODE:

the Ricci flow equation (2.67) is a kind of nonlinear geometric generalization of the standard linear *heat equation*[50]

$$\partial_t u = \Delta u. \tag{2.68}$$

Like the heat equation (2.68), the Ricci flow equation (2.67) is well behaved in forward time and acts as a kind of smoothing operator (but is usually impossible to solve in backward time). If some parts of a solid object are hot and others are cold, then, under the heat equation, heat will flow from hot to cold, so that the object gradually attains a uniform temperature. To some extent the Ricci flow behaves similarly, so that the Ricci curvature 'tries' to become more uniform [Mil03], thus resembling a monotonic *entropy growth*,[51] $\partial_t S \geq 0$, which is due to the positive definiteness of the metric $g_{ij} \geq 0$, and naturally implying the *arrow of time* [II07].

In a suitable local coordinate system, the Ricci flow equation (2.67) has a nonlinear heat-type form, as follows. At any time t, we can choose local harmonic coordinates so that the coordinate functions are locally defined harmonic functions in the metric $g(t)$. Then the Ricci flow takes the general form (see e.g., [And04])

$$\partial_t g_{ij} = \Delta_M g_{ij} + Q_{ij}(g, \partial g), \tag{2.69}$$

$$\dot{x} = -\lambda f(x),$$

for a positive function $f(x)$. We can get some insight into its solution from the simple exponential decay ODE,

$$\dot{x} = -\lambda x \qquad \text{with the solution} \qquad x(t) = x_0 \mathrm{e}^{-\lambda t},$$

(where $x = x(t)$ is the observed quantity with its initial value x_0 and λ is a positive decay constant), as well as the corresponding nth order rate equation (where $n > 1$ is an integer),

$$\dot{x} = -\lambda x^n \qquad \text{with the solution} \qquad \frac{1}{x^{n-1}} = \frac{1}{{x_0}^{n-1}} + (n-1)\,\lambda t.$$

[50] More precisely, the negative sign is to make the equation parabolic so that there is a theory of existence and uniqueness. Otherwise the equation would be backwards parabolic and not have any theory of existence, uniqueness, etc.

[51] Note that two different kinds of entropy functional have been introduced into the theory of the Ricci flow, both motivated by concepts of entropy in thermodynamics, statistical mechanics and information theory. One is Hamilton's entropy, the other is Perelman's entropy. While in Hamilton's entropy, the scalar curvature R of the metric g_{ij} is viewed as the leading quantity of the system and plays the role of a probability density, in Perelman's entropy the leading quantity describing the system is the metric g_{ij} itself. Hamilton established the monotonicity of his entropy along the volume-normalized Ricci flow on the 2-sphere S^2 [Ham88]. Perelman established the monotonicity of his entropy along the Ricci flow in all dimensions [Per02].

where Δ_M is the *Laplace-Beltrami operator* (2.98) and $Q = Q_{ij}(g, \partial g)$ is a lower-order term quadratic in g and its first order partial derivatives ∂g. From the analysis of nonlinear heat PDEs, one obtains existence and uniqueness of forward-time solutions to the Ricci flow on some time interval, starting at any smooth initial metric g_0.

The quadratic Ricci flow equation (2.69) is our geometric framework for general bio-reaction-diffusion systems, so that the spatio-temporal PDE (2.66) corresponds to the quadratic Ricci flow PDE [II11b]:

$$\begin{array}{ccccc} \partial_t \mathbf{u} & = & \mathbf{D}\Delta\mathbf{u} & + & \mathbf{R}(\mathbf{u}) \\ \updownarrow & & \updownarrow & & \updownarrow \\ \partial_t g_{ij} & = & \Delta_M g_{ij} & + & Q_{ij}(g, \partial g) \end{array}$$

with:

- the metric $g = g_{ij}$ on an $n-$manifold M corresponding to the $n-$dimensional (or $n-$component, or $n-$phase) concentration $\mathbf{u}(\mathbf{x}, t)$;
- the Laplace-Beltrami differential operator Δ_M, as defined on C^2-functions on an $n-$manifold M, with respect to the Riemannian metric g_{ij}, by

$$\Delta_M \equiv \frac{1}{\sqrt{\det(g)}} \frac{\partial}{\partial x^i} \left(\sqrt{\det(g)} g^{ij} \frac{\partial}{\partial x^j} \right) \tag{2.70}$$

  - corresponding to the $n-$dimensional bio-diffusion term $\mathbf{D}\Delta\mathbf{u}$; and
- the quadratic $n-$dimensional Ricci-term, $Q = Q_{ij}(g, \partial g)$, corresponding to the $n-$dimensional bio-reaction term, $\mathbf{R}(\mathbf{u})$.

As a simple example of the Ricci flow equations (2.67)-(2.69), consider a round spherical boundary S^2 of the 3-ball radius r. The metric tensor on S^2 takes the form

$$g_{ij} = r^2 \hat{g}_{ij},$$

where $\hat{g}_{ij}$ is the metric for a unit sphere, while the Ricci tensor

$$R_{ij} = (n-1)\hat{g}_{ij}$$

is independent of r. The Ricci flow equation on S^2 reduces to

$$\dot{r}^2 = -2(n-1), \qquad \text{with the solution} \qquad r^2(t) = r^2(0) - 2(n-1)t.$$

Thus the boundary sphere S^2 collapses to a point in finite time (see [Mil03]).

More generally, the geometrization conjecture [Thu82] holds for any 3-manifold M (see below). Suppose that we start with a compact initial 3-manifold M_0 whose Ricci tensor R_{ij} is everywhere positive definite. Then, as M_0 shrinks to a point under the Ricci flow (2.67), it becomes rounder and rounder. If we rescale the metric g_{ij} on M_0 so that the volume of M_0 remains constant, then M_0 converges towards another compact 3-manifold M_1 of constant positive curvature [Ham82].

In case of even more general 3−manifolds (outside the class of positive Ricci curvature metrics), the situation is much more complicated, as various singularities may arise. One way in which singularities may arise during the Ricci flow is that a spherical boundary $S^2 = \partial M$ of an 3−manifold M may collapse to a point in finite time. Such collapses can be eliminated by performing a kind of 'geometric surgery' on the 3-manifold M, that is a sophisticated sequence of cutting and pasting without accumulation of time errors[52] (see [Per03]). After a finite number of such surgeries, each component either: (i) converges towards a 3-manifold of constant positive Ricci curvature which shrinks to a point in finite time, or possibly (ii) converges towards an $S^2 \times S^1$ which shrinks to a circle S^1 in finite time, or (iii) admits a 'thin-thick' decomposition of [Thu82]. Therefore, one can choose the surgery parameters so that there is a well defined Ricci flow with surgery, that exists for all time [Per03].

In this section we use the evolving n−dimensional geometric machinery of the volume-decaying and entropy-growing Ricci flow $g(t)$, given by equations (2.67)-(2.69), for modeling various biological reaction-diffusion systems and dissipative solitons, defined by special cases of the general spatio-temporal model (2.66).

Basic Hamiltonian Machinery

We develop (autonomous) Hamiltonian dynamics on the configuration manifold M in three steps, following the standard symplectic geometry prescription (see [II06b, II07]):

Step A Find a symplectic *momentum phase-space* (P, ω).

Recall that a symplectic structure on a smooth manifold M is a nondegenerate closed[53] 2−form ω on M, i.e., for each $x \in M$, $\omega(x)$ is nondegenerate, and $d\omega = 0$.

[52] Hamilton's idea was to perform surgery to cut off the singularities and continue his flow after the surgery. If the flow develops singularities again, one repeats the process of performing surgery and continuing the flow. If one can prove there are only a finite number of surgeries in any finite time interval, and if the long-time behavior of solutions of the Ricci flow (2.67) with surgery is well understood, then one would be able to recognize the topological structure of the initial manifold. Thus Hamilton's program, when carried out successfully, would lead to a proof of the Poincaré Conjecture and Thurston's Geometrization Conjecture [Yau06].

[53] A p−form β on a smooth manifold M is called *closed* if its exterior derivative $d = \partial_i dx^i$ is equal to zero,

$$d\beta = 0.$$

From this condition one can see that the closed form (the *kernel* of the exterior derivative operator d) is conserved quantity. Therefore, closed p−forms possess certain invariant properties, physically corresponding to the *conservation laws*.

Also, a p−form β that is an exterior derivative of some $(p-1)$−form α,

$$\beta = d\alpha,$$

Let T_x^*M be a cotangent space to M at m. The cotangent bundle T^*M represents a union $\cup_{m\in M}T_x^*M$, together with the standard topology on T^*M and a natural smooth manifold structure, the dimension of which is twice the dimension of M. A 1−form θ on M represents a section $\theta : M \to T^*M$ of the cotangent bundle T^*M.

$P = T^*M$ is our momentum phase-space. On P there is a nondegenerate symplectic 2−form ω is defined in local joint coordinates $x^i, p_i \in U$, U open in P, as $\omega = dx^i \wedge dp_i$. In that case the coordinates $x^i, p_i \in U$ are called canonical. In a usual procedure the canonical 1−form θ is first defined as $\theta = p_i dx^i$, and then the canonical 2-form ω is defined as $\omega = -d\theta$.

A *symplectic phase-space manifold* is a pair (P, ω).

Step B Find a *Hamiltonian vector-field* X_H on (P, ω).

Let (P, ω) be a symplectic manifold. A vector-field $X : P \to TP$ is called *Hamiltonian* if there is a smooth function $F : P \longrightarrow \mathbb{R}$ such that $i_X\omega = dF$ ($i_X\omega$ denotes the *interior product* or *contraction* of the vector-field X and the 2-form ω). X is *locally Hamiltonian* if $i_X\omega$ is closed.

Let the smooth real-valued *Hamiltonian function* $H : P \to \mathbb{R}$, representing the total configuraton energy $H(x,p) = T(p) + V(x)$ (T and V denote kinetic and potential energy of the system, respectively), be given in local canonical coordinates $x^i, p_i \in U$, U open in P. The *Hamiltonian vector-field* X_H, condition by $i_{X_H}\omega = dH$, is actually defined via symplectic matrix J, in a local chart U, as

$$X_H = J\nabla H = (\partial_{p_i}H, -\partial_{x^i}H), \qquad J = \begin{pmatrix} 0 & I \\ -I & 0 \end{pmatrix}, \tag{2.71}$$

where I denotes the $n \times n$ identity matrix and ∇ is the gradient operator.

Step C Find a *Hamiltonian phase-flow* ϕ_t of X_H.

Let (P, ω) be a symplectic phase-space manifold and $X_H = J\nabla H$ a Hamiltonian vector-field corresponding to a smooth real-valued Hamiltonian function $H : P \to \mathbb{R}$, on it. If a unique one-parameter group of diffeomorphisms $\phi_t : P \to P$ exists so that $\frac{d}{dt}|_{t=0}\, \phi_t x = J\nabla H(x)$, it is called the *Hamiltonian phase-flow*.

---

is called *exact* (the *image* of the exterior derivative operator d). By *Poincaré lemma,* exact forms prove to be closed automatically,

$$d\beta = d(d\alpha) = 0.$$

Since $d^2 = 0$, *every exact form is closed.* The converse is only partially true, by Poincaré lemma: every closed form is *locally exact*.

Technically, this means that given a closed p−form $\alpha \in \Omega^p(U)$, defined on an open set U of a smooth manifold M any point $m \in U$ has a neighborhood on which there exists a $(p-1)$−form $\beta \in \Omega^{p-1}(U)$ such that $d\beta = \alpha|_U$. In particular, there is a Poincaré lemma for contractible manifolds: Any closed form on a smoothly contractible manifold is exact.

A smooth curve $t \mapsto (x^i(t), p_i(t))$ on (P, ω) represents an *integral curve* of the Hamiltonian vector-field $X_H = J\nabla H$, if in the local canonical coordinates $x^i, p_i \in U$, U open in P, *Hamiltonian canonical equations* hold:

$$\dot{q}^i = \partial_{p_i} H, \qquad \dot{p}_i = -\partial_{q^i} H. \tag{2.72}$$

An integral curve is said to be *maximal* if it is not a restriction of an integral curve defined on a larger interval of $\mathbb{R}$. It follows from the standard theorem on the *existence* and *uniqueness* of the solution of a system of ODEs with smooth r.h.s, that if the manifold (P, ω) is Hausdorff, then for any point $x = (x^i, p_i) \in U$, U open in P, there exists a maximal integral curve of $X_H = J\nabla H$, passing for $t = 0$, through point x. In case X_H is complete, i.e., X_H is C^p and (P, ω) is compact, the maximal integral curve of X_H is the Hamiltonian phase-flow $\phi_t : U \to U$.

The phase-flow ϕ_t is *symplectic* if ω is constant along ϕ_t, i.e., $\phi_t^* \omega = \omega$
($\phi_t^* \omega$ denotes the *pull-back*[54] of ω by ϕ_t),
iff $\mathfrak{L}_{X_H} \omega = 0$
($\mathfrak{L}_{X_H} \omega$ denotes the *Lie derivative*[55] of ω upon X_H).

Symplectic phase-flow ϕ_t consists of canonical transformations on (P, ω), i.e., diffeomorphisms in canonical coordinates $x^i, p_i \in U$, U open on all (P, ω) which leave ω invariant. In this case the *Liouville theorem* is valid: ϕ_t *preserves* the *phase volume* on (P, ω). Also, the system's total energy H is conserved along ϕ_t, i.e., $H \circ \phi_t = \phi_t$.

---

54 Given a map $f : X \longrightarrow X'$ between the two manifolds, the *pullback* on X of a form α on X' by f is denoted by $f^* \alpha$. The pullback satisfies the relations

$$f^*(\alpha \wedge \beta) = f^* \alpha \wedge f^* \beta, \qquad df^* \alpha = f^*(d\alpha),$$

for any two forms $\alpha, \beta \in \Omega^p(X)$.

55 The *Lie derivative* $\mathfrak{L}_u \alpha$ of p–form α along a vector-field u is defined by Cartan's 'magic' formula (see [II06b, II07]):

$$\mathfrak{L}_u \alpha = u \rfloor d\alpha + d(u \rfloor \alpha).$$

It satisfies the *Leibnitz relation*

$$\mathfrak{L}_u (\alpha \wedge \beta) = \mathfrak{L}_u \alpha \wedge \beta + \alpha \wedge \mathfrak{L}_u \beta.$$

Here, the *contraction* $\rfloor$ of a vector-field $u = u^\mu \partial_\mu$ and a p–form $\alpha = \alpha_{\lambda_1 \dots \lambda_p} dx^{\lambda_1} \wedge \dots \wedge dx^{\lambda_p}$ on a configuraton manifold X is given in local coordinates on X by

$$u \rfloor \alpha = u^\mu \alpha_{\mu \lambda_1 \dots \lambda_{p-1}} dx^{\lambda_1} \wedge \dots \wedge dx^{\lambda_{p-1}}.$$

It satisfies the following relation

$$u \rfloor (\alpha \wedge \beta) = u \rfloor \alpha \wedge \beta + (-1)^{|\alpha|} \alpha \wedge u \rfloor \beta.$$

Recall that the Riemannian metrics $g = <,>$ on the configuration manifold M is a positive-definite quadratic form $g : TM \to \mathbb{R}$, in local coordinates $x^i \in U$, U open in M, given by (2.56) above. Given the metrics g_{ij}, the system's Hamiltonian function represents a momentum p-dependent quadratic form $H : T^*M \to \mathbb{R}$ - the system's kinetic energy $H(p) = T(p) = \frac{1}{2} < p, p >$, in local canonical coordinates $x^i, p_i \in U_p$, U_p open in T^*M, given by

$$H(p) = \frac{1}{2} g^{ij}(x, m)\, p_i p_j, \tag{2.73}$$

where $g^{ij}(x, m) = g_{ij}^{-1}(x, m)$ denotes the *inverse* (contravariant) material *metric tensor*

$$g^{ij}(x, m) = \sum_{\chi=1}^{n} m_\chi \delta_{rs} \frac{\partial x^i}{\partial x^r} \frac{\partial x^j}{\partial x^s}.$$

T^*M is an *orientable* manifold, admitting the standard *volume form*

$$\Omega_{\omega_H} = \frac{(-1)^{\frac{N(N+1)}{2}}}{N!} \omega_H^N.$$

For Hamiltonian vector-field, X_H on M, there is a base integral curve $\gamma_0(t) = \big(x^i(t), p_i(t)\big)$ iff $\gamma_0(t)$ is a *geodesic*, given by the one-form *force equation*

$$\dot{\bar{p}}_i \equiv \dot{p}_i + \Gamma^i_{jk}\, g^{jl} g^{km}\, p_l p_m = 0, \qquad \text{with} \qquad \dot{x}^k = g^{ki} p_i. \tag{2.74}$$

The l.h.s $\dot{\bar{p}}_i$ of the covariant momentum equation (2.74) represents the intrinsic or Bianchi covariant derivative of the momentum with respect to time t. Basic relation $\dot{\bar{p}}_i = 0$ defines the *parallel transport* on T^N, the simplest form of human-motion dynamics. In that case Hamiltonian vector-field X_H is called the *geodesic spray* and its phase-flow is called the *geodesic flow*.

For Earthly dynamics in the gravitational *potential* field $V : M \to \mathbb{R}$, the Hamiltonian $H : T^*M \to \mathbb{R}$ (2.73) extends into potential form

$$H(p, x) = \frac{1}{2} g^{ij} p_i p_j + V(x),$$

with Hamiltonian vector-field $X_H = J\nabla H$ still defined by canonical equations (2.72).

A general form of a *driven*, non-conservative Hamiltonian equations reads:

$$\dot{x}^i = \partial_{p_i} H, \qquad \dot{p}_i = F_i - \partial_{x^i} H, \tag{2.75}$$

where $F_i = F_i(t, x, p)$ represent any kind of joint-driving *covariant torques*, including active neuro-muscular-like controls, as functions of time, angles and momenta, as well as passive dissipative and elastic joint torques. In the covariant momentum formulation (2.74), the non-conservative Hamiltonian equations (2.75) become:

$$\dot{\bar{p}}_i \equiv \dot{p}_i + \Gamma^i_{jk}\, g^{jl} g^{km}\, p_l p_m = F_i, \qquad \text{with} \qquad \dot{x}^k = g^{ki} p_i.$$

The general form of (autonomous) Hamiltonian dynamics is given by dissipative, driven Hamiltonian equations on T^*M:

$$\dot{x}^i = \frac{\partial H}{\partial p_i} + \frac{\partial R}{\partial p_i}, \tag{2.76}$$

$$\dot{p}_i = F_i - \frac{\partial H}{\partial x^i} + \frac{\partial R}{\partial x^i}, \tag{2.77}$$

$$x^i(0) = x^i_0, \qquad p_i(0) = p^0_i, \tag{2.78}$$

including *contravariant equation* (2.76) - the *velocity vector-field*, and *covariant equation* (2.77) - the *force 1-form* (field), together with initial joint angles and momenta (2.78). Here $R = R(x,p)$ denotes the Raileigh nonlinear (biquadratic) dissipation function, and $F_i = F_i(t,x,p)$ are covariant driving torques of *equivalent muscular actuators*, resembling muscular excitation and contraction dynamics in rotational form. The velocity vector-field (2.76) and the force 1−form (2.77) together define the generalized Hamiltonian vector-field X_H; the Hamiltonian energy function $H = H(x,p)$ is its generating function.

As a Lie group, the configuration manifold $M = \prod_j SE(3)^j$ is Hausdorff.[56] Therefore, for $x = (x^i, p_i) \in U_p$, where U_p is an open coordinate chart in T^*M, there exists a unique one-parameter group of diffeomorphisms $\phi_t : T^*M \to T^*M$, that is the *autonomous Hamiltonian phase-flow:*

$$\phi_t : T^*M \to T^*M : (p(0), x(0)) \mapsto (p(t), x(t)), \tag{2.79}$$
$$(\phi_t \circ \phi_s = \phi_{t+s}, \quad \phi_0 = \text{identity}),$$

given by (2.76-2.78) such that

$$\frac{d}{dt}|_{t=0}\, \phi_t x = J\nabla H(x).$$

Pontryagin's Maximum Principle

Pontryagin's Maximum Principle (MP) applies to a general optimization problem called a *Bolza problem* (see [PBK62]). To apply MP to optimal control, we need to define Hamiltonian function:

$$H(\psi, x, u) = (\psi, f(x,u)) = \psi_i f^i(x,u), \qquad (i = 1, ..., n). \tag{2.80}$$

Then in order for a control $u(t)$ and a trajectory $x(t)$ to be *optimal*, it is necessary that there exist a nonzero absolutely continuous vector function $\psi(t) = (\psi_0(t), \psi_1(t), ..., \psi_n(t))$ corresponding to the functions $u(t)$ and $x(t)$ such that:

[56] That is, for every pair of points $x_1, x_2 \in M$, there are disjoint open subsets (charts) $U_1, U_2 \subset M$ such that $x_1 \in U_1$ and $x_2 \in U_2$.

1. The function $H(\psi(t), x(t), u(t))$ attains its maximum at the point $u = u(t)$ almost everywhere in the interval $t_0 \leq t \leq T$,
$$H(\psi(t), x(t), u(t)) = \max_{u \in U} H(\psi(t), x(t), u(t)).$$
2. At the *terminal time* T, the relations $\psi_0(T) \leq 0$ and $H(\psi(T), x(T), u(T))$ are satisfied.

MP states the following algorithm: To maximize the set of *steering functions* $\gamma_i x^i(t)$ (with n constants γ_i) for controlling the changes in the *state variables*

$$\dot{x}^i(t) = f^i(x^i, u_k), \qquad (i = 0, 1, ..., n, \quad k = 1, ..., m),$$

we maximize at each instant the Hamiltonian function (2.80), where

$$\dot{\psi}_i = -\psi_j \frac{\partial f^j}{\partial x^i} \qquad \text{and} \qquad \psi_i(T) = \gamma_i.$$

Affine Control Systems

Now, let us look at MP as applied to the *affine control system* (see [Lew00b])

$$\dot{y}(t) = f_0(\gamma(t)) + u^a(t)\, f_a(\gamma(t)),$$

with $\gamma(t) \in M$, u taking values in $U \subset \mathbb{R}^m$, and objective function $L(x, u)$.
We need to have the control Hamiltonian on $U \times T^*M$:

$$H(\alpha_x, u) = \underbrace{\alpha_x(f_0(x))}_{H_1} + \underbrace{\alpha_x(u^a f_a(x))}_{H_2} - \underbrace{L(x, u)}_{H_3}.$$

One of several consequences of the MP is that if (u, γ) is a *minimizer* then there exists a 1−form field λ along γ with the property that $t \mapsto \lambda(t)$ is an integral curve for the time–dependent Hamiltonian $(\alpha_x, u) \mapsto H(\alpha_x, u)$. The Hamiltonian $H(\alpha_x, u)$ is a sum of three terms, and so too will be the Hamiltonian vector-field.

Let us look at the first term, that with (old) Hamiltonian $H_1 = \alpha_x(f_0(x))$. In local coordinates X_{H_1} is written as

$$\dot{x}^i = f_0^i(x), \qquad \dot{p}_i = -\frac{\partial f_0^j(x)}{\partial x^i} p_j. \tag{2.81}$$

X_{H_1} is the *cotangent lift* of f_0 and, following [Lew00b], we denote it $f_0^{T^*}$. So we want to understand $f_0^{T^*}$ on TM with $f_0 = Z$.

Let f_0 be a vector-field on a general manifold N with f_0^T its *tangent lift* defined by

$$f_0^T(v_x) = \left.\frac{d}{dt}\right|_{t=0} T_x F_t(v_x),$$

where F_t denotes the flow of f_0. Therefore, f_0^T is the 'linearisation' of f_0 and in local coordinates it is given by (compare with (2.81))

$$\dot{x}^i = f_0^i(x), \qquad \dot{v}^i = -\frac{\partial f_0^i(x)}{\partial x^j} v^j.$$

The flow of f_0^T measures how the integral curves of f_0 change as we change the initial condition in the direction of v_x.

Now, perhaps we can understand Z^T on TM with $f_0 = Z$ in the discussion of tangent lift. Let $\gamma(t)$ be a geodesic. By varying the initial condition for the geodesic we generate an 'infinitesimal variation' which satisfies the (extended) *Jacobi equation*,

$$\nabla^2_{\dot{y}(t)}\xi(t) + R(\xi(t), \dot{y}(t))\,\dot{y}(t) + \nabla_{\dot{y}(t)}\left(T(\xi(t), \dot{y}(t))\right) = 0. \tag{2.82}$$

To make the 'connection' between Z^T and the Jacobi equation, we perform constructions on the tangent bundle using the spray Z. ∇ comes from a linear connection on M which induces an *Ehresmann connection* on $\tau_M : TM \to M$. Thus we may write $T_{v_q}TM \simeq T_qM \oplus T_qM$. Now, if $I_M : TTM \to TTM$ is the canonical involution then $I_M^* Z^T$ is a spray. We use $I_M^* Z^T$ to induce an Ehresmann connection on $\tau_{TM} : TTM \to TM$. Thus,

$$T_{X_{v_q}}TTM \simeq T_{v_q}TM \oplus T_{v_q}TM \simeq \underbrace{T_qM \oplus T_qM}_{geodesic\ equations} \oplus \underbrace{T_qM \oplus T_qM}_{variation\ equations}.$$

One represents Z^T in this splitting and determines that the Jacobi equation sits 'inside' one of the four components. Now one applies similar constructions to T^*TM and Z^{T^*} to derive a 1–form version of the Jacobi equation (2.82), the so–called *adjoint Jacobi equation* [Lew00b]:

$$\nabla^2_{\dot{y}(t)}\lambda(t) + R^*(\lambda(t), \dot{y}(t))\,\dot{y}(t) - T^*\left(\nabla_{\dot{y}(t)}\lambda(t), \dot{y}(t)\right) = 0, \tag{2.83}$$

where we have used $\langle R^*(\alpha, u)v; \omega\rangle = \langle \alpha; R(\omega, u)v\rangle$, and $\langle T^*(\alpha, u); \omega\rangle = \langle \alpha; T(\omega, u)\rangle$.

The adjoint Jacobi equation forms the backbone of a general statement of the MP for affine connection control systems. When objective function is the Lagrangian $L(u, v_q) = \frac{1}{2}g(v_q, v_q)$, when ∇ is the Levi–Civita connection for the Riemannian metric g, and when the system is fully actuated, then we recover the equation of [NHP89]

$$\nabla^3_{\dot{y}(t)}\,\dot{y}(t) + R\left(\nabla_{\dot{y}(t)}\dot{y}(t), \dot{y}(t)\right) = 0.$$

Therefore, the adjoint Jacobi equation (2.83) captures the interesting part of the Hamiltonian vector-field Z^{T^*}, which comes from the MP, in terms of affine geometry, i.e., from Z^{T^*} follows

$$\nabla_{\dot{y}(t)}\dot{y}(t) = 0, \qquad \nabla^2_{\dot{y}(t)}\lambda(t) + R^*(\lambda(t), \dot{y}(t))\,\dot{y}(t) - T^*\left(\nabla_{\dot{y}(t)}\lambda(t), \dot{y}(t)\right) = 0.$$

The geometry of Z on TM provides a way of globally pulling out the adjoint Jacobi equation from the MP in an intrinsic manner, which is not generally possible in the MP [Lew00b].

Basic Hodge Machinery

Hodge star operator

The *Hodge star* operator $\star : \Omega^p(M) \to \Omega^{n-p}(M)$, which maps any $p-$form α into its *dual* $(n-p)-$form $\star\,\alpha$ on a smooth $n-$manifold M, is defined as (see, e.g. [Voi02])

$$\begin{aligned}
&\alpha \wedge \star\,\beta = \beta \wedge \star\,\alpha = \langle \alpha, \beta \rangle\, \mu, \qquad (\text{for } \alpha, \beta \in \Omega^p(M)),\\
&\star \star\,\alpha = (-1)^{p(n-p)}\alpha,\\
&\star\,(c_1\alpha + c_2\beta) = c_1(\star\,\alpha) + c_2(\star\,\beta),\\
&\alpha \wedge \star\,\alpha = 0 \Rightarrow \alpha \equiv 0.
\end{aligned}$$

The $\star$ operator depends on the Riemannian metric $g = g_{ij}$ on M and also on the orientation (reversing orientation will change the sign)[57] [II07].

The *volume form* μ is defined in local coordinates on an $n-$manifold M as (compare with Hodge inner product below)

$$\mu = \mathrm{vol} = \star\,(1) = \sqrt{\det(g_{ij})}\; dx^1 \wedge ... \wedge dx^n, \tag{2.84}$$

and the total volume on M is given by

$$\mathrm{vol}(M) = \int_M \star\,(1).$$

For example, in Euclidean $\mathbb{R}^3$ space with Cartesian (x, y, z) coordinates, we have:

$$\star\,dx = dy \wedge dz, \qquad \star\,dy = dz \wedge dx, \qquad \star\,dz = dx \wedge dy.$$

The Hodge dual in this case clearly corresponds to the 3D cross−product.

In the 4D−electrodynamics, the dual 2−form *Maxwell* $\star\,F$ satisfies the electric Maxwell equation with the source [MTW73],

$$\text{Dual Bianchi identity}: \quad d \star F = \star\,J,$$

where $\star\,J$ is the 3−form dual to the charge−current 1−form J.

[57] Note that in local coordinates on a smooth manifold M, the metric $g = g_{ij}$ is defined for any orthonormal basis $(\partial_i = \partial_{x^i})$ in M by $g_{ij} = g(\partial_i, \partial_j) = \delta_{ij}$, $\partial_k g_{ij} = 0$.

The Hodge $\star$ operator is defined locally in an orthonormal basis (coframe) of 1-forms $e_i dx^i$ on a smooth manifold M as: $\star(e_i \wedge e_j) = e_k$, $(\star)^2 = 1$.

Hodge inner product

For any two p–forms $\alpha, \beta \in \Omega^p(M)$ with compact support on an n–manifold M, we define bilinear and positive–definite Hodge L^2–inner product as

$$(\alpha, \beta) = \int_M \langle \alpha, \beta \rangle \star (1) = \int_M \alpha \wedge \star \beta, \tag{2.85}$$

where $\alpha \wedge \star \beta$ is an n–form. We can extend the product $(\cdot, \cdot)$ to $L^2(\Omega^p(M))$; it remains bilinear and positive–definite, because as usual in the definition of L^2, functions that differ only on a set of measure zero are identified. The inner product (2.85) is evidently linear in each variable and symmetric, $(\alpha, \beta) = (\beta, \alpha)$. We have: $(\alpha, \alpha) \geq 0$ and $(\alpha, \alpha) = 0$ iff $\alpha = 0$. Also, $(\star \alpha, \star \beta) = (\alpha, \beta)$. Thus, operation (2.85) turns the space $\Omega^p(M)$ into an infinite–dimensional inner–product space.

From (2.85) it follows that for every p–form $\alpha \in \Omega^p(M)$ we can define the *norm functional*

$$\|\alpha\|^2 = (\alpha, \alpha) = \int_M \langle \alpha, \alpha \rangle \star (1) = \int_M \alpha \wedge \star \alpha,$$

for which the *Euler–Lagrangian equation* becomes the Laplace equation (see Hodge Laplacian below),

$$\Delta \alpha = 0.$$

For example, the standard *Lagrangian* for the free Maxwell electromagnetic field,
$F = dA$ (where $A = A_i dx^i$ is the electromagnetic potential 1–form), is given by [II06b, II07, II08b]

$$\mathcal{L}(A) = \frac{1}{2}(F \wedge \star F),$$

with the corresponding action

$$S(A) = \frac{1}{2} \int F \wedge \star F.$$

Using the Hodge L^2–inner product (2.85), we can rewrite this electrodynamic action as

$$S(A) = \frac{1}{2}(F, F). \tag{2.86}$$

Hodge codifferential operator

The Hodge dual (or, formal adjoint) to the exterior derivative $d : \Omega^p(M) \to \Omega^{p+1}(M)$ on a smooth manifold M is the *codifferential* δ, a linear map $\delta : \Omega^p(M) \to \Omega^{p-1}(M)$, which is a generalization of the divergence, defined by [Rha84, Voi02]

$$\delta = (-1)^{n(p+1)+1} \star d \star, \qquad \text{so that} \qquad d = (-1)^{np} \star \delta \star .$$

That is, if the dimension n of the manifold M is even, then $\delta = -\star d \star$.

Applied to any p–form $\omega \in \Omega^p(M)$, the codifferential δ gives

$$\delta\omega = (-1)^{n(p+1)+1} \star d \star \omega, \qquad \delta d\omega = (-1)^{np+1} \star d \star d\omega.$$

If $\omega = f$ is a 0–form, or function, then $\delta f = 0$. If a p–form α is a codifferential of a $(p+1)$–form β, that is $\alpha = \delta\beta$, then β is called the *coexact* form. A p–form α is *coclosed* if $\delta\alpha = 0$; then $\star\alpha$ is closed (i.e., $d \star \alpha = 0$) and conversely.

The Hodge codifferential δ satisfies the following set of rules:

- $\delta\delta = \delta^2 = 0$, the same as $dd = d^2 = 0$;
- $\delta\star = (-1)^{p+1} \star d$; $\star\delta = (-1)^p \star d$;
- $d\delta\star = \star\delta d$; $\star d\delta = \delta d\star$.

Standard example is classical electrodynamics, in which the *gauge field* is an electromagnetic potential 1–form (a *connection* on a $U(1)$–bundle),

$$A = A_\mu dx^\mu = A_\mu dx^\mu + df, \qquad (f = \text{ arbitrary scalar field}),$$

with the corresponding electromagnetic field 2–form (the *curvature* of the connection A)

$$F = dA, \qquad \text{in components given by}$$
$$F = \frac{1}{2} F_{\mu\nu}\, dx^\mu \wedge dx^\nu, \qquad \text{with} \qquad F_{\mu\nu} = \partial_\nu A_\mu - \partial_\mu A_\nu.$$

Electrodynamics is governed by the *Maxwell equations*,[58] which in exterior formulation read:

$$dF = 0, \qquad \delta F = -4\pi J, \qquad \text{or in components,}$$
$$F_{[\mu\nu,\eta]} = 0, \qquad F_{\mu\nu},^{\mu} = -4\pi J_\mu,$$

where comma denotes the partial derivative and the 1–form of electric current $J = J_\mu dx^\mu$ is conserved, by the electrical *continuity equation*,

$$\delta J = 0, \qquad \text{or in components,} \qquad J_\mu,^{\mu} = 0.$$

[58] Note that the first, sourceless Maxwell equation, $dF = 0$, gives vector magneto-statics and magneto-dynamics,

$$\text{Magnetic Gauss' law : } \operatorname{div} \mathbf{B} = 0,$$
$$\text{Faraday's law : } \partial_t \mathbf{B} + \operatorname{curl} \mathbf{E} = 0.$$

The second Maxwell equation with source, $\delta F = J$ (or, $d \star F = -\star J$), gives vector electro-statics and electro-dynamics,

$$\text{Electric Gauss' law : } \operatorname{div} \mathbf{E} = 4\pi\rho,$$
$$\text{Ampère's law : } \partial_t \mathbf{E} - \operatorname{curl} \mathbf{B} = -4\pi\mathbf{j}.$$

Hodge Laplacian operator

The codifferential δ can be coupled with the exterior derivative d to construct the *Hodge Laplacian* $\Delta : \Omega^p(M) \to \Omega^p(M)$, a harmonic generalization of the Laplace–Beltrami differential operator, given by[59]

$$\Delta = \delta d + d\delta = (d + \delta)^2.$$

Δ satisfies the following set of rules:

$$\delta\,\Delta = \Delta\,\delta = \delta d\delta; \qquad d\,\Delta = \Delta\,d = d\delta d; \qquad \star\,\Delta = \Delta\,\star\,.$$

A p–form α is called *harmonic* iff

$$\Delta\alpha = 0 \;\Leftrightarrow\; (d\alpha = 0, \delta\alpha = 0).$$

Thus, α is harmonic in a compact domain $D \subset M$ iff it is both closed and coclosed in D. Informally, every harmonic form is both closed and coclosed. [Note that a domain D is compact iff it is both closed and bounded; or, formally, iff every open cover of D has a finite subcover.]

As a proof, we have:

$$0 = (\alpha,\ \Delta\alpha) = (\alpha, d\delta\alpha) + (\alpha, \delta d\alpha) = (\delta\alpha, \delta\alpha) + (d\alpha, d\alpha)\,.$$

Since $(\beta,\ \beta) \geq 0$ for any form β, $(\delta\alpha, \delta\alpha)$ and $(d\alpha, d\alpha)$ must vanish separately. Thus, $d\alpha = 0$ and $\delta\alpha = 0$.

All harmonic p–forms on a smooth manifold M form the vector space $H^p_\Delta(M)$.

Also, given a p–form λ, there is another p–form η such that the equation: $\Delta\eta = \lambda$ –is satisfied iff for any harmonic p–form γ we have $(\gamma, \lambda) = 0$.

---

[59] Note that the difference $d - \delta = \partial_D$ is called the *Dirac operator*. Its square ∂_D^2 equals the Hodge Laplacian Δ. Also, in his QFT-based rewriting the Morse topology, E. Witten [Wit82] considered also the operators:

$$d_t = \mathrm{e}^{-tf} d e^{tf}, \qquad \text{their adjoints}: \quad d_t^* = \mathrm{e}^{tf} d e^{-tf},$$
$$\text{as well as their Laplacian:} \quad \Delta_t = d_t d_t^* + d_t^* d_t.$$

For $t = 0$, Δ_0 is the *Hodge Laplacian*, whereas for $t \to \infty$, one has the following expansion

$$\Delta_t = dd^* + d^*d + t^2\,\|df\|^2 + t \sum_{k,j} \frac{\partial^2 h}{\partial x^k \partial x^j} [i\,\partial_{x^k}, dx^j],$$

where $(\partial_{x^k})_{k=1,\dots,n}$ is an orthonormal frame at the point under consideration. This becomes very large for $t \to \infty$, except at the critical points of f, i.e., where $df = 0$. Therefore, the eigenvalues of Δ_t will concentrate near the critical points of f for $t \to \infty$, and we get an *interpolation* between De Rham cohomology and Morse cohomology.

For example, to translate notions from standard 3D vector calculus, we first identify scalar functions with 0–forms, field intensity vectors with 1–forms, flux vectors with 2–forms and scalar densities with 3–forms. We then have the following correspondence:

grad $\longrightarrow d$: on 0–forms; curl $\longrightarrow \star d$: on 1–forms;
div $\longrightarrow \delta$: on 1–forms; div grad $\longrightarrow \Delta$: on 0–forms;
curl curl − grad div $\longrightarrow \Delta$: on 1–forms.

We remark here that exact and coexact p–forms ($\alpha = d\beta$ and $\omega = \delta\beta$) are mutually orthogonal with respect to the L^2–inner product (2.85). The orthogonal complement consists of forms that are both closed and coclosed: that is, of harmonic forms ($\Delta\gamma = 0$).

Hodge adjoints and self–adjoints

If α is a p–form and β is a $(p+1)$–form then we have [Rha84]

$$(d\alpha, \beta) = (\alpha, \delta\beta) \qquad \text{and} \qquad (\delta\alpha, \beta) = (\alpha, d\beta). \tag{2.87}$$

This relation is usually interpreted as saying that the two exterior differentials, d and δ, are *adjoint* (or, dual) to each other. This identity follows from the fact that for the volume form μ given by (2.84) we have $d\mu = 0$ and thus

$$\int_M d(\alpha \wedge \star\beta) = 0.$$

Relation (2.87) also implies that the Hodge Laplacian Δ is *self–adjoint* (or, self–dual),

$$(\Delta\alpha, \beta) = (\alpha, \Delta\beta),$$

which is obvious as either side is $(d\alpha, d\beta) + (\delta\alpha, \delta\beta)$. Since $(\Delta\alpha, \alpha) \geq 0$, with $(\Delta\alpha, \alpha) = 0$ only when $\Delta\alpha = 0$, Δ is a positive–definite (elliptic) self–adjoint differential operator.

Hodge decomposition theorem

The celebrated *Hodge decomposition theorem* (HDT) states that, on a compact orientable smooth n–manifold M (with $n \geq p$), any exterior p–form can be written as a unique sum of an *exact* form, a *coexact* form, and a *harmonic* form. More precisely, for any form $\omega \in \Omega^p(M)$ there are unique forms $\alpha \in \Omega^{p-1}(M)$, $\beta \in \Omega^{p+1}(M)$ and a harmonic form $\gamma \in \Omega^p(M)$, such that the following result holds:

$$\text{HDT}: \quad \underset{\omega}{\text{any form}} = \underset{d\alpha}{\text{exact form}} + \underset{\delta\beta}{\text{coexact form}} + \underset{\gamma}{\text{harmonic form}}.$$

For the proof, see [Rha84, Voi02].

In physics community, the exact form $d\alpha$ is called *longitudinal*, while the coexact form $\delta\beta$ is called *transversal*, so that they are mutually orthogonal.

Thus, any form can be orthogonally decomposed into a harmonic, a longitudinal and transversal form. For example, in fluid dynamics, any vector-field v can be decomposed into the sum of two vector-fields, one of which is divergence–free, and the other is curl–free.

Since γ is harmonic, $d\gamma = 0$. Also, by Poincaré lemma, $d(d\alpha) = 0$. In case ω is a closed $p-$form, $d\omega = 0$, then the term $\delta\beta$ in HDT is absent, so we have the *short Hodge decomposition*,

$$\omega = d\alpha + \gamma, \tag{2.88}$$

thus ω and γ differ by $d\alpha$. In topological terminology, ω and γ belong to the same *cohomology class* $[\omega] \in H^p(M)$. Now, by the De Rham theorems it follows that if C is any $p-$cycle, then

$$\int_C \omega = \int_C \gamma,$$

that is, γ and ω have the same periods. More precisely, if ω is any closed $p-$form, then there exists a unique harmonic $p-$form γ with the same periods as those of ω (see [Rha84, Fla63]).

The *Hodge–Weyl theorem* [Rha84, Voi02] states that every De Rham cohomology class has a unique harmonic representative. In other words, the space $H^p_\Delta(M)$ of harmonic $p-$forms on a smooth manifold M is isomorphic to the De Rham cohomology group (2.58), or
$H^p_\Delta(M) \cong H^p_{DR}(M)$. That is, the harmonic part γ of HDT depends only on the global structure, i.e., the topology of M.

For example, in $(2+1)$D electrodynamics, $p-$form Maxwell equations in the Fourier domain Σ are written as [TC99]

$$\begin{aligned} dE &= i\omega B, \qquad & dB &= 0, \\ dH &= -i\omega D + J, \qquad & dD &= Q, \end{aligned}$$

where H is a $0-$form (magnetizing field), D (electric displacement field), J (electric current density) and E (electric field) are $1-$forms, while B (magnetic field) and Q (electric charge density) are $2-$forms. From $d^2 = 0$ it follows that the J and the Q satisfy the *continuity equation*

$$dJ = iwQ,$$

where $i = \sqrt{-1}$ and w is the field frequency. Constitutive equations, which include all metric information in this framework, are written in terms of Hodge star operators (that fix an isomorphism between p forms and $(2-p)$ forms in the $(2+1)$ case)

$$D = \star E, \qquad B = \star H.$$

Applying HDT to the electric field intensity $1-$form E, we get [HT05]

$$E = d\phi + \delta A + \chi,$$

where ϕ is a $0-$form (a scalar field) and A is a $2-$form; $d\phi$ represents the static field and δA represents the dynamic field, and χ represents the harmonic field component. If domain Σ is contractible, χ is identically zero and we have the short Hodge decomposition,

$$E = d\phi + \delta A.$$

Basic Applications to Quantum Field Theory

Feynman Path Integral

The 'driving engine' of quantum field theory is the Feynman path integral.[60] Very briefly, there are three basic forms of the path integral (see, e.g. [Zee03, II08b]):

1. *Sum−over−histories,* developed in Feynman's version of quantum mechanics (QM); [Fey48];

---

[60] Note that Feynman's *amplitude* is a space-time version of the Schrödinger's *wave-function* ψ, which describes how the (non-relativistic) quantum state of a physical system changes in space and time, i.e.,

$$\langle \mathrm{Out}_{t_{fin}} | \mathrm{In}_{t_{ini}} \rangle = \psi(\mathbf{x}, t), \qquad (\text{for } \mathbf{x} \in [\mathrm{In}, \mathrm{Out}],\ t \in [t_{ini}, t_{fin}]).$$

In particular, quantum wave-function ψ is a complex-valued function of real space variables $\mathbf{x} = (x_1, x_2, ..., x_n) \in \mathbb{R}^n$, which means that its domain is in $\mathbb{R}^n$ and its range is in the complex plane, formally $\psi(\mathbf{x}) : \mathbb{R}^n \to \mathbb{C}$. For example, the one-dimensional *stationary plane wave* with wave number k is defined as

$$\psi(x) = \mathrm{e}^{\mathrm{i}kx}, \qquad (\text{for } x \in \mathbb{R}),$$

where the real number k describes the wavelength, $\lambda = 2\pi/k$. In n dimensions, this becomes

$$\psi(x) = \mathrm{e}^{\mathrm{i}\mathbf{p}\cdot\mathbf{x}},$$

where the momentum vector $\mathbf{p} = \mathbf{k}$ is the vector of the wave numbers $\mathbf{k}$ in natural units (in which $\hbar = m = 1$).

More generally, quantum wave-function is also time dependent, $\psi = \psi(\mathbf{x}, \mathbf{t})$. The time-dependent plane wave is defined by

$$\psi(\mathbf{x}, \mathbf{t}) = \mathrm{e}^{\mathrm{i}\mathbf{p}\cdot\mathbf{x} - \mathrm{i}p^2 t/2}. \tag{2.89}$$

In general, $\psi(\mathbf{x}, \mathbf{t})$ is governed by the Schrödinger equation [Tha00, II08b] (in natural units $\hbar = m = 0$)

$$\mathrm{i}\frac{\partial}{\partial t}\psi(\mathbf{x}, t) = -\frac{1}{2}\Delta\psi(\mathbf{x}, t),$$

2. *Sum–over–fields,* started in Feynman's version of quantum electrodynamics (QED) [Fey98] and later improved by Fadeev–Popov [FS80]; and
3. *Sum–over–geometries/topologies* in quantum gravity (QG), initiated by S. Hawking and properly developed in the form of causal dynamical triangulations (see [ALW08]; for a 'softer' review, see [Lol08]).

In all three versions, Feynman's *action–amplitude formalism* includes two components:

- A real–valued, classical, *Hamilton's action functional,*

$$S[\Phi] := \int_{t_{ini}}^{t_{fin}} L[\Phi]\, dt,$$

with the Lagrangian energy function defined over the Lagrangian density $\mathcal{L}$,[61]

where Δ is the n–dimensional Laplacian. Its solution is given by the integral of the time-dependent plane wave (2.89),

$$\psi(\mathbf{x},t) = \frac{1}{(2\pi)^{n/2}} \int_{\mathbb{R}^n} \mathrm{e}^{\mathrm{i}\mathbf{p}\cdot\mathbf{x} - \mathrm{i}p^2 t/2} \hat{\psi}_0(\mathbf{p}) d^n p,$$

which means that $\psi(\mathbf{x},t)$ is the inverse Fourier transform of the function

$$\hat{\psi}(\mathbf{p},t) = \mathrm{e}^{-\mathrm{i}p^2 t/2} \hat{\psi}_0(\mathbf{p}),$$

where $\hat{\psi}_0(\mathbf{p})$ has to be calculated for each initial wave-function. For example, if initial wave-function is Gaussian,

$$f(x) = \exp(-a\frac{x^2}{2}), \qquad \text{with the Fourier transform} \qquad \hat{f}(p) = \frac{1}{\sqrt{a}}\exp(-\frac{p^2}{2a}).$$

then $\quad \hat{\psi}_0(p) = \frac{1}{\sqrt{a}}\exp(-\frac{p^2}{2a})$.

[61] Note that in Lagrangian field theory, the fundamental quantity is the action

$$S[\Phi] = \int_{t_{in}}^{t_{out}} L\, dt = \int_{\mathbb{R}^4} d^n x\, \mathcal{L}(\Phi, \partial_\mu \Phi)\,,$$

so that the least action principle, $\delta S[\Phi] = 0$, gives

$$\begin{aligned} 0 &= \int_{\mathbb{R}^4} d^n x \left\{ \frac{\partial \mathcal{L}}{\partial \Phi}\delta\Phi + \frac{\partial \mathcal{L}}{\partial(\partial_\mu \Phi)}\delta(\partial_\mu \Phi) \right\} \\ &= \int_{\mathbb{R}^4} d^n x \left\{ \frac{\partial \mathcal{L}}{\partial \Phi}\delta\Phi - \partial_\mu \left(\frac{\partial \mathcal{L}}{\partial(\partial_\mu \Phi)} \right) \delta\Phi + \partial_\mu \left(\frac{\partial \mathcal{L}}{\partial(\partial_\mu \Phi)}\delta\Phi \right) \right\}. \end{aligned}$$

The last term can be turned into a surface integral over the boundary of the $\mathbb{R}^4$ (4D space-time region of integration). Since the initial and final field configurations are assumed given, $\delta\Phi = 0$ at the temporal beginning t_{in} and end t_{out} of this region, which implies that the surface term is zero. Factoring out the $\delta\Phi$ from the first two terms, and since the integral must vanish for arbitrary $\delta\Phi$, we arrive at the Euler-lagrange equation of motion for a field,

$$L[\Phi] = \int d^n x \, \mathcal{L}(\Phi, \partial_\mu \Phi), \qquad (\partial_\mu \equiv \partial/\partial x^\mu),$$

while Φ is a common symbol denoting all three things to be summed upon (histories, fields and geometries). The action functional $S[\Phi]$ obeys the *Hamilton's least action principle,* $\delta S[\Phi] = 0$, and gives, using standard variational methods, the Euler–Lagrangian equations, which define the shortest path, the extreme field, and the geometry of minimal curvature (and without holes).

- A complex–valued, quantum *transition amplitude,*[62]

$$\partial_\mu \left(\frac{\partial \mathcal{L}}{\partial(\partial_\mu \Phi)} \right) - \frac{\partial \mathcal{L}}{\partial \Phi} = 0.$$

If the Lagrangian (density) $\mathcal{L}$ contains more fields, there is one such equation for each. The momentum density $\pi(x)$ of a field, conjugate to $\Phi(x)$ is defined as: $\pi(x) = \frac{\partial \mathcal{L}}{\partial_\mu \Phi(x)}$.

For example, the standard electromagnetic action

$$S = -\frac{1}{4} \int_{\mathbb{R}^4} d^4 x \, F_{\mu\nu} F^{\mu\nu}, \qquad \text{where} \qquad F_{\mu\nu} = \partial_\mu A_\nu - \partial_\nu A_\mu,$$

gives the sourceless Maxwell's equations:

$$\partial_\mu F^{\mu\nu} = 0, \qquad \epsilon^{\mu\nu\sigma\eta} \partial_\nu F_{\sigma\eta} = 0,$$

where the field strength tensor $F_{\mu\nu}$ and the Maxwell equations are invariant under the *gauge transformations,*

$$A_\mu \longrightarrow A_\mu + \partial_\mu \epsilon.$$

The equations of motion of charged particles are given by the Lorentz-force equation,

$$m \frac{du^\mu}{d\tau} = e F^{\mu\nu} u_\nu,$$

where e is the charge of the particle and $u^\mu(\tau)$ its four-velocity as a function of the proper time.

[62] Note that the transition amplitude (2.92) is closely related to *partition function* Z, which is a quantity that encodes the statistical properties of a system in thermodynamic equilibrium. It is a function of temperature and other parameters, such as the volume enclosing a gas. Other thermodynamic variables of the system, such as the total energy, free energy, entropy, and pressure, can be expressed in terms of the partition function or its derivatives. In particular, the partition function of a *canonical ensemble* is defined as a sum $Z(\beta) = \sum_j \mathrm{e}^{-\beta E_j}$, where $\beta = 1/(k_B T)$ is the 'inverse temperature', where T is an ordinary temperature and k_B is the Boltzmann's constant. However, as the position x^i and momentum p_i variables of an ith particle in a system can vary continuously, the set of microstates is actually uncountable. In this case, some form of *coarse-graining* procedure must be carried out, which essentially amounts to treating two mechanical states as the same microstate if the differences in their position and momentum variables are 'small enough'. The partition function then takes the form of an integral. For instance, the partition function of a gas consisting of N molecules is proportional

$$\langle \mathrm{Out}_{t_{fin}} | \mathrm{In}_{t_{ini}} \rangle := \int_{\Omega} \mathcal{D}[\Phi] \, \mathrm{e}^{\mathrm{i}S[\Phi]}, \tag{2.92}$$

where $\mathcal{D}[\Phi]$ is 'an appropriate' Lebesgue–type measure,

---

to the $6N$–dimensional phase-space integral,

$$Z(\beta) \sim \int_{\mathbb{R}^{6N}} d^3p_i \, d^3x^i \exp[-\beta H(p_i, x^i)],$$

where $H = H(p_i, x^i)$, $(i = 1, ..., N)$ is the classical Hamiltonian (total energy) function.

Given a set of random variables X_i taking on values x^i, and purely potential Hamiltonian function $H(x^i)$, the partition function is defined as

$$Z(\beta) = \sum_{x^i} \exp\left[-\beta H(x^i)\right].$$

The function H is understood to be a real-valued function on the space of states $\{X_1, X_2, \cdots\}$ while β is a real-valued free parameter (conventionally, the inverse temperature). The sum over the x^i is understood to be a sum over all possible values that the random variable X_i may take. Thus, the sum is to be replaced by an integral when the X_i are continuous, rather than discrete. Thus, one writes

$$Z(\beta) = \int dx^i \exp\left[-\beta H(x^i)\right],$$

for the case of continuously-varying random variables X_i.

Now, the number of variables X_i need not be countable, in which case the set of coordinates $\{x^i\}$ becomes a field $\phi = \phi(x)$, so the sum is to be replaced by the *Euclidean path integral* (that is a Wick-rotated Feynman transition amplitude (2.93) in imaginary time), as

$$Z(\phi) = \int \mathcal{D}[\phi] \exp\left[-H(\phi)\right].$$

More generally, in quantum field theory, instead of the field Hamiltonian $H(\phi)$ we have the action $S(\phi)$ of the theory. Both Euclidean path integral,

$$Z(\phi) = \int \mathcal{D}[\phi] \exp\left[-S(\phi)\right], \qquad \text{real path integral in imaginary time,} \tag{2.90}$$

and Lorentzian one,

$$Z(\phi) = \int \mathcal{D}[\phi] \exp\left[\mathrm{i}S(\phi)\right], \qquad \text{complex path integral in real time,} \tag{2.91}$$

are usually called 'partition functions'. While the Lorentzian path integral (2.91) represents a quantum-field theory-generalization of the Schrödinger equation, the Euclidean path integral (2.90) represents a statistical-field-theory generalization of the Fokker-Planck equation.

$$\mathcal{D}[\Phi] = \lim_{N\to\infty} \prod_{s=1}^{N} \Phi_s^i, \qquad (i = 1, ..., n),$$

so that we can 'safely integrate over a continuous spectrum and sum over a discrete spectrum of our problem domain Ω', of which the absolute square is the real–valued probability density function,

$$P \;:=\; |\langle \mathrm{Out}_{t_{fin}} | \mathrm{In}_{t_{ini}} \rangle\rangle|^2.$$

The above transition amplitude (2.92) procedure can be redefined in a mathematically cleaner way if we Wick–rotate the time variable t to imaginary values, $t \mapsto \tau = t$, thereby making all integrals real:

$$\int \mathcal{D}[\Phi]\, \mathrm{e}^{\mathrm{i}S[\Phi]} \xrightarrow{Wick} \int \mathcal{D}[\Phi]\, \mathrm{e}^{-S[\Phi]}. \tag{2.93}$$

For example, in non-relativistic quantum mechanics, the propagation amplitude from x_a to x_b is given by the *configuration path integral*[63]

$$U(x_a, x_b; T) = \langle x_b | x_a \rangle = \left\langle x_b \right| \mathrm{e}^{-\mathrm{i}HT} \left| x_a \right\rangle = \int \mathcal{D}[x(t)]\, \mathrm{e}^{\mathrm{i}S[x(t)]},$$

which satisfies the Schrödinger equation (in natural units)

$$i\frac{\partial}{\partial T} U(x_a, x_b; T) = \hat{H} U(x_a, x_b; T), \qquad \text{where} \qquad \hat{H} = -\frac{1}{2}\frac{\partial^2}{\partial x_b^2} + V(x_b).$$

[63] On the other hand, the *phase-space path integral* (without peculiar constants in the functional measure) reads

$$U(q_a, q_b; T) = \left(\prod_i \int \mathcal{D}[q(t)]\mathcal{D}[p(t)]\right) \exp\left[\mathrm{i}\int_0^T \left(p_i \dot{q}^i - H(q,p)\right) dt\right],$$

where the functions $q(t)$ (space coordinates) are constrained at the endpoints, but the functions $p(t)$ (canonically-conjugated momenta) are not. The functional measure is just the product of the standard integral over phase space at each point in time

$$\mathcal{D}[q(t)]\mathcal{D}[p(t)] = \prod_i \frac{1}{2\pi} \int dq^i dp_i.$$

Applied to a non-relativistic real scalar field $\phi(x,t)$, this path integral becomes

$$\left\langle \phi_b(x,t) \right| \mathrm{e}^{-\mathrm{i}HT} \left| \phi_a(x,t) \right\rangle = \int \mathcal{D}[\phi] \exp\left[\mathrm{i}\int_0^T \mathcal{L}(\phi)\, d^4x\right],$$

$$\text{with} \qquad \mathcal{L}(\phi) = \frac{1}{2}(\partial_\mu \phi)^2 - V(\phi).$$

Functional measure on the space of differential forms

The Hodge inner product (2.85) leads to a natural (metric–dependent) functional measure $\mathcal{D}\mu[\omega]$ on $\Omega^p(M)$, which normalizes the *Gaussian functional integral*

$$\int \mathcal{D}\mu[\omega]\, \mathrm{e}^{\mathrm{i}\langle\omega|\omega\rangle} = 1. \tag{2.94}$$

One can use the invariance of (2.94) to determine how the functional measure transforms under the Hodge decomposition. Using HDT and its orthogonality with respect to the inner product (2.85), it was shown in [GK94] that

$$\langle\omega,\omega\rangle = \langle\gamma,\gamma\rangle + \langle d\alpha, d\alpha\rangle + \langle\delta\beta,\delta\beta\rangle = \langle\gamma,\gamma\rangle + \langle\alpha,\delta d\alpha\rangle + \langle\beta, d\delta\beta\rangle\,, \tag{2.95}$$

where the following differential/conferential identities were used [CD82]

$$\langle d\alpha, d\alpha\rangle = \langle\alpha,\delta d\alpha\rangle \qquad \text{and} \qquad \langle\delta\beta,\delta\beta\rangle = \langle\beta, d\delta\beta\rangle.$$

Since, for any linear operator O, one has

$$\int \mathcal{D}\mu[\omega] \exp \mathrm{i}\langle\omega|O\omega\rangle = \det{}^{-1/2}(O),$$

(2.94) and (2.95) imply that

$$\mathcal{D}\mu[\omega] = \mathcal{D}\mu[\gamma]\mathcal{D}\mu[\alpha]\mathcal{D}\mu[\beta] \det{}^{1/2}(\delta d)\det{}^{1/2}(d\delta).$$

Abelian Chern–Simons theory

Recall that the classical action for an Abelian Chern–Simons theory,

$$S = \int_M A \wedge dA\,,$$

is invariant (up to a total divergence) under the gauge transformation:

$$A \longmapsto A + d\varphi. \tag{2.96}$$

We wish to compute the *partition function* for the theory

$$Z := \int \frac{1}{V_G} \mathcal{D}\mu[A]\, \mathrm{e}^{\mathrm{i}S[A]},$$

where V_G denotes the volume of the group of gauge transformations in (2.96), which must be factored out of the partition function in order to guarantee that the integration is performed only over physically distinct gauge fields. We can handle this by using the Hodge decomposition to parametrize the potential A

in terms of its gauge invariant, and gauge dependent parts, so that the volume of the group of gauge transformations can be explicitly factored out, leaving a functional integral over gauge invariant modes only [GK94].

We now transform the integration variables:

$$A \longmapsto \alpha, \beta, \gamma,$$

where α, β, γ parameterize respectively the exact, coexact, and harmonic parts of the connection A. Using the Jacobian (2.95) as well as the following identity on 0−forms $\Delta = \delta d$, we get [GK94]

$$Z = \int \frac{1}{V_G} \mathcal{D}\mu[\alpha]\mathcal{D}\mu[\beta]\mathcal{D}\mu[\gamma] \det{}^{1/2}(\Delta) \det{}^{1/2}(d\delta)\, \mathrm{e}^{\mathrm{i}S},$$

from which it follows that

$$V_G = \int \mathcal{D}\mu[\alpha], \tag{2.97}$$

while the classical action functional becomes, after integrating by parts, using the harmonic properties of γ and the nilpotency of the exterior derivative operators, and dropping surface terms:

$$S = -\langle \beta, \star\delta d\delta\beta \rangle \ .$$

Note that S depends only the coexact (transverse) part of A. Using (2.97) and integrating over β yields:

$$Z = \int \mathcal{D}\mu[\gamma] \det{}^{-1/2}(\star\delta d\delta) \det{}^{1/2}(\Delta) \det{}^{1/2}(d\delta)\,.$$

Also, it was proven in [GK94] that

$$\det(\star\delta d\delta) = \det{}^{1/2}((d\delta d)(\delta d\delta)) = \det{}^{\frac{3}{2}}(d\delta).$$

As a consequence of Hodge duality we have the identity

$$\det(\delta d) = \det(d\delta),$$

from which it follows that

$$Z = \int \mathcal{D}\mu[\gamma] \det{}^{-3/4}\left(\Delta^T_{(1)}\right) \det{}^{1/2}(\Delta) \det{}^{1/2}\left(\Delta^T_{(1)}\right).$$

The operator $\Delta^T_{(1)}$ is the transverse part of the Hodge Laplacian acting on 1−forms:

$$\Delta^T_{(1)} := (\delta d)_{(1)}.$$

Applying identity for the Hodge Laplacian $\Delta_{(p)}$ [GK94]:

$$\det\left(\Delta_{(p)}\right) = \det\left((\delta d)_{(p)}\right) \det\left((\delta d)_{(p-1)}\right),$$

we get

$$\det\left(\Delta_{(1)}^T\right) = \det\left(\Delta_{(1)}\right)/\det\left(\Delta\right)$$

and hence

$$Z = \int \mathcal{D}\mu[\gamma] \det{}^{-1/4}\left(\Delta_{(1)}\right) \det{}^{3/4}\left(\Delta\right).$$

The space of harmonic forms γ (of any order) is a finite set. Hence, the integration over harmonic forms (2.8.4) is a simple sum.

Lie Algebra-Valued Exterior Forms

Denote by $\Omega^p = \Omega^p(T^*M)$ the set of all smooth p-forms $\{\omega_i\}$ on a smooth m-manifold M (for all $p = 1, ..., m$). Let $\mathfrak{g}$ be a real nD Lie algebra (with $n \leq m$). A $\mathfrak{g}$-valued p-form $\widetilde{\omega}$ on M is an element of the tensor product $\Omega^p \otimes \mathfrak{g}$. If the basis for $\mathfrak{g}$ is given by the set of vectorfields $\{X^i\}$, (for $i = 1, ..., n$), then $\widetilde{\omega} \in \Omega^p \otimes \mathfrak{g}$ is given by the component tensor products:

$$\widetilde{\omega} = \omega_i \otimes X^i.$$

For any three Lie algebra-valued exterior forms ($\widetilde{\omega} \in \Omega^p \otimes \mathfrak{g}$, $\widetilde{\varphi} \in \Omega^q \otimes \mathfrak{g}$, $\widetilde{\rho} \in \Omega^r \otimes \mathfrak{g}$) the following relations hold.

Standard exterior operations:

1. Exterior derivative: $d\widetilde{\omega} = (d\omega_i) \otimes X^i$, so we have: $dd\widetilde{\omega} = 0$;
2. Hodge star: $*\widetilde{\omega} = (*\omega_i) \otimes X^i$;
3. Exterior product: $\alpha \wedge \widetilde{\omega} = (\alpha \wedge \omega_i) \otimes X^i$, for any p-form $\alpha \in \Omega^p$, so we have: $d(\alpha \wedge \widetilde{\omega}) = (d\alpha) \wedge \widetilde{\omega} + (-1)^p \alpha \wedge \widetilde{\omega}$;
4. Interior product: $Y \lrcorner \widetilde{\omega} = (Y \lrcorner \omega_i) \otimes X^i$, for any vectorfield $Y \in \mathfrak{g}$;
5. Pullback: $f^*\widetilde{\omega} = (f^*\omega_i) \otimes X^i$, for any map $f \in M$;
6. Lie derivative: $\mathcal{L}_Y \widetilde{\omega} = (\mathcal{L}_Y \omega_i) \otimes X^i$, for any vectorfield $Y \in \mathfrak{g}$.

Graded Lie bracket: $[\widetilde{\omega}, \widetilde{\varphi}] \in \Omega^p \wedge \Omega^q \otimes \mathfrak{g}$ is defined by:

$$[\widetilde{\omega}, \widetilde{\varphi}] = (\omega_i \wedge \varphi_j) \otimes [X^i, X^j], \tag{2.98}$$

where $[X^i, X^j] \in \mathfrak{g}$ is the standard Lie bracket.The graded Lie bracket (2.98) satisfies the following properties (inherited from the standard Lie bracket):

1. $\mathbb{Z}_2$-graded anticommutativity: $[\widetilde{\omega}, \widetilde{\varphi}] = (-1)^{pq+1}[\widetilde{\varphi}, \widetilde{\omega}]$;
2. Bilinearity: $[\widetilde{\omega}, \widetilde{\varphi} + \widetilde{\psi}] = [\widetilde{\omega}, \widetilde{\varphi}] + [\widetilde{\omega}, \widetilde{\psi}]$; and
3. $\mathbb{Z}_2$-graded Jacoby identity: $(-1)^{pr}[\widetilde{\omega}, [\widetilde{\varphi}, \widetilde{\rho}]] + (-1)^{pq}[\widetilde{\varphi}, [\widetilde{\rho}, \widetilde{\omega}]] + (-1)^{rq}[\widetilde{\rho}, [\widetilde{\omega}, \widetilde{\varphi}]] = 0$.

$\mathfrak{g}$-valued exterior product: $\widetilde{\omega} \wedge \widetilde{\varphi} \in \Omega^p \wedge \Omega^q \otimes \mathfrak{g}^2$ is the matrix-valued $(p+q)$-form, defined by:

$$\widetilde{\omega} \wedge \widetilde{\varphi} = (\omega_i \wedge \varphi_j) \otimes (X^i X^j),$$

where $X^i X^j$ is the standard matrix product of vectorfields X^i and X^j. In particular, $\widetilde{\omega} \wedge \widetilde{\omega} = (\omega_i \wedge \omega_j) \otimes (X^i X^j)$, and (from $\omega_i \wedge \omega_j = -\omega_j \wedge \omega_i$), we have:

$$\widetilde{\omega} \wedge \widetilde{\omega} = (\omega_i \wedge \omega_j) \otimes [X^i, X^j].$$

Exterior Connection, Curvature and Torsion

An *exterior covariant derivative*, or an *exterior connection* in a smooth vector bundle $E \to M$ over a smooth n-manifold M is a differential operator:

$$D : \Gamma(E) \to \Gamma(E \otimes T^* M),$$

where $\Gamma(E)$ denotes the space of smooth sections v of the bundle E. If f is a smooth function with exterior derivative df, then D is defined as:

$$D(fv) = v \otimes df + fDv.$$

The curvature 2-form Ω of a connection 1-form ω in the bundle E is defined by:

$$\Omega = D\omega = d\omega + \omega \wedge \omega.$$

In terms of the exterior connection D, the curvature is given by:

$$\Omega(v) = D(Dv) = D^2 v, \qquad \text{(for any } v \in E).$$

Thus the curvature measures the failure of the sequence:

$$\Gamma(E) \xrightarrow{D} \Gamma(E \otimes \Omega^1(M)) \xrightarrow{D} \Gamma(E \otimes \Omega^2(M)) \xrightarrow{D} \dots \xrightarrow{D} \Gamma(E \otimes \Omega^n(M))$$

(where $\Omega^1(M) = T^* M$) to be a de Rham chain complex.

If the fibre dimension of the vector bundle E is equal to $\dim M = n$,then E can be additionally equipped with a *soldering form*, a globally defined vector-valued 1-form $\theta \in \Gamma(T^* M)$ such that the mapping: $\theta_x : T^* M \to E_x$ is a linear isomorphism for all $x \in M$. Then we can define an E-valued 2-form on M called the *torsion* Θ of the connection ω, as:

$$\Theta = D\theta.$$

In the local frame $\{e_i\}$ on E, the components of the soldering form θ and torsion Θ are:

$$\theta = \theta^i e_i, \qquad \Theta^i = d\theta^i + \omega^i_j \wedge \theta^j.$$

Cartan Calculus

Main exterior forms

1. Basis 1-form: θ^a
2. Connection 1-form: ω_a^b
3. Curvature 2-form: Ω_a^b
4. Torsion 2-form: Θ^a

Structure equations

Curvature (or, second) structure equation:

$$\Omega_a^b = d\omega_a^b + \omega_a^c \wedge \omega_c^b$$

Torsion (or, first) structure equation:

$$\Theta^a = d\theta^a - \theta^b \wedge \omega_b^a$$

Bianchi identities

Taking the exterior derivative of the curvature structure equation gives:

$$d\Omega_a^b - \omega_a^c \wedge d\omega_c^b + \omega_c^b \wedge d\omega_a^c = 0,$$

and subsequently applying both structure equations, we obtain the second Bianchi identity:

$$d\Omega_a^b + \omega_p^b \wedge \Omega_a^p - \omega_a^p \wedge \Omega_p^b = 0.$$

Taking the exterior derivative of the torsion structure equation gives:

$$d\Theta^a - \theta^b \wedge d\omega_b^a + \omega_b^a \wedge d\theta^b = 0,$$

and subsequently applying both structure equations, we obtain the first Bianchi identity:

$$d\Theta^a - \theta^p \wedge \Omega_p^a + \omega_p^a \wedge \Theta^p = 0.$$

Cartan Equations, Gauge Potential and Field Strength

Let G be a Lie group with its Lie algebra $\mathfrak{g}$. The so-called *gauge potential* is a $\mathfrak{g}$-valued 1-form ω, a canonical *Maurer-Cartan form* defined by the equation:

$$\omega(X^g) = g^{-1}X^g, \qquad \text{(with } g \in G\text{)}.$$

The potential form ω is governed by the *Maurer-Cartan structure equation:*

$$d\omega + \frac{1}{2}[\omega, \omega] = O, \qquad \text{or} \qquad d\omega + \omega \wedge \omega = 0. \tag{2.99}$$

The corresponding *field strength* is its *exterior-covariant derivative:* $\Omega = D\omega \in G$, which is a $\mathfrak{g}$-valued 2-form given by:

$$\Omega = D\omega = d\omega + \frac{1}{2}[\omega, \omega] = d\omega + \omega \wedge \omega. \qquad (2.100)$$

Equation (2.100) means that the 2-form Ω measures the extent to which the 1-form ω fails to satisfy the Maurer-Cartan equation (2.99). In the language of *principal bundles*, the gauge potential ω corresponds to *connection*, while the field strength Ω corresponds to *curvature*.

The most important special case is when the group G is a Riemannian manifold with the tangent bundle TG and the cotangent bundle T^*G, with the *Levi-Civita connection* $\Gamma = \Gamma_{ij}^k(e)$ given by the *Christoffel symbols* $\Gamma_{ij}^k(e)$ (with $i, j, k = 1, 2, ..., \dim G$). In the *local frame* $e = \{e_i\} \subset TG$, the connection Γ is defined by the covariant derivative $\nabla = \nabla_{e_i}$ as:

$$\nabla e = \Gamma e, \qquad \text{in components:} \qquad \nabla_{e_i} e_j = \Gamma_{ij}^k e_k.$$

If $\theta = \{\theta^i\} \subset T^*G$ denotes the *local coframe,* i.e., the dual basis on T^*G, such that $\theta^i(e_j) = \delta^i_j$, then the connection 1-form $\omega = \omega_i^j(e)$ is defined in θ by:

$$\omega = \Gamma\theta, \qquad \text{in components:} \qquad \omega_i^j(e) = \Gamma_{ki}^j \theta^k(e).$$

The *curvature* 2-form $\Omega = \Omega_i^j(e)$, corresponding to the Levi-Civita connection Γ, still defined as the exterior-covariant derivative $\Omega = D\omega$ in (2.100), is in the local coframe θ given by:

$$\Omega = d\omega + \omega \wedge \omega, \qquad \text{in components:} \qquad \Omega_i^j(e) = d\omega_i^j(e) + \omega_k^j(e) \wedge \omega_i^k(e).$$

3

Global Categorical Framework for Complexity and Control

In this Chapter, we present a global structural framework for complexity and control, given by modern algebraic and computational machinery of categories, functors and derived natural structures. Category theory provides us with what amounts to a powerful abstraction that hides and unifies the details of various individual theories and models. The central concept is that of transformations between objects that preserve structure.

3.1 Introduction

In contrast with the local geometric machinery of the previous chapter, here we are concerned with establishing a global perspective on complexity and control using the theory of Categories. First, recall that a function (or, a map) f is a *rule* that assigns to each element x in a set A exactly one element, called $f(x)$, in a set B. A function could be also thought of as an *input-output system* $[[f]]$ with $x-$input (the *domain* of f is the set of all possible inputs) and $f(x)-$output (the *range* of f is the set of all possible outputs):

$$x \to [[f]] \to f(x).$$

Let $f : A \to B$ be a map between two sets: $A \equiv \operatorname{Dom} f$ and $B \equiv \operatorname{Cod} f$. Then:

$$\begin{aligned}
\operatorname{Ker} f &= f^{-1}(e_B) \text{ – is a kernel of } f; \\
\operatorname{Im} f &= f(A) \text{ – is an image of } f; \\
\operatorname{Coker} f &= \operatorname{Cod} f / \operatorname{Im} f \text{ – is a cokernel of } f; \\
\operatorname{Coim} f &= \operatorname{Dom} f / \operatorname{Ker} f \text{ – is a coimage of } f;
\end{aligned}$$

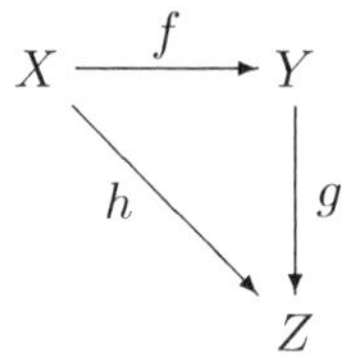

– is a commutative diagram, requiring $h = g \circ f$.

Let f and g be maps with domains A and B. Then the maps $f+g$, $f-g$, fg, and f/g are defined as follows (see, e.g. [II06b])

$$
\begin{aligned}
(f+g)(x) &= f(x)+g(x) && \text{domain} = A \cap B,\\
(f-g)(x) &= f(x)-g(x) && \text{domain} = A \cap B,\\
(fg)(x) &= f(x)\,g(x) && \text{domain} = A \cap B,\\
\left(\frac{f}{g}\right)(x) &= \frac{f(x)}{g(x)} && \text{domain} = \{x \in A \cap B : g(x) \neq 0\}.
\end{aligned}
$$

Given two maps f and g, the composite map $f \circ g$, called the *composition* of f and g, is defined by

$$(f \circ g)(x) = f(g(x)).$$

The $(f \circ g)-$machine is composed of the $g-$machine (first) and then the $f-$machine:

$$x \to [[g]] \to g(x) \to [[f]] \to f(g(x)).$$

For example, suppose that $y = f(u) = \sqrt{u}$ and $u = g(x) = x^2 + 1$. Since y is a function of u and u is a function of x, it follows that y is ultimately a function of x. We calculate this by substitution

$$y = f(u) = f \circ g = f(g(x)) = f(x^2+1) = \sqrt{x^2+1}.$$

Mathematical Example: Chain Homomorphism

Given two *chain complexes* C_* and D_* (see Appendix 3.8.1) with boundary operators ∂_C, ∂_D, a (homological) chain homomorphism is given by homomorphisms $\alpha_i : C_i \longrightarrow D_i$ such that the following diagram commutes:

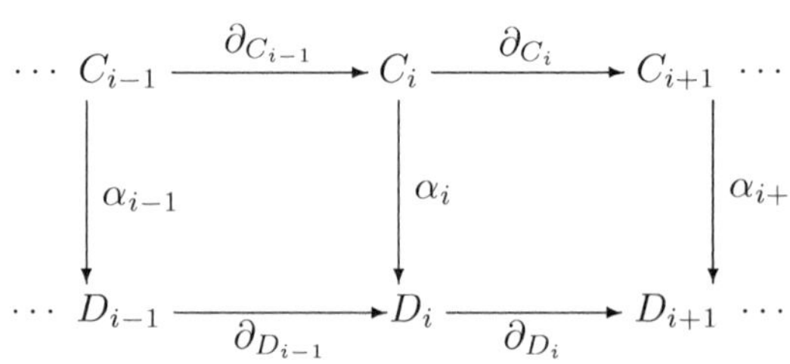

Biomedical Example of a Functional Composition

Many biochemical and biophysical reactions in human body represent compositions (or, sequences) of various biophysical processes. Here we present an elaborate example of *electrical muscular stimulation* EMS (see Figure 3.1), which is a nonsurgical biomedical rehabilitation procedure of externally stimulating weak human skeletal of face muscles. More precisely, the $EMS-$response mapping $\mathcal{F}$ of a skeletal muscle to the efferent functional stimulation from the Hodgkin-Huxley neural system (HH, see subsection 7.2.1 below) can be modeled as a *fifth-order electrical transmission cascade*

$$\mathcal{F} = (\mathcal{F}_1 \mapsto \mathcal{F}_2 \mapsto \mathcal{F}_3 \mapsto \mathcal{F}_4 \mapsto \mathcal{F}_5) = (\mathcal{F}_5 \circ \mathcal{F}_4 \circ \mathcal{F}_3 \circ \mathcal{F}_2 \circ \mathcal{F}_1),$$

where the maps $\mathcal{F}_i\,(i = 1, \ldots, 5)$ represent: *neural action potential, synaptic potential, muscular action potential, excitation-contraction coupling and muscle tension generating*, respectively. All transmission components of the cascade $\mathcal{F}$ represent (bio)*electrical diffusion-reaction processes*.

Formally, the mapping *cascade* $\mathcal{F}$ (for all included motor units in the particular muscular contraction) can be described by the following *recurrent electrical diffusion* system:

$$\mathcal{F}_i \,:\, C_i \frac{\partial V_i}{\partial t} = \frac{1}{R_i} \frac{\partial^2 V_{i-1}}{\partial z^2} - J_i(V_i),$$

with the boundary condition at $z = 0$,

$$V_1(0,t) = V_0 \sin(2\pi f t) \;=\; S(t), \qquad (i = 1, \ldots, 5).$$

The behavior of a single element $\mathcal{F}_4$ is now given by:

$$V_i(z,t) = V_0 \exp(-z_i/m) \sin(2\pi f(t - z_i/n)), \quad m = \frac{1}{R_i C_i f}, \quad n = \frac{4\pi f}{R_i C_i}.$$

3.2 Categories, Functors and Naturality

The properly defined mathematical framework for general functional compositions is the category theory, originated from the work in algebraic topology by S. Eilenberg and S. MacLane[1] [EM45, Mac71]. Here we give a short and intuitive brief on this abstract algebraic and topological theory.

[1] Sam Eilenberg and Sounders MacLane were American ex-members of the celebrated French Bourbaki group. Their best-known mathematical contribution is the *Eilenberg-MacLane space*, a special kind of topological space that is a building block for general homotopy theory.

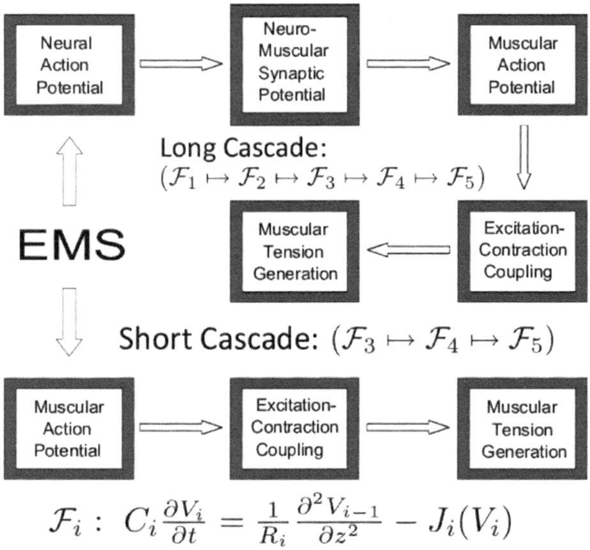

Fig. 3.1. *Basic diagram of the EMS−biophysics: a HH-based, fifth-order, electrical diffusion-reaction cascade: $(\mathcal{F}_1 \mapsto \mathcal{F}_2 \mapsto \mathcal{F}_3 \mapsto \mathcal{F}_4 \mapsto \mathcal{F}_5$: iff stimulating nerves) and a third-order cascade: $(\mathcal{F}_3 \mapsto \mathcal{F}_4 \mapsto \mathcal{F}_5$: iff stimulating muscles directly).*

3.2.1 Categories

Informally, a category is a generic mathematical structure consisting of a collection of *objects* (sets with possibly additional structure), with a corresponding collection of *arrows*, or *morphisms* (or just *maps*), between objects (agreeing with this additional structure). A category $\mathcal{K}$ is defined as a pair $(\mathtt{Ob}(\mathcal{K}), \mathtt{Mor}(\mathcal{K}))$ of generic objects $A, B, \dots$ in $\mathtt{Ob}(\mathcal{K})$ and generic arrows $f : A \to B$, $g : B \to C, \dots$ in $\mathtt{Mor}(\mathcal{K})$ between objects, with *associative composition* (see, e.g. [II06b, II07]):

$$A \xrightarrow{\;f\;} B \xrightarrow{\;g\;} C = A \xrightarrow{\;g \circ f\;} C,$$

and *identity* (*loop*) arrow. A category $\mathcal{K}$ is usually depicted as a commutative diagram:[2]

[2] Formally, we say that a *category* $\mathcal{K}$ is defined iff we have:

1. A *class of objects* $\{A, B, C, ...\}$ of $\mathcal{K}$, denoted by $\mathtt{Ob}(\mathcal{K})$;
2. A *set of morphisms*, or *morphisms* $\mathtt{Mor}_{\mathcal{K}}(A, B)$, with elements $f : A \to B$, defined for any *ordered pair* $(A, B) \in \mathcal{K}$, such that for two different pairs $(A, B) \neq (C, D)$ in $\mathcal{K}$, we have $\mathtt{Mor}_{\mathcal{K}}(A, B) \cap \mathtt{Mor}_{\mathcal{K}}(C, D) = \emptyset$;
3. For any *triplet* $(A, B, C) \in \mathcal{K}$ with $f : A \to B$ and $g : B \to C$, there is a *composition* of morphisms

$$\mathtt{Mor}_{\mathcal{K}}(B, C) \times \mathtt{Mor}_{\mathcal{K}}(A, B) \ni (g, f) \to g \circ f \in \mathtt{Mor}_{\mathcal{K}}(A, C),$$

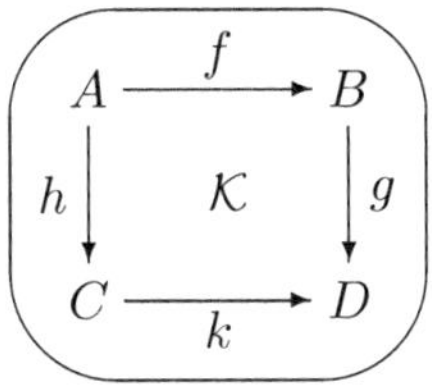

For example, a *groupoid* is a *category*[3] in which every morphism is invertible. A typical groupoid is the *fundamental groupoid* $\Pi_1(X)$ of a topological space X. An object of $\Pi_1(X)$ is a point $x \in X$, and a morphism $x \longrightarrow x^{prime}$ of $\Pi_1(X)$ is a homotopy class of paths f from x to x^{prime}. The *composition* of paths $g : x^{prime} \longrightarrow x^{primeprime}$ and $f : x \longrightarrow x^{prime}$ is the path h which is 'f followed by g'. Composition applies also to homotopy classes, and makes $\Pi_1(X)$ a category and a groupoid (the inverse of any path is the same path traced in the opposite direction).

Also, a *group* is a groupoid with one object, i.e., a *category with one object* in which *all morphisms are isomorphisms*. Therefore, if we try to generalize

---

written schematically as

$$\frac{f : A \to B, \qquad g : B \to C}{g \circ f : A \to C}.$$

[3] To make $\mathcal{K}$ a category, it must also fulfill the following two properties:

1. *Associativity of morphisms*: for all $f \in \mathtt{Mor}_{\mathcal{K}}(A, B)$, $g \in \mathtt{Mor}_{\mathcal{K}}(B, C)$, and $h \in \mathtt{Mor}_{\mathcal{K}}(C, D)$, we have $h \circ (g \circ f) = (h \circ g) \circ f$; in other words, the following diagram is commutative

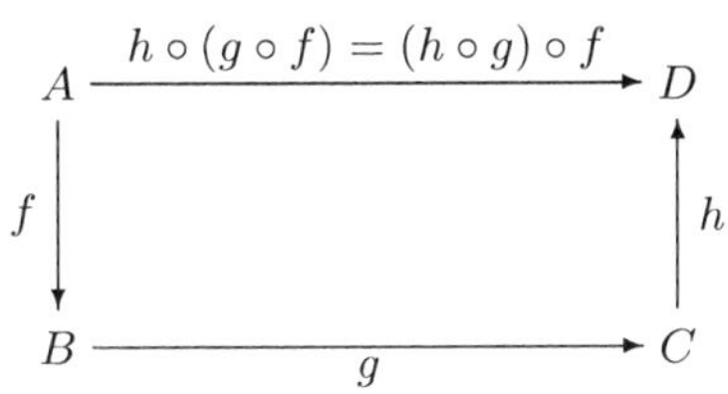

2. *Existence of identity morphism*: for every object $A \in \mathtt{Ob}(\mathcal{K})$ exists a unique identity morphism $1_A \in \mathtt{Mor}_{\mathcal{K}}(A, A)$; for any two morphisms $f \in \mathtt{Mor}_{\mathcal{K}}(A, B)$, and
$g \in \mathtt{Mor}_{\mathcal{K}}(B, C)$, compositions with identity morphism $1_B \in \mathtt{Mor}_{\mathcal{K}}(B, B)$ give $1_B \circ f = f$ and $g \circ 1_B = g$, i.e., the following diagram is commutative:

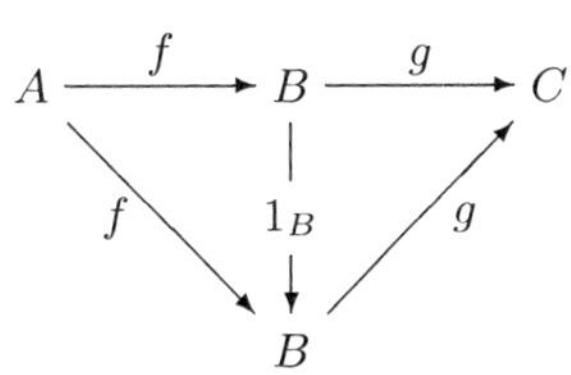

the concept of a group, keeping associativity as an essential property, we get the notion of a category.

3.2.2 Functors

A *functor* is a generic *picture* projecting (all objects and morphisms of) a source category into a target category. Let $\mathcal{K} = (\mathtt{Ob}(\mathcal{K}), \mathtt{Mor}(\mathcal{K}))$ be a *source* (or domain) *category* and $\mathcal{L} = (\mathtt{Ob}(\mathcal{L}), \mathtt{Mor}(\mathcal{L}))$ be a *target* (or codomain) category. A functor $\mathcal{F} = (\mathcal{F}_O, \mathcal{F}_M)$ is defined as a pair of maps, $\mathcal{F}_O : \mathtt{Ob}(\mathcal{K}) \to \mathtt{Ob}(\mathcal{L})$ and $\mathcal{F}_M : \mathtt{Mor}(\mathcal{K}) \to \mathtt{Mor}(\mathcal{L})$, preserving categorical symmetry (i.e., commutativity of all diagrams) of $\mathcal{K}$ in $\mathcal{L}$.[4]

Therefore, a *covariant functor*, or simply a *functor*, $\mathcal{F}_* : \mathcal{K} \to \mathcal{L}$ is a *picture* in the target category $\mathcal{L}$ of (all objects and morphisms of) the source category $\mathcal{K}$:

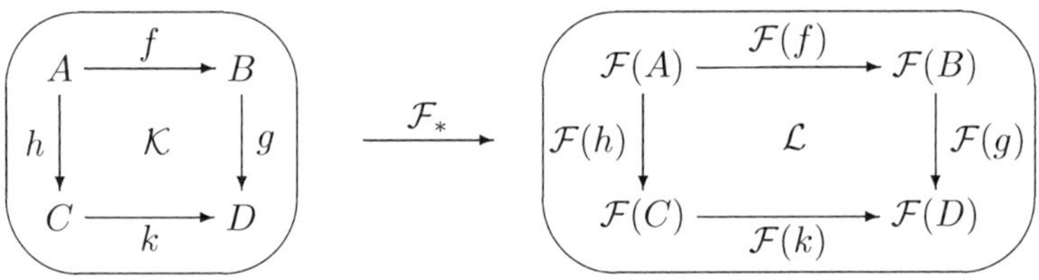

Similarly, a *contravariant functor*, or a *cofunctor*, $\mathcal{F}^* : \mathcal{K} \to \mathcal{L}$ is a *dual picture* with reversed arrows:[5]

[4] In algebraic topology, one attempts to assign to every topological space X some algebraic object $\mathcal{F}(X)$ in such a way that to every C^0–function $f : X \to Y$ there is assigned a homomorphism $\mathcal{F}(f) : \mathcal{F}(X) \longrightarrow \mathcal{F}(Y)$ (see [Swi75, II06b]). One advantage of this procedure is, e.g., that if one is trying to prove the non–existence of a C^0–function $f : X \to Y$ with certain properties, one may find it relatively easy to prove the non–existence of the corresponding algebraic function $\mathcal{F}(f)$ and hence deduce that f could not exist. In other words, $\mathcal{F}$ is to be a 'homomorphism' from one category (e.g., $\mathcal{T}$) to another (e.g., $\mathcal{G}$ or $\mathcal{A}$). Formalization of this notion is a *functor*.

[5] More formally, a *functor* $\mathcal{F} : \mathcal{K} \to \mathcal{L}$ from a *source* category $\mathcal{K}$ to a *target* category $\mathcal{L}$, is a pair $\mathcal{F} = (\mathcal{F}_O, \mathcal{F}_M)$ of maps $\mathcal{F}_O : \mathtt{Ob}(\mathcal{K}) \to \mathtt{Ob}(\mathcal{L})$, $\mathcal{F}_M : \mathtt{Mor}(\mathcal{K}) \to \mathtt{Mor}(\mathcal{L})$, such that:

1. If $f \in \mathtt{Mor}_{\mathcal{K}}(A, B)$ then $\mathcal{F}_M(f) \in \mathtt{Mor}_{\mathcal{L}}(\mathcal{F}_O(A), \mathcal{F}_O(B))$ in case of the *covariant* functor $\mathcal{F}_*$, and $\mathcal{F}_M(f) \in \mathtt{Mor}_{\mathcal{L}}(\mathcal{F}_O(B), \mathcal{F}_O(A))$ in case of the *contravariant* functor $\mathcal{F}^*$;
2. For all $A \in \mathtt{Ob}(\mathcal{K}) : \mathcal{F}_M(1_A) = 1_{\mathcal{F}_O(A)}$;
3. For all $f, g \in \mathtt{Mor}(\mathcal{K})$: if $\mathrm{cod}(f) = \mathrm{dom}(g)$, then $\mathcal{F}_M(g \circ f) = \mathcal{F}_M(g) \circ \mathcal{F}_M(f)$ in case of the *covariant* functor $\mathcal{F}_*$, and $\mathcal{F}_M(g \circ f) = \mathcal{F}_M(f) \circ \mathcal{F}_M(g)$ in case of the *contravariant* functor $\mathcal{F}^*$.

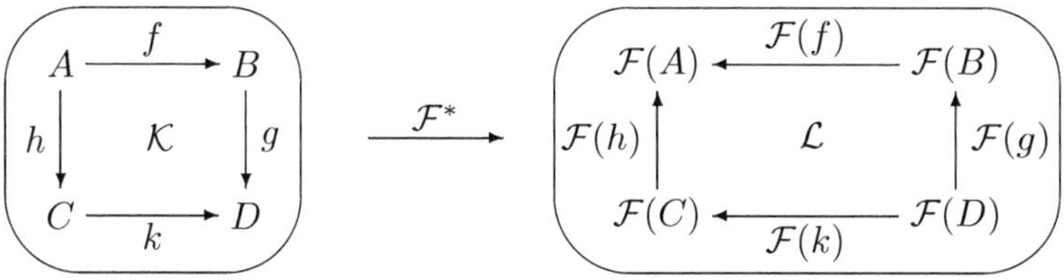

For example, in computer science, every *monad* is a functor, but not every functor is a monad. Monads are vitally important in creating higher-order functions that capture notions of computation such as sequence and control of flow. So, we can think of a functor as a structure that can be mapped over using a mapping that takes on its 'elements' to give a new structure.

As a standard topological example, the *fundamental group* π_1 is a functor.[6] Algebraic topology constructs a group called the *fundamental group* $\pi_1(X)$ from any topological space X, which keeps track of how many holes the space X has. But also, any map between topological spaces determines a homomorphism $\phi : \pi_1(X) \to \pi_1(Y)$ of the fundamental groups. So the fundamental group is really a functor $\pi_1 : \mathcal{T} \to \mathcal{G}$. This allows us to completely transpose any situation involving *spaces* and *continuous maps* between them to a parallel situation involving *groups* and *homomorphisms* between them, and thus reduce some topology problems to algebra problems.

Also, singular homology in a given dimension n assigns to each topological space X an Abelian group $H_n(X)$, its nth *homology group* of X, and also to each continuous map $f : X \to Y$ of spaces a corresponding homomorphism

[6] Another examples include covariant *forgetful* functors:

- From the category of topological spaces to the category of sets; it 'forgets' the topology–structure.
- From the category of metric spaces to the category of topological spaces with the topology induced by the metrics; it 'forgets' the metric.

For each category $\mathcal{K}$, the *identity functor* $I_{\mathcal{K}}$ takes every $\mathcal{K}$–object and every $\mathcal{K}$–morphism to itself.

Given a category $\mathcal{K}$ and its subcategory $\mathcal{L}$, we have an *inclusion functor* $\mathtt{In} : \mathcal{L} \to \mathcal{K}$.

Given a category $\mathcal{K}$, a *diagonal functor* $\Delta : \mathcal{K} \to \mathcal{K} \times \mathcal{K}$ takes each object $A \in \mathcal{K}$ to the object (A, A) in the product category $\mathcal{K} \times \mathcal{K}$.

Given a category $\mathcal{K}$ and a category of sets $\mathcal{S}$, each object $A \in \mathcal{K}$ determines a *covariant Hom–functor* $\mathcal{K}[A, \_] : \mathcal{K} \to \mathcal{S}$, a *contravariant Hom–functor* $\mathcal{K}[\_, A] : \mathcal{K} \to \mathcal{S}$, and a *Hom–bifunctor* $\mathcal{K}[\_, \_] : \mathcal{K}^{op} \times \mathcal{K} \to \mathcal{S}$.

A functor $\mathcal{F} : \mathcal{K} \to \mathcal{L}$ is a *faithful functor* if for all $A, B \in \mathtt{Ob}(\mathcal{K})$ and for all $f, g \in \mathtt{Mor}_{\mathcal{K}}(A, B)$, $\mathcal{F}(f) = \mathcal{F}(g)$ implies $f = g$; it is a *full functor* if for every $h \in \mathtt{Mor}_{\mathcal{L}}(\mathcal{F}(A), \mathcal{F}(B))$, there is $g \in \mathtt{Mor}_{\mathcal{K}}(A, B)$ such that $h = \mathcal{F}(g)$; it is a *full embedding* if it is both full and faithful.

A *representation of a group* is a functor $\mathcal{F} : \mathcal{G} \to \mathcal{V}$. Thus, a category is a generalization of a group and group representations are a special case of category representations.

$H_n(f) : H_n(X) \to H_n(Y)$ of groups, and this in such a way that $H_n(X)$ becomes a functor $H_n : \mathcal{T} \to \mathcal{A}$.

The leading idea in the *use of functors in topology* is that H_n or π_n gives an algebraic picture or image not just of the topological spaces X, Y but also of all the continuous maps $f : X \to Y$ between them.

Similarly, there is a functor $\Pi_1 : \mathcal{T} \to \mathcal{G}$, called the 'fundamental groupoid functor', which plays a very basic role in algebraic topology. Here's how we get from any space X its 'fundamental groupoid' $\Pi_1(X)$. To say what the groupoid $\Pi_1(X)$ is, we need to say what its objects and morphisms are. The objects in $\Pi_1(X)$ are just the *points* of X and the morphisms are just certain equivalence classes of *paths* in X. More precisely, a morphism $f : x \to y$ in $\Pi_1(X)$ is just an equivalence class of continuous paths from x to y, where two paths from x to y are decreed equivalent if one can be continuously deformed to the other while not moving the endpoints. (If this equivalence relation holds, we say the two paths are 'homotopic', and we call the equivalence classes 'homotopy classes of paths' [Swi75].

3.2.3 Natural Transformations

A *natural transformation* (i.e., a *functor morphism*, see Figure 3.2) $\boldsymbol{\tau} : \mathcal{F} \dot{\to} \mathcal{G}$ is a *map between two functors of the same variance*, $(\mathcal{F}, \mathcal{G}) : \mathcal{K} \rightrightarrows \mathcal{L}$, preserving categorical symmetry:

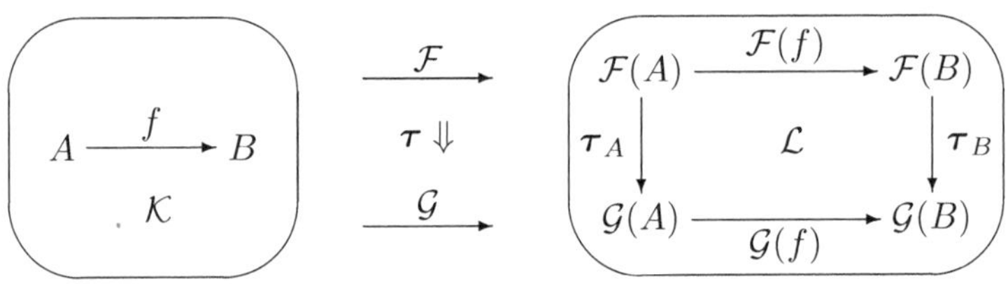

In other words, all functors of the same variance from a source category $\mathcal{K}$ to a target category $\mathcal{L}$ form themselves objects of the *functor category* $\mathcal{L}^{\mathcal{K}}$. Morphisms of $\mathcal{L}^{\mathcal{K}}$, called *natural transformations*, are defined as follows.

Let $\mathcal{F} : \mathcal{K} \to \mathcal{L}$ and $\mathcal{G} : \mathcal{K} \to \mathcal{L}$ be two functors of the same variance from a category $\mathcal{K}$ to a category $\mathcal{L}$. Natural transformation $\mathcal{F} \xrightarrow{\boldsymbol{\tau}} \mathcal{G}$ is a family of morphisms such that for all $f \in \mathtt{Mor}_{\mathcal{K}}(A, B)$ in the source category $\mathcal{K}$, we have $\mathcal{G}(f) \circ \boldsymbol{\tau}_A = \boldsymbol{\tau}_B \circ \mathcal{F}(f)$ in the target category $\mathcal{L}$. Then we say that the *component* $\boldsymbol{\tau}_A : \mathcal{F}(A) \to \mathcal{G}(A)$ *is natural in* A.

If we think of a functor $\mathcal{F}$ as giving a *picture* in the target category $\mathcal{L}$ of (all the objects and morphisms of) the source category $\mathcal{K}$, then a natural transformation $\boldsymbol{\tau}$ represents a set of morphisms mapping the picture $\mathcal{F}$ to another picture $\mathcal{G}$, preserving the categorial structure, that is, commutativity of all diagrams.

An invertible natural transformation, such that all components $\boldsymbol{\tau}_A$ are isomorphisms) is called a *natural equivalence* (or, *natural isomorphism*). In

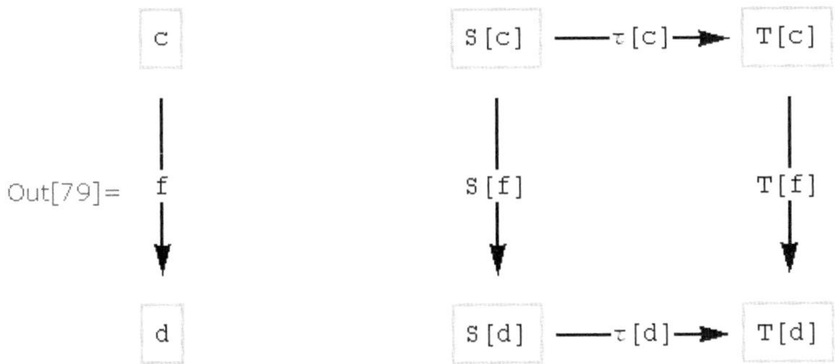

Fig. 3.2. *A sample natural transformation calculated and plotted in Mathematica®, using the WildCats package.*

this case, the inverses $(\boldsymbol{\tau}_A)^{-1}$ in $\mathcal{L}$ are the components of a natural isomorphism $(\boldsymbol{\tau})^{-1} : \mathcal{G} \xrightarrow{*} \mathcal{F}$. Natural equivalences are among the most important *metamathematical constructions* in algebraic topology (see [Swi75]).

As a mathematical example, let $\mathcal{B}$ be the category of Banach spaces over $\mathbb{R}$ and bounded linear maps. Define $D : \mathcal{B} \to \mathcal{B}$ by taking $D(X) = X^* =$ Banach space of bounded linear functionals on a space X and $D(f) = f^*$ for $f : X \to Y$ a bounded linear map. Then D is a cofunctor. $D^2 = D \circ D$ is also a functor. We also have the identity functor $1 : \mathcal{B} \to \mathcal{B}$. Define $T : 1 \to D \circ D$ as follows: for every $X \in \mathcal{B}$ let $T(X) : X \to D^2X = X^{**}$ be the *natural inclusion* – that is, for $x \in X$ we have $[T(X)(x)](f) = f(x)$ for every $f \in X^*$. T is a natural transformation. On the subcategory of nD Banach spaces T is even a natural equivalence. The largest subcategory of $\mathcal{B}$ on which T is a natural equivalence is called the category of reflexive Banach spaces [Swi75].

As a physical example, when we want to be able to conceive two physical systems A and B as one whole (see [Coe06, CP09]), we can denote this using a (symmetric) *monoidal tensor product* $A \otimes B$ (defined later in the section 3.6), and hence also need to consider the compound operations

$$A \otimes B \xrightarrow{f \otimes g} C \otimes D,$$

inherited from the operations on the individual systems. Now, a (symmetric) *monoidal category* is a category $\mathcal{K}$ defined as a pair $(\mathtt{Ob}(\mathcal{K}), \mathtt{Mor}(\mathcal{K}))$ of generic objects $A, B, \ldots$ in $\mathtt{Ob}(\mathcal{K})$ and generic morphisms $f : A \to B$, $g : B \to C, \ldots$ in $\mathtt{Mor}(\mathcal{K})$ between objects, defined using the symmetric monoidal tensor product:

$$\begin{aligned} \mathtt{Ob}(\mathcal{K}) &: \{A, B\} \mapsto A \otimes B, \\ \mathtt{Mor}(\mathcal{K}) &: \{A \xrightarrow{f} B, C \xrightarrow{g} D\} \mapsto A \otimes C \xrightarrow{f \otimes g} B \otimes D, \end{aligned}$$

with the additional notion of *bifunctoriality*: if we apply an operation f to one system and an operation g to another system, then the order in which we apply them does not matter; that is, the following diagram commutes:

$$\begin{array}{ccc} A_1 \otimes A_2 & \xrightarrow{f \otimes 1_{A_2}} & B_1 \otimes A_2 \\ {\scriptstyle 1_{A_1} \otimes g} \downarrow & & \uparrow {\scriptstyle 1_{B_1} \otimes g} \\ A_1 \otimes B_2 & \xrightarrow[f \otimes 1_{B_2}]{} & B_1 \otimes B_2 \end{array}$$

which shows that both paths yield the same result (see [Coe06, CP09] for technical details).

Natural transformations can be *composed* in two different ways. First, we have an 'ordinary' composition: if $\mathcal{F}, \mathcal{G}$ and $\mathcal{H}$ are three functors from the source category $\mathcal{A}$ to the target category $\mathcal{B}$, and then $\boldsymbol{\alpha} : \mathcal{F} \stackrel{\cdot}{\to} \mathcal{G}$, $\boldsymbol{\beta} : \mathcal{G} \stackrel{\cdot}{\to} \mathcal{H}$ are two natural transformations, then the formula

$$(\beta \circ \alpha)_A = \beta_A \circ \alpha_A, \qquad \text{(for all } A \in \mathcal{A}\text{)}, \tag{3.1}$$

defines a new natural transformation $\beta \circ \alpha : \mathcal{F} \stackrel{\cdot}{\to} \mathcal{H}$. This composition law is clearly associative and possesses a unit $1_{\mathcal{F}}$ at each functor $\mathcal{F}$, whose $\mathcal{A}$-component is $1_{\mathcal{F}A}$.

Second, we have the *Godement product* of natural transformations, usually denoted by $*$. Let $\mathcal{A}$, $\mathcal{B}$ and $\mathcal{C}$ be three categories, $\mathcal{F}, \mathcal{G}$, $\mathcal{H}$ and $\mathcal{K}$ be four functors such that $(\mathcal{F}, \mathcal{G}) : \mathcal{A} \rightrightarrows \mathcal{B}$ and $(\mathcal{H}, \mathcal{K}) : \mathcal{B} \rightrightarrows \mathcal{C}$, and $\boldsymbol{\alpha} : \mathcal{F} \stackrel{\cdot}{\to} \mathcal{G}$, $\boldsymbol{\beta} : \mathcal{H} \stackrel{\cdot}{\to} \mathcal{K}$ be two natural transformations. Now, instead of (3.1), the Godement composition is given by

$$(\beta * \alpha)_A = \beta_{GA} \circ H(\alpha_A) = K(\alpha_A) \circ \beta_{FA}, \qquad \text{(for all } A \in \mathcal{A}\text{)}, \tag{3.2}$$

which defines a new natural transformation $\beta * \alpha : \mathcal{H} \circ \mathcal{F} \stackrel{\cdot}{\to} \mathcal{K} \circ \mathcal{G}$.

Finally, the two compositions (3.1) and (3.2) of natural transformations can be combined as

$$(\delta * \gamma) \circ (\beta * \alpha) = (\delta \circ \beta) * (\gamma \circ \alpha),$$

where $\mathcal{A}$, $\mathcal{B}$ and $\mathcal{C}$ are three categories, $\mathcal{F}, \mathcal{G}$, $\mathcal{H}$, $\mathcal{K}$, $\mathcal{L}$, $\mathcal{M}$ are six functors, and $\boldsymbol{\alpha} : \mathcal{F} \stackrel{\cdot}{\to} \mathcal{H}$, $\boldsymbol{\beta} : \mathcal{G} \stackrel{\cdot}{\to} \mathcal{K}$, $\boldsymbol{\gamma} : \mathcal{H} \stackrel{\cdot}{\to} \mathcal{L}$, $\boldsymbol{\delta} : \mathcal{K} \stackrel{\cdot}{\to} \mathcal{M}$ are four natural transformations.

3.3 Adjunctions

The *adjunction* $\varphi : \mathcal{F} \dashv \mathcal{G}$ between two functors $(\mathcal{F}, \mathcal{G}) : \mathcal{K} \leftrightarrows \mathcal{L}$ of *opposite variance*, represents a *weak functorial inverse* (see, e.g. [II06b, II07])

$$\frac{f : \mathcal{F}(A) \to B}{\varphi(f) : A \to \mathcal{G}(B)},$$

forming a *natural equivalence* $\varphi : \texttt{Mor}_{\mathcal{K}}(\mathcal{F}(A), B) \xrightarrow{\varphi} \texttt{Mor}_{\mathcal{L}}(A, \mathcal{G}(B))$. The adjunction isomorphism is given by a *bijective correspondence* (a *1-1* and *onto* map on objects) $\varphi : \texttt{Mor}(\mathcal{K}) \ni f \to \varphi(f) \in \texttt{Mor}(\mathcal{L})$ of isomorphisms in the two categories, $\mathcal{K}$ (with a representative object A), and $\mathcal{L}$ (with a representative object B). It can be depicted as a *non-commutative diagram*

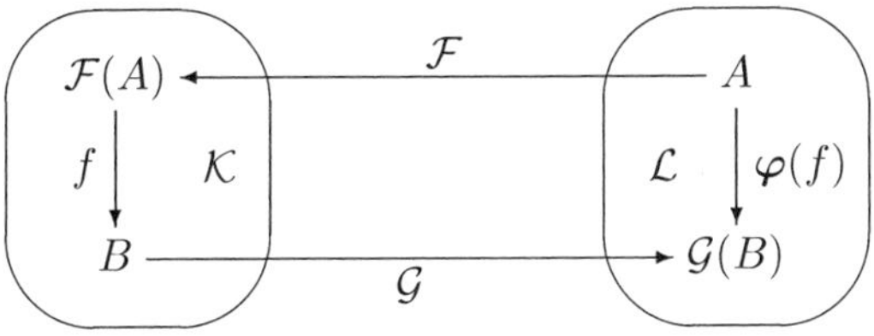

In this case $\mathcal{F}$ is called *left adjoint*, while $\mathcal{G}$ is called *right adjoint.*

In other words, an *adjunction* $F \dashv G$ *between two functors* $(\mathcal{F}, \mathcal{G})$ of opposite variance, from a source category $\mathcal{K}$ to a target category $\mathcal{L}$, is denoted by $(\mathcal{F}, \mathcal{G}, \boldsymbol{\eta}, \boldsymbol{\varepsilon}) : \mathcal{K} \leftrightarrows \mathcal{L}$. Here, $\mathcal{F} : \mathcal{L} \to \mathcal{K}$ is the *left (upper) adjoint functor*, $\mathcal{G} : \mathcal{K} \leftarrow \mathcal{L}$ is the *right (lower) adjoint functor*, $\boldsymbol{\eta} : 1_{\mathcal{L}} \to \mathcal{G} \circ \mathcal{F}$ is the *unit natural transformation* (or, *front adjunction*), and $\boldsymbol{\varepsilon} : \mathcal{F} \circ \mathcal{G} \to 1_{\mathcal{K}}$ is the *counit natural transformation* (or, *back adjunction*).[7]

3.3.1 Crowd/Team Dynamics Adjunction

The following unique dynamical adjunction, uncovering the natural equivalence between *geometrical* and *topological* structures of crowd/team dynamics has been established in [IR10c]:

---

[7] Closely related to adjunctions are *limits and colimits*. A *limit* of a covariant functor $\mathcal{F} : \mathcal{J} \to \mathcal{C}$ is an object L of $\mathcal{C}$, together with morphisms $\phi_X : L \to \mathcal{F}(X)$ for every object X of $\mathcal{J}$, such that for every morphism $f : X \to Y$ in $\mathcal{J}$, we have $\mathcal{F}(f)\phi_X = \phi_Y$, and such that the following *universal property* is satisfied: for any object N of $\mathcal{C}$ and any set of morphisms $\psi_X : N \to \mathcal{F}(X)$ such that for every morphism $f : X \to Y$ in $\mathcal{J}$, we have $\mathcal{F}(f)\psi_X = \psi_Y$, there exists precisely one morphism $u : N \to L$ such that $\phi_X u = \psi_X$ for all X. If $\mathcal{F}$ has a limit (which it need not), then the limit is defined up to a unique isomorphism, and is denoted by $\lim \mathcal{F}$.

Analogously, a *colimit* of the functor $\mathcal{F} : \mathcal{J} \to \mathcal{C}$ is an object L of $\mathcal{C}$, together with morphisms $\phi_X : \mathcal{F}(X) \to L$ for every object X of $\mathcal{J}$, such that for every morphism $f : X \to Y$ in $\mathcal{J}$, we have $\phi_Y \mathcal{F}(X) = \phi_X$, and such that the following universal property is satisfied: for any object N of $\mathcal{C}$ and any set of morphisms $\psi_X : \mathcal{F}(X) \to N$ such that for every morphism $f : X \to Y$ in $\mathcal{J}$, we have $\psi_Y \mathcal{F}(X) = \psi_X$, there exists precisely one morphism $u : L \to N$ such that $u\phi_X = \psi_X$ for all X. The colimit of $\mathcal{F}$, unique up to unique isomorphism if it exists, is denoted by $\operatorname{colim} \mathcal{F}$.

Limits and colimits are related as follows: A functor $\mathcal{F} : \mathcal{J} \to \mathcal{C}$ has a colimit iff for every object N of $\mathcal{C}$, the functor $X \longmapsto Mor_{\mathcal{C}}(\mathcal{F}(X), N)$ (which is a covariant functor on the dual category $\mathcal{J}^{op}$) has a limit. If that is the case, then $Mor_{\mathcal{C}}(\operatorname{colim} \mathcal{F}, N) = \lim Mor_{\mathcal{C}}(\mathcal{F}(-), N)$ for every object N of $\mathcal{C}$.

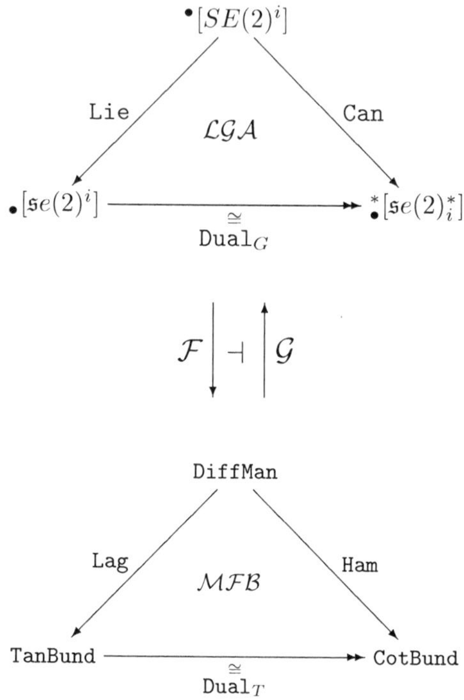

- by parallel development of Lagrangian (functor `Lag`) and Hamiltonian (functor `Ham`) formulations of crowd dynamics, both within the category $\mathcal{LGA}$ of Lie groups $(SE(2)^i)$ and their tangent and cotangent Lie algebras $(\mathfrak{se}(2)^i$ and $\mathfrak{se}(2)_i^*$, respectively) and in the category $\mathcal{MFB}$ of associated smooth manifolds `DiffMan` and their fiber bundles (tangent and cotangent bundles, `TanBund` and `CotBund`, respectively). In this way, a general crowd dynamics is formally defined as the adjunction:

$$(\mathcal{F}, \mathcal{G}) : \mathcal{LGA} \leftrightarrows \mathcal{MFB}.$$

3.3.2 Neurophysiological Sensory-Motor Adjunction

Now, recall that both human and animal sensations from the skin, muscles, and internal organs of the body, are transmitted to the central nervous system via axons that enter via spinal nerves. They are called *sensory pathways*. On the other hand, the motor system executes control over the skeletal muscles of the body via several major tracts (including pyramidal and extrapyramidal). They are called *motor pathways*. Sensory-motor (or, sensorimotor) control/coordination concerns relationships between sensation and movement or, more broadly, between perception and action. The interplay of sensory and motor processes provides the basis of observable human behavior. Anatomically, its top-level, association link can be visualized as a talk between sensory and motor Penfield's homunculi. This sensory-motor control system can be modeled as an adjunction between the afferent sensory functor $\mathcal{S} : \mathcal{BODY} \to \mathcal{BRAIN}$ and the efferent motor functor $\mathcal{M} : \mathcal{BRAIN} \to \mathcal{BODY}$. Thus, we have $\mathcal{SMC} : \mathcal{S} \dashv \mathcal{M}$, with $(\mathcal{S}, \mathcal{M}) : \mathcal{BRAIN} \leftrightarrows \mathcal{BODY}$ and depicted as:

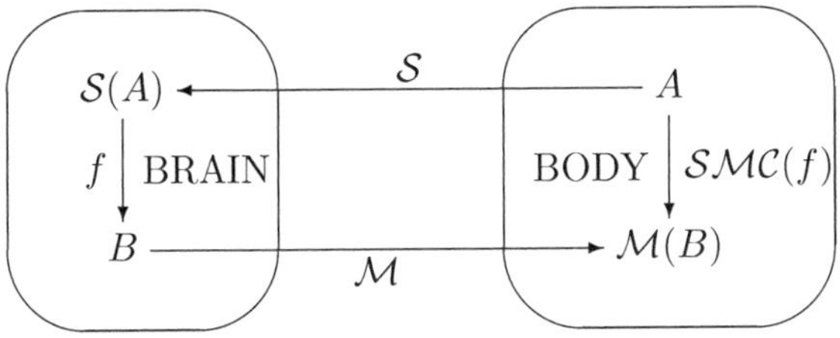

This adjunction offers a mathematical answer to the fundamental question: How would *Nature* solve a general biophysical control/coordination problem? *By using a weak functorial inverse of sensory neural pathways and motor neural pathways, Nature controls human behavior in general, and human motion in particular.*

More generally, normal functioning of human body is achieved through interplay of a number of physiological systems - Objects of the category BODY: musculoskeletal system, circulatory system, gastrointestinal system, integumentary system, urinary system, reproductive system, immune system and endocrine system. These systems are all interrelated, so one can say that the Morphisms between them make the proper functioning of the BODY as a whole. On the other hand, BRAIN contains the images of all above functional systems (Brain objects) and their interrelations (Brain morphisms), for the purpose of body control. This body-control performed by the brain is partly unconscious, through neuro-endocrine complex, and partly conscious, through neuro-muscular complex. A generalized sensory functor $\mathcal{SS}$ sends the information about the state of all Body objects (at any time instant) to their images in the Brain. A generalized motor functor $\mathcal{MM}$ responds to these upward sensory signals by sending downward corrective action-commands from the Brain's objects and morphisms to the Body's objects and morphisms.

For physiological details, see [II06a].

3.3.3 Quantum Teleportation Example

Recall that the celebrated *quantum teleportation* [BBC93, NKL98, FSB98, BPM97] represents a *quantum entanglement*[8] based process by which a quantum qubit can be transmitted from one location to another, without being

[8] Recall that in 1935, A. Einstein, B. Podolsky, and N. Rosen (*EPR*) published a seminal paper entitled "Can Quantum-Mechanical Description of Physical Reality Be Considered Complete?", which started a long lasting debate between Einstein and the so-called Copenhagen school of the interpretation of quantum mechanics, about the status of quantum theory [EPR35a] and the possible existence of additional 'hidden variables'. It was not until 1964, that J. Bell presented a way to empirically test the two competing hypotheses [Bel64, Bel66, Bel87]. Famous

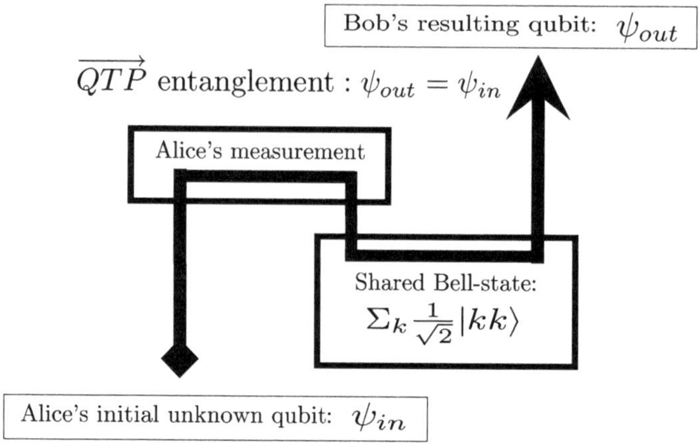

Fig. 3.3. *Quantum teleportation protocol $\overrightarrow{QTP}$: the incoming arrow represents Alice's initial unknown qubit and the outgoing arrow represents Bob's resulting qubit. The fact that the only labels it encounters are identities, implies that Bob's qubit is now in the state Alice's qubit was, formally defining the Bell-state entanglement.*

transmitted through the intervening space.[9] The *quantum teleportation protocol* ($\overrightarrow{QTP}$),[10] represents a *causal arrow* of quantum information which involves two agents, Alice (A) and Bob (B). Both Alice and Bob have their own qubits, as well as a shared entangled qubit in the particular Bell-state:

$$\overrightarrow{QTP}_{(\psi_{out}=\psi_{in})} = \sum_k \frac{1}{\sqrt{2}}|kk\rangle.$$

Alice's initial qubit is in an unknown quantum state. Alice performs a Bell-base measurement on her two qubits (a bipartite measurement which has the Bell-state as one of its eigenvectors, see Figure 3.3). If the measurement

Bell's theorem is not strictly about quantum mechanics. It is a statement concerning correlations of measurement outcomes at distant events that any physical theory may predict under the assumption of an underlying local classical model [Bal87, Mer93].

9 It is physically possible to transfer a quantum state to a different quantum system at a distant location without physically transmitting the actual quantum system in the particular state - provided that the parties initially share a pair of two-level systems in a maximally entangled Bell-state. This transfer of a quantum state was named *quantum teleportation*.

10 The $\overrightarrow{QTP}$ system consists of the following data: (i) a qubit that is to be teleported; (ii) a conventional communication channel capable of transmitting two classical bits (i.e., one of four states); and (iii) means of generating an entangled pair of qubits, performing a Bell measurement on the pair, and manipulating the quantum state of one of the pair. For the recent QTP-related paper, see [FB12].

outcome is the eigenvalue corresponding to the Bell-state, then Bob's qubit appears to be in the same unknown state in which Alice's qubit initially was (see [AC04, Coe05, Coe09] for details). We will continue our expose on *quantum protocols* in the section 3.6 below.

3.4 Hierarchical Recursive Categories

After about half-a-century of exploiting Eilenberg-MacLane's category theory in various mathematical disciplines, it has become increasingly obvious that the theory in its original settings as defined by MacLane's book [Mac71] is too narrow to meet the requirements of modern mathematical and computer sciences. So, in the last several decades a new idea of recursive higher-dimensional categories, or $n-$categories (or multicategories) has been developed (see, e.g. [Ben67, Bae97, BD98, Lei02, Lei03, Lei04]).

Informally, if we think of a point in geometrical space (either natural, or abstract) as an *object* (or, a $0-$cell), and a path between two points as an *arrow* (or, a $1-$*morphism*, or a $1-$cell), we could think of a 'path of paths' as a $2-$arrow (or, a $2-$morphism, or a $2-$cell), and a 'path of paths of paths' (or, a $3-$morphism, or a $3-$cell), etc. Here a 'path of paths' is just a continuous 1-parameter family of paths from between source and target points, which we can think of as tracing out a 2D surface, etc. In this way we get a 'skeleton' of an $n-$category, where a $1-$category operates with $0-$cells (objects) and $1-$cells (arrows, causally connecting *source* objects with *target* ones), a $2-$category operates with all the cells up to $2-$cells [Ben67], a $3-$category operates with all the cells up to $3-$cells, etc. This skeleton clearly demonstrates the *hierarchical self-similarity* of *n-categories*:

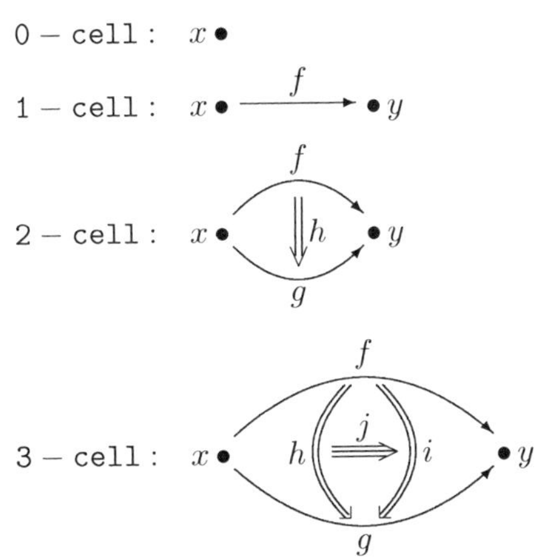

where triple arrow goes in the third direction, perpendicular to both single and double arrows. Categorical composition is defined by pasting arrows.

Thus, a $1-$category can be depicted as a commutative triangle:

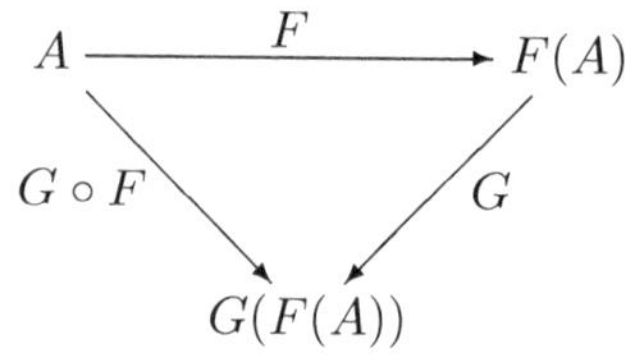

a 2−category is a commutative triangle:

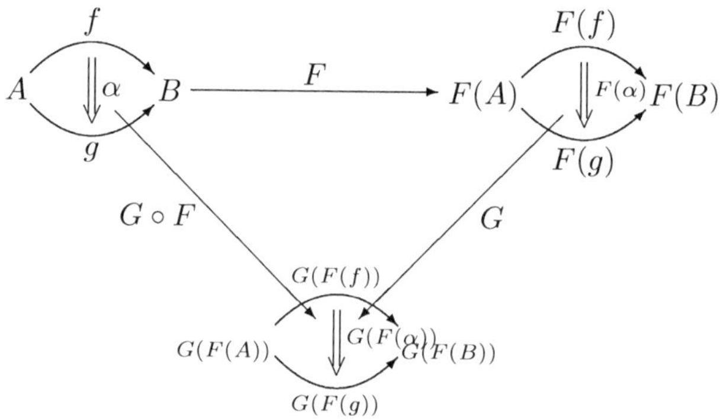

a 3−category is a commutative triangle:

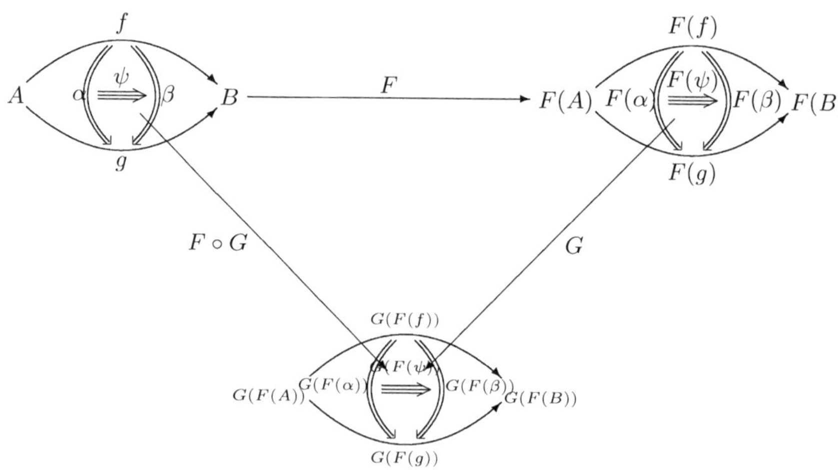

etc., up to n−categories.

Therefore, while an ordinary category theory uses $1D$ arrows: 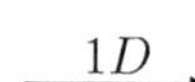

, a higher-dimensional category theory uses higher-dimensional arrows:

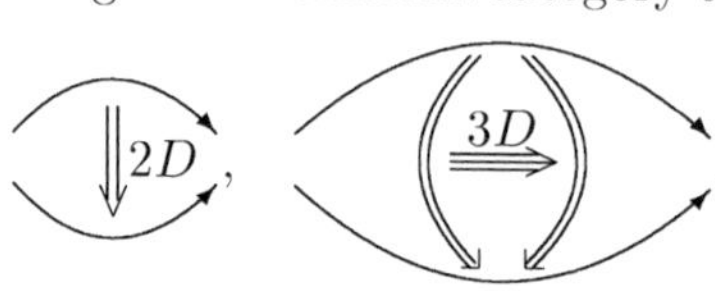

, etc. An n−category, or a *multicategory*, is

a generic mathematical structure consisting of a collection of objects, a collection of arrows between objects, a collection of 2−arrows between arrows [Ben67], a collection of 3−arrows between 2−arrows, and so on up to n [Bae97, BD98, Lei02, Lei03, Lei04].

Calculus of n−categories has been developed as follows. First, there is $\mathcal{K}_2$, the 2-category of all ordinary (small) categories. $\mathcal{K}_2$ has categories $\mathcal{K}, \mathcal{L}, ...$ as objects, functors $\mathcal{F}, \mathcal{G} : \mathcal{K} \rightrightarrows \mathcal{L}$ as arrows, and natural transformations, like $\tau : \mathcal{F} \stackrel{\cdot}{\to} \mathcal{G}$ as 2-arrows.

In a similar way, the arrows in a 3-category $\mathcal{K}_3$ are 2-functors $\mathcal{F}_2, \mathcal{G}_2, ...$ sending objects in $\mathcal{K}_2$ to objects in $\mathcal{L}_2$, arrows to arrows, and 2-arrows to 2-arrows, strictly preserving all the structure of $\mathcal{K}_2$

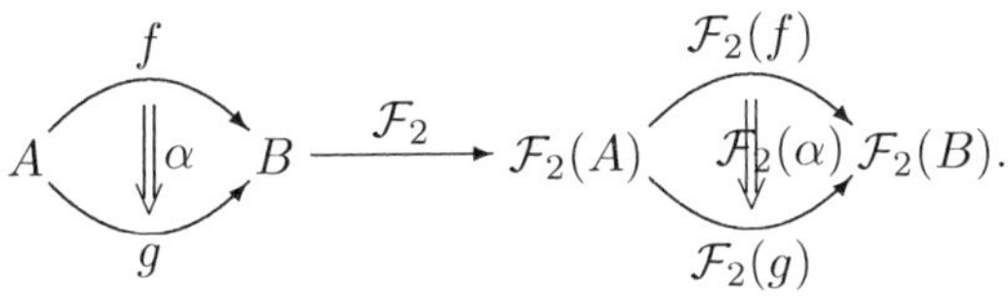

The 2-arrows in $\mathcal{K}_3$ are 2-natural transformations, like $\tau_2 : \mathcal{F}_2 \stackrel{2\cdot}{\Rightarrow} \mathcal{G}_2$ between 2-functors $\mathcal{F}_2, \mathcal{G}_2 : \mathcal{K}_2 \longrightarrow \mathcal{L}_2$ that sends each object in $\mathcal{K}_2$ to an arrow in $\mathcal{L}_2$ and each arrow in $\mathcal{K}_2$ to a 2-arrow in $\mathcal{L}_2$, and satisfies natural transformation-like conditions. We can visualize τ_2 as a prism going from one functorial picture of $\mathcal{K}_2$ in $\mathcal{L}_2$ to another, built using commutative squares:

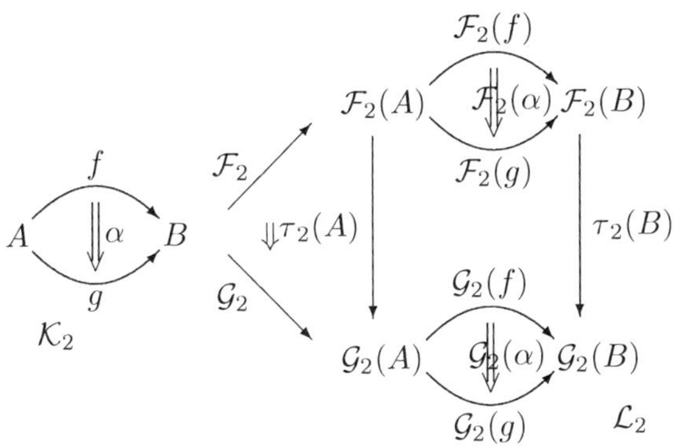

Similarly, the arrows in a 4-category $\mathcal{K}_4$ are 3-functors $\mathcal{F}_3, \mathcal{G}_3, ...$ sending objects in $\mathcal{K}_3$ to objects in $\mathcal{L}_3$, arrows to arrows, and 2-arrows to 2-arrows, strictly preserving all the structure of $\mathcal{K}_3$

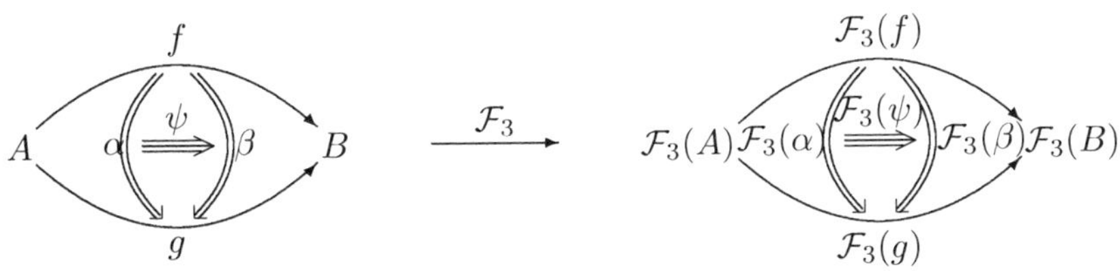

The 2-arrows in $\mathcal{K}_4$ are 3-natural transformations, like $\tau_3 : \mathcal{F} \overset{3:}{\Rightarrow} \mathcal{G}$ between 3-functors $\mathcal{F}_3, \mathcal{G}_3 : \mathcal{K}_3 \to \mathcal{L}_3$ that sends each object in $\mathcal{K}_3$ to a arrow in $\mathcal{L}_3$ and each arrow in $\mathcal{K}_3$ to a 2-arrow in $\mathcal{L}_3$, and satisfies natural transformation-like conditions. We can visualize τ_3 as a prism going from one picture of $\mathcal{K}_3$ in $\mathcal{L}_3$ to another, built using commutative squares:

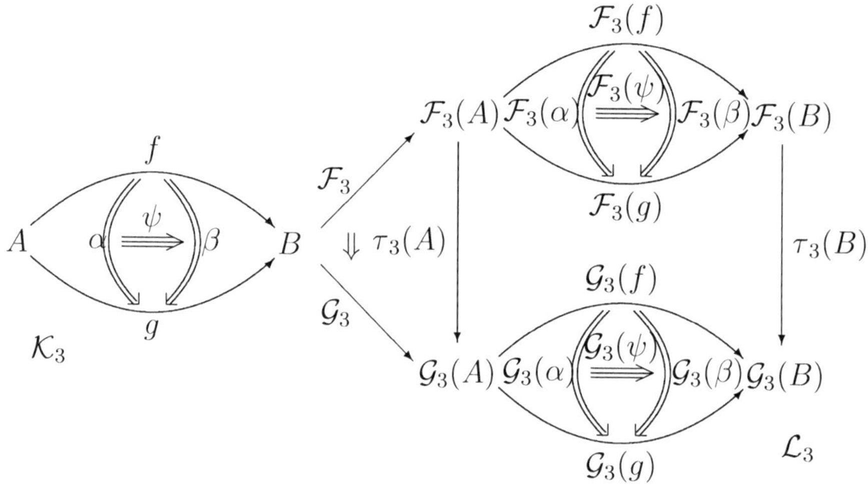

3.4.1 Topological Structure of $n-$Categories

We already emphasized the topological nature of ordinary category theory. This fact is even more obvious in the general case of $n-$categories (see [Lei02, Lei03, Lei04]).

Hierarchical Recursive Homotopy Theory

Any topological manifold M induces an $n-$*category* $\Pi_n(M)$ (its *fundamental groupoid*[11]), in which 0-cells are *points* in M; 1-cells are *paths* in M (i.e., parameterized continuous maps $f : [0,1] \to M$); 2-cells are *homotopies* (denoted by $\simeq$) *of paths* relative to endpoints (i.e., parameterized continuous maps $h : [0,1] \times [0,1] \to M$); 3-cells are *homotopies of homotopies* of paths in M (i.e., parameterized continuous maps $j : [0,1] \times [0,1] \times [0,1] \to M$); categorical *composition* is defined by *pasting* paths and homotopies. In this way the following 'homotopy skeleton' emerges:

[11] A *groupoid* is a category in which every morphism is invertible; its special case with only one object is a *group*.

$0-\texttt{cell}: x\bullet \qquad x \in M;$

$1-\texttt{cell}: x\bullet \xrightarrow{f} \bullet y \qquad f : x \simeq y \in M,$

$f : [0,1] \to M,\ f : x \mapsto y,\ y = f(x),\ f(0) = x,\ f(1) = y;$

e.g., linear path: $f(t) = (1-t)x + ty;$

$2-\texttt{cell}:$

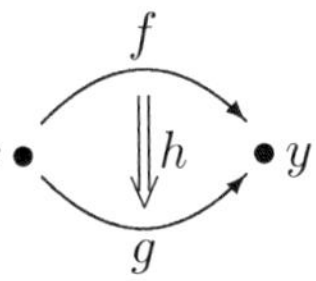

$h : f \simeq g \in M,$

$h : [0,1]\times[0,1] \to M,\ h : f \mapsto g,\ g = h(f(x)),$

$h(x,0) = f(x),\ h(x,1) = g(x),\ h(0,t) = x,\ h(1,t) = y$

e.g., linear homotopy: $h(x,t) = (1-t)f(x) + tg(x);$

$3-\texttt{cell}:$

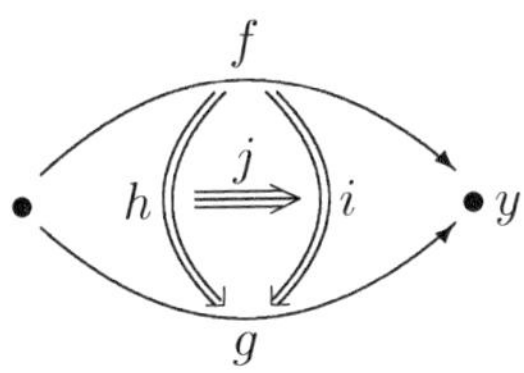

$j : h \simeq i \in M,$

$j : [0,1]\times[0,1]\times[0,1] \to M,\ j : h \mapsto i,\ i = j(h(f(x)))$

$j(x,t,0) = h(f(x)),\ j(x,t,1) = i(f(x)),$

$j(x,0,s) = f(x),\ j(x,1,s) = g(x),$

$j(0,t,s) = x,\ j(1,t,s) = y$

e.g., linear composite homotopy: $j(x,t,s) = (1-t)\,h(f(x)) + t\,i(f(x)).$

If M is a *smooth* manifold, then all included paths and homotopies need to be *smooth*.

Category $\mathcal{TT}$

Topological $n-$category $\mathcal{TT}$ has:

- 0-cells: topological spaces X
- 1-cells: continuous maps $X \xrightarrow{f} Y$
- 2-cells: homotopies h between f and g :

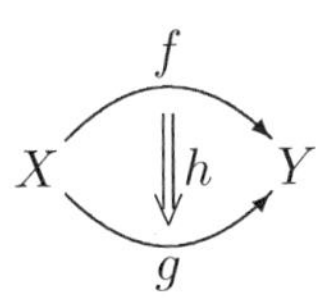

  i.e., continuous maps $h : X \times [0,1] \to Y$, such that $\forall x \in X$, $h(x,0) = f(x)$ and $h(x,1) = g(x)$

- 3-cells: homotopies between homotopies : 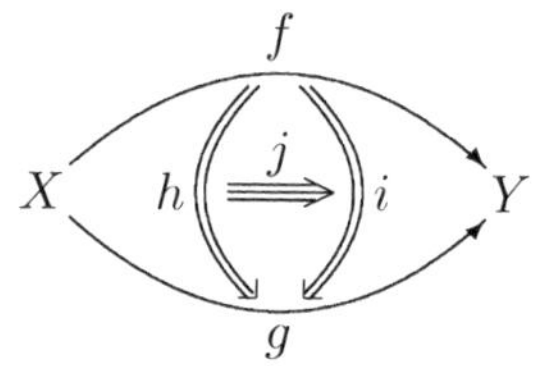

  i.e., continuous maps $j : X \times [0,1] \times [0,1] \to Y$.

Category $\mathcal{CK}$

Consider an $n-$category $\mathcal{CK}$, which has:

- 0-cells: chain complexes A (of Abelian groups, for example)
- 1-cells: chain maps $A \xrightarrow{f} B$
- 2-cells: chain homotopies 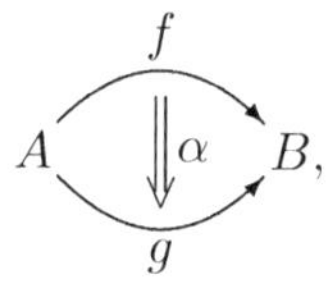

  i.e., maps $\alpha : A \to B$ of degree 1
- 3-cells 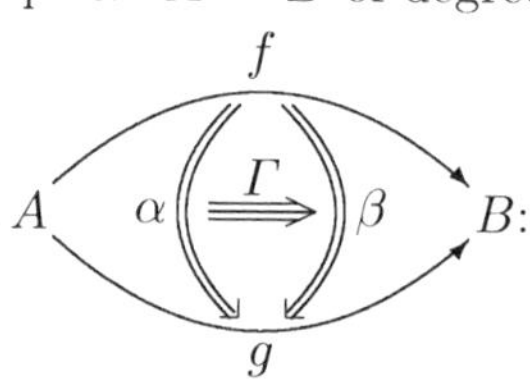

  : homotopies between homotopies,

  i.e., maps $\Gamma : A \to B$ of degree 2 such that $d\Gamma - \Gamma d = \beta - \alpha$.

There ought to be some kind of map $\mathcal{CC} : \mathcal{TT} \Rightarrow \mathcal{CK}$ (see [Lei02, Lei03, Lei04]).

3.4.2 Multicategorical Team/Group Dynamics

The *crowd manifold* M (see [IR10a, IR10b, IR12]) has quite a sophisticated topological structure [IR10c] defined by its macrostate Euler-Lagrangian dynamics. As a Riemannian smooth $n-$manifold, M gives rise to its fundamental $n-$groupoid, or $n-$category $\Pi_n(M)$. In $\Pi_n(M)$, 0-cells are points in M; 1-cells are paths in M (i.e., parameterized smooth maps $f : [0,1] \to M$); 2-cells are smooth homotopies (denoted by $\simeq$) of paths relative to endpoints (i.e., parameterized smooth maps $h : [0,1] \times [0,1] \to M$); 3-cells are smooth homotopies of homotopies of paths in M (i.e., parameterized smooth maps $j : [0,1] \times [0,1] \times [0,1] \to M$). Categorical composition is defined by pasting paths and homotopies. In this way, the following recursive homotopy dynamics emerges on the crowd $3n-$manifold M:

$\texttt{0 - cell} : x_0 \bullet \qquad x_0 \in M; \qquad$ in the higher cells below: $t, s \in [0, 1]$;

$\texttt{1 - cell} : x_0 \bullet \xrightarrow{\ f\ } \bullet x_1 \qquad f : x_0 \simeq x_1 \in M,$

$f : [0, 1] \to M,\ f : x_0 \mapsto x_1,\ x_1 = f(x_0),\ f(0) = x_0,\ f(1) = x_1;$

e.g., linear path: $f(t) = (1 - t)\, x_0 + t\, x_1;$ or

Euler-Lagrangian f – dynamics with endpoint conditions (x_0, x_1) :

$$\frac{d}{dt} f_{\dot{x}^i} = f_{x^i}, \quad \text{with} \quad x(0) = x_0, \quad x(1) = x_1, \quad (i = 1, ..., n);$$

$\texttt{2 - cell} : x_0$ 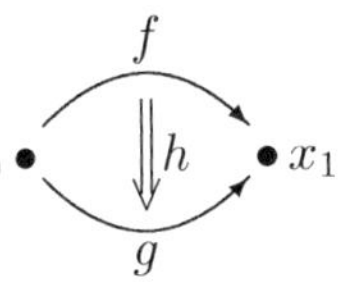 $x_1 \qquad h : f \simeq g \in M,$

$h : [0, 1] \times [0, 1] \to M,\ h : f \mapsto g,\ g = h(f(x_0)),$

$h(x_0, 0) = f(x_0),\ h(x_0, 1) = g(x_0),\ h(0, t) = x_0,\ h(1, t) = x_1$

e.g., linear homotopy: $h(x_0, t) = (1 - t)\, f(x_0) + t\, g(x_0);$ or

homotopy between two Euler-Lagrangian (f, g) – dynamics

with the same endpoint conditions (x_0, x_1) :

$$\frac{d}{dt} f_{\dot{x}^i} = f_{x^i}, \quad \text{and} \quad \frac{d}{dt} g_{\dot{x}^i} = g_{x^i} \quad \text{with} \quad x(0) = x_0, \quad x(1) = x_1;$$

$\texttt{3 - cell} : x_0$ 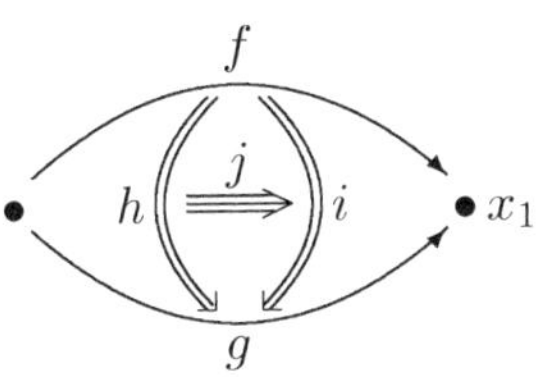 $x_1 \qquad j : h \simeq i \in M,$

$j : [0, 1] \times [0, 1] \times [0, 1] \to M,\ j : h \mapsto i,\ i = j(h(f(x_0)))$

$j(x_0, t, 0) = h(f(x_0)),\ j(x_0, t, 1) = i(f(x_0)),$

$j(x_0, 0, s) = f(x_0),\ j(x_0, 1, s) = g(x_0),$

$j(0, t, s) = x_0,\ j(1, t, s) = x_1$

e.g., linear composite homotopy: $j(x_0, t, s) = (1 - t)\, h(f(x_0)) + t\, i(f(x_0));$

or, homotopy between two homotopies between above two Euler-Lagrangian (f, g) – dynamics with the same endpoint conditions (x_0, x_1).

3.5 Crowd Symplectic Machine in a Category

Recall (from Chapter 2, subsection 2.5.2) that both Hamiltonian and Lagrangian formalisms on Kähler manifolds can be recast in a unique symplectic

formulation (see also [AM78, LR89, Tek05]). So, to start with a dual symplectic dynamics of general human crowds, given the crowd mD configuration Kähler manifold Q[12] with its cotangent bundle $M_H \equiv T^*Q$ (that is a $2m$D momentum phase-space manifold) and a regular Hamiltonian (kinetic plus potential energy) function $H : M_H \to \mathbb{R}$, then:
(i) there is a unique Hamiltonian vector-field $X_H \in M_H$, and
(ii) a unique symplectic 2-form $\omega_H \in M_H$,
such that their inner product (or, contraction) $i_{X_H}\omega_H \equiv \langle X_H, \omega_H\rangle$[13] is also unique and defines the crowd Hamiltonian equations of motion on M_H as the Hamiltonian 1-form:[14]

$$dH = i_{X_H}\omega_H. \tag{3.3}$$

The triple $\mathcal{H} = (M_H, X_H, \omega_H)$ defines the *crowd Hamiltonian system.*

Alternatively, if $M_L \equiv TQ$ is a tangent bundle of the crowd mD configuration Kähler manifold Q (which is a $2m$D velocity phase-space manifold) with a regular Lagrangian (kinetic minus potential energy) function $L : M_L \to \mathbb{R}$ and its associated total (kinetic plus potential energy) function E_L, then:
(i) there is a unique Lagrangian vector-field $X_L \in M_L$, and
(ii) a unique Lagrangian symplectic 2-form $\omega_L \in M_L$,[15]
such that their inner product $i_{X_L}\omega_L \equiv \langle X_L, \omega_L\rangle$ is also unique and defines the crowd Lagrangian equations of motion on M_L as the Lagrangian 1-form:[16]

$$dE_L = i_{X_L}\omega_L. \tag{3.4}$$

The triple $\mathcal{L} = (M_L, X_L, \omega_L)$ defines the *crowd Lagrangian system.*

The two dual crowd triple-systems, $\mathcal{H}$ and $\mathcal{L}$, represent two dual symplectic crowd categories based on the configuration Kähler manifold Q.[17] This symplectic crowd-dynamics duality has both geometric and topological origins (see [IR10c]). It is the starting point for the abstract crowd machine in a category, in the sense of [AM74, AM80], which can be defined as a pair of *adjoint crowd functors:*

$$(\mathfrak{C}_R, \mathfrak{C}_L) : \mathcal{H} \leftrightarrows \mathcal{L},$$

with the right crowd adjoint, $\mathfrak{C}_R : \mathcal{H} \longrightarrow \mathcal{L}$, and the left crowd adjoint, $\mathfrak{C}_L : \mathcal{L} \longrightarrow \mathcal{H}$.

---

12 Recall that each individual agent can be defined as a rigid body in the (complex, or Euclidean) plane, formally an $SE(2)$-group; so the whole crowd configuration m-manifold is defined as the product: $Q = \prod_{i=1}^m SE(2)^i$.

13 In the usual case of a holonomic crowd system, ω_H is closed (i.e., it has a vanishing Lie derivative: $\mathcal{L}_{X_H}\omega_H \equiv di_{X_H}\omega_H + i_{X_H}d\omega_H = 0$) which leads to classical conservation laws.

14 It can be shown that classical Hamilton's equations of motion can be derived from (3.3).

15 In the usual, holonomic crowd case, ω_L is closed.

16 It can be shown that classical Lagrange's equations of motion can be derived from (3.4).

17 It can be shown that both $\mathcal{H}$ and $\mathcal{L}$ satisfy the category axioms.

3.6 Quantum Categorical Structures

3.6.1 Monoidal Tensor Product

To combine two quantum systems A and B into one, we use a *symmetric monoidal tensor product* $A \otimes B$, together with its compound operations: $A \otimes B \xrightarrow{f \otimes g} C \otimes D$. Using the tensor product $\otimes$, we can define a symmetric *monoidal category* $\mathcal{K}$ as a pair $(\mathtt{Ob}(\mathcal{K}), \mathtt{Mor}(\mathcal{K}))$ of generic objects: $A, B, \ldots \in \mathtt{Ob}(\mathcal{K})$ and generic morphisms between objects: $f : A \to B$, $g : B \to C, \ldots \in \mathtt{Mor}(\mathcal{K})$. Formally, we have (see [Coe06, CP09]):

$$\begin{aligned} \mathtt{Ob}(\mathcal{K}) &: \{A, B\} \mapsto A \otimes B, \\ \mathtt{Mor}(\mathcal{K}) &: \{A \xrightarrow{f} B, C \xrightarrow{g} D\} \mapsto A \otimes C \xrightarrow{f \otimes g} B \otimes D, \end{aligned}$$

together with the commutative *bifunctoriality*: the order in which two operations/functions, f (applied to one quantum system) and g (applied to another system), does not matter, that is, the following diagram commutes:

$$\begin{array}{ccc} A_1 \otimes A_2 & \xrightarrow{f \otimes 1_{A_2}} & B_1 \otimes A_2 \\ {\scriptstyle 1_{A_1} \otimes g} \downarrow & & \uparrow {\scriptstyle 1_{B_1} \otimes g} \\ A_1 \otimes B_2 & \xrightarrow[f \otimes 1_{B_2}]{} & B_1 \otimes B_2 \end{array}$$

which shows that both paths from $A_1 \otimes A_2$ to $B_1 \otimes B_2$ yield the same result (see [Coe06, CP09] for technical details).

3.6.2 Snake Lemma and Tensor Products

One of the common theorem-proving tools in algebraic topology is the *snake lemma*, which concerns a commutative and exact diagram called a *snake diagram* [Lan02]:

$$\begin{array}{ccccccc} & & M' & \xrightarrow{f} & M & \xrightarrow{g} & M'' & \longrightarrow & 0 \\ & & \downarrow {\scriptstyle d'} & & \downarrow {\scriptstyle d} & & \downarrow {\scriptstyle d''} & & \\ 0 & \longrightarrow & N' & \xrightarrow[f]{} & N & \xrightarrow[g]{} & N'' & & \end{array}$$

Given a snake diagram as above, the map:

$$\delta : \operatorname{Ker} d'' \longrightarrow \operatorname{Coker} d'$$

is well defined and we have the following exact sequence [Lan02]:

$$\operatorname{Ker} d' \longrightarrow \operatorname{Ker} d \longrightarrow \operatorname{Ker} d'' \xrightarrow{\delta} \operatorname{Coker} d' \longrightarrow \operatorname{Coker} d \longrightarrow \operatorname{Coker} d''$$

where the maps besides δ are the natural ones. The *extended snake diagram* includes the following kernels and cokernels:

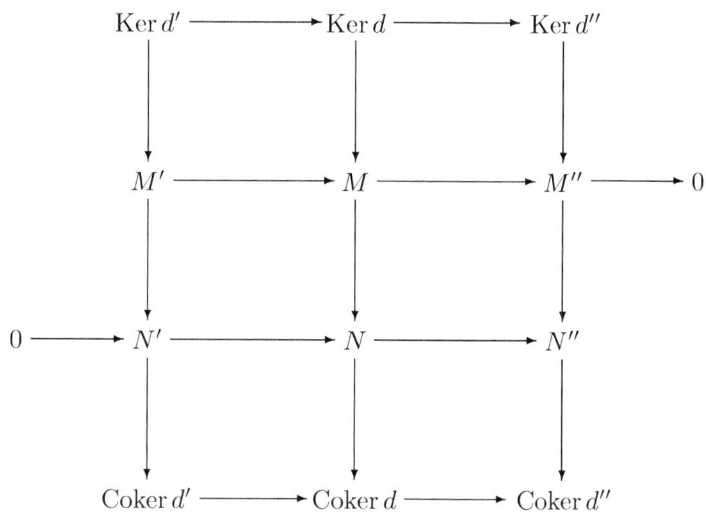

together with the connection map:

$$\delta : \operatorname{Ker} d'' \longrightarrow \operatorname{Coker} d'.$$

For example, consider a commutative diagram of $R-$modules and homomorphisms such that each row is exact:

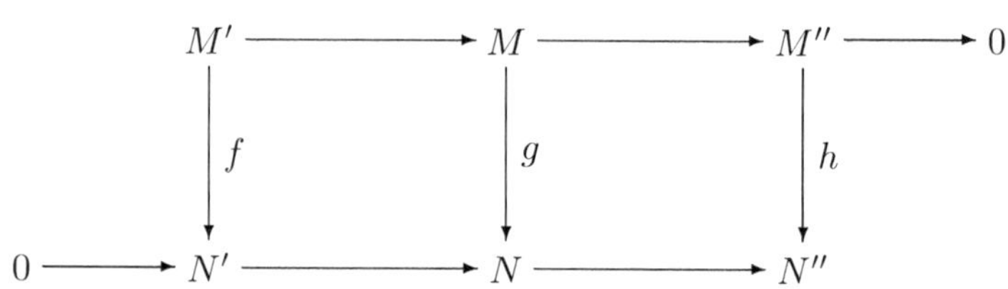

The following assertions about this diagram can be proved [Lan02]:

1. If f, h are monomorphisms, then g is a monomorphism.
2. If f, h are surjective, then g is surjective.
3. Assume in addition that $0 \longrightarrow M' \longrightarrow M$ is exact and that $N \longrightarrow N'' \longrightarrow 0$ is exact. If any two of f, g, h are isomorphisms, then so is the third.

Now, the following conditions are formally equivalent and define the *tensor exact module* F:

1. For every exact sequence

$$E' \longrightarrow E \longrightarrow E''$$

the following sequence is exact:

$$F \otimes E' \longrightarrow F \otimes E \longrightarrow F \otimes E'',$$

where $\otimes$ defines the tensor product operation, which will be used later for modelling crowd behavior in a topos.

2. For every short exact sequence

$$0 \longrightarrow E' \longrightarrow E \longrightarrow E'' \longrightarrow 0$$

the following sequence is exact:

$$0 \longrightarrow F \otimes E' \longrightarrow F \otimes E \longrightarrow F \otimes E'' \longrightarrow 0.$$

3. For every injection $0 \longrightarrow E' \longrightarrow E$ the following sequence is exact:

$$0 \longrightarrow F \otimes E' \longrightarrow F \otimes E.$$

3.6.3 Product of Hilbert Spaces and Quantum Entanglement

Recall that quantum-mechanical systems are properly described by Hermitian operators in complex Hilbert spaces[18] ($\mathcal{H}_A, \mathcal{H}_B, ...$). In the simplest quantum-interaction case, we have only two agent particles, Alice (A) and Bob (B), defined respectively by their state-vectors (or, wave-functions): $\psi_A = |\psi\rangle_A$ and $\psi_B = |\psi\rangle_B$, which live in their corresponding individual Hilbert state-spaces, $\mathcal{H}_A$ and $\mathcal{H}_B$. Therefore, the interaction between Alice and Bob can be formaly represented by a causal arrow:

$$Q : \mathcal{H}_A \ni A \mapsto B \in \mathcal{H}_B.$$

The Alice-Bob *compound quantum state*[19] $|\psi\rangle_A \otimes |\psi\rangle_B$ lives in the composite Hilbert state-space $\mathcal{H}_A \otimes \mathcal{H}_B$, which is the *monoidal tensor product* $\otimes$ of $\mathcal{H}_A$ and $\mathcal{H}_B$.

---

18 In this section we are talking about finitary quantum mechanics, in which all Hilbert spaces are finite-dimensional. The state-space of the system is represented as a space $\mathcal{H}$, i.e. a finite-dimensional complex vector space with a 'sesquilinear' inner-product written $\langle\phi|\psi\rangle$, which is conjugate-linear in the first argument and linear in the second. A state of a quantum system is standardly represented by a vector $\psi = |\psi\rangle \in \mathcal{H}$ of unit norm. For quantum informatics purposes, the basic type is that of qubits, namely 2-dimensional Hilbert space, equipped with a computational basis: $\{|0\rangle, |1\rangle\}$.

19 Quantum compound systems are described by tensor products of the component systems. Here, the key phenomenon of *quantum entanglement* arises, since the general form of a vector in $\mathcal{H}_A \otimes \mathcal{H}_B$ is: $\sum_k \alpha_k \cdot \phi_k \otimes \psi_k$ [AC04].

From categorical perspective, $\mathcal{H}_A$ and $\mathcal{H}_B$ are objects of a (symmetric) monoidal category $\mathfrak{Hilb}$ defined by:

$$\begin{aligned}
\mathtt{Ob}(\mathfrak{Hilb}) &: \{\mathcal{H}_A, \mathcal{H}_B\} \mapsto \mathcal{H}_A \otimes \mathcal{H}_B, \\
\mathtt{Mor}(\mathcal{K}) &: \{\mathcal{H}_A \xrightarrow{f} \mathcal{H}_B, \mathcal{H}_C \xrightarrow{g} \mathcal{H}_D\} \\
&\mapsto \mathcal{H}_A \otimes \mathcal{H}_C \xrightarrow{f \otimes g} \mathcal{H}_B \otimes \mathcal{H}_D.
\end{aligned}$$

From standard quantum-mechanical perspective, $\mathcal{H}_A \otimes \mathcal{H}_B = \mathbb{C}^2 \otimes \mathbb{C}^2$ consists of all weighted vectors of the form

$$|\psi\rangle_A \otimes |\psi\rangle_B = \sum_{k,j} \omega_{kj} \cdot |k\rangle_A \otimes |j\rangle_B,$$

with complex weights $\omega_{kj} \in \mathbb{C}$.[20]

More precisely, if $\{|k\rangle\}_k$ is a basis for the Hilbert space $\mathcal{H}_A$ and $\{|j\rangle\}_j$ is a basis for the Hilbert space $\mathcal{H}_B$, then their tensor product, containing all composite states $|\psi\rangle_A \otimes |\psi\rangle_B$, is defined by the following comprehension set[21] (see [AC04, Coe05, Coe09]):

$$\mathcal{H}_A \otimes \mathcal{H}_B = \left\{ \sum_{k,j} \omega_{kj} \cdot |k\rangle_A \otimes |j\rangle_B \; : \; (\forall i, j), \; \omega_{kj} \in \mathbb{C} \right\}. \tag{3.5}$$

Two such composite vectors,

$$\psi = \sum_{k,j} \omega_{kj} \cdot |k\rangle \otimes |j\rangle \qquad \text{and} \qquad \psi' = \sum_{k,j} \omega'_{kj} \cdot |k\rangle \otimes |j\rangle,$$

are equal iff their complex weights ω_{kj} coincide for all k, j, that is, if their matrices $(\omega)_{kj}$ and $(\omega)'_{kj}$ are equal. As each matrix $(\omega)_{kj}$ is the matrix of some

---

20 This 2-qubit state represents the so-called *EPR pair* (a pair of qubits which jointly are in a *Bell-state*, i.e., entangled with each other), given by:

$$b_1 = |\psi^+\rangle_{AB} = \frac{1}{\sqrt{2}}(|00\rangle_{AB} + |11\rangle_{AB}) \in \mathcal{H}_A \otimes \mathcal{H}_B,$$

or, projected on one of 1D subspaces of $\mathcal{H}_A \otimes \mathcal{H}_B$ spanned by a vector in the so-called *Bell basis*:

$$\begin{aligned}
b_1 &:= \tfrac{1}{\sqrt{2}} \cdot (|00\rangle + |11\rangle), & b_2 &:= \tfrac{1}{\sqrt{2}} \cdot (|01\rangle + |10\rangle), \\
b_3 &:= \tfrac{1}{\sqrt{2}} \cdot (|00\rangle - |11\rangle), & b_4 &:= \tfrac{1}{\sqrt{2}} \cdot (|01\rangle - |10\rangle).
\end{aligned}$$

The EPR pair has the following properties: if qubit A is measured in the computational basis $\{|0\rangle, |1\rangle\}$ a uniformly random bit $x \in \{0, 1\}$ is observed and qubit B collapses to $|x\rangle$.

21 We remark here of the natural relation between the interaction tensor product (3.5) and complex-valued neural networks (see Appendix, section 3.8.5).

linear operator $\omega : \mathcal{H}_A \to \mathcal{H}_B$, namely the one for which we have $\omega(|j\rangle) = \sum_k \omega_{kj} \cdot |k\rangle$, this implies that there is a bijective correspondence between linear operators $\omega : \mathcal{H}_A \to \mathcal{H}_B$ and the composite vectors $\psi \in \mathcal{H}_A \otimes \mathcal{H}_B = \mathbb{C}^2 \otimes \mathbb{C}^2$.

The compound Alice-Bob quantum state:

$$|\psi\rangle_{AB} = |\psi\rangle_A \otimes |\psi\rangle_B = \sum_{k,j} \omega_{kj} \cdot |k\rangle_A \otimes |j\rangle_B \tag{3.6}$$

is called *separable quantum state* if the weight matrix $(\omega)_{kj}$ can be decomposed as:

$$\omega_{kj} = \omega_k^A \omega_j^B, \qquad \text{which implies :}$$
$$|\psi\rangle_A = \sum_k \omega_k^A \cdot |k\rangle_A \qquad \text{and} \qquad |\psi\rangle_B = \sum_j \omega_j^B \cdot |j\rangle_B \,.$$

Otherwise (when $\omega_{kj} \neq \omega_k^A \omega_j^B$), the compound Alice-Bob state $|\psi\rangle_{AB}$ given by (3.6) is called inseparable, or *entangled state*. In particular, the so-called *Bell-state* represents the simplest example of quantum entanglement, defined as a maximally entangled state of two qubits.

3.7 Quantum Protocols

Earlier (in the subsection 3.3.3), we introduced the basic quantum teleportation protocol. Here we mention three new quantum protocols: (i) *logic gate teleportation protocol* ($\overrightarrow{GTP}$), (ii) *entanglement swapping protocol* ($\overrightarrow{ESP}$), and (iii) *quantum gambling protocol* (QGP). For technical details of more general protocols, see [AC04, Coe05, Coe09].

3.7.1 Gate Teleportation Protocol

The quantum (logic) gate teleportation protocol ($\overrightarrow{GTP}$) (see Figure 3.4) is an extension of the $\overrightarrow{QTP}$, in which Alice and Bob are initially not necessarily an EPR-pair but may be in some other entangled state. $\overrightarrow{QTP}$ not only transfers a state ψ_{in} from Alice to Bob, but at the same time applies a linear operator ω to it: $\psi_{out} = \omega(\psi_{in})$. Because of this, $\overrightarrow{QTP}$ is important in applications since it enables robust information processing.

The basic functionality of the $\overrightarrow{GTP}$ was described in [KSW05] as:

$$\overrightarrow{GTP} = \begin{cases} |HH\rangle \to \ |HH\rangle \\ |HV\rangle \to \ |HV\rangle \\ |VH\rangle \to \ |VH\rangle \\ |VV\rangle \to -|VV\rangle, \end{cases} \tag{3.7}$$

where the Boolean states 0 and 1 are represented by the linear horizontal (H) and vertical (V) polarization states of a photon, respectively.

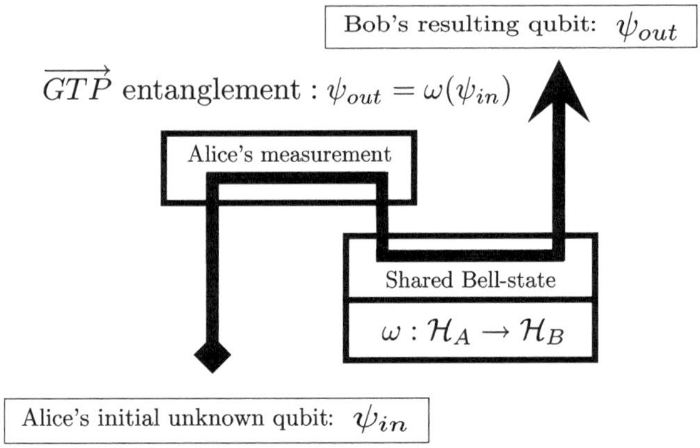

Fig. 3.4. *Gate teleportation protocol* GTP *is an extension of the quantum teleportation protocol* QTP *mentioned earlier.*

The $\overrightarrow{GTP}$-gate can be used to perform a complete Bell-state projection measurement by mapping four Bell states onto four orthogonal product states. In our case of the maximally entangled Bell-state:

$$|\widetilde{\psi^+}\rangle = \frac{1}{\sqrt{2}}(|H+\rangle + |V-\rangle),$$

(where $+(-)$ denotes $+45\,^\circ\,(-45\,^\circ)$ linear polarization), the $\overrightarrow{GTP}$-gate performs the following bilateral operation [SKW09]:

$$\begin{aligned} |\widetilde{\psi^+}\rangle &= \frac{1}{2}\left(|HH\rangle + |HV\rangle + |VH\rangle - |VV\rangle\right) \\ &\overleftrightarrow{GTP}\ \frac{1}{2}\left(|HH\rangle + |HV\rangle + |VH\rangle + |VV\rangle\right) \\ &= \frac{1}{2}\left(|H\rangle + |V\rangle\right) \otimes \left(|H\rangle + |V\rangle\right) = |{+}{+}\rangle. \end{aligned}$$

This means, the gate transforms between the product state $|{+}{+}\rangle$ and the maximally entangled Bell state $|\widetilde{\psi^+}\rangle$. Therefore, by detecting one of these four product states *behind* the phase gate, we know that the photons have been in the corresponding Bell state *before* the phase gate (see [SKW09] for technical details). This fact demonstrates the *causality* of the $\overrightarrow{GTP}$-gate.

3.7.2 Entanglement Swapping Protocol

The entanglement swapping protocol ($\overrightarrow{ESP}$) is another variation/extension of the $\overrightarrow{QTP}$. We start with two Bell-states, then apply a Bell-base measurement to one qubit in each pair (see Figure 3.5). The inherent quantum features of the teleportation process are best seen by performing the $\overrightarrow{ESP}$. In this quantum communication method, two photons that have never interacted in the past, become entangled by teleporting the state of one photon of an entangled pair onto one photon of another entangled pair (see [SKW09] for technical details).

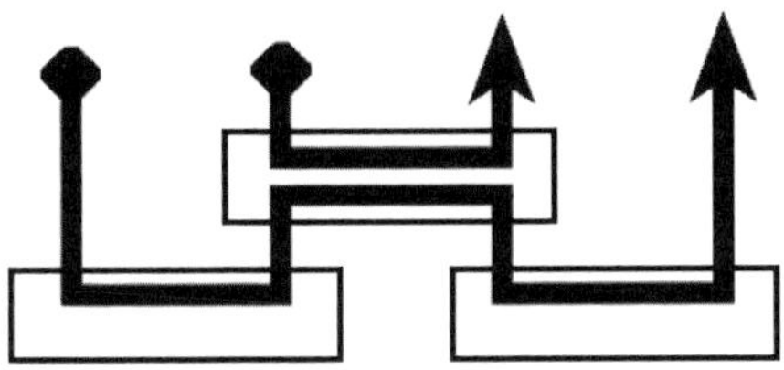

Fig. 3.5. *Entanglement swapping protocol* ESP *is another extension of the* QTP, *which maps two Alices into their corresponding Bobs.*

For derivation of all three quantum communication protocols in terms of special *quantum projectors* of the form, see [Coe03]:

$$P = |\psi\rangle\langle\psi| \qquad \text{with} \qquad |\psi\rangle = \sum_{k,j} \omega_{kj} \cdot |kj\rangle \in \mathcal{H} \otimes \mathcal{H},$$

where the main theorem is that any four linear operators satisfy the following *bifunctorial equation*:

$$(g \circ f) \otimes (k \circ h) = (g \otimes k) \circ (f \otimes h). \tag{3.8}$$

3.7.3 Quantum Gambling Protocol

Recall that *coin tossing* is defined as a method of generating a random bit over a communication channel between two distant parties. In case of quantum protocols, the two parties, Alice and Bob, do not trust each other, or a third party. They create the random bit by exchanging quantum and classical information. At the end of the protocol, the generated bit is known to both of them. If a party cheats, i.e., changes the occurrence probability of an outcome, the other party should be able to detect the cheating. We would consider a coin tossing protocol to be secure if it defines a parameter such that when it goes to infinity the probability to detect any finite change of probabilities goes to 1. Using a secure protocol, the parties can make certain decisions depending on the value of the random bit, without being afraid that the opponent may have some advantage. For instance, Alice and Bob can play a game $\overrightarrow{QGP}$ in which Alice wins if the outcome is '0' and Bob wins if it is '1'. Note that if bit commitment were secure, it could be used to implement coin tossing trivially: Alice commits a bit a to Bob; Bob tells Alice the value of a bit b; the random bit is the parity bit of a and b.

The gambling task $\overrightarrow{QGP} : A \to B$ can be precisely defined as follows [GVW99]: the casino (Alice) and the player (Bob) are physically separated, communicating via quantum and classical channels. The bet of Bob in a single game is taken for simplicity to be 1 coin. At the end of a game the player wins 1 or R coins, or loses 1 coin (his bet), depending on the result of the game. We have found a protocol which implement this game while respecting two requirements: First, the player can ensure that, irrespective of what the casino does, his expected gain is not less than δ coins, where δ is a negative function of R which goes to zero when R goes to infinity. The exact form of $\delta(R)$ will be specified below. Second, the casino can ensure that, irrespective of what the player does, its expected gain is not less than 0 coins.

In order to define the protocol rigorously, we will first present the rules of the game, then the strategies of the players which ensure the outcomes quoted above and finally we will prove the security of the method.

The Rules of the Game: Alice has two boxes, A and B, which can store a particle [GVW99]. The quantum states of the particle in the boxes are denoted by $|a\rangle$ and $|b\rangle$, respectively. Alice prepares the particle in some state and sends box B to Bob.

Bob wins in one of the two cases:

1. If he finds the particle in box B, then Alice pays him 1 coin (after checking that box A is empty).
2. If he asks Alice to send him box A for verification and he finds that she initially prepared a state different from

$$|\psi_0\rangle = \frac{1}{\sqrt{2}}(|a\rangle + |b\rangle), \tag{3.9}$$

then Alice pays him R coins.

In any other case Alice wins, and Bob pays her 1 coin.

The players' strategies which ensure (independently) an expectation value of Alice's gain $G_A \geq 0$ (irrespective of Bob's actions) and an expectation value of Bob's gain $G_B \geq \delta$ (irrespective of Alice's actions) are as follows:
Alice's Strategy: Alice prepares the equally distributed state $|\psi_0\rangle$ (given in (3.9)).
Bob's Strategy: After receiving box B, Bob splits the particle in two parts; specifically, he performs the following unitary operation:

$$|b\rangle \to \sqrt{1-\eta}\,|b\rangle + \sqrt{\eta}\,|b'\rangle, \tag{3.10}$$

where $\langle b'|b\rangle = 0$. The particular splitting parameter η he uses is $\eta = \tilde{\eta}(R)$ (to be specified below). After the splitting Bob measures the projection operator on the state $|b\rangle$, and then

I. If the measurement yields a positive result, i.e. he finds the particle, he announces Alice that he won.
II. If the measurement yields a negative result, he asks Alice for box A and verifies the preparation.

This completes the formal definition of the gaming protocol (for more technical details, see [GVW99]).

3.8 Appendix

3.8.1 Abelian Category of Chain and Cochain Complexes

Chain and Cochain Complexes on a Smooth Manifold

Recall that the central concept in (co)homology theory is the *category* $\mathbf{S}^\bullet(\mathcal{C})$ of *generalized (co)chain complexes* in an Abelian category $\mathcal{C}$ [Die88, II07]. The *objects* of the category $\mathbf{S}^\bullet(\mathcal{C})$ are infinite sequences

$$A^\bullet : \cdots \longrightarrow A^{n-1} \xrightarrow{d^{n-1}} A^n \xrightarrow{d^n} A^{n+1} \longrightarrow \cdots$$

where, for each $n \in \mathbb{Z}$, A^n is an object of $\mathcal{C}$ and d^n a morphism of $\mathcal{C}$, with the conditions

$$d^{n-1} \circ d^n = 0$$

for every $n \in \mathbb{Z}$. When $A^n = 0$ for $n < 0$, one speaks of *cochain complexes*. The d^n are called *coboundary operators*.

The *morphisms* of the category $\mathbf{S}^\bullet(\mathcal{C})$ are sequences $f^\bullet = (f^n) : A^\bullet \to B^\bullet$ where, for each $n \in \mathbb{Z}$, $f^n : A^n \to B^n$ is a morphism of $\mathcal{C}$, and in the diagram

$$\begin{array}{ccccccccc} \cdots \longrightarrow & A^{n-1} & \xrightarrow{d^{n-1}} & A^n & \xrightarrow{d^n} & A^{n+1} & \longrightarrow \cdots \\ & f^{n-1} \downarrow & & f^n \downarrow & & f^{n+1} \downarrow & \\ \cdots \longrightarrow & B^{n-1} & \xrightarrow{d^{n-1}} & B^n & \xrightarrow{d^n} & B^{n+1} & \longrightarrow \cdots \end{array} \tag{3.11}$$

all squares are commutative; one says the f^n commute with the coboundary operators. One has $\operatorname{Im} d^{n+1} \subset \operatorname{Ker} d^n \subset A^n$ for every $n \in \mathbb{Z}$; the quotient $H^n(A^\bullet) = \operatorname{Ker} d^n / \operatorname{Im} d^{n+1}$ is called the nth cohomology object of $A^\bullet$. From (3.11) it follows that there is a morphism

$$H^n(f^\bullet) : H^n(A^\bullet) \to H^n(B^\bullet)$$

deduced canonically from $f^\bullet$, and

$$(A^\bullet, f^\bullet) \Rightarrow (H^n(A^\bullet), H^n(f^\bullet))$$

is a *covariant functor* from $\mathbf{S}^\bullet(\mathcal{C})$ to $\mathcal{C}$.

The *cohomology exact sequence*: if three cochain complexes $A^\bullet, B^\bullet, C^\bullet$ are elements of a short exact sequence of morphisms

$$0 \longrightarrow A^\bullet \longrightarrow B^\bullet \longrightarrow C^\bullet \longrightarrow 0$$

then there exists an infinite sequence of canonically defined morphisms d^n : $H^n(C^\bullet) \to H^{n-1}(A^\bullet)$ such that the sequence

$$\cdots \longrightarrow H^n(A^\bullet) \longrightarrow H^n(B^\bullet) \longrightarrow H^n(C^\bullet) \longrightarrow H^{n-1}(A^\bullet) \longrightarrow \cdots$$

is *exact*, that is the *image* of each homomorphism in the sequence is exactly the *kernel* of the next one.

The *dual* to the category $\mathbf{S}^\bullet(\mathcal{C})$ is the *category* of $\mathbf{S}_\bullet(\mathcal{C})$ of *generalized chain complexes*. Its objects and morphisms are obtained by formal inversion of all arrows and lowering all indices.

(Co)Homologies in Abelian Categories Related to the Smooth (Team) Manifold

Let $\mathcal{M}^\bullet$ denote the Abelian category of cochains, (i.e., p-forms) on the crowd configuration manifold M. When $\mathcal{C} = \mathcal{M}^\bullet$, we have the category $\mathcal{S}^\bullet(\mathcal{M}^\bullet)$ of generalized cochain complexes $A^\bullet$ in $\mathcal{M}^\bullet$, and if $A' = 0$ for $n < 0$ we have a subcategory $\mathcal{S}^\bullet_{\mathcal{DR}}(\mathcal{M}^\bullet)$ of the *De Rham differential complexes* in $\mathcal{M}^\bullet$:

$$A^\bullet_{DR} : 0 \to \Omega^0(M) \xrightarrow{d} \Omega^1(M) \xrightarrow{d} \Omega^2(M) \cdots \xrightarrow{d} \Omega^n(M) \xrightarrow{d} \cdots . \quad (3.12)$$

Here $A' = \Omega^n(M)$ is the vector space over $\mathbb{R}$ of all p-forms ω on M (for $p = 0$ the smooth functions on M) and $d_n = d : \Omega^{n-1}(M) \to \Omega^n(M)$ is the exterior differential. A form $\omega \in \Omega^n(M)$ such that $d\omega = 0$ is a closed form or n-cocycle. A form $\omega \in \Omega^n(M)$ such that $\omega = d\theta$, where $\theta \in \Omega^{n-1}(M)$, is an exact form or n-coboundary. Let $Z^n(M) = \operatorname{Ker} d$ (resp. $B^n(M) = \operatorname{Im} d$ denote a real vector space of cocycles (resp. coboundaries) of degree n. Since $d_{n+1} \circ d_n = d^2 = 0$, we have $B^n(M) \subset Z^n(M)$. The quotient vector space,

$$H^n_{DR}(M) = \operatorname{Ker} d / \operatorname{Im} d = Z^n(M)/B^n(M),$$

is the *De Rham cohomology group*. The elements of $H^n_{DR}(M)$ represent equivalence sets of cocycles. Two cocycles ω_1, ω_2 belong to the same equivalence set, or are cohomologous (written $\omega_1 \sim \omega_2$) iff they differ by a coboundary $\omega_1 - \omega_2 = d\theta$. The de Rham's cohomology class of any form $\omega \in \Omega^n(M)$ is $[\omega] \in H^n_{DR}(M)$. The De Rham differential complex (3.12) can be considered as a system of second-order DEs: $d^2\theta = 0$, $\theta \in \Omega^{n-1}(M)$, having a solution represented by $Z^n(M) = \operatorname{Ker} d$.

Analogously, let $\mathcal{M}_\bullet$ denote the Abelian category of chains on the configuration manifold M. When $\mathcal{C} = \mathcal{M}_\bullet$, we have the category $\mathcal{S}_\bullet(\mathcal{M}_\bullet)$ of generalized chain complexes $A_\bullet$ in $\mathcal{M}_\bullet$, and if $A_n = 0$ for $n < 0$ we have a subcategory $\mathcal{S}^C_\bullet(\mathcal{M}_\bullet)$ of chain complexes in $\mathcal{M}_\bullet$:

$$A_\bullet : 0 \leftarrow C^0(M) \xleftarrow{\partial} C^1(M) \xleftarrow{\partial} C^2(M) \cdots \xleftarrow{\partial} C^n(M) \xleftarrow{\partial} \cdots .$$

Here $A_n = C^n(M)$ is the vector space over $\mathbb{R}$ of all finite chains C on the manifold M and $\partial_n = \partial : C^{n+1}(M) \to C^n(M)$. A finite chain C such that $\partial C = 0$ is an $n-$cycle. A finite chain C such that $C = \partial B$ is an $n-$boundary. Let $Z_n(M) = \operatorname{Ker}\partial$ (resp. $B_n(M) = \operatorname{Im}\partial$) denote a real vector space of cycles (resp. boundaries) of degree n. Since $\partial_{n+1} \circ \partial_n = \partial^2 = 0$, we have $B_n(M) \subset Z_n(M)$. The quotient vector space,

$$H^C_n(M) = \operatorname{Ker}\partial / \operatorname{Im}\partial = Z_n(M)/B_n(M),$$

is the $n-$homology group. The elements of $H^C_n(M)$ are equivalence sets of cycles. Two cycles C_1, C_2 belong to the same equivalence set, or are homologous (written $C_1 \sim C_2$), iff they differ by a boundary $C_1 - C_2 = \partial B$). The homology class of a finite chain $C \in C^n(M)$ is $[C] \in H^C_n(M)$.

The dimension of the $n-$cohomology (resp. $n-$homology) group equals the nth *Betti number* b^n (resp. b_n) of the manifold M. *Poincaré lemma* says that on an open set $U \in M$ diffeomorphic to $\mathbb{R}^N$, all closed forms (cycles) of degree $p \geq 1$ are exact (boundaries). That is, the *Betti numbers* satisfy $b^p = 0$ (resp. $b = 0$), for $p = 1, \ldots, n$.

The *De Rham theorem* states the following. The map Φ: $H_n \times H^n \to \mathbb{R}$ given by $([C], [\omega]) \to \langle C, \omega \rangle$ for $C \in Z_n, \omega \in Z^n$ is a bilinear nondegenerate map which establishes the duality of the groups (vector spaces) H_n and H^n and the equality $b_n = b^n$.

3.8.2 A Brief on Categorical Logic

In this subsection, we give a brief on advanced categorial logic. For the basic, first-order logic, see subsection 4.3.6 below.

Now we are almost ready to embark on our journey into topos theory. Before that, in this subsection we will make a brief excursion into related area of logic in *coherent Cartesian closed categories.*[22]

A *category with multiplication* is a category $\mathcal{C}$ together with a bifunctor $\cdot : \mathcal{C} \times \mathcal{C} \to \mathcal{C}$ and a special object I (a propositional constant, see below).

[22] The term "coherence" covers in category theory what from a logical point of view would be called problems of *completeness*, *axiomatizability* and *decidability*. A coherence condition, or coherence theorem expresses the statement that two or more morphisms between two given objects, the existence of which is given or follows from general properties, are equal. As different authors put stress on different things related to coherence, we stick to MacLane's usage of the term in [Mac63], the primordial paper on coherence. Basically, MacLane has shown that monoidal and symmetric monoidal categories are coherent.

In particular, a *category with binary products* is a category with a binary operation (Cartesian product $\times$) on objects, *projection morphisms* [DP01, DP04]

$$\mathbf{k}^1_{A,B} : A \times B \to A, \qquad \mathbf{k}^2_{A,B} : A \times B \to B,$$

and the *pairing operation* on morphisms $\langle , \rangle$ given by

$$\frac{f : C \to A \qquad g : C \to B}{\langle f, g \rangle : C \to A \times B}. \tag{3.13}$$

The morphisms must satisfy the following set of equations:

$$\mathbf{k}^1_{A,B} \circ \langle f, g \rangle = f, \qquad \mathbf{k}^2_{A,B} \circ \langle f, g \rangle = g,$$
$$\langle f, g \rangle \circ h = \langle f \circ h, g \circ h \rangle, \qquad \langle \mathbf{k}^1_{A,B}, \mathbf{k}^2_{A,B} \rangle = \mathbf{1}_{A \times B}.$$

A category has a *terminal object* T iff it has the special morphisms: $\mathbf{k}_A : A \to \mathrm{T}$, which satisfy the equation: for $f : A \to \mathrm{T}, \quad f = \mathbf{k}_A$.

A *cartesian category* is a category with binary products and a terminal object.

In particular, standard equational axiomatization of Cartesian categories (see [LS86]) is based on the universality of the Cartesian product and uses as primitives the following morphisms: $\mathbf{1}_A : A \to A$, $\boldsymbol{\pi}_{A,B} : A{\times}B \to A$, $\boldsymbol{\pi}'_{A,B} : A{\times}B \to B$ and $\mathbf{k}_A : A \to \mathrm{I}$ for all objects A and B, and a partial binary operation on morphisms (3.13). The following equations hold:

$$f = \mathbf{k}_A, \quad \text{for every} \quad f : A \to \mathrm{I};$$
$$\boldsymbol{\pi}_{A,B}\langle f, g \rangle = f; \quad \text{for } f : C \to A \text{ and } g : C \to B;$$
$$\boldsymbol{\pi}^{prime}_{A,B}\langle f, g \rangle = g; \quad \text{for } f : C \to A \text{ and } g : C \to B;$$
$$\langle \boldsymbol{\pi}_{A,B} h, \boldsymbol{\pi}^{prime}_{A,B} h \rangle = h, \quad \text{for } h : C \to A{\times}B,$$

together with the standard categorial equations:

$$\text{(cat 1)} \quad \mathbf{1}_B \circ f = f \circ \mathbf{1}_A = f, \qquad \text{(cat 2)} \quad h \circ (g \circ f) = (h \circ g) \circ f. \tag{3.14}$$

Now, we can define the propositional language $\mathcal{P}$ as generated from a set of *propositional letters* $\mathcal{L}$ with the nullary connectives, i.e. propositional constants, I and O, and the binary connectives $\times$ and $+$. The fragments $\mathcal{P}_{\times,+,\mathrm{I}}$, $\mathcal{P}_{\times,+}$ etc. of $\mathcal{P}$ are obtained by keeping only those formulae of $\mathcal{P}$ that contain the connectives in the index. For the propositional letters of $\mathcal{P}$, i.e., for the members of $\mathcal{L}$, we use the schematic letters $p, q, \dots, p_1, \dots$, and for the formulae of $\mathcal{P}$, or of its fragments, we use the schematic letters $A, B, \dots, A_1, \dots$ (see [DP04]).

Next we define inductively the *terms* that will stand for the morphisms of the free bicartesian category $\mathcal{C}$ generated by $\mathcal{L}$. Every term has a *type*, which is a pair (A, B) of formulae of $\mathcal{P}$. That a term f is of type (A, B) is written $f : A \to B$. The *atomic* terms of $\mathcal{C}$ are for every A of $\mathcal{P}$

$$\mathbf{1}_A : A \to A,$$
$$k_A : A \to \mathrm{I}, \qquad\qquad l_A : \mathrm{O} \to A.$$

The terms $\mathbf{1}_A$ are called *identities*. The other terms of $\mathcal{C}$ are generated with the following operations on terms, which we present by rules so that from the terms in the premises we obtain the terms in the conclusion (using $f, g, \ldots, f_1, \ldots$ as schematic letters for terms of $\mathcal{C}$) :

$$\frac{f : A \to B \qquad g : B \to C}{g \circ f : A \to C},$$
$$\frac{f : A \to C}{K_B^1 f : A \times B \to C}, \qquad \frac{f : C \to A}{L_B^1 f : C \to A + B},$$
$$\frac{f : B \to C}{K_A^2 f : A \times B \to C}, \qquad \frac{f : C \to B}{L_A^2 f : C \to A + B},$$
$$\frac{f : C \to A \qquad g : C \to B}{\langle f, g \rangle : C \to A \times B}, \qquad \frac{f : A \to C \qquad g : B \to C}{[f, g] : A + B \to C}.$$

The category $\mathcal{C}$ has as objects the formulae of $\mathcal{P}$ and as morphisms equivalence classes of terms[23] so that the both (3.14) and the following equations are satisfied for $i \in \{1, 2\}$ [DP02]

$$
\begin{array}{llll}
(K1) & g \circ K_A^i f = K_A^i (g \circ f), & (L1) & L_A^i g \circ f = L_A^i (g \circ f), \\
(K2) & K_A^i g \circ \langle f_1, f_2 \rangle = g \circ f_i, & (L2) & [g_1, g_2] \circ L_A^i f = g_i \circ f, \\
(K3) & \langle g_1, g_2 \rangle \circ f = \langle g_1 \circ f, g_2 \circ f \rangle, & (L3) & g \circ [f_1, f_2] = [g \circ f_1, g \circ f_2], \\
(K4) & \langle K_B^1 \mathbf{1}_A, K_A^2 \mathbf{1}_B \rangle = \mathbf{1}_{A \times B}, & (L4) & [L_B^1 \mathbf{1}_A, L_A^2 \mathbf{1}_B] = \mathbf{1}_{A+B}, \\
(k) & \text{for } f : A \to \mathrm{I},\ f = k_A, & (l) & \text{for } f : \mathrm{O} \to A,\ f = l_A.
\end{array}
$$

For more technical details on categorical logic, an interested reader might consult J. Lambek's categorial proof-theoretical program [Lam68, Lam69, Lam72, LS86].

3.8.3 Natural Geometrical Operations on Kähler Manifolds

Lie Derivative on a Kähler Manifold

To define how vector-fields operate on functions on a Kähler m-manifold M, we will use the *Lie derivative*.[24] Let $f : M \to \mathbb{R}$ so $Tf : TM \to T\mathbb{R} = \mathbb{R} \times \mathbb{R}$.

---

[23] Equivalence between proofs in intuitionistic logic is axiomatized independently of these diagrams in the *typed lambda calculus* and in various sorts of categories, like bicartesian closed categories. There, proofs are coded by typed lambda terms or by arrow terms, and two proofs are considered equivalent iff the coding terms are equal as lambda terms or as arrow terms in categories [DP02, DP04].

[24] Recall that the *Lie derivative* is popularly called the '*fisherman's derivative*'. In continuum mechanics it is called *Liouville operator*. This is a central differential operator in modern differential geometry and its physical and control applications.

By generalizing the notions of [AMR88] to the Kähler geometry, we write Tf acting on a vector $v \in T_m M$ in the form

$$Tf \cdot v = (f(m), df(m) \cdot v)\,.$$

This defines, for each point $m \in M$, the element $df(m) \in T_m^* M$. Thus df is a section of the cotangent bundle T^*M, i.e., a 1−form. The 1−form $df : M \to T^*M$ defined this way is called the *differential* of f. If f is C^∞, then df is C^{k-1}.

If $\phi : U \subset M \to V \subset E$ is a local chart for a Kähler m-manifold M, then the local representative of $f \in C^\infty(M, \mathbb{R})$ is the map $f : V \to \mathbb{R}$ defined by $f = f \circ \phi^{-1}$. The local representative of Tf is the tangent map for local manifolds,

$$Tf(x, v) = (f(x), Df(x) \cdot v)\,.$$

Thus the local representative of df is the derivative of the local representative of f. In particular, if $(x^1, ..., x^n)$ are local coordinates on M, then the local components of df are

$$(df)^i = \partial_{x^i} f.$$

The introduction of df leads to the following definition of the Lie derivative. The *directional* or *Lie derivative* $\mathcal{L}_X : C^\infty(M, \mathbb{R}) \to C^{k-1}(M, \mathbb{R})$ of a function $f \in C^\infty(M, \mathbb{R})$ along a vector-field X is defined by

$$\mathcal{L}_X f(m) = X[f](m) = df(m) \cdot X(m),$$

for any $m \in M$. Denote by $X[f] = df(X)$ the map $M \ni m \mapsto X[f](m) \in \mathbb{R}$. If f is F−valued, the same definition is used, but now $X[f]$ is F−valued.

If a local chart (U, ϕ) on a Kähler m-manifold M has local coordinates $(x^1, ..., x^n)$, the local representative of $X[f]$ is given by the function

$$\mathcal{L}_X f = X[f] = X^i\, \partial_{x^i} f.$$

Evidently if f is C^∞ and X is C^{k-1} then $X[f]$ is C^{k-1}.

Let $\varphi : M \to N$ be a diffeomorphism between two Kähler manifolds M and N. Then $\mathcal{L}_X$ is natural with respect to push–forward by φ. That is, for each $f \in C^\infty(M, \mathbb{R})$,

$$\mathcal{L}_{\varphi_* X}(\varphi_* f) = \varphi_* \mathcal{L}_X f,$$

i.e., the following diagram commutes:

$$\begin{array}{ccc}
C^\infty(M, \mathbb{R}) & \xrightarrow{\varphi_*} & C^\infty(N, \mathbb{R}) \\
\mathcal{L}_X \downarrow & & \downarrow \mathcal{L}_{\varphi_* X} \\
C^\infty(M, \mathbb{R}) & \xrightarrow[\varphi_*]{} & C^\infty(N, \mathbb{R})
\end{array}$$

Also, $\mathcal{L}_X$ is natural with respect to restrictions. That is, for U open in a Kähler m-manifold M and $f \in C^\infty(M, \mathbb{R})$,

$$\mathcal{L}_{X|U}(f|U) = (\mathcal{L}_X f)|U,$$

where $|U : C^\infty(M,\mathbb{R}) \to C^\infty(U,\mathbb{R})$ denotes restriction to U, i.e., the following diagram commutes:

$$\begin{array}{ccc} C^\infty(M,\mathbb{R}) & \xrightarrow{|U} & C^\infty(U,\mathbb{R}) \\ \Big\downarrow \mathcal{L}_X & & \Big\downarrow \mathcal{L}_{X|U} \\ C^\infty(M,\mathbb{R}) & \xrightarrow[|U]{} & C^\infty(U,\mathbb{R}) \end{array}$$

Since $\varphi^* = (\varphi^{-1})_*$ the Lie derivative is also natural with respect to pull–back by φ. This has a generalization to φ–related vector-fields as follows: Let $\varphi : M \to N$ (between two Kähler manifolds M and N) be a C^∞–map, $X \in \mathcal{X}^{k-1}(M)$ and $Y \in \mathcal{X}^{k-1}(N)$, $k \geq 1$. If $X \sim_\varphi Y$, then

$$\mathcal{L}_X(\varphi^* f) = \varphi^* \mathcal{L}_Y f$$

for all $f \in C^\infty(N,\mathbb{R})$, i.e., the following diagram commutes:

$$\begin{array}{ccc} C^\infty(N,\mathbb{R}) & \xrightarrow{\varphi^*} & C^\infty(M,\mathbb{R}) \\ \Big\downarrow \mathcal{L}_Y & & \Big\downarrow \mathcal{L}_X \\ C^\infty(N,\mathbb{R}) & \xrightarrow[\varphi^*]{} & C^\infty(M,\mathbb{R}) \end{array}$$

The Lie derivative map $\mathcal{L}_X : C^\infty(M,\mathbb{R}) \to C^{k-1}(M,\mathbb{R})$ is a *derivation*, i.e., for two functions $f, g \in C^\infty(M,\mathbb{R})$ the *Leibniz rule* is satisfied

$$\mathcal{L}_X(fg) = g\mathcal{L}_X f + f\mathcal{L}_X g;$$

Also, Lie derivative of a constant function is zero, $\mathcal{L}_X(\text{const}) = 0$.

The connection between the Lie derivative $\mathcal{L}_X f$ of a function $f \in C^\infty(M,\mathbb{R})$ and the flow F_t of a vector-field $X \in \mathcal{X}^{k-1}(M)$ is given as:

$$\frac{d}{dt}\left(F_t^* f\right) = F_t^*\left(\mathcal{L}_X f\right).$$

Lie Bracket on a Kähler Manifold

If $X, Y \in \mathcal{X}^k(M)$, $k \geq 1$ are two vector-fields on a Kähler m-manifold M, then

$$[\mathcal{L}_X, \mathcal{L}_Y] = \mathcal{L}_X \circ \mathcal{L}_Y - \mathcal{L}_Y \circ \mathcal{L}_X$$

is a derivation map from $C^{k+1}(M,\mathbb{R})$ to $C^{k-1}(M,\mathbb{R})$. Then there is a unique vector-field, $[X,Y] \in \mathcal{X}^k(M)$ of X and Y such that $\mathcal{L}_{[X,Y]} = [\mathcal{L}_X, \mathcal{L}_Y]$ and

$[X,Y](f) = X(Y(f)) - Y(X(f))$ holds for all functions $f \in C^\infty(M,\mathbb{R})$. This vector-field is also denoted $\mathcal{L}_X Y$ and is called the Lie derivative of Y with respect to X, or the *Lie bracket* of X and Y. In a local chart (U,ϕ) at a point $m \in M$ with coordinates $(x^1,...,x^n)$, for $X|_U = X^i \partial_{x^i}$ and $Y|_U = Y^i \partial_{x^i}$ we have

$$\left[X^i \partial_{x^i}, Y^j \partial_{x^j}\right] = \left(X^i \left(\partial_{x^i} Y^j\right) - Y^i \left(\partial_{x^i} X^j\right)\right) \partial_{x^j},$$

since second partials commute. If, also X has flow F_t, then [AMR88]

$$\frac{d}{dt}\left(F_t^* Y\right) = F_t^*\left(\mathcal{L}_X Y\right).$$

In particular, if $t = 0$, this formula becomes

$$\frac{d}{dt}|_{t=0}\left(F_t^* Y\right) = \mathcal{L}_X Y.$$

Then the unique C^{k-1} vector-field $\mathcal{L}_X Y = [X,Y]$ on M defined by

$$[X,Y] = \frac{d}{dt}|_{t=0}\left(F_t^* Y\right),$$

is called the Lie derivative of Y with respect to X, or the Lie bracket of X and Y, and can be interpreted as the leading order term that results from the sequence of flows

$$F_t^{-Y} \circ F_t^{-X} \circ F_t^{Y} \circ F_t^{-X}(m) = \epsilon^2 [X,Y](m) + \mathcal{O}(\epsilon^3), \qquad (3.15)$$

for some real $\epsilon > 0$. Therefore a Lie bracket can be interpreted as a 'new direction' in which the system can flow, by executing the sequence of flows (3.15).

Lie bracket satisfies the following property:

$$[X,Y][f] = X[Y[f]] - Y[X[f]],$$

for all $f \in C^{k+1}(U,\mathbb{R})$, where U is open in M.

An important relationship between flows of vector-fields is given by the *Campbell–Baker–Hausdorff formula*:

$$F_t^Y \circ F_t^X = F_t^{X+Y+\frac{1}{2}[X,Y]+\frac{1}{12}([X,[X,Y]]-[Y,[X,Y]])+\ldots} \qquad (3.16)$$

Essentially, if given the composition of multiple flows along multiple vector-fields, this formula gives the one flow along one vector-field which results in the same net flow. One way to prove the Campbell–Baker–Hausdorff formula (3.16) is to expand the product of two formal exponentials and equate terms in the resulting formal power series.

Lie bracket is the $\mathbb{R}-$bilinear map $[,] : \mathcal{X}^k(M) \times \mathcal{X}^k(M) \to \mathcal{X}^k(M)$ with the following properties:

1. $[X, Y] = -[Y, X]$, i.e., $\mathcal{L}_X Y = -\mathcal{L}_Y X$ for all $X, Y \in \mathcal{X}^k(M)$ – skew–symmetry;
2. $[X, X] = 0$ for all $X \in \mathcal{X}^k(M)$;
3. $[X, [Y, Z]] + [Y, [Z, X]] + [Z, [X, Y]] = 0$ for all $X, Y, Z \in \mathcal{X}^k(M)$ – the Jacobi identity;
4. $[fX, Y] = f[X, Y] - (Yf)X$, i.e., $\mathcal{L}_{fX}(Y) = f(\mathcal{L}_X Y) - (\mathcal{L}_Y f)X$ for all $X, Y \in \mathcal{X}^k(M)$ and $f \in C^\infty(M, \mathbb{R})$;
5. $[X, fY] = f[X, Y] + (Xf)Y$, i.e., $\mathcal{L}_X(fY) = f(\mathcal{L}_X Y) + (\mathcal{L}_X f)Y$ for all $X, Y \in \mathcal{X}^k(M)$ and $f \in C^\infty(M, \mathbb{R})$;
6. $[\mathcal{L}_X, \mathcal{L}_Y] = \mathcal{L}_{[x,y]}$ for all $X, Y \in \mathcal{X}^k(M)$.

The pair $(\mathcal{X}^k(M), [,])$ is the prototype of a *Lie algebra* [KMS93]. In more general case of a *general linear Lie algebra* $\mathfrak{gl}(n)$, which is the Lie algebra associated to the Lie group $GL(n)$, Lie bracket is given by a *matrix commutator*

$$[A, B] = AB - BA,$$

for any two matrices $A, B \in \mathfrak{gl}(n)$.

Let $\varphi : M \to N$ be a diffeomorphism. Then $\mathcal{L}_X : \mathcal{X}^k(M) \to \mathcal{X}^k(M)$ is natural with respect to push–forward by φ. That is, for each $f \in C^\infty(M, \mathbb{R})$,

$$\mathcal{L}_{\varphi_* X}(\varphi_* Y) = \varphi_* \mathcal{L}_X Y,$$

i.e., the following diagram commutes:

$$\begin{array}{ccc} \mathcal{X}^k(M) & \xrightarrow{\varphi_*} & \mathcal{X}^k(N) \\ {\scriptstyle \mathcal{L}_X}\downarrow & & \downarrow{\scriptstyle \mathcal{L}_{\varphi_* X}} \\ \mathcal{X}^k(M) & \xrightarrow[\varphi_*]{} & \mathcal{X}^k(N) \end{array}$$

Also, $\mathcal{L}_X$ is natural with respect to restrictions. That is, for U open in a Kähler m-manifold M and $f \in C^\infty(M, \mathbb{R})$,

$$[X|U, Y|U] = [X, Y]|U,$$

where $U : C^\infty(M, \mathbb{R}) \to C^\infty(U, \mathbb{R})$ denotes restriction to U, i.e., the following diagram commutes [AMR88]:

$$\begin{array}{ccc} \mathcal{X}^k(M) & \xrightarrow{|U} & \mathcal{X}^k(U) \\ {\scriptstyle \mathcal{L}_X}\downarrow & & \downarrow{\scriptstyle \mathcal{L}_{X|U}} \\ \mathcal{X}^k(M) & \xrightarrow[|U]{} & \mathcal{X}^k(U) \end{array}$$

If a local chart (U, ϕ) on a Kähler m-manifold M has local coordinates $(x^1, ..., x^n)$, then the local components of a Lie bracket are

$$[X, Y]^j = X^i \partial_{x^i} Y^j - Y^i \partial_{x^i} X^j,$$

that is, $[X, Y] = (X \cdot \nabla)Y - (Y \cdot \nabla)X$.

Let $\varphi : M \to N$ be a $C^\infty-$map, $X \in \mathcal{X}^{k-1}(M)$ and $Y \in \mathcal{X}^{k-1}(N)$, $k \geq 1$. Then $X \sim_\varphi Y$, iff

$$(Y[f]) \circ \varphi = X[f \circ \varphi]$$

for all $f \in C^\infty(V, \mathbb{R})$, where V is open in N.

For every $X \in \mathcal{X}^k(M)$, the operator $\mathcal{L}_X$ is a derivation on $\big(C^\infty(M, \mathbb{R}), \mathcal{X}^k(M)\big)$, i.e., $\mathcal{L}_X$ is $\mathbb{R}-$linear.

For any two vector-fields $X \in \mathcal{X}^k(M)$ and $Y \in \mathcal{X}^k(N)$, $k \geq 1$ with flows F_t and G_t, respectively, if $[X, Y] = 0$ then $F_t^* Y = Y$ and $G_t^* X = X$.

Lie Groups and Lie Algebras on Kähler Manifolds

In the setting of Kähler geometry, a *Lie group* is a Kähler m-manifold M that has at the same time a group $G-$structure consistent with its manifold $M-$structure in the sense that *group multiplication*

$$\mu : G \times G \to G, \qquad (g, h) \mapsto gh \tag{3.17}$$

and the *group inversion*

$$\nu : G \to G, \qquad g \mapsto g^{-1} \tag{3.18}$$

are $C^\infty-$maps (compare with [Che46, AMR88, MR99, Put93] for the case of Banach manifolds). A point $e \in G$ is called the *group identity element*.

Let G and H be two *Kähler Lie groups*. A map $G \to H$ is said to be a *morphism* of Lie groups (or their *smooth homomorphism*) if it is their homomorphism as abstract groups and their smooth map as manifolds [Pos86].

All Kähler Lie groups and all their morphisms form the category $\mathcal{LG}$. Formally, there is a countable family of categories $\mathcal{LG}$ depending on C^k-smoothness of the corresponding Kähler manifolds.

Similarly, a group G defined on a Kähler m-manifold M, which is at the same time a topological space is said to be a *topological group* if maps (3.17–3.18) are continuous, i.e., C^0-maps for it. The homomorphism $G \to H$ of topological groups is said to be continuous if it is a continuous map.[25] Topological groups and their continuous homomorphisms form the category $\mathcal{TG}$.

For every g in a Kähler Lie group G, the following two maps:

$$L_g : G \to G, \qquad h \mapsto gh, \qquad \text{and}$$
$$R_h : G \to G, \qquad g \mapsto gh,$$

[25] A topological group (as well as a smooth manifold) is not necessarily Hausdorff. A topological group G is Hausdorff iff its identity is closed. As a corollary we have that every Lie group is a Hausdorff topological group (see [Pos86]).

are called *left* and *right translation* maps. Since $L_g \circ L_h = L_{gh}$, and $R_g \circ R_h = R_{gh}$, it follows that $(L_g)^{-1} = L_{g^{-1}}$ and $(R_g)^{-1} = R_{g^{-1}}$, so both L_g and R_g are diffeomorphisms. Moreover $L_g \circ R_h = R_h \circ L_g$, i.e., left and right translation commute.

A vector-field X on Kähler Lie group G is called *left–invariant vector-field* if for every $g \in G$, $L_g^* X = X$, that is, if $(T_h L_g) X(h) = X(gh)$ for all $h \in G$, i.e., the following diagram commutes:

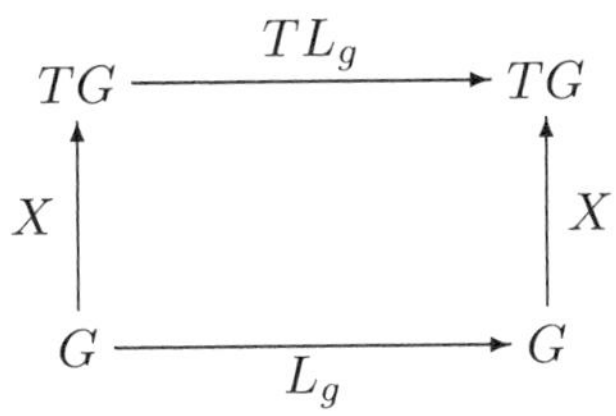

The correspondences $G \to TG$ and $L_g \to TL_g$ obviously define a functor $\mathcal{F} : \mathcal{LG} \Rightarrow \mathcal{LG}$ from the category G of Lie groups to itself. $\mathcal{F}$ is a special case of the vector bundle functor.

Let $\mathcal{X}_L(G)$ denote the set of left–invariant vector-fields on G; it is a Lie subalgebra of $\mathcal{X}(G)$, the set of all vector-fields on G, since $L_g^*[X, Y] = [L_g^* X, L_g^* Y] = [X, Y]$, so the Lie bracket $[X, Y] \in \mathcal{X}_L(G)$.

Let e be the identity element of G. Then for each ξ on the tangent space $T_e G$ we define a vector-field X_ξ on G by

$$X_\xi(g) = T_e L_g(\xi).$$

$\mathcal{X}_L(G)$ and $T_e G$ are isomorphic as vector spaces. Define the Lie bracket on $T_e G$ by

$$[\xi, \eta] = [X_\xi, X_\eta]\,(e),$$

for all $\xi, \eta \in T_e G$. This makes $T_e G$ into a Lie algebra. Also, by construction, we have

$$[X_\xi, X_\eta] = X_{[\xi,\eta]},$$

this defines a bracket in $T_e G$ via *left extension*. The vector space $T_e G$ with the above algebra structure is called the Lie algebra of the Lie group G and is denoted $\mathfrak{g}$.

For example, let V be a nD vector space. Then $T_e V \simeq V$ and the left–invariant vector-field defined by $\xi \in T_e V$ is the constant vector-field $X_\xi(\eta) = \xi$, for all $\eta \in V$. The Lie algebra of V is V itself.

Since any two elements of an Abelian Lie group G commute, it follows that all adjoint operators Ad_g, $g \in G$, equal the identity. Therefore, the Lie algebra g is Abelian; that is, $[\xi, \eta] = 0$ for all $\xi, \eta \in \mathfrak{g}$ [MR99].

Lie algebras and their smooth homomorphisms form the category $\mathcal{LAL}$. We can now introduce the fundamental *Lie functor*, $\mathcal{F} : \mathcal{LG} \Rightarrow \mathcal{LAL}$, from the category of Lie groups to the category of Lie algebras [Pos86].

Let X_ξ be a left–invariant vector-field on a Kähler Lie group G corresponding to ξ in $\mathfrak{g}$. Then there is a unique integral curve $\gamma_\xi : \mathbb{R} \to G$ of X_ξ starting at e, i.e.,

$$\dot{\gamma}_\xi(t) = X_\xi\left(\gamma_\xi(t)\right), \qquad \gamma_\xi(0) = e.$$

$\gamma_\xi(t)$ is a smooth *one–parameter subgroup* of G, i.e.,

$$\gamma_\xi(t+s) = \gamma_\xi(t) \cdot \gamma_\xi(s),$$

since, as functions of t both sides equal $\gamma_\xi(s)$ at $t = 0$ and both satisfy differential equation

$$\dot{\gamma}(t) = X_\xi\left(\gamma_\xi(t)\right)$$

by left invariance of X_ξ, so they are equal. Left invariance can be also used to show that $\gamma_\xi(t)$ is defined for all $t \in \mathbb{R}$. Moreover, if $\phi : \mathbb{R} \to G$ is a one–parameter subgroup of G, i.e., a *smooth homomorphism* of the additive group $\mathbb{R}$ into G, then $\phi = \gamma_\xi$ with $\xi = \dot{\phi}(0)$, since taking derivative at $s = 0$ in the relation

$$\phi(t+s) = \phi(t) \cdot \phi(s) \qquad \text{gives} \qquad \dot{\phi}(t) = X_{\dot{\phi}(0)}\left(\phi(t)\right),$$

so $\phi = \gamma_\xi$ since both equal e at $t = 0$. Therefore, all one–parameter subgroups of G are of the form $\gamma_\xi(t)$ for some $\xi \in \mathfrak{g}$.

The map $\exp : \mathfrak{g} \to G$, given by

$$\exp(\xi) = \gamma_\xi(1), \qquad \exp(0) = e, \tag{3.19}$$

is called the *exponential map* of the Lie algebra $\mathfrak{g}$ of G into G. exp is a C^∞–map, similar to the projection π of tangent and cotangent bundles; exp is locally a diffeomorphism from a neighborhood of zero in $\mathfrak{g}$ onto a neighborhood of e in G; if $f : G \to H$ is a smooth homomorphism of Lie groups, then

$$f \circ \exp_G = \exp_H \circ T_e f\,.$$

Also, in this case (see [Che46, MR99, Pos86])

$$\exp(s\xi) = \gamma_\xi(s).$$

Indeed, for fixed $s \in \mathbb{R}$, the curve $t \mapsto \gamma_\xi(ts)$, which at $t = 0$ passes through e, satisfies the differential equation

$$\frac{d}{dt}\gamma_\xi(ts) = sX_\xi\left(\gamma_\xi(ts)\right) = X_{s\xi}\left(\gamma_\xi(ts)\right).$$

Since $\gamma_{s\xi}(t)$ satisfies the same differential equation and passes through e at $t = 0$, it follows that $\gamma_{s\xi}(t) = \gamma_\xi(st)$. Putting $t = 1$ induces $\exp(s\xi) = \gamma_\xi(s)$ [MR99].

Hence exp maps the line $s\xi$ in $\mathfrak{g}$ onto the one–parameter subgroup $\gamma_\xi(s)$ of G, which is tangent to ξ at e. It follows from left invariance that the flow F_t^ξ of X satisfies $F_t^\xi(g) = g\exp(s\xi)$.

Globally, the exponential map exp, as given by (3.19), is a natural operation, i.e., for any morphism $\varphi : G \to H$ of Lie groups G and H and a Lie functor $\mathcal{F}$, the following diagram commutes [Pos86]:

$$\begin{array}{ccc} \mathcal{F}(G) & \xrightarrow{\mathcal{F}(\varphi)} & \mathcal{F}(H) \\ \downarrow \exp & & \downarrow \exp \\ G & \xrightarrow[\varphi]{} & H \end{array}$$

Let G_1 and G_2 be Kähler Lie groups with Lie algebras $\mathfrak{g}_1$ and $\mathfrak{g}_2$. Then $G_1 \times G_2$ is a Lie group with Lie algebra $\mathfrak{g}_1 \times \mathfrak{g}_2$, and the exponential map is given by [MR99].

$$\exp : \mathfrak{g}_1 \times \mathfrak{g}_2 \to G_1 \times G_2, \qquad (\xi_1, \xi_2) \mapsto (\exp_1(\xi_1), \exp_2(\xi_2)) \,.$$

The unit circle in the complex plane $S^1 = \{z \in \mathbb{C} : |z| = 1\}$ is an Abelian Lie group under multiplication. The tangent space T_eS^1 is the imaginary axis, and we identify $\mathbb{R}$ with T_eS^1 by $t \mapsto 2\pi i t$. With this identification, the exponential map $\exp : \mathbb{R} \to S^1$ is given by $\exp(t) = e^{2\pi i t}$.

The nD torus $T^n = S^1 \times \cdots \times S^1$ (n times) is an Abelian Lie group. The exponential map $\exp : \mathbb{R}^n \to T^n$ is given by

$$\exp(t_1, ..., t_n) = (e^{2\pi i t_1}, ..., e^{2\pi i t_n}).$$

Since $S^1 = \mathbb{R}/\mathbb{Z}$, it follows that

$$T^n = \mathbb{R}^n / \mathbb{Z}^n,$$

the projection $\mathbb{R}^n \to T^n$ being given by the exp map (see [MR99, Pos86]).

For every $g \in G$, the map

$$Ad_g = T_e \left(R_{g^{-1}} \circ L_g \right) : \mathfrak{g} \to \mathfrak{g}$$

is called the *adjoint map* (or *operator*) associated with g.

For each $\xi \in \mathfrak{g}$ and $g \in G$ we have

$$\exp\left(Ad_g \xi\right) = g \left(\exp \xi\right) g^{-1}.$$

The relation between the adjoint map and the Lie bracket is the following: For all $\xi, \eta \in \mathfrak{g}$ we have

$$\left.\frac{d}{dt}\right|_{t=0} Ad_{\exp(t\xi)} \eta = [\xi, \eta].$$

A *Kähler Lie subgroup* H of G is a subgroup H of G which is also a Kähler submanifold of G. Then $\mathfrak{h}$ is a Lie subalgebra of $\mathfrak{g}$ and moreover $\mathfrak{h} = \{\xi \in \mathfrak{g} | \exp(t\xi) \in H, \text{ for all } t \in \mathbb{R}\}$.

Finally, recall that one can characterize *Lebesgue measure* up to a multiplicative constant on $\mathbb{R}^n$ by its invariance under translations. Similarly, on a locally compact group there is a unique (up to a nonzero multiplicative constant) left–invariant measure, called *Haar measure*. For Kähler Lie groups the existence of such measures is especially simple [MR99]: Let G be a Kähler Lie group; then there is a volume form $Ub5$, unique up to nonzero multiplicative constants, that is left–invariant. If G is compact, $Ub5$ is right invariant as well.

Vector Bundles of Kähler Manifolds

Maps of fibre bundles (or, *bundle maps*), by definition, *preserve their fibrations*, i.e., send a fibre to a fibre. Namely, a *bundle map* of a fibre bundle $\pi : Y \to X$ to a fibre bundle $\pi' : Y' \to X'$ is defined as a pair (Φ, f) of *Kähler-manifold maps* such that the following diagram commutes:

$$\begin{array}{ccc} Y & \xrightarrow{\quad\Phi\quad} & Y' \\ {\scriptstyle\pi}\big\downarrow & & \big\downarrow{\scriptstyle\pi'} \\ X & \xrightarrow[\quad f\quad]{} & X' \end{array}$$

i.e., Φ is a fibrewise map over f which sends a fibre Y_x, (for all $x \in X$), to a fibre $Y'_{f(x)}$, (for all $f(x) \in X'$). A *bundle diffeomorphism* is called an *automorphism* if it is an *isomorphism* to itself. In field theory, any automorphism of a fibre bundle is treated as a *gauge transformation*.

Given a bundle $Y \to X$, every map $f : X' \to X$ induces a bundle $Y' = f^*Y$ over X' which is called the *pull–back* of the bundle Y by f, such that the following diagram commutes

$$\begin{array}{ccc} Y & \xleftarrow{\quad f^*\quad} & Y' \\ {\scriptstyle\pi}\big\downarrow & & \big\downarrow{\scriptstyle\pi'} \\ X & \xleftarrow[\quad f\quad]{} & X' \end{array}$$

In particular, the product $Y \times Y'$ over X of bundles $\pi : Y \to X$ and $\pi' : Y' \to X$ is the pull–back

$$Y \times Y' = \pi^*Y' = \pi'^*Y.$$

The most important fibre bundles are vector and affine bundles, which give a standard framework in both classical and quantum dynamics and field theory (e.g., matter fields are sections of vector bundles, while gauge potentials are sections of affine bundles).

Recall that both the tangent bundle (TM, π_M, M) and the cotangent bundle (T^*M, π^*_M, M) are examples of a more general notion of *vector bundle*

(E, π, M) of a Kähler manifold M, which consists of manifolds E (the total space) and M (the base), as well as a smooth map $\pi : E \to M$ (the projection) together with an equivalence class of vector bundle atlases (see [KMS93]). A vector bundle atlas $(U_\alpha, \phi_\alpha)_{\alpha \in A}$ for (E, π, M) is a set of pairwise compatible vector bundle charts (U_α, ϕ_α) such that $(U_\alpha)_{\alpha \in A}$ is an open cover of a Kähler m-manifold M. Two vector bundle atlases are called equivalent, if their union is again a vector bundle atlas.

On each fibre $E_m = \pi^{-1}(m)$ corresponding to the point $m \in M$ there is a unique structure of a real vector space, induced from any vector bundle chart (U_α, ϕ_α) with $m \in U_\alpha$. A section u of (E, π, M) is a smooth map $u : M \to E$ with $\pi \circ u = Id_M$.

Let (E, π_M, M) and (F, π_N, N) be vector bundles. A *vector bundle homomorphism* $\Phi : E \to F$ is a fibre respecting, fibre linear smooth map induced by the smooth map $\varphi : M \to N$ between the base Kähler manifolds M and N, i.e., the following diagram commutes:

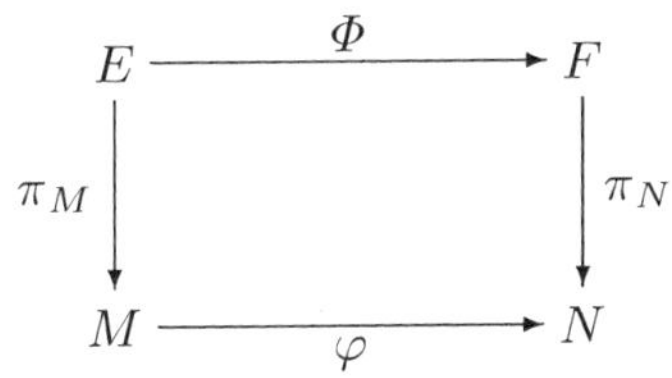

We say that Φ *covers* φ. If Φ is invertible, it is called a *vector bundle isomorphism*.

All smooth vector bundles together with their homomorphisms form a category $\mathcal{VB}$.

If (E, π, M) is a vector bundle which admits a vector bundle atlas $(U_\alpha, \phi_\alpha)_{\alpha \in A}$ with the given open cover, then, we have $\phi_\alpha \circ \phi_\beta^{-1}(m, v) = (m, \phi_{\alpha\beta}(m)v)$ for C^∞–transition functions $\phi_{\alpha\beta} : U_{\alpha\beta} = U_\alpha \cap U_\beta \to GL(V)$ (where we have fixed a standard fibre V). This family of transition maps satisfies the *cocycle condition*

$$\begin{cases} \phi_{\alpha\beta}(m) \cdot \phi_{\beta\gamma}(m) = \phi_{\alpha\gamma}(m) & \text{for each } m \in U_{\alpha\beta\gamma} = U_\alpha \cap U_\beta \cap U_\gamma, \\ \phi_{\alpha\alpha}(m) = e & \text{for all } m \in U_\alpha. \end{cases}$$

The family $(\phi_{\alpha\beta})$ is called the *cocycle* of transition maps for the vector bundle atlas (U_α, ϕ_α).

Now, let us suppose that the same vector bundle (E, π, M) is described by an equivalent vector bundle atlas $(U_\alpha, \psi_\alpha)_{\alpha \in A}$ with the same open cover (U_α). Then the vector bundle charts (U_α, ϕ_α) and (U_α, ψ_α) are compatible for each α, so $\psi_\alpha \circ \phi_\beta^{-1}(m, v) = (m, \tau_\alpha(m)v)$ for some $\tau_\alpha : U_\alpha \to GL(V)$. We get

$$\tau_\alpha(m)\, \phi_{\alpha\beta}(m) = \phi_{\alpha\beta}(m)\, \tau_\beta(m) \quad \text{for all } m \in U_{\alpha\beta},$$

and we say that the two cocycles $(\phi_{\alpha\beta})$ and $(\psi_{\alpha\beta})$ of transition maps over the cover (U_α) are *cohomologous*. If $GL(V)$ is an Abelian group, i.e., if the standard fibre V is of real or complex dimension 1, then the cohomology classes of cocycles $(\phi_{\alpha\beta})$ over the open cover (U_α) form a usual cohomology group $H^1(M, GL(V))$ with coefficients in the sheaf $GL(V)$ [KMS93].

Let (E, π, M) be a vector bundle and let $\varphi : N \to M$ be a smooth map between the base Kähler manifolds N and M. Then there exists the *pull-back vector bundle* $(\varphi^*E, \varphi^*\pi, \varphi^*N)$ with the same typical fibre and a vector bundle homomorphism, given by the commutative diagram [KMS93]:

$$\begin{array}{ccc} \varphi^*E & \xrightarrow{\pi^*\varphi} & E \\ {\scriptstyle \varphi^*\pi}\downarrow & & \downarrow{\scriptstyle \pi} \\ N & \xrightarrow[\varphi]{} & M \end{array}$$

The vector bundle $(\varphi^*E, \varphi^*\pi, \varphi^*N)$ is constructed as follows. Let $E = VB(\phi_{\alpha\beta})$ denote that E is described by a cocycle $(\phi_{\alpha\beta})$ of transition maps over an open cover (U_α) of M. Then $(\phi_{\alpha\beta} \circ \varphi)$ is a cocycle of transition maps over the open cover $\left(\varphi^{-1}(U_\alpha)\right)$ of N and the bundle is given by $\varphi^*E = VB(\phi_{\alpha\beta} \circ \varphi)$.

In other words, a vector bundle is a fibre bundle which admits an atlas of linear bundle coordinates. Typical fibres of a smooth vector bundle $\pi : Y \to X$ are *vector spaces* of some finite dimension (called the fibre dimension, $\mathrm{fdim}\, Y$ of Y), and Y admits a bundle atlas Ψ_Y, where trivialization maps $\psi_\xi(x)$ and transition functions $\rho_{\xi\zeta}(x)$ are linear isomorphisms of vector spaces. The corresponding bundle coordinates (y^i) obey a *linear* coordinate transformation law

$$y'^i = \rho^i_j(x) y^j.$$

We have the decomposition $y = y^i e_i(\pi(y))$, where

$$\{e_i(x)\} = \psi_\xi^{-1}(x)\{v_i\}, \qquad x = \pi(y) \in U_\xi,$$

are fibre bases (or frames) for fibres Y_x of Y and $\{v_i\}$ is a fixed basis for the typical fibre V of Y.

There are several standard constructions of new vector bundles from old ones:

- Given two vector bundles Y and Y' over the same base X, their *Whitney sum* $Y \oplus Y'$ is a vector bundle over X whose fibres are the direct sums of those of the vector bundles Y and Y'.
- Given two vector bundles Y and Y' over the same base X, their tensor product $Y \otimes Y'$ is a vector bundle over X whose fibres are the tensor products of those of the vector bundles Y and Y'. In a similar way the exterior product $Y \wedge Y$ of vector bundles is defined, so that the *exterior bundle* of Y is defined as

$$\wedge Y = X \times \mathbb{R} \oplus Y \oplus \wedge^2 Y \oplus \cdots \oplus \wedge^m Y, \qquad (m = \mathrm{fdim}\, Y).$$

- Let $Y \to X$ be a vector bundle. By $Y^* \to X$ is denoted the *dual vector bundle* whose fibres are the duals of those of Y. The *interior product* (or *contraction*) of Y and Y^* is defined as a bundle map

$$\rfloor : Y \otimes Y^* \to X \times \mathbb{R}.$$

Given a linear bundle map $\Phi : Y' \to Y$ of vector bundles over X, its kernel $\operatorname{Ker}\Phi$ is defined as the inverse image $\Phi^{-1}(\widehat{0}(X))$ of the canonical zero section $\widehat{0}(X)$ of Y. If Φ is of constant rank, its kernel $\operatorname{Ker}\Phi$ and its image $\operatorname{Im}\Phi$ are subbundles of the vector bundles Y' and Y, respectively. For example, monomorphisms and epimorphisms of vector bundles fulfil this condition. If Y' is a subbundle of the vector bundle $Y \to X$, the factor bundle Y/Y' over X is defined as a vector bundle whose fibres are the quotients Y_x/Y'_x, $x \in X$.

Consider the *short exact sequence* of vector bundles over X,

$$0 \to Y' \xrightarrow{i} Y \xrightarrow{j} Y'' \to 0, \tag{3.20}$$

which means that i is a bundle monomorphism, j is a bundle epimorphism, and $\operatorname{Ker} j = \operatorname{Im} i$. Then Y'' is the factor bundle Y/Y'. One says that the short exact sequence (3.20) admits a *splitting* if there exists a bundle monomorphism $s : Y'' \to Y$ such that $j \circ s = \mathrm{Id}_{Y''}$, i.e.,

$$Y = i(Y') \oplus s(Y'') \cong Y' \oplus Y''.$$

Vector bundles of rank 1 are called *line bundles*.

Multivector-fields and Tangent–Valued Exterior Forms

Recall that a vector-field on a Kähler m-manifold M is defined as a global section of the tangent bundle $TM \to M$. The set $\mathcal{V}^1(M)$ of vector-fields on M is a real Lie algebra with respect to the *Lie bracket* (see [GMS97]):

$$[v,u] = (v^\alpha \partial_\alpha u^\mu - u^\alpha \partial_\alpha v^\mu)\partial_\mu, \qquad v = v^\alpha \partial_\alpha, \qquad u = u^\alpha \partial_\alpha. \tag{3.21}$$

Every vector-field on a Kähler m-manifold M can be seen as an infinitesimal generator of a local 1–parameter Lie group of diffeomorphisms of M as follows [KN63/9]. Given an open subset $U \subset M$ and an interval $(-\epsilon, \epsilon) \in \mathbb{R}$, by a local 1–parameter group of diffeomorphisms of M defined on $(-\epsilon, \epsilon) \times U$ is denoted a map

$$G \to M, \qquad (t,x) \mapsto G_t(x)$$

such that:

1. for each $t \in (-\epsilon, \epsilon)$, the map G_t is a diffeomorphism of U onto the open subset $G_t(U) \subset M$; and
2. $G_{t+t'}(x) = (G_t \circ G_{t'})(x)$ if $t, t', t+t' \in (-\epsilon, \epsilon)$ and $G_{t'}(x), x \in U$.

Any local 1–parameter group of diffeomorphisms G on $U \subset M$ defines a local vector-field u on U by setting $u(x)$ to be the tangent vector to the curve $x(t) = G_t(x)$ at $t = 0$. Conversely, if u is a vector-field on a manifold M, there exists a unique local 1–parameter group G_u of diffeomorphisms on a neighborhood of every point $x \in M$ which defines u. We call G_u a *flow* of the vector-field u. A vector-field u on a manifold M is called *complete* if its flow is a 1–parameter group of diffeomorphisms of M. In particular, every vector-field on a compact manifold is complete [KN63/9].

A vector-field u on a fibre bundle $Y \longrightarrow X$ is an infinitesimal generator of a local 1–parameter group G_u of isomorphisms of $Y \longrightarrow X$ iff it is a projectable vector-field on Y. A vector-field u on a fibre bundle $Y \longrightarrow X$ is called projectable if it projects onto a vector-field on X, i.e., there exists a vector-field τ on X such that the following diagram commutes:

$$\begin{array}{ccc} Y & \xrightarrow{\quad u \quad} & TY \\ {\scriptstyle \pi}\downarrow & & \downarrow{\scriptstyle T\pi} \\ X & \xrightarrow[\quad \tau \quad]{} & TX \end{array}$$

A projectable vector-field has the coordinate expression:

$$u = u^\alpha(x^\mu)\partial_\alpha + u^i(x^\mu, y^j)\partial_i,$$

where u^α are local functions on X. A projectable vector-field is said to be vertical if it projects onto the zero vector-field $\tau = 0$ on X, i.e., $u = u^i\partial_i$ takes its values in the vertical tangent bundle VY.

For example, in field theory, projectable vector-fields on fibre bundles play a role of infinitesimal generators of local 1–parameter groups of gauge transformations.

In general, a vector-field $\tau = \tau^\alpha\partial_\alpha$ on a base X of a fibre bundle $Y \to X$ induces a vector-field on Y by means of a *connection* on this fibre bundle. Nevertheless, every natural fibre bundle $Y \to X$ admits the *canonical lift* $\widetilde{\tau}$ onto Y of any vector-field τ on X. For example, if Y is the *tensor bundle*, the above canonical lift reads:

$$\widetilde{\tau} = \tau^\mu\partial_\mu + [\partial_\nu\tau^{\alpha_1}\dot{x}^{\nu\alpha_2\cdots\alpha_m}_{\beta_1\cdots\beta_k} + \ldots - \partial_{\beta_1}\tau^\nu\dot{x}^{\alpha_1\cdots\alpha_m}_{\nu\beta_2\cdots\beta_k} - \ldots]\frac{\partial}{\partial\dot{x}^{\alpha_1\cdots\alpha_m}_{\beta_1\cdots\beta_k}}. \qquad (3.22)$$

In particular, we have the canonical lift onto the tangent bundle TX,

$$\widetilde{\tau} = \tau^\mu\partial_\mu + \partial_\nu\tau^\alpha\dot{x}^\nu\frac{\partial}{\partial\dot{x}^\alpha} \qquad (3.23)$$

and another one onto the cotangent bundle T^*X,

$$\widetilde{\tau} = \tau^\mu\partial_\mu - \partial_\beta\tau^\nu\dot{x}_\nu\frac{\partial}{\partial\dot{x}_\beta}. \qquad (3.24)$$

A *multivector-field* ϑ of degree r (or simply a r-vector-field) on a Kähler m-manifold M, by definition, is a global section of the bundle $\wedge^r TM \to M$. It is given by the coordinate expression

$$\vartheta = \vartheta^{\alpha_1\ldots\alpha_r}\partial_{\alpha_1}\wedge\cdots\wedge\partial_{\alpha_r}, \qquad |\vartheta| = r,$$

where summation is over all ordered collections $(\lambda_1, ..., \lambda_r)$.

Similarly, an *exterior* r*–form* on a Kähler m-manifold M with local coordinates x^α, by definition, is a global section of the skew–symmetric tensor bundle (exterior product) $\wedge^r T^*M \to M$,

$$\phi = \frac{1}{r!}\phi_{\alpha_1\ldots\alpha_r}dx^{\alpha_1}\wedge\cdots\wedge dx^{\alpha_r}, \qquad |\phi| = r.$$

The 1–forms are also called the *Pfaffian forms*.

The vector space $\mathcal{V}^r(M)$ of $r-$vector-fields on a Kähler m-manifold M admits the *Schouten–Nijenhuis bracket* (or, SN bracket)

$$[.,.]_{SN} : \mathcal{V}^r(M)\times\mathcal{V}^s(M) \to \mathcal{V}^{r+s-1}(M)$$

which generalizes the Lie bracket of vector-fields (3.21). The SN–bracket has the coordinate expression:

$$\begin{aligned}
&\vartheta = \vartheta^{\alpha_1\ldots\alpha_r}\partial_{\alpha_1}\wedge\cdots\wedge\partial_{\alpha_r}, \qquad \upsilon = \upsilon^{\alpha_1\ldots\alpha_s}\partial_{\alpha_1}\wedge\cdots\wedge\partial_{\alpha_s},\\
&[\vartheta,\upsilon]_{SN} = \vartheta\star\upsilon + (-1)^{|\vartheta||\upsilon|}\upsilon\star\vartheta, \qquad \text{where}\\
&\vartheta\star\upsilon = \vartheta^{\mu\alpha_1\ldots\alpha_{r-1}}\partial_\mu\upsilon^{\alpha_1\ldots\alpha_s}\partial_{\alpha_1}\wedge\cdots\wedge\partial_{\alpha_{r-1}}\wedge\partial_{\alpha_1}\wedge\cdots\wedge\partial_{\alpha_s}.
\end{aligned}$$

The following relations hold for the SN–bracket:

$$\begin{aligned}
&[\vartheta,\upsilon]_{SN} = (-1)^{|\vartheta||\upsilon|}[\upsilon,\vartheta]_{SN},\\
&[\nu,\vartheta\wedge\upsilon]_{SN} = [\nu,\vartheta]_{SN}\wedge\upsilon + (-1)^{|\nu||\vartheta|+|\vartheta|}\vartheta\wedge[\nu,\upsilon]_{SN},\\
&(-1)^{|\nu||\vartheta|+|\nu|}[\nu,\vartheta\wedge\upsilon]_{SN} + (-1)^{|\vartheta||\nu|+|\vartheta|}[\vartheta,\upsilon\wedge\nu]_{SN}\\
&+ (-1)^{|\upsilon||\vartheta|+|\upsilon|}[\upsilon,\nu\wedge\vartheta]_{SN} = 0.
\end{aligned}$$

In particular, let $w = w^{\mu\nu}\partial_\mu\wedge\partial_\nu$ be a *bivector-field.* We have

$$[w,w]_{SN} = w^{\mu\alpha_1}\partial_\mu w^{\alpha_2\alpha_3}\partial_{\alpha_1}\wedge\partial_{\alpha_2}\wedge\partial_{\alpha_3}. \tag{3.25}$$

Every bivector-field w on a Kähler m-manifold M induces the 'sharp' bundle map $w^\sharp : T^*M \to TM$ defined by

$$w^\sharp(p)\rfloor q := w(x)(p,q), \qquad w^\sharp(p) = w^{\mu\nu}(x)p_\mu\partial_\nu, \qquad (p,q\in T_x^*M). \tag{3.26}$$

A bivector-field w whose bracket (3.25) vanishes is called the *Poisson bivector-field*.

Let $\wedge^r(M)$ denote the vector space of exterior $r-$forms on a Kähler m-manifold M. By definition, $\wedge^0(M) = C^\infty(M)$ is the ring of smooth real functions on M. All exterior forms on M constitute the $\mathbb{N}$-graded exterior algebra

$\wedge^*(M)$ of global sections of the exterior bundle $\wedge T^*M$ with respect to the exterior product $\wedge$. This algebra admits the *exterior differential*

$$d : \wedge^r(M) \to \wedge^{r+1}(M),$$
$$d\phi = dx^\mu \wedge \partial_\mu \phi = \frac{1}{r!}\partial_\mu \phi_{\alpha_1 \dots \alpha_r} dx^\mu \wedge dx^{\alpha_1} \wedge \cdots dx^{\alpha_r},$$

which is *nilpotent*, i.e., $d \circ d = 0$, and obeys the relation

$$d(\phi \wedge \sigma) = d(\phi) \wedge \sigma + (-1)^{|\phi|} \phi \wedge d(\sigma).$$

The *interior product* (or, *contraction*) of a vector-field $u = u^\mu \partial_\mu$ and an exterior $r-$form ϕ on a Kähler m-manifold M is given by the coordinate expression

$$u\rfloor\phi = \sum_{k=1}^{r} \frac{(-1)^{k-1}}{r!} u^{\alpha_k} \phi_{\alpha_1 \dots \alpha_k \dots \alpha_r} dx^{\alpha_1} \wedge \cdots \wedge \widehat{dx}^{\alpha_k} \wedge \cdots \wedge dx^{\alpha_r} \tag{3.27}$$
$$= \frac{1}{(r-1)!} u^\mu \phi_{\mu \alpha_2 \dots \alpha_r} dx^{\alpha_2} \wedge \cdots \wedge dx^{\alpha_r},$$

where the caret $\widehat{\cdot}$ denotes omission. The following relations hold:

$$\phi(u_1, \dots, u_r) = u_r\rfloor \cdots u_1\rfloor\phi, \tag{3.28}$$
$$u\rfloor(\phi \wedge \sigma) = u\rfloor\phi \wedge \sigma + (-1)^{|\phi|} \phi \wedge u\rfloor\sigma, \tag{3.29}$$
$$[u, u']\rfloor\phi = u\rfloor d(u'\rfloor\phi) - u'\rfloor d(u\rfloor\phi) - u'\rfloor u\rfloor d\phi, \qquad (\phi \in \wedge^1(M)). \tag{3.30}$$

Recall from section 3.8.3 above, that the *Lie derivative* $\mathfrak{L}_u\sigma$ of an exterior form σ along a vector-field u is defined by the *Cartan relation*:

$$\mathfrak{L}_u\sigma = u\rfloor d\sigma + d(u\rfloor\sigma).$$

It satisfies the relation

$$\mathfrak{L}_u(\phi \wedge \sigma) = \mathfrak{L}_u\phi \wedge \sigma + \phi \wedge \mathfrak{L}_u\sigma.$$

In particular, if f is a function, then

$$\mathfrak{L}_u f = u(f) = u\rfloor df.$$

It is important for dynamical applications that an exterior form ϕ is invariant under a local 1–parameter group of diffeomorphisms G_t of M (i.e., $G_t^*\phi = \phi$) iff its Lie derivative $\mathfrak{L}_u\phi$ along the vector-field u, generating G_t, vanishes.

Let Ω be a two–form on a Kähler m-manifold M. It defines the 'flat' bundle map $\Omega^\flat$, as

$$\Omega^\flat : TM \to T^*M, \qquad \Omega^\flat(v) = -v\rfloor\Omega(x), \qquad (v \in T_xM). \tag{3.31}$$

In coordinates, if $\Omega = \Omega_{\mu\nu} dx^\mu \wedge dx^\nu$ and $v = v^\mu \partial_\mu$, then

$$\Omega^\flat(v) = -\Omega_{\mu\nu} v^\mu dx^\nu.$$

One says that Ω is of *constant rank* k if the corresponding map (3.31) is of constant rank k (i.e., k is the greatest integer n such that Ω^n is not the zero form). The rank of a *nondegenerate* two–form is equal to $\dim M$. A nondegenerate closed two–form is called the *symplectic form*.

Given a manifold map $f : M \to M'$, any exterior k-form ϕ on M' induces the pull–back exterior form $f^*\phi$ on M by the condition

$$f^*\phi(v^1, \ldots, v^k)(x) = \phi(Tf(v^1), \ldots, Tf(v^k))(f(x))$$

for an arbitrary collection of tangent vectors $v^1, \cdots, v^k \in T_xM$. The following relations hold:

$$f^*(\phi \wedge \sigma) = f^*\phi \wedge f^*\sigma, \qquad df^*\phi = f^*(d\phi).$$

In particular, given a fibre bundle $\pi : Y \to X$, the pull–back onto Y of exterior forms on X by π gives the monomorphism of exterior algebras

$$\pi^* : \wedge^*(X) \to \wedge^*(Y).$$

Elements of its image $\pi^* \wedge^* (X)$ are called basic forms. Exterior forms on Y such that $u\rfloor\phi = 0$ for an arbitrary vertical vector-field u on Y are said to be *horizontal forms*. They are generated by horizontal 1–forms $\{dx^\alpha\}$. For example, basic forms are horizontal forms with coefficients in $C^\infty(X) \subset C^\infty(Y)$. A horizontal form of degree $n = \dim X$ is called a *density*. For example, Lagrangians in field theory are densities.

Elements of the tensor product $\wedge^r(M) \otimes \mathcal{V}^1(M)$ are called the *tangent–valued r-forms* on M. They are sections

$$\phi = \frac{1}{r!}\phi^\mu_{\alpha_1 \ldots \alpha_r} dx^{\alpha_1} \wedge \cdots \wedge dx^{\alpha_r} \otimes \partial_\mu$$

of the tensor bundle

$$\wedge^r T^*M \otimes TM \to M.$$

Tangent-valued 1–forms are usually called the (1,1) *tensor fields*.

In particular, there is the 1–1 correspondence between the tangent–valued 1–forms on a Kähler m-manifold M and the linear bundle maps over M,

$$\phi : TM \to TM, \qquad \phi : T_xM \ni v \mapsto v\rfloor\phi(x) \in T_xM. \tag{3.32}$$

In particular, the *canonical tangent–valued one–form* $\theta_M = dx^\alpha \otimes \partial_\alpha$ defines the identity map of TM.

Tangent-valued forms play a prominent role in jet formalism and theory of connections on fibre bundles. In particular, tangent–valued 0-forms are vector-fields on M. Also, there is 1–1 correspondence between the tangent–valued 1–forms ϕ on a Kähler m-manifold M and the linear bundle endomorphisms

$$\widehat{\phi} : TM \to TM, \qquad \widehat{\phi} : T_xM \ni v \mapsto v\rfloor\phi(x) \in T_xM, \tag{3.33}$$

$$\widehat{\phi}^* : T^*M \to T^*M, \qquad \widehat{\phi}^* : T_x^*M \ni v^* \mapsto \phi(x)\rfloor v^* \in T_x^*M, \tag{3.34}$$

over M. For example, the *canonical tangent–valued 1–form* on M,

$$\theta_M = dx^\alpha \otimes \partial_\alpha\,, \tag{3.35}$$

corresponds to the identity maps (3.33) and (3.34).

The most important are the following types of vector-fields and differential forms on a bundle $Y \longrightarrow X$ (see [GMS97]):

- a *projectable vector-field* on Y,

$$u = u^\mu(x)\partial_\mu + u^i(y)\partial_i,$$

  which covers a vector-field $\tau_u = u^\mu(x)\partial_\mu$ on the base X such that the following diagram commutes:

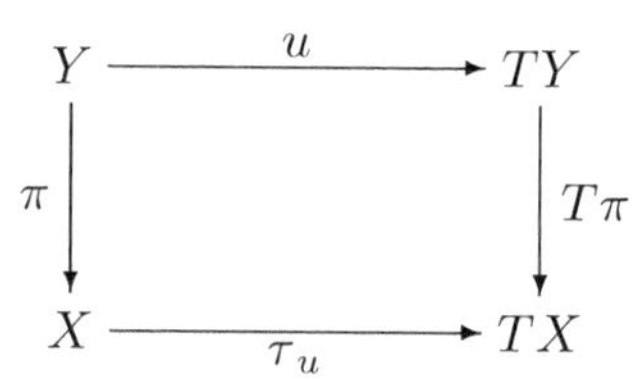

- a *vertical vector-field*, $u : Y \to VY$, given by $u = u^i(y)\partial_i$, is a projectable vector-field which covers $\tau_u = 0$;
- an *exterior horizontal form*, $\phi : Y \to \wedge^r T^*X$, given by

$$\phi = \frac{1}{r!}\phi_{\alpha_1\dots\alpha_r}(y)dx^{\alpha_1} \wedge \cdots \wedge dx^{\alpha_r};$$

- a *tangent–valued horizontal form*, $\phi : Y \to \wedge^r T^*X \otimes TY$, given by

$$\phi = \frac{1}{r!}dx^{\alpha_1} \wedge \cdots \wedge dx^{\alpha_r} \otimes [\phi^\mu_{\alpha_1\dots\alpha_r}(y)\partial_\mu + \phi^i_{\alpha_1\dots\alpha_r}(y)\partial_i];$$

- a *vertical–valued horizontal form*, $\phi : Y \to \wedge^r T^*X \otimes VY$, given by

$$\phi = \frac{1}{r!}\phi^i_{\alpha_1\dots\alpha_r}(y)dx^{\alpha_1} \wedge \cdots \wedge dx^{\alpha_r} \otimes \partial_i.$$

- a vertical-valued *soldering form*, $\sigma : Y \to T^*X \otimes VY$, given by

$$\sigma = \sigma^i_\alpha(y)dx^\alpha \otimes \partial_i \tag{3.36}$$

  and, in particular, the *canonical soldering form* on TX,

$$\theta_X = dx^\alpha \otimes \partial_\alpha.$$

The *pull–back–valued forms* on a bundle $Y \to X$ are the following two maps:[26]

$$Y \to \wedge^r T^*Y \otimes TX, \qquad \phi = \frac{1}{r!}\phi^\mu_{\alpha_1\dots\alpha_r}(y)dx^{\alpha_1}\wedge\cdots\wedge dx^{\alpha_r}\otimes\partial_\mu,$$
$$\text{and} \qquad (3.37)$$
$$Y \to \wedge^r T^*Y \otimes V^*X, \qquad \phi = \frac{1}{r!}\phi_{\alpha_1\dots\alpha_r i}(y)dx^{\alpha_1}\wedge\cdots\wedge dx^{\alpha_r}\otimes\overline{d}y^i.$$

The pull–back-valued forms (3.37) are exemplified by the canonical bundle monomorphism

$$\wedge^n T^*X \otimes V^*Y \hookrightarrow \wedge^{n+1}T^*Y, \qquad \omega\otimes\overline{d}y^i \mapsto \omega\wedge dy^i.$$

All horizontal $n-$forms on a bundle $Y \longrightarrow X$ are called *horizontal densities.*
For any vector-field τ on X, we can define its pull–back on Y,

$$\pi^*\tau = \tau\circ\pi : Y \longrightarrow TX.$$

This is not a vector-field on Y, for the tangent bundle TX of X fails to be a subbundle of the tangent bundle TY of Y. One needs a connection on $Y \longrightarrow X$ in order to set the imbedding $TX \hookrightarrow TY$.

The space $\wedge^*(M)\otimes\mathcal{V}^1(M)$ of tangent–valued forms admits the *Frölicher–Nijenhuis bracket* (or, FN bracket)

$$[.,.]_{FN} : \wedge^r(M)\otimes\mathcal{V}^1(M)\times\wedge^s(M)\otimes\mathcal{V}^1(M)\to\wedge^{r+s}(M)\otimes\mathcal{V}^1(M),$$
$$[\phi,\sigma]_{FN} = \frac{1}{r!s!}(\phi^\nu_{\alpha_1\dots\alpha_r}\partial_\nu\sigma^\mu_{\alpha_{r+1}\dots\alpha_{r+s}} - \sigma^\nu_{\alpha_{r+1}\dots\alpha_{r+s}}\partial_\nu\phi^\mu_{\alpha_1\dots\alpha_r} - \qquad (3.38)$$
$$r\phi^\mu_{\alpha_1\dots\alpha_{r-1}\nu}\partial_{\alpha_r}\sigma^\nu_{\alpha_{r+1}\dots\alpha_{r+s}} + s\sigma^\mu_{\nu\alpha_{r+2}\dots\alpha_{r+s}}\partial_{\alpha_{r+1}}\phi^\nu_{\alpha_1\dots\alpha_r})dx^{\alpha_1}$$
$$\wedge\cdots\wedge dx^{\alpha_{r+s}}\otimes\partial_\mu.$$

The following relations hold for the FN–bracket:

$$[\phi,\psi]_{FN} = (-1)^{|\phi||\psi|+1}[\psi,\phi]_{FN}, \qquad (3.39)$$
$$[\phi,[\psi,\theta]_{FN}]_{FN} = [[\phi,\psi]_{FN},\theta]_{FN} + (-1)^{|\phi||\psi|}[\psi,[\phi,\theta]_{FN}]_{FN}.$$

---

[26] The forms (3.37) are not tangent–valued forms. The pull–backs

$$\phi = \frac{1}{r!}\phi^\mu_{\alpha_1\dots\alpha_r}(x)dx^{\alpha_1}\wedge\cdots\wedge dx^{\alpha_r}\otimes\partial_\mu$$

of tangent–valued forms on X onto Y by π exemplify the pull–back-valued forms (3.37). In particular, we shall refer to the pull–back $\pi^*\theta_X$ of the canonical form θ_X on the base X onto Y by π. This is a pull–back-valued horizontal one–form on Y which we denote by the same symbol

$$\theta_X : Y \to T^*X\otimes TX, \qquad \theta_X = dx^\alpha\otimes\partial_\alpha.$$

Given a tangent–valued form θ, the *Nijenhuis differential*, $d_\theta\sigma$, along θ on $\wedge^*(M)\otimes\mathcal{V}^1(M)$ is defined as

$$d_\theta\sigma = [\theta,\sigma]_{FN}. \tag{3.40}$$

By virtue of the relation (3.39), it has the property

$$d_\phi[\psi,\theta]_{FN} = [d_\phi\psi,\theta]_{FN} + (-1)^{|\phi||\psi|}[\psi,d_\phi\theta]_{FN}.$$

In particular, if $\theta = u$ is a vector-field, the Nijenhuis differential becomes the Lie derivative of tangent–valued forms

$$\begin{aligned}\mathfrak{L}_u\sigma = \; & d_u\sigma = [u,\sigma]_{FN} = (u^\nu\partial_\nu\sigma^\mu_{\alpha_1\dots\alpha_s} - \sigma^\nu_{\alpha_1\dots\alpha_s}\partial_\nu u^\mu \\ & + s\sigma^\mu_{\nu\alpha_2\dots\alpha_s}\partial_{\alpha_1}u^\nu)dx^{\alpha_1}\wedge\dots\wedge dx^{\alpha_s}\otimes\partial_\mu, \qquad (\sigma\in\wedge^s(M)\otimes\mathcal{V}(M)).\end{aligned} \tag{3.41}$$

3.8.4 Tensor-Product State-Space for n Quantum Particles

Recall that classical state-space for the system of n particles is its $6N$D phase-space $\mathcal{P}$, including all position and momentum vectors, $\mathbf{r}_i = (x,y,z)_i$ and $\mathbf{p}_i = (p_x,p_y,p_z)_i$ respectively (for $i = 1,...,n$). The *quantization* is performed as a *linear representation* of the real Lie algebra $\mathcal{L}_P$ of the symplectic phase-space $\mathcal{P}$, defined by the Poisson bracket $\{A,B\}$ of classical variables A,B - into the corresponding real Lie algebra $\mathcal{L}_H$ of the Hilbert space $\mathcal{H}$, defined by the commutator $[\hat{A},\hat{B}]$ of skew-Hermitian operators $\hat{A},\hat{B}$ [II08b, II09].

We start with the *Hilbert space* $\mathcal{H}_x$ for a single 1D (non-relativistic) quantum particle, which is composed of all vectors $|\psi_x\rangle$ of the form

$$|\psi_x\rangle = \int_{-\infty}^{+\infty}\psi(x)\;|x\rangle\;dx,$$

where $\psi(x) = \langle x|\psi\rangle$ are square integrable Fourier coefficients,

$$\int_{-\infty}^{+\infty}|\psi(x)|^{\,2}\,dx < +\infty.$$

The position and momentum Hermitian operators, $\hat{x}$ and $\hat{p}$, respectively, act on the vectors $|\psi_x\rangle\in\mathcal{H}_x$ in the following way:

$$\hat{x}|\psi_x\rangle = \int_{-\infty}^{+\infty}\hat{x}\,\psi(x)\;|x\rangle\;dx, \qquad \int_{-\infty}^{+\infty}|x\,\psi(x)|^{\,2}\,dx < +\infty,$$

$$\hat{p}|\psi_x\rangle = \int_{-\infty}^{+\infty}-\mathrm{i}\hbar\partial_{\hat{x}}\psi(x)\;|x\rangle\;dx, \qquad \int_{-\infty}^{+\infty}|-\mathrm{i}\hbar\partial_x\psi(x)|^2\,dx < +\infty.$$

The *orbit Hilbert space* $\mathcal{H}_1^o$ for a single 3D quantum particle with the full set of compatible observable $\hat{\mathbf{r}} = (\hat{x},\hat{y},\hat{z}), \hat{\mathbf{p}} = (\hat{p}_x,\hat{p}_y,\hat{p}_z)$, is defined as

$$\mathcal{H}_1^o = \mathcal{H}_x\otimes\mathcal{H}_y\otimes\mathcal{H}_z,$$

where $\hat{\mathbf{r}}$ has the common generalized eigenvectors of the form

$$|\hat{\mathbf{r}}\rangle = |x\rangle \times |y\rangle \times |z\rangle .$$

$\mathcal{H}_1^o$ is composed of all vectors $|\psi_r\rangle$ of the form

$$|\psi_r\rangle = \int_{\mathcal{H}^o} \psi(\mathbf{r}) \; |\mathbf{r}\rangle \; d\mathbf{r} = \int_{-\infty}^{+\infty} \int_{-\infty}^{+\infty} \int_{-\infty}^{+\infty} \psi(x,y,z) \; |x\rangle \times |y\rangle \times |z\rangle \; dxdydz,$$

where $\psi(\mathbf{r}) = \langle \mathbf{r}|\psi_r\rangle$ are square integrable Fourier coefficients,

$$\int_{-\infty}^{+\infty} |\psi(\mathbf{r})|^{\,2} \, d\mathbf{r} < +\infty.$$

The position and momentum operators, $\hat{\mathbf{r}}$ and $\hat{\mathbf{p}}$, respectively, act on the vectors $|\psi_r\rangle \in \mathcal{H}_1^o$ in the following way:

$$\hat{\mathbf{r}}|\psi_r\rangle = \int_{\mathcal{H}_1^o} \hat{\mathbf{r}}\,\psi(\mathbf{r}) \; |\mathbf{r}\rangle \; d\mathbf{r}, \qquad \int_{\mathcal{H}_1^o} |\mathbf{r}\,\psi(\mathbf{r})|^{\,2} \, d\mathbf{r} < +\infty,$$

$$\hat{\mathbf{p}}|\psi_r\rangle = \int_{\mathcal{H}_1^o} -\mathrm{i}\hbar\partial_{\hat{\mathbf{r}}}\psi(\mathbf{r}) \; |\mathbf{r}\rangle \; d\mathbf{r}, \qquad \int_{\mathcal{H}_1^o} |-\mathrm{i}\hbar\partial_{\mathbf{r}}\psi(\mathbf{r})|^2 \, d\mathbf{r} < +\infty.$$

Now, if we have a system of n 3D particles, let $\mathcal{H}_i^o$ denote the orbit Hilbert space of the ith particle. Then the composite orbit state-space $\mathcal{H}_n^o$ of the whole system is defined as a direct product

$$\mathcal{H}_n^o = \mathcal{H}_1^o \otimes \mathcal{H}_2^o \otimes ... \otimes \mathcal{H}_n^o.$$

$\mathcal{H}_n^o$ is composed of all vectors

$$|\psi_r^n\rangle = \int_{\mathcal{H}_n^o} \psi(\mathbf{r}_1, \mathbf{r}_2, ..., \mathbf{r}_n) \; |\mathbf{r}_1\rangle \times |\mathbf{r}_2\rangle \times ... \times |\mathbf{r}_n\rangle \; d\mathbf{r}_1 d\mathbf{r}_2 ... d\mathbf{r}_n$$

where $\psi(\mathbf{r}_1, \mathbf{r}_2, ..., \mathbf{r}_n) = \langle \mathbf{r}_1, \mathbf{r}_2, ..., \mathbf{r}_n|\psi_r^n\rangle$ are square integrable Fourier coefficients

$$\int_{\mathcal{H}_n^o} |\psi(\mathbf{r}_1, \mathbf{r}_2, ..., \mathbf{r}_n)|^2 \, d\mathbf{r}_1 d\mathbf{r}_2 ... d\mathbf{r}_n < +\infty.$$

The position and momentum operators $\hat{\mathbf{r}}_i$ and $\hat{\mathbf{p}}_i$ act on the vectors $|\psi_r^n\rangle \in \mathcal{H}_n^o$ in the following way:

$$\hat{\mathbf{r}}_i|\psi_r^n\rangle = \int_{\mathcal{H}_n^o} \{\hat{\mathbf{r}}_i\} \, \psi(\mathbf{r}_1, \mathbf{r}_2, ..., \mathbf{r}_n) \; |\mathbf{r}_1\rangle \times |\mathbf{r}_2\rangle \times ... \times |\mathbf{r}_n\rangle \; d\mathbf{r}_1 d\mathbf{r}_2 ... d\mathbf{r}_n,$$

$$\hat{\mathbf{p}}_i|\psi_r^n\rangle = \int_{\mathcal{H}_n^o} \{-\mathrm{i}\hbar\partial_{\hat{\mathbf{r}}_i}\} \, \psi(\mathbf{r}_1, \mathbf{r}_2, ..., \mathbf{r}_n) \; |\mathbf{r}_1\rangle \times |\mathbf{r}_2\rangle \times ... \times |\mathbf{r}_n\rangle \; d\mathbf{r}_1 d\mathbf{r}_2 ... d\mathbf{r}_n,$$

with the square integrable Fourier coefficients

$$\int_{\mathcal{H}_n^o} |\{\hat{\mathbf{r}}_i\}\,\psi\,(\mathbf{r}_1, \mathbf{r}_2, ..., \mathbf{r}_n)|^2\, d\mathbf{r}_1 d\mathbf{r}_2 ... d\mathbf{r}_n < +\infty,$$

$$\int_{\mathcal{H}_n^o} |\{-\mathrm{i}\hbar\partial_{\mathbf{r}_i}\}\,\psi\,(\mathbf{r}_1, \mathbf{r}_2, ..., \mathbf{r}_n)|^2\, d\mathbf{r}_1 d\mathbf{r}_2 ... d\mathbf{r}_n < +\infty,$$

respectively. In general, any set of vector Hermitian operators $\{\hat{\mathbf{A}}_i\}$ corresponding to all the particles, act on the vectors $|\psi_r^n\rangle \in \mathcal{H}_n^o$ in the following way:

$$\hat{\mathbf{A}}_i|\psi_r^n\rangle = \int_{\mathcal{H}_n^o} \{\hat{\mathbf{A}}_i\}\psi\,(\mathbf{r}_1, \mathbf{r}_2, ..., \mathbf{r}_n)\ |\mathbf{r}_1\rangle\times|\mathbf{r}_2\rangle\times...\times|\mathbf{r}_n\rangle\ d\mathbf{r}_1 d\mathbf{r}_2 ... d\mathbf{r}_n,$$

with the square integrable Fourier coefficients

$$\int_{\mathcal{H}_n^o} \left|\left\{\hat{\mathbf{A}}_i\right\}\psi\,(\mathbf{r}_1, \mathbf{r}_2, ..., \mathbf{r}_n)\right|^2 d\mathbf{r}_1 d\mathbf{r}_2 ... d\mathbf{r}_n < +\infty.$$

3.8.5 Complex-Valued Neural Networks

The main characteristics of complex-valued, feedforward, multi-layered, back-propagation neural networks (NNs)[27] are (see [Nit97, II07] and references therein):

(a) the properties greatly different from those of the real-valued back-propagation (BP) network, including 2D motion structure of weights and the orthogonality of the decision boundary of a complex-valued neuron;

(b) the learning property superior to the real-valued back-propagation;

(c) the inherent 2D motion learning ability (an ability to transform geometric figures); and

(d) the ability to solve the XOR problem and detection of symmetry problem with a single complex-valued neuron.

Complex-valued NNs consist of the complex-valued neurons. Their input signals, weights, thresholds and output signals are all complex numbers. The net input U_n to a complex-valued neuron n is defined as

$$U_n = W_{mn}X_m + V_n,$$

where W_{mn} is the (complex-valued) weight connecting the complex-valued neurons m and n, V_n is the (complex-valued) threshold value of the complex-valued neuron n, and X_m is the (complex-valued) input signal from the complex-valued neuron m. To get the (complex-valued) output signal, convert

[27] It is expected that complex-valued neural networks, whose parameters (weights and threshold values) are all complex numbers, would have applications in all the fields dealing with complex numbers (e.g., quantum physics/computation, telecommunications, electronics, etc).

the net input U_n into its real and imaginary parts as follows: $U_n = x + \mathrm{i}y = z$ (where $\mathrm{i} = \sqrt{-1}$). The (complex-valued) output signal is defined to be:[28]

$$\sigma(z) = \tanh(x) + \mathrm{i}\tanh(y), \qquad \text{where} \qquad (\text{for all } u \in \mathbb{R})$$
$$\tanh(u) = (\exp(u) - \exp(-u)) = (\exp(u) + \exp(-u)).$$

A complex-valued NN consists of such complex-valued neurons. A typical network has three layers: $m \to n \to 1$, with $\omega_{ij} \in \mathbb{C}$ being the weight between the input neuron i and the hidden neuron j; $\omega_{0j} \in \mathbb{C}$ is the threshold of the hidden neuron j; $c_j \in \mathbb{C}$ is the weight between the hidden neuron j and the output neuron $(1 \le i \le m; 1 \le j \le n)$, and $c_0 \in \mathbb{C}$ is the threshold of the output neuron. Let $y_j(z), h(z)$ denote the output values of the hidden neuron j, and the output neuron for the input pattern $z = [z_1, ..., z_m]^t \in \mathbb{C}^m$, respectively. Let also $\nu_j(z)$ and $\mu(z)$ denote the net inputs to the hidden neuron j and the output neuron for the input pattern $z \in \mathbb{C}^m$, respectively. Then the following relations hold:

$$\nu_j(z) = \omega_{ij} z_i + \omega_{0j}, \qquad \mu(z) = c_j y_j(z) + c_0,$$
$$y_j(z) = \sigma(\nu_j(z)), \qquad h(z) = \sigma(\mu(z)).$$

The set of all $m \to n \to 1$ complex-valued NNs is usually denoted by $N_{m,n}$. In particular, the complex-BP learning rule for the $N_{m,n}$ has been obtained in [Nit97] by using a steepest-descent method.

[28] Note [illegible] $<$ Re[σ], Im[σ] < 1 and also that σ is not regular as a complex func[illegible]se the Cauchy-Riemann equations do not hold.

4

Dynamics of Crowd Behaviors: From Complex Plane to Quantum Random Fields

4.1 Complex Plane Dynamics of Crowds and Groups

In this section, we present a unique modelling framework for dynamics, simulation and control of groups and crowds behaviors in the complex plane. First, we formulate a general conceptual framework for representing human behaviour in the language of commutative diagrams. Second, we derive a nonlinear, attractor, Langevin-type dynamics in the complex plane from the previously published crowd dynamics model [IR12]. This previous model was based on the nonlinear Schrödinger equation; the development presented here is essentially a restatement using formal geometrical representation in terms of Kähler manifolds and the associated Kähler-Ricci flow. We simulate this Langevin-type crowd dynamics in the C-sharp visual environment, using a complexified version of the Runge-Kutta-Cash-Karp numerical integrator. Finally, we use our approach to perform finite-type control in the complex Hilbert space of hundreds of agents moving around in the complex plane.

4.1.1 Introduction

As a motivation, consider a local map of a sea with a fleet of vessels moving on it; the sea can be naturally modeled by a region in the complex-plane, so that dynamics of an ND ship fleet sailing on that local map is governed by a set of complex-valued ordinary differential equations (ODEs). Then all individual agents behaviors comprise a complex ND Hilbert control-space.

The above motivational example can be abstracted to formal conceptual modelling of a wide class of human behaviors, which can be externally observed as some form of two-dimensional (2D) dynamics, each consisting of a deterministic drift and random fluctuations. Such class of human behaviors can be efficiently modeled by Langevin-type dynamical equations in the complex plane. Some examples of such behaviors include:

(i) Movements of a group/crowd of people on any horizontal surface (e.g. see, soccer field, swimming pool, street, corridor, ship deck, gym, or any large hall); and

(ii) Movements of any kind of vehicles (land, sea, amphibian), or a fleet of N such vehicles, moving on a flat surface, etc.

As a more formal example, here is a functional definition of the general concept of *human behaviors*. Suppose that we are given the following data:

- A set of N nonlinear attractor ODEs governing individual agents' motions in the complex plane $\mathbb{C}$, or formally, a map:

$$\mathrm{Dyn} : \mathbb{R} \to \mathbb{C}, \quad \text{given by} \quad t \mapsto \dot{z}_k(t), \quad (\text{for } k = 1, ..., N); \quad \text{and}$$

- A set of the corresponding individual agents' controllers, formally given by a map:

$$\mathrm{Con} : \mathbb{C} \to \mathcal{H}(\mathbb{C}) \simeq \mathbb{C}^N, \quad \text{given by} \quad \dot{z}_k(t) \mapsto \mathrm{con}_k(t),$$

  where $\mathcal{H}(\mathbb{C})$ is the Hilbert space constructed over $\mathbb{C}$, by a double least-squares method and the inner product between them.[1]

Then the functional set *Behav* of agent behaviors is defined as a *functional composition* of the set *Dyn* of functional dynamics and the set *Con* of functional controls:

$$\mathrm{Behav} = \mathrm{Dyn} \circ \mathrm{Con} \; : \; \mathbb{R} \to \mathcal{H}(\mathbb{C}), \quad \text{given by} \quad t \mapsto \mathrm{con}_k(t).$$

---

[1] Consider a common practical problem of general data-fitting with a set of arbitrary functions, using an efficient Levenberg-Marquardt (LM) optimizer. This nonlinear *least-squares* problem does not have a closed-form solution, so it can be only iteratively solved, by minimizing a certain L^2-norm (i.e., setting its generalized gradient to zero). If the associated Cauchy L^2-sequence converges (to zero), we have created a real N-dimensional (ND) Banach space. Slightly more generally, if we can successfully fit two data-sets simultaneously, we will create two real ND Banach spaces. Then their "dot-product" is a complex ND Hilbert space $\mathcal{H}$, which represents a rigorous framework for analyzing lattice-ODE dynamics in the complex plane. [More precisely, this is a tensor product of two real Banach spaces V_1^N and V_2^N together with an inner product defined on it in such a way that for each pair of N-vectors $\left(\mathbf{v}_1 \in V_1^N, \mathbf{v}_2 \in V_2^N\right)$ there exists an inner (dot) product, a complex number $z = \mathbf{v}_1 \cdot \mathbf{v}_2 \in \mathcal{H}(\mathbb{C})$.] This lattice dynamics of N agents, each governed by its own wave-like descriptor $\psi(t)$, has some kind of a complex chain-form (or, a discrete nonlinear Schrödinger equation). Finally, if we replace the fitting sum (of-squares) with the integral (of square-integrable functions) – we create an infinite-dimensional Hilbert space, which represents a rigorous framework for analyzing fluid-PDE dynamics in the complex plane. Alternatively, by taking the continuous limit ($N \to \infty$) of an lattice-ODE dynamics, we come to the same 'fluid'-PDE dynamics with its infinite-dimensional Hilbert control-space. From computational perspective, however, we will be always constrained to finite-dimensional ND lattice dynamics and its associated ND Hilbert control-space. For example, consider a local map of a sea. It can be most-naturally modeled as a complex-plane, so that dynamics of an ND ship-fleet sailing on that local map is governed by the ODE-lattice dynamics. Then all individual behaviors of the ships in the fleet (i.e., all individual agents) comprise a complex ND Hilbert control-space.

Furthermore, suppose that we are given the following two functional sets:

- **Dyn** : adaptive dynamics $\psi_k(t)$ of N agents (for $k = 1, ..., N$) in the complex plane $\mathbb{C}$, for small N (say $N < 100$) given by a set of nonlinear attractor ODEs, otherwise (for $N \geq 100$) given by the *adaptive Schrödinger lattice*[2], with random initial conditions and parameter values:

$$\dot{\psi}_k = \mathrm{i}\left(\alpha_k \psi_{k-1} + 2\psi_k + \beta_k \psi_{k+1}\right), \qquad (\mathrm{i} = \sqrt{-1}). \tag{4.1}$$

  Here, α_k and β_k are the best-fit parameters, to be determined by the LM-least-squares from some externally observed data-set. The *quantum random field* (4.1) formally defines the map:

$$\mathbf{Dyn} : \mathbb{R} \longrightarrow \mathbb{C}^N, \quad \text{given by} \quad t \mapsto \dot{\psi}_k(t).$$

- **Con** : a set of the corresponding individual agents' controllers, formally given by a map:

$$\mathbf{Con} : \mathbb{C}^N \longrightarrow \mathcal{H}(\mathbb{C}), \quad \text{given by} \quad \dot{\psi}_k(t) \mapsto \mathrm{con}_k(t),$$

  where $\mathcal{H}(\mathbb{C})$ is the complex Hilbert space constructed over $\mathbb{C}$ by a double least-squares method and the inner product of vectors from the two real Banach spaces.

If the maps **Dyn** and **Con** are properly defined, then the following diagram commutes:

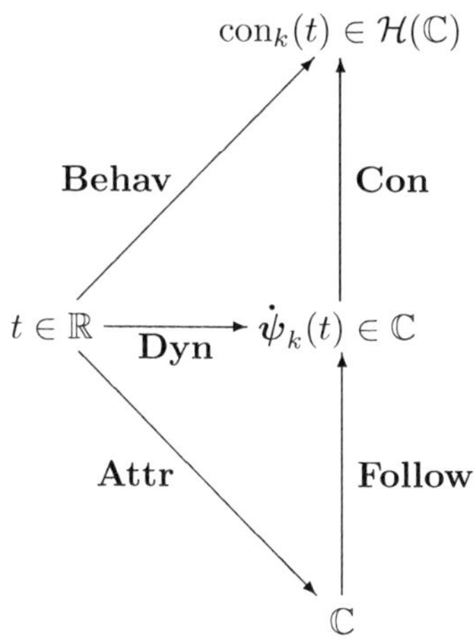

[2] This is a discretized (and adaptive) version of the standard linear Schrödinger equation (see, e.g. [II09]):

$$\mathrm{i}\partial_t \psi = -\frac{1}{2}\partial_{xx}\psi, \qquad (\partial_x \equiv \partial/\partial x)$$

governing the time evolution of the complex-valued wave function $\psi = \psi(x, t)$, describing a ('quantum cloud' around a) free elementary particle.

where **Attr** is an attractor in the complex plane $\mathbb{C}$, while **Follow** is the set of the path-following maps.

That is, the set **Behav** of dynamic agent behaviors is defined by the functional composition:

$$\mathbf{Behav} = \mathbf{Con} \circ \mathbf{Dyn}, \quad \text{given by} \quad t \mapsto \mathrm{con}_k(t). \tag{4.2}$$

For example, using the *Arrows* package in the *Haskell language* (see [OGS08]), the composition (4.2) can be implemented as:

$$\text{Behav} == \text{Con} \cdot \text{Dyn} == \text{Con(Dyn t))} == \text{(Con<<<Dyn) t},$$

by calling the two core C-functions RKF and LM; RFK implements the adaptive dynamics **Dyn**, and LM implements the agents' controllers **Con**.

More generally, *combining different behaviors* is defined by *composing maps* (e.g. functions, processes, signals, systems), which is like following directed paths from one object to another. Compositions along successive paths or arrows comprise a *commutative flow.*

As an illustration, consider a **path** $\overrightarrow{\mathbf{AB}}$ on a street directory, or a ship deck and corridors, to be followed by a crew or a special unit. $\overrightarrow{\mathbf{AB}}$ is given by the commutative flow below, i.e., the following set of compositional rules, where each arrow represents *behavioral dynamics* given by equation (4.1):

$$\begin{aligned} g \circ e &= h \circ f \\ &= k \circ c \circ a \circ i \\ &= k \circ d \circ b \circ i \\ &= k \circ c \circ j \circ e \\ &= k \circ d \circ l \circ f; \end{aligned}$$

e.g. imagine a map from a street directory (or, a building floor, a ship's deck, etc.) underneath this diagram and a special unit (or person of interest) moving from **A** to **B**, where each path is defined by (4.1).

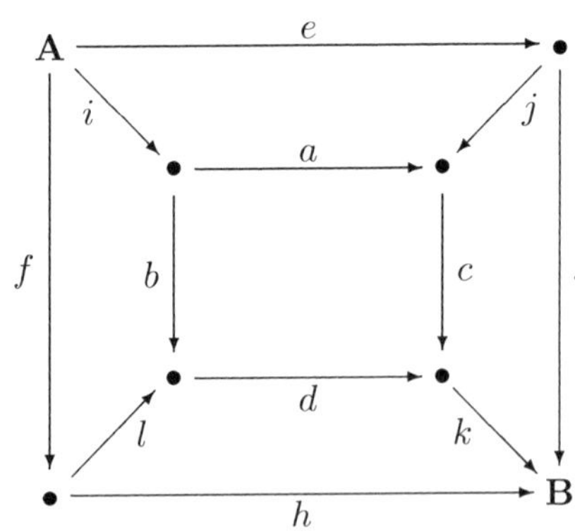

The objective of this problem might be to implement an abstract commutative-flow crowd framework in a functional programming language like Haskell and apply it to modelling and reasoning about general dynamics of individuals, teams, groups and crowds [IR10a, IR10b, IR10c, IR12], on various hierarchical levels of nonlinear complexity [II08a].

4.1.2 Complex-Valued Dynamics of Crowd and Group Behaviors

A *nonlinear Schrödinger equation* (NLS) of the cubic focusing form (see [II09] and references therein):

$$\mathrm{i}\partial_t\psi = -\frac{\Phi}{2}\partial_{xx}\psi + V|\psi|^2\psi, \qquad (4.3)$$

governing the time evolution of the complex-valued wave function $\psi = \psi(x,t)$ with the strength (scaling factor) Φ, describing a free elementary particle moving in adaptive potential well $V = V(x)$ has been recently used for financial option pricing [Iva10, Iva11a] as well as for general crowd modelling (see [IR12] and references therein). As a crowd dynamics simulator, the NLS equation (4.3) has been numerically solved with the *method of lines* (MOL), using the second-order central finite difference scheme (for technical details, see [IR12] and references therein). MOL-lines effectively but automatically define and solve the crowd velocity vector-field in the complex plane $\mathbb{C}$, thus implicitly governing motions of N agents.

In the present paper, we further develop crowd dynamics in the complex plane $\mathbb{C}$, by explicitly defining all *individual velocity controllers* in the complex plane $\mathbb{C}$, by the set of 1st-order ODEs:[3]

$$\dot{z}_k(t) = \Phi\left[A - f(z)\right], \qquad (\text{for } k = 1, ..., N) \qquad (4.4)$$

where $z = z(t)$ is a time-dependent complex number ($z = z + \mathrm{i}y$, with z as its real part and y as its imaginary part); $A = A(z)$ is an attractor function (trajectory) in the complex plane $\mathbb{C}$, while $f(z)$ is some nonlinear function of z, yet to be defined. Numerical solution of the system (4.4) gives the time evolution of N agents of $SE(2)$-kinematic type.[4]

To encapsulate the main characteristics of both linear Schrödinger quantum mechanics and solitonic quantum media given by the cubic NLS (4.3) the

---

[3] For comparison, we remark here that a direct discretizaton of the NLS equation (4.3) gives the following lattice NLS model (see the Kerr-cubic DNLS case in [OB07]):

$$\dot{\psi}_k(t) = \mathrm{i}\Phi\left[\psi_{k+1}(t) + \psi_{k-1}(t) + V\psi_k^3(t)\right], \qquad (\text{for } k = 1, ..., N).$$

[4] The Euclidean group $SE(2) \equiv SO(2) \times \mathbb{R}$, giving motion of a rigid-body in a plane, represents a set of all $3 \times 3-$matrices of the form:

$$\begin{bmatrix} \cos\theta & \sin\theta & x \\ -\sin\theta & \cos\theta & y \\ 0 & 0 & 1 \end{bmatrix},$$

including both rigid translations (i.e., Cartesian $x, y-$coordinates) and rotation matrix $\begin{bmatrix} \cos\theta & \sin\theta \\ -\sin\theta & \cos\theta \end{bmatrix}$ in Euclidean plane $\mathbb{R}^2 \approx \mathbb{C}$, see [II12].

function $f(z)$ is defined as: $f(z) = \alpha z_k + \beta z_k^3$. Further more, for the sake of more realistic dynamical simulations, the notion of 'quantum cloud' is added in the form of both additive and multiplicative Wiener-type zero-mean random noise, i.e. Brownian motion: $W(t) = \beta z_k(t)(\zeta - 0.5)$, so that the system (4.4) becomes the set of complex-valued Langevin-type equations:

$$\dot{z}_k = \Phi\left[A(t) - \alpha z_k - \beta z_k^3 + \beta z_k(\zeta - 0.5)\right], \tag{4.5}$$

where ζ is the pseudo-random number from the range [0,1], generated at each integration time step.

4.1.3 Kähler Geometry of Crowd and Group Dynamics

Crowd dynamics model from the previous section can be further developed using sophisticated machinery from Kähler differential geometry, which combines Riemannian (Lagrangian) and symplectic (Hamiltonian) formalisms with the quantum-like complex structure (see Appendix), as follows.

A Kähler manifold is a symplectic manifold that has an integrable almost-complex structure that maps with the symplectic form. Formally, the ODE system (4.4) defines the *crowd vector-field* on its own configuration Kähler N-manifold $\mathcal{K} = (M, \omega) = (M, g)$ (with the above-mentioned special case $\mathcal{K} = \mathbb{C}^N$), which has the following characteristics:

1. Crowd's local holomorphic coordinates: $\{z_1, \cdots, z_N\} \in U \subset M$, with the corresponding complex-valued crowd differentials:
$$dz_k = dx_k + \mathrm{i}dy_k, \qquad d\bar{z}_k = dx_k - \mathrm{i}dy_k;$$
2. Crowd's Hermitian metric tensor:[5] $g_{i\bar{j}} = g_{i\bar{j}}(z^i, z^{\bar{j}})$, with the corresponding Kähler metric g as a positive and symmetric (1,1)-form:
$$g = \mathrm{i}g_{i\bar{j}}\, dz^i \otimes dz^{\bar{j}},$$
and the associated main Kähler form ω as a closed ($d\omega = 0$) and positive (1,1)-form:
$$\omega = \mathrm{i}g_{i\bar{j}}\, dz^i \wedge dz^{\bar{j}}.$$
In the particular case of $\mathcal{K} = \mathbb{C}^N$ used before as a crowd manifold example, the Kähler form ω can be written as:
$$\begin{aligned}\omega &= \mathrm{i}\left(dz_1 \wedge \overline{dz_1} + dz_2 \wedge \overline{dz_2} + ... + dz_N \wedge \overline{dz_N}\right)\\ &= -2\left(dx_1 \wedge dy_1 + dx_2 \wedge dy_2 + ... + dx_N \wedge dy_N\right)\\ &= \sum_{k=1}^{N} dx_k \wedge dy_k.\end{aligned}$$

[5] The metric tensor $g_{i\bar{j}}(z^i, z^{\bar{j}})$ gives both distances between individual agents and their total mass-distribution within the crowd.

while the corresponding Kähler metric g is given by:

$$\begin{aligned} g &= \mathrm{i}\left(dz_1 \otimes \overline{dz_1} + dz_2 \otimes \overline{dz_2} + ... + dz_N \otimes \overline{dz_N}\right) \\ &= -2\left(dx_1 \otimes dy_1 + dx_2 \otimes dy_2 + ... + dx_N \otimes dy_N\right) \\ &= \sum_{k=1}^{N} dx_k \otimes dy_k = \delta^{ik} dx_k dy_k\,, \qquad (\delta^{ik} = \text{Kronecker delta}). \end{aligned}$$

3. Dolbeault cohomology group for the crowd:

$$H_{\overline{\partial}}^{1,1}(M,\mathbb{R}) = \frac{\{\overline{\partial}\text{-closed real (1,1)-forms}\}}{\{\overline{\partial}\text{-exact real (1,1)-forms}\}};$$

4. Main functional crowd relation:[6]

$$\text{Kähler crowd form} \longrightarrow \omega_\varphi = \omega + \mathrm{i}\partial\overline{\partial}\varphi \longleftarrow \text{Kähler crowd potential.}$$

5. Functional space of Kähler crowd potentials:

$$\mathcal{P} = \{\varphi \mid \omega_\varphi = \omega + \mathrm{i}\partial\overline{\partial}\varphi > 0\}.$$

Once the Kähler configuration manifold is properly defined, the so-called normalised Kähler-Ricci flow is usually written as:

$$\partial_t g = g - \mathrm{Ric}(g), \tag{4.6}$$

where $\mathrm{Ric}(g)$ is the Ricci form. In a local open chart on the Kähler crowd (configuration) manifold $\mathcal{K} = (M, g)$, starting from some smooth initial Kähler metric $g_0 = g_{i\bar{j}}(0)$, the Kähler-Ricci flow (4.6) is given (for $i, j = 1, \cdots, n$) by:

$$\partial_t g_{i\bar{j}}(t) = g_{i\bar{j}}(t) - R_{i\bar{j}}(t). \tag{4.7}$$

During the Kähler-Ricci flow (4.6)-(4.7), the corresponding evolutions of the Ricci curvature $R_{i\bar{j}} = R_{i\bar{j}}(t)$ and the scalar curvature $R = R(t)$ on the Kähler crowd manifold $\mathcal{K}$ are governed by:

$$\begin{aligned} \partial_t R_{i\bar{j}} &= \Delta_M R_{i\bar{j}} + R_{i\bar{j}p\bar{q}} R_{q\bar{p}} - R_{i\bar{p}} R_{p\bar{j}} \quad \text{and} \\ \partial_t R &= \Delta_M R + R_{i\bar{j}} R_{j\bar{i}} - R, \quad \text{respectively,} \end{aligned} \tag{4.8}$$

starting from some smooth initial Ricci and scalar curvatures, $R_{i\bar{j}}(0)$ and $R(0)$, respectively.

The Kähler-Ricci flow (4.7)–(4.8) is a geometrical representation of the basic crowd dynamics model (4.4). It was proved by [Cao85, CC99], following

[6] The Kähler crowd potential φ is a real–valued function φ on a Kähler configuration manifold for which the Kähler form ω can be written as $\omega = \mathrm{i}\partial\bar{\partial}\varphi$, where $\partial = \partial_{z_k} dz_k$ and $\bar{\partial} = \partial_{\bar{z}_k} d\bar{z}_k$.

the pioneering work of [Yau06], that the Kähler-Ricci flow (4.7) exists globally for any smooth initial Kähler metric $g_0 = g_{i\bar{j}}(0)$. For the most recent review, see [TZ13]. From the control-theory perspective, the most important characteristic of the Kähler-Ricci flow and the associated curvatures is the existence of its solitary solutions, which are shrinking, or decaying in time. This characteristic is associated to geometrical entropy decrease and gives the global Lyapunov stability to the crowd flow. This quality is particularly obvious in numerical crowd simulations driven by complex-plane attractors (see next section), where all initial randomness gradually vanishes during the simulation, while the attractor gradually becomes the most dominant dynamics factor.

4.1.4 Computer Simulations of Crowds and Croups Dynamics

We have built a dynamical simulator in the C# language for simulating attractor-driven Langevin-type crowd dynamics in the complex plane $\mathbb{C}$.

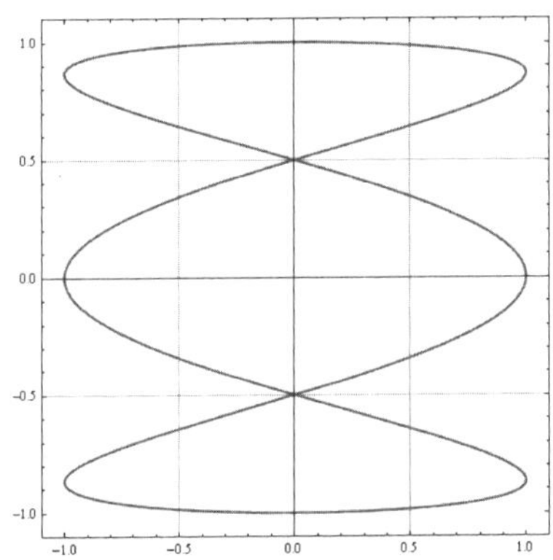

Fig. 4.1. *A sample complex attractor.*

We adopted the following approach to yield the illustrations here:

1. The set of complex-plane attractors, $A(t) \equiv \mathrm{Attr(t)}$, has been defined (see Figure 4.1) as:

$$\begin{aligned}
\sin(t) + \mathrm{i}\cos(t) &: \text{basic circle},\\
\sin(2t) + \mathrm{i}\cos(t) &: \text{vertical figure 8},\\
\sin(t) - \mathrm{i}\sin(2t) &: \text{horizontal figure 8},\\
\sin(3t) + \mathrm{i}\cos(t) &: \text{double figure 8},\\
\sin(4t) + \mathrm{i}\cos(t) &: \text{triple figure 8},\\
\sin(t) \pm \mathrm{i}\cos(2t) &: \text{convez/concave parabolas},
\end{aligned}$$

$$\begin{aligned}
\cos(3t) \pm \mathrm{i}\cos(t) &: \text{figures S/Z},\\
0 + \mathrm{i}\cos(t) &: \text{central vertical line},\\
\pm 1 + \mathrm{i}\cos(t) &: \text{left/right vertical lines},\\
\sin(t) + 0\,\mathrm{i} &: \text{central horizontal line},\\
\sin(t) \pm \mathrm{i} &: \text{top/bottom horizontal lines},\\
\sin(t) \pm \gamma\,\mathrm{i}\sin(t) &: \text{various diagonals (for } \gamma = 1,2,3)
\end{aligned}$$

using (real and imaginary) pairs of C# λ-functions of the form:

```
Func <double, double, double> f1 = (t, a) =>
(t < 0.5*tFin) ? 2 * Math.Sin(a * t) : 2 * Math.Cos(a * t);
```

– etc.

2. The set of complex-valued Langevin-type ODEs, one ODE per agent, has been defined, including the previously-defined attractors, as:

$$\dot{z}_k = \Phi\left[A(t) - \alpha z_k - \beta z_k^3 + \beta z_k(\zeta - 0.5)\right], \qquad \text{which in C\# reads :}$$

```
dz[k] = FldStrn * (Attr(t) - alpha * z[k] - beta * Math.Pow(z[k], 3)
        + beta * z[k] * (rand.NextDouble() - 0.5));
```

where $\Phi \equiv$ FldStrn is the field strength, $\alpha \equiv$ alpha and $\beta \equiv$ beta are coefficients, $\zeta \equiv$ rand.NextDouble() is the pseudo-random number from the range [0,1], generated at each integration time step;
3. Complexified version of the Runge-Kutta-Cash-Karp integrator [CK90] has been implemented;
4. Dynamics of several hundred agents has been visualised on the standard C-sharp panel component;
5. Plots of the simulations are visualised on the standard MS-chart component (see Figures 4.2, 4.3 and 4.4).

4.1.5 Braids of Agents' Behaviors in the Complex Plane

There is a close relationship between braid groups and configuration spaces of distinct points in the complex plane $\mathbb{C}$, which can provide a useful model for various agents' dynamic behaviors in different planar environments (e.g. land, sea, ship deck/corridor, street, etc.).

Informally, assume that all individual agents are represented as points in $\mathbb{C}$. If the subset $O \subset \mathbb{C}$ represents the obstacles which the agents need to avoid, then the configuration space $\mathcal{C}$, representing a 'safe[7] configuration' of n non-colliding agents, is given by [Ghr01]:

$$\begin{aligned}
\mathcal{C}(\mathbb{C} - O) &= [(\mathbb{C} - O) \times (\mathbb{C} - O) \cdots \times (\mathbb{C} - O)] - \Delta, \qquad \text{where}\\
\Delta &= \{(x_1, x_2, \ldots, x_n : x_i = x_j \text{ for some } i \neq j\} \quad \text{(the pairwise diagonal)}.
\end{aligned}$$

[7] According to [Ghr01] there is an analytical measure of 'safety', that is the distance from each agent to the boundary of $\mathcal{C}(\mathbb{C} - O)$.

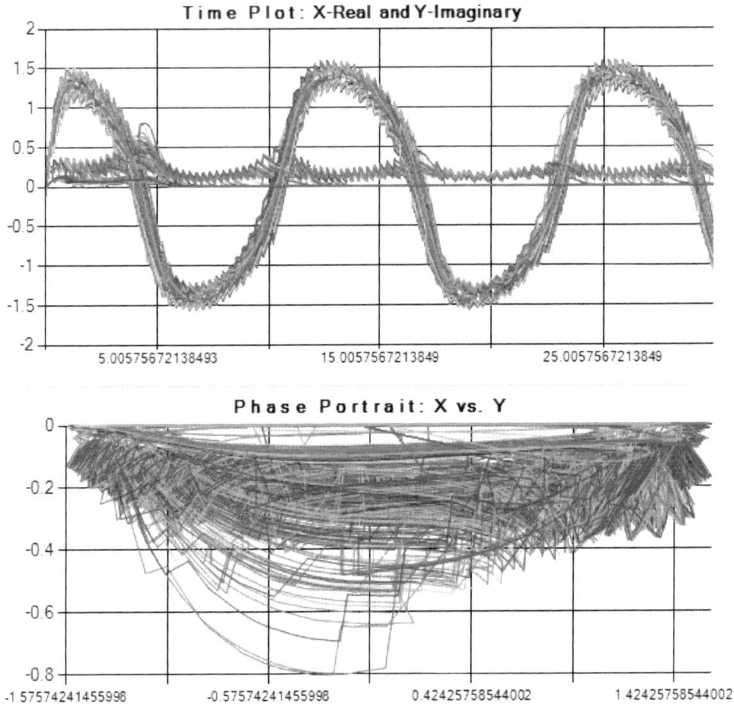

Fig. 4.2. *Simulating a crowd of 200 agents moving up and down the corridor.*

The unlabeled configuration space, $\mathcal{C}_{\mathrm{u}}(\mathbb{C}-O)$, is defined as the quotient $\mathcal{C}(\mathbb{C}-O)/S_n$, of $\mathcal{C}(\mathbb{C}-O)$ by the (natural action of) the symmetric group S_n that permutes the ordered points in $\mathbb{C}$. Now, the n-strand braid group of $(\mathbb{C}-O)$ is defined as:

$$B_n(\mathbb{C}-O) = \pi_1\left[\mathcal{C}_{\mathrm{u}}(\mathbb{C}-O)\right],$$

whereas the n-strand pure braid group of $(\mathbb{C}-O)$ is:

$$P_n(\mathbb{C}-O) = \pi_1\left[\mathcal{C}(\mathbb{C}-O)\right],$$

where $\pi_1(X)$ is the fundamental group of the topological space X. In case of $\mathbb{C}-O=\mathbb{R}^2$, these are classical Artin braid groups B_n and their pure relatives P_n.[8]

Formally, let E_n be an affine complex algebraic variety (i.e., the set of solutions of a system of polynomial equations over the field of complex numbers, singular if $n \geqslant 2$), given by the set of all unordered n-tuples of points of $\mathbb{C}$ [Rud83, Rud05]. A *discriminant locus* is a subset $\Delta \subset E_n$ of n-tuples with at

[8] Besides Artin's representation, there is also classical Burau's representation, given as a matrix (or, ring) of Laurent polynomials, and also related to Alexander's polynomial.

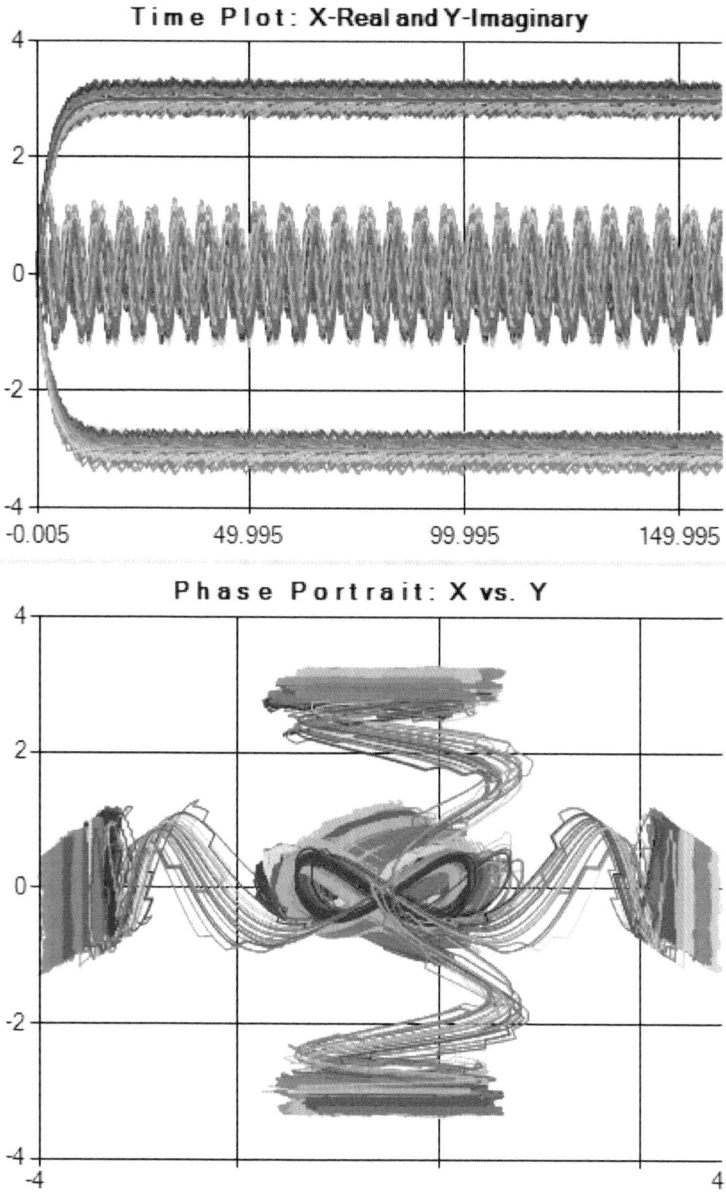

Fig. 4.3. *Simulating a fleet of 200 boats moving within the rectangular sea-region, driven by 'figure 8' attractors (both vertical and horizontal).*

least one duplication (representing an irreducible algebraic hypersurface). Its complement, $\mathcal{C} = E_n \backslash \Delta$, is the *configuration space* (of n distinct points in $\mathbb{C}$). Any function $F : X \to E_n$ defined on a topological space X is an n-valued

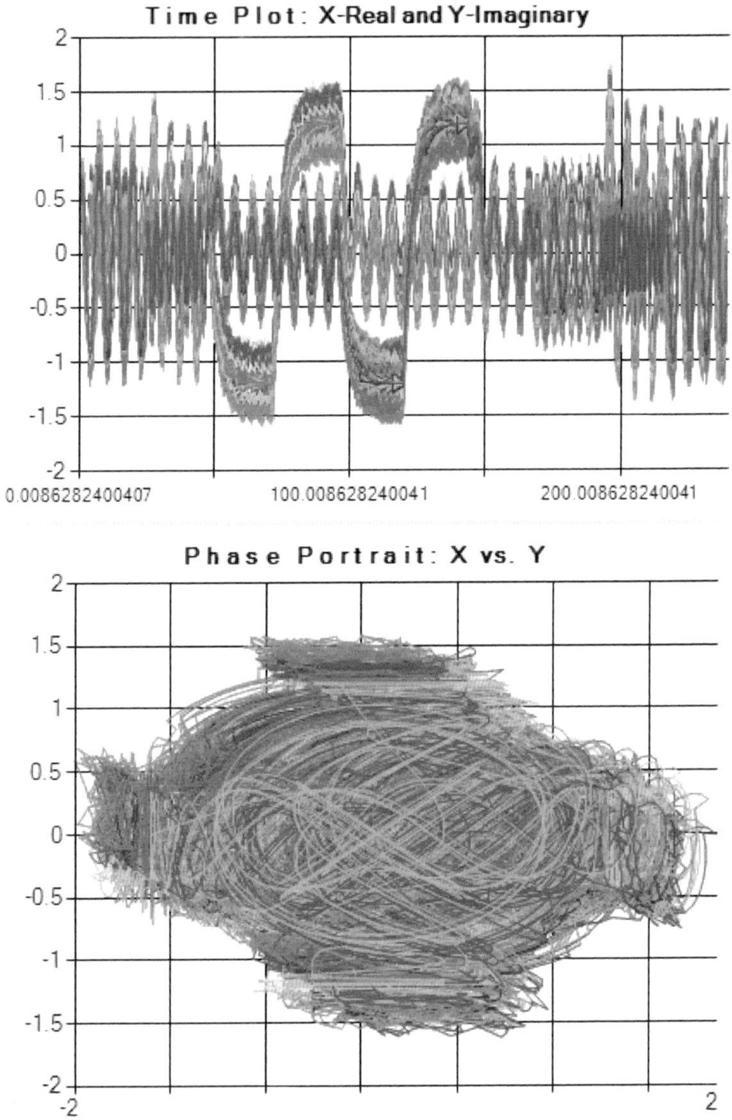

Fig. 4.4. *Simulating general behaviors of a crowd of 200 agents, including: up-and-down moves on both left and right sides, left-and-right on both top and bottom sides, left and right diagonals, clockwise and anticlockwise circles.*

analytic function (which is locally given by a convergent power series) on X (such that its graph is the subset of $X \times \mathbb{C}$).

The *braid group* B_n with n strands is the fundamental group $B_n = \pi_1(E_n \backslash \varDelta)$ of the configuration space $\mathcal{C} = E_n \backslash \varDelta$ [Bir75]. B_n acts (faithfully) as a group of automorphisms of the free group F_n of rank n. Explicitly, if $F_n = \pi_1(\mathbb{C}\setminus\{w_1, ..., w_n\}; w_0)$, the acting B_n is realized as $\pi_1(E_n - \Delta; \{w_1, \dots, w_n\})$; on standard free generators $x_1, \dots, x_n$ of F_n (positively oriented meridians around $w_1, \dots, w_n$), the action is:

$$x_i\sigma_i = x_i x_{i+1} x_i^{-1}, \qquad x_{i+1}\sigma_i = x_i, \qquad x_j\sigma_i = x_j \quad \text{for} \quad j \neq i, i+1.$$

There is a map from the set of strata of Q_n into a hierarchy of "types of expressions" of $\Delta^2 \in B_n$ as products $\beta(1)\cdots\beta(k)$ [Rud83, Rud05]. For example, $\Delta^2 = (\sigma_1\sigma^2\cdots\sigma_{n-1})^n$ is pure, and in terms of the standard generators of the pure braid group:

$$A_{ij} = (\sigma_i\cdots\sigma_{j-1})\sigma_j^2(\sigma_i\cdots\sigma_{j-1})^{-1}, \quad 1 \leqslant i \leqslant j \leqslant n-1,$$

Δ^2 is given by

$$\Delta^2 = A_{1,n-1}A_{1,n-2}\cdots A_{1,1}A_{2,n-1}\cdots A_{2,2}\cdots A_{n-1,n-1}.$$

More generally, let $\Delta^2 = \beta(1)\cdots\beta((n^2-n)/2)$, where $\beta(i) = A_{p,q}$. For each pair p, q the generators x_p, x_{q+1} commute, that is, e.g. the action of $A_{1,1} = \sigma_1^2$ on F_n is

$$x_1\sigma_1^2 = (x_1 x_2 x_1^{-1})\sigma_1 = x_1 x_2 x_1 x_2^{-1} x_1^{-1},$$
$$x_2\sigma_1^2 = x_1\sigma_1^2 = x_1 x^2 x_1^{-1}, \qquad x_k\sigma_1^2 = x_k, \qquad k \neq 1, 2;$$

and the relations $x_1 = x_1 x_2 x_1 x_2^{-1} x_1^{-1}$ and $x_2 = x_1 x_2 x_1^{-1}$ both say x_1 commutes with x_2. The group B_n is free Abelian of rank $n-1$.

4.1.6 Hilbert-Space Control of Crowds and Groups Dynamics

In the so-called *finite control problem* (e.g., output tracking and navigation, see [II12] and references therein) one wants the scalar system output $y(\mathbf{x}, t)$ to follow a continuous model (nominal) trajectory $\chi(t)$, for a given MD system-vector $\mathbf{x}$ and time t. This problem can be expressed as:

$$\min_{\mathbf{x}\in\mathbb{R}^M} \int_{t_0}^{t_1} [y(\mathbf{x}, t) - \chi(t)]^2\, dt, \tag{4.9}$$

and, upon time discretization (using a suitable quadrature scheme), becomes the following least-squares problem:

$$\min_{\mathbf{x}\in\mathbb{R}^M} f(\mathbf{x}) = \sum_{i=1}^{N} [\tilde{y}(\mathbf{x}, t_i) - \tilde{\chi}(t_i)]^2\, dt,$$

where both discretized functions, $\tilde{y}(\mathbf{x}, t_i)$ and $\tilde{\chi}(t_i)$, include the weights of the chosen quadrature scheme (see [Mat13]).

We remark here that the finite control problem (4.28) is formally a minimization of the square of the Banach metric (4.31), see Appendix. Similarly, a ship (or other vehicle or agent) navigation problem modeled in the complex plane is formally a minimization of the square of the Hilbert metric (4.34) (see Appendix).

To proceed, let us consider the vector space $V(f)$ of real-valued square-integrable functions f defined on an interval $[a,b] \subset \mathbb{R}$ (i.e., functions with the convergent integral: $I(f) = \int_a^b |f(x)|^2 dx < \infty$).

To be able to perform the analysis on Banach and Hilbert spaces, we first need to make $V(f)$ a normed space. However, a naive approach will not work here, because the most natural choice of a norm,

$$\|\cdot\| : f \mapsto \sqrt{\int_a^b |f(x)|^2 dx}, \tag{4.10}$$

is not a proper norm on $V(f)$, which can be seen, e.g. in case:

$$f(x) = \begin{cases} 0 \text{ for } & x \neq a \\ 1 \text{ for } & x = a \end{cases} \Longrightarrow \|f\| = 0, \text{ but } f \neq 0,$$

so it violates the **N1**-axiom (of positive definiteness). However, the problem is solved with a little help from measure theory, namely, (4.30) becomes a proper norm if we identify functions f which are equal almost everywhere (i.e., differ only on a set of measure zero in $[a,b]$). The resulting vector space of equivalence classes $[f]$ is (usually) denoted by $L^2[a,b]$ and is an infinite-dimensional Banach space with the norm of the equivalence class $[f]$ defined by:

$$L^2[a,b] = \|[f]\| = \sqrt{\int_a^b |f(x)|^2 dx}.$$

For the proof of the necessary completeness, see e.g. [Roy68].

The corresponding *Banach metric* (i.e., distance) between any two real-valued square-integrable functions f,g defined on an interval $[a,b] \subset \mathbb{R}$, associated with the norm $L^2[a,b]$ reads:

$$\mathbf{d}(f,g) = \sqrt{\int_a^b |f(x) - g(x)|^2 dx}. \tag{4.11}$$

For example, the finite control problem (4.28) is actually a minimization of the square of the Banach metric (4.31).

Next, if we define the inner product:

$$\langle f, g \rangle = \int_a^b f(x) g(x) dx,$$

$L^2[a,b]$ becomes the infinite-dimensional real Hilbert space $\mathcal{H}(\mathbb{R})$, which is used for statistical learning theory of *support vector machines.*

Finally, if we repeat this whole procedure (while instead of $f \in \mathbb{R}$) using complex-valued square-integrable wave-functions $\psi \in \mathbb{C}$ defined in a complex-plane region $[a, b] \subset \mathbb{C}$, with

$$\langle \psi, \varphi \rangle = \int_a^b \psi(x)\overline{\varphi(x)}dx, \tag{4.12}$$

we are landing at the infinite-dimensional quantum Hilbert space $L^2([a, b], \mathbb{C}) \equiv \mathcal{H}(\mathbb{C})$ of both QHO and AQO, with the L^2-norm defined by:

$$L^2([a, b], \mathbb{C}) = \|[\psi]\| = \sqrt{\int_a^b |\psi(x)|^2 dx}. \tag{4.13}$$

The corresponding *Hilbert metric* (i.e., distance) between any two complex-valued square-integrable functions f, g defined on an region $[a, b] \subset \mathbb{C}$, associated with the norm $L^2([a, b], \mathbb{C})$ reads:

$$\mathbf{d}(\psi, \varphi) = \sqrt{\int_a^b |\psi(x) - \varphi(x)|^2 dx}. \tag{4.14}$$

For example, the *ship-navigation problem* can be formulated in the complex-plane $\mathbb{C}$ as a minimization of the square of the Hilbert metric (4.34).

Summary

In this section we have proposed a unique modeling framework for dynamics, simulation and control of groups and crowds behaviors in the complex plane. The simulations with two-dimensional animations are performed in the C-sharp object-oriented environment, using the fast, complexified, Runge-Kutta-Cash-Karp numerical integrator. An efficient algorithm for simulating a variety of behaviors of several hundreds of agents was adopted on an ordinary PC, using a nonlinear, attractor, Langevin-type crowd dynamics. A finite-type control is performed in the complex Hilbert space.

4.2 Quantum Random Fields: A Unique Framework for Simulation, Optimization, Control and Learning

In this section, we generalize the complex-plane crowd behavioral dynamics into a unified modeling and simulation framework, of both rigorous theoretical and efficient computational nature, for developing/implementing five practical engineering problem-solving tools (simulation, optimization, learning, control, and logic) on a common physical ground of quantum random fields and within the common geometrical settings of infinite-dimensional Banach-Hilbert spaces.

4.2.1 Introduction

In our previous development of the Crowd Simulator in a powerful but not friendly C++ (see, [IR12]), we would use one crowd dynamics model (defined by two or more nonlinear Schrödinger equations) which would be put into different 3D graphics environments (a bush or a street) and we would apply to it different events (bushfire or car-explosion). So, we didn't mathematically define these 3D environments, they were just parts of the 3D graphics engine used (including the collision dynamics engine).

Recently, we started a different, much easier to implement, line of development in a 2D visual environment in a friendly C-sharp language. So far, we have developed various agents' dynamics $x_i = x_i(t)$, following different attractors in the complex plane $\mathbb{C}$, which are given *a priori* by the 'leaders' $:\to$ these could be called 'supervised dynamics.' We are also trying to make these dynamics adaptive, so that the attractors are not given *a priori* but rather learned or trained on the fly $:\to$ these could be called 'unsupervised or self-organized dynamics.' However, in both cases, these are only agents' dynamics, without mathematically defining any environments.

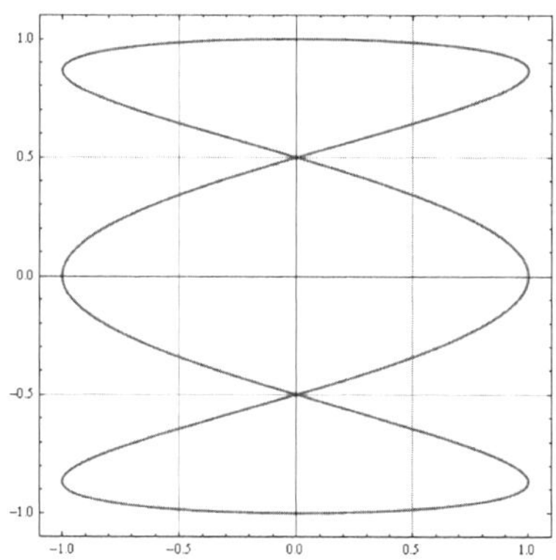

Fig. 4.5. *A sample line-attractor in the complex plane. In the background, we can see the chess-board like field lattice.*

Here is a field-theoretic proposal for modeling environments in the complex plane, which would affect both the attractors and the followers' dynamics. If we look carefully at the Figure 4.5, we will see not only the attractor line in the complex plane, but also the 'chess board' in the background, which can have N by N squares (for arbitrary N) and can be modeled as the *adaptive quantum field* (AQF). This AQF is our proposal for modeling environments in the complex plane $\mathbb{C}$, similar to the quantized (Maxwell) electrodynamics fields Φ_i (QED for short). In this context, the word 'quantized' can be interpreted as: 'each square has its own rules of behavior'; so, each square can represent a different environmental section Φ_i. So, we would have not only line or point attractors, but the whole area of the complex plane would be 'quantized' into many squares that would have their own, different rules of behavior. The line attractors (or, leaders) would be restricted by these rules of behavior, and all individual agents would strictly follow.

In this way, we would have the full dynamics of *agents' paths* $[x]$ + *environmental fields* $[\Phi]$:

$$\text{Full Complex Dynamics} = \oint \mathcal{D}^{\text{agents}'}_{\text{paths}}[x]\, \mathrm{e}^{\mathrm{i}S[x]} + \oint \mathcal{D}^{\text{environ}}_{\text{fields}}[\Phi]\, \mathrm{e}^{\mathrm{i}S[\Phi]}$$

4.2.2 Adaptive Quantum Oscillator

In this section, starting from a simple mechanical harmonic oscillator, we will derive our main physical object, the adaptive quantum oscillator (AQO), the generator of random fields, on which optimization, learning, control and logic. We will perform this AQO-derivation rather intuitively and using the necessary minimum of Schrödinger's quantum wave mechanics (leaving the rigorous quantization to Appendix) - yet with straightforward and clean descriptive brevity, thus setting up the scene for rigorous learning, control and logic development.

Quantum Harmonic Oscillator

The AQO-derivation starts with the *classical harmonic oscillator* (CHO), a simple mass-and-spring system which is a generator of sinusoidal oscillations (about the equilibrium point), and is formally given by the *Hamiltonian energy* (kinetic plus potential) function:

$$H(p,x) = \frac{1}{2}\left(\frac{p^2}{m} + kx^2\right), \tag{4.15}$$

where (x,p) is a pair of canonically-conjugated real *variables* (coordinate and momentum, respectively), while (m,k) is a pair of real *constants* (mass and spring, respectively). This is all that we need from classical physics for the moment. We will revisit the CHO later in the subsection 4.3.4 below, but now we quickly and intuitively move into the quantum world.

Standard quantum harmonic oscillator (QHO) is obtained from the classical one (4.15) by performing the process of the so-called *first quantization.* Informally, a time-dependent state of any quantum-mechanical system is determined by a normalized, complex-valued, wave psi-function $\psi = \psi(x,t)$. Semi-formally, in *Dirac's notation* (which is 'de facto' standard in quantum physics, see [Dir49]), $\psi(x,t)$ is a unit *ket*-vector $|\psi\rangle$, which is an element of the complex, infinite-dimensional *Hilbert space* of square-integrable wave ψ-functions $\mathcal{H}(\mathbb{C}) \equiv L^2(\mathbb{R}) = \{|\psi\rangle, |\phi\rangle, |\varphi\rangle, ...\}$ with the coordinate basis (x^i), for $i = 1, 2, ...\infty$.

Recall from Chapter 2 that the essential property of $\mathcal{H}(\mathbb{C})$ is the *Hermitian inner product* $\langle\psi,\phi\rangle$. In physics, it is usually denoted by Dirac's celebrated *bra-ket* $\langle\psi|\phi\rangle$, or Feynman's *probability amplitude* $\langle\text{Out}|\text{In}\rangle$, or statistical-mechanical *partition function* Z. Thus, all these description are more-or-less equivalent: $\langle\psi,\phi\rangle \equiv \langle\psi|\phi\rangle \equiv \langle\text{Out}|\text{In}\rangle \equiv Z$. This inner product can be applied to any pair of $\mathcal{H}$-vectors, $|\psi\rangle$ and $|\phi\rangle$, to produce a single complex number $\langle\psi,\phi\rangle \equiv \langle\psi|\phi\rangle \in \mathbb{C}$, which satisfies a number of simple algebraic properties:

$$\begin{aligned}
\langle\psi+\phi,\varphi\rangle &= \langle\psi,\varphi\rangle + \langle\phi,\varphi\rangle \\
\langle\psi,\phi+\varphi\rangle &= \langle\psi,\phi\rangle + \langle\psi,\varphi\rangle \\
\langle a\,\psi,\phi\rangle &= a\,\langle\psi,\phi\rangle\,, \qquad \langle\psi,a\,\phi\rangle = \bar{a}\,\langle\psi,\phi\rangle \\
\langle\psi,\phi\rangle &= \overline{\langle\phi,\psi\rangle} \qquad \text{(where bar denotes complex-conjugate)} \\
\langle\psi,\psi\rangle &\geq 0, \quad \langle\psi,\psi\rangle = 0 \text{ only if } \psi = 0 \text{ (positive definite)} \\
\langle\psi,\phi\rangle &= \sum_i \psi_i\bar{\phi}_i \in \mathbb{C}^n, \quad (\text{for } i = 1,n).
\end{aligned}$$

Probability of finding a quantum particle at a certain location, or more generally, a quantum system at a certain ket-state $|\psi\rangle$, is given by its *probability density function* (PDF), which is defined as the absolute square of the wave ψ-function:

$$PDF := |\psi|^2 \equiv \left||\psi\rangle\right|^2 \equiv \langle\psi|\psi\rangle \equiv \langle\psi,\psi\rangle,$$

which is simply the squared length of the ket-vector $|\psi\rangle$ in $\mathcal{H}(\mathbb{C})$ and a normalized state is given by a $\mathcal{H}$-vector $|1\rangle$ (whose squared length is unity). More generally, the PDF-description of any microscopic process, or even any (quantum-influenced) macroscopic process, is the In$\to$Out map determined by the absolute square of the probability amplitude:

$$PDF : \text{In} \mapsto \text{Out} := \left|\langle\text{Out}|\text{In}\rangle\right|^2 \equiv |\psi|^2.$$

Every quantum state-vector $|\psi\rangle \in \mathcal{H}(\mathbb{C})$ is subject to the action of *Hermitian operators* (i.e., self-adjoint operators with associated Hermitian matrices, real eigenvalues and orthogonal eigenfunctions) obtained by quantization of classical dynamical quantities, and whose real eigenvalues are being measured (see e.g. [Tha00, Gri05]). In other words, the quantization process generates Hermitian operators from the corresponding classical variables:

$$\begin{aligned}
x &\xrightarrow{\text{quant}} \hat{x} = x && \text{(coordinate operator)} \\
p &\xrightarrow{\text{quant}} \hat{p} = -\mathrm{i}\hbar\tfrac{d}{dx} && \text{(momentum operator)} \\
H &\xrightarrow{\text{quant}} \hat{H} = \hat{H}(x,\hat{p}) && \text{(Hamiltonian operator)},
\end{aligned}$$

where $\mathrm{i} = \sqrt{-1}$and $\hbar$ is Planck's constant. In general, Dirac's *canonical quantization*, by which classical dynamical variables$(f, g, ...)$ become quantum Hermitian operators $\left(\hat{f}, \hat{g}, ...\right)$ states:

$$\{f, g\} \longmapsto \frac{1}{\mathrm{i}\hbar}[\hat{f}, \hat{g}], \tag{4.16}$$

which means that the quantum commutator $[\hat{f}, \hat{g}]$ has the same values as the classical Poisson bracket $\{f, g\}$. See Appendix (subsection 4.3.4) for the rigorous geometric formulation.

Now, quantum mechanics can be described in three standard pictures (see, e.g. [II08b, II09]). In the most common *Schrödinger picture* (of the coordinate x-representation), under the action of the Hermitian *evolution operator* $\hat{S}(t)$, the quantum state-vector $|\psi(t)\rangle$ rotates, starting with the initial state-vector $|\psi(0)\rangle$:

$$|\psi(t)\rangle = \hat{S}(t)\,|\psi(0)\rangle,$$

while the infinite-dimensional coordinate basis (x^i) in the Hilbert space $\mathcal{H}(\mathbb{C})$ is fixed, so the Hermitian operators are constant in time:

$$\hat{F}(t) = \hat{F}(0) = \hat{F},$$

and the system evolution is determined by the time-dependent Schrödinger wave equation:[9]

$$\mathrm{i}\hbar\,\partial_t|\psi(t)\rangle = \hat{H}\,|\psi(t)\rangle, \qquad (\text{where } \partial_t = \partial/\partial_t), \tag{4.17}$$

where $\hat{H} = \hat{H}(x, \hat{p})$ is Schrödinger's Hamiltonian operator, which in our QHO-case reads:[10]

$$\hat{H} = \frac{1}{2}\left(\frac{\hat{p}^2}{m} + kx^2\right) \equiv -\frac{\hbar^2}{2m}\frac{d^2}{dx^2} + \frac{1}{2}kx^2 \quad (\text{where } \hat{p}^2 = -\hbar^2\frac{d^2}{dx^2}).$$

Alternatively, in the *Heisenberg picture*, under the action of the evolution operator $\hat{S}(t)$, the coordinate basis (x^i) rotates, and the system evolution is determined by the *Heisenberg equation* of motion:

---

9 The time-dependent Schrödinger equation (4.17) represents a complex-valued generalization of the real-valued Fokker-Planck equation for describing the spatio-temporal probability density function for the system exhibiting continuous-time Markov stochastic process.

10 If the Hamiltonian does not explicitly depend on time, or at $t = 0$, $\hat{H}(t) = \hat{H}$, which is the case with the absence of variables of macroscopic fields, the state vector $|\psi(t)\rangle$ can be presented in the form:

$$|\psi(t)\rangle = \exp\left(-\mathrm{i}\frac{E}{\hbar}t\right)|\psi\rangle,$$

where the stationary state vector $|\psi\rangle$ satisfies the time-independent (stationary) Schrödinger equation:

$$\hat{H}\,|\psi\rangle = E\,|\psi\rangle,$$

which gives the eigenvalues E_m and the corresponding eigenfunctions $|\psi_m\rangle$ of the Hamiltonian operator $\hat{H}$.

$$\mathrm{i}\hbar\,\partial_t\hat{F}(t) = [\hat{F}(t), \hat{H}(t)], \tag{4.18}$$

where $\hat{F}(t)$ denotes an arbitrary Hermitian operator of the system, while the quantum commutator $[,]$ is given by:

$$[\hat{F}(t), \hat{H}(t)] = \hat{F}(t)\,\hat{H}(t) - \hat{H}(t)\,\hat{F}(t).$$

In both pictures,[11] the evolution operator $\hat{S}(t)$ is determined by the Schrödinger-like equation:

$$\mathrm{i}\hbar\,\partial_t\hat{S}(t) = \hat{H}\,\hat{S}(t), \qquad \text{with } \hat{S}(0) = \hat{I}d.$$

This completes our 'intuitive quantization' of the classical harmonic oscillator (4.15), so that we now have two QHO-versions, the Schrödinger (4.17) one and the Heisenberg (4.18) one, and thus we can move on towards our main physical/computational object: the adaptive quantum oscillator. For mathematical completeness sake, a rigorous geometric CHO-quantization in the Heisenberg picture is provided in Appendix (section 4.3.4).

From Schrödinger to Cubic NLS to Lattice DNLS

We are now preparing to enter into the nonlinear world of quantum matter (or quantum media, or quantum super-fluids, like Bose-Einstein condensates, or quantum optics). So, we will remain in the complex Hilbert space $\mathcal{H}(\mathbb{C})$ (which is the natural stage for all complex PDEs), but we will not need Dirac's bras and kets nor Hermitian-operator caps. So, we firstly switch to dimensionless units in which ($\hbar = 1$), mass m becomes a dimensionless weight-parameter $\mathfrak{m}$ and spring constant k becomes a dimensionless spring-parameter $\mathfrak{k}$. In this way, the linear Schrödinger equation (4.17) simplifies into:

$$\mathrm{i}\partial_t\psi = H\,\psi \qquad \text{or} \qquad \mathrm{i}\partial_t\psi = -\frac{1}{2\mathfrak{m}}\partial_{xx}\psi + \frac{1}{2}\mathfrak{k}x^2\psi, \tag{4.19}$$

which generates an infinite-dimensional family of harmonic-waves as solutions (see e.g. [Tha00, Gri05]).

Now comes the critical shift from linear quantum wave mechanics to nonlinear quantum-matter waves, which comprise the macroscopic physical system with quantum origin and much more interesting wave mechanics that includes a zoo of solitons, kinks, breathers, shock waves and rogue waves (as

---

[11] Note that in the third, *Dirac interaction picture*, both the state vector $|\psi(t)\rangle$ and the coordinate basis (x^i) rotate, so the system evolution is determined by both the Schrödinger wave equation (4.17) and the Heisenberg equation of motion (4.18).

We remark here that there is also a number of other ways in which the CHO-quantization can be performed, including the well-known quantizations by Von Neumann (density matrix formalism), Feynman (path-integral formalism) and Wigner (phase-space formalism).

well as harmonic waves if nonlinearity approaches zero). The shift from linear to nonlinear wave mechanics is most easily performed by generalizing the spring-parameter's potential energy: $V_{\mathrm{spr}} = \frac{1}{2}\mathfrak{k}x^2$ of (4.19) into the following *wave-dependent, recursive potential energy*: $V_{\mathrm{rec}} = \frac{1}{2}\mathfrak{h}|\psi|^2$ with the *heat parameter* $\mathfrak{h}$. So, the spring-parameter $\mathfrak{k}$ becomes the heat-parameter $\mathfrak{h}$, while the square term x^2 becomes the absolute square term $|\psi|^2$. In this way, we come to the *nonlinear Schrödinger equation* (NLS), also called the *Gross-Pitaevskii equation* (GP), given by:[12]

$$\mathrm{i}\partial_t\psi = -\frac{1}{2\mathfrak{m}}\partial_{xx}\psi + \frac{1}{2}\mathfrak{h}|\psi|^2\psi. \tag{4.20}$$

The cubic NLS equation (4.20) is the one that has a before-mentioned zoo of solutions in the complex Hilbert space $\mathcal{H}(\mathbb{C})$.

We remark here that a slightly different form of the cubic NLS equation (4.20) was previously derived in a different way, from the linear Schrödinger equation governing a *free quantum particle*: $\mathrm{i}\partial_t\psi = -\frac{1}{2}\partial_{xx}\psi$, which was then put into a nonlinear feedback loop[13] that gave rise to the peculiar cubic NLS-term: $|\psi|^2\psi$ (see [IR12] and references therein). This NLS-form has been recently implemented[14] as a Crowd Dynamics Simulator, IP of Australian Defence.

Furthermore, the NLS (4.20) can be spatially-discretized to give the following lattice model,[15] or discrete NLS (DNLS), in the complex plane $\mathbb{C}$: (see, e.g. the Kerr (cubic) case in [OB07])

$$\dot{\psi}_i(t) = \mathrm{i}\,w_1\left[\psi_{i+1}(t) + \psi_{i-1}(t) + w_2\psi_i^3(t)\right], \qquad (i = 1,...N;\ \dot{\psi} = \partial_t\psi) \tag{4.21}$$

where the two weights (i.e., best fit parameters w_1, w_2) are defined by: $w_1 = 1/\mathfrak{m}$, $w_2 = 2\mathfrak{h}$.

The lattice DNLS model (4.21) represents a set of differential-difference equations, an N-dimensional (ND) chain in the complex plane $\mathbb{C}$. This is the basic model for the adaptive quantum oscillator (AQO). It is meant for numerical (iterative) solution, using the sophisticated Cash-Karp (RKF) numerical integrator, starting from random initial waves: $\psi_i(0)$ and random initial weights: (w_1, w_2). The numerical solution of (4.21) is meant to evolve along a certain time-period in such a way that at every adaptive time-step dt,

---

[12] Note that the standard form of the non-dimensional *cubic NLS* is:

$$\mathrm{i}\partial_t\psi = -\frac{1}{2}\Delta\psi \pm |\psi|^2\psi.$$

The sign + in front of the cubic nonlinearity $|\psi|^2\psi$ represents *defocusing NLS*, while the − sign represents *focusing NLS*.

[13] Every feedback loop is necessarily nonlinear.

[14] using the numerical *method of lines*, so that each single line became the controller for a single agent

[15] This is a simplified analog to the method of lines.

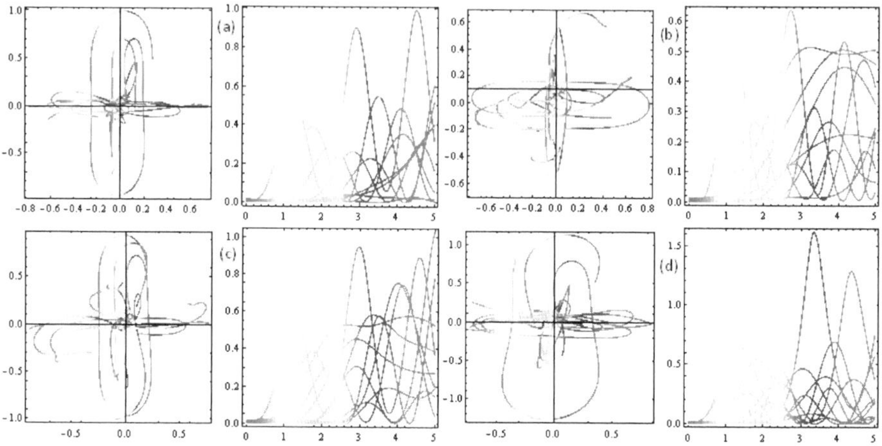

Fig. 4.6. Simulation of the lattice DNLS model (4.21) in Mathematica® gives different AQO-signatures of its quantum random field, *presented here in four sub-figures (a)-(d). The signatures are slightly different from each other, depending on random initial and boundary conditions. Fields are shown on the left, while the corresponding PDFs are on the right of each sub-figure. The code for this simulation is given below. While here we use NDSolve ODE-solver, we also have the same DNLS-implementation in several native and managed programming languages (using RKF (Cash-Karp) numerical ODE-solver).*

an efficient LM-optimizer (see next section) is called and the local minimum (w_1, w_2)-values are adopted for the continuation of the time evolution. In this way, different AQO-signatures (see Figure 4.6) are obtained, depending on numerical values of the weights (w_1, w_2).

Mathematica Code for the Adaptive Quantum Oscillator (Random Field)

```
tt := Table[n = 10; Tfin = 5; w1 = 1; w2 = 1;

DNLS = Table[{Derivative[1][Subscript[\[Psi], i]][t] ==
I*w1*(Subscript[\[Psi], i + 1][t] + Subscript[\[Psi], i-1][t]
+ w2*Subscript[\[Psi], i][t]^3),
Subscript[\[Psi], i][0] == 0.1*Random[Complex]}, {i,n}];
Subscript[\[Psi], 0][t] = Sin[3*Random[]*t];
Subscript[\[Psi], n + 1][t] = I*Sin[3*Random[]*t];

sol = Quiet[NDSolve[Flatten[DNLS],
Table[Subscript[\[Psi], i], {i, n}], {t, Tfin}]];

pl1 = Table[Subscript[g, i] = ParametricPlot[
    Evaluate[{Re[Subscript[\[Psi], i][t]],
```

```
    Im[Subscript[\[Psi], i][t]]} /. sol], {t, 0, Tfin},
    PlotRange -> All, PlotStyle -> AbsoluteThickness[1.5],
    Frame -> True, ColorFunction -> (Hue[1*#1] & ),
    DisplayFunction -> Identity], {i, n}];

pl2 = Table[Subscript[f, i] =
    Plot[Evaluate[Abs[Subscript[\[Psi], i][t]^2] /. sol],
    {t, 0, Tfin}, PlotRange -> All, PlotStyle ->
    AbsoluteThickness[1.5], Frame -> True,
    ColorFunction -> (Hue[1*#1] & ), DisplayFunction ->
    Identity], {i, n}]; {Show[pl1, AspectRatio -> 1,
    ImageSize -> 200], Show[pl2, AspectRatio -> 1,
    ImageSize -> 200]}, {j, 1, 1}];
```

Finally, our sought-for complex ND AQO is readily obtained from the basic AQO-model (4.21), using the following complex-valued optimization procedure. Namely, the pair of weights (w_1, w_2) will not be given *a priori,* but rather computed as a minimum of the *complex cost function* $\mathcal{C} \equiv (w_1, w_2) \in \mathbb{C}$, defined by:

$$\mathcal{C} = \min \frac{1}{2} \sum_{j=1}^{N} z_j^2 = \min \frac{1}{2} \sum_{j=1}^{N} \left(\alpha_j + \mathrm{i}\,\beta_j\right)^2, \tag{4.22}$$

where the components $\left(\alpha_j + \mathrm{i}\,\beta_j\right) \equiv (z_j) \in \mathbb{C}^N$ of the complex ND *weight vector* $\mathbf{z} = (z_j)$ are derived from a specific *domain knowledge,* presented in the form of two sets of weighted rules (one giving α_j-weights and another one giving β_j-weights) of some first-order logic implementation.

4.2.3 Optimization and Learning on Banach and Hilbert Spaces

Nonlinear Least-Squares Optimization

In this subsection, we give an informal and intuitive review of the most powerful algorithm for general least-squares optimization: *Levenberg-Marquardt (LM) method,* and briefly outline three main fields of its application: (i) general data fitting in nonlinear regression statistics, (ii) finite control (output tracking and navigation), and (iii) machine learning (function approximation in multilayered neural networks).

At the core of modern nonlinear optimization problems is the minimization of an ND *cost function* $\mathcal{C}(\varphi)$ defined as a *sum of squares* of the set (or, vector) $\mathbf{f}$ of *residuals:*

$$\mathcal{C}(\varphi) = \frac{1}{2} \sum_{i=1}^{N} f_i(\varphi)^2, \tag{4.23}$$

where $[f_i(\varphi)] = \mathbf{f} : \mathbb{R}^M \to \mathbb{R}^N$ is a nonlinear ND vector function of M parameters φ_m (where $N \geq M$). The values of φ_m-parameters that minimize $\mathcal{C}(\varphi)$

are known as the *best fit* parameters. Here is a C-code example of a highly-nonlinear vector of residuals $\mathbf{f}$, including 7 functions of 7 variables, which have been quickly and very accurately minimized using the LM-algorithm:

```
{
    fvec[0] = tanh(x[0]*x[1]) - 1.7*x[2]
        + log(1.5 + x[3]*x[4]);
    fvec[1] = sin(-4.0*x[0]) - 3.*tanh(x[1]*x[5])
        + 0.1*x[2]*x[4]+x[3]*x[6];
    fvec[2] = 0.5*x[1]*x[3] - tanh(x[2]*x[4])
        + sinh(x[5]*x[6]) + x[2]*x[1];
    fvec[3] = x[5]*x[1] + x[2]*x[3] + x[0]*x[2]
         - x[6]*x[4];
    fvec[4] = sinh(x[0]*x[1]) + x[2]*x[3]
         + x[6]*x[4] - x[3]*x[4];
    fvec[5] = 0.5*x[1]*x[4] - sinh(x[2]*x[3])
        + tanh(x[5]*x[6]) + x[2]*x[1];
    fvec[6] = x[0]*x[1] + x[3]*x[4]
        + x[2]*x[5] - atan(x[3]*x[6]);
}
```

As an illustration, Figure (4.7) shows a sample output of the LM-algorithm, performing an efficient optimization of the vector function $\mathbf{f}$, including 7 highly-nonlinear functions $f_i = \text{fvec}[i]$ of 7 variables $x[i]$, $(i = 0, ..., 6)$. The output, achieved in a split second on an ordinary desktop PC, shows an extremely strong convergence of the Cauchy sequence for the L^2-norm: starting from the initial norm $L^2 \approx 0.4$, it monotonically converges to the final $L^2 \approx 10^{-219}$. This means that the 7-dimensional Banach space $V^7(\mathbf{f})$ is effectively constructed. While next subsection gives rigorous details on optimization on real (Euclidean) Banach spaces and complex (quantum) Hilbert spaces, here we outline the main idea. Very briefly, suppose that we are given to optimize not one, but rather two nonlinear vector functions, $\mathbf{f}_1$ and $\mathbf{f}_2$. If, under the optimization by the LM-algorithm, their corresponding Cauchy sequences converge with their L^2-norm monotonically approaching zero, we have effectively constructed two real N-dimensional Banach spaces, $V_1^N(\mathbf{f}_1)$ and $V_2^N(\mathbf{f}_2)$. Then their inner dot-product $\langle V_1^N, V_2^N \rangle$ forms a complex N-dimensional Hilbert space $\mathcal{H}(\mathbb{C}^N)$, which naturally appears when minimizing complex cost functions (4.22) on the AQO-system (4.21).

For example, in general nonlinear data fitting, the vector $\mathbf{f}$ of residuals is given by:

$$f_i(\boldsymbol{\varphi}) = \frac{x(t_i, \boldsymbol{\varphi}) - x_i}{\sigma_i}, \tag{4.24}$$

where $x(t_i, \boldsymbol{\varphi})$ is a model of the observed data sequence $\{t_i, x_i\}$, which depends on a vector of unknown parameters $\boldsymbol{\varphi} = [\varphi_m]$. The weight vector $\boldsymbol{\sigma} = [\sigma_i]$ measures the uncertainty in observed data, that is the deviation of the model

```
          *** Levenberg-Marquardt Optimizer (V.I., July 2013) ***
Minimizes the sum-of-squares of M(=7) nonlinear functions of N(=7) vars
Effectively minimizes the Euclidean L2-norm:  norm2(f) = sqrt(sum(f(1:M)**2))
If L2 Cauchy minimization sequence is convergent, a Banach space is constructed

Sample initial solution x:
0.00000e+000 0.00000e+000 0.00000e+000 0.00000e+000 0.00000e+000 0.00000e+000 0.
00000e+000

Sample initial vector function f:
4.05465e-001 0.00000e+000 0.00000e+000 0.00000e+000 0.00000e+000 0.00000e+000 0.
00000e+000

--------------------------------
Banach space construction  -->
Cauchy sequence for the L2-norm:
--------------------------------
L2-norm = 4.054651081081644e-001
L2-norm = 3.054651080612439e-001
L2-norm = 1.054651080800816e-001
L2-norm = 4.954581189764459e-011
L2-norm = 2.966234025397137e-026
L2-norm = 2.332738056142589e-034
L2-norm = 2.301529560794739e-042
L2-norm = 3.898305307318532e-050
L2-norm = 2.305448647966667e-058
L2-norm = 2.000366512095712e-066
L2-norm = 1.111652579163495e-074
L2-norm = 3.973336476123791e-083
L2-norm = 1.129530926415996e-090
L2-norm = 2.144042488921030e-098
L2-norm = 4.249209513538372e-106
L2-norm = 3.234237918332130e-114
L2-norm = 1.716811219828818e-122
L2-norm = 1.980808886737351e-130
L2-norm = 3.471499571640520e-138
L2-norm = 1.283798978050475e-146
L2-norm = 3.022742168661423e-154
L2-norm = 3.695373351279775e-162
L2-norm = 5.257880559902580e-170
L2-norm = 1.639060684746655e-178
L2-norm = 2.033735678685008e-186
L2-norm = 8.611867574276758e-195
L2-norm = 8.290247140192018e-203
L2-norm = 9.046835632368352e-211

225 functional evalutations performed

Computed solution x:
6.74070e-220 -7.31495e-219 2.38509e-001 2.70377e-219 5.54778e-219 7.51335e-219 3
.99893e-001

Computed minimum of the vector function f:
0.00000e+000 -1.48274e-219 -6.33432e-221 -1.41287e-219 2.86339e-219 6.14978e-220
 7.10781e-220

Final L2-norm = 3.644332623711892e-219

 Extremely strong convergence : L2-norm <= 1.0e-150
====================================================
 :- Banach space is constructed
```

Fig. 4.7. *Output of the LM-algorithm in C on efficiently minimizing the vector function* $\mathbf{f}$, *including 7 highly-nonlinear functions* $f_i = \mathrm{fvec}[i]$ *of 7 variables* $x[i]$, $(i = 0, ..., 6)$.

$x(t_i, \boldsymbol{\varphi})$ from the observation sequence $\{t_i, x_i\}$ (see, e.g. [TS12] and references therein).

The LM-algorithm is a modification of the Gauss-Newton (GN) method (see [Lev44, Mar63, PTV07]). Recall that the standard GN algorithm is based on a local linearization of the residuals:

$$f_i(\boldsymbol{\varphi} + \delta\boldsymbol{\varphi}) \approx f_i(\boldsymbol{\varphi}) + J_{im}\delta\varphi^m, \tag{4.25}$$

where $\mathbf{J} = [J_{im}] = [\partial f_i/\partial \varphi_m]$ is the Jacobian matrix. The *GN-iteration* is defined by:

$$\delta\varphi = -(\mathbf{J}^T\mathbf{J})^{-1}\nabla\mathcal{C}(\varphi) = -(\mathbf{J}^T\mathbf{J})^{-1}\mathbf{J}^T\mathbf{f}, \qquad (4.26)$$

where $\nabla\mathcal{C}(\varphi) = \mathbf{J}^T\mathbf{f}$ is the gradient of the cost function. The GN-method usually converges quickly if it begins sufficiently near a minimum of $\mathcal{C}(\varphi)$. However, the matrix $\mathbf{J}^T\mathbf{J}$ is often ill-conditioned (with a very wide range of eigenvalues). So, unless the initial guess is very good, the GN method would fail to converge.

To correct the shortcomings of the GN-method, Levenberg and Marquardt suggested (separately) damping the $\mathbf{J}^T\mathbf{J}$ matrix by a diagonal cutoff [Lev44, Mar63]. The *LM-iteration* is defined by:

$$\delta\varphi = -\left(\mathbf{J}^T\mathbf{J} + \lambda\mathbf{D}^T\mathbf{D}\right)^{-1}\nabla\mathcal{C}(\varphi) = -\left(\mathbf{J}^T\mathbf{J} + \lambda\mathbf{D}^T\mathbf{D}\right)^{-1}\mathbf{J}^T\mathbf{f}, \qquad (4.27)$$

where $\mathbf{D}^T\mathbf{D}$ is a diagonal, positive-definite matrix (with the scaled parameters φ_n) and λ is a *damping parameter* adjusted by the LM-algorithm. When the LM-parameter λ is large, the method takes a small step in the gradient direction. As the algorithm comes closer to a solution, λ is chosen to be small and the method converges quickly via the GN-method [TS12]. The LM-method can interpolate between gradient descent and the GN-algorithm for quick and efficient convergence, by properly adjusting the damping parameter λ.

In finite control (i.e., output tracking and navigation) one wants the scalar system output $y(\mathbf{x}, t)$ to follow a continuous model (nominal) trajectory $\chi(t)$, for a given MD system-vector $\mathbf{x}$ and time t. This problem can be expressed as:

$$\min_{\mathbf{x}\in\mathbb{R}^M} \int_{t_0}^{t_1} [y(\mathbf{x}, t) - \chi(t)]^2\, dt, \qquad (4.28)$$

and, upon time discretization (using a suitable quadrature scheme), becomes the following least-squares problem:

$$\min_{\mathbf{x}\in\mathbb{R}^M} f(\mathbf{x}) = \sum_{i=1}^{N} [\tilde{y}(\mathbf{x}, t_i) - \tilde{\chi}(t_i)]^2\, dt,$$

where both discretized functions, $\tilde{y}(\mathbf{x}, t_i)$ and $\tilde{\chi}(t_i)$, include the weights of the chosen quadrature scheme (see [Mat13]). We remark here that the finite control problem (4.28) is formally a minimization of the square of the Banach metric (4.31), see subsection 4.2.3 below. Similarly, a ship navigation problem modeled in the complex plane is formally a minimization of the square of the Hilbert metric (4.34).

Finally, in machine learning (mainly supervised learning of multilayer feed-forward neural networks, see Appendix 4.3.5) the LM-algorithm is mostly suitable for *function approximation* problems[16] where the network has up to

[16] The *universal approximation theorem* of Kolmogorov states [Hay94]:
Let $\phi(\cdot)$ be a nonconstant, bounded, and monotone-increasing continuous (C^0)

a few hundred weights and the approximation must be very accurate; if this is the case, then on the average the LM-algorithm is over four times faster than any other known algorithm (see [BHD13]). In multilayer feedforward neural networks, given the weight-vector $\mathbf{w}(n) = [w_0(n), ..., w_D(n)]^T$, if we put

$$\nabla \mathbf{e} = \mathbf{J}^T \mathbf{e},$$

where $\mathbf{e}$ is a vector of network errors and $\nabla \mathbf{e} = \left[\frac{\partial e}{\partial w_0}, ..., \frac{\partial e}{\partial w_D}\right]^T$ is the MSE-gradient (see Appendix 4.3.5), we obtain the *LM-learning algorithm*:

$$\mathbf{w}(n+1) = \mathbf{w}(n) - [\mathbf{J}^T\mathbf{J} + \eta \mathbf{I}]^{-1}\mathbf{J}^T\mathbf{e}, \qquad (4.29)$$

where η is the learning rate (step size).

Optimization-Based Control on Complex Hilbert spaces

Now we will try and move our rigorous optimization, control and learning methodology back to the complex Hilbert space. To start width, let us consider the vector space $V(f)$ of real-valued square-integrable functions f defined on an interval $[a, b] \subset \mathbb{R}$ (i.e., functions with the convergent integral: $I(f) = \int_a^b |f(x)|^2 dx < \infty$).

To be able to perform the analysis on Banach and Hilbert spaces, we first need to make $V(f)$ a normed space. However, a naive approach will not work here, because the most natural choice of a norm,

$$\|\cdot\| : f \mapsto \sqrt{\int_a^b |f(x)|^2 dx}, \qquad (4.30)$$

is not a proper norm on $V(f)$, which can be seen, e.g. in case:

$$f(x) = \begin{cases} 0 \text{ for } & x \neq a \\ 1 \text{ for } & x = a \end{cases} \implies \|f\| = 0, \text{ but } f \neq 0,$$

so it violates the **N1**-axiom (of positive definiteness). However, the problem is solved with a little help from measure theory, namely, (4.30) becomes a

---

function. Let I^N denote an ND unit hypercube $[0,1]^N$. The space of C^0–functions on I^N is denoted by $C(I^N)$. Then, given any function $f \in C(I^N)$ and $\epsilon > 0$, there exist an integer M and sets of real constants $\alpha_i, \theta_i, \omega_{ij}$, $i = 1, \dots, M$; $j = 1, \dots, N$ such that we may define

$$F(x_1, \dots, x_N) = \alpha_i \phi(\omega_{ij} x_j - \theta_i),$$

as an approximate realization of the function $f(\cdot)$; that is

$$|F(x_1, \dots, x_N) - f(x_1, \dots, x_N)| < \epsilon \qquad \text{for all} \quad \{x_1, \dots, x_N\} \in I^N.$$

proper norm if we identify functions f which are equal almost everywhere (i.e., differ only on a set of measure zero in $[a, b]$). The resulting vector space of equivalence classes $[f]$ is (usually) denoted by $L^2[a, b]$ and is an infinite-dimensional Banach space with the norm of the equivalence class $[f]$ defined by:

$$L^2[a, b] = \|[f]\| = \sqrt{\int_a^b |f(x)|^2 dx}.$$

For the proof of the necessary completeness, see e.g. [Roy68].

The corresponding *Banach metric* (i.e., distance) between any two real-valued square-integrable functions f, g defined on an interval $[a, b] \subset \mathbb{R}$, associated with the norm $L^2[a, b]$ reads:

$$\mathbf{d}(f, g) = \sqrt{\int_a^b |f(x) - g(x)|^2 dx}. \tag{4.31}$$

For example, the finite control problem (4.28) is actually a minimization of the square of the Banach metric (4.31).

Next, if we define the inner product:

$$\langle f, g \rangle = \int_a^b f(x) g(x) dx,$$

$L^2[a, b]$ becomes the infinite-dimensional real Hilbert space $\mathcal{H}(\mathbb{R})$, which is used for statistical learning theory of *support vector machines*.

Finally, if we repeat this whole procedure (while instead of $f \in \mathbb{R}$) using complex-valued square-integrable wave-functions $\psi \in \mathbb{C}$ defined in a complex-plane region $[a, b] \subset \mathbb{C}$, with

$$\langle \psi, \varphi \rangle = \int_a^b \psi(x) \overline{\varphi(x)} dx, \tag{4.32}$$

we are landing at the infinite-dimensional quantum Hilbert space $L^2([a, b], \mathbb{C}) \equiv \mathcal{H}(\mathbb{C})$ of both QHO and AQO, with the L^2-norm defined by:

$$L^2([a, b], \mathbb{C}) = \|[\psi]\| = \sqrt{\int_a^b |\psi(x)|^2 dx}. \tag{4.33}$$

The corresponding *Hilbert metric* (i.e., distance) between any two complex-valued square-integrable functions f, g defined on an region $[a, b] \subset \mathbb{C}$, associated with the norm $L^2([a, b], \mathbb{C})$ reads:

$$\mathbf{d}(\psi, \varphi) = \sqrt{\int_a^b |\psi(x) - \varphi(x)|^2 dx}. \tag{4.34}$$

For example, the *ship-navigation problem* mentioned above can be formulated in the complex-plane $\mathbb{C}$ as a minimization of the square of the Hilbert metric (4.34).

Summary

In this section we have proposed a unified theoretical and computational framework for developing and implementing the following five practical engineering problem-solving tools: numerical simulation of ODEs and PDEs, general least-squares optimization, machine learning, finite navigation control, and first-order logic.

The rigorous theoretical framework for optimization-based learning is proposed in the form of both finite- and infinite dimensional real Banach spaces and complex Hilbert spaces.

The central physical/computational object is a two-core adaptive quantum oscillator (AQO, RKF-integrator + LM-optimizer), designed for simulation of quantum random fields, nonlinear optimization, learning and control, while an external first-order logic module is currently under implementation (in both Erlang and Haskell, to be called separately). Both AQO-cores are very fast, designed for high-dimensional, real-life problems, and can be called from any other languages.

4.3 Appendix

4.3.1 Complex-Valued Image Processing

Complex Fourier Transform Approach

A linear (albeit sophisticated and elegant) 'distant relative' of our nonlinear crowd dynamics in the complex plane is a *complex-valued image processing*, in which every pixel of a 2D image is represented by its own complex number $c \in \mathbb{C}$. For the sake of completeness, in this subsection we give a brief review of this approach that is commonly used in computer vision, as well as in satellite- and neuro-imaging.

Basically, given two real-valued square-integrable signals or functions $f, g \in L^2(\mathbb{R})$, their inner product is defined by:

$$\langle f, g \rangle = \int_{\mathbb{R}} f(x)\overline{g(x)}\, dx,$$

while the *Fourier transform* (FT) of f, denoted by $F = \mathcal{F}[f(x)]$, is given by:

$$\mathcal{F}[f(x)] = \hat{f}(\xi) = \int_{\mathbb{R}} f(x)\, \mathrm{e}^{-2\pi \mathrm{i}\xi x} dx, \qquad (4.35)$$

such that the original signal $f(x)$ is recovered by the *inverse Fourier transform* (IFT) of $\hat{f}(\xi)$:

$$f(x) = \mathcal{F}^{-1}[\hat{f}(\xi)] = \int_{\mathbb{R}} \hat{f}(\xi)\, \mathrm{e}^{2\pi \mathrm{i}\xi x} df.$$

Without going into technical details about image processing in the frequency space, based on either *discrete Fourier transform* (DFT) or *fast Fourier transform* (FFT)[17] (see [Wik13b, II12], or any textbook on digital image processing; e.g., common are those with applications in *Matlab*®, like [SB10]), we focus here on the basic DFT-algorithm[18] with complex Fourier components: $c_k = (a_k - ib_k)/2$, for $k = 0, ..., N-1$ (where N is the number of samples of the signal function $f(n)$, defined by the formula:

$$c_k = \frac{1}{N}\sum_{n=0}^{N-1}\left[f(n)\exp\left(-\frac{\mathrm{i}2\pi kn}{N}\right)\right], \qquad \text{where} \tag{4.36}$$

$$\exp\left(-\frac{\mathrm{i}2\pi kn}{N}\right) = \cos\left(\frac{2\pi kn}{N}\right) - \mathrm{i}\sin\left(\frac{2\pi kn}{N}\right).$$

According to (4.36), each complex Fourier component c_k is a complex sum of complex products of the signal $f(n)$ and complex vectors spinning around the unit circle $z \leq 1$ in the $\mathbb{C}$-plane. For example in C#, the complex DFT formula (4.36) can be given the following *naive* implementation (see [Mos13]):

```
public void ComplexDFT(int N) {
  int k, n;  TKomplex w;
  if (N > 0)
  {
    for (k = 0; k < N; k++)
    {
      c[k].real = 0;  c[k].imag = 0;
      for (n = 0; n < N; n++)
      {
        w.real = Math.Cos((double)(2.0 * Math.PI
            * (double)(k * n) / (double)(N)));
        w.imag = -Math.Sin((double)(2.0 * Math.PI
            * (double)(k * n) / (double)(N)));
        c[k] = ksum(c[k], kprod(w, y[n]));
      }
      c[k].real = c[k].real / (double)(N) * 2.0;
      c[k].imag = -c[k].imag / (double)(N) * 2.0;
    }
  }
  c[0].real = c[0].real / 2;
  c[0].imag = c[0].imag / 2;
```

[17] Recall that an FFT is an efficient algorithm to compute the discrete Fourier transform (DFT) and its inverse. The most commonly used FFT approach is the Cooley–Tukey algorithm [CT65], a *divide and conquer algorithm* that recursively breaks down a DFT of any composite size $N = N_1N_2$ into many smaller DFTs of sizes N_1 and N_2.

[18] A DFT decomposes a sequence of values into components of different frequencies.

```
}
```

A more efficient complex DFT implementation, firstly builds the look-up table consisting of the cosine (real) and sine (imaginary) terms:

```
for (k = 1; k < N; k++) {
   we[k].real = Math.Cos((2.0 * Math.PI
        * (double)(k) / (double)(N)));
   we[k].imag = -Math.Sin((2.0 * Math.PI
        * (double)(k) / (double)(N)));
}
```

– and secondly, reads the above look-up table and implements formula (4.36) using the following code (see [Mos13]):

```
public void ComplexDFT() {
  int k, n;
  if (N > 0)
  {
    for (k = 0; k < N; k++)
    {
      c[k].real = 0;  c[k].imag = 0;
      for (n = 0; n < (N - 1); n++)
      {
        c[k] = ksum(c[k], kprod(we[(k * n) % N], y[n]));
      }
      c[k].real = c[k].real / N * 2;
      c[k].imag = -c[k].imag / N * 2;
    }
    c[0].real = c[0].real / 2;
    c[0].imag = c[0].imag / 2;
  }
}
```

Using a similar (albeit more efficient) DFT code, for both forward and inverse Fourier transforms, one can perform *low-pass and high-pass filtering* of the complex-valued image, following the procedural prescription implemented in the popular AForge.NET library for complex-valued image processing[19] and machine learning (see [Kir07]):

```
// 1. Create a complex image from a bitmap
AForge.Imaging.ComplexImage cimage =
    AForge.Imaging.ComplexImage.FromBitmap( bitmap );
// 2. Perform the forward FFT
cimage.ForwardFourierTransform( );
```

[19] Here, a C# class is used to keep image represented in complex numbers suitable for FFT.

```
// 3. Get the frequency view
System.Drawing.Bitmap img = cimage.ToBitmap( );
// 4. Perform the lowpass or hipass filtering
cimage.FrequencyFilter( new Range( 0, 100 ) );
// 5. Perform the backward FFT (or, IFFT)
cimage.BackwardFourierTransform( );
// 6. Get the filtered image
System.Drawing.Bitmap img = cimage.ToBitmap( );
```

Complex Wavelet Transform Approach

Uncertainty principle of signal processing

Recall that the purpose of the *wavelet transform* (WT) is to provide information about both time t and (angular) frequency w of a signal, which satisfy the fundamental *uncertainty principle*[20] of signal processing:

$$\Delta t \Delta \omega \geq \frac{1}{2},$$

which means that the higher the required resolution of a signal in time, the lower is its resolution in frequency.[21]

The wavelet-transformed signal provides information about the time and the frequency. Therefore, wavelet-transformation contains information similar to the *short-time Fourier transform* (STFT),[22] but with additional special properties of the wavelets, which show up at the resolution in time at higher analysis frequencies of the basis function. In contrast to the FT, the WT offers the 2D expansion for a time–dependent signal with the scale and translation parameters which are interpreted physically one employs the so-called *mother*

[20] This is analogous to Heisenberg's *uncertainty principles of quantum mechanics*: (i) $\Delta q \Delta p \geq \frac{\hbar}{2}$ between coordinate (q) and momentum (p), and (2) $\Delta t \Delta E \geq \frac{\hbar}{2}$ between time (t) and energy (E).

[21] In case of large Δt, we have: (i) bad time resolution, (ii) good frequency resolution, and (iii) low frequency, large scaling factor. In case of small Δt, we have: (i) good time resolution, (ii) bad frequency resolution, and (3) high frequency, small scaling factor.

[22] Recall that Fourier transform (FT) requires that a signal to be examined is stationary, not giving the time evolution of the frequency pattern. However, most real-life (e.g., biological) signals are usually non-stationary. This FT-limitation can be partly resolved by using the STFT. Assuming that the signal is quasi–stationary in the narrow time period, the FT is applied with time–evolving narrow windows. Then STFT yields the time evolution of the frequency spectrum. The STFT, however, has a critical limitation violating the above *uncertainty principle*, which asserts that if the window is too narrow, the frequency resolution will be poor whereas if the window is too wide, the time resolution will be less precise. This limitation becomes serious for signals with much transient components, like spike signals.

wavelet which is localized in both frequency and time domains. The WT expansion is carried out in terms of a family of wavelets which is made by dilation and translation of the mother wavelet. The time evolution of frequency pattern can be followed with an optimal time–frequency resolution. The WT appears to be an ideal tool for analyzing signals of a non– stationary nature.

Briefly, there are two types of WTs: one is the *continuous wavelet transform* (CWT) and the other the *discrete wavelet transform* (DWT). In the former the parameters denoting the scale and translation are continuous variables while in the latter they are discrete variables.

CWT

The CWT[23] for a given regular function $f(t)$ can be defined, using the quantum-like notation, by the inner product in the complex Hilbert space $\mathcal{H}(\mathbb{C}) = L^2(\mathbb{C})$ (see [Has02]):

$$c(a,b) = \int \bar{\psi}_{ab}(t) f(t) dt \;\; = \;\; \langle \psi_{ab}(t), f(t) \rangle \,,$$

where $\psi_{ab}(t)$ is the family of wavelets generated by:

$$\psi_{ab}(t) = |a|^{-1/2} \psi(\frac{t-b}{a}), \tag{4.37}$$

in which $\psi(t)$ is the *mother wavelet*,[24] while a and b express the scale change and translation, respectively, and they physically stand for the inverse of the frequency and the time. Then the CWT transforms the time–dependent function $f(t)$ into the frequency- and time-dependent function $c(a,b)$. The mother wavelet (e.g., Mexican Hat wavelet, Meyer wavelet, or Morlet wavelet) is a smooth function with good localization in both frequency and time spaces. A wavelet family given by (4.37) plays a role of elementary function, representing the function $f(t)$ as a superposition of wavelets $\psi_{ab}(t)$.

The inverse wavelet transform (IWT) can then be given by:

$$f(t) = C_\psi^{-1} \int \frac{da}{a^2} \int c(a,b) \psi_{ab}(t) db,$$

when the mother wavelet satisfies the following two conditions:

(i) the admissibility condition given by

$$0 < C_\psi < \infty, \qquad \text{with} \qquad C_\psi = \int_{-\infty}^{\infty} |\hat{\Psi}(\omega)|^2 / |\omega| \, d\omega,$$

---

[23] For more introductory information about continuous wavelet transforms (starting from orthonormal wavelets that can be used to define a basis for the Hilbert space of (real or complex) square integrable functions), see [Wik13e, II12].

[24] One of the advantages of the WT over FT is that one can choose a proper mother wavelet among many mother wavelets, depending on a signal to be analyzed.

where $\hat{\Psi}(\omega)$ is the Fourier transform of $\psi(t)$, and

(ii) the zero mean of the mother wavelet:

$$\int_{-\infty}^{\infty} \psi(t)dt = \hat{\Psi}(0) = 0.$$

DWT

From computational perspective, much more interesting is the DWT, which is an invertible WT that permits sparse signal decompositions at a low computational cost (see [Mal09] for a general overview). Again using the quantum-like notation, the DWT can be defined for *discrete* values of $a = 2^j$ and $b = 2^j k$ $(j, k \in \mathbb{Z})$ as

$$c_{jk} \equiv c(2^j, 2^j k) = \left\langle \psi_{jk}(t),\ f(t) \right\rangle, \qquad \text{with} \qquad \psi_{jk}(t) = 2^{-j/2}\psi(2^{-j}t - k). \tag{4.38}$$

The orthonormal condition for the wavelet functions is given by (see [Has02])

$$\left\langle \psi_{jk}(t), \psi_{j'k'}(t) \right\rangle = \delta_{jj'}\delta_{kk'},$$

which leads to the *inverse DWT*:

$$f(t) = \sum_j \sum_k c_{jk}\psi_{jk}(t). \tag{4.39}$$

MRA

In the *multiresolution analysis* (MRA) of the DWT, one introduces a scaling function $\phi(t)$, which satisfies the recurrent relation with $2K$ masking coefficients, h_k, given by:

$$\phi(t) = \sqrt{2} \sum_{k=0}^{2K-1} h_k \phi(2t - k),$$

with the normalization condition for $\phi(t)$ given by:

$$\int \phi(t)dt = 1.$$

A family of wavelet functions is generated by:

$$\psi(t) = \sqrt{2} \sum_{k=0}^{2K-1} (-1)^k h_{2K-1-k} \phi(2t - k).$$

The scaling and wavelet functions satisfy the ortho–normal relations:

$$\langle \phi(t), \phi(t - m) \rangle = \delta_{m0}, \qquad \langle \psi(t), \psi(t - m) \rangle = \delta_{m0}, \qquad \langle \phi(t), \psi(t - m) \rangle = 0.$$

A set of masking coefficients h_j is chosen so as to satisfy the conditions shown above. Here, $g_k = (-1)^k h_{2M-1-k}$, which is derived from the orthonormal relations among wavelet and scaling functions.

The simplest wavelet function for $K = 1$ is the *Harr wavelet* for which we get $h_0 = h_1 = 1/\sqrt{2}$, and

$$\begin{aligned} \psi_{\mathrm{H}}(t) &= 1, && (\text{for } 0 \le t < 1/2) \\ &= -1, && (\text{for } 1/2 \le t < 1) \\ &= 0, && \text{otherwise.} \end{aligned}$$

In the more sophisticated wavelets like the *Daubechies wavelet*, an additional condition given by

$$\int t^\ell \psi(t) dt = 0, \quad (\text{for } \ell = 0, 1, 2, 3..., L-1)$$

is imposed for the smoothness of the wavelet function. Furthermore, e.g., in the *Coiflet wavelet*, a similar smoothing condition is imposed also for the scaling function as

$$\int t^\ell \phi(t) dt = 0, \quad (\text{for } \ell = 1, 2, 3..., L'-1) \,.$$

Once WT coefficients are obtained, we can calculate various quantities such as auto– and cross–correlations and SNR, as will be discussed shortly. In principle the expansion coefficients c_{jk} in DWT may be calculated by using (4.38)-(4.39) for a given function $f(t)$ and an adopted mother wavelet $\psi(t)$.

For example, a 1D forward and inverse Harr wavelets can be respectively implemented in C# as:

```
public void ForwHarr(double[] dat) {
    double[] tmp = new double[dat.Length];
    int h = dat.Length >> 1;
    for (int i = 0; i < h; i++) {
        int k = (i << 1);
        tmp[i] = dat[k] * s0 + dat[k + 1] * s1;
        tmp[i + h] = dat[k] * w0 + dat[k + 1] * w1;
    }
    for (int i = 0; i < dat.Length; i++)  dat[i] = tmp[i];
}

public void BackHaar(double[] dat) {
    double[] tmp = new double[dat.Length];
    int h = dat.Length >> 1;
    for (int i = 0; i < h; i++) {
        int k = (i << 1);
        tmp[k] = (dat[i] * s0 + dat[i + h] * w0) / w0;
```

```
            tmp[k + 1] = (dat[i] * s1 + dat[i + h] * w1) / s0;
        }
        for (int i = 0; i < dat.Length; i++)  dat[i] = tmp[i];
    }
```

DT-CWT

The DWT has been successfully employed in many applications, including image compression, noise reduction and speech recognition. However, in the area of *statistical signal processing*, the DWT is less effective due to the high *translation sensitivity* of the DWT[25] (see [BDS12] and references therein). In particular, for sophisticated complex-valued signal and image analysis, the best currently-available tool is the so-called *dual-tree complex wavelet transform* (DT-CWT), which exhibits better shift-invariance than the conventional DWT.[26] In this subsection, mainly following [BDS12], we briefly describe main features of DT-CWT, based on the *Hilbert transform* (HT) $\mathcal{H}$, which is itself defined by its relation to the Fourier transform (4.35):

$$\mathcal{H}\{\mathcal{F}[f(x)]\} = \widehat{\mathcal{H}f}(\xi) = -\mathrm{i}\,\mathrm{sign}(\xi)\hat{f}(\xi).$$

The Hilbert transform is orthogonal to the signal, commutes with translations and positive dilatations, and $\mathcal{H}^{-1} = -\mathcal{H}$.

Let $\{\psi_{j,k}\}_{j,k\in\mathbb{Z}}$ and $\{\psi'_{j,k}\}_{j,k\in\mathbb{Z}}$ be two real-valued bi-orthogonal wavelet systems that form a *Hilbert-transform pair*, i.e., $\psi' = \mathcal{H}\psi$. We define the wavelet coefficients of f with respect to these wavelet systems by (see [BDS12]):

$$a_j[k] = \langle f, \psi_{j,k}\rangle \quad \text{and} \quad b_j[k] = \langle f, \psi'_{j,k}\rangle, \qquad (\text{for all } j,k \in \mathbb{Z}),$$

from which the following wavelet identities follow:

$$f = \sum_{j,k\in\mathbb{Z}} a_j[k]\tilde{\psi}_{j,k} \quad \text{and} \quad f = \sum_{j,k\in\mathbb{Z}} b_j[k]\tilde{\psi}'_{j,k},$$

where $\tilde{\psi}_{j,k}$ and $\tilde{\psi}'_{j,k}$ represent the dual wavelets of $\psi_{j,k}$ and $\psi'_{j,k}$ respectively.

Now we can introduce the *complex wavelets:*

---

[25] Small shifts in the input signal may completely change the wavelet coefficient pattern; therefore, algorithms based on the DWT need to recognize a wide variety of different wavelet patterns.

[26] The near shift-invariance property of the DT-CWT has been extensively studied over the last decade (see [BDS12] and references therein). In particular, an amplitude-phase representation for dual-tree complex wavelet transforms that involve modulated wavelets was studied in [CU10], linking the multiresolution framework of the wavelet components to the frequency decomposition through Fourier analysis, which demonstrated the improved shiftability of the DT-CWT.

$$\Psi_{j,k} = \frac{\psi_{j,k} + i\psi'_{j,k}}{2} \quad \text{and} \quad \tilde{\Psi}_{j,k} = \frac{\tilde{\psi}_{j,k} + i\tilde{\psi}'_{j,k}}{2},$$

for which DT-CWT *coefficients* are given by (see [BDS12]):

$$c_j[k] = \langle f, \Psi_{j,k} \rangle = \tfrac{1}{2}(a_j[k] - ib_j[k]), \qquad (\text{for all } \ j, k \in \mathbb{Z}).$$

In particular, for *dyadic wavelet transforms*, the level j coefficients of a shifted signal $f(\cdot + s)$ with $s = 2^{-j}m$, $m \in \mathbb{Z}$, can be easily predicted from the coefficients of the reference signal. They satisfy the following relations:

$$c_j^s[k] = \big\langle f(\cdot + 2^{-j}m), \Psi_{j,k} \big\rangle = \big\langle f, \Psi_{j,k}(\cdot - 2^{-j}m) \big\rangle = \langle f, \Psi_{j,k+m} \rangle = c_j[k+m].$$

This well-known property can be adapted for arbitrary shifts s by decomposing s into a dyadic number $2^{-j}m$ and some remainder h with $|h| < 2^{-j}$: $s = 2^{-j}m + h$. Then we have:

$$c_j^s[k] = \langle f(\cdot + s), \Psi_{j,k} \rangle = \langle f(\cdot + h), \Psi_{j,k+m} \rangle = c_j^h[k+m].$$

The *adjusted shift error*, given by:

$$|c_j[k+m] - c_j^s[k]| = |c_j[k+m] - c_j^h[k+m]|,$$

is in general much smaller than the original shift error $|c_j[k] - c_j^s[k]|$. In order to further reduce the shift error $|c_j[k+m] - c_j^h[k+m]|$, we will perform a phase change of $c_j[k+m]$ over an angle ϕ_h that partially compensates for the small shift h ($|h| < 2^{-j}$), so that

$$c_j^s[k] = c_j^h[k+m] \approx e^{i\phi_h} c_j[k+m].$$

As suggested in [CU10], the assumption that the involved wavelet ψ is *modulated* is made here, i.e.,

$$\psi(x) = w(x)\cos(\omega_0 x + \xi_0)$$

for $\omega_0, \xi_0 > 0$ where the localization window w is bandlimited to $[-\Omega, \Omega]$ for some $\Omega < \omega_0$. Examples of modulated wavelets are the *Shannon wavelets* and *Gabor wavelets*. As the orthonormal spline, resp. B-spline, wavelets resemble the Shannon, resp. Gabor, wavelet, they can be seen as a kind of modulated wavelets. Using the so-called *Bedrosian identity* (see [CU10]), one can show that

$$\psi'(x) = w(x)\sin(\omega_0 x + \xi_0).$$

In this way, we obtain the identity

$$\Psi(x) = \frac{e^{i\xi_0}}{2} w(x) e^{i\omega_0 x}.$$

For more technical details on DT-CWT, including the study of the *phase-compensated shift error*, see [BDS12] and references therein.

Application: Handwritten signature identification

A method for handwritten signature identification, based on complex wavelet filters was proposed in [SK11]. Use of the rotated complex wavelet filters were proposed in conjunction with DT-CWT, to derive signature feature extraction, which captures information in twelve different directions. The proposed method proved to be superior over conventional DWT.

The following two algorithms were used for DWT and DT-CWT, respectively:

```
Algorithm 1: Feature datbase creation using DWT Input:
    Signature image datbase: DB
    1D filters : LF, HF
    Handwritten Signature : Si
Output:
    Feature datbase FV
Begin
    For each Si in DB do
        Decompose the Si by applying low pass LF and
        high pass HF filters up to 6th level
        Calculate energy E and standard deviation SD for
        each wavelet-decomposed subband
        respectively in each level
        Feature vector f = [E U SD]
        FV=FV U f
    End for
End

Algorithm 2: Feature datbase creation using DT-CWT Input:
    Signature image datbase: DB
    2D DT-CWT filters : F
    Handwritten Signature : Si
Output:
    Feature datbase : FV
Begin
    If DT-RCWF
        Rotate 2D filters F by 45 0
    End if
    For each Si in DB do
        Decompose the Si by applying 2D filters F up to 6th
        Level. Calculate energy E and standard deviation SD
        for each wavelet-decomposed subband respectively in
        each level
```

```
        Feature vector f = [E U SD]
        FV=FV U f
    End for
End
```

For more technical details on signature feature extraction using DWT and DT-CWT, see [SK11].

4.3.2 Linear Integral Equations

As a reference, we give here a short review of common integral equations. For more technical details, see [PZ04] and references therein.

Electro-Mechanical Systems

In electrical and control engineering (in particular in robotics and mechatronics) various electro-mechanical input-output systems are ubiquitous. Some of them are most naturally described by integro-differential equations, while others (usually overdamped ones) are usually given by integral equations.

Integro-Differential Equations

We start with several examples of linear integro-differential equations arising in electrical and mechanical engineering:

- Series RLC-circuit with voltage-input $u(t)$, resistance R, inductance L, capacitance C and current $i(t)$, is governed by the following integro-differential equation:[27]

  $$L\frac{di(t)}{dt} + R\,i(t) + C^{-1}\int_0^t i(t)\,dt = u(t).$$

  Applcation of the Laplace transform $\mathcal{L}$ to this equation (assuming zero initial conditions) gives:[28]

  $$Ls\,I(s) + R\,I(s) + (Cs)^{-1}\,I(s) = U(s),$$

  which implies the impedance of the series RLC-circuit:

  $$Z(s) = \frac{U(s)}{I(s)} = Ls + R + (Cs)^{-1}.$$

[27] In terms of electric charge $q(t)$, given by $i(t) = \dot{q}(t)$, this equation becomes the 2nd-order ODE:

$$L\,\ddot{q}(t) + R\,\dot{q}(t) + C^{-1}q(t) = u(t).$$

[28] We denote: $\mathcal{L}[g(t)] = G(s)$.

- Translational mass-spring-damper with force-input $f(t)$, mass M, damper viscosity B, spring stiffness K and velocity $v(t)$, is governed by the integro-differential equation:[29]

$$M\frac{dv(t)}{dt} + B\,v(t) + K\int_0^t v(t)\,dt = f(t).$$

In the Laplace transform, this equation reads:

$$M s\,V(s) + B\,V(s) + (K/s)\,V(s) = F(s),$$

which implies the mechanical impedance of the translational mass-spring-damper:

$$Z(s) = \frac{F(s)}{V(s)} = Ms + B + \frac{K}{s}\,.$$

- Rotational mass-spring-damper with torque-input $\tau(t)$, inertia-moment J, angular damper B, angular spring stiffness K and angular velocity $w(t)$, is governed by the integro-differential equation:[30]

$$J\frac{dw(t)}{dt} + D\,w(t) + \kappa\int_0^t w(t)\,dt = \tau(t).$$

In the Laplace transform, this equation reads:

$$Js\,W(s) + D\,W(s) + (\kappa/s)\,W(s) = T(s),$$

which implies the mechanical impedance of the rotational mass-spring-damper:

$$Z(s) = \frac{T(s)}{W(s)} = Js + D + \frac{\kappa}{s}\,.$$

- Linear electric RLC-network is governed by the following system of integro-differential equations:

$$\begin{aligned}
z_{11}i_1 + z_{12}i_2 + \dots + z_{1n}i_n &= u_1(t)\\
z_{21}i_1 + z_{22}i_2 + \dots + z_{2n}i_n &= u_2(t)\\
\dots \quad \dots \quad \dots \quad &\dots \quad \dots\\
z_{n1}i_1 + z_{n2}i_2 + \dots + z_{nn}i_n &= u_n(t),
\end{aligned}$$

[29] In terms of displacement $x(t)$, given by $v(t) = \dot{x}(t)$, this equation becomes the 2nd-order ODE:

$$M\,\ddot{x}(t) + B\,\dot{x}(t) + K\,x(t) = f(t).$$

[30] In terms of angle $\varphi(t)$, given by $w(t) = \dot{\varphi}(t)$, this equation becomes the 2nd-order ODE:

$$J\,\ddot{\varphi}(t) + D\,\dot{\varphi}(t) + \kappa\,\varphi(t) = \tau(t).$$

with integro-differential operators:

$$z_{jk}i_k = L_{jk}\frac{di_k(t)}{dt} + R_{jk}\, i_k(t) + C_{jk}^{-1}\int_0^t i_k(t)\, dt.$$

In the Laplace transform, this equation reads:

$$L_{jk}s\, I_k(s) + R_{jk}\, I_k(s) + (C_{jk}s)^{-1}\, I_k(s) = U_j(s),$$

which implies the impedance of the RLC-network:

$$Z_{jk}(s) = \frac{U_j(s)}{I_k(s)} = L_{jk}s + R_{jk} + (C_{jk}s)^{-1}.$$

Linear Integral Equations

If we remove derivative terms from the above integro-differential equations representing linear electrical and mechanical systems, we get the following Volterra integral equations:

- Series RC-circuit with voltage-input $u(t)$, resistance R, capacitance C and current $i(t)$, is governed by the following integral equation:

$$R\, i(t) + C^{-1}\int_0^t i(t)\, dt = u(t).$$

  Applying the Laplace transform to this equation (for zero initial conditions) gives:

$$R\, I(s) + (Cs)^{-1}\, I(s) = U(s),$$

  which implies the RC-circuit impedance:

$$Z(s) = \frac{U(s)}{I(s)} = R + (Cs)^{-1}.$$

- Translational overdamped spring with force-input $f(t)$, damper viscosity B, spring stiffness K and velocity $v(t)$, is governed by the integral equation:

$$B\, v(t) + K\int_0^t v(t)\, dt = f(t).$$

  In the Laplace transform, this equation reads:

$$B\, V(s) + (K/s)\, V(s) = F(s),$$

  which implies the mechanical impedance of the translational spring-damper:

$$Z(s) = \frac{F(s)}{V(s)} = B + \frac{K}{s}.$$

- Rotational overdamped spring with torque-input $\tau(t)$, angular damper B, angular spring stiffness K and angular velocity $w(t)$, is governed by the integral equation:
$$D\,w(t) + \kappa \int_0^t w(t)\,dt = \tau(t).$$
In the Laplace transform, this equation reads:
$$D\,W(s) + (\kappa/s)\,W(s) = T(s),$$
which implies the mechanical impedance of the rotational spring-damper:
$$Z(s) = \frac{T(s)}{W(s)} = D + \frac{\kappa}{s}\,.$$

- Linear electric RC-network is governed by the following system of integral equations:
$$\begin{aligned} z_{11}i_1 + z_{12}i_2 + \dots + z_{1n}i_n &= u_1(t) \\ z_{21}i_1 + z_{22}i_2 + \dots + z_{2n}i_n &= u_2(t) \\ \dots \quad \dots \quad \dots \quad \dots & \quad \dots \\ z_{n1}i_1 + z_{n2}i_2 + \dots + z_{nn}i_n &= u_n(t), \end{aligned}$$
with integral operators:
$$z_{jk}i_k = R_{jk}\,i_k(t) + C_{jk}^{-1} \int_0^t i_k(t)\,dt.$$
In the Laplace transform, this equation reads:
$$R_{jk}\,I_k(s) + (C_{jk}s)^{-1}\,I_k(s) = U_j(s),$$
which implies the impedance of the linear RC-network:
$$Z_{jk}(s) = \frac{U_j(s)}{I_k(s)} = R_{jk} + (C_{jk}s)^{-1}\,.$$

Solutions of Some Volterra Integral Equations

Basic objective of this paragraph is to solve for $y(x)$ a linear Volterra equation with *kernel* $K(x,t)$:
$$\int_a^x K(x,t)\,y(t)\,dt = f(x).$$
where $(a \le x \le b)$ and $f(x)$ is a given function. In particular, we have the *difference kernel:*
$$K(x,t) = K(x-t).$$

In the following, zero initial conditions are always assumed: $0 = f(a) = f_x(a) = f_{xx}(a) = f_{xxx}(a)$ ... Also, all integrals $\int_0^x K(x,t)y(t)\,dt$ are evaluated below for: $y(x) = \sin x$.

Simple kernels

1.

$$\begin{aligned}&\text{Eq}: \int_a^x y(t)\,dt = f(x),\\&\text{Sol}: y(x) = f_x(x), \qquad \text{for} \quad y(x) = \sin x,\\&\text{Int}: \int_0^x y(t)\,dt = 1 - \cos x.\end{aligned}$$

2.

$$\begin{aligned}&\text{Eq}: \int_a^x (x-t)\,y(t)\,dt = f(x),\\&\text{Sol}: y(x) = f_{xx}(x),\\&\text{Int}: \int_0^x (x-t)\,y(t)\,dt = x - \sin x.\end{aligned}$$

3.

$$\begin{aligned}&\text{Eq}: \int_a^x (x-t)^2\,y(t)\,dt = f(x),\\&\text{Sol}: y(x) = \frac{1}{2}f_{xxx}(x),\\&\text{Int}: \int_0^x (x-t)^2\,y(t)\,dt = x^2 + 2\cos x - 2.\end{aligned}$$

4.

$$\begin{aligned}&\text{Eq}: \int_a^x (x-t)^3\,y(t)\,dt = f(x),\\&\text{Sol}: y(x) = \frac{1}{3!}f_{xxxx}(x),\\&\text{Int}: \int_0^x (x-t)^3\,y(t)\,dt = x^3 - 6x + 6\sin x.\end{aligned}$$

5.

$$\begin{aligned}&\text{Eq}: \int_a^x (x-t)^n\,y(t)\,dt = f(x),\\&\text{Sol}: y(x) = \frac{1}{n!}f_x^{(n+1)}(x),\\&\text{Int}: \int_0^x (x-t)^4\,y(t)\,dt = x^4 - 12x^2 - 24\cos x + 24.\end{aligned}$$

6.

$$\begin{aligned}
\text{Eq}: & \int_a^x (x^2 - t^2)\, y(t)\, dt = f(x), \\
\text{Sol}: & \ y(x) = \frac{1}{2}\frac{d}{dx}\left[\frac{1}{x} f_x(x)\right], \\
\text{Int}: & \int_0^x (x^2 - t^2)\, y(t)\, dt = x^2 - 2x\sin x - 2\cos x + 2.
\end{aligned}$$

7.

$$\begin{aligned}
\text{Eq}: & \int_a^x (x^3 - t^3)\, y(t)\, dt = f(x), \\
\text{Sol}: & \ y(x) = \frac{1}{3}\frac{d}{dx}\left[\frac{1}{x^2} f_x(x)\right], \\
\text{Int}: & \int_0^x (x^3 - t^3)\, y(t)\, dt = x^3 - 3\left(x^2 - 2\right)\sin x - 6x\cos x.
\end{aligned}$$

8.

$$\begin{aligned}
\text{Eq}: & \int_a^x (x^n - t^n)\, y(t)\, dt = f(x), \qquad n - \text{integer} \\
\text{Sol}: & \ y(x) = \frac{1}{n}\frac{d}{dx}\left[\frac{1}{x^{n-1}} f_x(x)\right], \\
\text{Int}: & \int_0^x (x^4 - t^4)\, y(t)\, dt = x^4 - 4\left(x^2 - 6\right)\ x\sin x - 12\left(x^2 - 2\right)\cos x - 24.
\end{aligned}$$

9.

$$\begin{aligned}
\text{Eq}: & \int_a^x (x^\mu - t^\mu)\, y(t)\, dt = f(x), \qquad \mu - \text{noninteger} \\
\text{Sol}: & \ y(x) = \frac{1}{\mu}\frac{d}{dx}\left[x^{1-\mu} f_x(x)\right].
\end{aligned}$$

Exponential kernels

1.

$$\begin{aligned}
\text{Eq}: & \int_a^x \mathrm{e}^{\lambda(x-t)}\, y(t)\, dt = f(x), \\
\text{Sol}: & \ y(x) = f_x(x) - \lambda f(x), \\
\text{Int}: & \int_0^x \mathrm{e}^{\lambda(x-t)}\, y(t)\, dt = \frac{\mathrm{e}^{\lambda x} - \lambda\sin x - \cos x}{\lambda^2 + 1}.
\end{aligned}$$

In particular,

$$\begin{aligned}
\text{Eq}: & \int_0^x \mathrm{e}^{\lambda(x-t)}\, y(t)\, dt = Ax, \\
\text{Sol}: & \ y(x) = A(1 - \lambda x).
\end{aligned}$$

2.

$$\text{Eq}: \int_a^x \left[e^{\lambda(x-t)} - 1\right] y(t)\,dt = f(x),$$

$$\text{Sol}: y(x) = \frac{1}{\lambda} f_{xx}(x) - f_x(x),$$

$$\text{Int}: \int_0^x \left[e^{\lambda(x-t)} - 1\right] y(t)\,dt = \frac{e^{\lambda x} - \lambda \sin x - \cos x}{\lambda^2 + 1} + \cos x - 1.$$

3.

$$\text{Eq}: \int_a^x e^{\lambda x + \beta t}\, y(t)\,dt = f(x),$$

$$\text{Sol}: y(x) = e^{-(\lambda+\beta)x}\left[f_x(x) - \lambda f(x)\right],$$

$$\text{Int}: \int_0^x e^{\lambda x + \beta t}\, y(t)\,dt = \frac{e^{\lambda x}\left(e^{\beta x}(\beta \sin x - \cos x) + 1\right)}{\beta^2 + 1}.$$

In particular,

$$\text{Eq}: \int_0^x e^{\lambda x + \beta t}\, y(t)\,dt = A \sin(\gamma x),$$

$$\text{Sol}: y(x) = A e^{-(\lambda+\beta)x}\left[\gamma \cos(\gamma x) - \lambda \sin(\gamma x)\right].$$

4.

$$\text{Eq}: \int_a^x e^{\lambda(x^2 - t^2)}\, y(t)\,dt = f(x),$$

$$\text{Sol}: y(x) = f_x(x) - 2\lambda x f(x),$$

$$\text{Int}: \int_0^x e^{\lambda(x^2 - t^2)}\, y(t)\,dt =$$

$$\frac{\sqrt{\pi} e^{x^2\lambda - \frac{1}{4\lambda}}\left(-\mathrm{i}\,\mathrm{erf}\left(\frac{2\lambda x - i}{2\sqrt{\lambda}}\right) + \mathrm{i}\,\mathrm{erf}\left(\frac{2\lambda x + i}{2\sqrt{\lambda}}\right) + 2\mathrm{erfi}\left(\frac{1}{2\sqrt{\lambda}}\right)\right)}{4\sqrt{\lambda}}.$$

5.

$$\text{Eq}: \int_a^x e^{\lambda(x^\beta - t^\beta)}\, y(t)\,dt = f(x),$$

$$\text{Sol}: y(x) = f_x(x) - \lambda \beta x^{\beta - 1} f(x).$$

6.

$$\text{Eq}: \int_a^x \left(e^{\lambda x} - e^{\lambda t}\right) y(t)\,dt = f(x),$$

$$\text{Sol}: y(x) = e^{-\lambda x}\left[\frac{1}{\lambda} f_{xx}(x) - f_x(x)\right],$$

$$\text{Int}: \int_0^x \left(e^{\lambda x} - e^{\lambda t}\right) y(t)\,dt =$$

$$\frac{e^{\lambda x}(\cos x - \lambda \sin x) - 1}{\lambda^2 + 1} - e^{\lambda x}(\cos x - 1).$$

7.

$$\begin{aligned}
&\text{Eq} : \int_a^x \left(e^{\lambda x^2} - e^{\lambda t^2}\right) y(t)\, dt = f(x),\\
&\text{Sol} : y(x) = \frac{1}{2\lambda}\frac{d}{dx}\left[\frac{f_x(x)}{x\, e^{\lambda x^2}}\right],\\
&\text{Int} : \int_0^x \left(e^{\lambda x^2} - e^{\lambda t^2}\right) y(t)\, dt = -e^{\lambda x^2}(\cos x - 1)\\
&\quad - \frac{\sqrt{\pi} e^{\frac{1}{4\lambda}}\left(\text{erf}\left(\frac{1-2i\lambda x}{2\sqrt{\lambda}}\right) + \text{erf}\left(\frac{1+2i\lambda x}{2\sqrt{\lambda}}\right) - 2\text{erf}\left(\frac{1}{2\sqrt{\lambda}}\right)\right)}{4\sqrt{\lambda}}.
\end{aligned}$$

8.

$$\begin{aligned}
&\text{Eq} : \int_a^x (x-t)\, e^{\lambda(x-t)}\, y(t)\, dt = f(x),\\
&\text{Sol} : y(x) = f_{xx}(x) - 2\lambda f_x(x) + \lambda^2 f(x),\\
&\text{Int} : \int_0^x (x-t)\, e^{\lambda(x-t)}\, y(t)\, dt =\\
&\quad \frac{e^{\lambda x}\left(\lambda x^2 + x - 2\lambda\right) + \left(\lambda^2 - 1\right)\sin x + 2\lambda\cos x}{\left(\lambda^2+1\right)^2}.
\end{aligned}$$

9.

$$\begin{aligned}
&\text{Eq} : \int_a^x \left(e^{\lambda x + \mu t} - e^{\mu x + \lambda t}\right) y(t)\, dt = f(x),\\
&\text{Sol} : y(x) = \frac{f_{xx}(x) - (\lambda+\mu)\, f_x(x) + \lambda\mu\, f(x)}{(\lambda-\mu)\, e^{(\lambda+\mu)x}},\\
&\text{Int} : \int_0^x \left(e^{\lambda x + \mu t} - e^{\mu x + \lambda t}\right) y(t)\, dt = \frac{e^{\mu x}\left(e^{\lambda x}(\cos x - \lambda \sin x) - 1\right)}{\lambda^2+1}\\
&\quad + \frac{e^{\lambda x}\left(e^{\mu x}(\mu \sin x - \cos x) + 1\right)}{\mu^2+1}.
\end{aligned}$$

Hyperbolic kernels

1.

$$\begin{aligned}
&\text{Eq} : \int_a^x \cosh\left[\lambda(x-t)\right]\, y(t)\, dt = f(x),\\
&\text{Sol} : y(x) = f_x(x) - \lambda^2 \int_a^x f(t)\, dt,\\
&\text{Int} : \int_0^x \cosh\left[\lambda(x-t)\right]\, y(t)\, dt = \frac{\cosh(\lambda x) - \cos x}{\lambda^2+1}.
\end{aligned}$$

2.

$$\text{Eq}: \int_a^x (\cosh[\lambda(x-t)] - 1)\, y(t)\, dt = f(x),$$

$$\text{Sol}: y(x) = \frac{1}{\lambda^2} f_{xxx}(x) - f_x(x),$$

$$\text{Int}: \int_0^x (\cosh[\lambda(x-t)] - 1)\, y(t)\, dt = \frac{\cosh(\lambda x) - \cos x}{\lambda^2 + 1} + \cos x - 1.$$

3.

$$\text{Eq}: \int_a^x [\cosh(\lambda x) - \cosh(\lambda t)]\, y(t)\, dt = f(x),$$

$$\text{Sol}: y(x) = \frac{1}{\lambda} \frac{d}{dx} \left[\frac{f_x(x)}{\sinh(\lambda x)} \right],$$

$$\text{Int}: \int_0^x [\cosh(\lambda x) - \cosh(\lambda t)]\, y(t)\, dt = \frac{\left(\lambda^2(-\cos x) + \lambda^2 + 1\right)\cosh(\lambda x) - \lambda \sin x \sinh(\lambda x) - 1}{\lambda^2 + 1}.$$

4.

$$\text{Eq}: \int_a^x [\cosh(\lambda x) - \cosh(\lambda t)]^n\, y(t)\, dt = f(x),$$

$$\text{Sol}: y(x) = \frac{\sinh(\lambda x)}{\lambda^n n!} \left[\frac{1}{\sinh(\lambda x)} \frac{d}{dx} \right]^{n+1} f(x),$$

$$\text{Int}: \int_0^x [\cosh(\lambda x) - \cosh(\lambda t)]^2\, y(t)\, dt = \frac{1}{8\lambda^4 + 10\lambda^2 + 2} [-6\lambda^3 \sin x \sinh(2\lambda x) - 4\left(4\lambda^2 + 1\right)\cosh(\lambda x)$$
$$+ \left(\lambda^2 + 1\right)\left(\left(4\lambda^2 + 1\right)\cosh(2\lambda x) + 8\lambda^2 + 3\right)$$
$$+ 2\lambda^2 \cos x \left(\left(1 - 2\lambda^2\right)\cosh(2\lambda x) - 4\lambda^2 - 1\right)]$$

5.

$$\text{Eq}: \int_a^x \sinh[\lambda(x-t)]\, y(t)\, dt = f(x),$$

$$\text{Sol}: y(x) = \frac{1}{\lambda} f_{xx}(x) - \lambda f(x),$$

$$\text{Int}: \int_0^x \sinh[\lambda(x-t)]\, y(t)\, dt = \frac{\sinh(\lambda x) - \lambda \sin x}{\lambda^2 + 1}.$$

6.

$$\text{Eq}: \int_a^x [\sinh(\lambda x) - \sinh(\lambda t)]\, y(t)\, dt = f(x),$$

$$\text{Sol}: y(x) = \frac{1}{\lambda}\frac{d}{dx}\left[\frac{f_x(x)}{\cosh(\lambda x)}\right],$$

$$\text{Int}: \int_0^x [\sinh(\lambda x) - \sinh(\lambda t)]\, y(t)\, dt = \frac{\left(\lambda^2(-\cos x) + \lambda^2 + 1\right)\sinh(\lambda x) - \lambda \sin x \cosh(\lambda x)}{\lambda^2 + 1}.$$

7.

$$\text{Eq}: \int_a^x \left[\sinh^2(\lambda x) - \sinh^2(\lambda t)\right] y(t)\, dt = f(x),$$

$$\text{Sol}: y(x) = \frac{1}{\lambda}\frac{d}{dx}\left[\frac{f_x(x)}{\cosh(2\lambda x)}\right],$$

$$\text{Int}: \int_0^x \left[\sinh^2(\lambda x) - \sinh^2(\lambda t)\right] y(t)\, dt = \frac{\cos(x)\left(\cosh(2\lambda x) - 4\lambda^2 - 1\right) - 2\lambda \sin x \sinh(2\lambda x) + 4\lambda^2}{8\lambda^2 + 2}$$
$$+(1 - \cos x)\sinh^2(\lambda x).$$

8.

$$\text{Eq}: \int_a^x [\sinh^\mu x - \sinh^\mu t]\, y(t)\, dt = f(x),$$

$$\text{Sol}: y(x) = \frac{1}{\mu}\frac{d}{dx}\left[\frac{f_x(x)}{\cosh x \, \sinh^{\mu-1} x}\right].$$

9.

$$\text{Eq}: \int_a^x [\tanh(\lambda x) - \tanh(\lambda t)]\, y(t)\, dt = f(x),$$

$$\text{Sol}: y(x) = \frac{1}{\lambda}\frac{d}{dx}\left[\cosh^2(\lambda x)\; f_x(x)\right].$$

10.

$$\text{Eq}: \int_a^x \left[\tanh^2(\lambda x) - \tanh^2(\lambda t)\right] y(t)\, dt = f(x),$$

$$\text{Sol}: y(x) = \frac{d}{dx}\left[\frac{\cosh^3(\lambda x)\; f_x(x)}{2\lambda \sinh(\lambda x)}\right].$$

11.

$$\text{Eq}: \int_a^x \left[\tanh^\mu x - \tanh^\mu t\right] y(t)\,dt = f(x),$$

$$\text{Sol}: y(x) = \frac{1}{\mu}\frac{d}{dx}\left[\frac{\cosh^{\mu+1}x\ f_x(x)}{\sinh^{\mu-1}x}\right].$$

12.

$$\text{Eq}: \int_a^x \left[\tanh(\lambda x) - \tanh(\lambda t)\right]^n y(t)\,dt = f(x),$$

$$\text{Sol}: y(x) = \frac{1}{\lambda^n n!\cosh^2(\lambda x)}\left[\cosh^2(\lambda x)\frac{d}{dx}\right]^{n+1} f(x).$$

13.

$$\text{Eq}: \int_a^x \left[\coth(\lambda x) - \coth(\lambda t)\right] y(t)\,dt = f(x),$$

$$\text{Sol}: y(x) = -\frac{1}{\lambda}\frac{d}{dx}\left[\sinh^2(\lambda x)\ f_x(x)\right].$$

14.

$$\text{Eq}: \int_a^x \left[\coth^2(\lambda x) - \coth^2(\lambda t)\right] y(t)\,dt = f(x),$$

$$\text{Sol}: y(x) = -\frac{d}{dx}\left[\frac{\sinh^3(\lambda x)\ f_x(x)}{2\lambda\cosh(\lambda x)}\right].$$

15.

$$\text{Eq}: \int_a^x \left[\coth^\mu x - \coth^\mu t\right] y(t)\,dt = f(x),$$

$$\text{Sol}: y(x) = -\frac{1}{\mu}\frac{d}{dx}\left[\frac{\sinh^{\mu+1}x\ f_x(x)}{\cosh^{\mu-1}x}\right].$$

16.

$$\text{Eq}: \int_a^x \left[\coth(\lambda x) - \coth(\lambda t)\right]^n y(t)\,dt = f(x),$$

$$\text{Sol}: y(x) = \frac{(-1)^n}{\lambda^n n!\sinh^2(\lambda x)}\left[\sinh^2(\lambda x)\frac{d}{dx}\right]^{n+1} f(x).$$

Trigonometric kernels

1.

$$\begin{aligned}
&\text{Eq}: \int_a^x \cos\left[\lambda(x-t)\right]\, y(t)\, dt = f(x),\\
&\text{Sol}: y(x) = f_x(x) + \lambda^2 \int_a^x f(x)\, dx,\\
&\text{Int}: \int_0^x \cos\left[\lambda(x-t)\right]\, y(t)\, dt = \frac{\cos x - \cos(\lambda x)}{\lambda^2 - 1}.
\end{aligned}$$

2.

$$\begin{aligned}
&\text{Eq}: \int_a^x \left(\cos\left[\lambda(x-t)\right] - 1\right)\, y(t)\, dt = f(x),\\
&\text{Sol}: y(x) = -\frac{1}{\lambda^2} f_{xxx}(x) - f_x(x),\\
&\text{Int}: \int_0^x \left(\cos\left[\lambda(x-t)\right] - 1\right)\, y(t)\, dt = \frac{\cos x - \cos(\lambda x)}{\lambda^2 - 1} + \cos x - 1.
\end{aligned}$$

3.

$$\begin{aligned}
&\text{Eq}: \int_a^x \left[\cos(\lambda x) - \cos(\lambda t)\right]\, y(t)\, dt = f(x),\\
&\text{Sol}: y(x) = -\frac{1}{\lambda}\frac{d}{dx}\left[\frac{f_x(x)}{\sin(\lambda x)}\right],\\
&\text{Int}: \int_0^x \left[\cos(\lambda x) - \cos(\lambda t)\right]\, y(t)\, dt =\\
&\frac{\left(\lambda^2(-\cos x) + \lambda^2 - 1\right)\cos(\lambda x) - \lambda \sin x \sin(\lambda x) + 1}{\lambda^2 - 1}.
\end{aligned}$$

4.

$$\begin{aligned}
&\text{Eq}: \int_a^x \left[\cos^2(\lambda x) - \cos^2(\lambda t)\right]\, y(t)\, dt = f(x),\\
&\text{Sol}: y(x) = -\frac{1}{\lambda}\frac{d}{dx}\left[\frac{f_x(x)}{\sin(2\lambda x)}\right].
\end{aligned}$$

5.

$$\begin{aligned}
&\text{Eq}: \int_a^x \left(\cos^\mu x - \cos^\mu t\right)\, y(t)\, dt = f(x),\\
&\text{Sol}: y(x) = -\frac{1}{\mu}\frac{d}{dx}\left[\frac{f_x(x)}{\sin x \cos^{\mu-1} x}\right].
\end{aligned}$$

6.

$$\text{Eq}: \int_a^x [\cos(\lambda x) - \cos(\lambda t)]\ ^n y(t)\, dt = f(x),$$

$$\text{Sol}: y(x) = \frac{(-1)^n}{\lambda^n n!} \sin(\lambda x) \left[\frac{1}{\sin(\lambda x)} \frac{d}{dx} \right]^{n+1} f(x).$$

7.

$$\text{Eq}: \int_a^x \sin[\lambda(x-t)]\ y(t)\, dt = f(x),$$

$$\text{Sol}: y(x) = \frac{1}{\lambda} f_{xx}(x) + \lambda f(x),$$

$$\text{Int}: \int_0^x \sin[\lambda(x-t)]\ y(t)\, dt = \frac{\lambda \sin x - \sin(\lambda x)}{\lambda^2 - 1}.$$

8.

$$\text{Eq}: \int_a^x [\sin(\lambda x) - \sin(\lambda t)]\ y(t)\, dt = f(x),$$

$$\text{Sol}: y(x) = \frac{1}{\lambda} \frac{d}{dx} \left[\frac{f_x(x)}{\cos(\lambda x)} \right],$$

$$\text{Int}: \int_0^x [\sin(\lambda x) - \sin(\lambda t)]\ y(t)\, dt = \frac{\sin(\lambda x)\left(\lambda^2(-\cos x) + \lambda^2 - 1\right) + \lambda \sin x \cos(\lambda x)}{\lambda^2 - 1}.$$

9.

$$\text{Eq}: \int_a^x \left[\sin^2(\lambda x) - \sin^2(\lambda t)\right]\ y(t)\, dt = f(x),$$

$$\text{Sol}: y(x) = \frac{1}{\lambda} \frac{d}{dx} \left[\frac{f_x(x)}{\sin(2\lambda x)} \right],$$

$$\text{Int}: \int_0^x \left[\sin^2(\lambda x) - \sin^2(\lambda t)\right]\ y(t)\, dt = \frac{\left(4\lambda^2 \cos x - 4\lambda^2 + 1\right) \cos(2\lambda x) + 2\lambda \sin x \sin(2\lambda x) - 1}{8\lambda^2 - 2}.$$

10.

$$\text{Eq}: \int_a^x (\sin^\mu x - \sin^\mu t)\ y(t)\, dt = f(x),$$

$$\text{Sol}: y(x) = \frac{1}{\mu} \frac{d}{dx} \left[\frac{f_x(x)}{\cos x \sin^{\mu-1} x} \right].$$

11.

$$\text{Eq}: \int_a^x [\tan(\lambda x) - \tan(\lambda t)]\, y(t)\, dt = f(x),$$
$$\text{Sol}: y(x) = \frac{1}{\lambda}\frac{d}{dx}\left[\cos^2(\lambda x)\ f_x(x)\right].$$

12.

$$\text{Eq}: \int_a^x \left[\tan^2(\lambda x) - \tan^2(\lambda t)\right]\, y(t)\, dt = f(x),$$
$$\text{Sol}: y(x) = \frac{d}{dx}\left[\frac{\cos^3(\lambda x)\ f_x(x)}{2\lambda \sin(\lambda x)}\right].$$

13.

$$\text{Eq}: \int_a^x (\tan^\mu x - \tan^\mu t)\, y(t)\, dt = f(x),$$
$$\text{Sol}: y(x) = \frac{1}{\mu}\frac{d}{dx}\left[\frac{\cos^{\mu+1} x\ f_x(x)}{\sin^{\mu-1} x}\right].$$

14.

$$\text{Eq}: \int_a^x [\tan(\lambda x) - \tan(\lambda t)]^n\, y(t)\, dt = f(x),$$
$$\text{Sol}: y(x) = \frac{1}{\lambda^n n! \cos^2(\lambda x)}\left[\cos^2(\lambda x)\frac{d}{dx}\right]^{n+1} f(x).$$

15.

$$\text{Eq}: \int_a^x [\cot(\lambda x) - \cot(\lambda t)]\, y(t)\, dt = f(x),$$
$$\text{Sol}: y(x) = -\frac{1}{\lambda}\frac{d}{dx}\left[\sin^2(\lambda x)\ f_x(x)\right].$$

16.

$$\text{Eq}: \int_a^x \left[\cot^2(\lambda x) - \cot^2(\lambda t)\right]\, y(t)\, dt = f(x),$$
$$\text{Sol}: y(x) = -\frac{d}{dx}\left[\frac{\sin^3(\lambda x)\ f_x(x)}{2\lambda \cos(\lambda x)}\right].$$

17.

$$\text{Eq}: \int_a^x (\cot^\mu x - \cot^\mu t)\, y(t)\, dt = f(x),$$
$$\text{Sol}: y(x) = -\frac{1}{\mu}\frac{d}{dx}\left[\frac{\sin^{\mu+1} x\ f_x(x)}{\cos^{\mu-1} x}\right].$$

18.

$$\text{Eq}: \int_a^x [\cot(\lambda x) - \cot(\lambda t)]^n \, y(t)\, dt = f(x),$$

$$\text{Sol}: y(x) = \frac{(-1)^n}{\lambda^n n! \sin^2(\lambda x)} \left[\sin^2(\lambda x) \frac{d}{dx}\right]^{n+1} f(x).$$

Inverse trigonometric kernels

1.

$$\text{Eq}: \int_a^x [\arccos(\lambda x) - \arccos(\lambda t)] \, y(t)\, dt = f(x),$$

$$\text{Sol}: y(x) = -\frac{1}{\lambda} \frac{d}{dx} \left[\sqrt{1 - \lambda^2 x^2}\, f_x(x)\right].$$

2.

$$\text{Eq}: \int_a^x [\arccos^\mu(\lambda x) - \arccos^\mu(\lambda t)] \, y(t)\, dt = f(x),$$

$$\text{Sol}: y(x) = -\frac{1}{\lambda\mu} \frac{d}{dx} \left[\frac{\sqrt{1 - \lambda^2 x^2}\, f_x(x)}{\arccos^{\mu-1}(\lambda x)}\right].$$

3.

$$\text{Eq}: \int_a^x [\arccos(\lambda x) - \arccos(\lambda t)]^{\,n} y(t)\, dt = f(x),$$

$$\text{Sol}: y(x) = \frac{(-1)^n}{\lambda^n n! \sqrt{1 - \lambda^2 x^2}} \left[\sqrt{1 - \lambda^2 x^2} \frac{d}{dx}\right]^{n+1} f(x).$$

4.

$$\text{Eq}: \int_a^x [\arcsin(\lambda x) - \arcsin(\lambda t)] \, y(t)\, dt = f(x),$$

$$\text{Sol}: y(x) = \frac{1}{\lambda} \frac{d}{dx} \left[\sqrt{1 - \lambda^2 x^2}\, f_x(x)\right].$$

5.

$$\text{Eq}: \int_a^x [\arcsin^\mu(\lambda x) - \arcsin^\mu(\lambda t)] \, y(t)\, dt = f(x),$$

$$\text{Sol}: y(x) = \frac{1}{\lambda\mu} \frac{d}{dx} \left[\frac{\sqrt{1 - \lambda^2 x^2}\, f_x(x)}{\arcsin^{\mu-1}(\lambda x)}\right].$$

6.

$$\text{Eq}: \int_a^x [\arcsin(\lambda x) - \arcsin(\lambda t)]^{\,n} \, y(t)\, dt = f(x),$$

$$\text{Sol}: y(x) = \frac{1}{\lambda^n n! \sqrt{1 - \lambda^2 x^2}} \left[\sqrt{1 - \lambda^2 x^2} \frac{d}{dx}\right]^{n+1} f(x).$$

7.

$$\text{Eq}: \int_a^x [\arctan(\lambda x) - \arctan(\lambda t)]\, y(t)\, dt = f(x),$$

$$\text{Sol}: y(x) = \frac{1}{\lambda}\frac{d}{dx}\left[\left(1+\lambda^2 x^2\right)\ f_x(x)\right].$$

8.

$$\text{Eq}: \int_a^x [\arctan(\lambda x) - \arctan(\lambda t)]\ ^n y(t)\, dt = f(x),$$

$$\text{Sol}: y(x) = \frac{1}{\lambda^n n!\left(1+\lambda^2 x^2\right)}\left[\left(1+\lambda^2 x^2\right)\frac{d}{dx}\right]^{n+1} f(x).$$

9.

$$\text{Eq}: \int_a^x [\arctan^\mu(\lambda x) - \arctan^\mu(\lambda t)]\, y(t)\, dt = f(x),$$

$$\text{Sol}: y(x) = \frac{1}{\lambda\mu}\frac{d}{dx}\left[\frac{\left(1+\lambda^2 x^2\right)\ f_x(x)}{\arctan^{\mu-1}(\lambda x)}\right].$$

10.

$$\text{Eq}: \int_a^x [\operatorname{arccot}(\lambda x) - \operatorname{arccot}(\lambda t)]\, y(t)\, dt = f(x),$$

$$\text{Sol}: y(x) = -\frac{1}{\lambda}\frac{d}{dx}\left[\left(1+\lambda^2 x^2\right)\ f_x(x)\right].$$

11.

$$\text{Eq}: \int_a^x [\operatorname{arccot}(\lambda x) - \operatorname{arccot}(\lambda t)]\ ^n y(t)\, dt = f(x),$$

$$\text{Sol}: y(x) = \frac{(-1)^n}{\lambda^n n!\left(1+\lambda^2 x^2\right)}\left[\left(1+\lambda^2 x^2\right)\frac{d}{dx}\right]^{n+1} f(x).$$

12.

$$\text{Eq}: \int_a^x [\operatorname{arccot}^\mu(\lambda x) - \operatorname{arccot}^\mu(\lambda t)]\, y(t)\, dt = f(x),$$

$$\text{Sol}: y(x) = -\frac{1}{\lambda\mu}\frac{d}{dx}\left[\frac{\left(1+\lambda^2 x^2\right)\ f_x(x)}{\operatorname{arccot}^{\mu-1}(\lambda x)}\right].$$

Combined elementary kernels

1.

$$\text{Eq}: \int_a^x e^{\lambda(x-t)}\, [\ln(x/t)]\ ^n y(t)\, dt = f(x),$$

$$\text{Sol}: y(x) = \frac{1}{n!x} e^{\lambda x}\left(x\frac{d}{dx}\right)^{n+1} e^{-\lambda x} f(x).$$

2.

$$\text{Eq}: \int_a^x \mathrm{e}^{\lambda(x-t)} (\ln x - \ln t)\ y(t)\, dt = f(x),$$
$$\text{Sol}: y(x) = \mathrm{e}^{\lambda x} \left[x\varphi_{xx}(x) + \varphi_x(x)\right], \qquad \varphi_x(x) = \mathrm{e}^{-\lambda x} f(x).$$

3.

$$\text{Eq}: \int_a^x \mathrm{e}^{\mu(x-t)} \left[\cos(\lambda x) - \cos(\lambda t)\right]^{\,n} y(t)\, dt = f(x),$$
$$\text{Sol}: y(x) = \frac{(-1)^n}{\lambda^n n!} \mathrm{e}^{\mu x} \sin(\lambda x) \left[\frac{1}{\sin(\lambda x)} \frac{d}{dx}\right]^{n+1} \mathrm{e}^{-\mu x} f(x).$$

4.

$$\text{Eq}: \int_a^x \mathrm{e}^{\mu(x-t)} \cos\left[\lambda(x-t)\right]\, y(t)\, dt = f(x),$$
$$\text{Sol}: y(x) = f_x(x) - \mu f(x) + \lambda^2 \int_a^x \mathrm{e}^{\mu(x-t)} f(t)\, dt.$$

5.

$$\text{Eq}: \int_a^x \mathrm{e}^{\mu(x-t)} \sin\left[\lambda(x-t)\right]\, y(t)\, dt = f(x),$$
$$\text{Sol}: y(x) = \frac{1}{\lambda} \left[f_{xx}(x) - 2\mu f_x(x) + (\lambda^2 + \mu^2) f(x)\right].$$

6.

$$\text{Eq}: \int_a^x \left(\sinh\left[\lambda(x-t)\right] - \sin\left[\lambda(x-t)\right]\right)\, y(t)\, dt = f(x),$$
$$\text{Sol}: y(x) = \frac{1}{2\lambda^3} \left(\frac{d^4}{dx^4} - \lambda^4\right) f(x).$$

Logarithmic kernels

1.

$$\text{Eq}: \int_a^x (\ln x - \ln t)\ y(t)\, dt = f(x),$$
$$\text{Sol}: y(x) = x f_{xx}(x) + f_x(x),$$
$$\text{Int}: \int_0^x (\ln x - \ln t)\ y(t)\, dt = \ln x - \mathrm{Ci}(x) + \gamma(\simeq 0.577216).$$

2.

$$\text{Eq}: \int_a^x \left[\ln^2(\lambda x) - \ln^2(\lambda t)\right]\ y(t)\, dt = f(x),$$
$$\text{Sol}: y(x) = \frac{d}{dx} \left[\frac{x\ f_x(x)}{2\ln(\lambda x)}\right].$$

3.

$$\text{Eq}: \int_a^x \left[\ln^\mu(\lambda x) - \ln^\mu(\lambda t)\right] y(t)\, dt = f(x),$$

$$\text{Sol}: y(x) = \frac{1}{\mu}\frac{d}{dx}\left[x \ln^{1-\mu}(\lambda x)\, f_x(x)\right].$$

4.

$$\text{Eq}: \int_a^x \left[\ln(x/t)\right]^n y(t)\, dt = f(x),$$

$$\text{Sol}: y(x) = \frac{1}{n!x}\left(x\frac{d}{dx}\right)^{n+1} f(x).$$

5.

$$\text{Eq}: \int_a^x \ln\left(\frac{x+b}{t+b}\right) y(t)\, dt = f(x),$$

$$\text{Sol}: y(x) = (x+b)\, f_{xx}(x) + f_x(x).$$

Bessel kernels

1.

$$\text{Eq}: \int_a^x \left[J_0(\lambda x) - J_0(\lambda t)\right] y(t)\, dt = f(x),$$

$$\text{Sol}: y(x) = -\frac{d}{dx}\left[\frac{f_x(x)}{\lambda J_1(\lambda x)}\right].$$

2.

$$\text{Eq}: \int_a^x \left[J_\nu(\lambda x) - J_\nu(\lambda t)\right] y(t)\, dt = f(x),$$

$$\text{Sol}: y(x) = \frac{d}{dx}\left[\frac{x\, f_x(x)}{\nu J_\nu(\lambda x) - \lambda x J_{\nu+1}(\lambda x)}\right].$$

3.

$$\text{Eq}: \int_a^x (x-t)^{1/2} J_{1/2}\left[\lambda(x-t)\right] y(t)\, dt = f(x),$$

$$\text{Sol}: y(x) = \sqrt{\frac{\pi}{2\lambda}}\left[f_{xx}(x) + \lambda^2 f(x)\right].$$

4.

$$\text{Eq}: \int_a^x (x-t)^{3/2} J_{3/2}\left[\lambda(x-t)\right] y(t)\, dt = f(x),$$

$$\text{Sol}: y(x) = \frac{\sqrt{\pi}}{(2\lambda)^{3/2}}\left(\frac{d^2}{dx^2} + \lambda^2\right)^2 f(x).$$

5.

$$\text{Eq}: \int_a^x (x-t)^{\frac{2n-1}{2}} J_{\frac{2n-1}{2}}\left[\lambda(x-t)\right] y(t)\,dt = f(x),$$

$$\text{Sol}: y(x) = \frac{\sqrt{\pi}}{(2\lambda)^{\frac{2n-1}{2}}(n-1)!}\left(\frac{d^2}{dx^2}+\lambda^2\right)^n f(x).$$

6.

$$\text{Eq}: \int_a^x \left[Y_0(\lambda x) - Y_0(\lambda t)\right] y(t)\,dt = f(x),$$

$$\text{Sol}: y(x) = -\frac{d}{dx}\left[\frac{f_x(x)}{\lambda Y_1(\lambda x)}\right].$$

7.

$$\text{Eq}: \int_a^x \left[Y_\nu(\lambda x) - Y_\nu(\lambda t)\right] y(t)\,dt = f(x),$$

$$\text{Sol}: y(x) = \frac{d}{dx}\left[\frac{x\, f_x(x)}{\nu Y_\nu(\lambda x) - \lambda x Y_{\nu+1}(\lambda x)}\right].$$

8.

$$\text{Eq}: \int_a^x (x-t)^{1/2} I_{1/2}\left[\lambda(x-t)\right] y(t)\,dt = f(x),$$

$$\text{Sol}: y(x) = \sqrt{\frac{\pi}{2\lambda}}\left[f_{xx}(x) - \lambda^2 f(x)\right].$$

9.

$$\text{Eq}: \int_a^x (x-t)^{3/2} I_{3/2}\left[\lambda(x-t)\right] y(t)\,dt = f(x),$$

$$\text{Sol}: y(x) = \frac{\sqrt{\pi}}{(2\lambda)^{3/2}}\left(\frac{d^2}{dx^2}-\lambda^2\right)^2 f(x).$$

10.

$$\text{Eq}: \int_a^x (x-t)^{\frac{2n-1}{2}} I_{\frac{2n-1}{2}}\left[\lambda(x-t)\right] y(t)\,dt = f(x),$$

$$\text{Sol}: y(x) = \frac{\sqrt{\pi}}{(2\lambda)^{\frac{2n-1}{2}}(n-1)!}\left(\frac{d^2}{dx^2}-\lambda^2\right)^n f(x).$$

11.

$$\text{Eq}: \int_a^x \left[K_0(\lambda x) - K_0(\lambda t)\right] y(t)\,dt = f(x),$$

$$\text{Sol}: y(x) = -\frac{d}{dx}\left[\frac{f_x(x)}{\lambda K_1(\lambda x)}\right].$$

Arbitrary kernels

1.

$$\text{Eq}: \int_a^x g(x)\, h(t)\, y(t)\, dt = f(x),$$

$$\text{Sol}: y(x) = \frac{1}{h(x)} \frac{d}{dx}\left[\frac{f(x)}{g(x)}\right].$$

2.

$$\text{Eq}: \int_a^x [g(x) - g(t)]\; y(t)\, dt = f(x),$$

$$\text{Sol}: y(x) = \frac{d}{dx}\left[\frac{f_x(x)}{g_x(x)}\right].$$

3.

$$\text{Eq}: \int_a^x [g(x) - g(t)]^n\; y(t)\, dt = f(x),$$

$$\text{Sol}: y(x) = \text{Sol}: y(x) = \frac{1}{n!} g_x(x) \left(\frac{1}{g_x(x)} \frac{d}{dx}\right)^{n+1} f(x).$$

4.

$$\text{Eq}: \int_a^x [g(x)\, h(t) - h(x) g(t)]\; y(t)\, dt = f(x),$$

$$\text{Sol}: y(x) = \frac{1}{h(x)} \frac{d}{dx}\left(\frac{\frac{d}{dx}\,[f(x)/h(x)]}{\frac{d}{dx}\,[g(x)/h(x)]}\right).$$

Difference kernel: $K(x,t) = K(x-t)$

1.

$$\text{Eq}: \int_{-\infty}^x K(x-t)\, y(t)\, dt = f(x).$$

$$\text{Sol1}: \text{If} \quad f(x) = \sum_{k=0}^n A_k x^k, \quad y(x) = \sum_{k=0}^n B_k x^k,$$

$(A_k, B_k$ undetermined coefficients$)$.

$$\text{Sol2}: \text{If} \quad f(x) = \mathrm{e}^{\lambda x} \sum_{k=0}^n A_k x^k, \quad y(x) = \mathrm{e}^{\lambda x} \sum_{k=0}^n B_k x^k.$$

$$\text{Sol3}: \text{If} \quad f(x) = \cos(\lambda x) \sum_{k=0}^n A_k x^k,$$

$$y(x) = \cos(\lambda x) \sum_{k=0}^{n} B_k x^k + \sin(\lambda x) \sum_{k=0}^{n} C_k x^k.$$

Sol4 : If $\quad f(x) = \sin(\lambda x) \sum_{k=0}^{n} A_k x^k,$

$$y(x) = \cos(\lambda x) \sum_{k=0}^{n} B_k x^k + \sin(\lambda x) \sum_{k=0}^{n} C_k x^k.$$

2.

Eq : $\displaystyle\int_x^{\infty} K(x-t)\, y(t)\, dt = f(x)$: Laplace transforms.

Sol1 : If $\quad f(x) = \sum_{k=0}^{n} A_k x^k, \quad y(x) = \sum_{k=0}^{n} B_k x^k, \quad (A_k, B_k$ undetermined coefficients).

Sol2 : If $\quad f(x) = e^{\lambda x} \sum_{k=0}^{n} A_k x^k, \quad y(x) = e^{\lambda x} \sum_{k=0}^{n} B_k x^k.$

Sol3 : If $\quad f(x) = \cos(\lambda x) \sum_{k=0}^{n} A_k x^k,$

$$y(x) = \cos(\lambda x) \sum_{k=0}^{n} B_k x^k + \sin(\lambda x) \sum_{k=0}^{n} C_k x^k.$$

Sol4 : If $\quad f(x) = \sin(\lambda x) \sum_{k=0}^{n} A_k x^k,$

$$y(x) = \cos(\lambda x) \sum_{k=0}^{n} B_k x^k + \sin(\lambda x) \sum_{k=0}^{n} C_k x^k.$$

For arbitrary right-hand side $f(x)$, the solution of this integral equation can be calculated by the inverse Laplace transform

$$y(x) = \frac{1}{2\pi i} \int_{c-i\infty}^{c+i\infty} \frac{F(s)}{k(-s)} e^{sx}\, ds, \qquad \text{with}$$

$$F(s) = \mathcal{L}[f(x)] = \int_0^{\infty} f(x) e^{-sx}\, dx,$$

$$k(-s) = \mathcal{L}[K(-z)] = \int_0^{\infty} K(-z) e^{sz}\, dz.$$

Some transformations

Let the solution of the integral equation

$$\int_a^x K(x,t)\,y(t)\,dt = f(x)$$

have the form

$$y(x) = \mathcal{F}\left[f(x)\right],$$

where $\mathcal{F}$ is some linear integro-differential operator. Then the solution of the more complicated integral equation

$$\int_a^x K(x,t)\,g(x)\,h(t)\,y(t)\,dt = f(x)$$

has the form

$$y(x) = \frac{1}{h(x)}\mathcal{F}\left[\frac{f(x)}{g(x)}\right].$$

1.

$$\text{Eq}: \int_a^x K(x,t)\,(x/t)^{\lambda} y(t)\,dt = f(x),$$
$$\text{Sol}: y(x) = x^{\lambda}\mathcal{F}\left[x^{-\lambda} f(x)\right].$$

2.

$$\text{Eq}: \int_a^x K(x,t)\,\mathrm{e}^{\lambda(x-t)}\,y(t)\,dt = f(x),$$
$$\text{Sol}: y(x) = \mathrm{e}^{\lambda x}\mathcal{F}\left[\mathrm{e}^{-\lambda x} f(x)\right].$$

4.3.3 Riemann-Liouville Fractional Calculus

Here we give a brief overview of Riemann-Liouville fractional calculus: (i) integral, (ii) differential, (iii) integral equations, and (iv) diffusion-wave PDE (for additional details see references listed in [FC13]).

Fractional Integral

Cauchy formula: an $n-$fold primitive $f_n(t)$ of a function $f(t)$ is calculated by a convolution integral:

$$J^n f(t) := f_n(t) = \frac{1}{(n-1)!}\int_0^t (t-\tau)^{n-1} f(\tau)\,d\tau\,, \quad (\text{for } t>0\,,\ n\in\mathbb{N}). \tag{4.40}$$

Riemann-Liouville fractional integral of order $\alpha > 0$ is given by:

$$J^{\alpha} f(t) := \frac{1}{\Gamma(\alpha)}\int_0^t (t-\tau)^{\alpha-1} f(\tau)\,d\tau\,, \quad (\text{for } t>0\,,\ \alpha\in\mathbb{R}^+), \tag{4.41}$$
$$J^0 f(t) := f(t), \qquad J^{\alpha} f(0^+) = \lim_{t\to 0^+} J^{\alpha} f(t).$$

Semigroup property:

$$J^\alpha J^\beta = J^{\alpha+\beta}\,, \quad (\text{for } \alpha\,,\, \beta \geq 0\,), \tag{4.42}$$

implies the *commutative property* $J^\beta J^\alpha = J^\alpha J^\beta$, and the effect of our operators J^α on the power functions (t^γ with $\gamma > -1,\ t > 0\,$,)

$$J^\alpha t^\gamma = \frac{\Gamma(\gamma+1)}{\Gamma(\gamma+1+\alpha)}\, t^{\gamma+\alpha}\,, \quad \alpha > 0\,, \quad \gamma > -1\,, \quad t > 0\,. \tag{4.43}$$

The properties (4.42-4.43) are natural generalization of those known when the order is a positive integer. The proofs are based on the properties of the two Eulerian integrals, i.e., the *Gamma function* and *Beta function*:

$$\Gamma(z) := \int_0^\infty \mathrm{e}^{-u}\, u^{z-1}\, du\,, \quad \Gamma(z+1) = z\,\Gamma(z)\,, \quad \{z\} > 0\,, \tag{4.44}$$

$$B(p,q) := \int_0^1 (1-u)^{p-1}\, u^{q-1}\, du \; = \frac{\Gamma(p)\,\Gamma(q)}{\Gamma(p+q)} = B(q,p)\,, \quad \{p\,,\, q\} > 0\,. \tag{4.45}$$

where $\mathrm{Re}\,\{p\,,\, q\} > 0$.

Let us now recall the notion of *Laplace convolution*, i.e., the convolution integral with two causal functions, which reads in a standard notation:

$$f(t) \, * \, g(t) := \int_0^t f(t-\tau)\, g(\tau)\, d\tau = g(t) \, * \, f(t)\,.$$

Also, it is convenient to introduce the following causal function

$$\Phi_\alpha(t) := \frac{t_+^{\alpha-1}}{\Gamma(\alpha)}\,, \quad \alpha > 0\,, \tag{4.46}$$

where the suffix $+$ is just denoting that the function is vanishing for $t < 0$. Being $\alpha > 0$, this function turns out to be *locally* absolutely integrable in $\mathbb{R}^+$.

Then we note from (4.41) and (4.46) that the fractional integral (of a function $f(t)$) of order $\alpha > 0$ can be considered as the Laplace convolution between $\Phi_\alpha(t)$ and $f(t)$, i.e.,

$$J^\alpha\, f(t) = \Phi_\alpha(t) \, * \, f(t)\,, \quad \alpha > 0\,. \tag{4.47}$$

Furthermore, based on the Eulerian integrals, one proves the *composition rule*

$$\Phi_\alpha(t) \, * \, \Phi_\beta(t) = \Phi_{\alpha+\beta}(t)\,, \quad \alpha\,,\, \beta > 0\,, \tag{4.48}$$

which can be used to re-obtain (4.42) and (4.43).

Introducing the Laplace transform by the notation:

$$\mathcal{L}\{f(t)\} := \int_0^{\infty} \mathrm{e}^{-st} f(t)\, dt = F(s)\,, \quad s \in \mathbb{C}\,,$$

and using the sign $\div$ to denote a Laplace transform pair: $f(t) \div F(s)$, we note the following rule for the Laplace transform of the fractional integral,

$$J^{\alpha} f(t) \div \frac{F(s)}{s^{\alpha}}\,, \quad \alpha > 0\,, \tag{4.49}$$

which is the straightforward generalization of the case with an n-fold repeated integral ($\alpha = n$). (For the proof it is sufficient to recall the convolution theorem for Laplace transforms and note the pair $\Phi_{\alpha}(t) \div 1/s^{\alpha}$, with $\alpha > 0$.)

Fractional Derivative

Riemann-Liouville Fractional Derivative

Denoting by D^n with $n \in \mathbb{N}$, the operator of the derivative of order n, we first note that

$$D^n\, J^n = I\,, \quad J^n\, D^n \neq I\,, \quad n \in \mathbb{N}\,, \tag{4.50}$$

i.e., D^n is left-inverse (and not right-inverse) to the corresponding integral operator J^n. In fact we easily recognize from (4.40) that

$$J^n\, D^n\, f(t) = f(t) - \sum_{k=0}^{n-1} f^{(k)}(0^+)\, \frac{t^k}{k!}\,, \quad t > 0\,. \tag{4.51}$$

As a consequence we expect that D^{α} is defined as left-inverse to J^{α}. For this purpose, introducing the positive integer m such that $m - 1 < \alpha \leq m$, one defines the *Riemann-Liouville fractional derivative of order* $\alpha > 0$:

$$D^{\alpha} f(t) := D^m\, J^{m-\alpha} f(t)\,, \qquad \text{namely}$$

$$D^{\alpha} f(t) := \begin{cases} \frac{d^m}{dt^m}\left[\frac{1}{\Gamma(m-\alpha)} \int_0^t \frac{f(\tau)}{(t-\tau)^{\alpha+1-m}}\, d\tau\right], & m-1 < \alpha \leq m\,, \\ \frac{d^m}{dt^m} f(t), \quad \alpha = m. & \end{cases} \tag{4.52}$$

Defining for complementation $D^0 = J^0 = I$, then we easily recognize that

$$D^{\alpha}\, J^{\alpha} = I\,, \quad \alpha \geq 0\,, \tag{4.53}$$

and

$$D^{\alpha}\, t^{\gamma} = \frac{\Gamma(\gamma+1)}{\Gamma(\gamma+1-\alpha)}\, t^{\gamma-\alpha}\,, \quad \alpha > 0\,, \quad \gamma > -1\,, \quad t > 0\,. \tag{4.54}$$

The properties (4.53-4.54) are a natural generalization of those known when the order is a positive integer.

Caputo Fractional Derivative

We now observe that an alternative definition of fractional derivative, originally introduced by Caputo in the late sixties and adopted by Caputo and Mainardi in the framework of the theory of *linear viscoelasticity*, is the so-called *Caputo fractional derivative* of order $\alpha > 0 : D_*^\alpha f(t) := J^{m-\alpha} D^m f(t)$ with $m-1 < \alpha \le m$, namely

$$D_*^\alpha f(t) := \begin{cases} \frac{1}{\Gamma(m-\alpha)} \int_0^t \frac{f^{(m)}(\tau)}{(t-\tau)^{\alpha+1-m}} d\tau \,, & m-1 < \alpha \le m \,, \\ \frac{d^m}{dt^m} f(t), & \alpha = m. \end{cases} \tag{4.55}$$

This definition is more restrictive than (4.52), in that requires the absolute integrability of the derivative of order m. Whenever we use the operator D_*^α we (tacitly) assume that this condition is met.

We easily recognize that in general

$$D^\alpha f(t) := D^m J^{m-\alpha} f(t) \neq J^{m-\alpha} D^m f(t) := D_*^\alpha f(t) \,, \tag{4.56}$$

unless the function $f(t)$ along with its first $m-1$ derivatives vanishes at $t = 0^+$. In fact, assuming that the passage of the m-derivative under the integral is legitimate, one recognizes that, for $m-1 < \alpha < m$ and $t > 0$,

$$D^\alpha f(t) = D_*^\alpha f(t) + \sum_{k=0}^{m-1} \frac{t^{k-\alpha}}{\Gamma(k-\alpha+1)} f^{(k)}(0^+) \,, \tag{4.57}$$

and therefore, recalling the fractional derivative of the power functions (4.54),

$$D^\alpha \left[f(t) - \sum_{k=0}^{m-1} \frac{t^k}{k!} f^{(k)}(0^+) \right] = D_*^\alpha f(t) \,. \tag{4.58}$$

The Caputo definition (4.55) for the fractional derivative thus incorporates the initial values of the function and of its integer derivatives of lower order. The subtraction of the Taylor polynomial of degree $m-1$ at $t = 0^+$ from $f(t)$ means a sort of regularization of the fractional derivative. In particular, according to this definition, the relevant property for which the fractional derivative of a constant is still zero, i.e.,

$$D_*^\alpha 1 \equiv 0 \,, \quad \alpha > 0 \,. \tag{4.59}$$

can be easily recognized.

Fractional Integral Equations

Abel's equation of the first kind

Let us consider the *Abel integral equation* of the first kind

$$\frac{1}{\Gamma(\alpha)} \int_0^t \frac{u(\tau)}{(t-\tau)^{1-\alpha}}\, d\tau = f(t)\,, \quad 0 < \alpha < 1\,, \tag{4.60}$$

where $f(t)$ is a given function. We easily recognize that this equation can be expressed in terms of a fractional integral, i.e.,

$$J^\alpha\, u(t) = f(t)\,, \tag{4.61}$$

and consequently solved in terms of a fractional derivative, according to

$$u(t) = D^\alpha\, f(t)\,. \tag{4.62}$$

To this end we need to recall the definition (4.41) and the property (4.53) $D^\alpha\, J^\alpha = I$.

Let us now solve (4.60) using the Laplace transform. Noting from (4.46-4.47) and (4.49) that

$$J^\alpha\, u(t) = \Phi_\alpha(t) * u(t) \;\div\; U(s)/s^\alpha\,,$$

we then obtain

$$\frac{U(s)}{s^\alpha} = F(s) \Longrightarrow U(s) = s^\alpha\, F(s)\,. \tag{4.63}$$

Now we can choose two different ways to get the inverse Laplace transform from (4.63), according to the standard rules. Writing (4.63) as

$$U(s) = s \left[\frac{F(s)}{s^{1-\alpha}}\right]\,, \tag{4.64}$$

we obtain

$$u(t) = \frac{1}{\Gamma(1-\alpha)} \frac{d}{dt} \int_0^t \frac{f(\tau)}{(t-\tau)^\alpha}\, d\tau\,. \tag{4.65}$$

On the other hand, writing (4.63) as

$$U(s) = \frac{1}{s^{1-\alpha}} \left[sF(s) - f(0^+)\right] + \frac{f(0^+)}{s^{1-\alpha}}\,, \tag{4.66}$$

we obtain

$$u(t) = \frac{1}{\Gamma(1-\alpha)} \int_0^t \frac{f'(\tau)}{(t-\tau)^\alpha}\, d\tau + f(0^+)\, \frac{t^{-\alpha}}{\Gamma(1-a)}\,. \tag{4.67}$$

Thus, the solutions (4.65) and (4.67) are expressed in terms of the fractional derivatives D^α and D_*^α, respectively, according to (4.52), (4.55) and (4.57) with $m = 1$.

The way *b*) requires that $f(t)$ be differentiable with $\mathcal{L}$-transformable derivative; consequently $0 \le |f(0^+)| < \infty$. Then it turns out from (4.67) that $u(0^+)$ can be infinite if $f(0^+) \ne 0$, being $u(t) = O(t^{-\alpha})$, as $t \to 0^+$.

The way a) requires weaker conditions in that the integral at the R.H.S. of (4.65) must vanish as $t \to 0^+$; consequently $f(0^+)$ could be infinite but with $f(t) = O(t^{-\nu})$, $0 < \nu < 1 - \alpha$ as $t \to 0^+$. To this end keep in mind that $\Phi_{1-\alpha} * \Phi_{1-\nu} = \Phi_{2-\alpha-\nu}$. Then it turns out from (4.65) that $u(0^+)$ can be infinite if $f(0^+)$ is infinite, being $u(t) = O(t^{-(\alpha+\nu)})$, as $t \to 0^+$.

Finally, let us remark that we can analogously treat the case of equation (4.60) with $0 < \alpha < 1$ replaced by $\alpha > 0$. If $m-1 < \alpha \le m$ with $m \in \mathbb{N}$, then again we have (4.61), now with $D^\alpha f(t)$ given by the formula (4.52) which can also be obtained by the Laplace transform method.

Abel's equation of the second kind

Let us now consider the Abel equation of the second kind

$$u(t) + \frac{\lambda}{\Gamma(\alpha)} \int_0^t \frac{u(\tau)}{(t-\tau)^{1-\alpha}}\, d\tau = f(t)\,, \quad \alpha > 0\,, \quad \lambda \in \mathbb{C}\,. \tag{4.68}$$

In terms of the fractional integral operator such equation reads

$$(1 + \lambda J^\alpha)\, u(t) = f(t)\,, \tag{4.69}$$

and consequently can be formally solved as follows:

$$u(t) = (1 + \lambda J^\alpha)^{-1} f(t) = \left(1 + \sum_{n=1}^{\infty} (-\lambda)^n J^{\alpha n}\right) f(t)\,. \tag{4.70}$$

Noting by (4.46-4.47) that

$$J^{\alpha n} f(t) = \Phi_{\alpha n}(t) * f(t) = \frac{t_+^{\alpha n-1}}{\Gamma(\alpha n)} * f(t)$$

the formal solution reads

$$u(t) = f(t) + \left(\sum_{n=1}^{\infty} (-\lambda)^n \frac{t_+^{\alpha n-1}}{\Gamma(\alpha n)}\right) * f(t)\,. \tag{4.71}$$

Using the definition of the *Mittag-Leffler function*,[31]

[31] The Mittag-Leffler function $E_\alpha(z)$ with $\alpha > 0$ is defined by the following series representation, valid in the whole complex plane,

$$E_\alpha(z) := \sum_{n=0}^{\infty} \frac{z^n}{\Gamma(\alpha n + 1)}\,, \quad \alpha > 0\,, \quad z \in \mathbb{C}\,.$$

It turns out that $E_\alpha(z)$ is an *entire function* of order $\rho = 1/\alpha$ and type 1. This property is still valid but with $\rho = 1/\mathrm{Re}\{\alpha\}$, if $\alpha \in \mathbb{C}$ with *positive real part*.The Mittag-Leffler function provides a simple generalization of the exponential function because of the substitution of $n! = \Gamma(n+1)$ with $(\alpha n)! = \Gamma(\alpha n + 1)$ in the denominator of the terms of the exponential series.

$$e_\alpha(t;\lambda) := E_\alpha(-\lambda\, t^\alpha) = \sum_{n=0}^{\infty} \frac{(-\lambda\, t^\alpha)^n}{\Gamma(\alpha n + 1)}, \quad t > 0, \quad \alpha > 0, \quad \lambda \in \mathbb{C}, \tag{4.72}$$

where E_α denotes the Mittag-Leffler function of order α, we note that

$$\sum_{n=1}^{\infty} (-\lambda)^n \frac{t_+^{\alpha n-1}}{\Gamma(\alpha n)} = \frac{d}{dt} E_\alpha(-\lambda t^\alpha) = e'_\alpha(t;\lambda), \quad t > 0. \tag{4.73}$$

Finally, the solution reads

$$u(t) = f(t) + e'_\alpha(t;\lambda) \, * \, f(t). \tag{4.74}$$

The above formal proof based on the series development of the operator $(1 + \lambda J^\alpha)^{-1}$,can be made rigorous. Simply observe that because of the rapid growth of the gamma function the infinite series in (4.71) and (4.73) are uniformly convergent in every bounded interval of the variable t so that term-wise integrations and differentiations are allowed. However, we prefer to use the alternative technique of Laplace transforms, which will allow us to obtain the solution in different forms, including the result (4.74).

Applying the Laplace transform to (4.68) we obtain

$$\left(1 + \frac{\lambda}{s^\alpha}\right) U(s) = F(s) \Longrightarrow U(s) = \frac{s^\alpha}{s^\alpha + \lambda} F(s). \tag{4.75}$$

Now, let us proceed to obtain the inverse Laplace transform of (4.75) using the following Laplace transform pair:

$$e_\alpha(t;\lambda) := E_\alpha(-\lambda\, t^\alpha) \div \frac{s^{\alpha-1}}{s^\alpha + \lambda}. \tag{4.76}$$

As for the Abel equation of the first kind, we can choose two different ways to get the inverse Laplace transforms from (4.75), according to the standard rules.

Writing (4.75) as

$$\bar{u}(s) = s \left[\frac{s^{\alpha-1}}{s^\alpha + \lambda} F(s) \right],$$

we obtain

$$u(t) = \frac{d}{dt} \int_0^t f(t-\tau)\, e_\alpha(\tau;\lambda)\, d\tau.$$

If we write (4.75) as

$$U(s) = \frac{s^{\alpha-1}}{s^\alpha + \lambda} \left[sF(s) - f(0^+) \right] + f(0^+) \frac{s^{\alpha-1}}{s^\alpha + \lambda},$$

we obtain

$$u(t) = \int_0^t f'(t-\tau)\, e_\alpha(\tau;\lambda)\, d\tau + f(0^+)\, e_\alpha(t;\lambda).$$

We also note that, $e_\alpha(t;\lambda)$ being a function differentiable with respect to t with $e_\alpha(0^+;\lambda) = E_\alpha(0^+) = 1$, there exists another possibility to re-write (4.75), namely

$$U(s) = \left(s\,\frac{s^{\alpha-1}}{s^\alpha+\lambda} - 1\right)F(s) + F(s)\,.$$

Then we obtain

$$u(t) = \int_0^t f(t-\tau)\,e'_\alpha(\tau;\lambda)\,d\tau + f(t)\,,$$

in agreement with (4.74).

We see that the way *b*) is more restrictive than the ways *a*) and *c*) since it requires that $f(t)$ be differentiable with $\mathcal{L}$-transformable derivative.

Newtonian heating problem

Consider the *equation of a heat flow*:

$$u_t - u_{xx} = 0\,, \quad u = u(x,t)\,, \tag{4.77}$$

in the semi-infinite intervals $0 < x < \infty$ and $0 < t < \infty$ of space and time, respectively. In this dimensionless equation $u = u(x,t)$ means temperature. Assume vanishing initial temperature, i.e., $u(x,0) = 0$ for $0 < x < \infty$ and given influx across the boundary $x = 0$ from $x < 0$ to $x > 0$,

$$-\,u_x(0,t) = p(t)\,. \tag{4.78}$$

Then, under appropriate regularity conditions, $u(x,t)$ is given by the formula:

$$u(x,t) = \frac{1}{\sqrt{\pi}}\int_0^t \frac{p(\tau)}{\sqrt{t-\tau}}\,\mathrm{e}^{-x^2/[4(t-\tau)]}\,d\tau\,, \quad x > 0\,,\ t > 0\,. \tag{4.79}$$

We turn our special interest to the interior boundary temperature $\phi(t) := u(0^+,t)\,,\ t > 0$, which by (4.79) is represented as

$$\frac{1}{\sqrt{\pi}}\int_0^t \frac{p(\tau)}{\sqrt{t-\tau}}\,d\tau = J^{1/2}\,p(t) = \phi(t)\,, \quad t > 0\,. \tag{4.80}$$

We recognize (4.80) as an Abel integral equation of first kind for *determination of an unknown influx $p(t)$ if the interior boundary temperature $\phi(t)$ is given* by measurements, or intended to be achieved by controlling the influx. Its solution is given by formula (4.52) with $m = 1$, $\alpha = 1/2$, as

$$p(t) = D^{1/2}\,\phi(t) = \frac{1}{\sqrt{\pi}}\frac{d}{dt}\int_0^t \frac{\phi(\tau)}{\sqrt{t-\tau}}\,d\tau\,. \tag{4.81}$$

We now modify our problem to obtain an Abel integral equation of *second kind*. Assume that the rod $x > 0$is bordered at $x = 0$by a bath of liquid in $x < 0$ with controlled exterior boundary temperature $u(0^-,t) := \psi(t)$.

Assuming Newton's radiation law, we have an influx of heat from 0^- to 0^+ proportional to the difference of exterior and interior temperature,

$$p(t) = \lambda\, [\psi(t) - \phi(t)]\,, \quad \lambda > 0\,. \tag{4.82}$$

Inserting (4.82) into (4.80) we obtain

$$\phi(t) = \frac{\lambda}{\sqrt{\pi}} \int_0^t \frac{\psi(\tau) - \phi(\tau)}{\sqrt{t-\tau}}\, d\tau\,,$$

namely, in operator notation,

$$\left(1 + \lambda\, J^{1/2}\right) \phi(t) = \lambda\, J^{1/2}\, \psi(t)\,. \tag{4.83}$$

If we *now assume the exterior boundary temperature* $\psi(t)$ *as given and the evolution in time of the interior boundary temperature* $\phi(t)$ *as unknown*, then (4.83) is an Abel integral equation of *second kind* for determination of $\phi(t)$.

Fractional Diffusion-Wave Equation

In this section, we deal with the family of evolution equations obtained from the standard diffusion equation (or the D'Alembert wave equation) by replacing the first-order (or the second-order) time derivative by a fractional derivative of order α with $1 \le \alpha \le 2$, namely

$$\frac{\partial^\alpha u}{\partial t^\alpha} = \frac{\partial^2 u}{\partial x^2}, \tag{4.84}$$

where $x \in \mathbb{R}$, $t \in \mathbb{R}^+$ denote the space and time variables, respectively.

In (4.84), $u = u(x,t)$ represents the response field variable and the fractional derivative of order α, $n-1 < \alpha < n$, $n \in \mathbb{N}$ is defined in the Caputo sense (compare with (4.55)):

$$\frac{\partial^\alpha u}{\partial t^\alpha} = \frac{1}{\Gamma(n-\alpha)} \int_0^t (t-\tau)^{n-\alpha-1} \frac{\partial^n u(\tau)}{\partial \tau^n}\, d\tau. \tag{4.85}$$

For $\alpha = n$, $n \in \mathbb{N}$, the Caputo fractional derivative is defined as the standard derivative of order n.

In order to guarantee existence and uniqueness of a solution, we must add to (4.84) some initial and boundary conditions. Denoting by $f(x)$, $x \in \mathbb{R}$ and $h(t)$, $t \in \mathbb{R}^+$ sufficiently well-behaved functions, the Cauchy problem for the time-fractional diffusion-wave equation with $1 \le \alpha \le 2$ is formulated as follows:

$$\begin{cases} u(x,0) = f(x)\,, & (\text{for } -\infty < x < +\infty), \\ u(\mp\infty, t) = 0\,, & (\text{for } t > 0). \end{cases} \tag{4.86}$$

If $1 < \alpha \le 2$, we must add to (4.86) the initial value of the first time derivative of the field variable, $u_t(x,0)$, since in this case the Caputo fractional derivative is expressed in terms of the second order time derivative. To ensure continuous dependence of the solution with respect to the parameter α we agree to assume

$$u_t(x,0) = 0\,, \qquad (\text{for } \; 1 < \alpha \le 2)\,.$$

In view of our subsequent analysis we find it convenient to set $\nu := \alpha/2$, so that $1/2 \le \nu \le 1$ for $1 \le \alpha \le 2$.

For the Cauchy problem, we introduce the so-called Green function $\mathcal{G}_c$ $(x, t; \nu)$, which represents the respective fundamental solution, obtained when $f(x) = \delta(x)$, δ being the Dirac δ-function. As a consequence, the solution of the Cauchy problem is obtained by a space convolution according to

$$u(x,t;\nu) = \int_{-\infty}^{+\infty} \mathcal{G}_c(x-\xi, t; \nu)\, f(\xi)\, d\xi\,.$$

It should be noted that $\mathcal{G}_c(x,t;\nu) = \mathcal{G}_c(|x|,t;\nu)$ since the Green function of the Cauchy problem turns out to be an even function of x. This means that we can restrict our investigation of the function $\mathcal{G}_c$ to non-negative values $x \ge 0$.

For the standard diffusion equation ($\nu = 1/2$) it is well known that

$$\mathcal{G}_c(x,t;1/2) := \mathcal{G}_c^d(x,t) = \frac{t^{-1/2}}{2\sqrt{\pi}}\, \mathrm{e}^{-x^2/4t}\,.$$

In the limiting case $\nu = 1$ we recover the standard wave equation, for which we get

$$\mathcal{G}_c(x,t;1) := \mathcal{G}_c^w(x,t) = \frac{1}{2}\left[\delta(x-t) + \delta(x+t)\right]\,.$$

In the case $1/2 < \nu < 1$, the Green function $\mathcal{G}_c$ can be determined by using the technique of the Laplace and the Fourier transforms.

Here we are interested in investigation of some important characteristics of the Green function $\mathcal{G}_c$ including location of its maximum point, its propagation speed, and its maximum value. Consider an integral representation of the Green function $\mathcal{G}_c$:

$$\mathcal{G}_c(x,t;\nu) = \frac{1}{\pi}\int_0^{\infty} E_{2\nu}\left(-\kappa^2 t^{2\nu}\right)\cos(x\kappa)\, d\kappa, \tag{4.87}$$

where $E_\alpha(z)$ is the *Mittag-Leffler function* defined by the series

$$E_\alpha(z) = \sum_{n=0}^{\infty} \frac{z^n}{\Gamma(\alpha n + 1)}, \qquad \alpha > 0. \tag{4.88}$$

The representation (4.87) can be easily obtained by transforming the Cauchy problem for the equation (4.84) into the Laplace-Fourier domain using the known formula

$$\mathcal{L}\left\{\frac{d^\alpha u(t)}{dt^\alpha}; s\right\} = s^\alpha \mathcal{L}\left\{u(t); s\right\} - \sum_{k=0}^{n-1} u^{(k)}(0^+) s^{\alpha-1-k}, \; n-1 < \alpha \le n, \tag{4.89}$$

with $n \in \mathbb{N}$, for the Laplace transform of the Caputo fractional derivative. This formula together with the standard formulas for the Fourier transform of the second derivative and of the Dirac δ-function lead to the representation

$$\widehat{\widetilde{\mathcal{G}}}_c(\kappa, s, \nu) = \frac{s^{2\nu-1}}{s^{2\nu} + \kappa^2}, \quad \nu = \alpha/2, \tag{4.90}$$

of the Laplace-Fourier transform $\widehat{\widetilde{\mathcal{G}_c}}$ of the Green function $\mathcal{G}_c$. Using the well-known Laplace transform formula:

$$\mathcal{L}\{E_\alpha(-t^\alpha); s\} = \frac{s^{\alpha-1}}{s^\alpha + 1},$$

and applying to the R.H.S of the formula (4.90) first the inverse Laplace transform and then the inverse Fourier transform we obtain the integral representation (4.87) if we take into consideration the fact that the Green function of the Cauchy problem is an even function of x that follows from the formula (4.90).

4.3.4 Rigorous Geometric Quantization

Now we perform the rigorous CHO-quantization in the Heisenberg picture using machinery from symplectic mechanics. Recall from section 4.2.2 that Dirac's canonical quantization (4.16) means that for any pair of dynamical variables (f, g), the quantum commutator $[\hat{f}, \hat{g}]$ has the same values as the classical Poisson bracket $\{f, g\}$. In this section, we will look at Dirac's quantization from symplectic geometric perspective.

Also, recall (see, e.g. [Put93, II06b]) that a *symplectic structure* on a smooth manifold M is a nondegenerate *closed* 2-form ω on M, i.e.,

$$(\forall x \in M) \quad d\omega = 0 \quad \text{and} \quad \omega(x) \text{ is nondegenerate.}$$

Let T_x^*Q be a cotangent space to a smooth configuration manifold Q of some classical system at the point $x \in Q$ that determines its state. The cotangent bundle $M = T^*Q$ represents a union $\cup_{x\in Q} T_x^*Q$,[32] the dimension of which is twice the dimension of Q. A *canonical* 1-*form* θ on M, such that $\omega = d\theta$, represents a section $\theta : M \to T^*M$ of the cotangent bundle M.[33] A symplectic phase-space manifold is a pair (M, ω). In our case of the CHO (4.15), we have (p, x) as canonical coordinates on $\mathbb{R}^2$ and the following symplectic dynamics (with unit mass and spring, for simplicity):

$$M = T^*\mathbb{R} \simeq \mathbb{R}^2, \quad \omega = dp \wedge dx, \quad H(p, x) = \frac{1}{2}\left(p^2 + x\right), \tag{4.91}$$

---

[32] together with the standard topology on T^*Q and a natural smooth manifold structure (see, e.g. [II06b])

[33] In the general case of an arbitrary symplectic manifold (M, ω) (not necessarily the cotangent bundle) we can find only a locally exact 1-form θ (such that $\omega = d\theta$).

from which the Hamiltonian equations of CHO-motion read:

$$\dot{x} = p, \qquad \dot{p} = -x.$$

In general, for any pair of smooth functions $f, g \in C^\infty(M, \mathbb{R})$ the Poisson bracket is given by

$$\{f, g\}_\omega = \frac{\partial f}{\partial x}\frac{\partial g}{\partial p} - \frac{\partial f}{\partial p}\frac{\partial g}{\partial x}.$$

Dirac's quantization can be formulated in the language of symplectic geometry in the following way: we can associate smooth functions $f, g, h, ...$ of any classical system (like the CHO), defined on their symplectic phase-space manifold (M, ω), with operators on a Hilbert space $\mathcal{H}(\mathbb{C})$, with the inner product (4.32), norm (4.33) and metric (4.34), in such a way that their Poisson brackets correspond to quantum commutators. Abstractly speaking, there is a *quantization functor*[34] from the category **Symplec** (of symplectic manifolds and smooth functions defined on them) to the category **Hilbert** (of complex Hilbert spaces and quantum Hermitian operators defined on them).

Formally, a symplectic manifold $(M, \omega = d\theta)$ is *quantizable* (i.e., we can define the Hilbert representation space $\mathcal{H}(\mathbb{C})$ and the Hermitian quantum operators δ_f in a globally consistent way) if ω defines an *integral cohomology class* (that is, it integrates to an integer on any 1D submanifold of M, i.e., an algebraic curve). If (M, ω) is a quantizable manifold, then the pair $(\mathcal{H}, \delta)$ defines its quantization (see [Put93]).

Let the Hamiltonian structure of the CHO be given by the symplectic dynamics (4.91). If the canonical 1-form θ on M (also called the *symplectic potential* of ω) is given by: $\theta = \frac{1}{2}(pdx - xdp)$, then the spectrum (i.e., set of eigenvalues) of the Hamiltonian operator:

$$\delta_H \equiv \hat{H} = \mathrm{i}\hbar\left(x\frac{\partial}{\partial p} - p\frac{\partial}{\partial x}\right) \quad \text{is} \quad \{..., -2\hbar, -\hbar, 0, \hbar, 2\hbar, ...\},$$

where each eigenvalue occurs with infinite multiplicity.

Furthermore, let $\mathfrak{g}$ be the vector space of the CHO, spanned by the operators:

$$\delta_x = \mathrm{i}\hbar\frac{\partial}{\partial p} + x, \quad \delta_p = -\mathrm{i}\hbar\frac{\partial}{\partial x}, \quad \delta_H = \mathrm{i}\hbar\left(x\frac{\partial}{\partial p} - p\frac{\partial}{\partial x}\right), \text{ and } Id. \tag{4.92}$$

Then $\mathfrak{g}$ is a *solvable Lie algebra*,[35] given by (see [Put93]):

---

[34] Strictly speaking, in this section we are performing the *(pre)quantization functor*, but for our application this subtle difference can be neglected.

[35] A Lie algebra $\mathfrak{g}$ is a vector space V provided with the commutator $[,] : \mathfrak{g} \times \mathfrak{g} \to \mathfrak{g}$, which is bilinear, skew-symmetric, and satisfies the *Jacobi identity* (for any three vector-fields $X, Y, Z \in \mathfrak{g}$):

$$[[X, Y], Z] = [X, [Y, Z]] - [X, [Y, Z]].$$

$$[\delta_p, \delta_x] = \mathrm{i}\hbar\delta_{\{p,x\}_\omega} = \mathrm{i}\hbar\, Id,$$
$$[\delta_H, \delta_x] = \mathrm{i}\hbar\delta_{\{H,x\}_\omega} = -\mathrm{i}\hbar\delta_p,$$
$$[\delta_H, \delta_p] = \mathrm{i}\hbar\delta_{\{H,p\}_\omega} = \mathrm{i}\hbar\delta_x,$$

such that the *commutator of Lie algebras* $[\mathfrak{g}, \mathfrak{g}]$ is itself spanned by (4.92), or equivalently, $\mathfrak{g}$ is called the *Heisenberg Lie algebra.*

4.3.5 Supervised Machine-Learning Methods

Given the $(n+1)$D vector of synaptic weights $\mathbf{w}(n) = [w_0(n), ..., w_D(n)]^T$ (with $w_0 = bias$), and the correspondent Mean Square Error (MSE) gradient (including partial derivatives of MSE w.r.t. weights)

$$\nabla\mathbf{e} = \left[\frac{\partial e}{\partial w_0}, ..., \frac{\partial e}{\partial w_D}\right]^T,$$

and the learning rate (step size) η, we have the following learning algorithms.

Basic gradient-descent method:

$$\mathbf{w}(n+1) = \mathbf{w}(n) - \eta\nabla\mathbf{e}(n),$$

which in index form reads

$$w_i(n+1) = w_i(n) - \eta\nabla e_i(n).$$

Least mean square (LMS) algorithm:

$$\mathbf{w}(n+1) = \mathbf{w}(n) + \eta\varepsilon(n)\, x(n),$$

where x is an input (measurement) vector, and ε is a zero-mean Gaussian noise vector uncorrelated with input, or

$$w_i(n+1) = w_i(n) + \eta\varepsilon(n)\, x^i(n).$$

Newton's method:

$$\mathbf{w}(n+1) = \mathbf{w}(n) - \eta\,\mathbf{R}^{-1}\mathbf{e}(n),$$

where $\mathbf{R}$ is input (auto)correlation matrix, or

$$\mathbf{w}(n+1) = \mathbf{w}(n) + \eta\,\mathbf{R}^{-1}\varepsilon(n)\, x(n),$$

Lie algebra $\mathfrak{g}$ is *solvable* if the derived series of commutators:

$$\mathfrak{g} > [\mathfrak{g}, \mathfrak{g}] > [[\mathfrak{g}, \mathfrak{g}], [\mathfrak{g}, \mathfrak{g}]] > [[[\mathfrak{g}, \mathfrak{g}], [\mathfrak{g}, \mathfrak{g}]], [[\mathfrak{g}, \mathfrak{g}], [\mathfrak{g}, \mathfrak{g}]]] > \cdots$$

eventually converges to zero.

Conjugate-gradient method:

$$\begin{aligned}\mathbf{w}(n+1) &= \mathbf{w}(n) + \eta\,\mathbf{p}(n),\\ \mathbf{p}(n) &= -\nabla\mathbf{e}(n) + \beta(n)\mathbf{p}(n-1),\\ \beta(n) &= \frac{\nabla\mathbf{e}(n)^T\nabla\mathbf{e}(n)}{\nabla\mathbf{e}(n-1)^T\nabla\mathbf{e}(n-1)}.\end{aligned}$$

4.3.6 First-Order Logic and Quantum Random Fields

First-order theories

Here, we are interested in the implementation of first-order theories for the purpose of knowledge representation and automated reasoning; towards this end, we present a condensed overview of some fundamental concepts in first-order logic. For more comprehensive treatments. Sentences in a first-order language L are constructed using disjoint countable sets of constants Σ^C, variables Σ^V, functions Σ^F and relations Σ^R, as well as logical connective symbols and quantifier symbols $\Omega = \{\neg, \wedge, \vee, \Rightarrow, \Leftrightarrow, \exists, \forall\}$. Constant symbols Σ^C name objects in the application domain of interest. Function symbols Σ^F map ordered tuples of constants to constants. Variable symbols Σ^V range over constants. Relations symbols Σ^R represent attributes of objects or describe relationships between objects in the domain. The non-logical symbols comprising the symbols in Σ^C, Σ^F and Σ^R, together with functions $v^F : \Sigma^F \longrightarrow \mathbb{N}$ and $v^R : \Sigma^R \longrightarrow \mathbb{N}$ giving the valence of function symbols and relation symbols, form the signature, say σ, of L.

A term is an expression that represents any object in the domain, and can be a constant, a variable, or a function of a tuple of terms; note also that a function may itself take a function term as one of its arguments. A ground term is any term that contains no variables. An atomic sentence in L is then any relation over terms, and a literal is an atomic sentence or a negated atomic sentence. Ground atoms are atomic sentences over atomic terms, and the (possibly countably infinite) set of all possible ground atoms generated by the signature σ of L is its Herbrand base, say $\mathcal{H}(\sigma)$. Compound sentences in L can also be constructed from atomic sentences in L using logical connective symbols and quantifier symbols in Ω. If ρ and ϕ are a sentences in L, then the following hold:

Negation: $\neg\rho \in L$ which evaluates to true iff ρ is false.
Conjunction: $\rho \wedge \phi \in L$, which is true iff both ρ and ϕ are true.
Disjunction: $\rho \vee \phi \in L$, which is true iff either ρ or ϕ are true.
Material Implication: $\rho \Rightarrow \phi \in L$, which is true iff either ρ is false or ϕ is true.
Material Equivalence: $\rho \Leftrightarrow \phi \in L$, which is true iff either both ρ and ϕ are false or both ρ and ϕ are true.
Universal Quantification: $\forall x\rho \in L$, which is true iff ρ is true for every assignment of constants to x.

Existential Quantification: $\exists x\rho \in L$, which is true iff ρ is true for some assignment of a constant to x.

Note that a *quantifier symbol* is not an operator, but rather can be considered as a unary operator in combination with the variable it binds.

We are specifically interested in first-order theories, which constitute knowledge about domains of interest: a first-order theory T is simply a set $T = \{\rho_1, ..., \rho_n\}$ of sentences ρ_i in some first-order language L [36]. The sentences in the set T are considered to be in conjunction, so T logically corresponds to the sentence $\rho_1 \wedge ... \wedge \rho_n \in L$. The sentences $\rho_i \in T$ are sometimes said to be axioms of T. For convenience, we will also directly associate the signature σ of language L with the theory T.

An interpretation is a mapping from a domain to symbols, and a Herbrand interpretation assigns true or false to every possible ground atom (noting that they may be countably infinite in number). Any sentence in L is said to be satisfiable if there is at least one possible Herbrand interpretation in which it evaluates to true; otherwise, the sentence is unsatisfiable. A sentence is said to be a tautology if it evaluates to true for all possible Herbrand interpretations. Any Herbrand interpretation for which all sentences ρ_i in a theory T evaluate to true is said to be a model of that theory.

It is often computationally convenient to convert sentences to equivalent sentences having regular structures, such as Disjunctive Normal Form (DNF) and Conjunctive Normal Form (CNF). The former is satisfied by all disjunctions over conjunctions of literals, the latter by all conjunctions of disjunctions of literals. Sentences in normal forms are not necessarily equivalent to those they replace, but are equisatisfiable with them. Existential quantifiers can be removed by Skolemization, which replaces existentially quantified variables with constant symbols or function symbols that are new (i.e. not in Σ^C or Σ^F, respectively). In finite domains, existentially quantified sentences can be replaced with disjunctions. Universal quantifiers can be similarly handled by

[36] There are numerous variations on this construction in the literature. Function-free first-order languages do not permit function symbols. We could also do away with constant symbols by admitting function symbols of valence 0 instead. The assumption implicit here that each relation symbol has unique valence can be relaxed in the obvious manner. Different references use various different symbols for the logical connectives, and we could reduce or extend the set of logical connectives in various ways. An important variation is the introduction of equality ($=$), also called identity, either as a relation symbol conventionally written in infix form, or as a part of the logic itself with the addition of a couple of extra rules in the definition of a term. Partial order relations may similarly appear in some constructions, either as relation symbols or directly in the logic with additional rules in the definition of terms. Finally, it can be useful to introduce a uniqueness modifier symbol on existential quantifiers, usually written as $\exists! x\rho$, which evaluates to true iff there is precisely one value of x in the domain that for which ρ is true.

some form of Skolemization with the observation that $\forall x \rho = \neg \exists x \neg \rho$, or, in finite domains, with conjunctions.

Logical inference in first-order logic is about deciding logical entailment: A first-order theory T is said to entail a sentence ρ when, in all interpretations in which all sentences in T are true, ρ is also true. Like satisfiability, this problem is only semi-decidable [37], which results in the widespread use crafted subsets of first-order logic (the most common of which is Horn Clauses, each of which contains at most one positive literal) to make inference on first-order theories implemented by knowledge bases decidable.

First-order logic and quantum random fields

The basic motivation behind using Markov Random Fields in automated reasoning and knowledge representation is to encode first-order theories as Markov Random fields. The effect is to wrap the first-order theory T in a kind of Bayesian shell: instead of a Herbrand interpretation either being a model of the theory or not, we regard it as more or less likely depending on how many sentences of the theory evaluate to false under it and how relatively important we regard satisfaction of each sentence of the theory. We associate with each sentence of T a weight that describes its relative importance. Typically, these weights are real-valued; with our proposed Quantum Random Fields, these weights will be complex-valued instead.

Each sentence in T is represented by a single clique (a completely connected component of the graph).

[37] A decision problem is semidecidable when there is an effective procedure that always halts with the answer "yes" on problem instances that are members of the set under question, but may map some instances that are not a member of the set to non-terminating computations rather than halting with "no". Equivalently, the set is said to be "recursively enumerable".

5

Hierarchical Self-Similarity in Group and Crowd Behaviors

In this Chapter,[1] a nonlinear, complex, Hamiltonian description of socio–cognio–physical dynamics at the macroscopic, classical, inter-personal crowd level and microscopic, quantum, intra-personal agent level, is presented, uniquely, in the form of the *open Liouville equation.* At the microscopic level, this can be considered to be a nonlinear extension of the linear correlation and factor dynamics. This implies the arrow of time in both microscopic and macroscopic processes and shows the existence of the formal crowd-agent space-time self-similarity. This in itself shows the *existence of a unique control law*, which acts on different scales of agent functioning. This self-similar socio–cognio–physical control law enables us to use the crowd dynamics simulator (previously developed at Defence Science & Technology Organisation, Australia), for recursive simulation of individual agents' representation spaces on a cluster of computers.

5.1 Introduction

There are a number of reasons why an understanding of *nonlinear complexity* is needed for modeling and simulation of multi-component systems, both in the general case, and for human *behavioral group dynamics,* in particular. In the general case, it is well-known that a system of interacting forces will be interdependent, due to nonlinear interactions, and this is the quality which is lost by decoupling and/or linearization (see, e.g. [Hak83, Hak93]. For *behavioral group dynamics*[2], it is also not possible to look at individual behaviors of

[1] The work presented in this Chapter has been developed in collaboration with Mr. Wayne Johnson, a senior defence scientist.

[2] Here we mean to imply that the resultant path and an individual agent will follow is due to a resultant *field* which has contributions from a social-field, a cognitive-field and a physical-field.

agents, in a crowded street scene for instance, in complete isolation[3], without breaking the whole web of interdependencies [Gol99].

We have recently shown that both individual and group/crowd behavioral dynamics exists both at the macroscopic, inter–personal, classical level and at the microscopic, intra–personal, quantum level (see [IR12] and references therein). In the present Chapter we try to describe group/crowd behavioral *socio–cognio–physics* on both levels by the *unique form* of a single equation, namely *open Liouville equation:* crowd–dynamics using its classical form, and individual agent dynamics using its quantum form in the Heisenberg picture. If this formulation is consistent, that would prove the *existence* of the formal macro–micro socio–cognio–physical crowd self-similarity. This proof implies the *existence of a unique behavioral control law*, which acts on different scales of human agent functioning. Such a self-similar control law would enable us to use the CDS, including the NLS effects, for recursive simulation of individual agents' representation spaces on a cluster of computers.

As this is an interdisciplinary study we will address technical preliminaries, concerning some nonlinear extensions of linear factor analysis first. Subsequently we will return to the more specify technical points concerning the *open Liouville equation framework*.

5.1.1 From Correlation Dynamics to Quantum Dynamics

To develop a linear correlation–factor dynamics model, we are using geometrical analogy with Schrödinger quantum mechanics (see [II08b, II09] for technical details). A time-dependent state of a quantum system is determined by a normalized (complex), time–dependent, wave psi–function $\psi = \psi(t)$, i.e. a unit Dirac's 'ket' vector $|\psi(t)\rangle$, an element of the Hilbert space $L^2(\psi)$ with a coordinate basis (q^i), under the action of the Hermitian operators, obtained by the procedure of quantization of classical mechanical quantities, for which real eigenvalues are measured. The state–vector $|\psi(t)\rangle$, describing the motion of L. de Broglie's waves, has a statistical interpretation as the probability amplitude of the quantum system, for the square of its magnitude determines the density of the probability of the system detected at various points of space. The summation over the entire space must yield unity and this is the normalization condition for the psi–function, determining the unit length of the state vector $|\psi(t)\rangle$.

In the coordinate q–representation and the Schrödinger S–picture we consider an action of an evolution operator (in normal units Planck constant $\hbar = 1$)

$$\hat{S} \equiv \hat{S}(t, t_0) = \exp[-\mathrm{i}\hat{H}(t - t_0)],$$

i.e., a one–parameter Lie–group of unitary transformations evolving a quantum system. The action represents an exponential map of the system's total

[3] As per a classic behavioral stimulus-response (S-R) controlled laboratory psychology experiment.

energy operator – Hamiltonian $\hat{H} = \hat{H}(t)$. It moves the quantum system from one instant of time, t_0, to some future time t, on the state–vector $|\psi(t)\rangle$, rotating it: $|\psi(t)\rangle = \hat{S}(t, t_0)|\psi(t_0)\rangle$. In this case the Hilbert coordinate basis (q^i) is fixed, so the system operators do not evolve in time, and the system evolution is determined exclusively by the time–dependent Schrödinger equation, in Dirac's notation given by:

$$\mathrm{i}\partial_t|\psi(t)\rangle = \hat{H}(t)|\psi(t)\rangle, \tag{5.1}$$

with initial condition given at one instant of time t_0 as $|\psi(t_0)\rangle = |\psi\rangle$.

If the Hamiltonian $\hat{H} = \hat{H}(t)$ does not explicitly depend on time (which is the case with the absence of variables of macroscopic fields), the state vector reduces to the exponential of the system energy:

$$|\psi(t)\rangle = \exp(-\mathrm{i}E(t - t_0)|\psi\rangle,$$

satisfying the time–independent (i.e., stationary) Schrödinger equation

$$\hat{H}|\psi\rangle = E|\psi\rangle, \tag{5.2}$$

which represents the characteristic equation for the Hamiltonian operator $\hat{H}$ and gives its real eigenvalues (stationary energy states) E_n and corresponding orthonormal eigenfunctions (i.e., probability amplitudes) $|\psi_n\rangle$.

To model the correlation and factor dynamics we start with the characteristic equation for the *correlation matrix*.

$$\mathbf{R}\mathbf{x} = \lambda\mathbf{x},$$

making heuristic analogy with the stationary Schrödinger equation (5.2). This analogy allows a 'physical' interpretation of the correlation matrix $\mathbf{R}$ as an operator of the 'total correlation or covariation energy' of the statistical system (the simulator–test data matrix $\mathbf{X} = \{x_{i\alpha}\}$), eigenvalues λ_n corresponding to the 'stationary energy states', and eigenvectors $\mathbf{x}_n$ resembling 'probability amplitudes' of the system.

So far we have considered one instant of time t_0. Including the time–flow into the stationary Schrödinger equation (5.2) we get the time–dependent Schrödinger equation (5.1) and returning back with our heuristic analogy, we get the basic equation of the n–dimensional correlation dynamics

$$\partial_t\mathbf{x}(t) = \mathbf{R}(t)\,\mathbf{x}_k(t), \tag{5.3}$$

with initial condition at time t_0 given as a stationary manifest–vectors $\mathbf{x}_k(t_0) = \mathbf{x}_k$ $(k = 1, \ldots, n)$.

In a more realistic case of 'many' observables (i.e., very big n), instead of the correlation dynamics (5.3), we can use the reduced–dimension factor dynamics, represented by analogous equation in the factor space spanned by the extracted (oblique) factors $\mathbf{F} = \mathbf{f}_i$, defined by the *oblique factor model* (see

[CL71, And84, Har75]) with inter–factor–correlation matrix $\mathbf{C} = c_{ij}$ $(i, j = 1, \ldots, \text{no. of factors})$[4]

$$\partial_t \mathbf{f}_i(t) = \mathbf{C}(t)\, \mathbf{f}_i(t), \tag{5.4}$$

subject to initial condition at time t_0 given as stationary vectors $\mathbf{f}_i(t_0) = \mathbf{f}_i$.

Now, according to the fundamental existence and uniqueness theorem for linear autonomous ordinary differential equations, if $A = A(t)$ is an $n \times n$ real matrix, then the initial value problem

$$\partial_t \mathbf{x}(t) = A\mathbf{x}(t), \qquad \mathbf{x}(0) = \mathbf{x}_0 \in \mathbb{R}^n,$$

has the unique solution

$$\mathbf{x}(t) = \mathbf{x}_0 \mathrm{e}^{tA}, \qquad \text{for all } t \in \mathbb{R}.$$

Therefore, analytical solutions of our correlation and factor–correlation dynamics equations (5.3) and (5.4) are given respectively by exponential maps

$$\mathbf{x}_k(t) = \mathbf{x}_k \exp[t\,\mathbf{R}],$$
$$\mathbf{f}_i(t) = \mathbf{f}_i \exp[t\,\mathbf{C}].$$

Thus, for each $t \in \mathbb{R}$, the matrix $\mathbf{x} \exp[t\,\mathbf{R}]$, respectively the matrix $\mathbf{f} \exp[t\,\mathbf{C}]$, maps

$$\mathbf{x}_k \mapsto \mathbf{x}_k \exp[t\,\mathbf{R}], \qquad \text{respectively} \qquad \mathbf{f}_i \mapsto \mathbf{f}_i \exp[t\,\mathbf{C}].$$

The sets $g^t_{corr} = \{\exp[t\,\mathbf{R}]\}_{t\in\mathbb{R}}$ and $g^t_{fact} = \{\exp[t\,\mathbf{C}]\}_{t\in\mathbb{R}}$ are 1–parameter families (groups) of linear maps of $\mathbb{R}^n$ into $\mathbb{R}^n$, representing the *correlation flow*, respectively the *factor–correlation flow* of simulator–tests. The linear flows g^t (representing both g^t_{corr} and g^t_{fact}) have two essential properties:

1. identity map: $g^0 = I$, and
2. composition: $g^{t_1+t_2} = g^{t_1} \circ g^{t_2}$.

They partition the state space $\mathbb{R}^n$ into subsets that we call 'correlation orbits', respectively 'factor–correlation orbits', through the initial states $\mathbf{x}_k$, and $\mathbf{f}_i$, of simulator tests, defined respectively by

$$\gamma(\mathbf{x}_k) = \{\mathbf{x}_k g^t | t \in \mathbb{R}\} \qquad \text{and} \qquad \gamma(\mathbf{f}_i) = \{\mathbf{f}_i g^t | t \in \mathbb{R}\}.$$

The correlation orbits can be classified as:

1. If $g^t \mathbf{x}_k = \mathbf{x}_k$ for all $t \in \mathbb{R}$, then $\gamma(\mathbf{x}_k) = \{\mathbf{x}_k\}$ and it is called a *point orbit*. Point orbits correspond to equilibrium points in the manifest and the factor space, respectively.

---

[4] The oblique factor model is a higher-order dimensionality reduction technique derivired from principal component analysis, which is itself derived from correlation matrix analysis. The overall aim is to distill statistically the small number of most important socio–cognitive descriptors and their interrelationships, without making any initial assumptions about which descriptors are important.

2. If there exists a $T > 0$ such that $g^T \mathbf{x}_k = \mathbf{x}_k$, then $\gamma(\mathbf{x}_k)$ is called a *periodic orbit*. Periodic orbits describe a system that evolves periodically in time in the manifest and the factor space, respectively.
3. If $g^t \mathbf{x}_k \neq \mathbf{x}_k$ for all $t \neq 0$, then $\gamma(\mathbf{x}_k)$ is called a *non-periodic orbit*.

Analogously, the factor–correlation orbits can be classified as:

1. If $g^t \mathbf{f}_i = \mathbf{f}_i$ for all $t \in \mathbb{R}$, then $\gamma(\mathbf{f}_i) = \{\mathbf{f}_i\}$ and it is called a point orbit. Point orbits correspond to equilibrium points in the manifest and the factor space, respectively.
2. If there exists a $T > 0$ such that $g^T \mathbf{f}_i = \mathbf{f}_i$, then $\gamma(\mathbf{f}_i)$ is called a periodic orbit. Periodic orbits describe a system that evolves periodically in time in the manifest and the factor space, respectively.
3. If $g^t \mathbf{f}_i \neq \mathbf{f}_i$ for all $t \neq 0$, then $\gamma(\mathbf{f}_i)$ is called a non–periodic orbit.

Now, to interpret properly the meaning of (really discrete) time in the correlation matrix $\mathbf{R} = \mathbf{R}(t)$ and factor–correlation matrix $\mathbf{C} = \mathbf{C}(t)$, we can perform a successive time–series $\{t, t + \Delta t, t + 2\Delta t, t + k\Delta t, \cdots\}$ of simulator tests (and subsequent correlation and factor analysis), and discretize our correlation (respectively, factor–correlation) dynamics, to get

$$\mathbf{x}_k(t + \Delta t) = \mathbf{x}_k(0) + \mathbf{R}(t)\,\mathbf{x}_k(t)\Delta t, \qquad \text{and}$$
$$\mathbf{f}_i(t + \Delta t) = \mathbf{f}_i(0) + \mathbf{C}(t)\,\mathbf{f}_i(t)\Delta t,$$

respectively. Finally we can represent the discrete correlation and factor–correlation dynamics in the form of the (computationally applicable) *three–point iterative dynamics equation*, respectively in the manifest space

$$\mathbf{x}_k^{s+1} = \mathbf{x}_k^{s-1} + \mathbf{R}_k^s\,\mathbf{x}_k^s,$$

and in the factor space

$$\mathbf{f}_i^{s+1} = \mathbf{f}_i^{s-1} + \mathbf{C}_i^s\,\mathbf{f}_i^s,$$

in which the time–iteration variable s labels the time occurrence of the simulator tests (and subsequent correlation and factor analysis), starting with the initial state, labelled $s = 0$.

5.2 Modeling Framework: Open Liouville Equation

Now we move into the general framework of the *open Liouville equation*, in which we will demonstrate the self-similarity between macro and micro crowd levels.

5.2.1 Hamiltonian Formalism

Suppose that on the macroscopic crowd level we have a conservative Hamiltonian system acting in a $2N$-dimensional symplectic phase space $M = T^*Q = \{q^i(t), p_i(t)\}$, $i = 1 \dots N$ (which is the cotangent bundle of the crowd–configuration manifold $Q = \{q^i\}$), with a Hamiltonian function $H = H(q^i, p_i, t) : T^*Q \times \mathbb{R} \to \mathbb{R}$. The conservative dynamics is defined by classical Hamiltonian canonical equations:

$$\begin{aligned} \dot{q}^i &= \partial_{p_i} H \quad \text{– contravariant velocity equation}, \\ \dot{p}_i &= -\partial_{q^i} H \quad \text{– covariant force equation}, \end{aligned} \tag{5.5}$$

(here and henceforth overdot denotes the total time derivative). Within the framework of the conservative Hamiltonian system (5.5) we can apply the formalism of classical Poisson brackets: for any two functions $A = A(q^i, p_i, t)$ and $B = B(q^i, p_i, t)$ their Poisson bracket is (using the summation convention) defined as:

$$[A, B] = (\partial_{q^i} A\, \partial_{p_i} B - \partial_{p_i} A\, \partial_{q^i} B).$$

5.2.2 Conservative Classical Dynamics

Any function $A(q^i, p_i, t)$ is called a constant (or, integral) of motion of the conservative system (5.5) if the following equation holds [II06b, II07]:

$$\dot{A} \equiv \partial_t A + [A, H] = 0, \qquad \text{which implies} \qquad \partial_t A = -[A, H]. \tag{5.6}$$

For example, if $A = \rho(q^i, p_i, t)$ is a density function of ensemble phase–points (or, a probability density to see a state $\mathbf{x}(t) = (q^i(t), p_i(t))$ of an ensemble at a moment t), then equation

$$\partial_t \rho = -[\rho, H] \tag{5.7}$$

represents the *Liouville theorem*, which is usually derived from the continuity equation

$$\partial_t \rho + div(\rho\, \dot{\mathbf{x}}) = 0.$$

Conserved quantity here is the Hamiltonian function $H = H(q^i, p_i, t)$, which is the sum of kinetic and potential energy. For example, in case of an ND harmonic oscillator, we have the phase space $M = T^*\mathbb{R}^N \simeq \mathbb{R}^{2N}$, with the symplectic form $\omega = dp_i \wedge dq^i$ and the Hamiltonian (total energy) function:

$$H = \frac{1}{2} \sum_{i=1}^{N} \left[p_i^2 + (q^i)^2\right]. \tag{5.8}$$

The corresponding Hamiltonian vector-field X_H is given by:

$$X_H = p_i \partial_{q^i} - q^i \partial_{p_i},$$

which gives canonical equations:

$$\dot{q}^i = p_i, \qquad \dot{p}_i = -\delta_{ij} q^j, \qquad \text{(where } \delta_{ij} \text{ is the Kronecker symbol).} \qquad (5.9)$$

In addition, for any two smooth ND functions $f, g : \mathbb{R}^{2N} \to \mathbb{R}$, the Poisson bracket is given by:

$$[f, g]_\omega = \frac{\partial f}{\partial q^i}\frac{\partial g}{\partial p_i} - \frac{\partial f}{\partial p_i}\frac{\partial g}{\partial q^i},$$

which implies that the particular functions $f = p_i p_j + q^i q^j$ and $g = p_i q^j + p_j q^i$ (for $i, j = 1, ..., N$) – are constants of motion. This system is integrable in an open set of $T^*\mathbb{R}^n$ with N integrability functions:

$$K_1 = H, \qquad K_2 = p_2^2 + (q^2)^2, \quad ..., \quad K_N = p_N^2 + (q^N)^2.$$

5.2.3 Conservative Quantum Dynamics

Here we perform the formal canonical quantization of the conservative equation (5.7) in the Heisenberg picture: all variables become Hermitian operators (denoted by '$\hat{(\cdot)}$'), the symplectic phase space $T^*Q = \{q^i, p_i\}$ becomes the combined Hilbert state space for N particles: $\mathcal{H} = \mathcal{H}_{\hat{q}^i} \otimes \mathcal{H}_{\hat{p}_i}$ (where $\mathcal{H}_{\hat{q}^i} = \mathcal{H}_{\hat{q}^1} \otimes ... \otimes \mathcal{H}_{\hat{q}^N}$ and $\mathcal{H}_{\hat{p}_i} = \mathcal{H}_{\hat{p}_1} \otimes ... \otimes \mathcal{H}_{\hat{p}_N}$), while the classical Poisson bracket $[\,,]$ becomes the quantum commutator $\{\,,\}$ multiplied by $-\mathrm{i}/\hbar$ [II08a]:

$$[\,,] \longrightarrow -\mathrm{i}\{\,,\} \qquad (\hbar = 1 \text{ in normal units}). \qquad (5.10)$$

In this way the classical Liouville equation (5.7) becomes the *quantum Liouville equation* [II08b]

$$\partial_t \hat{\rho} = \mathrm{i}\{\hat{\rho}, \hat{H}\}, \qquad (5.11)$$

where $\hat{H} = \hat{H}(\hat{q}^i, \hat{p}_i, t)$ is the Hamiltonian evolution operator, while

$$\hat{\rho} = \sum_a P(a)|\Psi_a><\Psi_a|, \quad \text{with} \quad \mathrm{Tr}(\hat{\rho}) = 1$$

denotes the von Neumann *density matrix operator*, where each quantum state $|\Psi_a>$ occurs with probability $P(a)$; $\hat{\rho} = \hat{\rho}(\hat{q}^i, \hat{p}_i, t)$ is closely related to another von Neumann concept – entropy:

$$S = -\mathrm{Tr}(\hat{\rho}[\ln \hat{\rho}]).$$

5.2.4 Open Classical Dynamics: Hamiltonian Crowd Model

We now move to the open (nonconservative) system: on the macroscopic crowd level the opening operation is equivalent to adding a covariant vector-field of external (dissipative and/or motor) forces $F_i = F_i(q^i, p_i, t)$ to (the right-hand-side of) the covariant Hamiltonian force equation, so that crowd Hamiltonian equations[5] obtain the open (dissipative and/or forced) form [II06b, II07]:

[5] For equivalent Lagrangian crowd model, see [IR10c].

$$\dot{q}^i = \partial_{p_i} H, \qquad \dot{p}_i = -\partial_{q^i} H + F_i \,. \tag{5.12}$$

In the framework of the open Hamiltonian system (5.12), dynamics of any function $A(q^i, p_i, t)$ is defined by the open (dissipative and/or forced) evolution equation:

$$\partial_t A = -[A, H] + F_i[A, q^i] \,, \qquad ([A, q^i] = -\partial_{p_i} A) \,. \tag{5.13}$$

In particular, if $A = \rho(q^i, p_i, t)$ represents the density function of ensemble phase-points then its dynamics is given by the *open Liouville equation* (dissipative and/or forced) [II06b, II07]:

$$\partial_t \rho = -[\rho, H] + F_i[\rho, q^i] \,. \tag{5.14}$$

Equation (5.14) represents the open classical model of our microscopic crowd dynamics.

For example, in case of our ND oscillator, Hamiltonian function (5.8) is not conserved any more, the canonical equations (5.9) become

$$\dot{q}^i = p_i, \qquad \dot{p}_i = F_i - \delta_{ij} q^j,$$

and the system is not integrable any more.

5.2.5 Neural Crowd System

Using analogy with neuronal biophysics (see [II06a] and neural references therein), we will consider two different types of agents: (i) 'male' agents of *graded-response type* (GRA), and (ii) 'female' agents of *coupled-oscillators type* (COA). Behavior of both types of agents can be presented in the form of a *Langevin equation* (see, e.g. [II08a])

$$\dot{\sigma}_i = f_i + \eta_i(t), \tag{5.15}$$

where $\sigma_i = \sigma_i(t)$ are the continual personal variables of the ith agent (representing either action potentials in case of GRA, or oscillator phases in case of COA); J_{ij} are individual synaptic weights; $f_i = f_i(\sigma_i, J_{ij})$ are the deterministic forces (given, in GRA-case, by

$$f_i = \sum_j J_{ij} \tanh[\gamma \sigma_j] - \sigma_i + \theta_i, \qquad \text{with} \quad \gamma > 0$$

and with the θ_i representing input functions, and in COA-case, by

$$f_i = \sum_j J_{ij} \sin(\sigma_j - \sigma_i) + \omega_i,$$

with ω_i representing the natural frequencies of the individual oscillators); the noise variables are given as

$$\eta_i(t) = \lim_{\Delta \to 0} \zeta_i(t)\sqrt{2T/\Delta},$$

where $\zeta_i(t)$ denote uncorrelated Gaussian distributed random forces and the parameter T controls the amount of noise in the system, ranging from $T = 0$ (deterministic dynamics) to $T = \infty$ (completely random dynamics).

More convenient, probabilistic description of the random process (5.15) is provided by the Fokker-Planck equation, describing the time evolution of the probability density $P(\sigma_i)$ [II08a]

$$\partial_t P(\sigma_i) = -\sum_i \partial_{\sigma_i}[f_i P(\sigma_i)] + T\sum_i \partial_{\sigma_i^2} P(\sigma_i). \tag{5.16}$$

In the case of deterministic dynamics $T = 0$, equation (5.16) can be easily put into the form of the conservative Liouville equation (5.7), by making the substitutions:

$$P(\sigma_i) \to \rho, \quad f_i = \dot{\sigma}_i, \qquad \text{and} \qquad [\rho, H] = \mathrm{div}(\rho\,\dot{\sigma}_i) \equiv \sum_i \partial_{\sigma_i}(\rho\,\dot{\sigma}_i),$$

where $H = H(\sigma_i, J_{ij})$. Further, we can formally identify the stochastic forces, i.e., the second-order noise-term $T\sum_i \partial_{\sigma_i^2}\rho$ with $F^i[\rho, \sigma_i]$, to get the open Liouville equation (5.14).

Therefore, on the crowd level deterministic dynamics corresponds to the conservative system (5.7). Inclusion of stochastic forces corresponds to the system opening (5.14), implying the macroscopic arrow of time.

5.2.6 Open Quantum System

By formal quantization of equation (5.14), we obtain the *quantum open Liouville equation* [II08a, II08b]:

$$\partial_t \hat{\rho} = \mathrm{i}\{\hat{\rho}, \hat{H}\} - \mathrm{i}\hat{F}_i\{\hat{\rho}, \hat{q}^i\}, \tag{5.17}$$

where $\hat{F}_i = \hat{F}_i(\hat{q}^i, \hat{p}_i, t)$ represents the covariant quantum operator of external friction forces in the Hilbert state space $\mathcal{H} = \mathcal{H}_{\hat{q}^i} \otimes \mathcal{H}_{\hat{p}_i}$.

Equation (5.17) represents the open quantum-decoherence model for our microscopic agent–dynamics. In EMN–language of *non-critical bosonic strings*[6] [EMN99, MN97, Nan95], the general agent–dynamics equation (5.17) expands into *bosonic string equation of motion* [II08a, II08b]:

[6] In order for any string theory to be consistent, its world-sheet (a 2D manifold which describes the embedding of a string in space-time) must be conformally invariant. For the bosonic string this can be accomplished by a world-sheet theory consisting of 26 free bosons. Since each boson is interpreted as a flat space-time dimension, the critical dimension of the bosonic string is 26. The non-critical string theory describes the relativistic string without enforcing the critical dimension. Therefore, a non-critical bosonic string consists of N bosons, and thus has N dimensions.

$$\partial_t \hat{\rho} = \mathrm{i}\{\hat{\rho}, \hat{H}\} - \mathrm{i}\hat{g}_{ij}\{\hat{\rho}, \hat{q}^i\}\hat{\dot{q}}^j, \qquad (i, j = 1, ..., N), \tag{5.18}$$

where the target-space density matrix $\hat{\rho}(\hat{q}^i, \hat{p}_i)$ is viewed as a function of coordinates $\hat{q}^i$ that parameterize the couplings of the composite oriented string world-sheet (see Figure 5.1), and their conjugate momenta $\hat{p}_i$, while $\hat{g}_{ij} = \hat{g}_{ij}(\hat{q}^i)$ is the quantum operator of the positive definite metric in the space of bosonic couplings. Therefore, the covariant quantum operator of external friction forces is in EMN–formulation given as $\hat{F}_i(\hat{q}^i, \hat{\dot{q}}^i) = \hat{g}_{ij}\,\hat{\dot{q}}^j$.

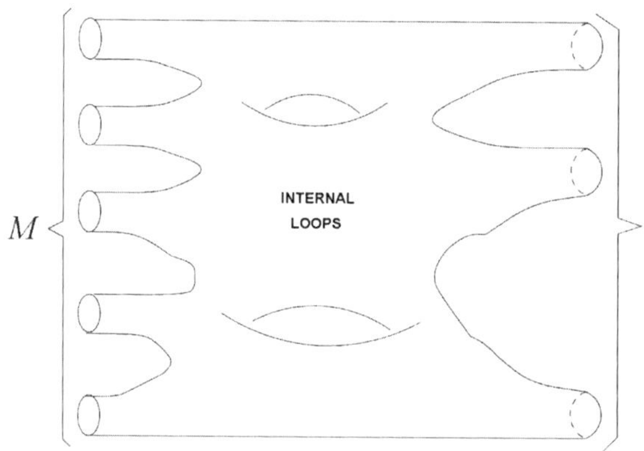

Fig. 5.1. *A composite oriented string world-sheet with M incoming strings (inputs), interacting through several internal loops (state feedbacks), to produce N outgoing strings (outputs). The arrow of time goes from left to right. Note the striking similarity with MIMO–systems of (nonlinear) control theory.*

From statistical point of view, equation (5.18) can be considered to be a sophisticated nonlinear extension of a linear N-dimensional *factor dynamics equation* (5.4). Physically speaking, the string equation 5.18) establishes the conditions under which a large–scale coherent state appears in the agent–network, which can be considered responsible for highly efficient, loss–free, energy transfer within the agent–networks.

The system-independent properties of equation (5.18), are:
(i) Conservation of probability P

$$\partial_t P = \partial_t[\mathrm{Tr}(\hat{\rho})] = 0. \tag{5.19}$$

(ii) Conservation of energy E, on the average

$$\partial_t \langle\langle E\rangle\rangle \equiv \partial_t[\mathrm{Tr}(\hat{\rho}E)] = 0. \tag{5.20}$$

(iii) Monotonic increase in entropy

$$\partial_t S = \partial_t[-\mathrm{Tr}(\hat{\rho}\ln\hat{\rho})] = (\hat{\dot{q}}^i\hat{g}_{ij}\hat{\dot{q}}^j)S \geq 0, \tag{5.21}$$

due to the positive definiteness of the metric $\hat{g}_{ij}$, and thus automatically and naturally implying a microscopic arrow of time [EMN99].

5.2.7 Equivalence of Hierarchical Models

Both the macroscopic crowd–equation (5.14) and the microscopic agent–equation (5.17) have the same open Liouville form, which implies the arrow of time [II08b]. These demonstrates the existence of the formal macro–micro socio–cognio–physical crowd self-similarity.

5.3 Computational Compositions for Crowd Actions

In our future work we will focus on computational aspects of a human/crowd modeling, in which the crowd is perceived as a *construction of individual computational actions,* which is itself a collective computational action.[7] Such computational actions can be effectively represented in the functional language *Haskell* using *Arrows* package [Hug00, Hug04],[8] which are (in our view) the most efficient way of structuring computations. The basic arrow combinators, for both serial and parallel processing of crowd actions, are:

$$\begin{aligned} (f <<< g)x &== f(gx)) == f.g, && \text{(actional composition)}, \\ (f >>> g)x &== g(fx) == g.f, && \text{(inverse composition)}, \\ (f^{***}g)(x,y) &== (fx, gy), && \text{(split into two actions)}, \\ (f\&\&\&g)x &== (fx, gx), && \text{(fusion of two actions)}. \end{aligned} \tag{5.22}$$

– so, they enable parallelism and therefore also time–flow modeling of crowd actions.

A similar computational framework has already been applied in the so-called *arrowized functional reactive programming* (AFRP, see e.g. [Nil02]), and implemented in the Haskell package *Yampa.*[9]

5.3.1 Haskell Example Code

For example, on a closed time-domain interval: $t_s = [0.0,\ 1.5] \subset \mathbb{R}$ we can define the following two crowd action signals:

$$f : \mathbb{R} \supset t_s \ni t \mapsto f(t) = 100(1 - \exp(-5t^3)) \in \mathbb{R}, \quad \text{end} \tag{5.23}$$

$$g : \mathbb{R} \supset t_s \ni t \mapsto g(t) = 100(1 - \exp(-5t^2)) \in \mathbb{R}, \tag{5.24}$$

and apply to them the four above-defined actional functions (5.22) within the Haskell *ghci* environment (interactive mode), as follows:

---

[7] Our idea is that human crowd has a 'collective unconscious mind' including some basic 'goal-driven actions'.

[8] For the Haskell *Arrows* package, see link: `http://www.haskell.org/ghc/docs/latest/html/libraries/base/Control-Arrow.html`

[9] For the Haskell *Yampa* package, see link: `http://www.haskell.org/haskellwiki/Yampa`.

```
ghci> import Control.Arrow -- import the package
-- define the time domain [0.0,0.1,0.2,0.3,0.4,0.5,0.6,0.7,
    ...,1.5]:
ghci> let ts = [0.0,0.1..1.5]
-- define two crowd actions:
ghci> let f t = 100*(1-exp(-5*t^3)) -- one way
ghci> let g = \t -> 100*(1-exp(-5*t^2)) -- another way
-- map them to the time domain:
ghci> map f ts -- gives the output:
[0.0,0.498752080731768,3.9210560847676823,12.628408831196559,
    ...,99.99999530883598]
ghci> map g ts -- gives the output:
[0.0,4.877057549928598,18.12692469220182,36.23718483782268,
    ...,99.9986992702346]
-- now map the four actional combinations:
ghci> map (f<<<g) ts -- composition gives the output:
[0.0,100.0,100.0,100.0,100.0,100.0,100.0,100.0,100.0,100.0,
    100.0,...,100.0]
ghci> map (f>>>g) ts -- inv. composition gives the output:
[0.0,71.17041848844417,100.0,100.0,100.0,100.0,100.0,100.0,
    100.0,...,100.0]
ghci> map (f&&&g) ts -- splitting gives the output:
[(0.0,0.0),(0.498752080731,4.877057549928),...,
    (99.999995308836,99.9986992702)]
ghci> map (f***g) (map (f&&&g) ts) -- fusion gives the
output:
[(0.0,0.0),(46.22341156852793,100.0),(100.0,100.0),
    (100.0,100.0),...,(100.0,100.0)]
```

Summary

In this Chapter we have proposed a nonlinear, complex, Hamiltonian model of a general socio–cognio–physical dynamics at the macroscopic, classical, inter-personal *crowd* level and microscopic, quantum, intra-personal *agent* level, is presented in the unique form of the open Liouville equation. At the microscopic level, this can be considered as a nonlinear extension of the linear correlation and factor dynamics. The *open* Liouville equation implies the arrow of time in both socio–cognio–physical dynamic processes and shows the existence of the formal crowd-agent space-time self-similarity. This shows the *existence of a unique behavioral control law*, which acts on different hierarchical levels of agents functioning.

6

Hybrid Topological Lie-Hamiltonian Learning in Evolving Energy Landscapes

In this Chapter, a novel bidirectional algorithm for hybrid (discrete + continuous-time) Lie-Hamiltonian evolution in adaptive energy landscape-manifold is designed and its topological representation is proposed. The algorithm is developed within a geometrically and topologically extended framework of Hopfield's neural nets and Haken's synergetics (it is currently designed in *Mathematica*, although with small changes it could be implemented in Symbolic C++ or any other computer algebra system). The adaptive energy manifold is determined by the Hamiltonian multivariate cost function H, based on the user-defined vehicle-fleet configuration matrix W, which represents the pseudo-Riemannian metric tensor of the energy manifold. Search for the global minimum of H is performed using random signal differential Hebbian adaptation. This stochastic gradient evolution is driven (or, pulled-down) by 'gravitational forces' defined by the 2nd Lie derivatives of H. Topological changes of the fleet matrix W are observed during the evolution and its topological invariant is established. The evolution stops when the $W-$topology breaks down into several connectivity-components, followed by topology-breaking instability sequence (i.e., a series of phase transitions).

6.1 Introduction

The present Chapter addresses the general problem of designing a cognitive-type topological evolution of adaptable complex systems (see [II08a]) within the framework of Hopfield's neural networks [Hop82), Hop84, HT85, DHS91, II07] and Haken's synergetics [Hak83, Hak93, Hak91, Hak96, Hak00, Hak02] (see Appendix, 6.3.1), using the case of a vehicle fleet configuration matrix. We propose a novel, Lie-derivative based generalization of a classical stochastic gradient-descent algorithm, for the hybrid topological evolution of an adaptive fleet of n vehicles along its energy landscape which geometrically represents the Hilbert n-manifold. The hybrid evolution consists of number of discrete generations, each having a short continuous-time flow, suitable for defining symbolic derivatives. The proposed adaptive evolution is 'topological' in two ways: (i) it is visually depicted as a graph-theoretic evolution, and (ii) it has a special kind of a 'topological invariant'.

6.2 The Hybrid Evolution Model

The hybrid evolution model within an adaptive energy landscape-manifold, proposed in this article, is summarized in Figure 6.1, which visualizes a bidirectional hybrid evolution of an energy landscape cost function H formulated from the *fleet configuration matrix* $W(t)$. Adaptive evolution in the $\langle Q \cdot P \rangle$ −Hilbert space starts with the initial binary configuration matrix $W_0 = W(0)$; it proceeds with discrete time-steps of sudden changes, followed by continuous time-steps of their smooth modifications. Search for the global minimum of H (the *white* path) is performed using random signal differential Hebbian adaptation (RSDH).[1] The evolution is pulled-down by 'gravitational forces' defined by the 2nd Lie derivatives of H. Topological changes of the fleet matrix W are observed during this hybrid evolution and its topological invariant is established. The evolution stops when the W−topology breaks down into several connectivity-components followed by topology-breaking instability sequence (i.e., a series of phase transitions).

6.2.1 Hamiltonian Cost-Function Model

The Hamiltonian cost-function is given as a negative-definite bilinear form:

$$H(\omega_{\alpha\beta}) = -\frac{1}{2}\sum_{\alpha=1}^{n}\sum_{\beta=1}^{n}\left[\omega_{\alpha\beta}(t)q_\alpha(t)q_\beta(t) + \omega_{\alpha\beta}(t)p_\alpha(t)p_\beta(t)\right]. \tag{6.1}$$

The *adaptive evolution*, in the Hilbert space determined by the inner product $\langle Q \cdot P\rangle = \sum_\alpha q_\alpha(t)p_\alpha(t)$, starts from the initial binary fleet configuration matrix $W = \omega_{\alpha\beta}(0)$, which is either predefined or random (see Figure 6.5), and its dynamics is governed by the following *'field dynamics equations'*:

$$\begin{aligned}
\dot q_\alpha(t) &= q_\alpha(t) - \mathcal{L}\mathcal{L}_F Q, \qquad q_\alpha(0) = \eta, \\
\dot\omega_{\alpha\beta}(t) &= -\omega_{\alpha\beta}(t) + Sq_\alpha Sp_\beta + S\dot q_\alpha S\dot p_\beta + \eta, \qquad \omega_{\alpha\beta}(0) = W_{\alpha\beta}, \\
\dot p_\beta(t) &= p_\beta(t) - \mathcal{L}\mathcal{L}_G P, \qquad p_\beta(0) = \eta,
\end{aligned} \tag{6.2}$$

where $\alpha, \beta = 1, ..., n \equiv$ number of vehicles in the fleet.

In (6.2) η represents the standard Gaussian random noise (see Appendix 6.3.1); $Sq_\alpha = \tanh[q_\alpha(t)]$ and $Sp_\beta = \tanh[p_\beta(t)]$ are sigmoid activation

[1] Recall that basic Hebbian adaptation or learning reflects the basic principle of neurobiology: 'Cells that fire together, wire together'. Here we are using its random signal differential formulation.

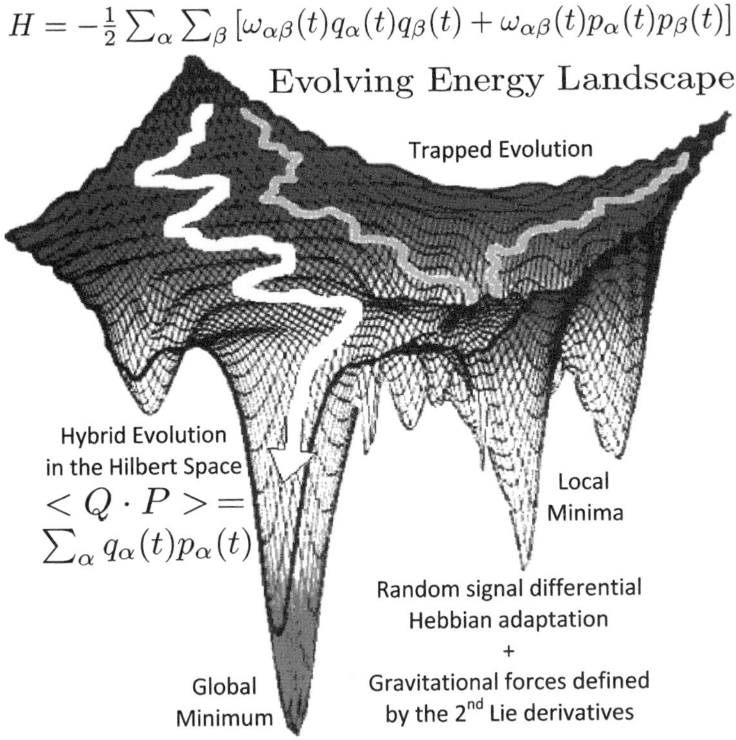

Fig. 6.1. *A hybrid evolution within the evolving energy landscape defined by the Hamiltonian cost function* $H = H(w, q, p)$.

functions and $S\dot{q}_\alpha$ and $S\dot{p}_\beta$ are the corresponding 'signal velocities'. $\mathcal{L}\mathcal{L}_F Q$ and $\mathcal{L}\mathcal{L}_G P$ are the 2nd-order (iterated) Lie derivatives[2] of the Hamiltonian

[2] Let $F(M)$ denote the set of all C^∞−smooth real-valued scalar functions $f : M \to \mathbb{R}$ on a smooth manifold M, $V(M)$ - the set of all smooth vector-fields on M, and $V^*(M)$ - the set of all differential one-forms on M. Also, let the vector-field $\zeta \in V(M)$ be given with its local flow $\phi_t : M \to M$ such that at a point $x \in M$, $\frac{d}{dt}|_{t=0}\, \phi_t x = \zeta(x)$, and ϕ_t^* representing the pull-back by ϕ_t. The Lie derivative differential operator $\mathcal{L}_\zeta$ is defined [II06b, II07]:

(i) on a scalar function $f \in F(M)$ as

$$\mathcal{L}_\zeta : F(M) \to F(M), \qquad \mathcal{L}_\zeta f = \frac{d}{dt}(\phi_t^* f)|_{t=0} = \zeta[f] = \sum_{\alpha=1}^{n} \frac{\partial f}{\partial x^\alpha} \zeta^\alpha;$$

(ii) on a vector-field $X \in V(M)$ as

$$\mathcal{L}_\zeta : V(M) \to V(M), \qquad \mathcal{L}_\zeta X = \frac{d}{dt}(\phi_t^* X)|_{t=0} \equiv [\zeta, X];$$

- that is the Lie bracket, and

$H(\omega_{\alpha\beta})$ with respect to the signal vector-fields: $F(q) = \{Sq_\alpha, \alpha = 1, ..., n\}$ and $G(p) = \{Sp_\beta, \beta = 1, ..., n\}$.

The basic structure (without Lie derivatives and signal functions) of the hybrid field dynamics equations (6.2) can be derived as a real-valued discretization of the two-party crowd dynamics model (see [IR12] and references therein), defined by a pair of strongly-coupled nonlinear Schr ödinger equations (NLS, see e.g. [II09]) with the simple Hebbian adaptation:

$$\begin{aligned}
\text{BLUE}: \quad i\partial_t\psi_\text{B} &= -\frac{a_\text{B}}{2}|\phi_\text{R}|^2\partial_{xx}\psi_\text{B} + V(w)|\psi_\text{B}|^2\psi_\text{B}, \\
\text{HEBB}: \quad \dot{w}_i &= -w_i + c_\text{H}|\psi_\text{B}|g_i|\phi_\text{R}|, \qquad V(w) = \sum_{i=1}^n w_i g_i, \\
\text{RED}: \quad i\partial_t\phi_\text{R} &= -\frac{b_\text{R}}{2}|\psi_\text{B}|^2\,\partial_{xx}\phi_\text{R} + V(w)|\phi_\text{R}|^2\phi_\text{R},
\end{aligned}$$

where, $a_\text{B}, b_\text{R}, c_\text{H}$ are parameters related to Red, Blue and Hebb equations, respectively, while g_i are Gaussian kernel functions. Besides, a similar NLS equation has been used for an option-pricing model in [Iva10b]. Spatial discretization:

$$\psi_\text{B}(x,t) \longmapsto q_\alpha(t), \qquad \phi_\text{R}(x,t) \longmapsto p_\beta(t),$$

can be performed using the standard *method of lines*, that is, the 2nd order central finite difference approximation for $\partial_{xx}\psi$ and $\partial_{xx}\phi$, defined as:

$$\partial_{xx}\psi \equiv \frac{\partial^2\psi}{\partial x^2} \approx \frac{\psi_{k+1} - 2\psi_k + \psi_{k-1}}{\Delta x^2},$$

where k is an index designating a position along a grid in x which has n points, corresponding to the number n of the vehicles in the fleet, and Δx is the spacing in x along the grid.

In each generation, the field equations (6.2) are numerically solved for a short time and the resulting fleet matrix $\omega_{\alpha\beta}(t_i)$ gives an initial fleet configuration for the next generation $\omega_{\alpha\beta}(t_{i+1})$.

6.2.2 Synergetics Interpretation of the Hybrid Model

In terms of Haken's synergetics (see Appendix, 6.3.1), the two vector evolution equations in the hybrid model (6.2):

(iii) on a one-form $\alpha \in V^*(M)$ as

$$\mathcal{L}_\zeta : V^*(M) \to V^*(M), \qquad \mathcal{L}_\zeta\alpha = \frac{d}{dt}(\phi_t^*\alpha)|_{t=0}.$$

In general, for any smooth tensor field $\mathbf{T}$ on the manifold M, the Lie derivative $\mathcal{L}_\zeta\mathbf{T}$ geometrically represents a directional derivative of $\mathbf{T}$ along the flow ϕ_t of the vector-field ζ.

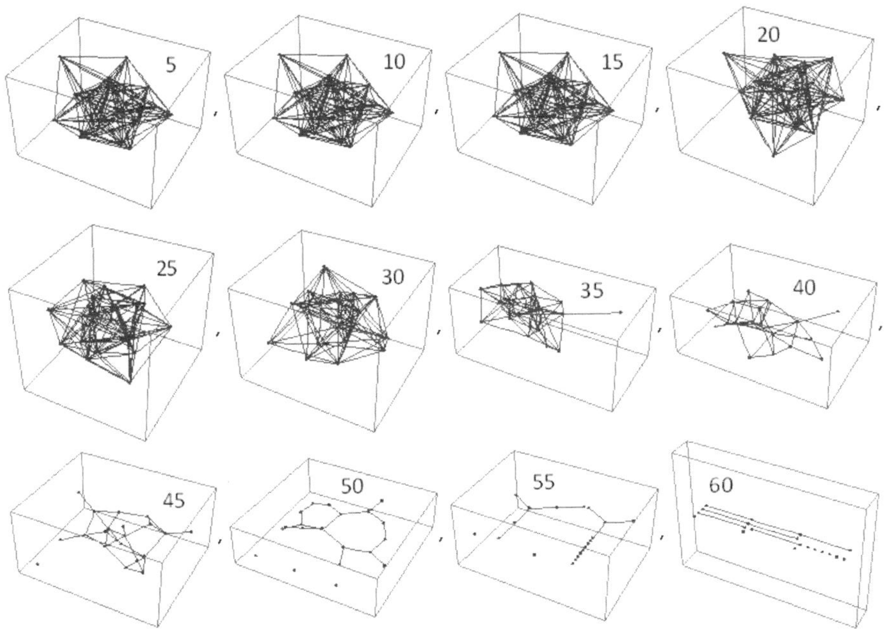

Fig. 6.2. *Adaptive hybrid evolution of a fleet of* $n = 20$ *vehicles, performed in* 60 *discrete generations with* 0.02*sec of continuous evolution within each generation (starting with the time-step of* 10^{-7}*). The evolution is represented as a series of 3D graphs, showing topology-breaking instability sequence ranging from the 45th to the 60th generation.*

$$\dot{q}_\alpha(t) = q_\alpha(t) - \mathcal{L}\mathcal{L}_F Q, \qquad q_\alpha(0) = \eta,$$
$$\dot{p}_\beta(t) = p_\beta(t) - \mathcal{L}\mathcal{L}_G P, \qquad p_\beta(0) = \eta,$$

represent the weak-noised Langevin equations driven by $\mathcal{L}\mathcal{L}$ -derivatives, while the matrix evolution equation:

$$\dot{\omega}_{\alpha\beta}(t) = -\omega_{\alpha\beta}(t) + Sq_\alpha Sp_\beta + S\dot{q}_\alpha S\dot{p}_\beta + \eta, \qquad \omega_{\alpha\beta}(0) = W_{\alpha\beta},$$

is a *stochastic gradient system* that can be interpreted as the overdamped motion of a *representative particle* in a scalar potential field $H(\omega_{\alpha\beta})$, subject to fluctuating forces: $F_{\omega_{\alpha\beta}}(t) = Sq_\alpha Sp_\beta + S\dot{q}_\alpha S\dot{p}_\beta$. During the hybrid evolution, the particle undergoes a series of non-equilibrium phase transitions caused by *topology-breaking instability sequence* (see Figures 6.2 and 6.3).

6.2.3 Geometry and Topology of the Hybrid Model

The evolving energy landscape geometrically represents a real nD *Hilbert manifold*, that is a pseudo-Riemannian n-dimensional configuration manifold M, with an evolving "$\langle bra \cdot ket\rangle$"-like inner product:

$$\langle Q \cdot P \rangle = \sum_{\alpha=1}^{n} q_\alpha(t) p_\alpha(t),$$

and the negative-definite, bilinear metric form (6.1), in which the fleet matrix $W = \omega_{\alpha\beta}(t)$ represents the adaptive *pseudo-Riemannian metric tensor* of M. The adaptive evolution on M geometrically represents a generalized gradient flow Φ_t, which is a one-parameter group of diffeomorphisms (smooth time-dependent homeomorphisms) of M.

The *evolution-flow* Φ_t can be formally geometrically modeled by the Ricci-flow (see [II11b] and references therein):

$$\partial_t \omega_{\alpha\beta} = -2\mathcal{R}_{\alpha\beta} \tag{6.3}$$

on the configuration manifold M, where $\mathcal{R}_{\alpha\beta}$ is the *Ricci curvature tensor* tensor of M. It states that the fleet matrix changes are governed by the changing curvature of M. The negative sign in (6.3) is essential to insure a global *volume exponential decay*,[3] since the Ricci flow equation (6.3) is a nonlinear geometric generalization of the standard linear *heat equation*

$$\partial_t \Psi = \Delta \Psi.$$

Topological analysis

Topology changes (see [II08a]) start with the 20th generation. Topology-braking instability sequence starts with the 45th generation (see Figures 6.2 and 6.3).

Hamiltonian cycle is a closed path that visits each vertex (vehicle) exactly once. It is a topological invariant for the adaptive hybrid evolution (see Figure 6.4).

---

[3] This complex geometric process is globally similar to a generic exponential decay:

$$\dot{x} = -\lambda f(x),$$

for a positive function $f(x)$. We can get some insight into its solution from the simple exponential decay ODE,

$$\dot{x} = -\lambda x \qquad \text{with the solution} \qquad x(t) = x_0 \mathrm{e}^{-\lambda t},$$

(where $x = x(t)$ is the observed quantity with its initial value x_0 and λ is a positive decay constant), as well as the corresponding nth order rate equation (where $n > 1$ is an integer),

$$\dot{x} = -\lambda x^n \qquad \text{with the solution} \qquad \frac{1}{x^{n-1}} = \frac{1}{{x_0}^{n-1}} + (n-1)\,\lambda t.$$

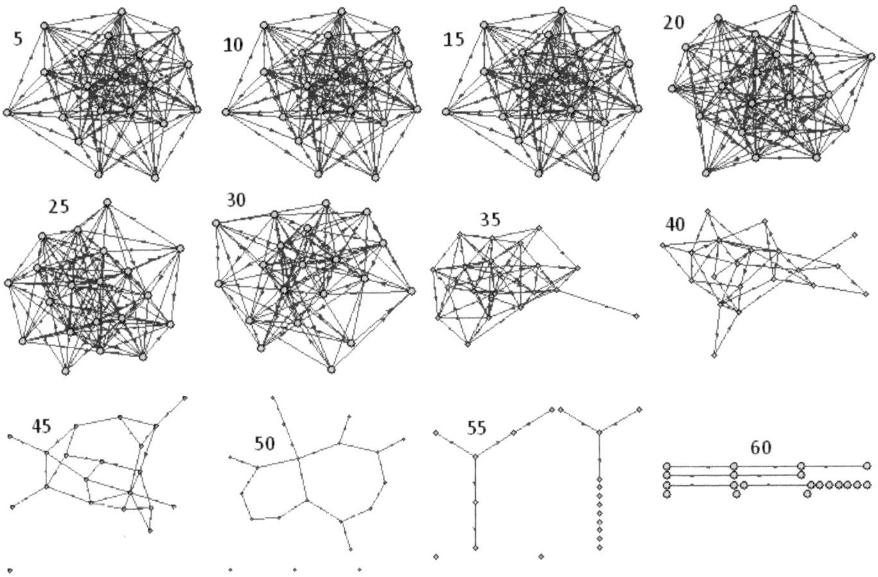

Fig. 6.3. *Adaptive hybrid evolution (performed under the conditions defined in Figure 6.2) represented as a series of 2D graphs, showing topology-breaking instability sequence ranging from the 45th to the 60th generation.*

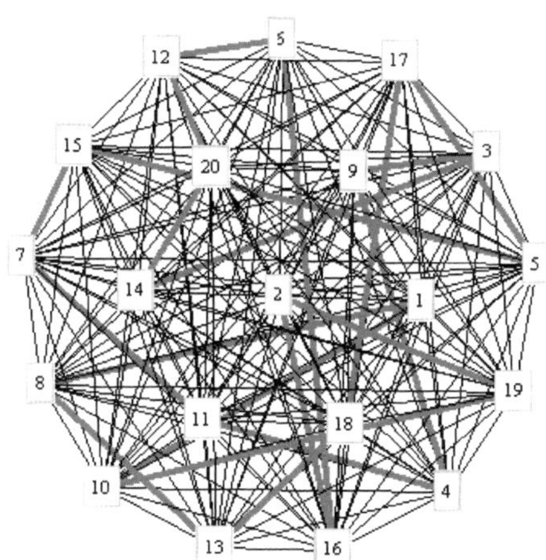

Fig. 6.4. *Hamiltonian cycle is a topological invariant of the hybrid adaptive evolution.*

Summary

We have presented a new algorithm for bidirectional hybrid (discrete + continuous-time) topological evolution in adaptive energy Hilbert n -manifold

M, currently developed in $Mathematica^{TM}$ (although with small changes it could be implemented in Symbolic C++ or any other computer algebra system). The algorithm is developed in a geometrically and topologically generalized Hopfield-Haken framework. The energy landscape manifold is determined by the Hamiltonian multivariate cost function H, based on the user-defined vehicle-fleet configuration matrix W. The Hamiltonian H is the negative-definite, bilinear metric form for the manifold M, in which the fleet matrix $W = \omega_{\alpha\beta}$ represents the pseudo-Riemannian metric tensor of M. Search for the global minimum of H is performed using random signal differential Hebbian adaptation. The evolution is driven (or, pulled-down) by 'gravitational forces' defined by the 2nd Lie derivatives of H. In terms of Haken's synergetics this stochastic gradient system can be interpreted as the overdamped motion of a representative particle in a scalar potential field $H(\omega_{\alpha\beta})$, subject to fluctuating forces, such that during the adaptive evolution, the particle undergoes a series of non-equilibrium phase transitions caused by topology-breaking instability sequence. The adaptive evolution geometrically represents a generalized gradient flow: Φ_t, that can be formally modeled by the Ricci-flow on the configuration manifold M, where $\mathcal{R}_{\alpha\beta}$ is the Ricci curvature tensor of M. This equation states that the fleet matrix changes are governed by the curvature of M. Topological changes of the fleet matrix W are observed during the evolution and its topological invariant is established. The evolution stops when the $W-$topology breaks down into several connectivity-components, followed by topology-breaking instability sequence.

6.3 Appendix

6.3.1 Haken's Synergetics and Hopfield's Overlaps

Recall that *Haken synergetics* deals with complex systems (far from thermal equilibrium) that are composed of many individual components that interact with each other and are able to produce spatial, temporal or functional structures by self-organization (see [Hak83, Hak93, Hak91, Hak96, Hak00, Hak02]).

The aim of synergetics is to describe processes of spontaneous self-organization and cooperation in such complex systems of various bio-psycho-socio-physical nature. General properties of the subsystems are their own nonlinear/chaotic dynamics as well as mutual nonlinear/chaotic interactions. Furthermore, the systems of synergetics are open. The influence from outside is measured by a small set of control parameters. Synergetic processes are described by a small set of order parameters, similar to those in Landau's *phase-transition theory* of physical systems in *thermal equilibrium*.

Synergetics originated in Haken's modeling of optical lasers using modified *Lorenz attractor equations*, where the control parameter was the laser pump rate. At low pump rate, the laser waves are entirely uncorrelated as in a usual lamp. When the pump rate is increased to a first critical value, the noise disappears and is replaced by a *pure focused signal*. This means that the atoms emit a pure sinusoidal light wave which means that they become *self-organized*. When the pump rate is increased beyond a second critical value, the laser may periodically emit very intense and *short pulses*. Under different conditions the frequency spectrum becomes broadened and the light emission may become *chaotic*. In this way the following *instability sequence of phase transitions* occurs:

NO OSCILLATION $\mapsto$ FIRST FREQUENCY $\mapsto$ SECOND FREQUENCY $\mapsto$ CHAOS

A similar instability sequence occurs in our hybrid topology-changing evolution (see Figures 6.2 and 6.3).

In particular, Haken's *brain synergetics* considers two types of agents (for an overview, see [H07, H08a]): (i) *graded-response type* (GR), and (ii) *coupled-oscillators type*(CO), both governed by the *vector Langevin equation*:

$$\dot{\sigma}_i = f_i(\sigma, W) + \eta_i,$$

where $\sigma_i = \sigma_i(t)$ are order-parameters: neural action potentials in case of GR and oscillator phases in case of CO, while η_i is the noise. $f_i = f_i(\sigma, W)$ are deterministic forces given by:

$$GR : f_i = \sum_j W_{ij} \tanh[\sigma_j] - \sigma_i + \theta_i,$$

$$CO : f_i = \sum_j W_{ij} \sin(\sigma_j - \sigma_i) + \omega_i,$$

where W_{ij} are synaptic weights, θ_i are input GR-functions, and ω_i are natural CO-frequencies.

The noise variables η_i in the Langevin equation are given by:

$$\eta_i(t) = \lim_{h\to 0} \zeta_i(t)\sqrt{T/h},$$

where $\zeta_i(t)$ denote uncorrelated Gaussian-distributed random forces and the control parameter T controls the amount of noise in the system, ranging from $T = 0$ (deterministic dynamics) to $T = \infty$ (completely random dynamics).

An alternative, *probabilistic description* of the same Langevin random process is given by the Fokker-Planck equation, describing the time evolution of the probability density function (PDF):

$$\partial_t P(\sigma_i) = -\sum_i \partial_{\sigma_i}[f_i P(\sigma_i)] + \frac{T}{2}\sum_i \partial_{\sigma_i^2} P(\sigma_i). \tag{6.4}$$

Now, recall that the similarity between two different n-dimensional patterns ξ_i^μ and ξ_i^ν is in Hopfield theory [Hop84, DHS91] measured by their *mutual overlap* or *cross-overlap* $m_{\mu\nu}$,[4] equal

$$m_{\mu\nu} = n^{-1}\xi_i^{\mu}\xi_i^{\nu}. \qquad (6.5)$$

For similar patterns the cross-overlap is close to unity whereas for uncorrelated patterns it is random variable with zero mean and small $(1/\sqrt{N})$ variance. The dynamics of overlaps is governed by the nonlinear ordinary differential equation (ODE, generalized from [DHS91], pp. 23):

$$\dot{m}_{\mu\nu}(t) = -m_{\mu\nu}(t) + \left\langle \xi_{\mu}(t)\xi_{\nu}(t)\tanh[\beta m_{\mu\nu}(t)\xi^{\mu}(t)\xi^{\nu}(t)]\right\rangle, \qquad (6.6)$$

(where the angular brackets denote an average over the q patterns $\xi^{\mu}(t)$),[5] which has the fixed-point stationary solution:

$$m_{\mu\nu} = \left\langle \xi_{\mu}\xi_{\nu}\tanh[\beta m_{\mu\nu}\xi^{\mu}\xi^{\nu}]\right\rangle.$$

In terms of synergetics, the ODE (6.6) represents an *order parameter equation*. If we introduce an overlap-dependent scalar quadratic potential field:

$$V(m_{\mu\nu}) = -\frac{1}{2}\sum_{\mu=1}^{q} m_{\mu\nu}^2,$$

the ODE (6.6) becomes the *stochastic-gradient* order parameter equation

$$\dot{m}_{\mu\nu}(t) = -\frac{\partial V(m_{\mu\nu})}{\partial m_{\mu\nu}(t)} + F_{\mu\nu}(t), \qquad (6.9)$$

where $F_{\mu\nu}(t)$ in (6.9) denotes a tensor fluctuating force, with average (over the underlying random process):

$$\langle F_{\mu\nu}(t)\rangle = \left\langle \xi_{\mu}(t)\xi_{\nu}(t)\tanh[\beta m_{\mu\nu}(t)\xi^{\mu}(t)\xi^{\nu}(t)]\right\rangle,$$

---

4 In other parlance, Hopfield's cross-overlap is called *Karhunen-Loeve covariance matrix* that extracts the principal components from a data set. It resembles the cross-correlation function of two time-series, with several distinct peaks, indicating that the two series are very similar at each point in time where the peaks occur.

5 The ODE (6.6) corresponds to the average, self-organizing Hebbian-type adaptation scheme (see [II07]):

$$\dot{m}_{\mu\nu}(t) = -m_{\mu\nu}(t) + \langle \mathcal{I}_{\mu\nu}\rangle, \qquad (6.7)$$

with random signal *innovation*:

$$\mathcal{I}_{\mu\nu} = f_{\mu}[\xi^{\mu}(t)]\, f_{\nu}[\xi^{\nu}(t)] + \eta_{\mu\nu}(t), \qquad (6.8)$$

where $\eta_{\mu\nu}$, denotes the tensorial, additive, zero-mean, Gaussian white noise, independent of the main innovation function $I_{\mu\nu}$, while $f_{\mu,\nu}[\cdot]$ represent the sigmoid (tanh) activation functions.

and variation

$$\langle F_{\mu\nu}(t)F_{\mu\nu}(t')\rangle = Q_{\mu\nu}\delta(t-t'), \tag{6.10}$$

in which the coefficient $Q_{\mu\nu}$ denotes the strength of the random processes, while the Dirac δ-function $\delta(t-t')$ gives its short-term memory.

The non-equilibrium phase transitions of the overlap $m_{\mu\nu}(t)$ can be described in terms of the corresponding PDF defined by the Fokker-Planck equation similar to (6.4):

$$\dot{p}(m_{\mu\nu},t) = p(m_{\mu\nu},t) + \frac{1}{2}Q_{\mu\nu}\frac{\partial^2 p(m_{\mu\nu},t)}{\partial m_{\mu\nu}^2}.$$

6.3.2 The Hybrid-Evolution Algorithm Design

Functional-type definitions of the gradient and Lie derivative of a scalar field:

$$\text{Gradient} : \text{Grad}[s\_, x\_\text{List}] := \partial_{\#}s@x;$$
$$\text{Lie derivative} : \text{LieDer}[v\_\text{List}\,, s\_, x\_\text{List}] := \text{Grad}[s,x].v;$$

Set up (DOF: $n = 20$ vehicles in the fleet) with two sets of time-dependent coordinates:

$$Q = \text{Table}\left[q_\alpha(t), \{\alpha, n\}\right];$$
$$P = \text{Table}\left[p_\beta(t), \{\beta, n\}\right];$$

Initial fleet state (configuration) given by random adjacency matrix:

$$\text{Wo} = \text{Table}\left[W_{\alpha,\beta} = \text{RandomInteger}[], \{\alpha, n\}, \{\beta, n\}\right]; \quad \text{Wo//MatrixForm}$$

Initial fleet graph/network plots in 2D and 3D

See Figure 6.5.

$$\{\text{GraphPlot}[\text{Wo}, \text{VertexLabeling} \to \text{True}],$$
$$\text{GraphPlot3D}[\text{Wo}, \text{VertexLabeling} \to \text{True}]\}$$

Set up activation functions and directional vector-fields:

Sigmoid threshold functions and their time derivatives (for RSDH adaptation) are defined as:

$$\text{Table}\left[\text{Sq}_\alpha = \text{Tanh}\left[q_\alpha(t)\right], \{\alpha, n\}\right];$$
$$\text{Table}\left[\text{Sp}_\beta = \text{Tanh}\left[p_\beta(t)\right], \{\beta, n\}\right];$$
$$\text{Table}\left[\text{Sdq}_\alpha = 1 - \text{Sq}_\alpha^2, \{\alpha, n\}\right];$$
$$\text{Table}\left[\text{Sdp}_\beta = 1 - \text{Sp}_\beta^2, \{\beta, n\}\right];$$

Directional vector-fields $F(q)$ and $G(p)$:

$$F = \text{Table}\left[\text{Sq}_\alpha, \{\alpha, n\}\right]; \quad G = \text{Table}\left[\text{Sp}_\beta, \{\beta, n\}\right];$$

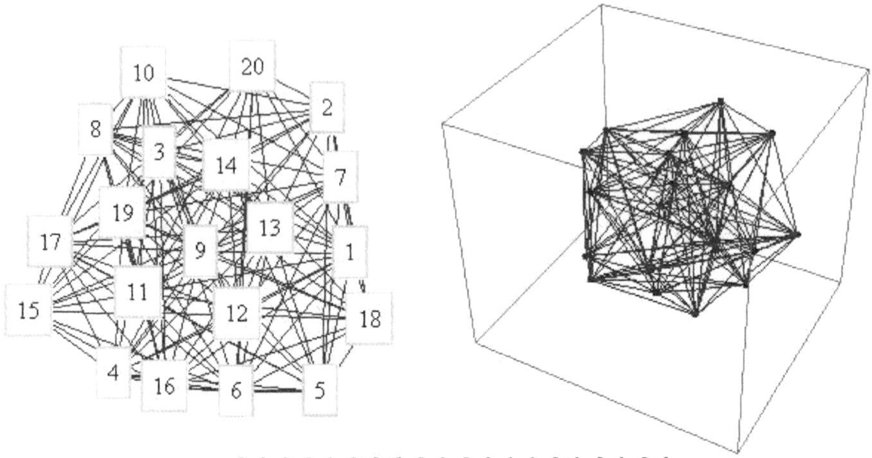

0	1	1	0	1	0	0	0	0	1	0	1	1	1	0	1	0	1	0	1
0	0	0	1	1	0	1	1	1	0	0	0	1	1	0	0	0	1	0	1
0	1	1	0	0	1	0	1	1	1	0	0	0	1	1	1	0	0	1	1
1	1	1	0	1	0	0	0	0	0	0	1	1	0	0	1	0	0	1	0
0	0	0	0	1	0	0	0	0	0	1	0	1	0	0	0	1	0	0	0
1	0	1	1	1	0	1	0	0	0	0	0	0	0	1	0	1	1	1	0
0	0	0	0	1	0	0	1	1	1	0	1	1	0	1	0	0	1	0	1
0	1	1	0	0	0	0	1	1	1	1	1	1	0	0	0	1	1	1	0
1	0	0	1	1	0	0	1	1	1	1	0	1	1	0	1	0	1	0	0
0	1	1	0	0	0	1	0	1	1	1	0	0	1	1	0	0	0	0	0
1	0	0	0	0	1	0	1	0	0	0	0	1	1	1	0	1	0	1	0
0	0	1	1	1	1	0	1	1	0	0	1	0	1	1	1	1	0	0	1
1	1	1	0	0	1	0	1	1	1	1	1	1	1	0	1	1	0	0	1
1	0	1	1	1	0	0	1	0	1	1	0	1	0	1	0	0	1	1	1
0	0	1	1	0	0	1	0	1	1	0	0	0	1	0	1	0	0	1	0
1	0	1	1	1	1	0	1	1	1	0	1	1	0	0	1	1	0	1	0
0	0	0	1	1	1	0	0	1	1	1	1	1	1	1	0	0	0	1	0
0	1	0	0	1	1	1	0	1	0	1	1	0	1	0	1	0	0	0	0
0	1	0	0	0	1	1	1	1	1	1	1	0	0	1	0	0	1	1	0
1	1	0	0	0	1	1	0	0	0	1	1	1	0	0	0	1	0	1	0

Fig. 6.5. *Initial fleet state (configuration) given by the random adjacency matrix (down) and depicted as a 2D graph (left-up) and 3D graph (right-up).*

Define (negative) Energy Landscape (Hamiltonian) H as a negative-definite $(Q,P)-$bilinear form:

$$H = -\frac{1}{2}\sum_{\alpha=1}^{n}\sum_{\beta=1}^{n}[\omega_{\alpha,\beta}[t]q_{\alpha}[t]q_{\beta}[t] + \omega_{\alpha,\beta}[t]p_{\alpha}[t]p_{\beta}[t]];$$

Driving 'gravitational' forces for the evolution defined by the second Lie derivatives of the Hamiltonian wrt. $F(q)$ and $G(p)$:

$$\mathrm{LQ}_F = \mathrm{LieDer}[F, H, Q]; \qquad \mathrm{LLQ}_F = \mathrm{LieDer}\ [F, \mathrm{LQ}_F, Q];$$
$$\mathrm{LP}_G = \mathrm{LieDer}[G, H, P]; \qquad \mathrm{LLP}_G = \mathrm{LieDer}\ [G, \mathrm{LP}_G, P];$$

Adaptive field equations defined:

$$
\begin{aligned}
q'_\alpha[t] &== q_\alpha[t] - \mathrm{LQ}_F - \mathrm{LLQ}_F, & q_\alpha[0] &== \eta, \\
p'_\beta[t] &== p_\beta[t] - \mathrm{LP}_G - \mathrm{LLP}_G, & p_\beta[0] &== \eta, \\
\omega'_{\alpha,\beta}[t] &== -\omega_{\alpha,\beta}[t] + \ \mathrm{Sq}_\alpha \mathrm{Sp}_\beta + \mathrm{Sdq}_\alpha \mathrm{Sdp}_\beta + \eta, & \omega_{\alpha,\beta}[0] &== W_{\alpha,\beta}
\end{aligned}
$$

Main loop (hybrid simulations for all generations/iterations):

$$
\begin{aligned}
\mathrm{Tfin} &= 0.02; \quad \mathfrak{M} = \mathrm{NumGener} = 60; \\
\mathrm{For[gen} &= 1, \quad \mathrm{gen} \le \mathfrak{M}, \qquad (\text{* Main loop *}) \\
\mathrm{Eqns} &= \mathrm{Flatten[Join[}(\text{* Field equations defined *}) \\
&\mathrm{Table}[\{(q_\alpha)'(t) == q_\alpha(t) - \mathrm{LQ}_F - \mathrm{LLQ}_F, q_\alpha[0] \\
&\qquad\qquad == \mathrm{RR}[\{-10, 10\}]\}, \{\alpha, n\}], \\
&\mathrm{Table}[\{(p_\beta)'(t) == p_\beta(t) - \mathrm{LP}_G - \mathrm{LLP}_G, p_\beta[0] \\
&\qquad\qquad == \mathrm{RR}[\{-10, 10\}]\}, \{\beta, n\}], \\
&\mathrm{Table}[\{(\omega_{\alpha,\beta})'(t) == -\omega_{\alpha,\beta}(t) + \mathrm{Sq}_\alpha \mathrm{Sp}_\beta \\
&\qquad\qquad + \mathrm{Sdq}_\alpha \mathrm{Sdp}_\beta + \mathrm{RR}[\{-1, 1\}], \\
&\omega_{\alpha,\beta}[0] == W_{\alpha,\beta}\}, \{\alpha, n\}, \{\beta, n\}]]]; \\
\mathrm{Sol} &= \mathrm{NDSolve[Eqns}, \qquad (\text{* Field equations numerically solved *}) \\
&\mathrm{Flatten}\left[\mathrm{Join}\left[\mathrm{Table}\left[q_\alpha, \{\alpha, n\}\right], \mathrm{Table}\left[p_\beta, \{\beta, n\}\right],\right.\right. \\
&\left.\left.\mathrm{Table}\left[\omega_{\alpha,\beta}, \{\alpha, n\}, \{\beta, n\}\right]\right]\right], \ \{t, 0, \mathrm{Tfin}\}, \\
&\left.\mathrm{StartingStepSize} \to \frac{1}{10^7}\right]; \\
&\mathrm{Do}\left[W_{\alpha,\beta} = \mathrm{Evaluate}\left[\omega_{\alpha,\beta}[\mathrm{Tfin}]/.\mathrm{Sol}\right][[1]], \{\alpha, n\}, \{\beta, n\}\right]; \\
&\mathrm{Print}[\{\mathrm{gen}, \qquad (\text{* Printout results for each generation *}) \\
&\mathrm{MatrixForm}\left[\mathrm{WW[gen]} = \mathrm{Table}\left[\ \mathrm{Evaluate}\left[\omega_{\alpha,\beta}[\mathrm{Tfin}]/.\mathrm{Sol}\right][[1]],\right.\right. \\
&\left.\left.\{\alpha, n\}, \{\beta, n\}\right]\right]\}] \\
&\mathrm{Print}\left[\mathrm{MatrixForm}\left[\{\mathrm{Table}\left[\mathrm{Evaluate}\left[q_\alpha[\mathrm{Tfin}]/.\mathrm{Sol}\right][[1]], \{\alpha, n\}\right],\right.\right. \\
&\left.\left.\mathrm{Table}\left[\mathrm{Evaluate}\left[p_\beta[\mathrm{Tfin}]/.\mathrm{Sol}\right][[1]], \{\beta, n\}\right]\}\right]\right] \\
\mathrm{gen} &++ \,] \qquad (\text{* End of main loop *})
\end{aligned}
$$

Here *RR* is an abbreviation for *RandomReal*.

For this particular simulation of the hybrid fleet evolution we used 60 discrete generations and the short continuous time for each generation is chosen to be 0.02 sec.

6.3.3 Topological Analysis of the Hybrid Evolution

Fleet-matrix Plots in the i-th generations (see Figure 6.6)

$$\mathrm{Table}[\mathrm{ArrayPlot}[\mathrm{WW}[i], \mathrm{ColorFunction} \to \text{"Rainbow"}], \{i, 5, 60, 5\}]$$

Fig. 6.6. *The evolution of the fleet-matrix (plots in every fifth generation) - the lighter the color, the weaker the connection.*

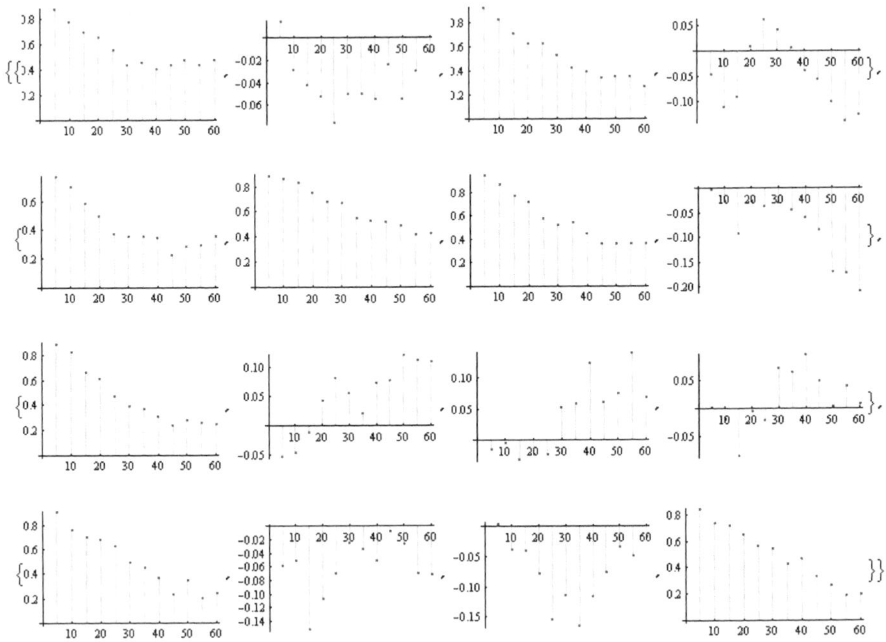

Fig. 6.7. *Plotting (some of) the matrix values* $j, k = 1, ..., 4$ *in every fifth generation.*

Plotting (some of) the matrix values (j,k) in the i-th generations (see Figure 6.7)

$$\text{Table}[\text{DiscretePlot}[\text{WW}[i][[j,k]], \{i, 5, 60, 5\}], \{j, 4\}, \{k, 4\}]$$

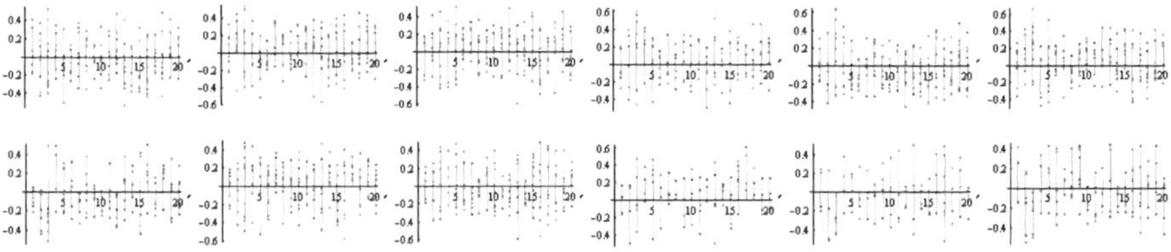

Fig. 6.8. *Plotting the fleet-matrix eigenvectors in every fifth generation.*

Plotting the Fleet-matrix Eigenvectors in the i-th generations (see Figure 6.8)

Evolution of the Fleet Network Topology in 2D (Figure 6.3) and in 3D (Figure 6.2)

```
Table[GraphPlot[Round[WW[i]], DirectedEdges → Truc,
VertexRenderingFunction → ({Green, EdgeForm[ Black], Disk[#, .05]}&),
ImageSize → {300, 260}], {i, 5, 60, 5}]
Table[GraphPlot3D[Round[WW[i]]], {i, 5, 60, 5}]
```

A Topological Invariant of adaptive fleet evolution (Figure 6.4)

Hamiltonian Cycle doesn't change during the hybrid fleet evolution: so it is a topological invariant for this adaptive evolution.

```
Needs[GraphUtilities`]
c2 = FindHamiltonianCycle[WW[2]];
cs2 = Transpose[{c2, RotateRight[c2]}];
GraphPlot[WW[2], EdgeRenderingFunction →
(If[MemberQ[cs2, #2]||MemberQ[cs2 , Reverse[#2]],
{Red, Thickness[.01], Line[#1]}, {Black , Line[#1]}]&),
VertexLabeling → True, MultiedgeStyle → False]
c38 = FindHamiltonianCycle[WW[38]];
cs38 = Transpose[{c38, RotateRight[c38]}];
GraphPlot[WW[38],EdgeRenderingFunction →
(If[MemberQ[cs38, #2]||MemberQ[cs38, Reverse[#2]],
{Red, Thickness[.01], Line[#1]}, {Black , Line[#1]}]&),
VertexLabeling → True, MultiedgeStyle → False]
```

7

Complexity and Control in Solitary Conductive PDEs

In this Chapter, we review and analyze models of controlled complexity in nonlinear pulse conduction, ranging from the Hodgkin-Huxley action potentials propagating along neural fibers to rogue waves in optical fibers. The novel model proposed is an alternative to the Hodgkin-Huxley neural model in the form of the sine-Gordon wave equation. This new alternative explains pulse conduction in terms of general wave phenomena (such as kinks, solitons and breathers).

7.1 Introduction

In this multidisciplinary Chapter, we explore new models of controlled complexity (see [II12, II08a]) in general pulse conduction. Towards this end, we review previously existing models, starting with the Hodgkin-Huxley [HH52, Hod64] action potentials propagating along a neural fiber. The electrical behavior of neurons is studied in terms of electrically equivalent circuits: the Hodgkin-Huxley model is loosely based on the lossy transmission line (see Figure 7.1 as well as [II07]). Resistors model various types of ion channels in membranes, and capacitors model charge storage of cell membranes. The power source models electrochemical potentials between regions of different ion concentrations.

Here, Rs are resistances [in which the *Ohm law* between the voltage $V(t)$ and current $I(t)$ holds: $V(t) = R\,I(t)$], Gs are the conductances (inverse resistances: $G = I^{-1}$), Cs are capacitances [defined by: $I(t) = C\,\dot{V}(t)$], and Ls are inductances [defined by: $V(t) = L\,\dot{I}(t)$]). The velocity of a signal $x(t)$ propagating along the line at the point x is (assuming small loss due to resistance) given by: $\dot{x}(t) = 1/\sqrt{L(x)C(x)}$, while the characteristic impedance at the same point is given by: $z(x) = \sqrt{L(x)/C(x)}$.

Later, we also review the development of solitons, positons, breathers and rogue waves in optical fibers (see, e.g. [MB99]). While much of the text is a review of both traditional and recent theories, the novel model is a second

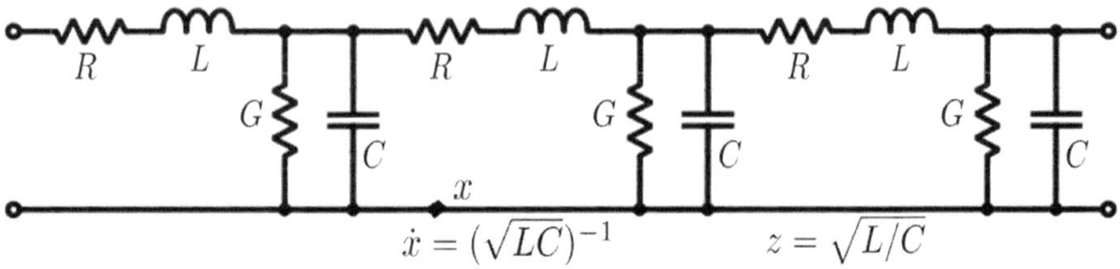

Fig. 7.1. *An equivalent electric-circuit representation (or, a lumped-parameter model) of a lossy transmission line.*

alternative to the Hodgkin-Huxley neural model in the form of sine-Gordon solitons, kinks and breathers (initiated in [II13]).

Ultimately, we are interested in broad questions concerning observability and controllability in highly complex environments, such as the exercise of authority over assigned forces in accomplishing military goals. The present subject of controlled complexity in biophysical systems is a case study in which to develop an understanding of certain phenomena, particularly traveling waves (solitons, breathers and kinks-antikinks).

7.2 Neural Action-Potential Solitons

The foundation of both electro-physiology and modern biophysics is given by electrical theory of neural impulse conduction (or, action potential propagation), proposed in 1952 by A.L. Hodgkin and A.F. Huxley in the form of their celebrated *HH-model*, for which they were awarded the 1963 Nobel Prize in Physiology or Medicine. More generally, the Hodgkin-Huxley electrical theory of neural action potentials is the accepted textbook basis of modern neurophysiology; for a recent review, see [GWP12, II09]. The HH-model describes the flow of membrane currents through the membrane capacitor and ion channels (see Figure 7.2).

The total membrane current, $I(t) = I_C + I_{\text{ion}}$, is the sum of the capacitive current $I_C = C_m \dot{V}(t)$ and the ionic current I_{ion} that is associated with the conductances g_{Na}, g_K, g_L due to different concentrations of the sodium Na and potassium K ions (while the leak current I_L approximates the passive properties of the neuron). Formally, it is written:

$$I_{\text{ion}} = I_{Na} + I_K + I_L = g_{Na}\,(V - V_{Na}) + g_K\,(V - V_K) + g_L\,(V - V_L)\,.$$

7.2.1 Hodgkin-Huxley Theory

In its basic form (without any of many modern extensions), the HH-model consists of four coupled nonlinear first-order ODEs, including the *cable equation* for the neural membrane potential V, together with m, h and n equations for the gating variables of Na and K channels and leakage [HH52, Hod64]:

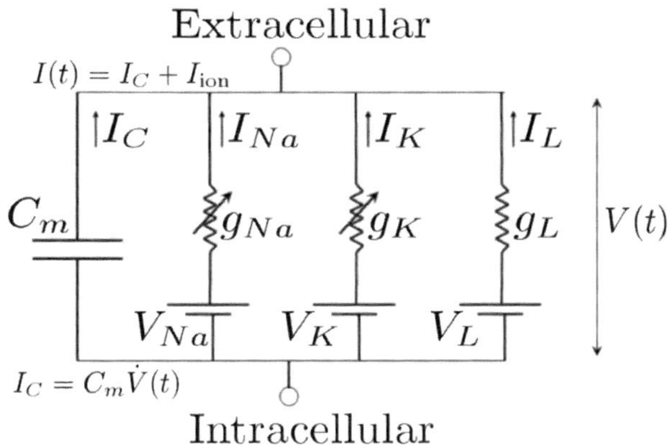

Fig. 7.2. *An equivalent electric-circuit representation of an excited (depolarized) neural membrane, pioneered by Hodgkin and Huxley in 1952 [HH52, Hod64]*

$$\begin{aligned} C_m\dot{V} &= -g_{Na}m^3h[V(t)-V_{Na}] - g_K n^4[V(t)-V_K] - g_L[V(t)-V_L] + I_j^{\text{ext}}, \\ \dot{m} &= -(a_m+b_m)\,m + a_m, \qquad \dot{h} = -(a_h+b_h)\,h + a_h, \\ \dot{n} &= -(a_n+b_n)\,n + a_n, \end{aligned} \tag{7.1}$$

where

$$\begin{aligned} a_m &= 0.1\,(V+40)/[1-\mathrm{e}^{-(V+40)/10}], \qquad b_m = 4\,\mathrm{e}^{-(V+65)/18}, \\ a_n &= 0.01\,(V+55)/[1-\mathrm{e}^{-(V+55)/10}], \qquad b_n = 0.125\,\mathrm{e}^{-(V+65)/80}, \\ a_n &= 0.07\,\mathrm{e}^{-(V+65)/20}, \qquad b_n = 1/[1+\mathrm{e}^{-(V+35)/10}]. \end{aligned}$$

In general, conductances are dependent on other factors such as membrane potential and concentrations of other ions in the intracellular medium. Hodgkin and Huxley assumed that g_K and g_{Na} are dependent on voltage, while they took leakage g_L to be constant, in order to resolve their model with their experimental data.

Here the reversal potentials of Na and K channels and leakage are: $V_{Na} = 50\text{mV}$, $V_K = -77\text{mV}$ and $V_L = -54.5\text{mV}$; the maximum values of corresponding conductivities are: $g_{Na} = 120\text{mS/cm}^2$, $g_K = 36\text{mS/cm}^2$ and $g_L = 0.3\text{mS/cm}^2$; the capacity of the membrane is: $C_m = 1\ \mu\text{F/cm}^2$. The external, input current is given by:

$$I_j^{\text{ext}} = g_{\text{syn}}(V_a - V_c)\sum_n \alpha(t - t_{in}), \tag{7.2}$$

which is induced by the pre-synaptic spike-train input applied to the neuron i, given by: $U_i(t) = V_a \sum_n \delta(t - t_{in})$. In (7.2), t_{in} is the nth firing time of the spike-train inputs, g_{syn} and V_c denote the conductance and the reversal potential, respectively, of the synapse, τ_{s} is the time constant relevant to the synapse conduction, and $\alpha(t) = (t/\tau_{\text{s}})\,\mathrm{e}^{-t/\tau_{\text{s}}}\Theta(t)$, where $\Theta(t)$ is the Heaviside function. Numerical simulation (using the fast fourth-fifth order Runge-Kutta-Cash-Karp integrator [CK90]) of the system (7.1)-(7.2) is given in Figure 7.3.

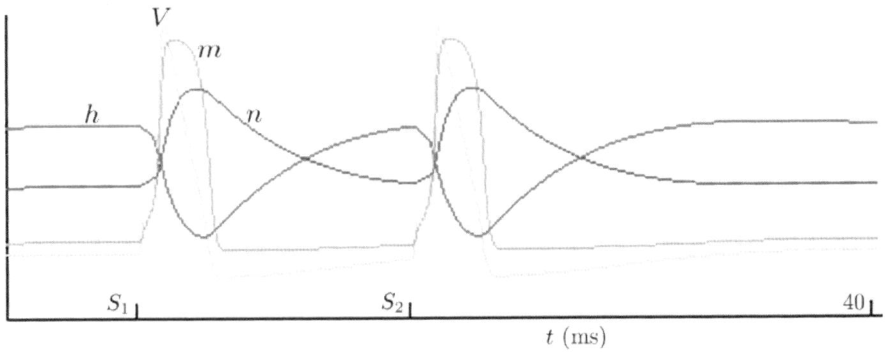

Fig. 7.3. *Numerical simulation of the HH-model (7.1)-(7.2), using the parameter-values defined in the text.*

In addition, Hodgkin and Huxley postulated that the total current is the sum of the trans-membrane current and the current along the axon, and that a propagating solution exists that fulfills a wave equation. The simple cable ODE figuring in the basic HH-model (7.1) was further expanded on this basis into the following PDE for the propagating nerve impulse, depending on the axon radius a:

$$\frac{a}{2R_i}V_{xx} = C_m V_t + g_K(V - E_K) + g_{Na}(V - E_{Na}).$$

Here, R_i is the resistance of the cytosol within the nerve (see [HJ07] for technical review).

The HH-model was originally proposed to account for the property of squid giant axons [HH52, Hod64] and it has been generalized with modifications of ion conductances. More generally, the so-called HH-type models (including the Morris-Lecar model, see [SchMor13]) have been widely adopted for a study on activities of *transducer neurons* such as motor and thalamus relay neurons, which transform the amplitude-modulated input to spike-train outputs.

For the attempts to relate the HH-model with propagation of solitons in neural cell membranes, see [DS95] and the references therein. The simplified form, namely the FitzHugh-Nagumo (FN) model [Fit55, Fit61, NAY62], with the tunnel-diode equivalent neural circuit (see [SchFit13]), is also discussed. Among a number of forms of the FN-model, the simplest one (similar to the Van der Pol oscillator) is suggested by FitzHugh himself in [Fit55]:

$$\epsilon\dot{x} = x - x^3 - y, \qquad \dot{y} = \gamma x - y + b,$$

where $x = x(t)$ is the fast voltage variable and $y = y(t)$ is the slow recovery variable, while γ, b, ϵ $(0 \leq \epsilon \ll 1)$ are parameters.

More generally, dynamics of an N-dimensional ensemble of FN neurons is given by the following set of nonlinear ODEs (see, e.g. [RT96, TR98, TP01]):

$$\dot{x}_i = F(x) - cy_i + I_i^{(c)} + I^{(e)} + \xi_i, \qquad \dot{y}_i = bx_i - dy_i, \quad (\text{for } i = 1, ..., N), \quad (7.3)$$

where $x_i = x_i(t)$ and $y_i = y_i(t)$ denote the fast (voltage) variable and slow (recovery) variable, respectively; $F(x) = kx\,(x-a)\,(1-x)$ is the nonlinear spring-like forcing term (with positive parameters k, a, b, c, d); and $\xi_i = \xi_i(t)$ is the Gaussian white noise with

$$\langle \xi_i(t) \rangle = 0 \quad \text{and} \quad \langle \xi_i(t)\, \xi_j(t') \rangle = \beta_i^2 \delta_{ij}\, \delta(t-t'), \qquad (7.4)$$

where $\langle \cdot \rangle$ means the average over random variables.

The inputs, $I_i^{(c)} = I_i^{(c)}(t)$ and $I^{(e)} = I^{(e)}(t)$, represent N coupling terms and an external (single-spike) input applied to all neurons, respectively. They are given by:

$$I_i^{(c)} = \frac{w}{N} \sum_{j(\neq i)} G(x_j) \quad \text{with} \quad G(x) = \left(1 + \mathrm{e}^{-(x-\theta)/\alpha}\right)^{-1},$$

where w is the coupling strength and $G(x)$ is the sigmoid function with the threshold θ and the width α, and

$$I^{(e)} = A\Theta(t - t_{in})\Theta(t_{in} + T_w - t), \qquad (7.5)$$

where $\Theta = \Theta(x)$ is the Heaviside step (threshold) function with the input time t_{in}, width T_w and amplitude A.

Even more generally, a network of noisy FitzHugh-Nagumo neurons is governed by the following set of nonlinear stochastic ODEs (see [KB93]):

$$dx_i = \left[\phi\,(x_i, y_i) + I_i(t) + \sum_{k=1}^{N} J_{ik}\Theta(x_k)\right] dt + \xi_i dW_i\,, \qquad (7.6)$$
$$dy_i = \psi\,(x_i, y_i)\, dt, \qquad (\text{for } i, k = 1, ..., N),$$

where ϕ and ψ are nonlinear functions, while J_{ik} are synaptic weights for the connection from neuron i to neuron k. The Ito-type stochastic analysis of the system (7.6) was performed in [RT96].

7.2.2 Wave Equation Alternative

The discussion of previous subsections represents compressed history of neural biophysics, as is written in advanced neurophysiological textbooks. Recently, two biophysicists from the Niels Bohr Institute in Copenhagen, T. Heimburg and A. Jackson (see [HJ05, HJ07]), challenged the half-a-century old Hodgkin-Huxley theory. Their main claim is that the HH-model fails to explain a number of features of the propagating nerve pulse, including the reversible release and reabsorption of heat, as well as the accompanying mechanical, fluorescence, and turbidity changes [HJ05, HJ07]. In short, electrical currents through resistors generate heat, independent of the direction of the ion flux; the heat production in the HH-model should always be positive, while the heat dissipation should be related to the power of a circuit through the resistor, i.e. $\dot{Q} = P = V \cdot I > 0$ (for each of the conducting objects in all phases of the action potential).

> "The most striking feature of the isothermal and isentropic compression modulus is its significant undershoot and striking recovery. These features lead generically to the conclusions (i) that there is a minimum velocity of a soliton and (ii) that the soliton profiles are remarkably stable as a function of the soliton velocity. There is a maximum amplitude and a minimum velocity of the solitons that is close to the propagation velocity in myelinated nerves..."

The previous neural-conduction related work of Englishman Archibald V. Hill (a Nobel Laureate in Physiology or Medicine, 1922), specifically regarding heat production in nerves [AHH58]. This was based on his prior work on heat production in contracting muscles [Hil38], which was actually reviewed by Hodgkin in [Hod64]. However, Hodgkin noted that the heat release and absorption response during the neural action potential "is important but is not understood," after which the thermodynamic path of neural biophysics research was effectively shut down for the next half-century.[1]

Based on the thermodynamic relation between heat capacity and membrane area compressibility, Heimburg and Jackson considered a (1+1) hydrodynamic PDE for the dispersive sound propagation in a cylindrical membrane [HJ05, HJ07]. A density-pulse, governing the changes $\Delta\rho^A$ (along the x-axis) of the lateral membrane density ρ^A, is defined by: $\Delta\rho^A(x,t) = \rho^A(x,t) - \rho_0^A$, where $\rho_0^A = 4.035 \cdot 10^{-3}\, g/m^2$ is the equilibrium lateral area density in the fluid phase of the membrane slightly above the melting point. The related sound velocity c can be expanded into a power series (close to the lipid melting transition) as:

$$c^2 = c_0^2 + p(\Delta\rho^A) + q(\Delta\rho^A)^2 + \ldots, \tag{7.7}$$

where $c_0 = 176.6$ m/s is the velocity of small amplitude sound, while p and q are parameters ($p = -16.6\, c_0^2/\rho_0^A$ and $q = 79.5\, c_0^2/(\rho_0^A)^2$).

In the standard $\phi-$notation, with $\phi(x,t) \equiv \Delta\rho^A(x,t)$, the *dispersive wave equation* of [HJ05, HJ07] can be rewritten as a driven wave equation with the driving force $f(\phi)$:

[1] An analogous example in 20th Century physics is A. Einstein's shutting down of the Kaluza-Klein five-dimensional relativity theory, which was recently revived by the advent of string theory.

$$\phi_{tt} = c^2 \phi_{xx} - f(\phi)\,. \tag{7.8}$$

Here we make two remarks regarding the dispersive wave equation (7.8):

1. If the compressibility is approximately constant and if $\Delta\rho^A \ll \rho_0^A$, then $f(\phi) = 0$ (that is, small changes in compressibility means small deviation of $f(\phi)$ from 0); (7.8) then reduces to the standard wave equation:

$$\phi_{tt} = c_0^2 \phi_{xx}\,.$$

2. Higher sound frequencies result in higher propagation velocities, since the isentropic compressibility is a decreasing function of frequency. If higher frequencies become dominant, the dispersive forcing function $f(\phi)$ in (7.8) needs to be defined, or *ad-hoc chosen* [HJ05, HJ07] to mimic the linear frequency-dependence of the sound velocity with a positive parameter h: $f(\phi) = h\phi_{xxxx}$. Furthermore, by introducing the sound propagation velocity v the wave equation (7.8) can be recast into a time-independent form, describing the *shape* of a propagating density excitation (in original Heimburg-Jackson form), giving

$$v^2 \frac{\partial^2}{\partial z^2}\Delta\rho^A = \frac{\partial}{\partial z}\left[\left(c_0^2 + p(\Delta\rho^A) + q(\Delta\rho^A)^2\right)\frac{\partial}{\partial z}\Delta\rho^A\right] - h\frac{\partial^4}{\partial z^4}\Delta\rho^A,$$

which has a localized (stationary) solution [LAJ11]:

$$\Delta\rho^A(z) = \frac{p}{q}\cdot\frac{1-\left(\frac{v^2-v_{min}^2}{c_0^2-v_{min}^2}\right)}{1+\left(1+2\sqrt{\frac{v^2-v_{min}^2}{c_0^2-v_{min}^2}}\cosh\left(\frac{c_0}{h}z\sqrt{1-\frac{v^2}{c_0^2}}\right)\right)}.$$

This solution represents a *sech-type soliton,* a typical solution for nonlinear PDEs (e.g. KdV and NLS).

Instead of arguing either pro- or contra- Heimburg-Jackson theory of *neural sound propagation*, as an alternative to Hodgkin-Huxley *neural electrical theory*, we simply accept the natural solitary explanation of the nerve impulse conduction, regardless of the physical medium that is carrying it (sound, or heat, or electrical, or smectic liquid crystal [DS95], or possibly quantum-mechanical [MN95a, MN95b, MN97]). Under this standpoint, we are still free to chose a different form for the dispersive force term $f(\phi)$ in the perturbed wave equation (7.8). Therefore, instead of the Heimburg-Jackson's ad-hoc choice of the forth-derivative term, we can, following [STZ93] and subsequent studies of neural micro-tubules, choose a double-well quartic dispersive potential:

$$V(\phi_n) = -\frac{1}{2}A\phi_n^2 + \frac{1}{4}B\phi_n^4\,, \tag{7.9}$$

where A and B are real parameters, with A being a linear function of temperature. If we now plug-in the (first two terms of the) Taylor-series expansion of the sine term, we have

$$\sin\phi = \phi - \frac{1}{6}\phi^3 + O(\phi^4) \approx A\phi - B\phi^3. \tag{7.10}$$

The quartic dispersive potential (7.9) would result in another novel alternative to the Hodgkin-Huxley model, in the form of standard sine-Gordon equation (see equation (7.21) in the following subsection), which generates analytical solutions in the form of traveling waves. That is, we have solitons, kinks-antikinks and breathers (see [II13]).

7.2.3 Sine-Gordon Alternative

The sine-Gordon equation (SGE), representing a new alternative to the Hodgkin-Huxley neural model, is the following real-valued, hyperbolic, spatiotemporal, nonlinear wave-type PDE, defined on the (1+1) space-time $\mathbb{R}^{1,1}$ (see Figure 7.4):

$$\phi_{tt} = \phi_{xx} - \sin\phi, \qquad \text{or} \qquad \phi_{tt}(x,t) = \phi_{xx}(x,t) - \sin\phi(x,t). \tag{7.11}$$

A basic solution of the SGE (7.21) is:

$$\phi(x,t) = 4\arctan\left[\exp\left(\pm\frac{x-vt}{\sqrt{1-v^2}}\right)\right], \tag{7.12}$$

which describes a *soliton* moving with velocity $0 \le v < 1$ and changing the phase either from 0 to 2π or from 2π to 0. The first case is a *kink*, when the sign is positive, and the second, when the sign is negative, is an *anti-kink*. Each traveling-wave solution of the SGE has the corresponding surface in $\mathbb{R}^3$ (see [TU00]).

The first *1-soliton* solution of the SGE (7.21) was given by [AKN73, AC91] in the form:

$$\phi(x,t) = 4\arctan\left[\frac{\sqrt{1-\omega^2}\cos(\omega t)}{\omega\cosh(x\sqrt{1-\omega^2})}\right],$$

which, for $\omega < 1$, is periodic in time t and decays exponentially when moving away from $x = 0$.

There is a well-known traveling solitary wave solution with velocity v (see [Tab89]), given by the following generalization of (7.22):

$$\phi(x,t) = 4\arctan\left[\exp\frac{\pm 2(z-z_0)}{\sqrt{1-v^2}}\right], \qquad \text{with} \qquad (z = \mu(x+vt)), \tag{7.13}$$

and the center at z_0. In (7.31), as with (7.22), the case when the sign is positive describes kink, while the case when the sign is negative represents an anti-kink.

The *stationary kink* with the center at x_0 is defined by

$$\phi(x) = 2\arctan\left[\exp(x - x_0)\right],$$

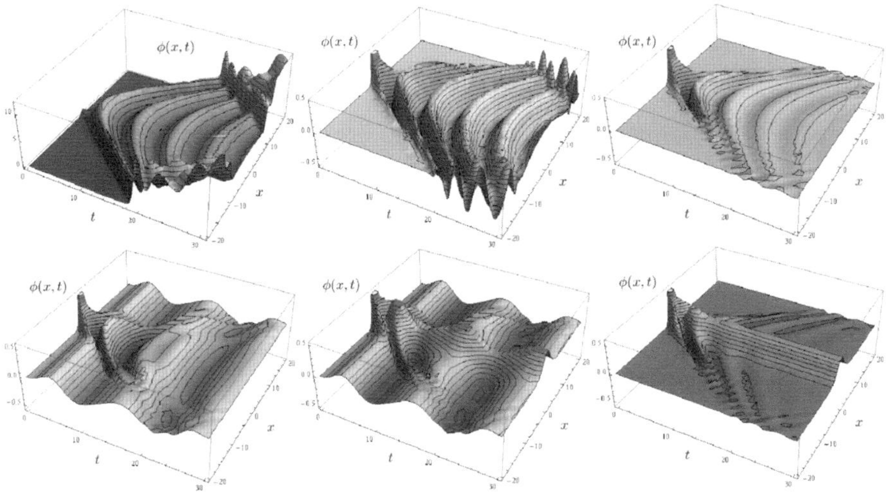

Fig. 7.4. *Mathematica® simulations of the (1+1) sine-Gordon PDE (7.21) for different initial and boundary conditions (for technical details, see [II13]).*

in which the position of the center x_0 can be varied continuously ($-\infty < x_0 < \infty$) and represents the solution of the first-order ODE $\phi_x(x) = \sin\phi(x)$.

Regarding solutions of the slightly more general three-parameter SGE

$$\phi_{tt} = a\phi_{xx} + b\sin(\lambda\phi), \tag{7.14}$$

the following cases were established in the literature (see [PZ04, II13] and references therein):

1. If a function $w = \phi(x,t)$ is a solution of (7.32), then so are also the following functions:

$$w_1 = \frac{2\pi n}{b} \pm \phi(C_1 \pm x, C_2 \pm t) \qquad \text{for} \quad (n = 0, \pm1, \pm2, ...),$$
$$w_2 = \pm\phi\left(x\cosh C_3 + t\sqrt{a}\sinh C_3,\ x\frac{\sinh C_3}{\sqrt{a}} + t\cosh C_3\right),$$

where C_1, C_2, and C_3 are arbitrary constants.

2. Traveling-wave solutions:

$$\phi(x,t) = \frac{4}{\lambda}\arctan\left[\exp\left(\pm\frac{b\lambda(C_1x + C_2t + C_3)}{\sqrt{b\lambda(C_2^2 - aC_1^2)}}\right)\right] \tag{7.15}$$

if $b\lambda(C_2^2 - aC_1^2) > 0$,

$$\phi(x,t) = -\frac{\pi}{\lambda} + \frac{4}{\lambda}\arctan\left[\exp\left(\pm\frac{b\lambda(C_1x + C_2t + C_3)}{\sqrt{b\lambda(aC_1^2 - C_2^2)}}\right)\right]$$

if $b\lambda(C_2^2 - aC_1^2) < 0$,

where the first expression (for $b\lambda(C_2^2 - aC_1^2) > 0$) represents another 1-soliton solution, which is kink in case of $\exp\left(\frac{b\lambda(C_1x+C_2t+C_3)}{\sqrt{b\lambda(C_2^2-aC_1^2)}}\right)$ and antikink in case of $\exp\left(-\frac{b\lambda(C_1x+C_2t+C_3)}{\sqrt{b\lambda(C_2^2-aC_1^2)}}\right)$. In case of the standard SGE (7.21), this kink-antikink expression specializes to the Lorentz-invariant solution similar to (7.31):

$$\phi^{\mathrm{K}}(x,t) = 4\arctan\left[\exp\left(\frac{\pm(x-x_c)-vt}{\sqrt{1-v^2}}\right)\right], \qquad (7.16)$$

where the velocity v $(0 < v < 1)$ and the soliton-center x_c are real-valued constants. The kink solution has the following physical (EM) characteristics:
(i) Energy:

$$E[\phi^{\mathrm{K}}(x,t)] = \int T^{00}dx = \frac{8}{\sqrt{1-v^2}};$$

(ii) Momentum:

$$P[\phi^{\mathrm{K}}(x,t)] = \int T^{10}dx = -\frac{8v}{\sqrt{1-v^2}}.$$

3. Functional separable solution:

$$w(x,t) = \frac{4}{\lambda}\arctan\left[f(x)g(t)\right],$$

where the functions $f = f(x)$ and $g = g(t)$ are determined by the first-order autonomous separable ODEs:

$$f_x^2 = Af^4 + Bf^2 + C, \qquad g_t^2 = -aCg^4 + (aB + b\lambda)g^2 - aA,$$

where A, B, and C are arbitrary constants. In particular, for $A = 0, B = k^2 > 0$, and $C > 0$, we have the 2-*soliton* solution [PS62]:

$$w(x,t) = \frac{4}{\lambda}\arctan\left[\frac{\eta\sin(kx+A_1)}{k\sqrt{a}\cosh(\eta t + B_1)}\right], \qquad \text{with } \left(\eta^2 = ak^2 + b\lambda > 0\right),$$

where k, A_1, and B_1 are arbitrary constants.

The only stable traveling wave SGE-solutions, meaning localized solutions with identical boundary conditions $\phi = 0$ and $\phi = 2\pi$, for a scalar field ϕ are 2π-kinks [OS83, Das85] . However, easier to follow experimentally are non-localized π-kinks [KGS92], where separate regions have different values of the field ϕ (see also [ZML98] and references therein).

On the other hand, a *breather* is spatially localized, time periodic, oscillatory SGE-solution (see, e.g. [HS98]). It represents a field which is periodically oscillating in time and decays exponentially in space as the distance from the center $x = 0$ is increased. This oscillatory solution of (7.21) is characterized by some phase that depends on the breather's evolution history. This could be, in particular, a bound state of vortex with an antivortex in a Josephson junction. In this case, breather may appear as a result of collision of a *fluxon* (a propagating magnetic flux-quantum) with an *antifluxon*, or even in the process of measurements of switching current characteristics. stationary breather solutions form one-parameter families of solutions. An example of a breather-solution of (7.21) is given by [GK06]:

$$\phi = 4 \arctan \left(\frac{\sin T}{u \cosh \left(g(u)x \right)} \right),$$

with parameters $u = u(t)$ and $T = T(t)$, such that

$$g(u) = 1/\sqrt{1+u^2} \quad \text{and} \quad T(t) = \int_0^t g(u(t'))\, u(t')\, dt'.$$

For more technical details on SGE, including its solitary solutions as well as various generalizations and discretizations, see Appendix, subsection 7.4.1.

7.3 Fiber-Optics Solitons

One half of the 2009 Nobel Prize in Physics was awarded to Charles K. Kao "for ground-breaking achievements concerning the transmission of light in fibers for optical communication," mainly for his pioneering work [KH66] published in July 1966, "the date now regarded as the birth of optical fiber communication." The importance of this discovery for today's world can be illustrated by the following quote from Kao's Nobel lecture:

> "The news of the Nobel Prize reached me in the middle of the night at 3 am in California, through a telephone call from Stockholm (then in their morning) no doubt carried on optical fibers; congratulations came literally minutes later from friends in Asia (for whom it was evening), again through messages carried on optical fibers."

For the present Chapter, it is important that there are two kinds of solitons traveling along optical fibers [MB99, GTW12]. The first kind is governed by the nonlinear Schrödinger (NLS) equation,[2] describing the balance between the 2nd-order dispersion and Kerr nonlinearity [Agr01]. The second is governed by the Maxwell-Bloch equation, describing the self-induced transparency soliton in the two-level resonance medium [AE75, XTA12].

[2] Note that there is a number of other NLS-applications, including financial analysis [Iva10, Iva11a] and crowd dynamics [IR12, IJ12].

7.3.1 NLS-Maxwell-Bloch System

More generally, in erbium doped fibers[3] [GH13], those two types of solitons usually coexist, which is governed by the coupled NLS-Maxwell-Bloch system [AE75, GH13]:

$$q_z = \mathrm{i}\,(\frac{1}{2}\,q_{tt} + |q|^2 q) + 2p, \tag{7.17}$$
$$p_t = 2\mathrm{i}\omega p + 2q\eta, \qquad \eta_t = -(q\bar{p} + p\bar{q}),$$

where q is the (slowly-varying) amplitude of the complex-valued pulse-field envelope, ω is the frequency, $p = \psi_1\bar{\psi}_2$ is the polarization and $\eta = |\psi_2|^2$ is the population inversion, while ψ_1 and ψ_2 are the wave functions in a two-level atomic system. For more technical details on this nonlinear PDE system, see [AE75].

7.3.2 Hirota-Maxwell-Bloch System

Although representing a common pulse-conduction model, the NLS-Maxwell-Bloch system (7.17) does not include the higher-order effects (including the 3rd-order dispersion, self-steepening and nonlinear response effects), which are important for modeling the ultra-short pulse propagation in nonlinear optical fibres. A more realistic optical pulse-conduction model is given by the following Hirota-Maxwell-Bloch system (see [AE75, GH13]):

$$q_z = \mathrm{i}\alpha(\frac{1}{2}q_{tt} + 2|q|^2 q) + \beta(q_{ttt} + 6|q|^2 q_t) + 2p, \tag{7.18}$$
$$p_t = 2\mathrm{i}\omega p + 2q\eta, \qquad \eta_t = -(q\bar{p} + p\bar{q}).$$

The symbols have the same meaning as in (7.17); in addition, α comprises both the 2nd-order dispersion and Kerr nonlinearity, while β represents the strength of the higher order nonlinear effects [AE75]. The Hirota-Maxwell-Bloch system (7.18) has been recently analyzed from different aspects. For example, conservation laws and multi-soliton solutions (on the zero background) were derived in [XtZ09] using Darboux transformation, Painlevé properties were defined in [PN96], while rouge wave solutions (on continuous-wave backgrounds) were constructed in [LHP13].

Lax pair representation

It was demonstrated in [PN95] that the system (7.18) is integrable and also admits a Lax pair representation, as well as other required properties of complete integrability. The linear eigenvalue problem is expressed in the form of the Lax pair (U, V), defined by the linear equation:

---

[3] Erbium (Er) is a chemical element belonging to the group of rare earth metals. It is widely used in the form of the trivalent ion Er3+ as the laser-active dopant of gain media (including crystals and glasses), which can amplify the power of light, which is required in a laser to compensate for the resonator losses.

$$\Phi_t = U\Phi, \qquad \Phi_z = V\Phi. \tag{7.19}$$

In (7.19), the eigenfunctions U and V (corresponding to the eigenvalue λ) are defined as follows:

$$U = \lambda\begin{pmatrix} -\mathrm{i} & 0 \\ 0 & \mathrm{i} \end{pmatrix} + \begin{pmatrix} 0 & q \\ -\bar{q} & 0 \end{pmatrix} = -\mathrm{i}\lambda\sigma + U_0, \qquad \text{where} \qquad \sigma = \begin{pmatrix} 1 & 0 \\ 0 & -1 \end{pmatrix}.$$

$$V = \lambda^3 V_3 + \lambda^2 V_2 + \lambda V_1 + V_0 + \mathrm{i}\frac{1}{\lambda+\omega}V_{-1}, \qquad \text{where}$$

$$V_3 = \begin{pmatrix} 4\mathrm{i}\beta & 0 \\ 0 & -4\mathrm{i}\beta \end{pmatrix}, \quad V_2 = \begin{pmatrix} -\alpha\mathrm{i} & -4\beta q \\ 4\beta\bar{q} & \alpha\mathrm{i} \end{pmatrix}, \quad V_1 = \begin{pmatrix} -2\beta\mathrm{i}|q|^2 & \alpha q - 2\beta\mathrm{i}q_t \\ -\alpha\bar{q} - 2\beta\mathrm{i}q_t^* & 2\beta\mathrm{i}|q|^2 \end{pmatrix}$$

$$V_0 = \begin{pmatrix} \frac{\alpha}{2}\mathrm{i}|q|^2 - \beta(q\bar{q}_t - q_t\bar{q}) & 2\beta|q|^2 q + \frac{\alpha}{2}\mathrm{i}q_t + \beta q_{tt} \\ -2\beta|q|^2\bar{q} + \frac{\alpha}{2}\mathrm{i}\bar{q}_t - \beta\bar{q}_{tt} & -\frac{\alpha}{2}\mathrm{i}|q|^2 + \beta(q\bar{q}_t - q_t\bar{q}) \end{pmatrix}, \quad V_{-1} = \begin{pmatrix} \eta & -p \\ -\bar{p} & -\eta \end{pmatrix}.$$

From the Lax pair (U, V), the system (7.18) is obtained by setting $\alpha = 2, \beta = -1$.

Method of successive Darboux transformations

The so-called Darboux transformation is an efficient method for constructing the soliton solutions for integrable nonlinear PDEs [MS92]. To apply this method to the Lax pair (7.19) of the the Hirota-Maxwell-Bloch model (7.18), consider the following transformation [LHP13]:

$$\Phi' = T\Phi = (\lambda A - S)\Phi, \qquad \text{where}$$

$$A = \begin{pmatrix} a_{11} & a_{12} \\ a_{21} & a_{22} \end{pmatrix}, \qquad S = \begin{pmatrix} s_{11} & s_{12} \\ s_{21} & s_{22} \end{pmatrix}$$

about linear function $\Phi = \Phi(\lambda)$, in which the new function Φ' satisfies:

$$\Phi'_t = U'\Phi', \qquad \Phi'_z = V'\Phi', \tag{7.20}$$

such that the matrix T satisfies the following identities:

$$T_t + TU = U'T, \qquad T_z + TV = V'T.$$

Substituting the matrices A and S into (7.20) and comparing the coefficients of both sides gives the following conditions:

$$a_{12} = a_{21} = 0, \qquad (a_{11})_t = (a_{22})_t = 0.$$

Of particular interest is the choice: $A = I$ and $T = (\lambda I - S)$. The relation between old solutions (q, p, η) and new solutions (q', p', η') is called Darboux transformation and can be obtained via (7.20).

Details of the one-fold and n-fold Darboux transformations performed on the above Lax pair representation of the Hirota-Maxwell-Bloch model are shown in [LHP13], from which a variety of soliton solutions of the Hirota-Maxwell-Bloch system (7.18) has been constructed (by assuming suitable seed solutions). In particular, assuming trivial seed solutions ($q = 0, p = 0, \eta = 1$) the linear system (7.19) becomes [LHP13]:

$$\Phi_t = U\Phi, \qquad \Phi_z = V\Phi, \qquad \text{with}$$

$$\Phi = \begin{pmatrix} \Phi_1 \\ \Phi_2 \end{pmatrix}, \qquad U = \begin{pmatrix} -\mathrm{i}\lambda & 0 \\ 0 & \mathrm{i}\lambda \end{pmatrix},$$
$$V = \begin{pmatrix} 4\mathrm{i}\beta\lambda^3 - \alpha\mathrm{i}\lambda^2 & 0 \\ 0 & -4\beta\mathrm{i}\lambda^3 + \alpha\mathrm{i}\lambda^2 \end{pmatrix} + \frac{\mathrm{i}}{\lambda + \omega}\begin{pmatrix} 1 & 0 \\ 0 & -1 \end{pmatrix}.$$

and the explicit eigenfunctions (with arbitrarily fixed real constants x_0, y_0, θ) given by:

$$\Phi_1 = \mathrm{e}^{-\mathrm{i}\lambda t + (4\beta\mathrm{i}\lambda^3 - \alpha\mathrm{i}\lambda^2 + \frac{i}{\lambda+\omega})z + \frac{x_0 + \mathrm{i}y_0}{2}},$$
$$\Phi_2 = \mathrm{e}^{\mathrm{i}\lambda t + (-4\beta\mathrm{i}\lambda^3 + \alpha\mathrm{i}\lambda^2 - \frac{i}{\lambda+\omega})z - \frac{x_0 + \mathrm{i}y_0}{2} + \mathrm{i}\theta}.$$

Their substituton into the one-fold Darboux transformation (while choosing $\lambda = \alpha_1 + I\beta_1$, $x_0 = 0$, $y_0 = 0, \theta = 0$), has given the following soliton solutions:

$$q = 2\beta_1 \mathrm{e}^{-2\mathrm{i}\frac{C}{B}} \operatorname{sech}(2\beta_1 \frac{A}{B}),$$
$$p = \frac{\mathrm{i}\beta_1 \left[(\alpha_1 - \mathrm{i}\beta_1 + \omega)\mathrm{e}^{-2\frac{D}{B}} + (\alpha_1 + \mathrm{i}\beta_1 + \omega)\mathrm{e}^{-2\frac{F}{B}}\right]}{B} \operatorname{sech}^2(2\beta_1 \frac{A}{B}),$$
$$\eta = 1 - \frac{2\beta_1^2}{B}\operatorname{sech}^2(2\beta_1 \frac{A}{B}).$$

Similarly, the construction of the *two-soliton* solutions of the Hirota-Maxwell-Bloch system (7.18) was performed in [LHP13] by using two spectral parameters $\lambda_1 = \alpha_1 + i\beta_1$ and$\lambda_2 = \alpha_2 + i\beta_2$ and the second Darboux transformation. In the case of two-soliton solutions, if λ_2 is assumed to be close to λ_1, performing the Taylor expansion of wave function to first order up to λ_1 leads to a new kind of solution which is called smooth *positon* solution.[4] Bright and dark positon solutions of the system (7.18) have been constructed in [LHP13], using four linear functions and second Darboux transformation. An example of such constructed positon solutions q_p has the form:

$$q_p = 8\mathrm{i}\mathrm{e}^{14.8\mathrm{i}z - \mathrm{i}t}[1582z\cosh(2t + 2.40z) - 688\mathrm{i}z\sinh(2t + 2.4z)$$
$$- 50\mathrm{i}\cosh(2t + 2.4z) + 100\mathrm{i}\, t\sinh(2t + 2.4z)]/$$
$$[400t^2 + 119296z^2 - 5504tz + 50 + 50\cosh(4.8z + 4t)].$$

4 A term *positon* was coined by Matveev [MS92, Mat92b, Mat92a, Mat02] for the Korteweg-de Vries (KdV) equation by the same limiting approach.

Next, by formulating (through the second Darboux transformation) the complex modified Korteweg-de Vries-Maxwell-Bloch system out of the Hirota-Maxwell-Bloch system (7.18), both dark and bright *breather* solutions were constructed in the form:

$$
\begin{aligned}
q_b = {} & [d^2(\mathrm{e}^{\mathrm{i}\frac{A_b(w)}{Y}} + \mathrm{e}^{\frac{A_b(-w)}{Y}\mathrm{i}}) + 2d\beta_1(\mathrm{e}^{\mathrm{i}\frac{\bar{C}_b}{Y}} + \mathrm{e}^{\frac{C_b}{Y}\mathrm{i}}) - d\beta_1(\mathrm{e}^{\frac{D_b}{X}\mathrm{i}} + \mathrm{e}^{\frac{\bar{D}_b}{X}\mathrm{i}}) \\
& + 2\beta_1(w-\beta_1)\mathrm{e}^{\frac{F_b(w)}{Y}\mathrm{i}} - 2\beta_1(w+\beta_1)\mathrm{e}^{\frac{F_b(-w)}{Y}\mathrm{i}}]/[2d\cosh(2\mathrm{i}w\beta_1 z\frac{G_b}{Y}) \\
& - 2\beta_1\cosh(2w\frac{H_b}{Y})], \qquad \text{where} \\
& w = \sqrt{\beta_1^2 - d^2}, \qquad X = (\alpha_1 + \beta_1\mathrm{i} + \omega)(-\alpha_1 + \beta_1\mathrm{i} - \omega), \\
& Y = (\alpha_1 + \omega)(\alpha_1 + \beta_1\mathrm{i} + \omega)(-\alpha_1 + \beta_1\mathrm{i} - \omega),
\end{aligned}
$$

while $[A_b(w), C_b, D_b, F_b(w)]$ and G_b, H_b are polynomials of: $(t, z, \omega, w, a, b, d, \alpha, \beta, \alpha_1, \beta_1)$ defined in [LHP13].

Finally, using the limit method of the NLS equation [HZW12], the bright and dark *rogue-wave* solutions of the system (7.18) were constructed in [LHP13], with the specific rogue-wave solution q_r:

$$
\begin{aligned}
q_r = {} & \mathrm{e}^{-\frac{\mathrm{i}}{4}(8+20\beta+7\alpha)}[40\alpha - 17\alpha^2 + 156\beta - 40 - 585\beta^2 - 192\alpha\beta \\
& + \mathrm{i}(32 + 8\alpha + 96\beta)]/(1170\beta^2 - 80\alpha + 34\alpha^2 - 312\beta + 112 + 384\alpha\beta).
\end{aligned}
$$

Numerical simulations of positons, 2-solitons, breathers and rogue-waves are given in Figure 7.5 (for more technical details, see [LHP13]). The figure illustrates some of the wide range of behaviors for which our generalized system can account.

Summary

We have looked to models of controlled complexity in nonlinear pulse conduction and proposed a novel new alternative, in the form of a sine-Gordon wave equation, to the traditional Hodgkin-Huxley neural model. Our proposed model explains pulse conduction in terms of more general traveling wave phenomena, such as kinks, solitons, breathers, positons and rogue waves.

What about our broader questions concerning controlling complexity? Our investigations here represent one case study in biological control from which we - eventually - hope to be able to formulate useful models for significantly broader contexts, such as in military command and control. We conclude that there are two significant barriers which we must tackle in pursuing this broader goal: such problems do not enjoy well established and stable problem statements, and discrepancies in their conceptualization may particularly strongly limit the scope and usefulness of a model. Unlike the pulse conduction problem, we do not more broadly expect to have a range of solutions to a single stable problem, but rather multiple solutions to multiple overlapping – and possibly contradictory - problems.

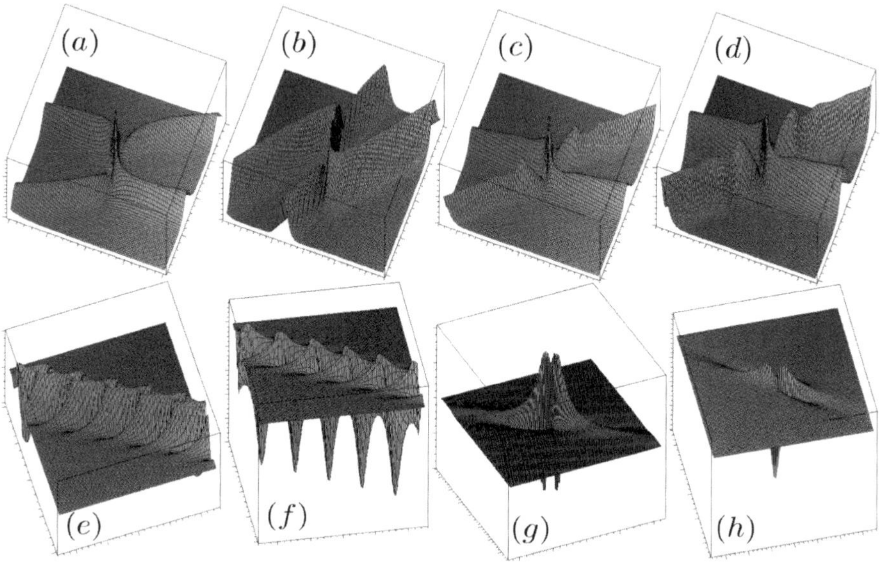

Fig. 7.5. *Numerical simulations of positons (a − b), 2-solitons (c − d), breathers (e − f) and rogue-waves (g − h), for various boundary and initial conditions; modified and adapted from [LHP13].*

7.4 Appendix

7.4.1 A 'Zoo' of Sine–Gordon Solitons

In spatio-temporal dynamics of complex nonlinear systems (see, e.g. [II08a, IR12]), the *Sine–Gordon equation* (SGE) is, together with the Korteweg–deVries (KdV) and the nonlinear Schrödinger (NLS) equations, one of the celebrated nonlinear-yet-integrable partial differential equations (PDEs), with a variety of traveling solitary waves as solutions (see Figures 7.6 and 7.7).

Briefly, a *solitary wave* is a traveling wave (with velocity v) of the form: $\phi(x,t) = f(x - vt)$, for a smooth function f that decays rapidly at infinity; e.g., a nonlinear wave equation:

$$\phi_{tt} - \phi_{xx} = \phi(2\phi^2 - 1)$$

has a family of solitary–wave solutions:

$$\phi(x,t) = \mathrm{sech}(x\mathrm{cosh}\mu + t\mathrm{sinh}\mu),$$

parameterized by $\mu \in \mathbb{R}$. In complex physical systems, SGE solitons, kinks and breathers appear in various situations, including propagation of magnetic flux (fluxons) in long Josephson junctions, dislocations in crystals, nonlinear spin waves in superfluids, and waves in ferromagnetic and anti-ferromagnetic materials.

In this subsection we give a brief overview of the SGE; for further technical details, see [II13] and references therein.

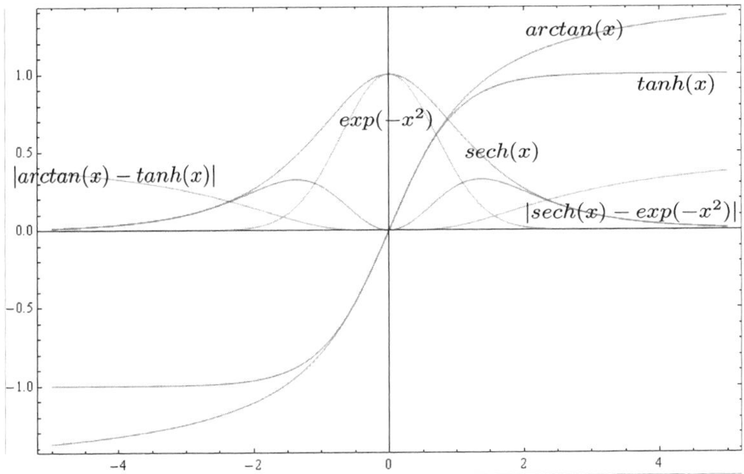

Fig. 7.6. *Basic static examples of kinks:* $\tanh(x)$, $\arctan(x)$ *and bell–shaped solitons:* $\operatorname{sech}(x)$, $\exp(-x^2)$, *together with their (absolute) differences; plotted in* $Mathematica^{TM}$.

The Sine–Gordon equation (SGE)

SGE is a real-valued, hyperbolic, nonlinear wave equation defined on $\mathbb{R}^{1,1}$, which appears in two equivalent forms (using standard indicial notation for partial derivatives: $\phi_{zz} = \partial_z^2\phi = \partial^2\phi/\partial x^z$):

- In the (1+1) space-time $(x,t)-$coordinates, the SGE reads:

$$\phi_{tt} = \phi_{xx} - \sin\phi, \qquad \text{or} \qquad \phi_{tt}(x,t) = \phi_{xx}(x,t) - \sin\phi(x,t), \tag{7.21}$$

  which shows that it is a nonlinear extension of the standard linear wave equation: $\phi_{tt} = \phi_{xx}$. The solutions $\phi(x,t)$ of (7.21) determine the internal Riemannian geometry of surfaces of constant negative scalar curvature $R = -2$, given by the line-element:

$$ds^2 = \sin^2\left(\frac{\phi}{2}\right)dt^2 + \cos^2\left(\frac{\phi}{2}\right)dx^2,$$

  where the angle ϕ describes the embedding of the surface into Euclidean space $\mathbb{R}^3$. A basic solution of the SGE (7.21) is:

$$\phi(x,t) = 4\arctan\left[\exp\left(\pm\frac{x-vt}{\sqrt{1-v^2}}\right)\right], \tag{7.22}$$

describing a *soliton* moving with velocity $0 \leq v < 1$ and changing the phase from 0 to 2π (*kink,* the case of + sign) or from 2π to 0 (*anti-kink,* the case of − sign). Each traveling soliton solution of the SGE has the corresponding surface in $\mathbb{R}^3$.

- In the (1+1) light-cone (u, v)−coordinates, defined by: $u = (x + t)/2$, $v = (x + t)/2$, in which the line-element (depending on the angle ϕ between two asymptotic lines: $u = \text{const}, v = \text{const}$) is given by:

$$ds^2 = du^2 + 2\cos\phi \, du \, dv + dv^2,$$

the SGE describes a family of pseudo-spherical surfaces with constant Gaussian curvature $K = -1$, and reads:

$$\phi_{uv} = \sin\phi, \qquad \text{or} \qquad \phi_{uv}(u, v) = \sin\phi(u, v). \tag{7.23}$$

SGE (7.23) is the single Codazzi–Mainardi compatibility equation between the first (I_G) and second (II_C) fundamental forms of a surface, defined by the Gauss and Codazzi equations, respectively:

$$I_G = du^2 + 2\cos\phi \, du \, dv + dv^2, \qquad II_C = 2\sin\phi \, du \, dv.$$

A typical, spatially-symmetric, boundary-value problem for (7.21) is defined by:

$$x \in [-L, L] \subset \mathbb{R}, \qquad (t \in \mathbb{R}^+),$$
$$\phi(x, 0) = f(x), \qquad \phi_t(x, 0) = 0, \qquad \phi(-L, t) = \phi(L, t),$$

where $f(x) \in \mathbb{R}$ is an axially-symmetric function (e.g., Gaussian or sech, see Figure 7.8).

Bäcklund transformations (BT) for the SGE (7.21) were devised in 1880s in Riemannian geometry of surfaces and are attributed to Bianchi and Bäcklund. [In 1883, A. Bäcklund showed that if $L : M \to M'$ is a pseudo-spherical line congruence between two surfaces M, M', then both M and M' are pseudo-spherical and L maps asymptotic lines on M to asymptotic lines on M'. Analytically, this is equivalent to the statement that if ϕ is a solution of the SGE (7.21), then so are also the solutions of the ODE system (7.24). BT have the form:

$$\frac{1}{2}(\phi + \varphi)_\xi = \alpha \sin\frac{\phi - \varphi}{2}, \qquad \frac{1}{2}(\phi - \varphi)_\eta = \frac{1}{\alpha}\sin\frac{\phi + \varphi}{2}, \tag{7.24}$$

where both ϕ and φ are solutions of the SGE (7.21), and can be viewed as a transformation of the SGE into itself. BT (7.24) allows one to find a 2-parameter family of solutions, given a particular solution ϕ_0 of (7.21). For example, consider the trivial solution $\phi = 0$ that, substituted into (7.24), gives:

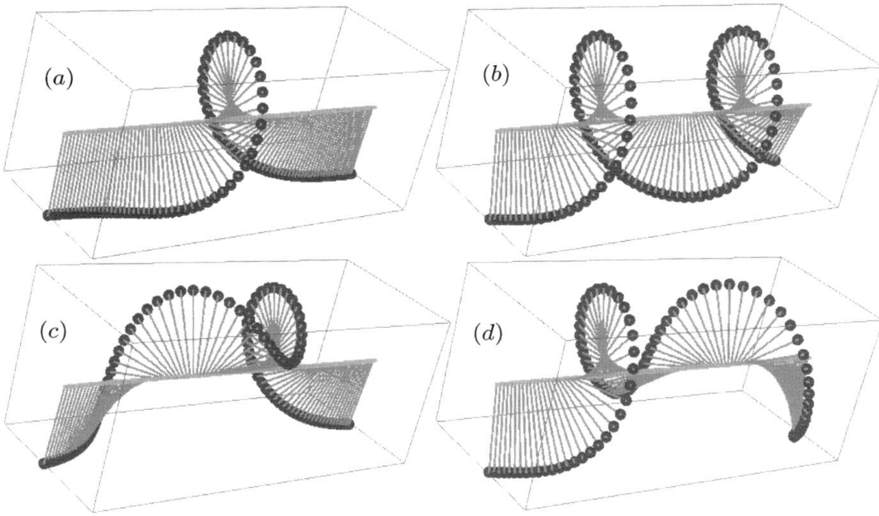

Fig. 7.7. *Basic solitary SGE–solutions, simulated in Mathematica as systems of spring-coupled torsional pendula: (a) single soliton:* $\phi(x,t) = 4\arctan\left(\exp\frac{x-vt}{1-v^2}\right)$; *(b) soliton–soliton collision:* $\phi(x,t) = 4\arctan\left(\frac{v\sinh\frac{x}{1-v^2}}{\cosh\frac{vt}{1-v^2}}\right)$; *(c) soliton–antisoliton collision:* $\phi(x,t) = 4\arctan\left(\frac{\sinh\frac{vt}{1-v^2}}{v\cosh\frac{x}{1-v^2}}\right)$; *and (d) single breather:* $\phi(x,t) = 4\arctan\left(\frac{\sin\frac{vt}{1-v^2}}{v\cosh\frac{x}{1-v^2}}\right)$.

$$\varphi_\xi = -2\alpha\sin\frac{\varphi}{2}, \qquad \varphi_\eta = -\frac{2}{\alpha}\sin\frac{\varphi}{2},$$

which, by integration, gives:

$$2\alpha\xi = -2\ln(\tan\frac{\varphi}{4}) + p(\eta), \qquad \frac{2}{\alpha}\eta = -2\ln(\tan\frac{\varphi}{4}) + p(\xi),$$

from which the following new solution is generated:

$$\varphi = 4\arctan\left[\exp(-\alpha\xi - \frac{1}{\alpha}\eta + \mathrm{const})\right].$$

The sine–forcing term in the SGE can be viewed as a *nonlinear deformation:* $\phi \to \sin\phi$, of the linear forcing term in the Klein–Gordon equation (KGE, a vacuum linearization of the SGE), which is commonly used for describing scalar fields (quantum) field theory:

$$\phi_{tt} = \phi_{xx} - \phi, \tag{7.25}$$

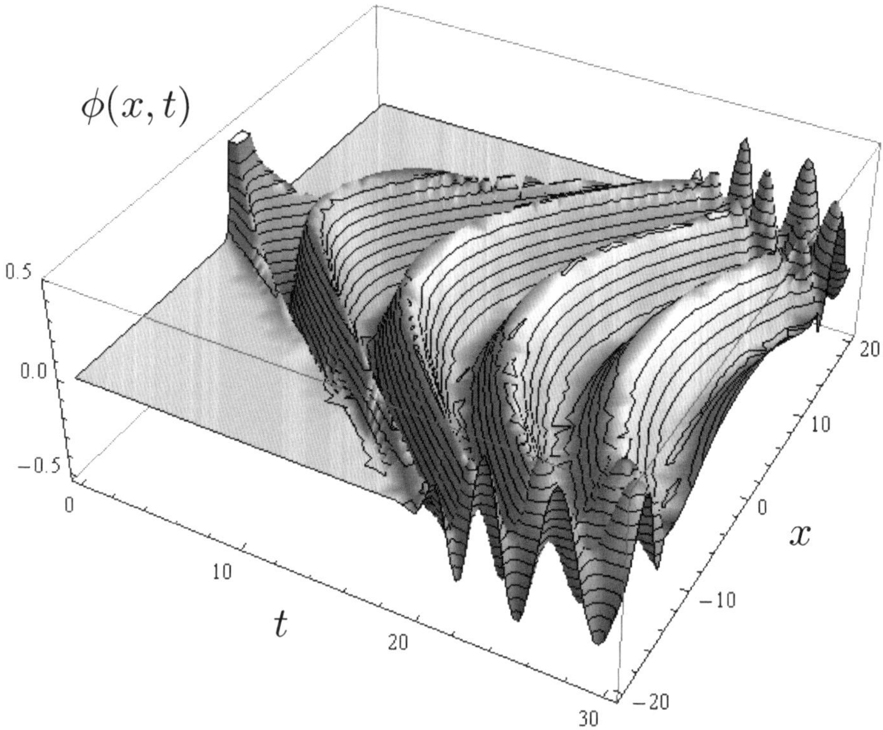

Fig. 7.8. *Numerical solution of the SGE (7.21) in Mathematica, using numerical ODE/PDE integrator NDSolve, with the following data (including the Gaussian initial state, zero initial velocity and symmetric boundary condition):* $x \in [-20, 20]$, $t \in [0, 30]$, $\phi(x, 0) = \exp(-x^2)$, $\phi_t(x, 0) = 0$, $\phi(-20, t) = \phi(20, t)$. *The waves oscillate around the zero plane and increase their width with time. Both near-periodicity and nonlinearity of the time evolution are apparent.*

This, in turn, implies that (as a field equation) SGE can be derived as an Euler–Lagrangian equation from the Lagrangian density:

$$\mathcal{L}_{\mathrm{SG}}(\phi) = \frac{1}{2}(\phi_t^2 - \phi_x^2) - 1 + \cos\phi. \tag{7.26}$$

It could be expected that $\mathcal{L}_{\mathrm{SG}}(\phi)$ is a 'deformation' of the KG Lagrangian:

$$\mathcal{L}_{\mathrm{KG}}(\phi) = \frac{1}{2}(\phi_t^2 - \phi_x^2) - \frac{\phi^2}{2}. \tag{7.27}$$

That can be demonstrated by the Taylor–series expansion of the cosine term:

$$\cos\phi = \sum_{n=0}^{\infty} \frac{(-\phi^2)^n}{(2n)!},$$

so that we have the following relationship between the two Lagrangians:

$$\mathcal{L}_{\mathrm{SG}}(\phi) = \mathcal{L}_{\mathrm{KG}}(\phi) + \sum_{n=2}^{\infty} \frac{(-\phi^2)^n}{(2n)!}.$$

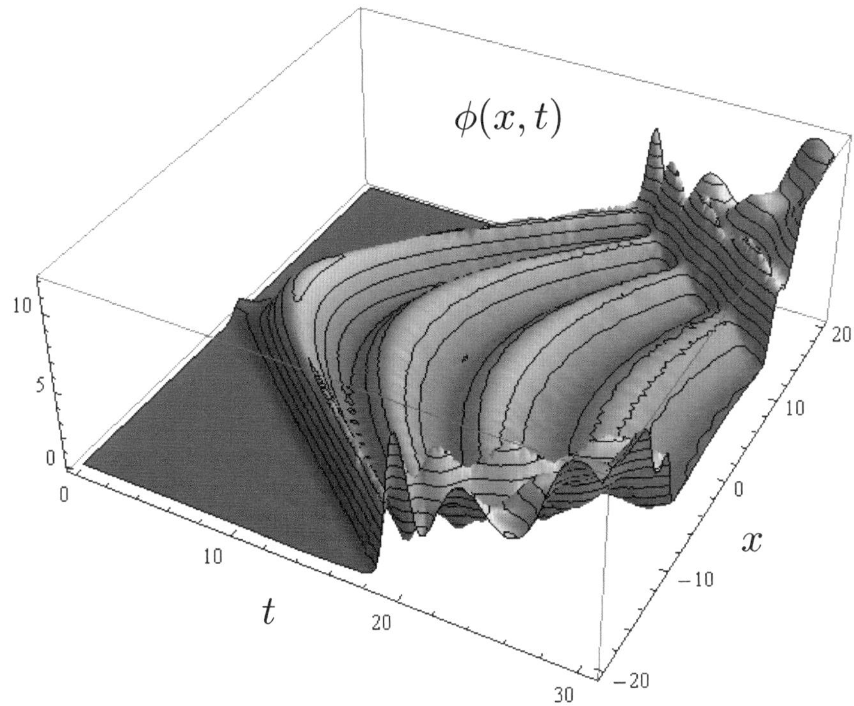

Fig. 7.9. *Numerical solution of the SGE (7.37) in Mathematica, with the following data (including the Gaussian initial state, zero initial velocity and symmetric boundary condition):* $x \in [-20, 20]$, $t \in [0, 30]$, $\phi(x, 0) = \exp(-x^2)$, $\phi_t(x, 0) = 0$, $\phi(-20, t) = \phi(20, t)$. *Under the same boundary conditions, the SGE with the plus sine gives about 20 times higher amplitude waves, which are all above the zero plane and decrease their width with time. Again, both near-periodicity and nonlinearity of the time evolution are apparent.*

The corresponding Hamiltonian densities, of kinetic plus potential energy type, are given in terms of canonically–conjugated coordinate and momentum fields by:

$$\mathcal{H}_{\text{SG}}(\phi,\pi) = \pi\phi_t - \mathcal{L}_{\text{SG}}(\phi) = \frac{1}{2}(\pi^2 + \phi_x^2) + 1 - \cos\phi,$$
$$\mathcal{H}_{\text{KG}}(\phi,\pi) = \pi\phi_t - \mathcal{L}_{\text{KG}}(\phi) = \frac{1}{2}(\pi^2 + \phi_x^2) + \phi^2.$$

Both SGE and KGE are infinite–dimensional Hamiltonian systems, with Poisson brackets given by:

$$\{F,G\} = \int_{-\infty}^{\infty} \left[\frac{\delta F}{\delta\phi(x)}\frac{\delta G}{\delta\pi(x)} - \frac{\delta F}{\delta\pi(x)}\frac{\delta G}{\delta\phi(x)}\right] dx, \tag{7.28}$$

so that both (7.21) and (7.25) follow from Hamilton's equations with Hamiltonian H and symplectic form ω:

$$\phi_t = \{H,\phi\}, \qquad \pi_t = \{H,\pi\},$$
$$\text{with} \quad H = \int_{-\infty}^{\infty} \mathcal{H}(\phi,\pi)\,dx, \qquad \omega = \int_{-\infty}^{\infty} d\pi\wedge d\phi\,dx. \tag{7.29}$$

The Hamiltonian (7.29) is conserved by the flow of both SGE (7.21) and KGE (7.25), with an infinite number of commuting constants of motion (common level sets of these constants of motion are generically infinite-dimensional tori of maximal dimension). Both SGE and KGE admit their own infinite families of conserved functionals in involution with respect to their Poisson bracket (7.28). This fact allows them both to be solved with the *inverse scattering transform*.

Momentum and energy of SGE–solitons

SGE is Lorentz–covariant (i.e., invariant with respect to special–relativistic Lorentz transformations; each SGE–soliton behaves as a relativistic object and contracts when $v \to c \equiv$ the speed of light), and for this fact it has been used in (quantum) field theory.

In both forms (7.21) and (7.23), the SGE has the following symmetries:

$$\begin{array}{llll}
t \to t+t_0, & x \to x, & \phi \to \phi & \text{(shift in } t\text{)},\\
t \to t, & x \to x + x_0, & \phi \to \phi & \text{(shift in } x\text{)},\\
t \to t, & x \to x, & \phi \to \phi + 2\pi n & \text{(discrete shifts in } \phi\text{)},\\
t \to -t, & x \to x, & \phi \to \phi & \text{(reflection in } t\text{)},\\
t \to t, & x \to -x, & \phi \to \phi & \text{(reflection in } x\text{)},\\
t \to t, & x \to x, & \phi \to -\phi & \text{(reflection in } \phi\text{)},
\end{array}$$
$$t \to \tfrac{t-vx}{\sqrt{1-v^2}},\ x \to \tfrac{x-vx}{\sqrt{1-v^2}},\ \phi \to \phi \text{ (Lorentz transformations with velocity } v\text{)},$$

where e.g. reflection in ϕ means: if ϕ is a solution then so is $-\phi$, etc.

In Minkowski (1+1) space-time coordinates ($x^\mu \in \mathbb{R}^{1,1}$, $x^0 = t$, $x^1 = x$) with metric tensor $\eta_{\mu\nu}$ ($\mu,\nu = 0,1; \eta_{11} = -\eta_{22} = 1, \eta_{11} = \eta_{11} = 0$), the SG–Lagrangian density has the following 'massive form' of kinetic minus potential energy, with mass m and coupling constant λ:

$$\mathcal{L}_{\mathrm{SG}}^{\mathrm{Mink}}(\phi) = \frac{1}{2}(\phi_t^2 - \phi_x^2) - \frac{m^4}{\lambda}\left[1 - \cos\left(\frac{\sqrt{\lambda}}{m}\phi\right)\right],$$

which reduces to the dimensionless form (7.26) by re-scaling the fields and coordinates:

$$\frac{\sqrt{\lambda}}{m}\phi \to \phi, \qquad mx^\mu \to x^\mu. \tag{7.30}$$

The SG–Lagrangian density $\mathcal{L}_{\mathrm{SG}}^{\mathrm{Mink}}(\phi) \equiv \frac{m^4}{\lambda}\mathcal{L}_{\mathrm{SG}}(\phi)$ obeys the conservation law and admits topological (i.e., not sensitive to local degrees of freedom) *Noether current* with respect to (7.30):

$$j^\mu = \frac{1}{2\pi}\varepsilon^{\mu\nu}\partial_\nu\phi \quad \text{with zero-divergence:} \quad \partial_\mu j^\mu = 0,$$

where $\varepsilon^{\mu\nu}$ is the $\mathbb{R}^{1,1}$–Levi–Civita tensor. The corresponding topological *Noether charge* is given by:

$$Q = \int \left|\partial_t j^0(x,t)\right| dx = \frac{1}{2\pi}\left|\phi(+\infty,t) - \phi(-\infty,t)\right|,$$

$$\text{with} \quad Q_t = \frac{1}{2\pi}\left|\phi_t(-\infty,t) - \phi_t(+\infty,t)\right| = 0.$$

The most important physical quality of SGE is its energy–momentum (EM) tensor $T_{\mu\nu}$, which is the Noether current corresponding to spacetime–translation symmetry: $x^\mu \to x^\mu + \xi^\mu$; this conserved quantity is derived from the Lagrangian (7.26) as:

$$T_{\mu\nu} = \partial_\mu\phi\partial_\nu\phi - \eta_{\mu\nu}\mathcal{L}_{\mathrm{SG}}(\phi).$$

$T_{\mu\nu}$ has the following components:

$$T_{00} = \frac{1}{2}(\phi_t^2 + \phi_x^2) + 1 - \cos\phi, \qquad T_{10} = \phi_{xt} = T_{01}.$$
$$T_{11} = \frac{1}{2}(\phi_t^2 + \phi_x^2) - 1 + \cos\phi,$$

EM's contravariant form $T^{\mu\nu}$ has the following components:

$$T^{00} = T_{00}, \qquad T^{11} = T_{11}, \qquad T^{10} = -T_{01},$$

obtained by raising the indices of $T_{\mu\nu}$ using the inverse metric tensor $\eta^{\mu\nu} = 1/(\eta)_{\mu\nu}$.

EM's conserved quantities are: *momentum* $P = \int T^{10}dx$, which is the Noether charge with respect to space–translation symmetry, and *energy* $E = \int T^{00}dx$, which is the Noether charge with respect to time–translation symmetry. Energy and momentum follow from EM's zero divergence:

$$\partial_\mu T^{\mu\nu} = 0 \Longrightarrow \begin{cases} \partial_t T^{00} - \partial_x T^{10} = 0 \\ \partial_t T^{01} - \partial_x T^{11} = 0 \end{cases} \Longrightarrow \begin{cases} \partial_t E = \partial_t \int T^{00}dx = 0 \\ \partial_t P = \partial_t \int T^{10}dx = 0 \end{cases}.$$

SGE solutions and integrability

SGE solitons, kinks and breathers

The first *one-soliton* solution of the SGE (7.21) was given by [AKN73, AC91] in the form:

$$\phi(x,t) = 4\arctan\left[\frac{\sqrt{1-\omega^2}\cos(\omega t)}{\omega\cosh(x\sqrt{1-\omega^2})}\right],$$

which, for $\omega < 1$, is periodic in time t and decays exponentially when moving away from $x = 0$.

There is a well-known traveling solitary wave solution with velocity v, given by the following generalization of (7.22):

$$\phi(x,t) = 4\arctan\left[\exp\frac{\pm 2(z-z_0)}{\sqrt{1-v^2}}\right], \qquad \text{with} \quad (z = \mu(x+vt)), \tag{7.31}$$

and the center at z_0. In (7.31), the case $+2$ describes kink, while the case -2 corresponds to antikink.

The *stationary kink* with the center at x_0 is defined by:

$$\phi(x) = 2\arctan\left[\exp(x-x_0)\right],$$

(in which the position of the center x_0 can be varied continuously: $-\infty < x_0 < \infty$) and represents the solution of the first-order ODE: $\phi_x(x) = \sin\phi(x)$.

Regarding solutions of the slightly more general, three-parameter SGE:

$$\phi_{tt} = a\phi_{xx} + b\sin(\lambda\phi), \tag{7.32}$$

the following cases were established in the literature (see [II13] and references therein):

1. If a function $w = \phi(x,t)$ is a solution of (7.32), then so are also the following functions:

$$w_1 = \frac{2\pi n}{b} \pm \phi(C_1 \pm x, C_2 \pm t) \qquad \text{for} \quad (n = 0, \pm 1, \pm 2, ...),$$

$$w_2 = \pm\phi\left(x\cosh C_3 + t\sqrt{a}\sinh C_3,\ x\frac{\sinh C_3}{\sqrt{a}} + t\cosh C_3\right),$$

   where C_1, C_2, and C_3 are arbitrary constants.

2. Traveling-wave solutions:

$$\phi(x,t) = \frac{4}{\lambda}\arctan\left[\exp\left(\pm\frac{b\lambda(C_1x + C_2t + C_3)}{\sqrt{b\lambda(C_2^2 - aC_1^2)}}\right)\right] \tag{7.33}$$

   if $b\lambda(C_2^2 - aC_1^2) > 0$,

$$\phi(x,t) = -\frac{\pi}{\lambda} + \frac{4}{\lambda}\arctan\left[\exp\left(\pm\frac{b\lambda(C_1x + C_2t + C_3)}{\sqrt{b\lambda(aC_1^2 - C_2^2)}}\right)\right]$$

   if $b\lambda(C_2^2 - aC_1^2) < 0$,

where the first expression (for $b\lambda(C_2^2 - aC_1^2) > 0$) represents another 1-soliton solution, which is kink in case of $\exp\left(\frac{b\lambda(C_1x+C_2t+C_3)}{\sqrt{b\lambda(C_2^2-aC_1^2)}}\right)$ and antikink in case of $\exp\left(-\frac{b\lambda(C_1x+C_2t+C_3)}{\sqrt{b\lambda(C_2^2-aC_1^2)}}\right)$. In case of the standard SGE (7.21), this kink–antikink expression specializes to the Lorentz-invariant solution similar to:

$$\phi^{\mathrm{K}}(x,t) = 4\arctan\left[\exp\left(\frac{\pm(x-x_c)-vt}{\sqrt{1-v^2}}\right)\right], \tag{7.34}$$

where the velocity v $(0 < v < 1)$ and the soliton-center x_c are real-valued constants. The kink solution has the following physical (EM) characteristics:
(i) Energy:

$$E[\phi^{\mathrm{K}}(x,t)] = \int T^{00}dx = \frac{8}{\sqrt{1-v^2}};$$

(ii) Momentum:

$$P[\phi^{\mathrm{K}}(x,t)] = \int T^{10}dx = -\frac{8v}{\sqrt{1-v^2}}.$$

3. Functional separable solution:

$$w(x,t) = \frac{4}{\lambda}\arctan\left[f(x)g(t)\right],$$

where the functions $f = f(x)$ and $g = g(t)$ are determined by the first-order autonomous separable ODEs:

$$f_x^2 = Af^4 + Bf^2 + C, \qquad g_t^2 = -aCg^4 + (aB + b\lambda)g^2 - aA,$$

where A, B, and C are arbitrary constants. In particular, for $A = 0, B = k^2 > 0$, and $C > 0$, we have the *two-soliton* solution:

$$w(x,t) = \frac{4}{\lambda}\arctan\left[\frac{\eta\sin(kx + A_1)}{k\sqrt{a}\cosh(\eta t + B_1)}\right], \qquad \text{with } \left(\eta^2 = ak^2 + b\lambda > 0\right),$$

where k, A_1, and B_1 are arbitrary constants.

On the other hand, a *breather* is spatially localized, time periodic, oscillatory SGE–solution . It represents a field which is periodically oscillating in time and decays exponentially in space as the distance from the center $x = 0$ is increased. This oscillatory solution of (7.21) is characterized by some phase that depends on the breather's evolution history. This could be, in particular, a bound state of vortex with an antivortex in a Josephson junction. In this case, breather may appear as a result of collision of a *fluxon* (a propagating magnetic flux-quantum) with an *antifluxon*, or even in the process of measurements of switching current characteristics. stationary breather solutions form one-parameter families of solutions. An example of a breather–solution of (7.21) is given by:

$$\phi = 4 \arctan \left(\frac{\sin T}{u \cosh \left(g(u) x \right)} \right),$$

with parameters $u = u(t)$ and $T = T(t)$, such that

$$g(u) = 1/\sqrt{1+u^2} \quad \text{and} \quad T(t) = \int_0^t g(u(t'))\, u(t')\, dt'.$$

Lax–pair and general SGE integrability

In both cases (7.21) and (7.23), the SGE admits a *Lax–pair formulation*:[5]

$$\dot{L} = [L, M], \tag{7.35}$$

where overdot means time derivative, L and M are linear differential operators and $[L, M] \equiv LM - ML$ is their commutator (or, Lie bracket).

For example, the SGE (7.21) is integrable through the following Lax pair:

$$\phi_t = L\phi, \qquad \phi_x = M\phi, \qquad \text{where} \tag{7.36}$$

$$L = \begin{pmatrix} \frac{\mathrm{i}}{4}(\phi_x + \phi_t) & -\frac{1}{16\lambda}\mathrm{e}^{\mathrm{i}\phi} + \lambda \\ \frac{1}{16\lambda}\mathrm{e}^{-\mathrm{i}\phi} - \lambda & -\frac{\mathrm{i}}{4}(\phi_x + \phi_t) \end{pmatrix}, \qquad (\mathrm{i} = \sqrt{-1})$$

$$M = \begin{pmatrix} \frac{\mathrm{i}}{4}(\phi_x + \phi_t) & \frac{1}{16\lambda}\mathrm{e}^{\mathrm{i}\phi} + \lambda \\ -\frac{1}{16\lambda}\mathrm{e}^{-\mathrm{i}\phi} - \lambda & -\frac{\mathrm{i}}{4}(\phi_x + \phi_t) \end{pmatrix}, \qquad (\lambda \in \mathbb{R}).$$

---

[5] Historically, the first Lax-pair for a nonlinear PDE was found by P. Lax in 1968 consisting of the following two operators [Lax68]:

$$L = \frac{d^2}{dx^2} - u, \qquad M = 4\frac{d^3}{dx^3} - 6u\frac{d}{dx} - 3u_x,$$

such that their Lax formulation (7.35) gives the *KdV equation*:

$$u_t - 6uu_x + u_{xxx} = 0, \qquad \text{by}$$

$$L_t = -u_t, \quad LM - ML = u_{xxx} - 6uu_x.$$

The Lax-pair form of the KdV–PDE immediately shows that the eigenvalues of L are independent of t. The key importance of Lax's observation is that any PDE that can be cast into such a framework for other operators L and M, automatically obtains many of the features of the KdV–PDE, including an infinite number of local conservation laws.

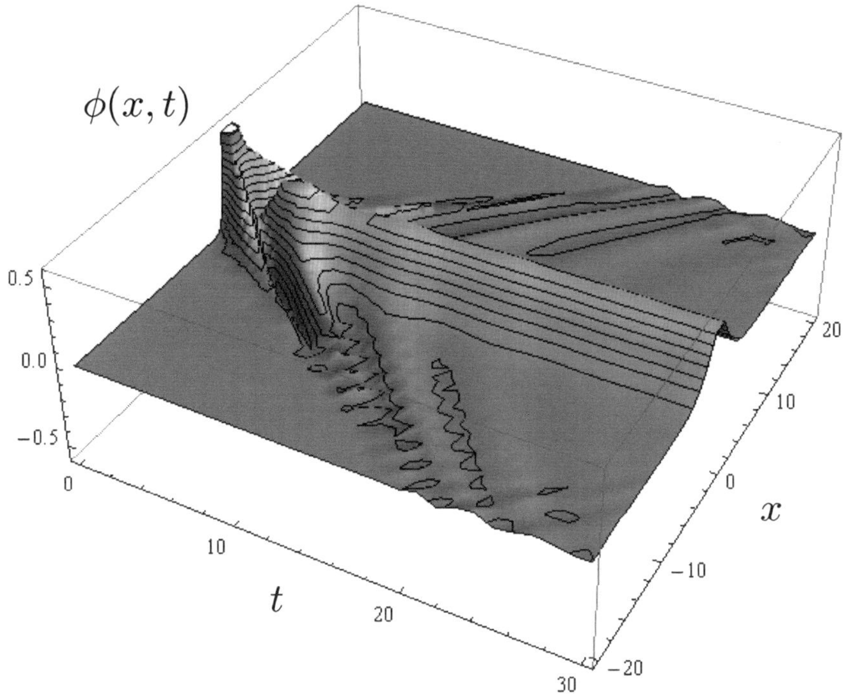

Fig. 7.10. *Numerical solution of the damped and spatially-forced SGE (7.38) in Mathematica, with the following data (including the Gaussian initial state, zero initial velocity and symmetric boundary condition):* $x \in [-20, 20]$, $t \in [0, 30]$, $\phi(x, 0) = \exp(-x^2)$, $\phi_t(x, 0) = 0$, $\phi(-20, t) = \phi(20, t)$, $\gamma = 0.2$, $F(x) = 0.5\text{sech}(x)$. *We can see the central sech-forcing along all time axis. Damping of the SG-waves is also apparent.*

The Lax pair (7.36) possesses the following complex-conjugate symmetry: if $\phi = \begin{pmatrix} \phi_1 \\ \phi_2 \end{pmatrix}$ solves the Lax pair (7.36) at (λ, ϕ), then $\begin{pmatrix} \overline{\phi_2} \\ \overline{\phi_1} \end{pmatrix}$ solves the Lax pair (7.36) at $(-\bar{\lambda}, \phi)$. In addition, there is a Darboux transformation for the Lax pair (7.36) as follows: let

$$u = \phi + 2\mathrm{i} \ln \left[\frac{\mathrm{i}\phi_2}{\phi_1} \right], \qquad (u \in \mathbb{R}).$$

If $\phi = \phi|_{\lambda=\nu}$ for some $\nu \in \mathbb{R}$, then

$$\psi = \begin{pmatrix} -\nu\phi_2/\phi_1 & \lambda \\ -\lambda & \nu\phi_1/\phi_2 \end{pmatrix} \phi.$$

solves the Lax pair (7.36) at (λ, u). Also, from its spatial part: $\phi_x = M\phi$, a complete Floquet theory can be developed.

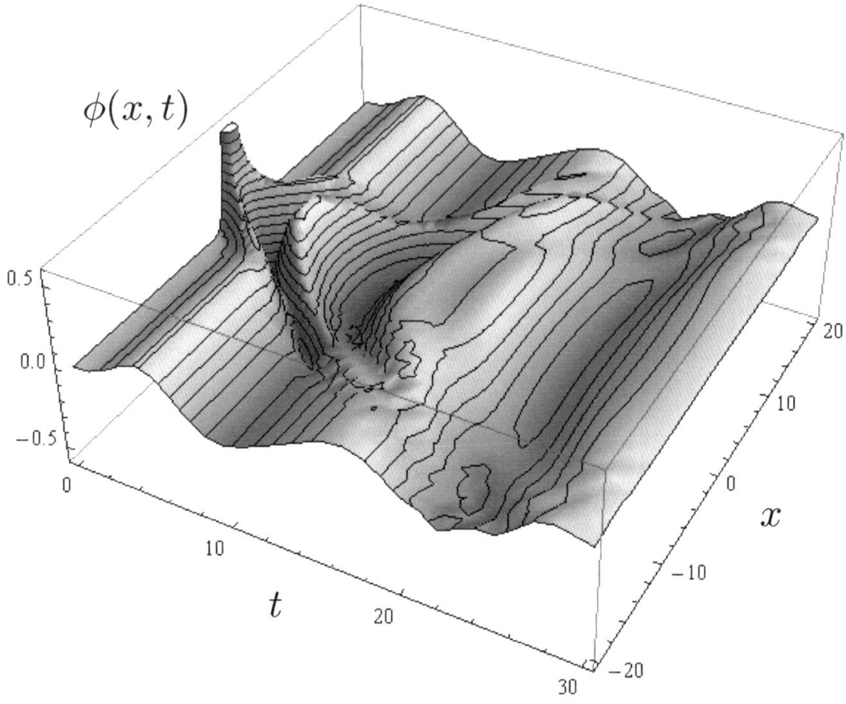

Fig. 7.11. *Numerical solution of the damped and temporally-forced SGE (7.38) in Mathematica, with the following data (including the Gaussian initial state, zero initial velocity and symmetric boundary condition):* $x \in [-20,20]$, $t \in [0,30]$, $\phi(x,0) = \exp(-x^2)$, $\phi_t(x,0) = 0$, $\phi(-20,t) = \phi(20,t)$, $\gamma = 0.2$, $F(x) = 0.1\sin(t/2)$. *We can see the sine-forcing along all time axis. Damping of the SG-waves is also apparent.*

SGE modifications

SGE with the positive sine term

The simplest SGE modification is to replace the minus sine term with the plus sine:

$$\phi_{tt} = \phi_{xx} + \sin\phi, \qquad \text{or} \qquad \phi_{tt}(x,t) = \phi_{xx}(x,t) + \sin\phi(x,t). \tag{7.37}$$

Again, a typical, spatially-symmetric, boundary-value problem for (7.37) is defined by:

$$x \in [-L,L] \subset \mathbb{R}, \qquad \left(t \in \mathbb{R}^+\right),$$
$$\phi(x,0) = f(x), \quad \phi_t(x,0) = 0, \quad \phi(-L,t) = \phi(L,t),$$

where $f(x) \in \mathbb{R}$ is an axially-symmetric function (see Figure 7.9).

Perturbed SGE and π-kinks

As we have seen above, the (1+1) SGE is integrable. In general though, the perturbations to this equation associated with the external forces and inhomogeneities spoil its integrability and the equation can not be solved exactly. Nevertheless, if the influence of these perturbations is small, the solution can be found perturbatively (see [II13] and references therein).

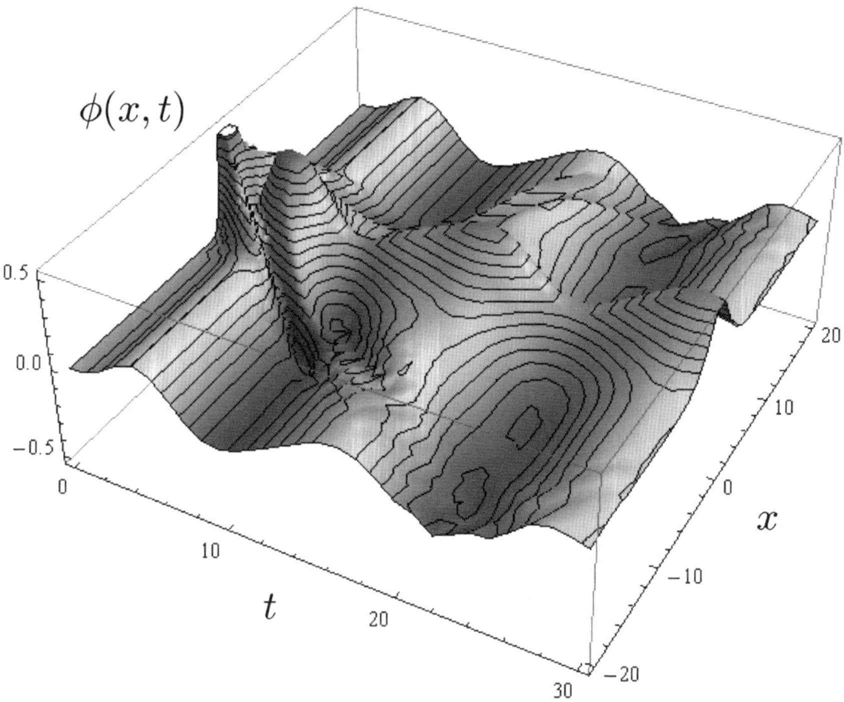

Fig. 7.12. *Numerical solution of the damped and both spatially and temporally forced SGE (7.38) in Mathematica, with the following data (including the Gaussian initial state, zero initial velocity and symmetric boundary condition):* $x \in [-20, 20]$, $t \in [0, 30]$, $\phi(x, 0) = \exp(-x^2)$, $\phi_t(x, 0) = 0$, $\phi(-20, t) = \phi(20, t)$, $\gamma = 0.2$, $F(x, t) = 0.1 \sin(t/2) + 0.5\,\mathrm{sech}(x)$. *We can see both temporal sine-forcing and spatial sech-forcing along all time axis. Damping of the SG-waves is still visible.*

One common form is a damped and driven SGE:

$$\phi_{tt} + \gamma\phi_t - \phi_{xx} + \sin\phi = F, \tag{7.38}$$

where $\gamma\phi_t$ is the damping term and $F(x, t)$ is the spatiotemporal forcing. Special cases of the forcing term $F = F(x, t)$ in (7.38) are: (i) purely temporal

$F = F(t)$ (e.g., periodic, see Figure 7.10); (ii) purely spatial $F = F(x)$ (e.g., central-symmetric, see Figure 7.11); and (iii) spatiotemporal $F = F(x,t)$ (e.g., temporally-periodic and spatially central-symmetric, see Figure 7.12).

Considering (for simplicity) purely spatial forcing: $F(x,t) = F(x)$, if $F(x_0) = 0$ for some point $x_0 \in \mathbb{R}$, this can be an equilibrium position for the soliton. If there is only one zero, in case of a soliton this is a stable equilibrium position if $\left(\frac{\partial F(x)}{\partial x}\right)_{x_0} > 0$; in case of an antisoliton, this is a stable equilibrium position if $\left(\frac{\partial F(x)}{\partial x}\right)_{x_0} < 0$ (see and references therein).

In particular if

$$F(x) = 2(\beta^2 - 1)\sinh(\beta x)/\cosh^2(\beta x), \qquad (\beta \in \mathbb{R}),$$

the exact stationary kink-solution of (7.38) is:

$$\phi_k = 4\arctan\left[\exp\left(\beta x\right)\right].$$

The stability analysis, which considers small amplitude oscillations around $\phi_k \left[\phi(k,x) = \phi_k(x) + f(x)\mathrm{e}^{\lambda t}\right]$, leads to the following eigenvalue problem:

$$\widehat{L}f = \Gamma f, \quad \text{where} \quad \widehat{L} = -\partial_x^2 + \left[1 - 2\cosh^{-2}(\beta x)\right] \quad \text{and} \quad \Gamma = -\lambda^2 - \gamma\lambda\,.$$

The eigenvalues of the discrete spectrum are given by the formula

$$\Gamma_n = \beta^2(\Lambda + 2\Lambda n - n^2) - 1,$$

where $\Lambda(\Lambda+1) = 2/\beta^2$. The integer part of Λ, yields the number of eigenvalues in the discrete spectrum, which correspond to the soliton modes (this includes the translational mode Γ_0, and the internal or shape modes Γ_n with $n > 0$.

In case of a function F defined in such a way that it possesses many zeroes, maxima and minima, perturbed SGE (7.38) describes an array of inhomogeneities. For example,

$$F(x) = \sum_{n=-q}^{q} 4\left(1-\beta^2\right) \frac{\mathrm{e}^{\beta(x+x_n)} - \mathrm{e}^{3\beta(x+x_n)}}{\left(\mathrm{e}^{2\beta(x+x_n)}+1\right)^2},$$

where $x_n = (n+2)\log\left(\sqrt{2}+1\right)/\beta$ $(n = -q, -q+1\cdots, q-1, q)$, and $q+2$ is the number of extrema points of $F(x)$. When the soliton is moving over intervals where $\frac{dF(x)}{dx} < 0$, its internal mode can be excited. The points x_i where $F(x_i) = 0$ and $\frac{dF(x_i)}{dx} < 0$, are 'barriers' which the soliton can overcome due to its kinetic energy.

Study of non-localized π-kinks in parametrically forced SGE (PSGE):

$$\phi_{tt} = \phi_{xx} - a(t/\epsilon)\sin\phi, \tag{7.39}$$

(over the fast time scale ϵ, where a is a mean-zero periodic function with a unit amplitude), has been performed via 2π-kinks as approximate solutions.

In particular, a finite-dimensional counterpart of the phenomenon of π-kinks in PSGE is the stabilization of the inverted *Kapitza pendulum* by periodic vibration of its suspension point. The so-called *geometric averaging technique* (see [II13] and references therein; here, the averaged forces in a rapidly forced system) was applied as a series of canonical near-identical transformations via V. Arnold's normal form technique, as follows.

Starting with the Hamiltonian of PSGE (7.39), given by:

$$H(\phi) = \int_{-\infty}^{+\infty} \left(\frac{p^2}{2} + \frac{\phi_x^2}{2} - a\cos\phi \right) dx, \qquad \text{where} \quad \left(p \equiv \phi_t \equiv \dot{\phi} \right),$$

a series of canonical transformations was performed by Zharnitsky with the aim to kill all rapidly-oscillating terms, the following slightly-perturbed Hamiltonian was obtained:

$$H_{\text{per}} = \int_{-\infty}^{+\infty} \left(\frac{p_3^2}{2} + \frac{\phi_{3x}^2}{2} + \frac{1}{2}\epsilon^2 \langle a_{-1}^2 \rangle \sin^2\phi_3 \right) dx + O(\epsilon^3),$$

which, after rescaling: $X = \epsilon x$, $T = \epsilon t$, $P = 2\epsilon^{-1} p_3$, $\Phi = 2\phi_3$, gave the following system of a slightly perturbed SGE with 2π-kinks as approximate solutions:

$$\Phi_T = P + O(\epsilon^2), \qquad P_T = \Phi_{XX} - \langle a_{-1}^2 \rangle \sin\Phi + O(\epsilon),$$

where a_{-1} is an anti-derivative with zero average. Finally, after rescaling back to variables (ϕ_3, p_3), approximate solutions $\phi_3 \approx \psi(x,t)$ in the form of π-kinks were obtained, with

$$\psi(x,t) = 2\arctan\left[\exp\left(\epsilon\sqrt{\langle a_{-1}^2 \rangle}\frac{x - ct}{\sqrt{1 - c^2}} \right) \right],$$

where c is the wave-propagation velocity.

In addition, he following two versions of the perturbed SGE have been studied by Zharnitsky:

1. Directly forced SGE:

$$\phi_{tt} - \phi_{xx} + \sin\phi = Mf(\omega t).$$

After shifting to the oscillating reference frame by the transformation:

$$\phi = \theta + M\omega^{-2}F(\omega t), \tag{7.40}$$

where F has zero mean and $F''(\tau) = f(\tau)$, the parametrically forced ODE is obtained:

$$\ddot{\theta} = -\sin\left(\theta + M\omega^{-2}F(\omega t)\right), \qquad \text{with} \tag{7.41}$$
$$H = \frac{p^2}{2} - A(\omega t)\cos(\theta) + B(\omega t)\sin(\theta),$$

where p is the momentum canonically conjugate to θ, and

$$A(\omega t) = \cos(M\omega^{-2}F(\omega t)), \qquad B(\omega t) = \sin(M\omega^{-2}F(\omega t)).$$

From (7.41), the corresponding evolution PDE (in canonical form) is obtained for a new phase θ on top of a rapidly oscillating background field:

$$\theta_t = p, \qquad p_t = \theta_{xx} - \sin\left(\theta + M\omega^{-2}F(\omega t)\right).$$

After retracing the identical transformation (7.40), the so-obtained (approximate) solutions become π-kinks.

2. Damped and driven SGE

$$\phi_{tt} - \phi_{xx} + \sin\phi = Mf(\omega t) - \alpha\phi_t + \eta, \tag{7.42}$$

which is frequently used to describe long Josephson junctions. [In (7.42), ϕ represents the phase-difference between the quantum-mechanical wave functions of the two superconductors defining the Josephson junction, t is the normalized time measured relative to the inverse plasma frequency, x is space normalized to the Josephson penetration depth, while $Mf(\omega t)$ represents tunneling of superconducting Cooper pairs (normalized to the critical current density).] Starting with a homogeneous transformation to the oscillating reference frame, analogous to (7.40) and designed to remove the free oscillatory term: $\phi = \theta + G(t)$, and substituting this transformation to (7.42), while choosing the function G so that it solves the following ODE:

$$\ddot{G} + \alpha\dot{G} = Mf(\omega t),$$

the following evolution PDE is obtined (in canonical form):

$$\theta_t = p, \qquad p_t = \theta_{xx} - \alpha p + \eta - \sin(\theta + G(\omega t)). \tag{7.43}$$

For the particular case of $f(\tau) = \sin\tau$, the the function G is found to be:

$$G(\tau) = -\frac{\alpha}{\omega}\frac{M}{\alpha^2+\omega^2}\cos\tau - \frac{M}{\alpha^2+\omega^2}\sin\tau.$$

SGE in (2+1) dimensions

The (2+1)D SGE with additional spatial coordinate[6] (y) is defined on $\mathbb{R}^{2,1}$as:

[6] In the case of a long Josephson junction, the soliton solutions of (7.44) describe Josephson vortices or *fluxons*. These excitations are associated with the distortion of a Josephson vortex line and their shapes can have an arbitrary profile, which is retained when propagating. In (7.44), φ denotes the superconducting phase difference across the Josephson junction; the coordinates x and y are normalized by the Josephson penetration length λ_J, and the time t is normalized by the inverse Josephson plasma frequency ω_p^{-1}.

$$\varphi_{tt} = \Delta\varphi - \sin\varphi = \varphi_{xx} + \varphi_{yy} - \sin\varphi. \tag{7.44}$$

A special class of solutions of (7.44) can be constructed by generalization of the solution of (7.21) which does not depend on one of the coordinates, or, obtained by Lorentz transforming the solutions of a stationary 2D SGE. However, there are numerical solutions of (7.44) which cannot be derived from the (7.21) or (7.23), e.g., radial breathers (or, *pulsons*).

A more general class of solutions of the (2+1)D SG equation has the following form:[7]

$$\varphi(x, y, t) = 4\ \arctan\exp\left[y - f(x \pm t)\right], \tag{7.45}$$

which exactly satisfies (7.44) with an arbitrary real-valued twice-differentiable function $f = f(x \pm t)$. The excitations, described by f are similar to elastic shear waves in solid mechanics.

Since the equation (7.44) is Lorentz-covariant, we can obtain other solutions performing Lorentz transformations on (7.45), which leads to a class of solutions of the form:

$$\varphi(x, y, t) = 4\ \arctan\exp\left[\frac{y - vt}{\sqrt{1-v^2}} - f\left(x \pm \frac{t - vy}{\sqrt{1-v^2}}\right)\right].$$

Sine–Gordon chain and discrete breathers

Frenkel–Kontorova model

The original Frenkel–Kontorova model [FK38, BK98, BK04] of stationary and moving crystal dislocations, was formulated historically decades before the continuous SGE. It consists of a chain of harmonically coupled atoms

[7] Because of the arbitrariness of f, solution (7.45) describes a variety of excitations of various shapes. Choosing f localized in a finite area, e.g., $f = A/\cosh(x - t)$, solution (7.45) describes an excitation, localized along x that keeps its shape when propagating, i.e., a soliton. For each solitary wave of this type, there exists an anti-partner with an f of opposite sign in (7.45). For solitary waves to be solitons, there is an additional important criterion: restoring their shapes after they collide.

Consider a trial function

$$\varphi(x, y, t) = 4\ \arctan\exp\left[y - f(x + t) \pm f(x - t)\right],$$

that, when $t \to -\infty$, describes the propagation of two solitary shape waves toward each other (minus sign) or a solitary wave and its anti-partner (plus sign). One can see that (7.45) can only *approximately* satisfy (7.44) when $|f'(x+t)f'(x-t)| \ll 1$ for all values of x and t. This suggests that, in general, the condition for restoring the shapes may not be satisfied. In general case, (7.44) can not be satisfied, that prompts that the collision of two solitary waves leads to distortion of the original excitations.

in a spatially periodic potential, governed by the set of differential-difference equations:

$$\ddot{\phi}_n + \frac{1}{\Delta x^2}\left[\phi_{n+1} - 2\phi_n + \phi_{n-1}\right] + \sin\phi_n = 0, \qquad (7.46)$$

where ϕ_n denotes the position of the nth atom in the chain. Alternatively, system (7.46) represents a chain of torsionally-coupled pendula (see Figure 7.7), where ϕ_n is the angle which the nth pendulum makes with the vertical.

Sine–Gordon chain

To derive dynamical equations of the sine–Gordon chain (SGC), consisting of anharmonic oscillators with the coupling constant μ, we start with the three-point, central, finite-difference approximation of the spatial derivative term ϕ_{xx} in the SGE:

$$\begin{aligned}\phi_{xx} &\approx \frac{1}{\Delta x^2}\left[\phi_{n+1} - 2\phi_n + \phi_{n-1}\right] + O(x^2)\\ &= -\frac{1}{\Delta x^2}\left[(\phi_n - \phi_{n-1}) - (\phi_{n+1} - \phi_n)\right] + O(x^2).\end{aligned}$$

Applying this finite-difference approximation to the SGE (7.21), and also performing the corresponding replacements: $\phi \to \phi_n$, $\phi_{tt} \to \ddot{\phi}_n$ and $\mu = 1/\Delta x^2$, we obtain the set of difference ODEs defining the SGC:

$$\ddot{\phi}_n + \mu\left[(\phi_n - \phi_{n-1}) - (\phi_{n+1} - \phi_n)\right] + \sin\phi_n = 0\,. \qquad (7.47)$$

The system (7.47) describes a chain of interacting particles subjected to a periodic on-site potential $V(x) = \sin(x)$. In the continuum limit, (7.47) becomes the standard SGE (7.21) and supports stable propagation of a kink-soliton of the form (7.34).

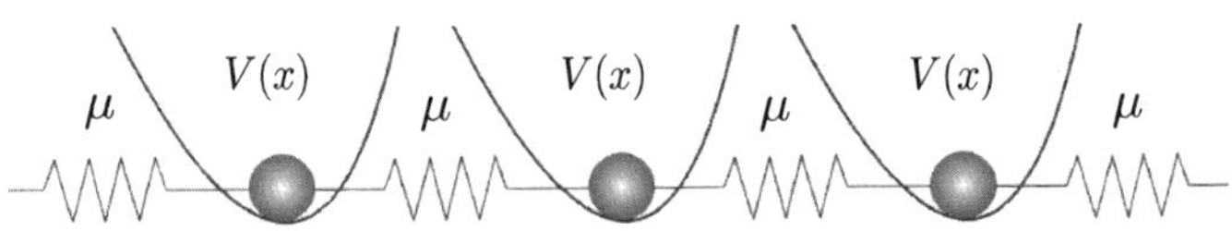

Fig. 7.13. *Simple sine–Gordon chain (SGC) with the coupling constant μ and the periodic on-site potential $V(x) = \sin(x)$.*

The linear-wave spectrum of (7.47) around a kink has either one or two localized modes (which depends on the value of μ). The frequencies of these modes lie inside the spectrum gap. The linear spectrum, with the linear frequency ω and the wave number k, is given by:

$$\omega^2 = 1 + 4\mu\sin^2\frac{k}{2}, \qquad (7.48)$$

while the gap edge frequency is $\omega = 1$.

The simplest example of (7.47), containing only two oscillators, is defined by:

$$\ddot{\phi}_{1,2} + \mu(\phi_{1,2} - \phi_{2,1}) + \sin\phi_{1,2} = 0\,. \tag{7.49}$$

A perturbed SGC, damped and driven by a large-amplitude ac-force of the form:

$$\ddot{\phi}_n - \mu\left[\phi_{n+1} - 2\phi_n + \phi_{n-1}\right] + \sin\phi_n = \chi + \alpha\sin\omega t - \gamma\dot{\phi}_n, \tag{7.50}$$

(where χ is a dc-force, α and ω are the normalized (large) amplitude and frequency of a periodic force, respectively, while γ is the normalized dissipative coefficient) – might support localized kink solitons. Without the forcing on the right-hand side, (7.50) reduces to (7.47).

Continuum limits

Perturbed SGEs have their corresponding perturbed SGCs. The following 0-π SGC:

$$\ddot{\phi}_n = \frac{\phi_{n-1} - 2\phi_n + \phi_{n+1}}{a^2} - \sin(\phi_n + \theta_n) + \gamma, \tag{7.51}$$

has been proposed as an equation of a phase ϕ_n-motion (of a 0-π array of Josephson junctions). Here, a is the lattice spacing parameter, $\gamma > 0$ is the applied bias current density, and $\theta_n = (0$ if $n \le 0$ and $-\pi$ if $n > 0)$ is the phase jump of π in ϕ_n. The SGC equation (7.51) is derived from the following discrete Lagrangian:

$$L_{\mathrm{D}} = \int \sum_{n\in\mathbb{Z}} \left[\frac{1}{2}\left(\frac{d\phi_n}{dt}\right)^2 - \frac{1}{2}\left(\frac{\phi_{n+1} - \phi_n}{a}\right)^2 - 1 + \cos(\phi_n + \theta_n) + \gamma\phi_n\right] dt. \tag{7.52}$$

In the continuum limit $a \ll 1$ Lagrangian (7.52) becomes

$$L_{\mathrm{C}} = \iint_{-\infty}^{\infty} \left[\frac{1}{2}(\phi_t)^2 - \frac{1}{2}\left(\widetilde{L}_a\phi_x\right)^2 - 1 + \cos(\phi + \theta) + \gamma\phi\right] dx\, dt\,,$$

from which, the continuum limit of (7.51) gives the following perturbed SGE:

$$\phi_{tt} = L_a\phi_{xx} - \sin(\phi + \theta) + \gamma,$$

where $\theta = (0$ if $x \le 0$ and $-\pi$ if $x > 0)$, while the differential operators $L_a\phi_{xx}$ and $\widetilde{L}_a\phi_x$ are given by the following Taylor expansions:

$$L_a\phi_{xx} = \frac{\phi_{n-1} - 2\phi_n + \phi_{n+1}}{a^2} = 2\sum_{k=0}^{\infty} \frac{a^{2k}}{(2k+2)!}\,\partial_{xx}^k\phi_{xx}(na),$$

$$\widetilde{L}_a\phi_x = \frac{\phi_{n+1} - \phi_n}{a} = \sum_{k=0}^{\infty} \frac{a^k}{(k+1)!}\,\partial_x^k\phi(na).$$

For more technical details, including several other continuum limits, see [DDG07].

Discrete breathers

Generally speaking, it is a well-known fact that different types of *excitations*, most notably *phonons* (propagating linear waves) and *discrete breathers* (DBs for short; they are time-periodic spatially localized excitations, also labeled intrinsic localized modes or discrete solitons) can occur as solutions of spatially-discrete nonlinear lattices. According to S. Flach *et al.* [FMF03, FW98, FG05]. DBs are caused by a specific interplay between the nonlinearity and discreteness of the lattice. The lattice nonlinearity provides with an amplitude-dependent tunability of oscillation or rotation frequencies of DBs, while its spatial discreteness leads to finite upper bounds of the frequency spectrum of small amplitude waves. Different types of DBs depending on the spectrum of linear waves propagating in the lattice, including: acoustic breathers, rotobreathers and optical breathers (see Figure 7.14).

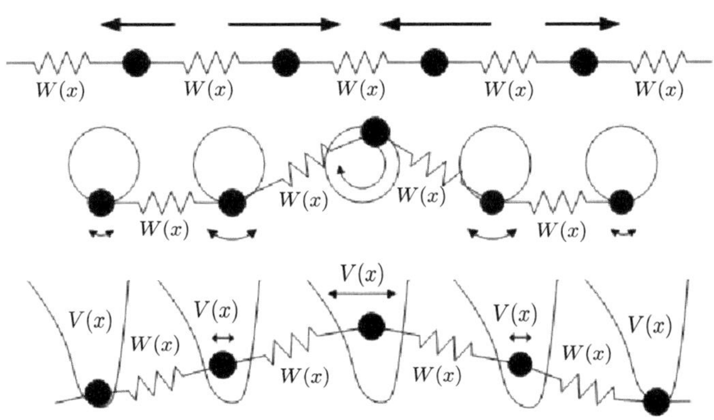

Fig. 7.14. *Different types of discrete breathers (DBs): acoustic breather (top), rotobreather (middle), and optical breather (bottom); modified and adapted from [FMF03, FW98, FG05]).*

A particular DB system is characterized by the following *lattice Hamiltonian*:

$$H = \sum_n \left(\frac{1}{2}\dot{x}_n^2 + W(x_n - x_{n-1}) + V(x_n)\right) \tag{7.53}$$
$$= \sum_n \left(\frac{1}{2}p_n^2 + W(x_n - x_{n-1}) + V(x_n)\right),$$

where $x_n = x_n(t)$ are time-dependent coordinates with canonically-conjugate momenta $p_n = \dot{x}_n(t)$, $W(x_n) = W(x)$ is the nearest neighbor interaction, and $V(x_n) = V(x)$ is an optional on-site (substrate) potential. From (7.53) the following equations of motion are derived:

$$\ddot{x}_n = -W'(x_n - x_{n-1}) + W'(x_{n+1} - x_n) - V'(x_n), \qquad \text{or}$$
$$\dot{x}_n = \dot{x}_n, \qquad \dot{p}_n = -W'(x_n - x_{n-1}) + W'(x_{n+1} - x_n) - V'(x_n),$$

where (for simplicity) the following zero initial conditions are assumed:

$$V(0) = W(0) = V'(0) = W'(0) = 0, \quad V''(0) \geq 0, \quad W''(0) > 0.$$

Hamiltonian (7.53) supports the excitation of small amplitude linear waves:

$$x_n(t) \sim \exp\left[\mathrm{i}(\omega_q t - qn)\right],$$

with the wave number q and the corresponding frequency spectrum ω_q^2 which, due to the underlying lattice, depends periodically on q:

$$\omega_q^2 = V''(0) + 4W''(0)\sin^2\left(\frac{q}{2}\right),$$

and its absolute value has always a finite upper bound. The maximum (Debye) frequency of small amplitude waves is:

$$\omega_q = \sqrt{V''(0) + 4W''(0)}.$$

For more technical details on SGE solitons, kinks and (discrete) breathers, see [II13] and references therein.

7.4.2 The Emperor's New Clothes: From Tesla's 'Æther' to Modern Quantum Turbulence

Much of Nikola Tesla's discovery (including over 800 patents only on AC electricity) can be attributed to some sort of 'deus ex machina', or 'spirit of technology'. At the same time, Tesla frequently used the term 'natural medium' through he was actually creating his machines. We argue that these two obscure (or rather mystical) terms denote one and the same thing, a controversial 'æther'.

Recall that classical Aristotelian concept of *æther*, which was operatively used in the fundamental electrodynamics research of *Michael Faraday*, *James Clerk Maxwell*, *Hendrik Lorentz* and *Nikola Tesla*,[8] can be naturally recast

[8] While Einstein's special relativity showed that Maxwell's electrodynamics equations do not require the æther for their existence, it also does not imply the non-existence of the
aether. In other words, mathematics does not require the æther; what about physics? Einstein even wrote that the special relativity theory "does not compel us to deny the æther. We may assume the existence of an æther". Einstein sometimes used the word 'æther' for the *gravitational field* within general relativity.

We argue that Tesla, who referenced the æther as a *natural medium,* used his deep knowledge of the æther to create his unique lightning strokes and plasma-balls.

into modern nonlinear physics of *quantum plasma*, an old research field with a new spin [Shu09], tightly related to *quantum turbulence*.

We propose here a simple and clear definition for Tesla's 'medium':

$$\text{Tesla's medium} \stackrel{def}{=} \text{æther} \stackrel{def}{=} \text{quantum plasma} .$$

This modern 'physical beast' is very different both from Aristotelian 'changeless quintessence' and from Newtonian 'absolute space'. Our æther is neither static, nor incompressible. It is a highly-nonlinear, yet fully predictable and controllable *quantum medium*[9] (e.g., a quantum superfluid), which exhibits *quantum turbulence* (see [IR12] and references therein), based on *nonlinear Schrödinger equation*, while its covariant (relativistic) generalizations are based on nonlinear Dirac equation (see, e.g. [II08b, II09]).

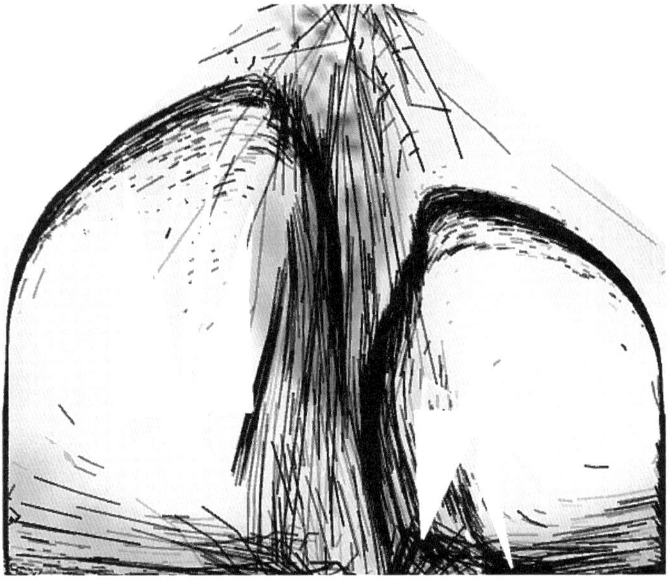

Fig. 7.15. *Using (a simplified numerical approximation of) the modern quantum turbulence to simulate Tesla's 'æther' dynamics.*

'Tesla's æther deus' is a nonlinear, adaptive and wave-form phenomenon. Its simplest model is the nonlinear Schrödinger equation (NLS), defining

[9] Note that standard quantum mechanics, both nonrelativistic and relativistic, as well as quantum field and gauge field theories are all *linear theories*, completely reversible in time. In our view, all these linear time-reversible theories are just first approximations of some more realistic future theories of Nature. From the shear vastness of Tesla's creation, it is quite clear that his æther-based intuition was far superior then all these linear theories.

the time-dependent complex-valued *macroscopic wave function* $\psi = \psi(x,t)$, whose absolute square $|\psi(x,t)|^2$ represents the *æther density function*. In natural quantum units ($\hbar = 1,\ m = 1$), with $\mathrm{i} = \sqrt{-1}$, the NLS equation can be written:

$$\mathrm{i}\partial_t\psi = -\frac{1}{2}\partial_{xx}\psi + V(x)|\psi|^2\psi, \qquad (\text{with } \partial_z\psi = \frac{\partial\psi}{\partial z}), \tag{7.54}$$

where $V(x)$ denotes the adaptive heat potential, trained either by the Hebbian or Levenberg-Marquardt learning (see [IR12] and references therein). Physically, the NLS equation (7.54) describes a nonlinear wave in a quantum matter (such as helium superfluid and Bose-Einstein condensates). It generates a whole 'zoo' of nonlinear waves (such solitons, kinks, shock-waves as and rogue-waves), representing a variety of dynamical forms of 'Tesla's æther deus'.

The main characteristic of 'Tesla's æther deus' is *turbulence*. Why? Because turbulence is necessary for any kind of creation, and Tesla was primarily a creator. However, the main characteristic of classical fluid turbulence (e.g., turbulence at the bottom of Niagara Falls) is total unpredictability. Clearly, such a wild turbulence is useless. What every creator needs is a fully predictable and controllable kind of turbulence. And luckily there is one: it is *quantum turbulence*, the main property of *quantum plasma*. Just for comparison, we will briefly analyze both classical and quantum turbulence, that is to say, both 'wild æther' and 'tamed æther'.

Classical Plasma $\Longrightarrow$ Classical Turbulence: Totaly Unpredictable and Uncontrollable

Classical plasma is usually described in terms of *Navier-Stokes equations* of fluid dynamics combined with *Maxwell's electromagnetic equations*. It exhibits totally unpredictable/uncontrollable classical turbulence. It is an ionized, electrically conducting gas/fluid of charged particles, which is a foundation of a number of attractive natural phenomena, such as lightning (both terrestrial and upper-atmospheric), ionosphere and polar aurorae, all of which fascinated Tesla, who tried to use and/or re-create them in his lab in Colorado Springs.

Formally, the so-called *viscous æther flows* evolve according to nonlinear Navier-Stokes equations:

$$\mathbf{u}\cdot\boldsymbol{\nabla}\mathbf{u} + \dot{\mathbf{u}} + \nabla p/\rho = \mathbf{f} + \nu\Delta\mathbf{u}, \tag{7.55}$$

where $\mathbf{u} = \mathbf{u}(\mathbf{x},t)$ is the 3D velocity of the æther flow, $\dot{\mathbf{u}} \equiv \partial_t\mathbf{u}$ is the 3D acceleration of the æther flow, $p = p(\mathbf{x},t)$ is the æther's pressure field, $\mathbf{f} = \mathbf{f}(\mathbf{x},t)$ is the external energy source to the æther, while ρ,ν are the æther's flow density and viscosity coefficient, respectively. Such an æther flow can be characterized by the ratio of the first term on the left-hand side of the equation (7.55), $\mathbf{u}\cdot\nabla\mathbf{u}$, referred to as the *æther inertial term*, and the second term on the right-hand side, $\nu\Delta\mathbf{u}$, that we call the *æther viscous term*.[10]

Classically, plasma dynamics is usually modeled using *magnetohydrodynamics* (MHD), which is formally defined by the following set of nonlinear PDEs for the *charge volume density* ρ, scalar pressure field p, electric conductivity σ, fluid velocity field $\mathbf{v}$ and magnetic field $\mathbf{B}$, given by:

$$\partial_t\rho = -\mathrm{div}(\rho\mathbf{v}), \quad \partial_t\mathbf{B} = \mathrm{curl}\,(\mathbf{v}\times\mathbf{B}) + \frac{1}{\mu_0\sigma}, \qquad \mathrm{div}\,\mathbf{B}=0,$$

$$\rho\dot{\mathbf{v}} = \rho\boldsymbol{\rho} + \mu\boldsymbol{\Delta}\mathbf{v} - \nabla p + \frac{1}{\mu_0}\mathrm{curl}\,\mathbf{B}\times\mathbf{B} + \left(\lambda + \frac{1}{3}\mu\right)\nabla(\nabla\cdot\mathbf{v}),$$

where $p=f(\rho)$, $\left(f\in C^0\right)$, while (λ,μ,μ_0) are *transport viscosity coefficients*.

Alternatively, a plasma consisting of particles with charge q and mass m moving in Euclidean 3D space $\mathbb{R}^3$ with positions $\mathbf{x}$ and velocities $\mathbf{v}$, can be modeled using the following *Maxwell-Vlasov* PDEs as follows. Let $\rho(\mathbf{x},\mathbf{v},t)$ be the *plasma density* at time t, while $\mathbf{E}(\mathbf{x},t)$ and $\mathbf{B}(\mathbf{x},t)$ be the electric and magnetic fields; then the Maxwell-Vlasov equations are defined as [AMR88]:

$$\partial_t\rho + v\cdot\partial_{\mathbf{x}}\rho + \frac{q}{m}\left(\mathbf{E}+\frac{\mathbf{v}\times\mathbf{B}}{c}\right)\cdot\partial_{\mathbf{v}}\rho = 0, \tag{7.56}$$

$$\frac{1}{c}\partial_t\mathbf{B} = -\mathrm{curl}\,\mathbf{E}, \quad \frac{1}{c}\partial_t\mathbf{E} = \mathrm{curl}\,\mathbf{B} - \frac{q}{c}\int\mathbf{v}\,\rho(\mathbf{x},\mathbf{v},t)\,d^3v, \tag{7.57}$$

together with the non-evolutionary equations:

$$\mathrm{div}\,\mathbf{E}=\rho, \quad \text{where} \quad \rho = q\int\mathbf{v}\,\rho(\mathbf{x},\mathbf{v},t)\,d^3v, \quad \mathrm{div}\,\mathbf{B}=0.$$

If we let the speed of light $c\to\infty$, equation (7.56) leads to the *Poisson-Vlasov equation* for the scalar potential φ_ρ:

$$\partial_t\rho + v\cdot\partial_{\mathbf{x}}\rho - \frac{q}{m}\partial_{\mathbf{x}}\varphi_\rho\cdot\partial_{\mathbf{v}}\rho = 0, \quad \text{where} \quad \Delta\varphi_\rho = -\rho.$$

In above natural units, the Hamiltonian for the Maxwell-Vlasov system (7.56)-(7.57) is given by:

$$H(\rho,\mathbf{E},\mathbf{B}) = \frac{1}{2}\int\|\mathbf{v}\|^2\,\rho(\mathbf{x},\mathbf{v},t)\,dxdv + \frac{1}{2}\int\|\mathbf{E}(\mathbf{x},t)\|^2 + \|\mathbf{B}(\mathbf{x},t)\|^2\,d^3v.$$

[10] This ratio defines the *Reynolds number* $Re=\bar{v}D/\nu$, where $\bar{v}$ and D are a characteristic velocity and length scale, respectively. When $\bar{v}$ increases and the Reynolds number Re exceeds a critical value, the æther changes from a *laminar state* to a *turbulent state*, in which the æther flow is complicated and contains eddies. To simplify the problem, the so-called *Reynolds condition*, $\langle\mathbf{f}\cdot\mathbf{u}\rangle=\varepsilon$, can be impose on the nergy source $\mathbf{f}$, where ε is the average rate of energy injection to the æther.

With this Hamiltonian H, the Maxwell-Vlasov system (7.56)-(7.57) is equivalent to the general dynamical equation in the *Poisson-bracket* form:

$$\dot{F} = \{F, H\},$$

where F is any locally defined function on the corresponding with functional derivatives (see [AMR88]).

Quantum Plasma $\Longrightarrow$ *Quantum Turbulence: Fully Predictable and Controllable*

The so-called *quantum plasma* (see [Shu06, SE10, SE11, ES11, ES12, SE12], which exhibits fully predictable and controllable *quantum turbulence*, is usually described in terms of the complex-valued NLS ψ-equation (7.54) combined with the real-valued Poisson φ-equation: $\nabla^2\varphi = \rho$.

In particular, for studying the formation and dynamics of electrostatic nano-structures in the so-called *dense quantum plasmas*, the NLS-*Poisson system* has been used (see [Shu06, SE10, SE11, ES11, ES12, SE12] and references therein) as follows:

$$\mathrm{i}\partial_t\psi + H\nabla^2\psi + \varphi\psi - |\psi|^{4/D}\psi = 0, \qquad \nabla^2\varphi = |\psi|^2 - 1, \tag{7.58}$$

where H is the system Hamiltonian, $\psi = \psi(x,t)$ is the normalized complex-valued wave function, while $\varphi = \varphi(x)$ is the normalized real-valued electrostatic potential.

The system (7.58) is usually supplemented by the *complex-valued Maxwell equation* (see [AMR88, Shu06, SE10, SE11, ES11, ES12, SE12]):[11]

$$\partial_t\mathbf{E} = \mathrm{i}H\left(\psi\nabla\psi^* - \psi^*\nabla\psi\right), \tag{7.59}$$

where the electric field is: $\mathbf{E} = -\nabla\phi$. The system (7.58)-(7.59) has the following conserved integrals:

1. The *number of electrons*:

$$N = \int |\psi|\, d^3x,$$

2. The *electron momentum*:

$$\mathbf{P} = -\mathrm{i}\int \psi^*\nabla\psi\, d^3x,$$

3. The *electron angular momentum*:

$$\mathbf{L} = -\mathrm{i}\int \psi^*\mathbf{r}\times\nabla\psi\, d^3x, \qquad \text{and}$$

---

[11] A slightly modified version of the nonlinear Schrödinger-Poisson system (7.58)-(7.59) has been used for exploring *quantum electron-fluid turbulence* both in 2D and in 3D (for more technical details, see [Shu06, SE10, SE11, ES11, ES12, SE12].

4. The *total energy*:

$$\mathcal{E} = \int [-\psi^* H \nabla^2 \psi + |\nabla \varphi|^2/2 + |\psi|^{2+4/D} D/(2+D)]\, d^3x.$$

The system (7.58)-(7.59) is our proposed 'Emperor's New Clothes': a quantum-turbulence model for Tesla's 'æther' dynamics (see Figure 7.15).

8

Quantum-Computation for Perceptual Control Architecture

In this Chapter, we propose a quantum-dynamical modeling approach to Perceptual Control Architecture, using large networks of Josephson Junctions and their category-theoretic generalizations with fuzzy associative functors.

Our approach provides the basis for composing modular multi-layered perceptual control architectures using Josephson Junction Networks, employing intuitively appealing category-theoretic abstractions to hide the algebraic details from the designer while nonetheless being able to rigorously ensure functional composition correctness. That is, our approach ensures that Josephson Junction Networks, as a modeling primitive, can be composed into formally correct multi-layered *perceptual control architectures*, while hiding the underlying algebraic systems of equations from the designer under a blanket of category-theoretic abstraction.

8.1 Introduction

8.1.1 From Brain Research to Perceptual Control Architecture

For the last few decades, a controversial issue in the research of biological neural networks has been how neurons communicate information (by firings or spikes, see [RWS96, Cha98, Egg98, UR99, CZ00, PDZ00]). In other words, there has been a continuous debate on the nature of the *neural code*, mainly focusing on the following two issues (see [Has03a]):

1. Whether information in the brain is encoded as the average firing rate of individual neurons (the *rate code* hypothesis), or in the precise firing times (the *temporal code* hypothesis). For example, firing activities of motor and sensory neurons in the brain have been reported to vary in response to applied stimuli, which was first noted in [Adr26], forms the basis of the rate-code hypothesis. In temporal-code hypothesis (see [SK93, KES96, SZ98]) detailed spike timings are assumed to play an important role in

information transmission (information is encoded in interspike intervals or in relative timings between firing times of spikes).

2. Whether information in the brain is encoded in the activity of a single neuron (or a small number of neurons), or in the activity of a large neural ensembles. The *ensemble rate code* hypothesis [AS98], proposes that information is coded in the relative firing rates of ensembles of neurons; this view has been adopted in the most of the theoretical analysis. The alternative, *ensemble temporal-code* hypothesis (see [Hop95, HL98, Rt01]), assumes that relative timings between spikes in ensemble neurons may be used as an encoding mechanism for *perceptual processing* - which is closely related to *perceptual control theory* (PCT), a system of hierarchical negative feedbacks, constituted by number of layers of perceptual and behavioral loops (including: intensities, sensations, configuration, transitions, events, relationship, category, principle and system concepts; see [Pow73a, Pow73b, Pow11, Pow05]).

Recently, a PCT-based architecture has been proposed in [WTW12] as a unified modeling approach to cyber-physical systems, inspired by the effective organization of living systems (accommodating heterogeneous information processing and environmental interaction) and applied to traffic incident management. The authors essentially propose a *model-free approach* for (the construction of) the PCT-architecture, based on Bayesian network reasoning.

In the present paper, we propose an alternative, quantum-dynamical *modeling* approach to perceptual control architecture, using large networks of Josephson junctions and their category-theoretic generalizations with fuzzy associative functors.

8.1.2 Josephson Junctions

A *Josephson junction* (JJ)[1] is a type of electronic circuit (an insulating barrier separating two superconducting materials and producing the *Josephson effect*,[2] see Appendix capable of switching at very high frequency ($10^{10}-10^{11}$ Hz) when operated at temperatures approaching absolute zero. It is the hardware for *qubits* (or, more generally, N-qubits, see Figure 8.1 below), which

[1] The Josephson Junction is named after Brian D. Josephson, 1973 Nobel Prize for Physics for the discovery of tunnelling supercurrents - also known as the Josephson effect - in 1962. See [Jos74].

[2] While researching superconductivity, B.D. Josephson studied the properties of a junction between two superconductors. Following up on earlier work by L. Esaki and I. Giaever (who both shared the Nobel Prize with Josephson), he demonstrated that in a situation when there is electron flow between two superconductors through an insulating layer, in the absence of an applied voltage, and a voltage is applied, the current stops flowing and oscillates at a high frequency. The Josephson effect is influenced by magnetic fields in the vicinity, a capacity that enables the Josephson junction to be used in devices that measure extremely weak magnetic fields, such as superconducting quantum interference devices (SQUIDs).

are basic building blocks of *quantum computers* (see [II09] and references therein).[3]

One of the most important JJs is the RCSJ model, a JJ with the critical current I_c, *shunted* by normal resistance R and capacitance C (see Appendix). RCSJ is governed by the 2nd-order ODE of motion (written here in normal units with $\hbar = 1$), similar to the equation for a driven and damped *pendulum* (that is potentially chaotic), derived from the Kirchoff's law (see, e.g. [CK08, II09] and references therein):

$$\frac{C}{2e}\ddot{\varphi} + \frac{1}{2eR}\dot{\varphi} + I_c \sin\varphi = I_{dc} + I_0 \cos(\omega t) \tag{8.1}$$

where, in accordance with the usual conventions, overdots refer to time derivatives, $2e$ and ω are the charge and frequency, φ is the *phase difference* of the Schrödinger wave ψ-function across the junction, I_{dc} is the dc-bias current and $I_0 \cos(\omega t)$ is the driving rf-field with the frequency ω.

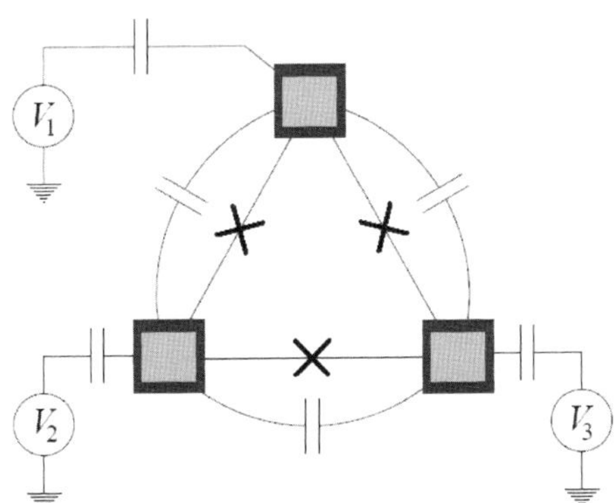

Fig. 8.1. *A superconducting-phase N-qubit is an inductance loop composed of $N = 3$ superconducting islands (squares) connected by Josephson junctions (crosses) and also capacitively coupled to each other and to the ground (modified and adapted from [IIG02]).*

8.2 Effective Josephson-Junction Networks

8.2.1 Qubits, Loops, Ladders, Arrays and Networks

By means of introduction to the quantum-mechanical zoo of loops, arrays, ladders and networks of JJs, let us consider its basic unit, a superconducting-

[3] The first *JJ-based quantum computer* was demonstrated in April 1999 by Nakamura *et al.* of NEC Fundamental Research Laboratories in Tsukuba [NPT99] and, independently, just a week later, by Ioffe *et al.* [IGF99].

phase *N-qubit* (see Figure 8.1, for the case $N = 3$), which is placed in an external magnetic field producing one half superconducting flux quantum $\Phi/2$ through the loop (around the islands). The N-qubit is governed by the following Hamiltonian operator[4] [IIG02]:

$$H = \frac{1}{2}\sum_{i,j=1}^{N} Q_i C_{ij}^{-1} Q_j + \sum_{i}^{N} V_i Q_i + \sum_{i=1}^{N} U_i(\varphi_{i+1} - \varphi_i - W_{i,i+1}), \qquad (8.2)$$

where φ_i and $Q_i = -\mathrm{i}\partial/\partial\varphi_i$ are the vectors of phases on the islands (represented as squares) and their conjugate charge operators, respectively; V_i and U_i are the vectors of gate-voltages applied to the islands and Josephson energies of the junctions (circles); C_{ij} is the nonsingular capacitance matrix (with the inverse C_{ij}^{-1}) and $W_{i,i+1}$ is the electromagnetic vector potential (induced by an external magnetic field).

We now introduce the Coulomb charging energy E_c of a single island and $E_J = I_c/2e$, where I_c is the critical current of a single JJ, for measuring the strength of the phase-coupling between the superconducting islands. In the charge-dominated limit [LPM96]: $E_c \gg E_J$ and assuming the restriction to tunneling JJs with $U_i(\varphi) = -E_J \cos(\varphi)$, the operators Q_i are diagonal and the Josephson term in the Hamiltonian operator (8.2) takes the simple linear form [IIG02]):

$$H_J = -\frac{E_J}{2}\sum_i (L_i^+ L_{i+1}^- \mathrm{e}^{\mathrm{i}\pi/n} + O^{(2)}), \qquad (8.3)$$

where L_i^+ is the charge raising operator and L_i^-) is the corresponding lowering operator on the ith island with $L_i^\pm |Q_i\rangle = |Q_i \pm 1\rangle$, and $O^{(2)}$ is the 2nd-order remainder term.

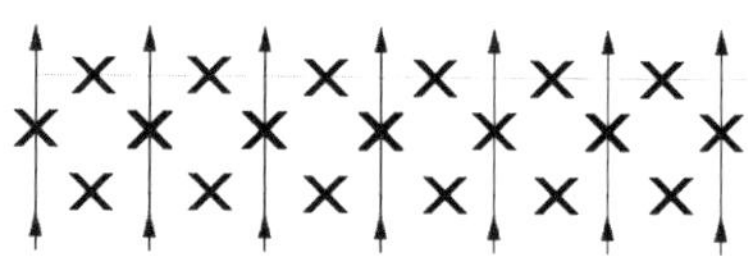

Fig. 8.2. *A schematic of a Josephson-junction ladder, with crosses denoting the junctions and arrows denoting the dc-bias current flow (abstracted from [MSF05]).*

N-qubits can be combined into ladders and/or arrays. A *Josephson-junction ladder* is a schematic depicted in Figure 8.2 and governed by the

[4] Note that, for simplicity reasons, we drop the usual 'hat' notation for Hermitian quantum-mechanical operators (self-adjoint complex-valued operators with real eigen-values). In particular, all Hamiltonians used in this paper are Hamiltonian operators.

following three coupled nonlinear pendulum-like ODEs [MSF05] (compare this with Appendix):

$$\begin{aligned}
\ddot{\varphi}_n^v + \alpha\dot{\varphi}_n^v + \sin\varphi_n^v &= \gamma + \frac{1}{\beta_L}(\Delta\varphi_n^v + \nabla\varphi_{n-1}^h - \nabla\tilde{\varphi}_{n-1}^h) \\
\ddot{\varphi}_n^h + \alpha\dot{\varphi}_n^h + \sin\varphi_n^h &= -\frac{1}{\eta\beta_L}(\nabla\varphi_n^v + \varphi_n^h - \tilde{\varphi}_n^h) \\
\ddot{\tilde{\varphi}}_n^h + \alpha\dot{\tilde{\varphi}}_n^h + \sin\tilde{\varphi}_n^h &= \frac{1}{\eta\beta_L}(\nabla\varphi_n^v + \varphi_n^h - \tilde{\varphi}_n^h)\,.
\end{aligned} \tag{8.4}$$

Here, $\nabla f_n \equiv f_{n+1} - f_n$ and $\Delta f_n \equiv f_{n+1} - 2f_n + f_{n-1}$ are the finite-difference approximations for the first and second spatial derivatives, respectively; φ_n^v, $\tilde{\varphi}_n^h$, φ_n^h denote the Josephson phases of vertical, upper- and lower-horizontal junctions in the nth cell, respectively; α and γ are the effective damping and the external dc-bias (across vertical junctions); β_L is the the discreteness inductance parameter (that characterizes the ratio of geometrical cell inductance and the Josephson inductance of vertical junctions), and $\eta = I_c^H/I_c^V$ denotes the anisotropy current parameter [here, I_c^V (resp. I_c^H) is the vertical (resp. horizontal) junction critical current]. For technical details, see [MSF05] and references therein.

By combining a number of ladders, a large-scale Josephson Junction array can be created. For large N, M such an array becomes a Josephson-junction network (JJN, see Figure 8.3). JJNs are the basic components of our PCT-architecture; they are described in the following subsections.

For a general information on Josephson-junction ladders, arrays and networks, consisting of small superconducting islands (squares) connected by JJs (crosses), see [GPG89, MWG90, ZFE92, ZEG96, DCH94, CDH98, TTH93, NLG00] and references therein.

8.2.2 Complexity: Breathers and Chaos Synchronization

JJNs exhibit several forms of nonlinear complexity, including breathers, chaos and synchronization.

JJNs support dynamic localized states which are common characteristics of all discrete nonlinear Hamiltonian systems called *discrete breathers* (DBs hereafter), which are time-periodic spatially localized excitations[5], also called intrinsic localized modes or discrete solitons. These can occur as solutions of spatially-discrete nonlinear lattices like JJNs (see [II13] and references therein). For example, the ladder system (8.4) represents three coupled damped and driven *sine-Gordon chains* of the form:

$$\ddot{\varphi}_n + \frac{1}{\Delta x^2}\left[\varphi_{n+1} - 2\varphi_n + \varphi_{n-1}\right] + \sin\varphi_n = \chi + \alpha\sin\omega t - \gamma\dot{\varphi}_n. \tag{8.5}$$

[5] According to S. Flach *et al.* [FW98, FMF03, FG05, FG08], DBs are caused by a specific interplay between the nonlinearity and discreteness of the lattice.

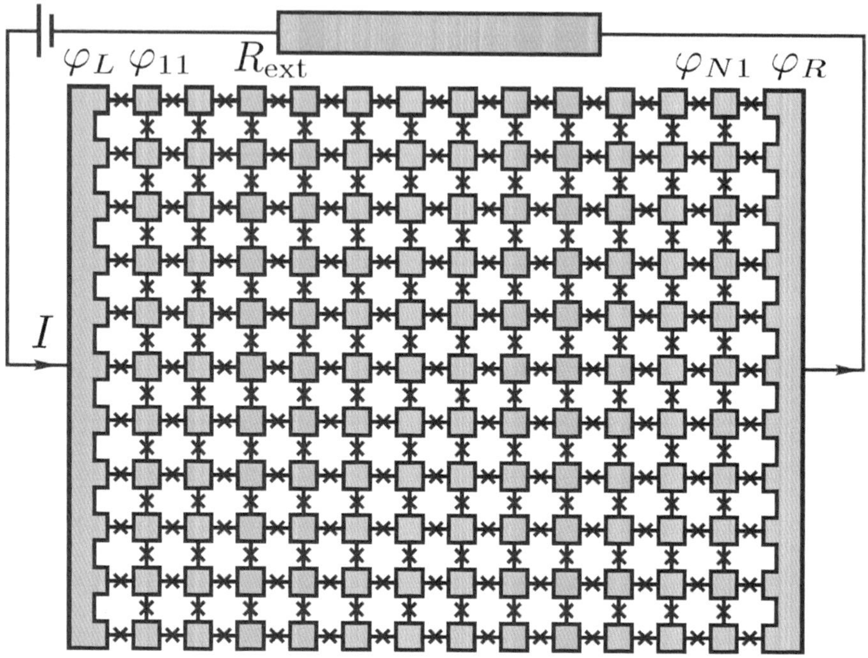

Fig. 8.3. *A Josephson-junction network (JJN) is given by a large-scale $N \times M$ array of superconducting islands (squares), each one connected by four neighboring JJs (crosses); an external current I is injected from the left through the electrode having the superconducting phase φ_L and extracted through the right electrode with the phase φ_R (modified and adapted from [FVB08]).*

Here, $\chi, \alpha, \omega, \gamma$ are respectively, the dc-force, the normalized amplitude and frequency of a periodic force, and the normalized dissipative coefficient. This perturbatively extends the classical Frenkel-Kontorova model [FK38, BK98, BK04, II13], which governs a chain of torsionally-coupled pendula:

$$\ddot{\varphi}_n + \frac{1}{\Delta x^2}\left[\varphi_{n+1} - 2\varphi_n + \varphi_{n-1}\right] + \sin\varphi_n = 0, .$$

The angle which the nth pendulum makes with the vertical is φ_n. The continuum limit of the sine-Gordon chain (8.5) gives the following perturbed sine-Gordon PDE:

$$\varphi_{tt} = \varphi_{xx} - \sin\varphi - \gamma\varphi_t + F, \qquad (F = \chi + \alpha\sin\omega t).$$

For example, the following 0-π sine-Gordon chain was analyzed in [DDG07] as an equation of a phase φ_n-motion of a 0-π array of JJs:

$$\ddot{\varphi}_n = \frac{1}{\Delta x^2}\left[\varphi_{n+1} - 2\varphi_n + \varphi_{n-1}\right] - \sin(\varphi_n + \theta_n) + \gamma. \qquad (8.6)$$

The applied bias current density is $\gamma > 0$, and $\theta_n = (0$ if $n \leq 0$ and $-\pi$ if $n > 0)$ is the phase jump of π in φ_n. The ODE (8.6) is derived from the following discrete Lagrangian:

$$L_{\mathrm{D}} = \int \sum_{n\in\mathbb{Z}} \left[\frac{1}{2}\left(\frac{d\varphi_n}{dt}\right)^2 - \frac{1}{2}\left(\frac{\varphi_{n+1}-\varphi_n}{a}\right)^2 - 1 + \cos(\varphi_n + \theta_n) + \gamma\varphi_n \right] dt.$$

Its continuum limit $a \ll 1$,

$$L_{\mathrm{C}} = \iint_{-\infty}^{\infty} \left[\frac{1}{2}(\varphi_t)^2 - \frac{1}{2}\left(\widetilde{L}_a\varphi_x\right)^2 - 1 + \cos(\varphi+\theta) + \gamma\varphi \right] dx\, dt\,,$$

gives the perturbed sine-Gordon PDE:

$$\varphi_{tt} = L_a\varphi_{xx} - \sin(\varphi+\theta) + \gamma,$$

where $\theta = (0$ if $x \leq 0$ and $-\pi$ if $x > 0)$. The differential operators $L_a\varphi_{xx}$ and $\widetilde{L}_a\varphi_x$ are given by the Taylor expansions [DDG07]:

$$L_a\varphi_{xx} = \frac{\varphi_{n-1} - 2\varphi_n + \varphi_{n+1}}{a^2} = 2\sum_{k=0}^{\infty} \frac{a^{2k}}{(2k+2)!}\,\partial_{xx}^k\varphi_{xx}(na),$$

$$\widetilde{L}_a\varphi_x = \frac{\varphi_{n+1}-\varphi_n}{a} = \sum_{k=0}^{\infty} \frac{a^k}{(k+1)!}\,\partial_x^k\varphi(na).$$

The solutions of the perturbed sine-Gordon PDEs produce a virtual zoo of solitons, kinks and breathers, as various combinations of kinks: $\tanh(x)$, $\arctan(x)$ and bell-shaped solitons: $\mathrm{sech}(x)$, $\exp(-x^2)$, see [II13].

Since it was first reported by [PC90, PC91] that synchronization of chaotic systems was possible, different types of synchronization - including complete, generalized and phase synchronization of chaotic oscillators - have been described theoretically and observed experimentally (see [RPK97a, RPK97b, PZR97, PRK00, PRK01, OPK02, GP05]). The presence of even a small phase-difference between the externally-applied fields was found to desynchronize a completely synchronized system; also, the phase difference could be applied for taming chaos in nonlinear systems.

An array of JJs was studied in [CK08] in the presence of a phase difference between the driving fields and its effect on synchronization and suppression of chaos. Formally, a JJ-array consisting of N coupled current-driven JJs of the type governed by (8.1) is described by the following set of 2nd-order ODEs of motion resembling the system of coupled, damped and driven pendula:

$$\begin{aligned} \ddot{\varphi}_1 + \beta\dot{\varphi}_1 + \sin\varphi_1 &= I_{dc} + I_0\cos(\Omega t) - \alpha_s\left[\dot{\varphi}_1 - \dot{\varphi}_2\right] \\ &\vdots \\ \ddot{\varphi}_i + \beta\dot{\varphi}_i + \sin\varphi_i &= \alpha_s\left[\dot{\varphi}_{i+1} + \dot{\varphi}_{i-1} - 2\dot{\varphi}_i\right] \\ &\vdots \\ \ddot{\varphi}_N + \beta\dot{\varphi}_N + \sin\varphi_N &= I_{dc} + I_0\cos(\Omega t) - \alpha_s\left[\dot{\varphi}_N - \dot{\varphi}_{N-1}\right] \end{aligned} \tag{8.7}$$

where $i = 2, ..., N-1$,. The dimensionless damping parameter is $\beta = \left(R\sqrt{2eI_c}\right)^{-1}$. The normalized time scale is used across the whole array, by weighting the single-JJ time by $\omega_{J1} = \sqrt{2eI_{c1}/C_1}$. The direct bias current I_{dc} and the RF amplitude I_0 are both normalized to the critical current I_{c1}. The actual frequency ω is re-scaled to $\Omega = \omega/\omega_{J1}$ and the coupling factor is defined as $\alpha_s = (R_1/R_s)\,\beta$. It was shown in [CK08] that the JJ-array governed by (8.7) is chaotic for the parameter values $\beta = 0.3, I_0 = 1.2, \omega = 0.6$ and $I_{dc} = 0.3$. The junctions were taken to be identical and for a coupling strength of $\alpha_s = 0.37$, the outer junctions synchronized while the inner junction remained uncorrelated with the two outer ones.

A particular case of (8.7) consists of three JJs linked in parallel. Assuming three identical junctions, the following set of first-order ODEs is given in [CK08]:

$$\begin{aligned}
\dot{\varphi}_1 &= \phi_1 \\
\dot{\phi}_1 &= -\beta\phi_1 - \sin\varphi_1 + I_{dc} + I_0\cos(\Omega t) - \alpha_s\left[\phi_1 - \phi_2\right] \\
\dot{\varphi}_2 &= \phi_2 \\
\dot{\phi}_2 &= -\beta\phi_2 - \sin\varphi_2 + \alpha_s\left[\phi_1 + \phi_3 - 2\phi_2\right] \\
\dot{\varphi}_3 &= \phi_3 \\
\dot{\phi}_3 &= -\beta\phi_3 - \sin\varphi_3 + I_{dc} + I_0\cos(\Omega t + \theta) - \alpha_s\left[\phi_3 - \phi_2\right].
\end{aligned}$$

The outer junctions are identical and symmetric to the interchange of variables in the absence of a phase-difference between the applied fields. Hence there exists an identical solution for the outer systems given by $\varphi_1 = \varphi_3 = \varphi(t)$, which demonstrates the complete synchronization. For further technical details, including stability analysis, see [CK08].

8.2.3 Formalism of Effective JNN Hamiltonians

In this subsection, we develop a general formalism of effective Hamiltonian operators (to be used in the next subsection), based on the following concerns:

1. A modern (geometric) version of the Rayleigh-Schrödinger *autonomous perturbation theory* [Ray894, Sch26] (for a review, see [LL77]),
2. *Adiabatic assumption*, which states that the eigen-energies $E_n = E_n(x^\mu)$ and eigen-states $|n\rangle = |n(x^\mu)\rangle$ of a multi-parameter Hamiltonian $H = H(x^\mu)$ are smooth functions of the parameter-set $\{x^\mu\}$, such that their values in the vicinity region of the parameter manifold $M = \{x^\mu\}$ can be expanded into a power series of x^μ (see, e.g. [Fra98, Mos97]);
3. Hellman-Feynman theorems [Hel37, Fey39] for calculating parametric derivatives of the Schrödinger Hamiltonian operator (for a computational review, see [Jen07]).

To start with, let us define a *linearly-parameterized Hamiltonian* $H = H(x^\mu)$, as a Hermitian operator $H : M \to \mathcal{H}$, which maps the *parameter*

manifold
$M := \{x^\mu : \mu = 1, ..., n\}$ onto the quantum Hilbert space $\mathcal{H}$, by the weak perturbation[6] [Sol81, RU06, Per13, II09]:

$$H = H_0 + \lambda H_{prt} \equiv H_0 + x^\mu F_\mu, \qquad \text{(using summation convention)}.$$

Here, H_0 is the unperturbed Schrödinger Hamiltonian,[7] H_{prt} is an external weak-disturbance potential-energy operator[8] with the associated perturbation parameter $\lambda \in [0, 1]$, while F_μ are the (generalized) force operators corresponding to the coordinate parameters $x^\mu \in M$ (e.g., if x^μ are magnetic-field coordinates, then F_μ are the corresponding magnetizations). Using the adiabatic assumption, we obtain the following power-series expansions[9]:

$$\begin{aligned} E_n(x^\mu) &= E_n^{(0)} + \lambda E_n^{(1)} + \lambda^2 E_n^{(2)} + O^{(3)} \\ &\equiv E_n + x^\mu \partial_\mu E_n + \frac{1}{2!} x^\mu x^\nu \partial_\mu \partial_\nu E_n + O^{(3)}, \\ |n(x^\mu)\rangle &= |n^{(0)}\rangle + \lambda |n^{(1)}\rangle + \lambda^2 |n^{(2)}\rangle + O^{(3)} \equiv |n\rangle + x^\mu |\partial_\mu n\rangle \\ &\quad + \frac{1}{2!} x^\mu x^\nu |\partial_\mu \partial_\nu n\rangle + O^{(3)}. \end{aligned}$$

Here, the 0th-order approximation corresponds to (8.9) from which we obtain the expectation value $\langle m|H|n\rangle$ of the unperturbed Hamiltonian H_0. In a similar way, consider the 1st-order approximation:

$$H_0|n^{(1)}\rangle + H_{prt}|n^{(0)}\rangle = E_n^{(0)}|n^{(1)}\rangle + E_n^{(1)}|n^{(0)}\rangle \equiv E_n|\partial_\mu n\rangle + \partial_\mu E_n|n\rangle$$

gives us the expectation value of the perturbed Hamiltonian $H(x^\mu)$:

$$E_n^{(1)} = \langle n^{(0)}|H_{prt}|n^{(0)}\rangle \equiv \langle n^{(0)}|\partial_\mu E_n|n^{(0)}\rangle \equiv \langle m|\partial_\mu H|n\rangle,$$

where we have used the first Hellman-Feynman theorem [Fey39] [derived by applying ∂_μ to both sides of (8.8)]:

---

[6] The energy levels E_n and eigen-states $|n\rangle$ of the perturbed Hamiltonian $H = H_0 + \lambda H_{prt} \equiv H(x^\mu)$ are given by the *perturbed Schrödinger equation:*

$$(H_0 + \lambda H_{prt})\,|n\rangle = E_n|n\rangle. \tag{8.8}$$

[7] H_0 satisfies the stationary Schrödinger equation:

$$H_0|n\rangle = E_n|n\rangle. \tag{8.9}$$

[8] Note that our potential-energy perturbation H_{prt} is commonly denoted as V (see, e.g. [RU06]). However, in this paper, symbol V is reserved for the voltage.

[9] Here, we are using the following notation:

$$\partial_\mu \equiv \frac{\partial}{\partial x^\mu}, \qquad E_n^{(m)} = \frac{1}{m!}\frac{d^m E_n}{d\lambda^m}, \qquad |n^{(m)}\rangle = \frac{1}{m!}\frac{d^m |n\rangle}{d\lambda^m}.$$

$$\partial_\mu E_n = \langle n|\partial_\mu H|n\rangle.$$

If we next apply the second Hellman-Feynman theorem [Fey39]:

$$\langle m|\partial_\mu n\rangle = \frac{\langle m|\partial_\mu H|n\rangle}{E_n - E_m} \qquad \text{and} \qquad \langle \partial_\mu m|n\rangle = \frac{\langle m|\partial_\mu H|n\rangle}{E_m - E_n},$$

we obtain the 2nd-order series expansion for the *effective Hamiltonian:*

$$H_{\text{eff}} = \langle m|H|n\rangle + \langle m|\partial_\mu H|n\rangle x^\mu +$$
$$\frac{1}{2!}\sum_{k\in\mathcal{H}}\left[\frac{\langle m|\partial_\mu H|k\rangle\langle k|\partial_\nu H|n\rangle}{E_m - E_k} + \frac{\langle m|\partial_\nu H|k\rangle\langle k|\partial_\mu H|n\rangle}{E_n - E_k}\right] x^\mu x^\nu + O^{(3)}.$$

8.2.4 Effective JJN Hamiltonian as Perturbative Sum

Using the effective Hamiltonian formalism and by simplifying and abstracting the dense quantum-mechanical calculations from [FVB08], we formulate the effective JJN-Hamiltonian operator (corresponding to Figure 8.3) in the form of the perturbative sum:

$$H_{\text{eff}} = H_0 + H_{\text{bath}} + H_{\text{int}} + H_{\text{ac}}\,. \tag{8.10}$$

This uses normal units (with $\hbar = 1$). The components are as follows:

- The unperturbed JJN-Hamiltonian, given by:

$$H_0 = \sum_{(i,j)}\left[\frac{(\dot\varphi_{ij} - \dot\varphi_{kl})^2}{4E_c} - E_J\cos\left(2etV_{ij-kl} + \varphi_{ij} - \varphi_{kl}\right)\right] + \sum_{i,j}\frac{\dot\varphi_{ij}^2}{4E_c}, \tag{8.11}$$

  where V_{ij-kl} are the dc-voltage drops on the junctions, while $\sum_{(i,j)}$ denotes the summation over pairs of adjacent junctions;
- The heat bath Hamiltonian H_{bath} (characterizing the thermal bath and modeled as a set of N harmonic oscillators with coordinates ξ_i, frequencies ω_i and masses m_i) is given by:

$$H_{\text{bath}} = \sum_{i=1}^{N}\left(\frac{\dot\xi_i^2}{2m_i} + m_i\omega_i^2\xi_i^2\right); \tag{8.12}$$

- The interaction Hamiltonian, given by:

$$H_{\text{int}} = \sum_{i=1}^{N} c_i\xi_i(\varphi_{\ L} - \varphi_{\ R}). \tag{8.13}$$

Here c_i denotes the coupling constants, so that $\sum_i c_i^2/(m_i\omega_i^2) \simeq R/R_{\text{ext}}$, where R_{ext} is the external resistance;

- The Hamiltonian H_{ac} for the AC-component I_{ac} of the Josephson current I [generated by the finite voltage V applied to JJN and oscillating with the frequency $\omega = 2e(V_j + V_N)$] is given by:

$$H_{\text{ac}} = \frac{2E_J^2}{E_c}\sum_{j=1}^{M}\cos\left[2e(V_{1j}+V_{Nj})t + \psi(t) - \varphi_{1j}(t) + \varphi_{Nj}(t)\right]. \qquad (8.14)$$

  The corresponding DC-component I_{dc} of the Josephson current I results from the time averaging of the AC-currents I_{ac}, and calculated by the correlations of the (time-dependent) fluctuations of the Josephson phases $\varphi_{ij}(t)$ across the JJN.

- The voltage-current JJN characteristics - that is, the JJN I-V curves - are defined by the quantum expectation value:

$$I(V) = \left\langle \frac{\partial H}{\partial[\varphi_R - \varphi_L]} \right\rangle. \qquad (8.15)$$

For the complete derivations of the above equations and other technical details, see [FVB08] and the references therein.

8.3 Commutative JJN Hierarchies

8.3.1 Fuzzy Associative Functors

Only Commutative Diagrams are Formally Valid

From a formal mathematical perspective, out of an infinite variety of possible action-control diagrams in any area of applied mathematics, only commutative diagrams are automatically valid: commutative diagrams define the general process of *functional composition,* thus implying both the existence and uniqueness of the corresponding algebraic equations. In modern mathematics, commutative diagrams are used as visual means for rigorous theorem proofs. They are illustratively appealing without sacrificing formal validity.

While commutative diagrams are particularly widespread in category theory [Mac71], which was born within algebraic topology [Die88], they are also heavily used in modern analysis [Die69] and its applications [AMR88], and in modern geometrical dynamics [II06b, II07].

JJN Categories and Functors

From the perspective of modern mathematics and computer science, a generic JJN, like the one in Figure 8.3, represents a *category*, in which the superconducting islands (squares) are *objects*, while their neighboring JJs (crosses) are *morphisms*, also called *arrows*. The corresponding maps between categories, the so-called *functors*, map squares (objects) to squares and crosses (morphisms) to crosses.

Fuzzy Associative Functors

Fuzzy associative functors are adaptive, or tuned, fuzzy inference systems (see [Zad65, Zad78, Kos93, Kos96, Kos99, II07]), which map objects and morphisms of a source category, say $\mathcal{A} = (\mathtt{Ob}(\mathcal{A}), \mathtt{Mor}(\mathcal{A}))$ into objects and morphisms of a target category, say $\mathcal{B} = (\mathtt{Ob}(\mathcal{B}), \mathtt{Mor}(\mathcal{B}))$. This is depicted in Figure 8.4.

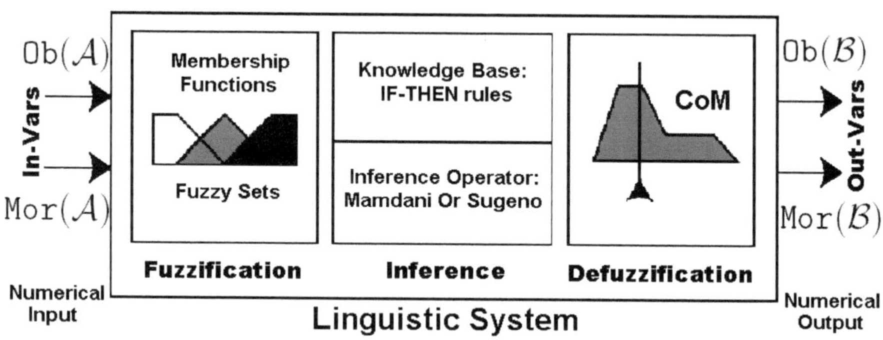

Fig. 8.4. *A fuzzy associative functor maps objects and morphisms of a source category* $\mathcal{A} = (\mathtt{Ob}(\mathcal{A}), \mathtt{Mor}(\mathcal{A}))$ *into objects and morphisms of a target category* $\mathcal{B} = (\mathtt{Ob}(\mathcal{B}), \mathtt{Mor}(\mathcal{B}))$. *It is a tuned (or, adaptive) fuzzy inference system, consisting of three modules: (i) Fuzzification module, which converts numerical inputs into membership functions (consisting of overlapping fuzzy sets); (ii) Inference module which is a fuzzy If-Then rule base (combined using Mamdani or Takagi-Sugeno operators), and (iii) Defuzzification module, converting fuzzy linguistic processes into numerical outputs, using CoM method (see, e.g. [II07]).*

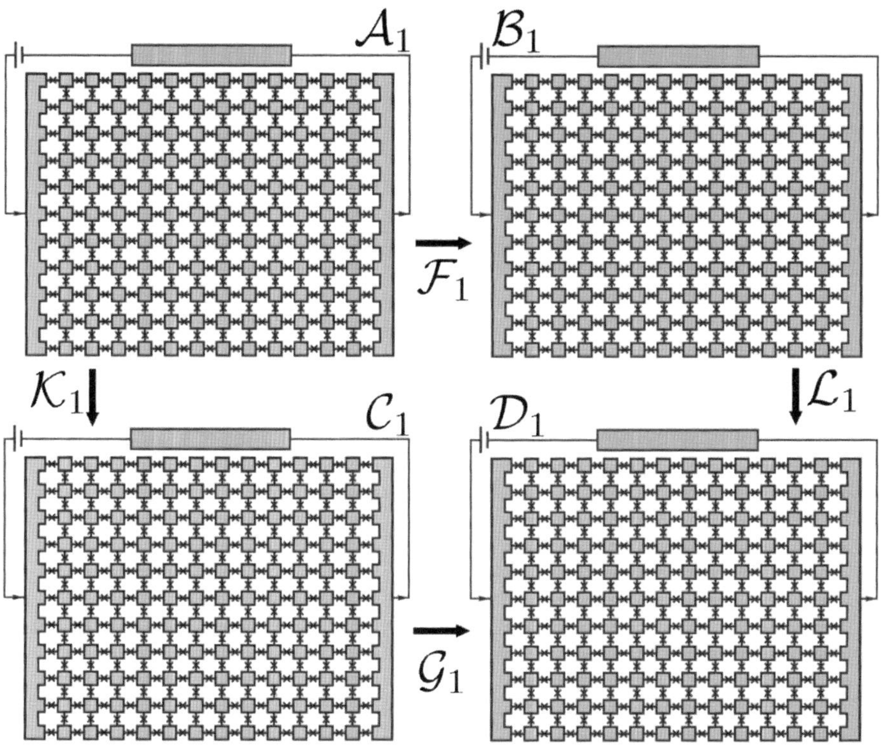

Fig. 8.5. *The 1st-level PCT architecture: a commutative square of JJNs.*

8.3.2 Hierarchy for JJN Architectures

The 1st-Level JJN Architecture

The 1st-level PCT architecture is defined by a commutative square of JJNs (see Figure 8.5). Here, $\mathcal{A}_1, \mathcal{B}_1, \mathcal{C}_1, \mathcal{D}_1$ are the JNN-categories, while $\mathcal{F}_1, \mathcal{G}_1, \mathcal{K}_1$, $\mathcal{L}_1$ are the corresponding fuzzy associative functors, mapping squares (islands) onto squares and crosses (junctions) onto crosses.

The 2nd-Level JJN Architecture

The 2nd-level PCT architecture is defined by a commutative square of the 1st-level architectures (see Figure 8.6). Here, $\mathcal{A}_2, \mathcal{B}_2, \mathcal{C}_2, \mathcal{D}_2$ are the 2-categories (each one consisting of four small JNN-categories), while $\mathcal{F}_2, \mathcal{G}_2, \mathcal{K}_2, \mathcal{L}_2$ are the corresponding fuzzy associative 2-functors.

The 3rd-Level JJN Architecture

The 3rd-level PCT architecture is defined by a commutative square of the 2nd-level architectures (see Figure 8.7). Here, $\mathcal{A}_3, \mathcal{B}_3, \mathcal{C}_3, \mathcal{D}_3$ are the 3-categories

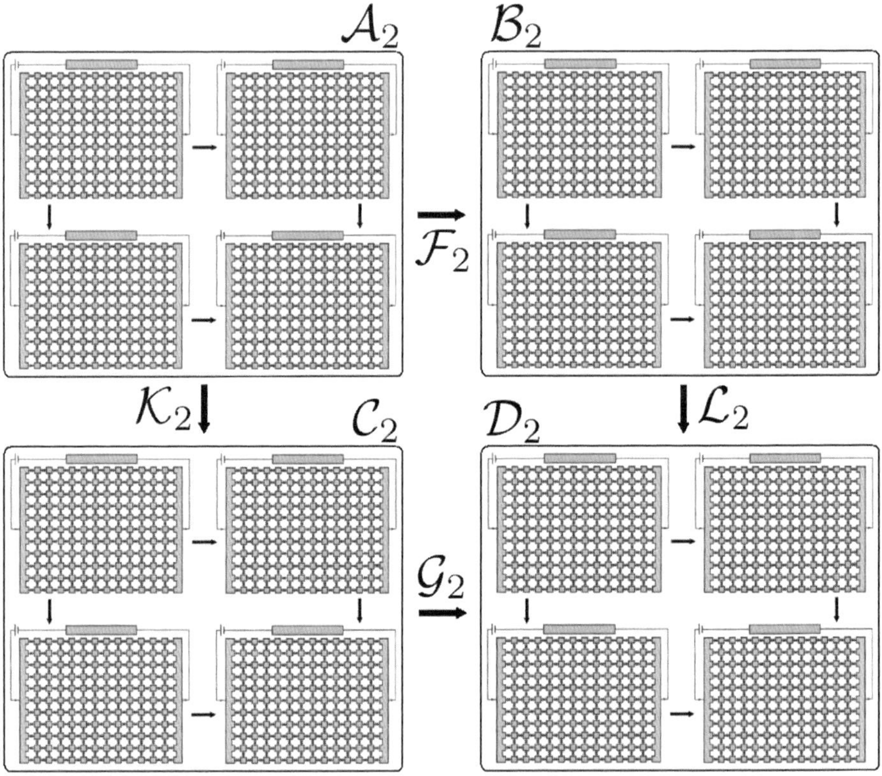

Fig. 8.6. *The 2nd-level PCT architecture: a commutative square of the 1st-level architectures.*

(each one consisting of four 2-categories), while $\mathcal{F}_3, \mathcal{G}_3, \mathcal{K}_3, \mathcal{L}_3$ are the corresponding fuzzy associative 3-functors.

Higher-Level JJN Architectures

By induction, the Nth-level PCT architecture is recursively defined by a commutative square of the (N-1)th-level architectures.

Summary

We have proposed a novel quantum-dynamical approach to Perceptual Control Architecture (PCT) modeling. In our conception, any such model will essentially consist of large networks of Josephson Junctions and their category theoretic generalizations, in a layered architecture. This contrasts with the essentially model-free PCT design mechanisms using reasoning in Bayesian networks elsewhere in the literature [WTW12].

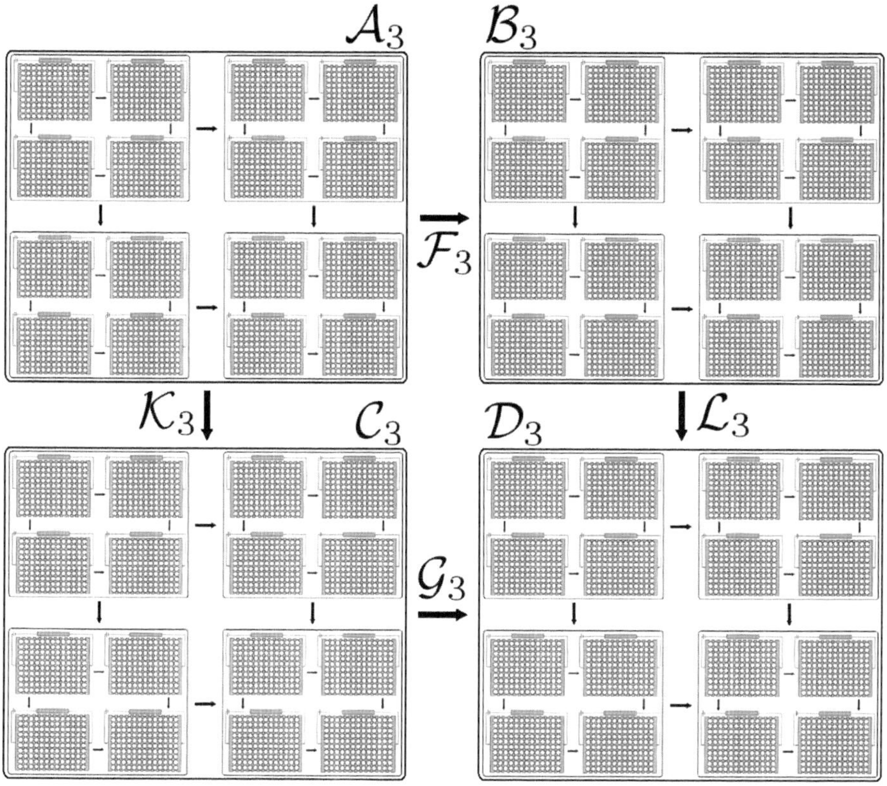

Fig. 8.7. *The 3rd-level PCT architecture: a commutative square of the 2nd-level architectures.*

Our approach allows complete PCT architectures to be composed from JJN units, mirroring the fundamental notion in theoretical computer science, where categories correspond to types. Thus we have the basis for composing PCT architectures while being able to rigorously prove the absence of undesirable outcomes; or, alternatively, to ensure that components fit together in ways that preserve the properties of proper functional compositions. Our proposed PCT composition is, in computer science parlance, strongly typed. That is, we have the basis for easily composing PCT architectures using intuitively appealing design methods while also ensuring formal validity: that the underlying algebraic equations exist and are unique without having we can guarantee the existence and uniqueness of underlying algebraic equations without having to deal with them directly.

Future research work lies in developing and extending the basis established here with practical machinery, to provide a new constructive modeling framework for PCT. We anticipate that this machinery will consist of tractable syntactic means for classifying PCT compositions according to the kinds of functions they represent.

8.4 Appendix

8.4.1 Hardware for Quantum Computers

A Josephson junction (JJ) is made up of two superconductors, separated by a weak coupling non-superconducting layer, so thin that electrons can cross through the insulating barrier. It can be conceptually represented as:

$$\begin{aligned} \text{Superconductor 1} &: \psi_1 \mathrm{e}^{\mathrm{i}\varphi_1} \\ \text{Weak Coupling} &\Updownarrow \\ \text{Superconductor 2} &: \psi_2 \mathrm{e}^{\mathrm{i}\varphi_2} \end{aligned}$$

where the two superconducting regions are characterized by simple *quantum-mechanical wave functions*, $\psi_1 \mathrm{e}^{\mathrm{i}\varphi_1}$ and $\psi_2 \mathrm{e}^{\mathrm{i}\varphi_2}$, respectively. One of JJ characteristics (part of the *Josephson effect*)[10] is that as the temperature is lowered, superconducting current flows through it even in the absence of voltage between the electrodes.

The basic equations governing the dynamics of the Josephson effect are (see, e.g. [BP82]):

$$U(t) = \frac{\hbar}{2e}\dot{\varphi}, \qquad I(t) = I_c \sin\varphi(t),$$

where overdot means time derivative, $\hbar$ is the Planck constant (h divided by 2π), $U(t)$ and $I(t)$ are the voltage and current across the JJ, $\varphi(t)$ is the phase difference between the wave functions in the two superconductors comprising the junction, and I_c is a constant, called the *critical current* of the junction. The critical current is an important phenomenological parameter of the device that can be affected by temperature as well as by an applied magnetic field. The physical constant $\hbar/2e$ is the magnetic flux quantum, the inverse of which is the *Josephson constant*.

The three main effects predicted by Josephson follow from these relations:

---

[10] The Josephson effect in particular results from two superconductors acting to preserve their long-range order across an insulating barrier. With a thin enough barrier, the phase of the electron wave-function in one superconductor maintains a fixed relationship with the phase of the wave-function in another superconductor. This linking up of phase is called phase coherence. It occurs throughout a single superconductor, and it occurs between the superconductors in a Josephson junction. The *phase coherence*, or *long-range order*, is the essence of the Josephson effect.

1. The *DC Josephson effect*. This refers to the phenomenon of a direct current crossing the insulator in the absence of any external electromagnetic field, owing to *Josephson tunnelling*. This DC Josephson current is proportional to the sine of the phase difference across the insulator, and may take values between $-I_c$ and I_c.

2. The *AC Josephson effect*. With a fixed voltage U_{DC} across the junctions, the phase will vary linearly with time and the current will be an AC current with amplitude I_c and frequency $2e/\hbar U_{DC}$. This means a JJ can act as a perfect *voltage-to-frequency converter*.

3. The *inverse AC Josephson effect*. If the phase takes the form

$$\varphi(t) = \varphi_0 + n\omega t + a\sin(\omega t),$$

the voltage and current will be

$$U(t) = \frac{\hbar}{2e}\omega[n + a\cos(\omega t)], \qquad I(t) = I_c \sum_{m=-\infty}^{\infty} J_n(a)\sin[\varphi_0 + (n+m)\omega t].$$

The DC components will then be

$$U_{DC} = n\frac{\hbar}{2e}\omega, \qquad I(t) = I_c J_{-n}(a)\sin\varphi_0.$$

Hence, for distinct DC voltages, the junction may carry a DC current and the junction acts like a perfect *frequency-to-voltage converter*.

To show a driven-damped pendulum analog of a microscopic description of a single JJ, we start with:

1. The *Josephson current-phase relation*

$$I = I_c \sin\varphi,$$

where I_c is the *critical current*, I is the bias current, and $\varphi = \varphi_2 - \varphi_1$ is the constant *phase difference* between the phases of the two superconductors that are weakly coupled; and

2. The *Josephson voltage-phase relation*

$$V = \frac{\hbar}{2e}\dot{\varphi},$$

where $V = V(t)$ is the instantaneous voltage across the junction and e is the charge on the electron.

Now, if we apply Kirhoff's voltage and current laws for the parallel RC-circuit with resistence R and capacitance C, we come to the first-order ODE

$$C\dot{V} + \frac{V}{R} + I_c\sin\varphi = I,$$

which can be recast solely in terms of the phase difference φ as the 2nd-order pendulum-like ODE:

$$\begin{aligned} \text{JJ}: &\qquad \frac{\hbar C}{2e}\ddot{\varphi} + \frac{\hbar}{2eR}\dot{\varphi} + I_c \sin\varphi = I, \\ \text{Pendulum}: &\qquad ml^2\ddot{\theta} + b\dot{\theta} + mgl\sin\theta = \tau. \end{aligned} \tag{8.16}$$

This mechanical analog has often proved useful in visualizing the dynamics of JJs [Str94]. If we divide (8.16) by I_c and define a dimensionless time

$$\tau = \frac{2eI_cR}{\hbar}t,$$

we get the (dimensionless) JJ-oscillator equation:

$$\beta\varphi'' + \varphi' + \sin\varphi = \frac{I}{I_c}, \tag{8.17}$$

where $\varphi' = d\varphi/d\tau$. The dimensionless group β, defined by

$$\beta = \frac{2eI_cR^2C}{\hbar},$$

is called the McCumber parameter and represents a dimensionless capacitance.

In a simple *overdamped limit* $\beta << 1$ with *resistive loading*, the 'inertial term' $\beta\varphi''$ may be neglected (as if oscillating in a highly-viscous medium), and so (8.17) reduces to a non-uniform oscillator

$$\varphi' = \frac{I}{I_c} - \sin\varphi, \tag{8.18}$$

with solutions approaching a stable fixed-point for $I < I_c$, and periodically varying for $I < I_c$. To find the current-voltage curve in the overdamped limit, we take the average voltage $\langle V \rangle$ as a function of the constant applied current I, assuming that all transients have decayed and the system has reached steady-state, and get

$$\langle V \rangle = I_cR\langle \varphi' \rangle.$$

An overdamped *array* of N JJs (8.18), parallel with a resistive load R, can be described by the system of 1st-order (dimensionless) ODEs [Str94]

$$\varphi'_k = \Omega + a\sin\varphi_k + \frac{1}{N}\sum_{j=1}^{N}\sin\varphi_j, \qquad (k = 1, ..., N), \tag{8.19}$$

where

$$\Omega = I_bR_0/I_cr, \qquad a = -(R_0 + r)/r, \qquad R_0 = R/N,$$

$$I_b = I_c\sin\varphi_k + \frac{1}{2eR}\left(\dot{\varphi}_k + \sum_{j=1}^{N}\dot{\varphi}_j\right).$$

8.4.2 Adaptive Fuzzy Inference Systems

In this subsection, we give a brief introduction to a *adaptive fuzzy inference system* (AFIS). For more technical details on this matter, see [Kos92, II07] and references therein.

Computational-Intelligence

Recall that term 'Computational Intelligence' (CI) was coined by IEEE as an alternative to classical AI.
IEEE CI-Society publishes three transactions journals:

- IEEE Transactions on Neural Networks;
- IEEE Transactions on Fuzzy Systems; and
- IEEE Transactions on Evolutionary Computation.

Thus, it is implicitly assumed that CI consists of these three areas and their combinations. Two main CI characteristics are: adaptation and knowledge-base.

Here, instead of trying to cover a very wide area of 'both worlds', neural networks and fuzzy logic (see, e.g. [Kos92, II07]), we focus on *adaptive fuzzy inference system* (AFIS), which is a superior MIMO-control technology already available in a digital chip. Besides, adaptivity (or, learning) is not necessarily neural; it can also be GA-evolutionary, or just a simple matrix iteration:

$$New\,Value\,(t+1) \;=\; Old\,Value\,(t) \;+\; Innovation\,(t+1)$$

$$\text{with}\quad Innovation = |desired\ output \;-\; achieved\ output|$$

$$\text{or}\quad innovation = |reward - penalty|$$

Motivational Fuzzy Examples

Easy Fuzzy-Control Example: Truck Backer-Upper

Let's start with an easy, textbook, fuzzy-control example: truck backer-upper steering control system. It has the following structure:

- Two input linguistic variables: position and direction of the truck, and one output variable: steering angle.
- Invars:
  1. position = {NL,NS,ZR,PS,PL} and
  2. direction = {NL,NM,NS,ZR,PS,PM,PL}

  where NL denotes Negative Large, NM is Negative Medium, NS is Negative Small, etc.
- Outvar: steering angle = {NL,NM,NS,ZR,PS,PM,PL}

Defining fuzzy sets (only output fuzzy sets are shown here):
defineSet [NB, -30, 1, -15, 0, -15, 0, -15, 0];
defineSet [NM, -25, 0, -15, 1, -15, 1, -5, 0];
defineSet [NS, -12, 0, -6, 1, -6, 1, 0, 0];
defineSet [ZE, -5, 0, 0, 1, 0, 1, 5, 0];
defineSet [PS, 0, 0, 6, 1, 6, 1, 12, 0];
defineSet [PM, 5, 0, 15, 1, 15, 1, 25, 0];
defineSet [PB, 18, 0, 30, 1, 30, 1, 30, 1];

Fuzzy IF-THEN rule-base:
IF direction is NL & position is NL, THEN steering angle is NL;
IF direction is NL & position is NS, THEN steering angle is NL;
IF direction is NL & position is ZE, THEN steering angle is PL;
IF direction is NL & position is PS, THEN steering angle is PL;
IF direction is NL & position is PL, THEN steering angle is PL;
IF direction is NM & position is NL, THEN steering angle is ZE;
. . . .
IF direction is PL & position is PL, THEN steering angle is PL.

Alternative representation: Kosko's FAM matrix
(works only for $2 \longrightarrow 1$ case)

$$\begin{pmatrix} \mathrm{PS}(\theta) & \mathrm{PM}(\theta) & \mathrm{PM}(\theta) & \mathrm{PM}(\theta) & \mathrm{PB}(\theta) \\ \mathrm{NS}(\theta) & \mathrm{PS}(\theta) & \mathrm{PM}(\theta) & \mathrm{PB}(\theta) & \mathrm{PB}(\theta) \\ \mathrm{NM}(\theta) & \mathrm{NS}(\theta) & \mathrm{PS}(\theta) & \mathrm{PM}(\theta) & \mathrm{PB}(\theta) \\ \mathrm{NM}(\theta) & \mathrm{NM}(\theta) & \mathrm{ZE}(\theta) & \mathrm{PM}(\theta) & \mathrm{PM}(\theta) \\ \mathrm{NB}(\theta) & \mathrm{NM}(\theta) & \mathrm{NS}(\theta) & \mathrm{PS}(\theta) & \mathrm{PM}(\theta) \\ \mathrm{NB}(\theta) & \mathrm{NB}(\theta) & \mathrm{NM}(\theta) & \mathrm{NS}(\theta) & \mathrm{PS}(\theta) \\ \mathrm{NB}(\theta) & \mathrm{NB}(\theta) & \mathrm{NM}(\theta) & \mathrm{NM}(\theta) & \mathrm{NS}(\theta) \end{pmatrix}$$

Harder Control Example: Truck-and-Trailer Backer-Upper Steering Control System

This is a nontrivial control problem (for a human operator).
Yet, a number of efficient controllers exist:
- sliding mode controllers;
- extended/unscented Kalman filters;
- fuzzy, ANN, GA, controllers;
- other nonlinear controllers, etc.

Very Hard Control problem: Truck-with-two (or more) Trailers Backer-Upper Steering Control System

To the best of our knowledge, only two solutions exist today:

- A *soft approach* of (adaptive) *fuzzy-logic controller*; and
- A *hard approach* of (adaptive) *Lie-derivative controller* (together with its exterior-differential dual).
- So, the only soft option is: **adaptive fuzzy logic**.

- Besides, many different fuzzy-chips already exist, starting from Motorola's 1996 pioneering one. Current Motorola's 68HC12 is the standard micro-controller family with a comprehensive fuzzy-logic instruction set and arbitrary number of inputs and outputs.

Fuzzy Logic Basics

Brief History of Fuzzy Logic

1965 Fuzzy Sets Theory founded by Lotfi Zadeh
(UC Berkeley); Zadeh subsequently founded both Fuzzy Logic and Possibility Theory.
1970 Fuzzy-Logic Control founded by E. Mamdani (Queen-Mary College, London) and
H. Zimmermann (RWTH Univ. Aachen, Germany);
1975 Introduction of Fuzzy Logic in Japan;
1985 Broad application of Fuzzy Logic in Japan;
1990 Broad application of Fuzzy Logic in Europe;
1995 Broad application of Fuzzy Logic in the US, in spite of two major obstacles: (i) wrong name "fuzzy", and (ii) strong Bayesian probability community, which did not like the possibility-theory competitor.

1996 IEEE Study of Fuzzy-Logic Applications

- Published over 1100 successful fuzzy-logic applications (an estimated 5% of those in existence);
- Almost all applications were MIMO supervisory control, rather than replacement of conventional SISO controllers;
- Applications range from Embedded Control (28%), Industrial Automation (62%) to Process Control (10%);
- In most cases, Fuzzy Logic has slashed Design time by more than half.

Fuzzy Logic Basics: Fuzzy Sets 1 A crisp set X is defined by a binary *characteristic function* $\mu_X(x)$ of its elements x:

$$\mu_X(x) = \begin{cases} 1, & \text{if } x \in X, \\ 0, & \text{if } x \notin X, \end{cases}$$

while a fuzzy set is defined by a continuous characteristic function (which is now called the *membership function*):

$$\mu_X(x) = [0,1]\,,$$

including all (possible) real values between the two crisp extremes 1 and 0, and including them as special cases.

Fuzzy Logic Basics: Fuzzy Sets 2 A fuzzy set X is a collection of ordered pairs

$$X = \{(x, \mu(x))\},$$

where $\mu(x)$ is the *membership function* representing the grade of membership of the element x in the set X.

A single pair is called a fuzzy *singleton* (for technical details, including possibility theory, see Zadeh, 1965; 1978).

Fuzzy Inference System

Fuzzy Inference System (FIS) is a MIMO-system that works as a nonlinear multivariate function approximator.
FIS *maps* a set of input linguistic variables ($IF-$part)
into a set of output linguistic variables ($THEN-$part).
FIS consists of three sequential modules:

$$Fuzzification \ \Rightarrow \ Inference \ \Rightarrow \ Defuzzification$$

Fuzzification Module In this module, numerical crisp input variables are fuzzified. This is performed as an overlapping partition of their universes of discourse by means of fuzzy membership functions $\mu(x)$, which can have various shapes, including triangular/trapezoidal (shown before), Gaussian:

$$\mu(x) = \exp\left[\frac{-(x-m)^2}{2\sigma^2}\right],$$

(with mean m and standard deviation σ), and sigmoid:

$$\mu(x) = \left[1 + \left(\frac{x-m}{\sigma}\right)^2\right]^{-1}.$$

Inference Module It has two submodules:

(i) The expert–knowledge base consisting of a set of fuzzy IF-THEN rules relating input and output variables, and

(ii) The inference method, or implication operator, that actually combines the rules to give the fuzzy output. Most common is *Mamdani Min–Max inference*, in which the membership functions for input variables are first combined inside the IF-THEN rules using AND ($\cap$, or Min) operator, and then the output fuzzy sets from different IF-THEN rules are combined using OR ($\cup$, or Max) operator to get the common fuzzy output.

Defuzzification Module In this module, fuzzy-linguistic outputs from the inference module are converted to numerical crisp values.
This is achieved by one of the several defuzzification algorithms. Most common is the *centroid*, or Center of Mass (CoM) method, in which the crisp

output value is calculated as the abscissa under the center of mass/gravity of the output fuzzy set.

Two Optional Modules In more complex technical applications of general function approximation, two optional blocks are usually added to the FIS:

- Preprocessor, preceding the fuzzification module, performing various kinds of normalization, scaling, filtering, averaging, differentiation or integration of input data.
- Postprocessor, succeeding the defuzzification module, performing the analog operations on output data.

Adaptive Fuzzy Inference System (AFIS)

In addition to all the FIS items, AFIS also includes:

- Weights on the fuzzy IF-THEN rules;
- Adjustable parameters in membership functions (e.g. mean and standard deviation in case of Gaussian fuzzy sets);
- Adjustable centroid (CoM) parameters.

Neuro-AFIS

Although FIS adaptivity can be achieved both by simple matrix iteration:

$$NewVal_{t+1} = OldVal_t + Innov_{t+1},$$

or by GA-evolution, the most common is a feedforward three-layer perceptron

Kosko's Standard Additive Model (SAM)

Fuzzy system F approximates a MIMO-function $f : \mathbb{R}^n \to \mathbb{R}^p$ by covering its graph with rule patches, which are *fuzzy Cartesian products* [Kos98]: $A_i \times B_i \subset \mathbb{R}^n \times \mathbb{R}^p$ (Kosko, IEEE-SMC, 1998). Almost all applied fuzzy systems use some form of SAM.
SAM is an additive map $F : \mathbb{R}^n \to \mathbb{R}^p$, which stores m rules:
'**If** $X = A_i$ **Then** $Y = B_i$', or formally, $A_i \times B_i \subset \mathbb{R}^n \times \mathbb{R}^p$,
and adds the 'fired' **Then**-parts $B_i^{'}(x)$, to give the output set $B(x)$ calculated as:

$$B(x) = \sum_{i=1}^{n} w_i B_i^{'}(x) = \sum_{i=1}^{n} w_i \mu_i(x) B_i(x), \qquad (i = 1, ..., n)$$

for a scalar rule weight $w_i > 0$. SAM map F computes its output $F(x)$ by taking the centroid CoM of the output set $B(x)$:

$$F(x) = CoM[B(x)]$$

Kosko's SAM Theorem:

$$F(x) = \sum_{i=1}^{n} p_i(x) c_i, \qquad p_i(x) = \frac{w_i \mu_i(x) V_i}{\sum_{k=1}^{n} w_k \mu_k(x) V_k}, \qquad (i = 1, ..., n)$$

where $p_i(x)$ are discrete probability weights, and V_i are finite positive volumes in the codomain space $\mathbb{R}^p$,

$$V_i = \int_{\mathbb{R}^p} b_i(y_1, ..., y_p) dy_1 ... dy_p > 0$$

while c_i is the centroid CoM of the **Then**-part set $B_i(x)$,

$$c_i = \frac{\int_{\mathbb{R}^p} y\, b_i(y_1, ..., y_p) dy_1 ... dy_p}{\int_{\mathbb{R}^p} b_i(y_1, ..., y_p) dy_1 ... dy_p}$$

Fuzzy Control Basics

There are four basics types of fuzzy-logic control:

- Direct Fuzzy Control (rare). Here, the FIS-output gives the *command variables* for direct control of an arbitrary plant. In other words, FIS works here like a conventional PID controller ("on steroids").
- Supervisory Fuzzy Control (usual). Here, FIS sets the values for a cluster of PID controllers.
  This is (clearly) a more appropriate FIS-application.
- Adaptive Fuzzy Control (by design, without NN). Here, by its own design, a non-adaptive FIS sets/adapts the (P,I,D) parameters of a conventional PID controller.
- Fuzzy Intervention Control. Here, FIS-controller and a conventional PID-controller are working in parallel.

Conclusion: When AFIS?

Informally: recall that there is a train in Japan (Sendai subway), which is controlled by fuzzy logic. The train pulls into the station within a few inches of its target and its timing is within a minute.
Comment by E. Cox [Cox92]: Fuzzy control is more accurate, but nevertheless replacing human expert-control.
And this is roughly a recipe when to use AFIS.

AFIS: Definition **Def:** *AFIS is a model-free strategy for human-like control of high-dimensional complex systems with uncertainty.*
So, AFIS should be used in the following 4 cases:

- When the problem is too complex (or, too hard) for one to be able to derive exact system's state/control equations.
- When there is a lot of practical (intuitive) expert knowledge available (e.g. pilot, driver, worker, player).
- When there are many inputs (e.g. 10) and outputs (e.g. 5).
- When the problem scenario is full of uncertainty.

9

Complexity and Control in Entropic and Stochastic Self-Organization

In this Chapter[1], we are primarily interested in developing advanced entropic and stochastic models of military *command and control (C2)*. The underpinning thesis here stands in sharp contrast to much grand historical theorizing based on the idea of discovering some inexorable laws of historical destiny; this belief shows up as efforts to predict the future by uncovering 'trends' and treating them as 'laws' as well as in narratives attributing causes to anything and everything except randomness that are constructed – of course – with the full benefits of hindsight. Instead, our approach expressly allows for the fact that significant combinations of events might arise purely randomly, and the observed realized outcome is merely one path of events among potentially innumerable unrealized possibilities. Entropy – an order parameter by which we may quantify uncertainty, information content, complexity, or relative inadequacy of historical determinism, depending on our point of view at the time – is therefore a central concern in our conceptualization.

9.1 A Path-Integral Model for Entropic Self-Organization

In this section, motivated by the notion of *perceptual error* (as a core of the perceptual control theory), we propose an action-amplitude model for *controlled entropic self-organization* (CESO). We present several aspects of this development that illustrate its explanatory power: (i) a physical view of partition functions and path integrals, as well as entropy and phase transitions; (ii) a global view of functional compositions and commutative diagrams; (iii) a local geometric view of the Kähler-Ricci flow and time-evolution of entropic action; and (iv) a computational view using various path-integral approximations.

[1] The work in this chapter has been developed in conjunction with Dr Jason Scholz, a Defence Research Leader.

Our basic C2-models we formulate as follows[2]. Our approach should also be applicable across a wide range of non-military organizational decision-making settings. Consider a group $\Gamma := \{\Gamma_i : i = 1, 2, ..., n\}$ of n agents, each characterized by its own Intent I_i and the set of m Capabilities $\{C_j : j = 1, 2, ..., J\}$. The basic modeling construction of the problem is that, according to *perceptual control theory* (see [Pow73a, Pow73b]), the *perceptual error* of an agent Γ_i is the difference between its intent and the consequences of its choice of a capability C_j:

$$E_i = |I_i - A_i(C_j)| \,. \tag{9.1}$$

Thus, under this construction, the optimal behavior Γ_i^{opt} of each agent Γ_i is given by minimizing the perceptual error E_i:

$$\Gamma_i^{opt} = \min_{C_j} (E_i) \,, \tag{9.2}$$

so that the probability $\mathrm{Pr}_i = \mathrm{Pr}_i(X_i = C_j)$ of a specific choice C_j is expected to be maximal when the perceptual error (9.2) is minimal.

The following geometrical interpretations can be given to the optimization problem (9.1)-(9.2):

1. The perceptual error E_i given by the simple absolute value (9.1) can be 'promoted' to the *Euclidean* L^2-norm:

$$E_i^{\mathrm{nrm}} = \|I, A\| = \sqrt{\sum_{i=1}^{n} |I_i - A_i(C_j)|^2}, \tag{9.3}$$

   or to the *Euclidean metric*:

$$E_i^{\mathrm{mtr}} = \mathbf{d}(I, A) = \sqrt{\sum_{i=1}^{n} [I_i^2 - A_i^2(C_j)]}. \tag{9.4}$$

2. Its continuous generalization, allowing for a continuum of capability choices, is given by the *Banach* L^2-norm:

$$B_i^{\mathrm{nrm}} = \|[I, A]\| = \sqrt{\int_a^b |I(x) - A\,[C(x)]\,|^2 dx}, \tag{9.5}$$

   and the associated *Banach metric* between any two real-valued square-integrable functions I, A defined on a real interval $[a, b] \subset \mathbb{R}$:

$$B_i^{\mathrm{mtr}} = \mathbf{d}(I, A) = \sqrt{\int_a^b |I(x)^2 - g(x)^2| dx}. \tag{9.6}$$

[2] This is the simplest formulation of the command and control problem.

For example, the *finite control problem*[3] is actually a minimization of the square of the Banach metric (9.6).

3. The optimization problem (9.2) can be rewritten in *variational formulation* as the stationary Hamilton action problem:

$$\delta E_i = 0,$$

with the following particular cases:

$$\delta \left| I_i - A_i(C_j) \right| = 0 \quad \text{or} \quad \delta \left| I(x) - A\left[C(x)\right] \right| = 0,$$

$$\delta \sqrt{\sum_{i=1}^{n} \left| I_i - A_i(C_j) \right|^2} = 0 \quad \text{or} \quad \delta \sqrt{\int_a^b \left| I(x) - A\left[C(x)\right] \right|^2 dx} = 0,$$

$$\delta \sqrt{\sum_{i=1}^{n} \left[I_i^2 - A_i^2(C_j) \right]} = 0 \quad \text{or} \quad \delta \sqrt{\int_a^b \left| I(x)^2 - g(x)^2 \right| dx} = 0.$$

From these cases, the set of discrete Euler–Lagrangian equations of motion on the group/graph Γ can be derived (see [IR14] as well as section 9.1.2 below). Geometrically, this means that the *perceptual error*, in all three forms – the absolute value (9.1), the L^2-norm (9.3)-(9.5), or the metric (9.4)-(9.6) – represents an *energy landscape*. The optimization problem is some kind of a gradient or Levenberg-Marquardt (LM)[4] descent along the Lagrangian geodesics of this energy landscape.

Next, we assume the probability $\Pr_i = \Pr_i(X_i = C_j)$ of a specific choice C_j in the optimization problem (9.1)-(9.2) to be a monotonically-decreasing function of the perceptual error E_i (or $E_i^{\rm nrm}$-$B_i^{\rm nrm}$, or $E_i^{\rm mtr}$-$B_i^{\rm mtr}$). If we consider

---

[3] In finite control (i.e. output tracking and navigation), we want the scalar system output $y(\mathbf{x}, t)$ to follow a continuous nominal objective trajectory, say $\chi(t)$, for a given MD system vector $\mathbf{x}$, and where t is time. This problem can be expressed as:

$$\min_{\mathbf{x}\in\mathbb{R}^M} \int_{t_0}^{t_1} \left[y(\mathbf{x}, t) - \chi(t) \right]^2 dt.$$

Upon time discretization using a suitable quadrature scheme, this becomes the following least-squares problem:

$$\min_{\mathbf{x}\in\mathbb{R}^M} f(\mathbf{x}) = \sum_{i=1}^{N} \left[\tilde{y}(\mathbf{x}, t_i) - \tilde{\chi}(t_i) \right]^2 dt.$$

Both discretized functions $\tilde{y}(\mathbf{x}, t_i)$ and $\tilde{\chi}(t_i)$ include the weights of the chosen quadrature scheme.

[4] The LM-algorithm, sometimes also known as *Damped Least Squares* (DLS), interpolates between the gradient descent and the Gauss-Newton algorithm for quick and efficient convergence, by properly adjusting the damping parameter. It is more robust than Gauss-Newton, meaning that it can often find an optimum even when initialised far from it.

only a finite set of capabilities, a useful example of the probability distribution function (PDF) can be formulated as an exponentially-weighted collection of discrete Dirac-delta functions:

$$\Pr_i(X_i = C_j) = \delta\,|I_i - A_i(C_j)|\,\mathrm{e}^{-\beta E_i}. \tag{9.7}$$

The PDF (9.7) can be easily generalized to the *Gibbs measure* of a random variable X_i having the set of corresponding coordinates $\{x^i\}$, defined by the PDF [Mou74]:

$$\Pr(X_i = x^i) = \frac{1}{Z(\beta)}\,\mathrm{e}^{-\beta H(x^i)} \equiv \frac{\exp\left[-\beta H(x^i)\right]}{\sum_{x^i}\exp\left[-\beta H(x^i)\right]},$$

where $H = H(x^i) \equiv E_i$ (or $H \equiv E_i^{\mathrm{mtr}}$, or $H \equiv E_i^{\mathrm{nrm}}$) is the Hamiltonian energy function of the above energy landscape with local coordinates $\{x^i\}$, and β is a free parameter (in thermodynamics, β would be inverse temperature). The corresponding *partition function* (see, e.g. [Lan41])

$$Z(\beta) = \sum_{x^i}\mathrm{e}^{-\beta H(x^i)} \tag{9.8}$$

provides the Gibbs measure on the system's state-space, which is a unique statistical distribution that maximizes the entropy for a fixed expectation value of the energy:

$$\langle H(x^i)\rangle = -\frac{\partial\log(Z(\beta))}{\partial\beta}.$$

The associated system's order parameter, *entropy,* is given by:

$$S = -\sum_{x^i}P(x^i)\ln P(x^i) = \beta\langle H(x^i)\rangle + \log Z(\beta).$$

Entropy describes both 'ignorance' – Heisenberg's uncertainty – and 'randomness' – also meaning 'complexity', 'incompressibility' or 'information content'.

A useful particular example of (9.8) is the partition function of a 3-dimensional ensemble of n 'molecules', given by the $6n$-dimensional phase–space integral:

$$Z(\beta) = \int_{\mathbb{R}^{6n}}\exp[-\beta H(x^i)]\,d^3p_i\,d^3x^i, \tag{9.9}$$

where $p_i = p_i(x^i)$ are generalized momenta.

More generally, we consider *Markov random fields/Markov networks*, which have a Markov property described by an undirected graph (see [Mou74]). In Markov networks with local vertex coordinates $\{x^i\}$, the PDF is usually given by the log-linear inner product model:

$$\Pr_i(X_i = x^i) = \frac{1}{Z}\exp\left(\sum_{i=1}^{n}w_i f_i\right), \tag{9.10}$$

where $f_i = f_i(x^i)$ are the feature functions and w_i are their corresponding weights (so that $\sum w_i f_i$ is their dot-product), while Z is the partition function. When the PDF (9.10) is strictly positive, the Markov network is often called the Gibbs random field.

We remark here that Markov random fields have been introduced as a Markovian framework for the *Ising spin-lattice model,* defined by the Hamiltonian energy function (given here in its simplest dot-product form):

$$H(\sigma) = -\sum_{i=1}^{n}\sum_{j=1}^{n} J_{ij}\sigma_i\sigma_j\,,$$

where $\sigma_i \in \{+1, -1\}$ are discrete spin-up and spin-down states (or, more generally, Pauli spin matrices), while J_{ij} is the interaction matrix with synaptic weights. The same Hamiltonian has also been used in *Hopfield neural networks* [Hop82)], in which case σ_i represents the state of the McCulloch-Pitts neuron [MP43].

In this paper we present several different views of an action-amplitude model for controlled entropic self-organization (CESO).

9.1.1 Physical Perspective

From partition function to Feynman's path integral

We have already seen from example (9.9) that the number of random variables X_i need not be countable; if we continue in this direction, the set of corresponding coordinates $\{x^i\}$ becomes a field $\phi = \phi(x) \in \mathbb{R}^n$. Consequently, the sum in (9.8) is replaced by the *path integral*[5] (see, e.g. [II08a]):

$$Z(\phi) = \int \mathcal{D}[\phi]\,\mathrm{e}^{-\beta H(\phi)}, \tag{9.11}$$

where $\mathcal{D}[\phi]$ represents the appropriate Lebesgue measure. We can interpret the path integral philosophically as the sum over possible histories.

More generally, in quantum field theory, instead of the field Hamiltonian $H(\phi)$ we have the classical (Lagrangian or Hamiltonian) action $A(\phi)$ of the theory. Both the real path integral in imaginary time (the so-called Euclidean path integral)

$$Z_{\mathrm{Euc}}(\phi) = \langle \mathrm{out}|\,\mathrm{in}\rangle_{\mathrm{Euc}} = \int_{\mathbb{R}^n} \mathcal{D}[\phi]\,\mathrm{e}^{-A[\phi]}, \tag{9.12}$$

and the complex path integral in real time (the so-called Lorentzian path integral)

[5] As any quantum system has both continuous spectrum (consisting of eigenfunctions) and discrete spectrum (consisting of eigenvectors), the path-integral symbol $\int \mathcal{D}[\phi]$ represents both the integration over the continuous spectrum and the summation over the discrete spectrum of the field system $\phi = \phi(x)$ [II08a].

$$Z_{\mathrm{Lor}}(\phi) = \langle \mathrm{out} | \, \mathrm{in} \rangle_{\mathrm{Lor}} = \int_{\mathbb{R}^n} \mathcal{D}[\phi] \, \mathrm{e}^{\mathrm{i}A[\phi]}, \tag{9.13}$$

represent partition functions of the quantum field theory in $\mathbb{R}^n$. In both cases, quantum probability $\mathrm{Pr}(\phi)$ is defined as the absolute square of the transition amplitude:

$$\mathrm{Pr}(\phi) = |\langle \mathrm{out} | \, \mathrm{in} \rangle|^2 .$$

Finally, we generalize our quantum theory of fields, from $\phi \equiv \phi(x) \in \mathbb{R}^n$ to $\Phi \equiv \Phi(x) \in M$ defined on an arbitrary nonlinear configuration n-manifold M, with its curved geometry and topology with a certain number of holes. In this way, we arrive at our main conceptual tool, the geometrical/topological path integral, in its Euclidean and Lorentzian versions, respectively given by (see [II09]):

$$\begin{aligned} Z_{\mathrm{Euc}}(\Phi) &= \langle \mathrm{out} | \, \mathrm{in} \rangle_{\mathrm{Euc}(M)} = \int_M \mathcal{D}[\Phi] \, \mathrm{e}^{-A[\Phi]}, \\ Z_{\mathrm{Lor}}(\Phi) &= \langle \mathrm{out} | \, \mathrm{in} \rangle_{\mathrm{Lor}(M)} = \int_M \mathcal{D}[\Phi] \, \mathrm{e}^{\mathrm{i}A[\Phi]}. \end{aligned} \tag{9.14}$$

Here, $A[\Phi]$ represents classical Lagrangian action, the integration is performed over all continuous $paths + fields + geometries$ defined on the configuration manifold M, while summation is performed along the edges of the corresponding discrete graph structure,[6] obtained either by putting the tram-tracks-like constraints along the manifold M, or by performing some form of triangulation discretization.

Three-phase entropic framework with transition amplitudes

Recall that *Prigogine's Extended Second Law of Thermodynamics* [NP77]

$$\partial_t S \geq 0, \qquad (\text{where } \ \partial_t S \equiv \partial S / \partial t), \tag{9.15}$$

considers open (i.e. non-isolated) irreversible systems or processes which exchange energy and matter with their environment, in such a way that the entropy change (or, entropy variation) is given by the sum of the internal and external entropy change:

$$\partial_t S = \frac{d_i S}{dt} + \frac{d_e S}{dt}, \tag{9.16}$$

where $d_i S$ denotes the internal entropy production within the system, while the $d_e S$ is the external entropy flux due to the exchanges with the environment; for an isolated system, $d_e S = 0$.

[6] Given any topological/smooth manifold M, one can always obtain not one but rather a family (or, a set) of discrete network/graph structures on it, defined either by fixing some tram-like constraints of motion, or by simplicial (or, Regge-type) triangulation. In terms of dynamics, this means spatio-temporal discretization (e.g., from standard continuous Laplacian (or, Laplace-Beltrami) operator $\triangle$ defined on a manifold M, one can obtain the discrete Laplacian $\triangle_d$.)

By further extending Prigogine's open Second Law (9.15) into the strict-control regime, we formulate a generic three-phase entropic framework for entropy S in any complex system described by one of the partition functions (9.11)-(9.14). the framework consists of the following ccomponents:

1. The phase of *Intent*, given by the monotonically increasing entropy: $\partial_t S > 0$;
2. The phase of *Action*, corresponding to conservation of information, described by the conserved entropy: $\partial_t S = 0$;
3. The phase of *Control*, described by the monotonically decreasing entropy: $\partial_t S < 0$.

The *phase transition* from one phase to another, caused by the system's topology change (see [II08a]), is described by the transition amplitude:

$$\langle \text{out phase}_{\,t_1} | \text{ in phase}_{\,t_0} \rangle := \int_{\mathrm{tch}^c_d(M)} \mathcal{D}[\Phi]\, \mathrm{e}^{\mathrm{i}A[\Phi]},$$

where '$\mathrm{tch}^c_d(M)$' denotes both continuous topology-change '$\mathrm{tch}^c(M)$' of the system's configuration manifold M and discrete topology-change '$\mathrm{tch}_d(M)$' of the system's configuration graph/network structure (e.g., removing or adding vertices of the graph).

9.1.2 Global Functional Perspective

The set – or, more appropriately, the category – of generic agents' behaviors, *Behav*, is defined by the functional composition of the following two successive multidimensional maps:

$$\begin{aligned} &Behav \equiv \mathcal{G} \circ \mathcal{F}, \qquad \text{where :} \\ &\quad \mathcal{F} \;:\; Intent_{\,t_0} \longrightarrow Action_{\,t_1} \qquad \text{and} \qquad \mathcal{G} : Action_{\,t_1} \longrightarrow Control_{\,t_2} \end{aligned}$$

such that the following diagram commutes:

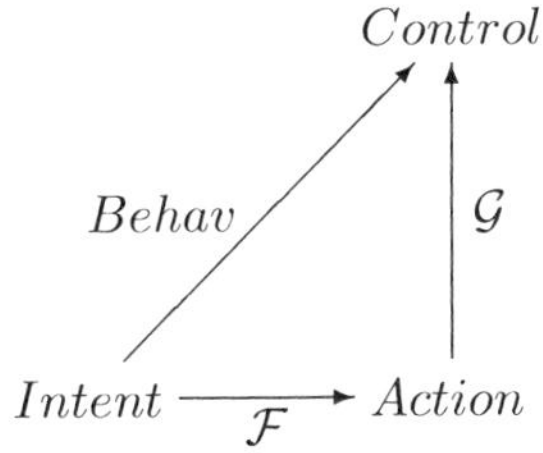

The maps $\mathcal{F}$ and $\mathcal{G}$ are given by their respective adaptive path integrals, meaning partition functions defined by the respective multi–phase and multi–path (multi–field and multi–geometry) transition amplitudes:

$$\mathcal{F} : Intent_{t_0} \longrightarrow Action_{t_1} := \langle \overset{\partial_t S=0}{Action}_{t_1} | \overset{\partial_t S>0}{Intent}_{t_0} \rangle = \int_{\mathrm{tch}(M)} \mathcal{D}[w\Phi]\, \mathrm{e}^{\mathrm{i}A_1[\Phi]}, \tag{9.17}$$

$$\mathcal{G} : Action_{t_1} \longrightarrow Control_{t_2} := \langle \overset{\partial_t S<0}{Control}_{t_2} | \overset{\partial_t S=0}{Action}_{t_1} \rangle = \int_{\mathrm{tch}(M)} \mathcal{D}[\omega\Psi]\, \mathrm{e}^{\mathrm{i}A_2[\Psi]}. \tag{9.18}$$

Here, the Lebesgue integration, in both integrals, is performed over all continuous
$(\Phi_c^i, \Psi_c^i) = paths + fields + geom/tops$. Summation is performed over all discrete random processes (i.e. Markov chains) and regional network-topologies $\left(\Phi_d^j, \Psi_d^j\right)$.

The path integrals (9.17) and (9.18) are constructed, respectively, from the classical *behavioral actions* (see [IR14]):

$$A_1[\Phi] = \int_{t_{ini}}^{t_{fin}} [\Phi(x^i) + \frac{1}{2} g_{ij}\, \dot{x}^i \dot{x}^j]\, dt, \qquad A_2[\Phi] = \int_{t_{ini}}^{t_{fin}} [\Psi(x^i) + \frac{1}{2} g_{ij}\, \dot{x}^i \dot{x}^j]\, dt, \tag{9.19}$$

where overdot denotes time derivative, and their corresponding action principles:

$$\delta A_1[\Phi] = 0 \qquad \text{and} \qquad \delta A_2[\Phi] = 0.$$

These correspond to classical Euler–Lagrangian equations of motion on the configuration manifold M, with the Riemannian metric tensor g_{ij}:

$$\frac{d}{dt} L_{\dot{x}^i} = L_{x^i} \qquad \text{and} \qquad \frac{d}{dt} \mathfrak{L}_{\dot{x}^i} = \mathfrak{L}_{x^i}$$

with Lagrangians

$$L = \frac{1}{2} g_{ij}\, \dot{x}^i \dot{x}^j - \Phi(x^i) \qquad \text{and} \qquad \mathfrak{L} = \frac{1}{2} g_{ij}\, \dot{x}^i \dot{x}^j - \Psi(x^i).$$

The symbolic differentials $\mathcal{D}[w\Phi]$ and $\mathcal{D}[\omega\Psi]$ in the path integrals (9.17) represent adaptive path measures, defined as the weighted products:

$$\mathcal{D}[w\Phi] = \lim_{N\to\infty} \prod_{s=1}^{N} w_s d\Phi_s^i, \qquad (i = 1, ..., N = con_i + dis_i),$$

$$\mathcal{D}[\omega\Psi] = \lim_{M\to\infty} \prod_{p=1}^{M} \omega_p d\Psi_s^k, \qquad (k = 1, ..., M = con_k + dis_k).$$

Both adaptive path integrals (9.17)-(9.18) represent arbitrary toplogy ∞-dimensional neural networks (see [II07]), with weights (w, ω) updated according to

$$w(t+1) = w(t) + Innov^w(t), \qquad \omega(t+1) = \omega(t) + Innov^\omega(t),$$

where $Innov^w(t)$ and $Innov^\omega(t)$ are the respective innovations.

9.1.3 Local Geometric Perspective

A complexified extension of the behavioral actions (9.19), which is more appropriate for general Lorentzian path integrals, is called the Kahler-Perelman entropic action.[7] In this section we give present its time-evolution along the complex geometric dynamics called the *Kähler-Ricci flow*. For this objective, we utilize the richest and most useful structure in the Kähler geometry[8], the so-called *Fano manifold*, which is a compact (i.e., closed and bounded) Kähler n-manifold (M, g) with positive first Chern class $c_1(M)$. For a recent review, see [TZ13] and references therein.

Consider the normalized Kähler-Ricci flow on a Fano n-manifold (M, g):

$$\partial_t g(t) = g(t) - \mathrm{Ric}\,[g(t)]) \qquad (\partial_t \equiv \partial/\partial t). \tag{9.20}$$

In a local open chart $U \subset (M, g)$, starting from some smooth initial Kähler metric $g_0 = g_{i\bar{j}}(0)$, the Ricci flow (9.20) is given by:

$$\partial_t g_{i\bar{j}}(t) = g_{i\bar{j}}(t) - R_{i\bar{j}}(t), \qquad (\text{for } i, j = 1, \cdots, n).$$

It was proved in [Cao85] that (9.20) has a global solution $g(t)$ in the case that g_0 has canonical Kähler class, i.e., $2\pi c_1(M)$ as its Kähler class. In particular, by the $\partial\bar{\partial}$-lemma,[9] there exists a family of real-valued functions

---

[7] The Kahler-Perelman entropic action $\mathcal{W}(g)$ is, roughly speaking, a complexified Einstein-Hilbert action from general relativity.

[8] Recall that a *Kähler manifold* (M, g) is a complex n-manifold that has the following basic characteristics:

1. A set of n local holomorphic coordinates: $\{z_1, \cdots, z_n\} \in U \subset M$, with the corresponding complex-valued differentials:
$$dz_k = dx_k + \mathrm{i}dy_k, \quad d\bar{z}_k = dx_k - \mathrm{i}dy_k;$$
2. Hermitian metric tensor: $g_{i\bar{j}} = g_{i\bar{j}}(z^i, z^{\bar{j}})$, with the corresponding Kähler metric g as a positive and symmetric (1,1)-form:
$$g = \mathrm{i}g_{i\bar{j}}\, dz^i \otimes dz^{\bar{j}},$$
and the associated Kähler form ω as a closed ($d\omega = 0$) and positive (1,1)-form:
$$\omega = \mathrm{i}g_{i\bar{j}}\, dz^i \wedge dz^{\bar{j}}.$$
3. Functional space of Kähler potentials:
$$\mathcal{P} = \{\varphi \mid \omega_\varphi = \omega + \mathrm{i}\partial\bar{\partial}\varphi > 0\}.$$

[9] $\partial \equiv \partial_j$ and $\bar{\partial} \equiv \partial_{\bar{j}}$ are the so-called Dolbeault differential operators. Any p-form α defined on the Kähler manifold (M, g) is called $\bar{\partial}$-*closed* iff $\bar{\partial}\alpha = 0$ and $\bar{\partial}$-*exact* iff $\alpha = \bar{\partial}\eta$ for some $(p-1)$-form η on (M, g). The *Dolbeault cohomology group*

$u(t)$, called *Ricci potentials* (see, e.g. [MT07]) of the metric $g(t)$, which are special Kähler potentials. They are determined by

$$g_{i\bar{j}} - R_{i\bar{j}} = \partial_i \partial_{\bar{j}} u, \qquad \frac{1}{V}\int_M \mathrm{e}^{-u(t)} dv_{g(t)} = 1,$$

where $V = \int dv_g$ denotes the volume of the Kähler-Ricci flow.

From the control-theory perspective, the most important characteristic of the Kähler-Ricci flow is the existence of its solitary solutions (solitons), which are shrinking, or decaying in time. This characteristic is associated to geometrical entropy decrease and gives the global Lyapunov stability to the flow.

Formally, a Riemannian manifold (M, g) represents a *shrinking Ricci soliton* iff (see, e.g. [MT07]):

$$\mathrm{Ric}(g) + \mathrm{Hess}(u) = \lambda g, \qquad (\lambda > 0).$$

In particular, if (M, g) is a Fano manifold with $g \in 2\pi c_1(M)$, it is a *shrinking Kähler-Ricci soliton* iff $\lambda = 1$ and $u \equiv u(t)$ is the Ricci potential; that is, iff (see [TZ13] and references therein):

$$\nabla\nabla u = 0,$$

or, applying the Bianchi identity, iff the following *Shur-identity* holds:

$$\Delta u - |\nabla u|^2 + u = a.$$

For any Kähler metric $g \in 2\pi c_1(M)$ with scalar curvature s and any smooth real time-function $u \equiv u(t)$, define the *Kähler-Perelman entropy* $\mathcal{E} \equiv \mathcal{E}(g)$ defined by the following infimum of sets of entropy functionals (compare with Perelman's original definition [Per02]):

$$\mathcal{E}(g) = \inf\left\{\mathcal{W}(g, u) : \int_M \mathrm{e}^{-u} dv = V\right\}, \qquad \text{where}$$
$$\mathcal{W}(g, u) = \frac{1}{V}\int_M (s + |\nabla u|^2 + u - n)\, \mathrm{e}^{-u} dv.$$

A smooth minimizer of the entropy $\mathcal{E}$ always exists, though it need not necessarily be unique (see [Rot81]). Entropy $\mathcal{E}$ admits a natural upper bound:

$$\mathcal{E}(g) \le \frac{1}{V}\int_M u\, \mathrm{e}^{-u} dv = a \le 0.$$

$H^{1,1}_{\bar{\partial}}(M, \mathbb{R})$ is a complexification of the standard de Rham cohomology group $H^2_d(M, \mathbb{R})$, defined on (M, g) as a quotient:

$$H^{1,1}_{\bar{\partial}}(M, \mathbb{R}) = \frac{\{\bar{\partial}\text{-closed real (1,1)-forms}\}}{\{\bar{\partial}\text{-exact real (1,1)-forms}\}}.$$

To see how the Kahler-Perelman entropic action $\mathcal{W}(g,u)$ evolves under the Kähler-Ricci flow (9.20), recall that for any solution $u(t)$ of the *backward heat equation* (compare with [Per02])

$$\partial_t u = -\triangle u + |\nabla u|^2 + \triangle u, \tag{9.21}$$

we have

$$\partial_t \mathcal{W}(g,u) = \frac{1}{V}\int_M \left[|\nabla\bar{\nabla}(u-u)|^2 + |\nabla\nabla u|^2\right] \mathrm{e}^{-u}dv,$$

which implies the *Kähler-Perelman monotonicity* condition on the geometric entropy [TZ13]:

$$(\forall t \geq 0), \qquad \mathcal{E}(g_0) \leq \mathcal{E}\left[g(t)\right] \leq 0.$$

Roughly speaking, application of control is trying to reduce the system's entropy. This is achieved through shrinking Kähler-Ricci solitons.

9.1.4 Computational Perspective

In this subsection we will outline a fast desktop simulation framework for controlled entropic self-organization, based on the preceding idea of functional composition of path-integrals (9.17)-(9.18).

In quantum field theory, there is both a theoretical and a numerical approach to solve a similar path integral as a sum-over-fields. A theoretical way consists of its perturbative expansion into a series of Feynman diagrams; although, there is a *Mathematica*® package 'FeynArts–FeynCalc' devoted to this, this approach does not scale well with respect to increasing numbers of agents, and therefore is not well suited to our task. A numerical way of handling the problem might be to discretize a path integral on a lattice (of dimension 2, 3, or 4) and use techniques of lattice gauge theory; again, although possible, this approach is not really feasible for the numbers of agents in which we are typically interested [10].

In non-relativistic quantum mechanics, the path integral can be numerically solved, either by a direct implementation of the Feynman formula (see, e.g. [II09]), or by Monte Carlo methods. Both these kinds of solvers are fairly slow, except for the Metropolis algorithm [MRR53] (see Figure 9.1). As shown in Feynman's first paper [Fey48], this path integral is equivalent to the linear Schrödinger equation. Its adaptive version is equivalent to the cubic Nonlinear Schrödinger equation (NLS, see [IR12] and references therein).

By analogy, in statistical mechanics, the real path integral in real time is equivalent to the linear Fokker-Planck equation, while its adaptive version is equivalent to the nonlinear Fokker-Planck equation. This approach is developed in the next subsection.

---

[10] Military Command and Control (C2) can involve hundreds of actors, and it is not unreasonable to expect that other non-military decision-making processes could involve similarly large numbers.

Simple Metropolis approach

For the sake of completeness, we include here Fortran 90 code simulating the ground state wave function using the Metropolis path-integral algorithm with discrete kinetic and potential energy given in Figure 9.1.

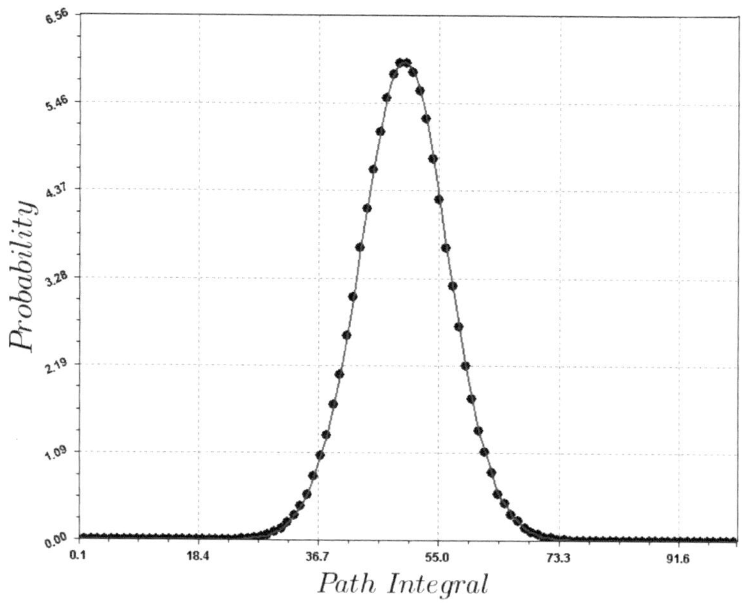

Fig. 9.1. *Simulating a non-relativistic quantum-mechanical path-integral using the Metropolis algorithm, with the probability of an uphill move give by the exponent of the difference between new energy and old energy:* $P = \exp\left[-(E_{\mathrm{new}} - E_{\mathrm{old}})\right]$.

```
program PathQmMetro ! Path integral: ground-state
wave-function
  implicit none      ! Metropolis algorithm; j=[1,100];
  i=[1,3.d5]
  integer :: i,j,N=100,element,prob(100),maxSteps=8.d5
  !=3.*10^5
  real*8 path(100),change,enerFun,newE,oldE,outPr
  do j=1,N  ! zero initial path and probability
    path(j) = 0.;  prob(j) = 0
  end do
  oldE = enerFun(path,N)    ! energy function of the initial
  path
  do i=1,maxSteps
```

```
    element = rand()*N + 1          ! pick random element
    change = 2*(rand()-0.5)         ! update path by rand +
    change
    path(element) = path(element) + change
    newE = enerFun(path,N)      ! enerFun of the current path
        ! Metropolis algorithm: uphill move =
        exp(-(newE-oldE))
    if ((newE > oldE) .and. (exp(oldE-newE) < rand())) then
      path(element) = path(element) - change     ! upd:
      path-change
    end if
    do j=1,N        ! summing probabilities [1,100]
      element = path(j)*10 + 50
      prob(element) = prob(element) + 1
    end do
    oldE = newE
  end do
  print *, 'Path integral: ground-state wave-function:';
  print *
  open(1,file='qPr.csv')
  write(1,*) 'j',',','P'; print *,'   j      P'; print *
  do j=1,N  ! printout to 'qPr.csv'
    outPr = prob(j); write(1,*) j,',',outPr/maxSteps
    print *, j, outPr/maxSteps;
  end do;   close(1)
end program

function enerFun(array,maxSteps)    ! total energy of the
system
  implicit none;  integer i,maxSteps
  real*8 enerFun,array(maxSteps)
  enerFun = 0.  ! initialize energy
  do i=1,(maxSteps-1)       ! kinetic plus potential energy
    enerFun = enerFun + (array(i+1) - array(i))**2 +
    array(i)**2
  end do
end function
```

Fokker-Planck-Ito approach

The Fokker-Planck equation, also known as the Kolmogorov forward equation

$$\begin{aligned}\partial_t P(x,t) &= -\partial_x\left[f(x,t)P(x,t)\right] + \frac{1}{2}\partial_{xx}\left[g^2(x,t)P(x,t)\right] \\ &= [-\partial_x f(x,t) + \frac{1}{2}\partial_{xx} g^2(x,t)\,]P(x,t)\end{aligned} \tag{9.22}$$

is a parabolic partial differential equation (PDE) that describes the time-forward evolution of the probability distribution $P(x,t)$, also called the probability density function (PDF, see Figure 9.2). The expression on the right-hand side of (9.22), including

$$\text{linear advective term}: -\partial_x P(x,t)\ f(x,t), \qquad \text{and}$$
$$\text{quadratic diffusive term}: \frac{1}{2} g^2(x,t)\, \partial_{xx} P(x,t),$$

defines the Fokker-Planck differential operator:

$$\mathcal{D}_{FP} := -\partial_x f(x,t) + \frac{1}{2} \partial_{xx} g^2(x,t).$$

Note that the quadratic diffusive term vanishes in case of zero noise $g(x,t) = 0$.

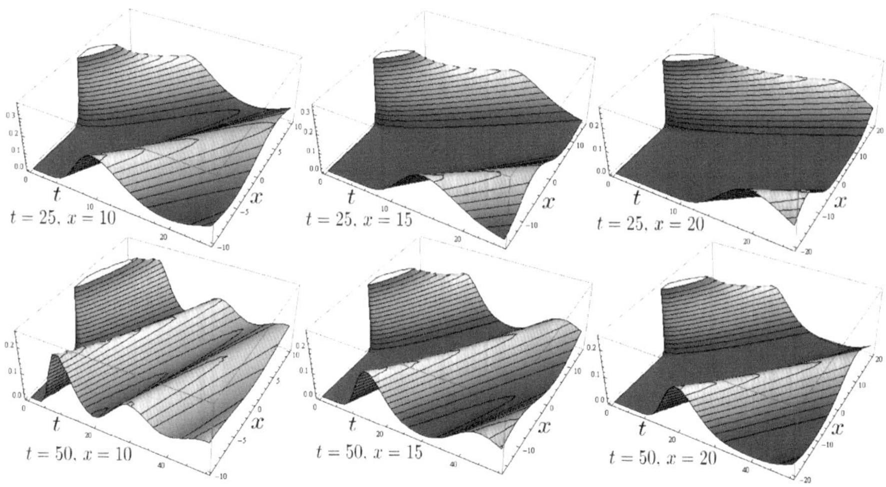

Fig. 9.2. *Illustrative simulation of a (1+1)D Fokker-Planck equation (9.22) in Mathematica*^{TM} *for the simple case of* $f(x,t) = g(x,t) = 1$*: depicting six PDF-snapshots for different time/space values.*

The PDE (9.22) satisfies the probability conservation law:[11]

[11] The backward Fokker-Planck equation, or the Kolmogorov backward equation:

$$\begin{aligned} \partial_t P(x,t) &= -f(x,t)\, \partial_x P(x,t) - \frac{1}{2} g^2(x,t)\, \partial_{xx} P(x,t) \\ &= -[f(x,t)\, \partial_x + \frac{1}{2} g^2(x,t)\, \partial_{xx}] P(x,t) \end{aligned} \tag{9.23}$$

describes the time-backward evolution of the PDF $P(x,t)$. It corresponds to the same SDE (9.24). Equation (9.23) is an evolution PDE with the formal adjoint

$$\partial_t P(x,t) + \partial_x J(x,t) = 0,$$

where $J(x,t)$ represents the probability current, defined as:

$$J(x,t) := f(x,t)P(x,t) - \frac{1}{2}\partial_x g(x,t)P(x,t).$$

The Fokker-Planck PDE (9.22) corresponds to the Ito stochastic differential equation [SDE, with deterministic drift $f(X_t,t)$ and random diffusion $g(X_t,t)$[12]]

$$dX_t = f(X_t,t)\,dt + g(X_t,t)\,dW_t, \tag{9.24}$$

where X_t is the Ito stochastic process (a solution to the SDE (9.24)). W_t is the Wiener process (or, Brownian motion), also known as the red noise, because its time derivative dW_t/dt represents the white noise.

As a simple demonstration case for the statistical path-integral simulation, we have implemented the following Ito-type SDE with the nonlinear drift: $f(X_t,t)$ and the vector Wiener process $g(X_t,t) = \sum_i g_i(X_t,t)\,dW_{i,t}$ that includes plane waves (*sine* and *cosine*) as well as soliton-type *sech* waves and shock-wave-type *tanh* waves (see Figure 9.3). Thus:

$$dX_t = f(X_t,t)\,dt + \sum_{i=1}^{4} g_i(X_t,t)\,dW_{i,t},$$

where

$$f(X_t,t) = -\frac{X_t}{\sqrt{X_t^2+1}}, \qquad g_1(X_t,t) = \sin(X_t), \qquad g_2(X_t,t) = \cos(X_t),$$
$$g_3(X_t,t) = \operatorname{sech}(0.5X_t), \qquad g_4(X_t,t) = \tanh(0.5X_t).$$

9.2 Self-Organization and Stochastic Delay Differential Equations

In this section, we review various forms of controlled complexity in neurodynamical feedback control systems defined by *delay differential equations* (DDEs) and *stochastic delay differential equations* (SDDEs).

To start with, as a motivation, we analyze several forms of complexity in dynamical and control systems defined by simple DDEs.

---

of the Fokker-Planck differential operator:

$$\mathcal{D}^*_{FP} := -[f(x,t)\,\partial_x + \frac{1}{2}g^2(x,t)\,\partial_{xx}].$$

[12] $g(X_t,t)$ describes the coupling of the drifting particle with the *heat bath*.

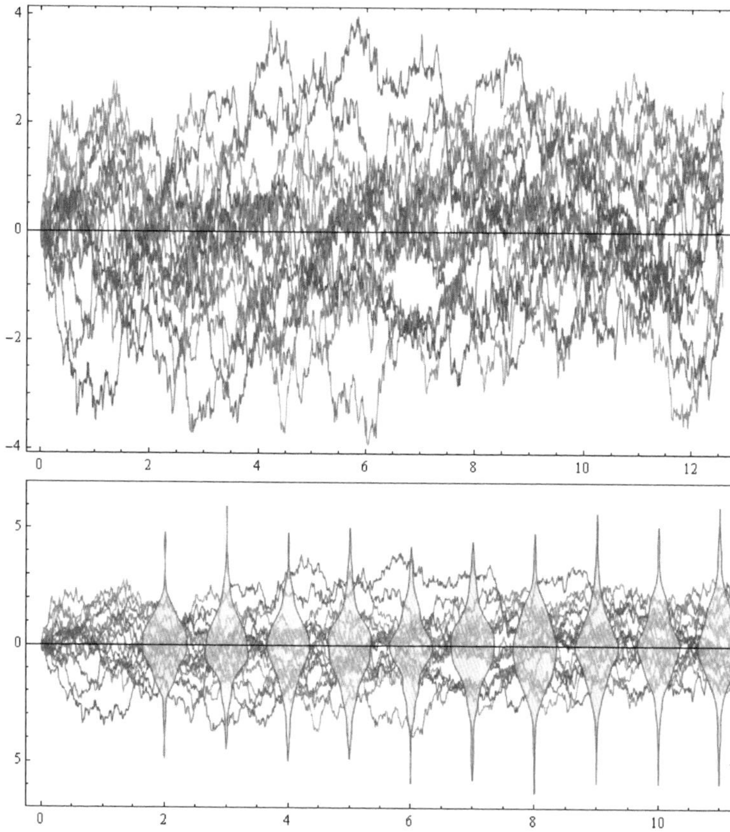

Fig. 9.3. *Illustrative simulation of the Ito stochastic process with a nonlinear drift and a vector Wiener process including both harmonic and nonlinear waves: 16 paths of this nonlinear random process are depicted (up) and also overlayed with a slice distribution of the process state at discrete time steps (down).*

Motivation 1: Inverted Pendulum Control System.

Dynamics of a damped, *inverted pendulum* with the mass $m > 0$, length $l > 0$ and damping $\gamma \geq 0$ (with gravity $g = 9.81\text{m/s}^2$ giving the only active external force) is governed by the 2nd-order ODE (see, e.g. [Jos]):

$$ml\ddot{\varphi} + \gamma l\dot{\varphi} - mg\sin\varphi = 0. \tag{9.25}$$

The act of balancing (i.e., control) of the pendulum defined by the equation of motion(9.25) means applying an external force $F = F(\varphi, \dot{\varphi}, \tau)$ on the right-hand side of (9.25), which depends on: (i) the deviation $\varphi = \varphi(t)$ of the pendulum from the vertical position $\varphi_0 = 0$, (ii) the angular velocity $\dot{\varphi} = \dot{\varphi}(t)$ of the pendulum, and (iii) the *delay* $\tau > 0$ (called the *reaction time*):

$$ml\ddot{\varphi} + \gamma l\dot{\varphi} - mg\sin\varphi = F[\varphi(t-\tau), \dot{\varphi}(t-\tau)], \tag{9.26}$$

thus defining a *delayed feedback control system.*

Formal definition of similar control problems is much easier (and more elegant) within the framework of LTI[13] systems [e.g., in case of (9.26), linearization is performed by the substitution: $\sin\varphi \to \varphi$]. In particular, in a SISO control system with the transfer function $H(s)$, a delay τ is incorporated by means of the exponential shift: $\mathrm{e}^{-s\tau}H(s)$, where s is the complex-valued Laplace-transform variable. More generaly, in a MIMO-control system, transport delays from given inputs to given outputs (i.e., I/O delays) are given by a transfer-function matrix. Here is an example of a 2×2 transfer function with four I/O delays $(\tau_1, \tau_2, \tau_3, \tau_4)$:

$$H(s) = \begin{bmatrix} \mathrm{e}^{-s\tau_1}H_1(s) & \mathrm{e}^{-s\tau_2}H_2(s) \\ \mathrm{e}^{-s\tau_3}H_3(s) & \mathrm{e}^{-s\tau_4}H_4(s) \end{bmatrix}, \qquad \text{with}$$

$$\begin{cases} H_1(s) = \frac{a}{s}, & H_2(s) = \frac{a}{s+b} \\ H_3(s) = \frac{s-1}{s+c}, & H_4(s) = \frac{s+1}{s+d} \end{cases}.$$

Motivation 2: Predator-Prey Models.

Given the sizes $x = x(t)$ and $y = y(t)$ of the prey (e.g., rabbits) and predator (e.g., foxes) populations at time t, with the corresponding growth rates $\dot{x} = \dot{x}(t)$ and $\dot{y} = \dot{y}(t)$ of the two populations over time, the standard Lotka-Volterra predator-prey model (see, e.g. [Mur03, LV13]) can be written as:

$$\dot{x} = r_1 x\,(1-x) - b_1 xy, \qquad \dot{y} = -r_2 y + b_2 xy, \tag{9.27}$$

where $r_1, r_2 > 0$ are the growth/death rates per capita, while $b_1, b_2 > 0$ are the predation coefficients.

A more realistic model [Cus77] includes a positive time delay: $\tau = t_1 - t_0$ (where t_0 is the time when the prey is eaten and $t_1 > t_0$ is the time when this food is converted into new biomass through births of predators) in the second equation of (9.27):

$$\begin{aligned} \dot{x} &= r_1 x\,(1-x) - b_1 xy, \\ \dot{y} &= -r_2 y(t) + b_2 x(t-\tau) y(t-\tau). \end{aligned} \tag{9.28}$$

An even more realistic model, using a continuous distribution of delays (including the fact that the gestation period is usually different for each individual), includes the convolution integral (with the probability-normalization condition: $\int_0^\infty G(\tau)d\tau = 1$) in the second equation [Cus77]:

$$\begin{aligned} \dot{x} &= r_1 x\,(1-x) - b_1 xy, \\ \dot{y} &= -r_2 y(t) + b_2 \int_0^\infty G(\tau) x(t-\tau) y(t-\tau) d\tau. \end{aligned} \tag{9.29}$$

In the particular case of the Dirac δ-function: $G(\tau) = \delta(t-\tau)$, the continuous delay-model (9.29) reduces to the discrete delay-model (9.28).

[13] LTI means 'linear time invariant'.

Motivation 3: Retarded Extension of the Logistic Map.

Recall that the classical route to chaos was first demonstrated in 1978 by M. Feigenbaum [Fei78], using the *logistic map:*[14]

$$x_{i+1} = rx_i(1 - x_i), \tag{9.30}$$

which shows successive bifurcations as one increases the parameter $r \geq 3$; the chaos occurs at $r = 3.5699$ (see Figure 9.4).

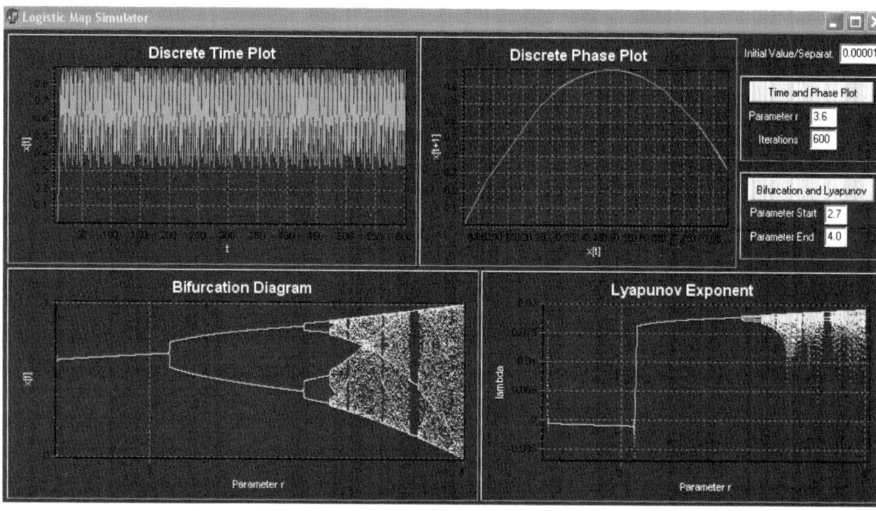

Fig. 9.4. *Output of the logistic-map simulator.*

One of interesting extensions of the basic logistic map (9.30) was proposed in [Ber01] by the τ-delayed continuous-time substitution: $x_i \to f(t - \tau)$ and addition of the inertia term $\sigma \dot{f}(t)$. In this way, the following retarded DDE was obtained:

$$\sigma \dot{f}(t) + f(t) = \alpha f(t - \tau)[1 - f(t - \tau)]. \tag{9.31}$$

DDE (9.31) reduces to the logistic map (9.30) for $\sigma = 0$ and $f(t - \tau) = x_i$. In general, such retarded DDEs can be quite complex and have been used to model both periodic and aperiodic dynamics of physiological systems [MG77]. Stability analysis of the system (9.31) was performed by [Mor03].

9.2.1 Mean-Field Neurodynamics

In this section, we review mean-field neurodynamics of an *ensemble* of FitzHugh-Nagumo neurons.

---

[14] Recall that logistic map was made very popular in 1970s by R. May [May76]; recently, its generalized version has been used in [SI10] for modeling human *situational awareness.*

From Brain Research to Perceptual Control Theory

For the last few decades, a controversial issue in the research of biological neural networks has been how neurons communicate information (by firings or spikes, see [RWS96, Cha98, Egg98, UR99, CZ00, PDZ00]). In other words, there has been a continuous debate on the nature of the *neural code*, mainly focusing on the following two issues (see [Has03a]):

1. Whether information in the brain is encoded in the *rate code* (i.e., the average firing rate of individual neurons), or in the precise firing times (*temporal code*). For example, firing activities of motor and sensory neurons in the brain have been reported to vary in response to applied stimuli, which was first noted in [Adr26], forms the basis of the rate-code hypothesis. In temporal-code hypothesis (see [SK93, KES96, SZ98]) detailed spike timings are assumed to play an important role in information transmission (information is encoded in interspike intervals or in relative timings between firing times of spikes).
2. Whether information in the brain is encoded in the activity of a single neuron (or a small number of neurons), or in the activity of a large neural ensembles. The *ensemble rate-code* hypothesis [AS98], assuming that information is coded in the relative firing rates of ensemble neurons, has been adopted in the most of the theoretical analysis. The alternative, *ensemble temporal-code* hypothesis (see [Hop95, HL98, Rt01]), assumes that relative timings between spikes in ensemble neurons may be used as an encoding mechanism for *perceptual processing* - which is closely related to *perceptual control theory* (PCT, see [Pow73a, Pow73b, Pow11, Pow05]).

Ensemble of FitzHugh-Nagumo Neurons

Dynamics of an N-dimensional ensemble of FitzHugh-Nagumo (FN) neurons is given by the following set of nonlinear ODEs (see, e.g. [RT96, TR98]):

$$\dot{x}_i = F(x) - cy_i + I_i^{(c)} + I^{(e)} + \xi_i, \qquad \dot{y}_i = bx_i - dy_i, \quad (\text{for } i = 1, ..., N), \tag{9.32}$$

where $x_i = x_i(t)$ and $y_i = y_i(t)$ denote the fast (voltage) variable and slow (recovery) variable, respectively; $F(x) = kx\,(x-a)\,(1-x)$ is the nonlinear spring-like forcing term (with positive parameters k, a, b, c, d); and $\xi_i = \xi_i(t)$ is the Gaussian white noise with

$$\langle \xi_i(t) \rangle = 0 \quad \text{and} \quad \langle \xi_i(t)\, \xi_j(t') \rangle = \beta_i^2 \delta_{ij}\, \delta(t-t'), \tag{9.33}$$

where $\langle \cdot \rangle$ means the average over random variables.

The inputs, $I_i^{(c)} = I_i^{(c)}(t)$ and $I^{(e)} = I^{(e)}(t)$, represent N coupling terms and an external (single-spike) input applied to all neurons, respectively. They are given by:

$$I_i^{(c)} = \frac{w}{N} \sum_{j(\neq i)} G(x_j) \quad \text{with} \quad G(x) = \left(1 + \mathrm{e}^{-(x-\theta)/\alpha}\right)^{-1}$$

where w is the coupling strength and $G(x)$ is the sigmoid function with the threshold θ and the width α, and

$$I^{(e)} = A\Theta(t - t_{in})\Theta(t_{in} + T_w - t), \tag{9.34}$$

where $\Theta = \Theta(x)$ is the Heaviside step (threshold) function with the input time t_{in}, width T_w and amplitude A.

Network of Stochastic FitzHugh-Nagumo Neurons

One of many possible generalizations of the neural ensemble (9.32) is represented by a network of noisy FitzHugh-Nagumo neurons, given by the following set of nonlinear stochastic ODEs (see [KB93]):

$$\begin{aligned} dx_i &= \left[\phi\left(x_i, y_i\right) + I_i(t) + \sum_{k=1}^{N} J_{ik}\Theta(x_k)\right] dt + \xi_i dW_i, \\ dy_i &= \psi\left(x_i, y_i\right) dt, \qquad (\text{for } \ i, k = 1, ..., N), \end{aligned} \tag{9.35}$$

where ϕ and ψ are nonlinear functions, while J_{ik} are synaptic weights for the connection from neuron i to neuron k. The Ito-type stochastic analysis of the system (9.35) is performed in [RT96].

Statistical Moment Method

In the *moment method* of [RT96, TR98, TP01], dynamics of the membrane potential of a neural ensemble stimulated by the white noise has been analyzed by replacing *stochastic* ODEs by *deterministic* ODEs described by moments of state variables, as follows (for a review see [Has03a]).

1. Global ensemble variables $[X = X(t),\ Y = Y(t)]$ and their averages $[\mu_1 = \mu_1(t),\ \mu_2 = \mu_2(t)]$ are defined as:

$$X = \frac{1}{N}\sum_{i=1}^{N} x_i, \qquad Y(t) = \frac{1}{N}\sum_{i=1}^{N} y_i, \qquad \mu_1 = \langle X \rangle, \qquad \mu_2 = \langle Y \rangle. \tag{9.36}$$

2. Using (9.36), the ensemble ODEs (9.32) can be rewritten as:

$$\begin{aligned} \dot{x}_i = {} & F(\mu_1) + F'(\mu_1)\delta x_i + \frac{1}{2}F^{(2)}(\mu_1)\delta x_i^2 \\ & + \frac{1}{6}F^{(3)}(\mu_1)\delta x_i^3 - c\mu_2 - c\delta y_i + I_i^{(c)} + I_i^{(e)} + \xi_i, \end{aligned}$$

$$\begin{aligned}
\dot{y}_i &= b\mu_1 - d\mu_2 + b\delta x_i - d\delta y_i + e, \qquad \text{with} \\
\delta x_i &= x_i - \mu_1, \qquad \delta y_i = y_i - \mu_2, \\
I_i^{(c)} &= w(1 - \frac{1}{N})G(\mu_1) + \frac{1}{N}\sum_{j(\neq i)}[G'(\mu_1)\delta x_j \\
&\quad + \frac{1}{2}G^{(2)}(\mu_1)\delta x_j^2 + \frac{1}{6}G^{(3)}(\mu_1)\delta x_j^3].
\end{aligned}$$

3. We define variances and covariances between local variables:

$$\gamma_{1,1} = \frac{1}{N}\sum_{i=1}^{N}\langle \delta x_i^2\rangle, \quad \gamma_{2,2} = \frac{1}{N}\sum_{i=1}^{N}\langle \delta y_i^2\rangle, \quad \gamma_{1,2} = \frac{1}{N}\sum_{i=1}^{N}\langle \delta x_i^2 \delta y_i^2\rangle,$$

and between global variables (using $\delta X = X - \mu_1$, $\delta Y = Y - \mu_2$):

$$\rho_{1,1} = \langle \delta X^2\rangle, \qquad \rho_{2,2} = \langle \delta Y^2\rangle, \qquad \rho_{1,2} = \langle \delta X\, \delta Y\rangle.$$

4. After some algebraic gymnastics (see [Has03a]), the following ODE-set can be derived:

$$\begin{aligned}
\dot{\mu}_1 &= f_0 + f_2\gamma_{1,1} - c\mu_2 + w\,(1 - \frac{1}{N})\,U_0 + I^{(e)}(t), \\
\dot{\mu}_2 &= b\mu_1 - d\mu_2 + e, \\
\dot{\gamma}_{1,1} &= 2[(f_1 + 3f_3\gamma_{1,1})\gamma_{1,1} - c\gamma_{1,2}] + 2w(\rho_{1,1} - \frac{\gamma_{1,1}}{N})\,U_1 + \beta^2, \\
\dot{\gamma}_{2,2} &= 2(b\gamma_{1,2} - d\gamma_{2,2}), \\
\dot{\gamma}_{1,2} &= b\gamma_{1,1} + (f_1 + 3f_3\gamma_{1,1} - d)\gamma_{1,2} - c\gamma_{2,2} + w\,(\rho_{1,2} - \frac{\gamma_{1,2}}{N})\,U_1, \\
\dot{\rho}_{1,1} &= 2[(f_1 + 3f_3\gamma_{1,1})\rho_{1,1} - c\rho_{1,2}] + 2w\,(1 - \frac{1}{N})\,\rho_{1,1}\,U_1 + \frac{\beta^2}{N}, \\
\dot{\rho}_{2,2} &= 2(b\rho_{1,2} - d\rho_{2,2}), \\
\dot{\rho}_{1,2} &= b\rho_{1,1} + (f_1 + 3f_3\gamma_{1,1} - d)\rho_{1,2} - c\rho_{2,2} + w\,(1 - \frac{1}{N})\,\rho_{1,2}\,U_1,
\end{aligned}$$

where $\beta^2 = \frac{1}{N}\sum_{i=1}^{N}\beta_i^2,$

$$\begin{aligned}
U_0 &= g_o + g_2\gamma_{1,1}, \qquad U_1 = g_1 + 3g_3\gamma_{1,1}, \\
f_k &= (1/k!)F^{(k)}(\mu_1), \qquad g_k = (1/k!)G^{(k)}(\mu_1).
\end{aligned}$$

Based on the *moment method*, a *dynamical mean-field approximation* (DMA) for an N-dimensional ensemble of FN-neurons (9.32) was proposed in [Has03a] and subsequently applied to Hodgkin-Huxley neurons in [Has03b].

9.2.2 Stochastic Neural DDEs

In real-life dynamical systems, we can often see the coexistence of both noises and time-delays. Their combined effect can be described by stochastic delay differential equations (SDDEs). For example, SDDEs have been used physiology [LMB90, CDK97] as well as in optics [GR96, Mas01] - for modeling of noise-driven systems with time-delayed feedbacks.

Langevin Ensembles

Dynamics of a Langevin ensemble with delayed couplings is given by the following set of SDDEs [Has04a]:

$$\dot{x}_i = F(x_i) + \frac{w}{N}\sum_{k=1}^{N} H\left[x_k(t-\tau)\right] + \xi_i + I^{(e)}, \qquad (\text{for } i = 1, ..., N), \quad (9.37)$$

where $F(x)$ and $H(x)$ are functions of x, $I^{(e)}$ is given by (9.34) and white noises ξ_i are given by (9.33).

In DMA [Has03a, Has03b], the global variable $X(t)$ and its mean $\mu(t)$ are given by (9.36), with correlation functions given by:

$$\gamma(t,t') = \frac{1}{N}\sum_{i=1}^{N}\langle\delta x_i(t)\,\delta x_i(t')\rangle, \qquad \rho(t,t') = \langle\delta X(t)\,\delta X(t')\rangle,$$

using $\delta x_i(t) = x_i(t) - \mu(t)$ and $\delta X(t) = X(t) - \mu(t)$.

The following three particular forms of the Langevin model (9.37) have been analyzed (see [Has04a]):

1. Linear Langevin model (with the positive parameter a), given by:

$$F(x) = -ax, \qquad H(x) = x,$$

giving the stochastic ODEs (for $m \geq 1$):

$$\begin{aligned}
\dot{\mu} &= -a\mu + w\mu(t-\tau) + I^{(e)}, \\
\dot{\gamma} &= -2a\gamma + 2w\rho(t,t-\tau) + \beta^2, \\
\dot{\rho} &= -2a\rho + 2w\rho(t,t-\tau) + \frac{\beta^2}{N}, \\
\dot{\rho}(t,t-m\tau) &= -2a\rho(t,t-m\tau) + w\rho\left[t,t-(m+1)\tau\right] \\
&\quad + w\rho(t-\tau,t-m\tau) + \left(\frac{\beta^2}{2}\right)\Delta(m\tau).
\end{aligned}$$

2. Cubic Langevin model (with $(a \geq 0,\ b > 0)$):

$$F(x) = -ax, \qquad H(x) = x - bx^3,$$

giving the stochastic ODEs (for $m \geq 1$):

$$\begin{aligned}
\dot{\mu} &= -a\mu(t) + wu_0(t-\tau) + I^{(e)}(t), \qquad (9.38)\\
\dot{\gamma} &= -2a\gamma(t,t) + 2wu_1(t-\tau)\rho(t,t-\tau) + \beta^2,\\
\dot{\rho} &= -2a\rho(t,t) + 2wu_1(t-\tau)\rho(t,t-\tau) + \frac{\beta^2}{N},\\
\dot{\rho}(t,t-m\tau) &= -2a\rho(t,t-m\tau) + wu_1(t-(m+1)\tau)\rho(t,t-(m+1)\tau)\\
&\quad + wu_1(t-\tau)\rho(t-\tau,t-m\tau) + \frac{\beta^2}{N}\Delta(m\tau), \qquad \text{using}
\end{aligned}$$

$$u_0(t) = \mu(t) - b\mu(t)^3 - 3b\mu(t)\gamma(t,t), \qquad u_1(t) = 1 - 3b\mu(t)^2 - 3b\gamma(t,t). \qquad (9.39)$$

3. Ikeda-Langevin model, using functions F and H previously used in [IDA80, IM87] for analyzing chaotic dynamics in time-delayed systems (for $a \geq 0$):

$$F(x) = -ax, \qquad H(x) = \sin(x),$$

as well as the relations:

$$H^{(2n)}(t) = (-1)^n \sin(x), \qquad H^{(2n+1)}(t) = (-1)^n \cos(x),$$

we get the ODEs (9.38), but with inputs given by:

$$u_0(t) = \sin[\mu(t)]\ \mathrm{e}^{-\gamma(t,t)/2}, \qquad u_1(t) = \cos[\mu(t)]\ \mathrm{e}^{-\gamma(t,t)/2},$$

instead of (9.39).

FitzHugh-Nagumo SDDEs

Dynamics of an N-unit ensemble of noisy FN neurons has been described in [Has04b] by the following set of SDDEs:

$$\begin{aligned}
\dot{x}_{1i} &= F[x_{1i}] - cx_{2i} + \frac{1}{N-1}\sum_{j(\neq i)} w_{ij} G\left[x_{1j}(t-\tau_{ij})\right] + \xi_i + I^{(e)},\\
\dot{x}_{2i} &= bx_{1i} - dx_{2i}, \qquad (9.40)
\end{aligned}$$

where $i = 1, ..., N$, $F(x) = kx[x-h][1-x]$, while the other terms have the same meaning as before.

After applying DMA [Has03a, Has03b], the system (9.40) reduces to the following set of DDEs [Has04b]:

$$\begin{aligned}
\dot{\mu}_1 &= f_0 + f_2\gamma_{1,1} - c\mu_2 + w\, u_0(t-\tau) + I^{(e)},\\
\dot{\mu}_2 &= b\mu_1 - d\mu_2 + e,\\
\dot{\gamma}_{1,1} &= 2[a\gamma_{1,1} - c\gamma_{1,2}] + 2wu_1(t-\tau)\,\zeta_{1,1}(t,t-\tau) + \beta^2,\\
\dot{\gamma}_{2,2} &= 2[b\gamma_{1,2} - d\gamma_{2,2}],\\
\dot{\gamma}_{1,2} &= b\gamma_{1,1} + [a-d]\gamma_{1,2} - c\gamma_{2,2} + wu_1(t-\tau)\,\zeta_{2,1}(t,t-\tau),
\end{aligned}$$

$$\dot{\rho}_{1,1} = 2[a\rho_{1,1} - c\rho_{1,2}] + 2wu_1(t-\tau)\rho_{1,1}(t,t-\tau) + \frac{\beta^2}{N},$$
$$\dot{\rho}_{2,2} = 2[b\rho_{1,2} - d\rho_{2,2}],$$
$$\dot{\rho}_{1,2} = b\rho_{1,1} + [a-d]\rho_{1,2} - c\rho_{2,2} + wu_1(t-\tau)\rho_{2,1}(t,t-\tau).$$

For the functions $a = a(t), u_0 = u_0(t), u_1 = u_1(t), f_k = f_k(t), g_k = g_k(t)$ and $\zeta_{\kappa,\nu}(t,t')$ and other technical details, see [Has04b].

9.3 Appendix

9.3.1 Adaptive Path-Integral Computation in Python/Cython

In this section, we demonstrate adaptive numerical Monte Carlo (MC) solution of quantum-mechanical (QM) path integrals as high-dimensional integrals with exponential integrands.

Class vegas.Integrator

A fast Python/Cython[15] class *vegas.Integrator* [Lep14] gives fast MC-estimates of arbitrary multidimensional integrals, using a recently enhanced version of Lepage's adaptive MC *Vegas algorithm*[16] [Lep78] (for a review, see [Ohl99]). The current vegas.Integrator implementation uses several adaptive strategies, the most important of which is the re-mapping of the integration variables in each direction, before it makes MC-estimates of the integral.[17]

Evaluation of multidimensional integrals

For example, consider a numerical solution of the following 4D integral:

$$I(x) = C\int_0^1 dx_0 \int_0^1 dx_1 \int_0^1 dx_2 \int_0^1 dx_3\; \mathrm{e}^{-100g(x)}, \qquad g(x) = \sum_d (x_d - 0.5)^2 \tag{9.41}$$

where the constant C is chosen so that the exact solution is: $I = 1$.

The integral (9.41) can be evaluated, using the vegas.Integrator class in several ways:

(i) The slow way is given by the following Python code):

[15] Cython is a compiled hybrid of Python and C, which approaches C and Fortran in speed.

[16] Classical Vegas algorithm, which was implemented in Fortran, C and C++, was a method for reducing error in MC-simulations by using a known (or approximate) PDF to concentrate the search in those areas of the integrand that make the greatest contribution to the total integral to be evaluated.

[17] This is equivalent to a standard MC-optimization technique called *importance sampling*.

```
import vegas
import numpy as np

C=1013.2118364296088 # const. chosen so that exact sol: I=1.
# define the integrand f(x) where x[d] is a point in
4D space.
def f(x):
    g=0.
    for d in range(4):
        g += (x[d]-0.5)**2       # exponent sum
    return np.exp(-100.*g)*C

# create a 4D integrator
integ = vegas.Integrator([[-1,1],[0,1],[0,1],[0,1]])

result = integ(f, nitn=10, neval=1e4)
print(result.summary())
print('result = %s    Q = %.2f' % (result, result.Q))
```

produces the following sample output:[18]

```
itn   integral        wgt average     chi2/dof        Q
-------------------------------------------------------
  1   1.03(13)        1.03(13)            0.00     1.00
  2   1.053(37)       1.052(36)           0.02     0.88
  3   1.003(17)       1.012(16)           0.77     0.46
  4   0.9970(89)      1.0007(78)          0.75     0.52
  5   1.0036(84)      1.0020(57)          0.58     0.68
  6   1.0048(75)      1.0031(45)          0.48     0.79
  7   1.0158(69)      1.0069(38)          0.80     0.57
  8   0.9943(59)      1.0032(32)          1.15     0.33
  9   0.9986(49)      1.0019(27)          1.08     0.37
 10   0.9970(44)      1.0005(23)          1.06     0.39

result = 1.0005(23)    Q = 0.39
```

The *result.summary*() returns a summary of results per iteration. Here, the first column is the iteration (ranges from 1 to $nitn = 10$), the second column is the integral value evaluated in the iteration, 'wgt average' is the weighted-average result per iteration (estimates are weighted by the inverse variance, which gives much less weight to the early iterations), which minimizes the following χ^2 measure:

$$\chi^2 \equiv \sum_i \frac{(I_i - \overline{I})^2}{\sigma_i^2},$$

[18] The result is slightly different in each simulation, due to random MC-fluctuations.

(where $I_i \pm \sigma_i$ are the estimates from individual iterations), while Q-value of the χ^2, which is the probability that a larger χ^2 could result from random fluctuations.

(ii) About an order-of-magnitude faster way, producing a similar quality result, is given by executing the following optimized Python (NumPy) code (which uses the vegas.batchintegrand() decorator[19] to a batch function for the integrand):

```
import vegas
import numpy as np

dim=4;  norm=1013.2118364296088**(dim/4.)

@vegas.batchintegrand
def f(x):
    g=0.# simultaneous integration at multiple points
    for d in range(dim):
        g += (x[:,d]-0.5)**2
    return np.exp(-100.*g)*norm

integ = vegas.Integrator(dim*[[0,1]],nhcube_batch=1000)

result = integ(f, nitn=10, neval=1e4)
print(result.summary())
print('result = %s    Q = %.2f' % (result,result.Q))
```

(iii) Even (a few times) faster way, producing a similar quality result, is given by executing the following Python/Cython code, which consists of two modules: the Cython integrand:

```
# file: cython_integrand.pyx

cimport vegas                      # for BatchIntegrand
from libc.math cimport exp         # use exp() from C library

import vegas
import numpy

cdef class f_cython(vegas.BatchIntegrand):
    cdef double norm
    cdef readonly int dim

    def __init__(self, dim):
        self.dim = dim
```

[19] This batch integrand is pretty fast because it is expressed in terms of NumPy-operators acting on entire arrays.

```
        self.norm = 1013.2118364296088**(dim/4.)

    def __call__(self, double[:, ::1] x):
        cdef int i, d
        cdef double g
        cdef double[::1] f = numpy.empty(x.shape[0], float)
        for i in range(f.shape[0]):
            g = 0.
            for d in range(self.dim):
                g += (x[i,d]-0.5)**2
            f[i] = exp(-100.*g)*self.norm
        return f
```

and the main Python code (that calls the Cython integrand):

```
import pyximport; pyximport.install()

import vegas
from cython_integrand import f_cython

f = f_cython(dim=4)
integ = vegas.Integrator(f.dim*[[0,1]],nhcube_batch=1000)

result = integ(f, nitn=10, neval=1e4)
print(result.summary())
print('result = %s    Q = %.2f' % (result,result.Q))
```

Evaluation of oscillatory QM path-integrals

Now, we are ready for adaptive MC-approximation of QM path-integrals (with the Hamiltonian operator $\hat{H}$):

$$\Psi(x,t) = \langle x_{\text{out}}|\exp\left[-\hat{H}\left(t_{\text{out}} - t_{\text{in}}\right)|x_{\text{in}}\right] = \int \mathcal{D}x(t)\exp\left[-S\left(x\right)\right]. \quad (9.42)$$

with the classical action $S(x)$, given as a time integral of the Lagrangian $L = L(x, \dot{x})$ [which is a kinetic energy $m\,\dot{x}^2/2$ (for a mechanical system with mass m) minus potential energy $V(x)$]:

$$S\left(x\right) \equiv \int_{t_{\text{in}}}^{t_{\text{out}}} L dt = \int_{t_{\text{in}}}^{t_{\text{out}}} \left[\frac{m\,\dot{x}^2}{2} + V(x)\right] dt, \quad (9.43)$$

evaluated for each possible path $x = x(t)$, in such a way that the sum of probabilities of all paths is normalized to 1.

A generic Cython integrand for 1D path-integrals

The numeric approximation of oscillatory 1D path-integrals of the form of (9.42), has been implemented in the generic Cython integrand, based on the following approximation:

$$\int_{t_j}^{t_{j+1}} Ldt \approx a\left[\frac{m}{2}\left(\frac{x_{j+1}-x_j}{a}\right)^2+\frac{1}{2}\left(V(x_{j+1})+V(x_j)\right)\right]$$

(where $a = (t_{\mathrm{out}} - t_{\mathrm{in}})/N$), from which the discretized action follows (see [Lep98]):

$$S_{\mathrm{dis}}(x) \equiv \sum_{j=0}^{N-1}\left[\frac{m}{2a}(x_{j+1}-x_j)^2 + aV(x_j)\right], \qquad (\text{where } x_0 = x_N = x).$$

Here is the Cython implementation:

```
# cython: profile=True
# Modified from 'vegas-2.1.4/path_integrand.pyx' by
G.P. Lepage,
# Ref: G.P. Lepage, Lattice QCD for Novices, Proc.HUGS98
(J.Goity ed),
# World Sci.(2000)
""" Cython code for the integrand used in various Python path
integrals.
Contains two classes: (i) class PathIntegral is a base class
for classes that do path integrals for 1-d systems; (ii)
class
Oscillator is derived from it and specifies a specific
potential.  """
cimport vegas                              # import Cython
description of vegas
from libc.math cimport exp, tan, pow
# import funs from C
import numpy as np
import vegas


cdef class PathIntegral(vegas.BatchIntegrand):
    """Computes the amplitude: < x0 | exp(-H*T) | x0 >
                    # for some Hamiltonian H
    Parameters:
        T ...... (Euclidean) time
        x0 ..... starting and ending position
        (if None=>integrate over x0)
        ndT .... total number of time steps
        m ...... mass
```

```
    xscale . typical length scale for ground state
    wavefunction
    neval .. number of evaluations per iteration (vegas)
    nitn ... number of iterations (vegas)
"""
cdef readonly int ndT, neval, nitn
cdef readonly double T, m, xscale, norm
cdef readonly double[::1] x
cdef readonly object integ

def __init__(self,T,ndT=10,m=1,xscale=1.,neval=25000,
nitn=10):
    self.ndT = ndT;  self.T = T;  self.neval = neval
    self.nitn = nitn;  self.m = m;  self.xscale = xscale
    self.x = np.empty(self.ndT+1, float)
    self.norm = (self.m*self.ndT/2./np.pi/T)**
    (self.ndT/2.)

cdef double V(self, double x):
    """ Derived classes needs to fill this in. """
    raise NotImplementedError('need to define V')

def __call__(self, theta):
    """ integrand for the path integral """
    cdef int i, j
    cdef double S, jac
    cdef double a = self.T/self.ndT
    cdef double m_2a = self.m/2./a
    cdef double[::1] f = np.empty(theta.shape[0],float)
    for i in range(len(f)):
        jac = self.norm  # compute Jacobian; map back to
        range
            -oo to +oo;
        for j in range(theta.shape[1]):
            self.x[j+1] = self.xscale*tan(theta[i,j])
            jac *= (self.xscale + self.x[j+1]**2/
            self.xscale)
        self.x[0] = self.x[-1]  # enforce periodic
        bound. cond.
        # compute the action according to eq.(17) from
        (Lepage,98)
        S = 0.
        for j in range(self.ndT):
            S += m_2a*(self.x[j+1]-self.x[j])**2 +
            a*self.V(self.x[j])
```

```
            f[i] = jac*exp(-S)
        return f

    def correlator(self, x0=None):
        """Compute the amplitude: <x0|exp(-H*T)|x0> for array
        of x0 values.
        If x0 is None, integrate over x0. """
        if x0 is None or len(x0) == 0:
            # integrate over endpoints -> exp(-E0*T)
            self.integ = vegas.Integrator(self.ndT*
            [[-np.pi/2,np.pi/2]])
            # train integrator
            self.integ(self, neval=self.neval,
            nitn=self.nitn)
            # final integral
            return self.integ(self,neval=self.neval,
                nitn=self.nitn,alpha=0.1)
        else:
            ans = [] # set endpoints equal to x0[i]
            -> psi(x0[i])^2 *
                exp(-E0*T)
            self.integ = vegas.Integrator((self.ndT-1)*
            [[-np.pi/2,np.pi/2]])
            # train integrator
            self.x[0] = x0[0];  self.x[-1] = x0[0]
            self.integ(self, neval=self.neval, nitn=self.
            nitn)
            for x0i in x0:  # do final integrals
                self.x[0] = x0i;  self.x[-1] = x0i
                ans.append(self.integ(self,neval=self.neval,
                    nitn=self.nitn,alpha=0.1))
            return np.array(ans)

cdef class Oscillator(PathIntegral):
    """ V(x) = x^2/2 + c*x^4 + c1*x^6
    Exact solution: E0=0.5 for c=0; E0=0.602405 for c=0.2 """
    cdef double c, c1
    def __init__(self, c=0.0, c1=0.0, c2=0.0, *args,
    **kargs):
        super(Oscillator, self).__init__(*args, **kargs)
        self.c = c;  self.c1 = c1
    cdef double V(self, double x):
        return x*x/2. + self.c*pow(x,4) + self.c1*pow(x,6)
```

Harmonic oscillator

We start with the path-integral (PI) code for the (exactly solvable) harmonic oscillator (HO) with the quadratic potential energy: $V(x) = x^2/2$ (see Figure 9.5):

```
# Path Integral for the Harmonic Oscillator: V = x^2/2
# Modified from 'vegas-2.1.4/path-integral.py' by G.P. Lepage
import pyximport; pyximport.install()   # compiles
PIO_integrand.pyx
import vegas
import numpy as np
import matplotlib.pyplot as plt
from PIO_integrand import Oscillator

def main():
    np.random.seed((1,2))   # seed random nums for
    repeatability
    T=4.;  ndT = 8;  neval = 30000;  nitn = 15   # initialize
    PI
    pathint = Oscillator(T=T, ndT=ndT, neval=neval,
    nitn=nitn)
    exp_E0T = pathint.correlator()  # compute ground-state
    energy
    E0 = -np.log(exp_E0T)/T
    x0 = np.linspace(0.,2.,6)        # compute psi^2 at
    points x0
    psi2 = pathint.correlator(x0=x0)/exp_E0T.mean
    exact_hosc = np.empty(psi2.shape,float)
    plot_results(E0,x0,psi2,T)

def plot_results(E0,x0,corr,T):
    def make_plot(x0=x0,E0=E0,corr=corr,T=T):
        corr_mean = np.array([z.mean for z in corr])
        corr_sdev = np.array([z.sdev for z in corr])
        plt.errorbar(x=x0,y=corr_mean,yerr=corr_sdev,
            fmt='bo',label='path integral')
        x = np.linspace(0,2.,100)
        y = np.exp(-x**2)/np.sqrt(np.pi)
        plt.plot(x,y,'r:',label='exact')
        plt.legend(('numeric','exact'), frameon=False)
        plt.xlabel('$x$')
        plt.ylabel('$|\psi(x)|^2$')
        plt.text(1.4,0.475,'$E_0 =$ %s'%E0)
        plt.title("Harmonic Oscillator: Psi^2")
        plt.draw()
```

```
    make_plot()
    plt.show()

if __name__ == '__main__':
    main()
```

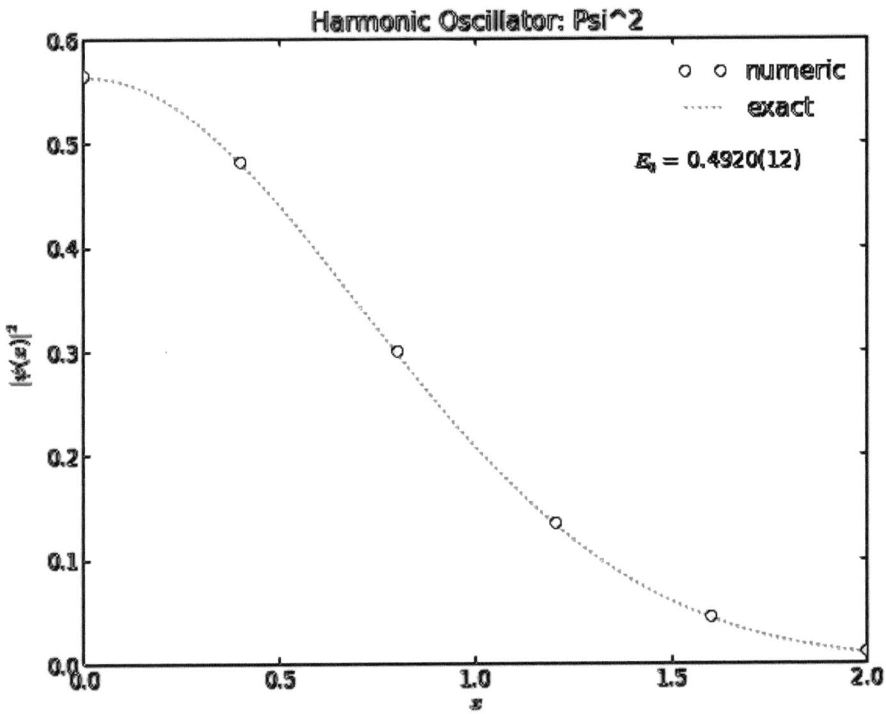

Fig. 9.5. *Path integral: harmonic oscillator (HO, dots), compared to the exact solution* E_0 *(line).*

Quartic anharmonic oscillator

Next, here is the PI-code for the weak quantum anharmonic oscillator with the quartic potential energy: $V(x) = x^2/2 + x^4/4$ (see Figure 9.6):

```
# Path Integral for the Anharmonic Oscillator:
V = x^2/2 + x^4/4
# Modified from 'vegas-2.1.4/path-integral.py' by G.P. Lepage
import pyximport; pyximport.install()   # compiles
PIO_integrand.pyx
import vegas
```

```
import numpy as np
import matplotlib.pyplot as plt
from PIO_integrand import Oscillator

def main():
    np.random.seed((1,2))   # seed random nums for
    repeatability
    T=4.;  ndT = 8;  neval = 30000;  nitn = 15   # initialize
    PI
    pathint = Oscillator(c=0.25,T=T,ndT=ndT,neval=neval,
    nitn=nitn)
    exp_E0T = pathint.correlator()  # compute ground-state
    energy
    E0 = -np.log(exp_E0T)/T
    x0 = np.linspace(0.,2.,6)       # compute psi^2 at
    points x0
    psi2 = pathint.correlator(x0=x0)/exp_E0T.mean
    plot_results(E0,x0,psi2,T)

def plot_results(E0,x0,corr,T):
    def make_plot(x0=x0,E0=E0,corr=corr,T=T):
        corr_mean = np.array([z.mean for z in corr])
        corr_sdev = np.array([z.sdev for z in corr])
        plt.errorbar(x=x0,y=corr_mean,yerr=corr_sdev,
            fmt='bo',label='path integral')
        x = np.linspace(0,2.,100)
        y = np.exp(-x**2)/np.sqrt(np.pi)
        plt.plot(x,y,'r:')
        plt.xlabel('$x$')
        plt.ylabel('$|\psi(x)|^2$')
        plt.text(1.4,0.475,'$E_0 =$ %s'%E0)
        plt.title("Anharmonic Oscillator: Psi^2")
        plt.draw()
    make_plot()
    plt.show()

if __name__ == '__main__':
    main()
```

Sixtic anharmonic oscillator

Finally, here is the PI-code for the strong quantum anharmonic oscillator with the sixtic potential energy: $V(x) = x^2/2 + x^4/4 + x^6/6$ (see Figure 9.7):

```
# Path Integral for the Anharmonic Oscillator 2:
```

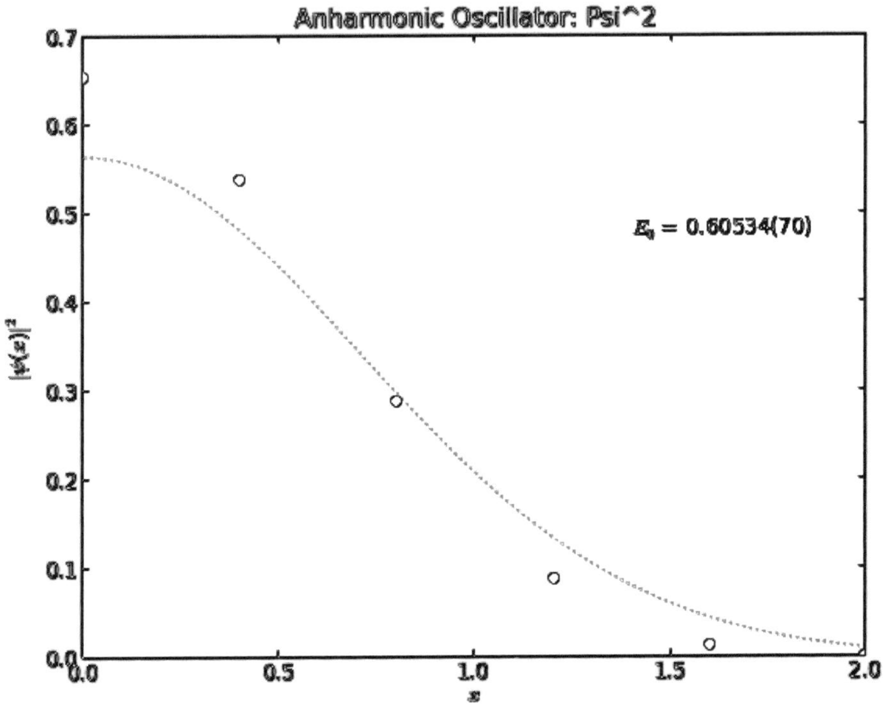

Fig. 9.6. *Path integral: quartic anharmonic oscillator (dots), compared to the HO (line). The exact solution E_0 is also given.*

```
    V = x^2/2 + x^4/4 + x^6/6
# Modified from 'vegas-2.1.4/path-integral.py' by G.P. Lepage
import pyximport; pyximport.install()   # compiles
PIO_integrand.pyx
import vegas
import numpy as np
import matplotlib.pyplot as plt
from PIO_integrand import Oscillator

def main():
    np.random.seed((1,2))   # seed random nums for
    repeatability
    T=4.;  ndT = 8;  neval = 30000;  nitn = 15   # initialize
    PI
    pathint = Oscillator(c=0.25,c1=0.166666667,T=T,ndT=ndT,
        neval=neval,nitn=nitn)
    exp_E0T = pathint.correlator()  # compute ground-state
    energy
```

```
    E0 = -np.log(exp_E0T)/T
    x0 = np.linspace(0.,2.,6)        # compute psi^2 at
    points x0
    psi2 = pathint.correlator(x0=x0)/exp_E0T.mean
    plot_results(E0,x0,psi2,T)

def plot_results(E0,x0,corr,T):
    def make_plot(x0=x0,E0=E0,corr=corr,T=T):
        corr_mean = np.array([z.mean for z in corr])
        corr_sdev = np.array([z.sdev for z in corr])
        plt.errorbar(x=x0,y=corr_mean,yerr=corr_sdev,
            fmt='bo',label='path integral')
        x = np.linspace(0,2.,100)
        y = np.exp(-x**2)/np.sqrt(np.pi)
        plt.plot(x,y,'r:')
        plt.xlabel('$x$')
        plt.ylabel('$|\psi(x)|^2$')
        plt.title("Anharmonic Oscillator 2: Psi^2")
        plt.draw()
    make_plot()
    plt.show()

if __name__ == '__main__':
    main()
```

9.3.2 Main Continuous Probability Distributions

Some Important Special Functions

The following special functions are frequently used in probability distributions (see [AS72]):

- The *Euler gamma function*:

$$\Gamma(\alpha) = \int_0^\infty t^{\alpha-1} \mathrm{e}^{-t} dt.$$

The *incomplete gamma function*:

$$\Gamma(\alpha, \beta) = \int_\beta^\infty t^{\alpha-1} \mathrm{e}^{-t} dt.$$

- The *Euler beta function*:

$$B(\alpha, \beta) = \frac{\Gamma(\alpha)\Gamma(\beta)}{\Gamma(\alpha+\beta)} = \int_0^1 t^{\alpha-1}(1-t)^{\beta-1} dt.$$

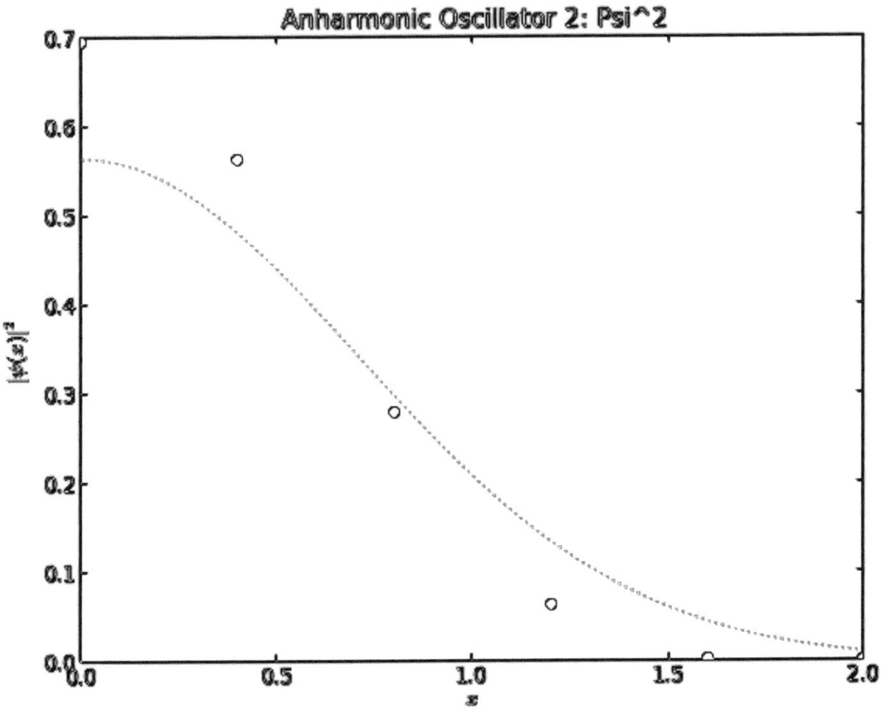

Fig. 9.7. *Path integral: sixtic anharmonic oscillator (dots), compared to the HO (line).*

The *incomplete beta function*:

$$B_t(\alpha, \beta) = \int_0^t t^{\alpha-1}(1-t)^{\beta-1} dt.$$

- The *error function*:

$$\mathrm{erf}(\alpha) = \frac{2}{\sqrt{\pi}} \int_0^\alpha \exp(-t^2)\, dt.$$

Uniform distribution

Uniform distribution **dunif(min, max)**, giving values between min and max, is given by the following PDF:

$$f(t, \min, \max) = \begin{cases} \frac{1}{\max - \min} & \min \le t \le \max \\ 0 & \text{otherwise} \end{cases},$$
$$(\text{with } -\infty \le \min \le t \le \max \le \infty).$$

The mean, variance and CDF of the log-normal distribution are given by:

$$
\begin{aligned}
Mean &= \frac{\max + \min}{2},\\
Variance &= \frac{1}{12}(\max - \min)^2,\\
CDF &= \begin{cases} \frac{t-\min}{\max - \min} & \min \le t \le \max \\ 1 & t > \max \end{cases}.
\end{aligned}
$$

Normal distribution

Normal (Gaussian) distribution **dnorm**(μ, σ), with mean μ and standard deviation σ, is given by the following PDF:

$$
f(t, \mu, \sigma) = \frac{\exp\left[-\frac{(t-\mu)^2}{2\sigma^2}\right]}{\sqrt{2\pi}\sigma}, \qquad [\text{with } \mu, \sigma > 0,\ t \in (-\infty, \infty)].
$$

Its CDF is given by:

$$
CDF = \frac{1}{2}\left(\operatorname{erf}\left(\frac{t-\mu}{\sqrt{2}\sigma}\right) + 1\right),
$$

where $\operatorname{erf}\left(\frac{t-\mu}{\sqrt{2}\sigma}\right)$ is the error function.

Log-Normal distribution

Log-normal distribution **dlnorm**(μ, σ), where (μ, σ) are the mean and the standard deviation of the corresponding normal distribution **dnorm**(μ, σ), is given by the following PDF:

$$
f(t, \mu, \sigma) = \frac{\exp\left[-\frac{(\log(t)-\mu)^2}{2\sigma^2}\right]}{\sqrt{2\pi}\sigma t}, \qquad [\text{with } \mu, \sigma > 0,\ t \in (-\infty, \infty)].
$$

The mean, variance and CDF of the log-normal distribution are given by:

$$
\begin{aligned}
Mean &= \exp\left(\mu + \frac{\sigma^2}{2}\right),\\
Variance &= \left[\exp\left(\sigma^2\right) - 1\right]\exp\left(2\mu + \sigma^2\right),\\
CDF &= \begin{cases} \frac{1}{2}\left[\operatorname{erf}\left(\frac{\log(t)-\mu}{\sqrt{2}\sigma}\right) + 1\right], & t > 0 \\ 0, & t \le 0 \end{cases},
\end{aligned}
$$

where $\operatorname{erf}\left(\frac{\log(t)-\mu}{\sqrt{2}\sigma}\right)$ is the error function.

Beta distribution

Log-normal distribution **dbeta($\boldsymbol{\alpha}, \boldsymbol{\beta}$)**, with shape parameters α and β, is given by the following PDF:

$$f(t, \alpha, \beta) = \frac{t^{\alpha-1}(1-t)^{\beta-1}}{B(\alpha, \beta)}, \qquad [\text{with } \alpha, \beta > 0,\ t \in (0,1)],$$

where $B(\alpha, \beta)$ is the Euler beta function.

The mean, variance and CDF of the log-normal distribution are given by:

$$Mean = \frac{\alpha}{\alpha+\beta}$$
$$Variance = \frac{\alpha\beta}{(\alpha+\beta)^2(\alpha+\beta+1)},$$
$$CDF = I_t(\alpha, \beta) \equiv \frac{B_t(\alpha, \beta)}{B(\alpha, \beta)},$$

where $I_t(\alpha, \beta)$ is the regularized incomplete beta function.

Chi-squared (χ^2) distribution

Chi-squared distribution **dchisqr(ν)**, with ν degrees of freedom, is given by the following PDF:

$$f(t, \nu) = \frac{2^{(-\nu/2)} t^{\left(\frac{\nu}{2}-1\right)} \exp(-\frac{t}{2})}{\Gamma\left(\frac{\nu}{2}\right)}, \qquad [\text{with } t, \nu > 0].$$

The mean, variance and CDF of the χ^2 distribution are given by:

$$Mean = \nu$$
$$Variance = 2\nu,$$
$$CDF = Q\left(\frac{\nu}{2}, 0, \frac{t}{2}\right) \equiv \frac{\Gamma\left(\frac{\nu}{2}, 0, \frac{t}{2}\right)}{\Gamma\left(\frac{\nu}{2}\right)},$$

where $Q\left(\frac{\nu}{2}, 0, \frac{t}{2}\right)$ is the generalized regularized incomplete gamma function.

Student-t distribution

Student distribution **dt($\boldsymbol{\mu}, \boldsymbol{\lambda}, \boldsymbol{\nu}$)**, with location parameter μ,scale parameter λ, and degrees of freedom ν, is given by the following PDF:

$$f(t, \mu, \lambda, \nu) = \frac{\left[\nu/(\frac{(t-\mu)^2}{\lambda^2} + \nu)\right]^{\frac{\nu+1}{2}}}{\lambda\sqrt{\nu}B\left(\frac{\nu}{2}, \frac{1}{2}\right)}, \qquad [\text{with } \lambda, \nu > 0;\ t, \mu \in (-\infty, \infty)].$$

The mean, variance and CDF of the χ^2 distribution are given by:

$$Mean = \begin{cases} \mu & \nu > 1 \\ \text{Indeterminate} & \nu \leq 1 \end{cases}$$

$$Variance = \begin{cases} \frac{\lambda^2 \nu}{\nu - 2} & \nu > 2 \\ \text{Indeterminate} & \nu \leq 2 \end{cases},$$

$$CDF = \frac{1}{2}\left[\text{sgn}(t-\mu)\, I_\alpha\left(\frac{\nu}{2}, \frac{1}{2}\right) + 1\right],$$

$$\text{with} \quad \alpha = \left(\frac{\nu}{\frac{(t-\mu)^2}{\lambda^2} + \nu}, 1\right), \quad I_\alpha\left(\frac{\nu}{2}, \frac{1}{2}\right) \equiv \frac{B_\alpha(\frac{\nu}{2}, \frac{1}{2})}{B(\frac{\nu}{2}, \frac{1}{2}\beta)},$$

where $I_\alpha\left(\frac{\nu}{2}, \frac{1}{2}\right)$ is the regularized incomplete beta function.

Weibull distribution

Weibull distribution **dweib**(α, λ) is given by the following PDF:

$$f(t, \alpha, \lambda) = \alpha t^{\alpha - 1} \lambda^{-\alpha} \exp\left[-\left(t/\lambda\right)^\alpha\right], \qquad (\text{with } t, \alpha, \lambda > 0),$$

where α is a shape parameter and λ is a scale parameter.

The mean, variance and CDF of the Weibull distribution are given by:

$$Mean = \lambda \Gamma\left(1 + \frac{1}{\alpha}\right),$$

$$Variance = \lambda^2 \left(\Gamma\left(1 + \frac{2}{\alpha}\right) - \Gamma\left(1 + \frac{1}{\alpha}\right)^2\right),$$

$$CDF = 1 - \exp\left[-\left(t/\lambda\right)^\alpha\right],$$

where $\Gamma(z)$ is the Euler gamma function.

10

Crash Simulator: Brain-and-Spine Injury Mechanics

Recently, the first author has proposed a new coupled loading-rate hypothesis as a unique cause of both brain and spinal injuries, which states that they are both caused by a Euclidean jolt, an impulsive loading that strikes head and spine (or, any other part of the human body) - in several coupled degrees-of-freedom simultaneously. Injury never happens in a single direction only, nor is it ever caused by a static force. It is always an impulsive translational plus rotational force. The Euclidean jolt causes two basic forms of brain, spine and other musculo-skeletal injuries: (i) localized translational dislocations; and (ii) localized rotational disclinations. In the present Chapter, we first review this unique mechanics of a general human mechanical injury, and then describe how it can be predicted and controlled by a crash simulator toolbox. This rigorous *Matlab* toolbox has been developed using an existing third-party toolbox *DiffMan*, for accurately solving differential equations on smooth manifolds and mechanical Lie groups. The present crash simulator toolbox performs prediction/control of brain and spinal injuries within the framework of the Euclidean group SE(3) of rigid motions in our natural 3-dimensional space.[1]

10.1 Introduction

Prediction and prevention of traumatic brain injury and spinal injury, as well as general musculo-skeletal injury, is a very important aspect of preventive medical science. In a series of papers [Iva09a, Iva09b, Iva09c, Iva10a], the first author of the present article proposed a new coupled loading-rate hypothesis as a unique cause of all above injuries. This new hypothesis states that the main cause of all mechanical injuries is a Euclidean Jolt, which is

[1] The work in this chapter has been developed in collaboration with Dr. Shady Mohamed, Centre for Intelligent Systems Research, Deakin University, Australia; e-mail:Shady.Mohamed@deakin.edu.au

an impulsive loading that strikes any part of the human body (head, spine or any bone/joint) - in several coupled degrees-of-freedom simultaneously. It never goes in a single direction only. Also, it is never a static force. It is always an impulsive translational and/or rotational force coupled to some mass eccentricity.

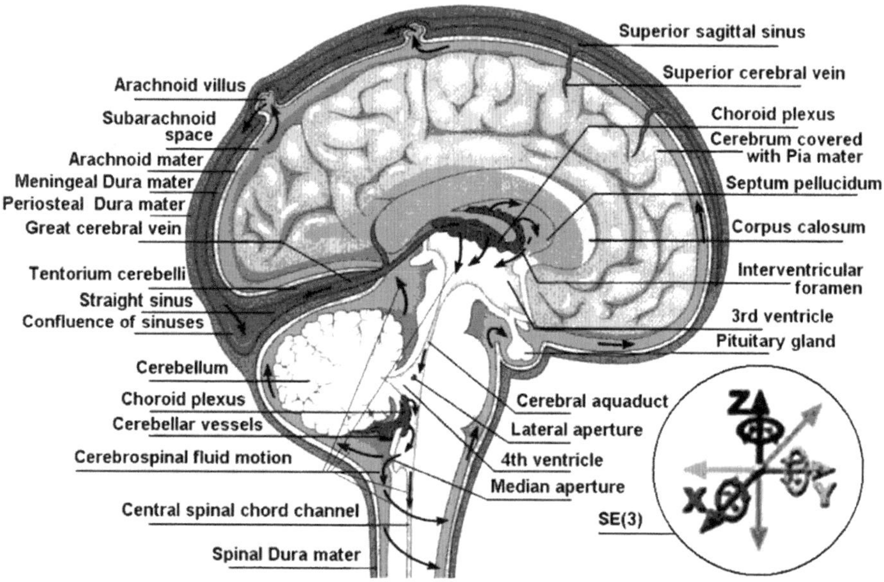

Fig. 10.1. *Human brain and its $SE(3)$-group of microscopic three-dimensional motions within the cerebrospinal fluid inside the cranial cavity.*

To show this, based on the previously defined *covariant force law* [II06a, II05, II06b], we have firstly formulated the fully coupled Newton-Euler dynamics of:

1. Brain's micro-motions within the cerebrospinal fluid inside the cranial cavity;
2. Any local inter-vertebral motions along the spine; and
3. Any major joint motions in the human musculo-skeletal system.

Then, from it, we have defined the essential concept of *Euclidean Jolt*, which is the main cause of all mechanical human injuries. The Euclidean Jolt has two main components:

- Sudden motion, caused either by an accidental impact or slightly distorted human movement; and

- Unnatural mass distribution of the human body (possibly with some added external masses), which causes some mass eccentricity from the natural physiological body state.

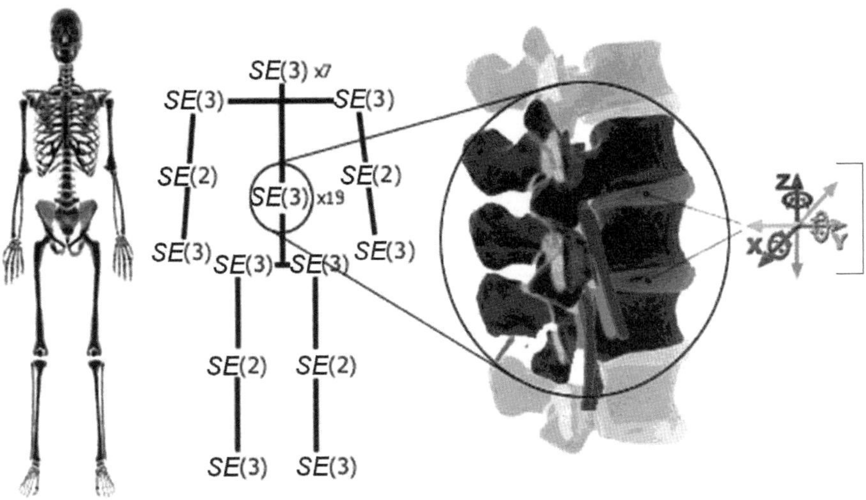

Fig. 10.2. *Human body representation in terms of $SE(3)/SE(2)$-groups of rigid-body motion, with the vertebral column represented as a chain of 26 flexibly-coupled $SE(3)$-groups.*

This can be intuitively (in "plain English") explained in the following way. As we live in a (Euclidean) 3D space, one could think that motion of any part of the human body, caused either by a voluntary human movement, or by an accidental impact, "simply obeys classical mechanics in 6 degrees-of-freedom: three translations and three rotations". However, these 6 degrees-of-freedom are not independent motions as it is suggested by the standard term "degrees-of-freedom". In reality, these six motions of any body in space are coupled. Firstly, three rotations are coupled in the so-called rotation group (or matrix, or quaternion). Secondly, three translations are coupled with the rotation group to give the full Euclidean group of rigid body motions in space. A simple way to see this is to observe someone throwing an object in the air or hitting a tennis ball: how far and where it will fly depends not only on the standard "projectile" mechanics, but also on its local "spin" around all three axes simultaneously. Every golf and tennis player knows this simple fact. Once the spin is properly defined we have a "fully coupled Newton-Euler dynamics" - to start with.

The covariant force law for any biodynamical system goes one step beyond the Newton-Euler dynamics. It states:

Euclidean Force covector field
$=$ **Body mass distribution** $\times$ **Euclidean Acceleration vector field**

This is a nontrivial biomechanical generalization of the fundamental Newton's definition of the force acting on a single particle. Unlike classical engineering mechanics of multi-body systems, this fundamental law of biomechanics proposes that forces acting on a multi-body system and causing its motions are fundamentally different physical quantities from the resulting accelerations. In simple words, forces are massive quantities while accelerations are massless quantities. More precisely, the acceleration vector-field includes all linear and angular accelerations of individual body segments. When we couple them all with the total body's mass-distribution matrix of all body segments (including all masses and inertia moments), we get the force covector-field, comprising all the forces and torques acting on the individual body segments. In this way, we have defined the 6-dimensional Euclidean force for an arbitrary biomechanical system.

Now, for prediction of injuries, we need to take the rate-of-change (or derivative, with respect to time) of the Euclidean biomechanical force defined above. In this way, we get the Euclidean Jolt, which is the sudden change (in time) of the 6-dimensional Euclidean force:

Euclidean Jolt covector field
$=$ **Body mass distribution** $\times$ **Euclidean Jerk vector field**

And again, it consists of two components: (i) massless linear and angular jerks (of all included body segments), and (ii) their mass distribution. For the sake of simplicity, we can say that the mass distribution matrix includes all involved segmental masses and inertia moments, as well as "eccentricities" or "pathological leverages" from the normal physiological state.

Therefore, the unique cause of all brain, spine and musculo-skeletal injuries has two components:

- Coupled linear and angular jerks; and
- Mass distribution with "eccentricities".

In other words, *there are no injuries in static conditions without any mass eccentricities; all injuries are caused by mutually coupled linear and angular jerks, which are also coupled with the involved mass distribution.*

The Euclidean Jolt causes two forms of discontinuous brain, spine or musculo-skeletal injury:

1. Mild rotational disclinations; and
2. Severe translational dislocations and/or bone fractures.

In the cited papers above, we have developed the soft-body dynamics of biomechanical disclinations and dislocations, caused by the Euclidean Jolt, using the Cosserat multipolar viscoelastic continuum model.

Implications of the new universal theory are various, as follows.

A. The research in traumatic brain injury (TBI, see Figure 10.1) has so far identified the rotation of the brain-stem as the main cause of the TBI due to various crashes/impacts. The contribution of my universal Jolt theory to the TBI research is the following:

1. Rigorously defined this brain rotation as a mechanical disclination of the brain-stem tissue modeled by the Cosserat multipolar soft-body model;
2. Showing that brain rotation is never uni-axial but always three-axial;
3. Showing that brain rotation is always coupled with translational dislocations. This is a straightforward consequence of my universal Jolt theory.

These apparently 'obvious' facts are actually *radically new:* we cannot separately analyze rapid brain's rotations from translations, because they are in reality always coupled.

One practical application of the brain Jolt theory is in design of helmets. Briefly, a 'hard' helmet saves the skull but not the brain; alternatively, a 'soft' helmet protects the brain from the collision jolt but does not protect the skull. A good helmet is both 'hard' and 'soft'. A proper helmet would need to have both a hard external shell (to protect the skull) and a soft internal part (that will dissipate the energy from the collision jolt by its own destruction, in the same way as a car saves its passengers from the collision jolt by its own destruction).

Similarly, in designing safer car air-bags, the two critical points will be (i) their placement within the car, and (ii) their "soft-hard characteristics", similar to the helmet characteristics described above.

B. In case of spinal injury (see Figure 10.2), the contribution of my universal Jolt theory is the following:

1. The spinal injury is always localized at the certain vertebral or inter-vertebral point;
2. In case of severe translational injuries (vertebral fractures or discus herniae) they can be identified using X-ray or other medical imaging scans; in case of microscopic rotational injuries (causing the back-pain syndrome) they cannot be identified using current medical imaging scans;
3. There is no spinal injury without one of the following two causes:
    a. Impulsive rotational + translational loading caused by either fast human movements or various crashes/impacts; and/or

b. Static eccentricity from the normal physiological spinal form, caused by external loading;

c. Any spinal injury is caused by a combination of the two points above: impulsive rotational + translational loading and static eccentricity.

This is a straightforward consequence of my universal Jolt theory. We cannot separately analyze translational and rotational spinal injuries. Also, there are no "static injuries" without eccentricity. Indian women have for centuries carried bulky loads on their heads without any spinal injuries; they just prevented any load eccentricities and any jerks in their motion.

The currently used "Principal loading hypothesis" that describes spinal injuries in terms of spinal tension, compression, bending, and shear, covers only a small subset of all spinal injuries covered by my universal Jolt theory. To prevent spinal injuries we need to develop spinal jolt awareness: ability to control all possible impulsive spinal loadings as well as static eccentricities.

C. In case of general musculo-skeletal injury, the contribution of my universal Jolt theory is the following:

1. The injury is always localized at the certain joint or bone and caused by an impulsive loading, which hits this particular joint/bone in several coupled degrees-of-freedom simultaneously;
2. Injury happens when most of the body mass is hanging on that joint; for example, in case of a knee injury, when most of the body mass is on one leg with a semi-flexed knee – and then, caused by some external shock, the knee suddenly "jerks" (this can happen in running, skiing, and ball games, as well as various crashes/impacts); or, in case of shoulder injury, when most of the body mass is hanging on one arm and then it suddenly jerks.

To prevent all these injuries we need to develop musculo-skeletal jolt awareness. For example, never overload a flexed knee and avoid any kind of uncontrolled motions (like slipping) or collisions with external objects.

In this Chapter, we propose two things: firstly, a universal theory of brain-and-spine injury prediction and prevention; and secondly, a $Matlab^{TM}$ crash-simulator toolbox for prediction of brain-and-spine injury.

10.2 Brain-and-Spine Injury

10.2.1 Traumatic Brain Injury Mechanics

Traumatic brain injury (TBI) is still a major health problem, with over a half-a-milion cases per year, mostly caused by motor-vehicle accidents (frequently involving alcohol use). TBI occurs when physical trauma causes brain damage, which can result from a closed head injury[2] or a penetrating head injury.[3] In both cases, TBI is caused by rapid deformation of the brain, resulting in a cascade of pathological events and ultimately neuro-degeneration.[4] Parts of the brain that can be damaged include the cerebral hemispheres, cerebellum, and brain stem. TBI can cause a host of physical, cognitive, emotional, and social effects. TBI is a frequent cause of major long-term disability in individuals surviving head injuries sustained in war zones. This is becoming an issue of growing concern in modern warfare in which rapid deployment of acute interventions are effective in saving the lives of combatants with significant head injuries. Traumatic brain injury has been identified as the 'signature injury' among wounded soldiers of military engagement. Rapid deformation of brain matter caused by skull acceleration is most likely the cause of concussion, as well as more severe TBI. The inability to measure deformation directly has led to disagreement and confusion about the biomechanics of concussion and TBI (see [Iva09a] and references therein).

TBI can be mild, moderate, or severe (depending on the extent of the damage to the brain), while the final outcome can be anything from complete recovery to permanent disability or death (see [CCC08]). Some symptoms are evident immediately, while others do not surface until several days or weeks after the injury[5] (see [NIH02]).

The natural cushion that protects the brain from trauma is the *cerebrospinal fluid* (CSF). It resides within cranial and spinal cavities and moves in a pulsatile fashion to and from the cranial cavity (see Figure 10.1). This motion can be measured by functional magnetic resonance imaging (fMRI, see [Sok08] for a review) and may be of clinical importance in the diagnosis of several brain and spinal cord disorders such as hydrocephalus, Chiari

[2] A closed injury occurs when the head suddenly and violently hits an object but the object does not break through the skull.

[3] A penetrating injury occurs when an object pierces the skull and enters brain tissue.

[4] In many cases of TBI widespread disruption of the axons occurs through a process known as diffuse axonal injury (DAI) or traumatic axonal injury (TAI).

[5] With mild TBI, the patient may remain conscious or may lose consciousness for a few seconds or minutes; the person may also feel dazed or not like him- or herself for several days or weeks after the initial injury; other symptoms include: headache, mental confusion, lightheadedness, dizziness, double vision, blurred vision (or tired eyes), ringing in the ears, bad taste in the mouth, fatigue or lethargy, a change in sleep patterns, behavioral or mood changes, trouble with memory/concentration/calculation. With moderate or severe TBI, the patient may show these same symptoms, but may also have: loss of consciousness, personality change, a severe/persistent/worsening headache, repeated vomiting/nausea, seizures, inability to awaken, dilation (widening) of one or both pupils, slurred speech, weakness/numbness in the extremities, loss of coordination, increased confusion, restlessness/agitation; vomiting and neurological deficit together are important indicators of prognosis and their presence may warrant early CT scanning and neurosurgical intervention.

malformation, and syringomyelia. It was found in [MHJ94] that brain and CSF of healthy volunteers exhibited periodic motion in the frequency range of normal heart rate. Both brain hemispheres showed periodic squeezing of the ventricles, with peak velocities up to 1 mm/sec followed by a slower recoil. Superimposed on the regular displacement of the brain stem was a slow, respiratory-related periodic shift of the neutral position. During the Valsalva maneuver, the brain stem showed initial caudal and subsequent cranial displacement of 2-3 mm. Coughing produced a short swing of CSF in the cephalic direction. The pressure gradient waveform of a linearized Navier-Stokes model of the pulsatile CSF flow was found in [LYA01] to be almost exclusively dependent on the flow waveform and cross-sectional area.

The microscopic motion of human brain within the skull is, in the language of modern dynamics [II06a, II05, II06b], governed by the Euclidean SE(3)-group of 3D motions. Within brain's SE(3)-group we have both SE(3)-kinematics (consisting of SE(3)-velocity and its two time derivatives: SE(3)-acceleration and SE(3)-jerk) and SE(3)-dynamics (consisting of SE(3)-momentum and its two time derivatives: SE(3)-force and SE(3)-jolt), which is brain's kinematics $\times$ brain's mass-inertia distribution.

As already explained, the external SE(3)-jolt[6] is a sharp and sudden change in the SE(3)-force acting on brain's mass-inertia distribution (given by brain's mass and inertia matrices). That is, a 'delta'-change in a 3D force-vector coupled to a 3D torque-vector, striking the head-shell with the brain immersed into the cerebrospinal fluid. In other words, the SE(3)-jolt is a sudden, sharp and discontinues shock in all 6 coupled dimensions of brain's continuous micro-motion within the cerebrospinal fluid (see Figure 10.1), namely within the three Cartesian (x, y, z)-translations and the three corresponding Euler angles around the Cartesian axes: roll, pitch and yaw. If the SE(3)-jolt produces a mild shock to the brain (e.g., strong head shake), it causes mild TBI, with temporary disabled associated sensory-motor and/or cognitive functions and affecting respiration and movement. If the SE(3)-jolt produces a hard shock (hitting the head with external mass), it causes severe TBI, with the total loss of gesture, speech and movement.

The SE(3)-jolt is rigorously defined in terms of differential geometry [II06b, II07]. Briefly, it is the absolute time-derivative of the covariant force 1-form (or, co-vector-field). As already stated, the fundamental law of biomechanics is the covariant force law:

[6] The mechanical SE(3)-jolt concept is based on the mathematical concept of higher-order tangency (rigorously defined in terms of jet bundles of the head's configuration manifold) [II06b, II07], as follows: When something hits the human head, or the head hits some external body, we have a collision. This is naturally described by the SE(3)-momentum, which is a nonlinear coupling of 3 linear Newtonian momenta with 3 angular Eulerian momenta. The tangent to the SE(3)-momentum, defined by the (absolute) time derivative, is the SE(3)-force. The second-order tangency is given by the SE(3)-jolt, which is the tangent to the SE(3)-force, also defined by the time derivative.

Force co-vector-field = Mass distribution × Acceleration vector-field,

which is formally written (using the Einstein summation convention, with indices labelling the three Cartesian translations and the three corresponding Euler angles):

$$F_\mu = m_{\mu\nu} a^\nu, \qquad (\mu, \nu = 1, ..., 6)$$

where F_μ denotes the 6 covariant components of the external "pushing" SE(3)-force co-vector-field, $m_{\mu\nu}$ represents the 6×6 covariant components of brain's inertia-metric tensor, while a^ν corresponds to the 6 contravariant components of brain's internal SE(3)-acceleration vector-field.

Now, the covariant (absolute, Bianchi) time-derivative $\frac{D}{dt}(\cdot)$ of the covariant SE(3)-force F_μ defines the corresponding external "striking" SE(3)-jolt co-vector-field:

$$\frac{D}{dt}(F_\mu) = m_{\mu\nu}\frac{D}{dt}(a^\nu) = m_{\mu\nu}\left(\dot{a}^\nu + \Gamma^\nu_{\mu\lambda} a^\mu a^\lambda\right), \qquad (10.1)$$

where $\frac{D}{dt}(a^\nu)$ denotes the 6 contravariant components of brain's internal SE(3)-jerk vector-field and overdot ($\dot{}$) denotes the time derivative. $\Gamma^\nu_{\mu\lambda}$ are the Christoffel's symbols of the Levi-Civita connection for the SE(3)-group, which are zero in case of pure Cartesian translations and nonzero in case of rotations as well as in the full-coupling of translations and rotations.

In the following, we elaborate on the SE(3)-jolt concept (using vector and tensor methods) and its biophysical TBI consequences in the form of brain's dislocations and disclinations.

$SE(3)$-group of brain's micro-motions within the CSF

The brain and the CSF together exhibit periodic microscopic translational and rotational motion in a pulsatile fashion to and from the cranial cavity, in the frequency range of normal heart rate (with associated periodic squeezing of brain's ventricles) [MHJ94]. This micro-motion is mathematically defined by the Euclidean (gauge) $SE(3)$-group. Briefly, the $SE(3)$-group is defined as a semidirect (noncommutative) product $\triangleright$ of 3D rotations and 3D translations:

$$SE(3) := SO(3) \triangleright \mathbb{R}^3.$$

Its most important subgroups are the following (see Appendix for technical details):

Subgroup	Definition
$SO(3)$, group of rotations in 3D (a spherical joint)	Set of all proper orthogonal 3×3 – rotational matrices
$SE(2)$, special Euclidean group in 2D (all planar motions)	Set of all 3×3 – matrices: $\begin{bmatrix} \cos\theta & \sin\theta & r_x \\ -\sin\theta & \cos\theta & r_y \\ 0 & 0 & 1 \end{bmatrix}$
$SO(2)$, group of rotations in 2D subgroup of $SE(2)$-group (a revolute joint)	Set of all proper orthogonal 2×2 – rotational matrices included in $SE(2)$ – group
$\mathbb{R}^3$, group of translations in 3D (all spatial displacements)	Euclidean 3D vector space

In other words, the gauge $SE(3)$-group of Euclidean micro-motions of the brain immersed in the cerebrospinal fluid within the cranial cavity, contains matrices of the form $\begin{pmatrix} \mathbf{R} & \mathbf{b} \\ 0 & 1 \end{pmatrix}$, where $\mathbf{b}$ is brain's 3D micro-translation vector and $\mathbf{R}$ is brain's 3D rotation matrix, given by the product $\mathbf{R} = R_\varphi \cdot R_\psi \cdot R_\theta$ of brain's three Eulerian micro-rotations, roll $= R_\varphi$, pitch $= R_\psi$, yaw $= R_\theta$, performed respectively about the $x-$axis by an angle φ, about the $y-$axis by an angle ψ, and about the $z-$axis by an angle θ [Iva04, PC05, II06a, II05, II06b]:

$$R_\varphi = \begin{bmatrix} 1 & 0 & 0 \\ 0 & \cos\varphi & -\sin\varphi \\ 0 & \sin\varphi & \cos\varphi \end{bmatrix}, \quad R_\psi = \begin{bmatrix} \cos\psi & 0 & \sin\psi \\ 0 & 1 & 0 \\ -\sin\psi & 0 & \cos\psi \end{bmatrix}, \quad R_\theta = \begin{bmatrix} \cos\theta & -\sin\theta & 0 \\ \sin\theta & \cos\theta & 0 \\ 0 & 0 & 1 \end{bmatrix}.$$

Therefore, brain's natural $SE(3)$-dynamics within the cerebrospinal fluid is given by the coupling of Newtonian (translational) and Eulerian (rotational) equations of micro-motion.

Brain's natural $SE(3)$-dynamics

To support our coupled loading-rate hypothesis, we formulate the coupled Newton-Euler dynamics of brain's micro-motions within the scull's $SE(3)$-group of motions. The forced Newton-Euler equations read in vector (boldface) form

$$\begin{aligned} \text{Newton}: \quad \dot{\mathbf{p}} &\equiv \mathbf{M}\dot{\mathbf{v}} = \mathbf{F} + \mathbf{p} \times \boldsymbol{\omega}, \\ \text{Euler}: \quad \dot{\boldsymbol{\pi}} &\equiv \mathbf{I}\dot{\boldsymbol{\omega}} = \mathbf{T} + \boldsymbol{\pi} \times \boldsymbol{\omega} + \mathbf{p} \times \mathbf{v}, \end{aligned} \tag{10.2}$$

where $\times$ denotes the vector cross product,[7]

[7] Recall that the cross product $\mathbf{u} \times \mathbf{v}$ of two vectors $\mathbf{u}$ and $\mathbf{v}$ equals $\mathbf{u} \times \mathbf{v} = uv\sin\theta\mathbf{n}$, where θ is the angle between $\mathbf{u}$ and $\mathbf{v}$, while $\mathbf{n}$ is a unit vector perpendicular to the plane of $\mathbf{u}$ and $\mathbf{v}$ such that $\mathbf{u}$ and $\mathbf{v}$ form a right-handed system.

$$\mathbf{M} \equiv M_{ij} = diag\{m_1, m_2, m_3\} \qquad \text{and} \qquad \mathbf{I} \equiv I_{ij} = diag\{I_1, I_2, I_3\}, \quad (i,j = 1,2,3)$$

are brain's (diagonal) mass and inertia matrices,[8] defining brain's mass-inertia distribution, with principal inertia moments given in Cartesian coordinates (x, y, z) by volume integrals

$$I_1 = \iiint \rho(z^2 + y^2)dxdydz, \quad I_2 = \iiint \rho(x^2 + z^2)dxdydz,$$
$$I_3 = \iiint \rho(x^2 + y^2)dxdydz,$$

dependent on brain's density $\rho = \rho(x, y, z)$,

$$\mathbf{v} \equiv v^i = [v_1, v_2, v_3]^t \qquad \text{and} \qquad \boldsymbol{\omega} \equiv \omega^i = [\omega_1, \omega_2, \omega_3]^t$$

(where $[\,]^t$ denotes the vector transpose) are brain's linear and angular velocity vectors (that is, column vectors),

$$\mathbf{F} \equiv F_i = [F_1, F_2, F_3] \qquad \text{and} \qquad \mathbf{T} \equiv T_i = [T_1, T_2, T_3]$$

are gravitational and other external force and torque co-vectors (that is, row vectors) acting on the brain within the scull,

$$\mathbf{p} \equiv p_i \equiv \mathbf{M}\mathbf{v} = [p_1, p_2, p_3] = [m_1 v_1, m_2 v_2, m_2 v_2] \qquad \text{and}$$
$$\boldsymbol{\pi} \equiv \pi_i \equiv \mathbf{I}\boldsymbol{\omega} = [\pi_1, \pi_2, \pi_3] = [I_1\omega_1, I_2\omega_2, I_3\omega_3]$$

are brain's linear and angular momentum co-vectors.

In tensor form, the forced Newton-Euler equations (10.2) read:

$$\dot{p}_i \equiv M_{ij}\dot{v}^j = F_i + \varepsilon^j_{ik} p_j \omega^k, \qquad (i,j,k = 1,2,3)$$
$$\dot{\pi}_i \equiv I_{ij}\dot{\omega}^j = T_i + \varepsilon^j_{ik} \pi_j \omega^k + \varepsilon^j_{ik} p_j v^k,$$

where the permutation symbol ε^j_{ik} is defined as:

[8] In reality, mass and inertia matrices $(\mathbf{M}, \mathbf{I})$ are not diagonal but rather full 3×3 positive-definite symmetric matrices with coupled mass- and inertia-products. Even more realistic, fully-coupled mass-inertial properties of a brain immersed in (incompressible, irrotational and inviscid) cerebrospinal fluid are defined by the single non-diagonal 6×6 positive-definite symmetric mass-inertia matrix $\mathcal{M}_{SE(3)}$, the so-called material metric tensor of the $SE(3)$-group, which has all nonzero mass-inertia coupling products. In other words, the 6×6 matrix $\mathcal{M}_{SE(3)}$ contains: (i) brain's own mass plus the added mass matrix associated with the fluid, (ii) brain's own inertia plus the added inertia matrix associated with the potential flow of the fluid, and (iii) all the coupling terms between linear and angular momenta. However, for simplicity, in this paper we shall consider only the simple case of two separate diagonal 3×3 matrices $(\mathbf{M}, \mathbf{I})$.

$$\varepsilon_{ik}^{j} = \begin{cases} +1 & \text{if } (i,j,k) \text{ is } (1,2,3),(3,1,2) \text{ or } (2,3,1), \\ -1 & \text{if } (i,j,k) \text{ is } (3,2,1),(1,3,2) \text{ or } (2,1,3), \\ 0 & \text{otherwise: } i=j \text{ or } j=k \text{ or } k=i. \end{cases}$$

In scalar form, the forced Newton-Euler equations (10.2) expand as

$$\text{Newton}: \begin{cases} \dot{p}_1 = F_1 - m_3 v_3 \omega_2 + m_2 v_2 \omega_3 \\ \dot{p}_2 = F_2 + m_3 v_3 \omega_1 - m_1 v_1 \omega_3 \\ \dot{p}_3 = F_3 - m_2 v_2 \omega_1 + m_1 v_1 \omega_2 \end{cases}, \tag{10.3}$$

$$\text{Euler}: \begin{cases} \dot{\pi}_1 = T_1 + (m_2 - m_3) v_2 v_3 + (I_2 - I_3)\omega_2\omega_3 \\ \dot{\pi}_2 = T_2 + (m_3 - m_1) v_1 v_3 + (I_3 - I_1)\omega_1\omega_3 \\ \dot{\pi}_3 = T_3 + (m_1 - m_2) v_1 v_2 + (I_1 - I_2)\omega_1\omega_2 \end{cases},$$

showing brain's individual mass and inertia couplings.

Equations (10.2)-(10.3) can be derived from the translational + rotational kinetic energy of the brain[9]

$$E_k = \frac{1}{2}\mathbf{v}^t \mathbf{M}\mathbf{v} + \frac{1}{2}\boldsymbol{\omega}^t \mathbf{I}\boldsymbol{\omega}, \tag{10.4}$$

or, in tensor form

$$E = \frac{1}{2} M_{ij} v^i v^j + \frac{1}{2} I_{ij}\omega^i \omega^j.$$

For this we use the *Kirchhoff-Lagrangian equations* (see [Iva09a] and references therein):

$$\frac{d}{dt}\partial_{\mathbf{v}} E_k = \partial_{\mathbf{v}} E_k \times \boldsymbol{\omega} + \mathbf{F}, \tag{10.5}$$
$$\frac{d}{dt}\partial_{\boldsymbol{\omega}} E_k = \partial_{\boldsymbol{\omega}} E_k \times \boldsymbol{\omega} + \partial_{\mathbf{v}} E_k \times \mathbf{v} + \mathbf{T},$$

where $\partial_{\mathbf{v}} E_k = \frac{\partial E_k}{\partial \mathbf{v}}$, $\partial_{\boldsymbol{\omega}} E_k = \frac{\partial E_k}{\partial \boldsymbol{\omega}}$; in tensor form these equations read

$$\frac{d}{dt}\partial_{v^i} E = \varepsilon_{ik}^{j}\left(\partial_{v^j} E\right)\omega^k + F_i,$$
$$\frac{d}{dt}\partial_{\omega^i} E = \varepsilon_{ik}^{j}\left(\partial_{\omega^j} E\right)\omega^k + \varepsilon_{ik}^{j}\left(\partial_{v^j} E\right) v^k + T_i.$$

Using (10.4)-(10.5), brain's linear and angular momentum co-vectors are defined as

$$\mathbf{p} = \partial_{\mathbf{v}} E_k, \qquad \boldsymbol{\pi} = \partial_{\boldsymbol{\omega}} E_k,$$

[9] In a fully-coupled Newton-Euler brain dynamics, instead of equation (10.4) we would have brain's kinetic energy defined by the inner product:

$$E_k = \frac{1}{2}\left[\begin{pmatrix}\mathbf{p}\\ \boldsymbol{\pi}\end{pmatrix} | \mathcal{M}_{SE(3)} \begin{pmatrix}\mathbf{p}\\ \boldsymbol{\pi}\end{pmatrix}\right].$$

or, in tensor form

$$p_i = \partial_{v^i} E, \qquad \pi_i = \partial_{\omega^i} E,$$

with their corresponding time derivatives, in vector form

$$\dot{\mathbf{p}} = \frac{d}{dt}\mathbf{p} = \frac{d}{dt}\partial_{\mathbf{v}} E, \qquad \dot{\boldsymbol{\pi}} = \frac{d}{dt}\boldsymbol{\pi} = \frac{d}{dt}\partial_{\boldsymbol{\omega}} E,$$

or, in tensor form

$$\dot{p}_i = \frac{d}{dt}p_i = \frac{d}{dt}\partial_{v^i} E, \qquad \dot{\pi}_i = \frac{d}{dt}\pi_i = \frac{d}{dt}\partial_{\omega^i} E,$$

or, in scalar form

$$\dot{\mathbf{p}} = [\dot{p}_1, \dot{p}_2, \dot{p}_3] = [m_1\dot{v}_1, m_2\dot{v}_2, m_3\dot{v}_3],$$
$$\dot{\boldsymbol{\pi}} = [\dot{\pi}_1, \dot{\pi}_2, \dot{\pi}_3] = [I_1\dot{\omega}_1, I_2\dot{\omega}_2, I_3\dot{\omega}_3].$$

While brain's healthy $SE(3)$-dynamics within the cerebrospinal fluid is given by the coupled Newton-Euler micro-dynamics, the TBI is actually caused by the sharp and discontinuous change in this natural $SE(3)$ micro-dynamics, in the form of the $SE(3)$-jolt, causing brain's discontinuous deformations.

Brain's traumatic dynamics: the $SE(3)$-jolt

The $SE(3)$-jolt, the actual cause of the TBI (in the form of the brain's plastic deformations), is defined as a coupled Newton+Euler jolt; in (co)vector form the $SE(3)$-jolt reads[10]

$$SE(3) - \text{jolt} : \begin{cases} \text{Newton jolt} : \dot{\mathbf{F}} = \ddot{\mathbf{p}} - \dot{\mathbf{p}} \times \boldsymbol{\omega} - \mathbf{p} \times \dot{\boldsymbol{\omega}} , \\ \text{Euler jolt} : \dot{\mathbf{T}} = \ddot{\boldsymbol{\pi}} - \dot{\boldsymbol{\pi}} \times \boldsymbol{\omega} - \boldsymbol{\pi} \times \dot{\boldsymbol{\omega}} - \dot{\mathbf{p}} \times \mathbf{v} - \mathbf{p} \times \dot{\mathbf{v}}, \end{cases}$$

where the linear and angular jolt co-vectors are

$$\dot{\mathbf{F}} \equiv \mathbf{M}\ddot{\mathbf{v}} = [\dot{F}_1, \dot{F}_2, \dot{F}_3], \qquad \dot{\mathbf{T}} \equiv \mathbf{I}\ddot{\boldsymbol{\omega}} = [\dot{T}_1, \dot{T}_2, \dot{T}_3],$$

where

$$\ddot{\mathbf{v}} = [\ddot{v}_1, \ddot{v}_2, \ddot{v}_3]^t, \qquad \ddot{\boldsymbol{\omega}} = [\ddot{\omega}_1, \ddot{\omega}_2, \ddot{\omega}_3]^t,$$

are linear and angular jerk vectors.

In tensor form, the $SE(3)$-jolt reads[11]

---

[10] Note that the derivative of the cross-product of two vectors follows the standard calculus product-rule: $\frac{d}{dt}(\mathbf{u} \times \mathbf{v}) = \dot{\mathbf{u}} \times \mathbf{v} + \mathbf{u} \times \dot{\mathbf{v}}$.

[11] In this paragraph the overdots actually denote the absolute Bianchi (covariant) time-derivative (10.1), so that the jolts retain the proper covector character, which would be lost if ordinary time derivatives are used. However, for the sake of simplicity and wider readability, we stick to the same overdot notation.

$$\dot{F}_i = \ddot{p}_i - \varepsilon^j_{ik}\dot{p}_j\omega^k - \varepsilon^j_{ik}p_j\dot{\omega}^k, \qquad (i,j,k=1,2,3)$$
$$\dot{T}_i = \ddot{\pi}_i \ - \varepsilon^j_{ik}\dot{\pi}_j\omega^k - \varepsilon^j_{ik}\pi_j\dot{\omega}^k - \varepsilon^j_{ik}\dot{p}_j v^k - \varepsilon^j_{ik}p_j\dot{v}^k,$$

in which the linear and angular jolt covectors are defined as:

$$\dot{\mathbf{F}} \equiv \dot{F}_i = \mathbf{M}\ddot{\mathbf{v}} \equiv M_{ij}\ddot{v}^j = [\dot{F}_1, \dot{F}_2, \dot{F}_3],$$
$$\dot{\mathbf{T}} \equiv \dot{T}_i = \mathbf{I}\ddot{\boldsymbol{\omega}} \equiv I_{ij}\ddot{\omega}^j = [\dot{T}_1, \dot{T}_2, \dot{T}_3],$$

where $\ddot{\mathbf{v}} = \ddot{v}^i$, and $\ddot{\boldsymbol{\omega}} = \ddot{\omega}^i$ are linear and angular jerk vectors.

In scalar form, the $SE(3)$-jolt expands as:

$$\text{Newton jolt}: \begin{cases} \dot{F}_1 = \ddot{p}_1 - m_2\omega_3\dot{v}_2 + m_3(\omega_2\dot{v}_3 + v_3\dot{\omega}_2) - m_2 v_2\dot{\omega}_3, \\ \dot{F}_2 = \ddot{p}_2 + m_1\omega_3\dot{v}_1 - m_3\omega_1\dot{v}_3 - m_3 v_3\dot{\omega}_1 + m_1 v_1\dot{\omega}_3, \\ \dot{F}_3 = \ddot{p}_3 - m_1\omega_2\dot{v}_1 + m_2\omega_1\dot{v}_2 - v_2\dot{\omega}_1 - m_1 v_1\dot{\omega}_2, \end{cases}$$

$$\text{Euler jolt}: \begin{cases} \dot{T}_1 = \ddot{\pi}_1 - (m_2 - m_3)(v_3\dot{v}_2 + v_2\dot{v}_3) - (I_2 - I_3)(\omega_3\dot{\omega}_2 + \omega_2\dot{\omega}_3), \\ \dot{T}_2 = \ddot{\pi}_2 + (m_1 - m_3)(v_3\dot{v}_1 + v_1\dot{v}_3) + (I_1 - I_3)(\omega_3\dot{\omega}_1 + \omega_1\dot{\omega}_3), \\ \dot{T}_3 = \ddot{\pi}_3 - (m_1 - m_2)(v_2\dot{v}_1 + v_1\dot{v}_2) - (I_1 - I_2)(\omega_2\dot{\omega}_1 + \omega_1\dot{\omega}_2). \end{cases}$$

We remark here that the linear and angular momenta $(\mathbf{p}, \boldsymbol{\pi})$, forces $(\mathbf{F}, \mathbf{T})$ and jolts $(\dot{\mathbf{F}}, \dot{\mathbf{T}})$ are co-vectors (row vectors), while the linear and angular velocities $(\mathbf{v}, \boldsymbol{\omega})$, accelerations $(\dot{\mathbf{v}}, \dot{\boldsymbol{\omega}})$ and jerks $(\ddot{\mathbf{v}}, \ddot{\boldsymbol{\omega}})$ are vectors (column vectors). This bio-physically means that the 'jerk' vector should not be confused with the 'jolt' co-vector. For example, the 'jerk' means shaking the head's own mass-inertia matrices (mainly in the atlanto-occipital and atlanto-axial joints), while the 'jolt'means actually hitting the head with some external mass-inertia matrices included in the 'hitting' SE(3)-jolt, or hitting some external static/massive body with the head (e.g., the ground - gravitational effect, or the wall - inertial effect). Consequently, the mass-less 'jerk' vector represents a (translational+rotational) *non-collision effect* that can cause only weaker brain injuries, while the inertial 'jolt' co-vector represents a (translational+rotational) *collision effect* that can cause hard brain injuries.

For example, while driving a car, the SE(3)-jerk of the head-neck system happens every time the driver brakes abruptly. On the other hand, the SE(3)-jolt means actual impact to the head. Similarly, the whiplash-jerk, caused by rear-end car collisions, is like a soft version of the high pitch-jolt caused by the boxing 'upper-cut'. Also, violently shaking the head left-right in the transverse plane is like a soft version of the high yaw-jolt caused by the sidewise, or hook punch.

Brain's dislocations and disclinations caused by the $SE(3)$-jolt

Recall from introduction that for mild TBI, the best injury predictor is considered to be the product of brain's strain and strain rate, which is the standard isotropic viscoelastic continuum concept. To improve this standard concept, in this subsection, we consider human brain as a 3D anisotropic multipolar *Cosserat viscoelastic continuum* (see [Iva09a] and references therein), exhibiting coupled-stress-strain elastic properties. This non-standard continuum model is suitable for analyzing plastic (irreversible) deformations and fracture mechanics in multi-layered materials with microstructure (in which slips and bending of layers introduces additional degrees of freedom, non-existent in the standard continuum models).

The $SE(3)$-jolt $(\dot{\mathbf{F}}, \dot{\mathbf{T}})$ causes two types of brain's rapid discontinuous deformations:

1. The Newton jolt $\dot{\mathbf{F}}$ can cause micro-translational *dislocations*, or discontinuities in the Cosserat translations;
2. The Euler jolt $\dot{\mathbf{T}}$ can cause micro-rotational *disclinations*, or discontinuities in the Cosserat rotations.

To precisely define brain's dislocations and disclinations, caused by the $SE(3)$-jolt $(\dot{\mathbf{F}}, \dot{\mathbf{T}})$, we first define the coordinate co-frame, i.e., the set of basis 1-forms $\{dx^i\}$, given in local coordinates $x^i = (x^1, x^2, x^3) = (x, y, z)$, attached to brain's center-of-mass. Then, in the coordinate co-frame $\{dx^i\}$ we introduce the following set of brain's plastic-deformation-related $SE(3)$-based differential $p-$forms[12] (see, e.g. [II06b, II07]):

---

[12] Differential $p-$forms are totally skew-symmetric covariant tensors, defined using the exterior wedge-product and exterior derivative. The proper definition of exterior derivative d for a $p-$form β on a smooth manifold M, includes the *Poincaré lemma* [II06b, II07]: $d(d\beta) = 0$, and validates the *general Stokes formula*

$$\int_{\partial M} \beta = \int_M d\beta,$$

where M is a $p-$dimensional *manifold with a boundary* and ∂M is its $(p-1)-$dimensional *boundary*, while the integrals have appropriate dimensions.

A $p-$form β is called *closed* if its exterior derivative is equal to zero,

$$d\beta = 0.$$

From this condition one can see that the closed form (the *kernel* of the exterior derivative operator d) is conserved quantity. Therefore, closed $p-$forms possess certain invariant properties, physically corresponding to the conservation laws.

A $p-$form β that is an exterior derivative of some $(p-1)-$form α,

$$\beta = d\alpha,$$

is called *exact* (the *image* of the exterior derivative operator d). By *Poincaré lemma*, exact forms prove to be closed automatically,

$$d\beta = d(d\alpha) = 0.$$

This lemma is the foundation of the de Rham cohomology theory.

the *dislocation current* 1-form, $\mathbf{J} = J_i\, dx^i$;
the *dislocation density* 2-form, $\boldsymbol{\alpha} = \frac{1}{2}\alpha_{ij}\, dx^i \wedge dx^j$;
the *disclination current* 2-form, $\mathbf{S} = \frac{1}{2}S_{ij}\, dx^i \wedge dx^j$; and
the *disclination density* 3-form, $\mathbf{Q} = \frac{1}{3!}Q_{ijk}\, dx^i \wedge dx^j \wedge dx^k$,

where $\wedge$ denotes the exterior wedge-product. According to Edelen [Ede80, KE83], these four $SE(3)$-based differential forms satisfy the following set of continuity equations:

$$\dot{\boldsymbol{\alpha}} = -\mathbf{dJ} - \mathbf{S}, \tag{10.6}$$

$$\dot{\mathbf{Q}} = -\mathbf{dS}, \tag{10.7}$$

$$\mathbf{d}\boldsymbol{\alpha} = \mathbf{Q}, \tag{10.8}$$

$$\mathbf{dQ} = \mathbf{0}, \tag{10.9}$$

where $\mathbf{d}$ denotes the exterior derivative.

In components, the simplest, fourth equation (10.9), representing the *Bianchi identity*, can be rewritten as

$$\mathbf{dQ} = \partial_l Q_{[ijk]}\, dx^l \wedge dx^i \wedge dx^j \wedge dx^k = 0,$$

where $\partial_i \equiv \partial/\partial x^i$, while $\theta_{[ij\ldots]}$ denotes the skew-symmetric part of $\theta_{ij\ldots}$.

Similarly, the third equation (10.8) in components reads

$$\frac{1}{3!}Q_{ijk}\, dx^i \wedge dx^j \wedge dx^k = \partial_k \alpha_{[ij]}\, dx^k \wedge dx^i \wedge dx^j, \qquad \text{or}$$
$$Q_{ijk} = -6\partial_k \alpha_{[ij]}.$$

The second equation (10.7) in components reads

$$\frac{1}{3!}\dot{Q}_{ijk}\, dx^i \wedge dx^j \wedge dx^k = -\partial_k S_{[ij]}\, dx^k \wedge dx^i \wedge dx^j, \qquad \text{or}$$
$$\dot{Q}_{ijk} = 6\partial_k S_{[ij]}.$$

Finally, the first equation (10.6) in components reads

$$\frac{1}{2}\dot{\alpha}_{ij}\, dx^i \wedge dx^j = (\partial_j J_i - \frac{1}{2}S_{ij})\, dx^i \wedge dx^j, \qquad \text{or}$$
$$\dot{\alpha}_{ij} = 2\partial_j J_i - S_{ij}.$$

In words, we have:

- The 2-form equation (10.6) defines the time derivative $\dot{\boldsymbol{\alpha}} = \frac{1}{2}\dot{\alpha}_{ij}\, dx^i \wedge dx^j$ of the dislocation density $\boldsymbol{\alpha}$ as the (negative) sum of the disclination current $\mathbf{S}$ and the curl of the dislocation current $\mathbf{J}$.
- The 3-form equation (10.7) states that the time derivative $\dot{\mathbf{Q}} = \frac{1}{3!}\dot{Q}_{ijk}\, dx^i \wedge dx^j \wedge dx^k$ of the disclination density $\mathbf{Q}$ is the (negative) divergence of the disclination current $\mathbf{S}$.

- The 3-form equation (10.8) defines the disclination density $\mathbf{Q}$ as the divergence of the dislocation density $\boldsymbol{\alpha}$, that is, $\mathbf{Q}$ is the *exact* 3-form.
- The Bianchi identity (10.9) follows from equation (10.8) by *Poincaré lemma* and states that the disclination density $\mathbf{Q}$ is conserved quantity, that is, $\mathbf{Q}$ is the *closed* 3-form. Also, every 4-form in 3D space is zero.

From these equations, we can derive two important conclusions:

1. Being the derivatives of the dislocations, brain's disclinations are higher-order tensors, and thus more complex quantities, which means that they present a higher risk for the severe TBI than dislocations – an old fact which *is* supported by the literature (see review of existing TBI-models given in Introduction).
2. Brain's dislocations and disclinations are mutually coupled by the underlaying $SE(3)$ group, which means that we cannot separately analyze translational and rotational TBIs – a new fact which *is not* supported by the literature.

10.2.2 Spinal Injury Mechanics

The traditional *principal loading hypothesis* [MM93, WZ98], which describes spinal injuries in terms of spinal tension, compression, bending, and shear, is insufficient to predict and prevent the cause of the back-pain syndrome. Its underlying mechanics is simply not accurate enough.

On the other hand, to be recurrent, musculo-skeletal injury must be associated with a histological change, i.e., the modification of associated tissues within the body. However, incidences of *functional* musculoskeletal injury, e.g., lower back pain, generally shows little evidence of *structural* damage [Wad98]. The incidence of injury is likely to be a continuum ranging from little or no evidence of structural damage through to the observable damage of muscles, joints or bones. The changes underlying functional injuries are likely to consist of torn muscle fibers, stretched ligaments, subtle erosion of join tissues, and/or the application of pressure to nerves, all amounting to a disruption of function to varying degrees and a tendency toward spasm.

For example, in a review of experimental studies on the role of mechanical stresses in the genesis of intervertebral disk degeneration and herniation [RCR01], the authors dismissed simple mechanical stimulations of functional vertebra as a cause of disk herniation, concluding instead that a complex mechanical stimulation combining forward and lateral bending of the spine followed by violent compression is needed to produce posterior herniation of the disk. Considering the use of models to estimate the risk of injury the authors emphasize the need to understand this complex interaction between the mechanical forces and the living body [SG01]. Compressive and shear loading increased significantly with exertion load, lifting velocity, and trunk asymmetry [GM95]. Also, it has been stated that up to two-thirds of all back injuries have been associated with trunk rotation [KN06]. In addition, load-lifting in awkward environment places a person at risk for low back pain and injury [RWM08]. These risks appear to be increased when facing up or down an inclined surface.

The above-mentioned safe spinal motions (flexion/extension, lateral flexion and rotation) *are* governed by standard Euler's rotational intervertebral dynamics coupled to Newton's micro-translational dynamics. On the other hand, the unsafe spinal events, the main cause of spinal injuries, are caused by intervertebral SE(3)-jolts, the sharp and sudden, "delta"- (forces + torques) combined, localized both in time and in space. These localized intervertebral SE(3)-jolts do not belong to the standard Newton-Euler dynamics. The only way to monitor them would be to measure "in vivo" the rate of the combined (forces + torques)- rise.

Ivancevic proposed in [Iva09b, Iva10a] a new locally-coupled loading-rate hypothesis, which states that the main cause of both soft- and hard-tissue spinal injury is a localized Euclidean jolt, or $SE(3)$-jolt, an impulsive loading that strikes a localized spine in several coupled degrees-of-freedom (DOF) simultaneously. To show this, based on the previously defined covariant force law, we formulate the coupled Newton-Euler dynamics of the local spinal motions and derive from it the corresponding coupled $SE(3)$-jolt dynamics. The $SE(3)$-jolt is the main cause of two forms of local discontinuous spinal injury: (i) hard-tissue injury of local translational dislocations; and (ii) soft-tissue injury of local rotational disclinations. Both the spinal dislocations and disclinations, as caused by the $SE(3)$-jolt, are described using the Cosserat multipolar viscoelastic continuum model.

While we can intuitively visualize the SE(3)-jolt, for the purpose of simulation we use the necessary simplified, decoupled approach (neglecting the 3D torque matrix and its coupling to the 3D force vector). Note that decoupling is a kind of linearization that prevents chaotic behavior, giving an illusion of full predictability. In this decoupled framework of reduced complexity, we define:

- The cause of hard spinal injuries (discus hernia) is a linear 3D-jolt vector hitting some intervertebral joint - the time rate-of-change of a 3D-force vector (linear jolt = mass × linear jerk); and
- The cause of soft spinal injuries (back-pain syndrome) is an angular 3-axial jolt hitting some intervertebral joint - the time rate-of-change of a 3-axial torque (angular jolt = inertia moment × angular jerk).

This decoupled framework has been implemented in the Human Biodynamics Engine, a world-class neuro-musculo-skeletal dynamics simulator (with 270 DOFs, the same number of equivalent muscular actuators and two-level neural reflex control), developed by the present author at Defence Science and Technology Organization, Australia. This kinematically validated human motion simulator has been described in a series of papers and books (see [Iva10c] and references therein).

As shown in [Iva10a], the mechanics of spinal (intervertebral) injury is essential the same as the mechanics of brain injury, described in the previous subsection. In particular, we can conclude that localized spinal dislocations and disclinations are mutually coupled by the underlaying $SE(3)$-group, which means that we cannot separately analyze translational and rotational spinal injuries – a new fact which *is not* supported by the literature.

10.3 Rigorous Crash Simulator Toolbox for *Matlab*®

A Matlab toolbox entitled *Rigorous Crash Simulator* (*RCS*) was recently developed jointly by Defence Science & Technology Organisation, Australia and the Centre for Intelligent Systems Research, Deakin University, Australia. This new toolbox is a spin-off of the Human Biodynamics Engine [Iva10c], based on two existing Matlab toolboxes: (i) the third-party toolbox *DiffMan* (for solving ODEs on manifolds), by K. Engø, A. Marthinsen and H. Munthe-Kaas, and (ii) the standard *Virtual Reality* (VR) toolbox for *Matlab* and Simulink.

Briefly, human spine with head and pelvis (see Figure 10.3), mechanically represents a chain of 27 rigid bodies, flexibly joined by 26 inter-vertebral joints. For rigorous prediction and prevention spinal injuries under various crash-impact situations, modern computational mechanics needs to be used. It is modeled as a chain of 26 Euclidean groups of motion and numerically solved by Lie-group integrators.

The *RCS* toolbox is developed around the main Lie-group integrator, called *Runge-Kutta-Munte-Kass* (RKMK) integrator (see next section).

10.3.1 Rigid Body Motion and ODEs on Smooth Manifolds

Recall from mechanics of brain-and-spine injury described in the previous section, that the special Euclidean group $SE(3)$ of rigid-body motions in our everyday Euclidean space $\mathbb{R}^3$, is a semidirect (non-commutative) product of the rotation group $SO(3)$ and the translation group $\mathbb{R}^3$. This practically means that the motion of a rigid body in a 3D space is given by a pair $(R, p) \in SE(3)$ of rotation matrix R and translation vector p, such that its angular velocity (attitude) matrix $\boldsymbol{\omega}$ and linear velocity vector $\mathbf{v}$ belong to its Lie algebra $\mathfrak{se}(3)$, that is: $(\boldsymbol{\omega}, \mathbf{v}) \in \mathfrak{se}(3) \approx \mathbb{R}^6$.

Kinematic equations of motion of a rigid body are:

$$\dot{p} = R\mathbf{v}, \qquad \dot{R} = R\boldsymbol{\omega}.$$

Kinetic energy of a rigid body has the symmetrical form:

$$E_k = \frac{1}{2}\mathbf{v}^T\mathbf{M}\mathbf{v} + \frac{1}{2}\boldsymbol{\omega}^T\mathbf{I}\boldsymbol{\omega}, \tag{10.10}$$

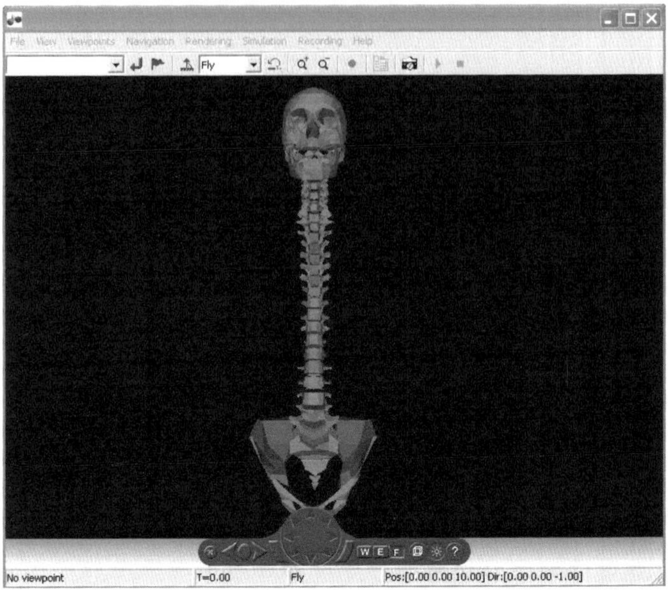

Fig. 10.3. *A 3D model of human spine implemented in the VR toolbox of Matlab.*

where (assuming uniform mass-distribution) mass and inertia matrices are diagonal:

$$\mathbf{M} = \mathrm{diag}\{m_1, m_2, m_3\},$$
$$\mathbf{I} = \mathrm{diag}\{I_1, I_2, I_3\}.$$

From the kinetic energy (10.10), dynamical *equations of motion* follow (these are coupled Newton-Euler equations, see my injury papers for the derivation):

$$\mathbf{M}\dot{\mathbf{v}} = \mathbf{M}\mathbf{v} \times \boldsymbol{\omega}, \qquad \mathbf{I}\dot{\boldsymbol{\omega}} = \mathbf{I}\boldsymbol{\omega} \times \boldsymbol{\omega} + \mathbf{M}\mathbf{v} \times \mathbf{v}.$$

Finally, by including the forces f_i and torques τ_i acting on the body ($i = 1, ..., n$), with input controls $u^i = u(t)$, the dynamical-control equations become:

$$\mathbf{M}\dot{\mathbf{v}} = \mathbf{M}\mathbf{v} \times \boldsymbol{\omega} + \mathbf{F}, \qquad (\text{with } \mathbf{F} = \sum_i f_i u^i), \tag{10.11}$$

$$\mathbf{I}\dot{\boldsymbol{\omega}} = \mathbf{I}\boldsymbol{\omega} \times \boldsymbol{\omega} + \mathbf{M}\mathbf{v} \times \mathbf{v} + \mathbf{T}, \qquad (\text{with } \mathbf{T} = \sum_i \tau_i u^i). \tag{10.12}$$

In the spinal crash-test model, the motion of the head, as well as of each individual vertebral body, is governed by the pair of vector equations (10.11)-(10.12). They are evolving on the smooth $SE(3)$-manifold.

Evolving ODEs on smooth manifolds

To give a brief description of the computational mechanics implemented in the *RCS* toolbox, consider the following ODE (ordinary differential equation) evolving in time ($t \geq 0$) on some configuration manifold M:[13]

$$\dot{x} = F(x), \qquad x(0) = x_0 \in M, \tag{10.13}$$

where $\dot{x} = dx/dt$, while $F(x) \in \mathfrak{X}(M)$ is the tangent *vector-field* on M passing through the points $x(t)$.

The solution of the ODE (10.13) is determined by its *flow* $\phi_{t,F}(x_0)$ that starts from the initial point $x(0) = x_0$, which is formally defined as:

$$x(t) = \phi_{t,F}(x_0), \qquad (\text{for} \;\; t \geq 0).$$

In general, any tangent vector-field is an *infinitesimal generator* of its flow. This means that the vector-field $F(x)$ is given by the time-derivative of the flow $\phi_{t,F}(x0)$ at the initial point:

$$F(x) = \frac{d}{dt}\phi_{t,F}(x_0)|t = 0.$$

The inverse of the time-derivative of the flow is something that plays the role of the time-integral, which is the *exponential map*, a nonlinear generalization of the matrix exponential. So, the flow $\phi_{t,F}$ is given by the exponential map of (tF):

$$\phi_{t,F} = \exp(tF).$$

In the special case of the linear ODE defined by some matrix A:

$$\dot{x} = Ax, \qquad \text{we have} \tag{10.14}$$
$$\phi_{t,A}(x) = \exp(tA)x_0$$

with the standard *matrix exponential*:

$$\exp(tA) = \sum_{n=0}^{\infty} \frac{1}{n!} t^n A^n.$$

In the case of linear ODEs, solved by the matrix exponentials, we can see that their flows do not commute: going first along the flow $\phi_{t,A}$ and then along some other flow $\phi_{t,B}$ is different from going first along the flow $\phi_{t,B}$ and then along the flow $\phi_{t,A}$. This is because matrix multiplication is not commutative, so it yields the *commutator*:

$$[A, B] = AB - BA \neq 0.$$

---

[13] For example, $M = SE(3)$, the configuration manifold of a rigid body.

This non-commutativity of flows is even more significant in a general case of nonlinear ODEs. Let us start from some point x_0 on the manifold M, and flow from x_0 first along $\phi_{t,F} = \exp(tF)$ and then along some other flow $\phi_{t,G} = \exp(tG)$, so that we come to some point x_1. If we now reverse the order of flows and starting from the same point x_0 we flow first along $\phi_{t,G}$ and then along $\phi_{t,F}$ - we will in general arrive at a different point $x_2 \neq x_1$. In terms of exponential maps this non-commutativity of flows can be written as:

$$\exp(sF) \circ \exp(tG) \circ \exp(-sF) \circ \exp(-tG) \neq 0.$$

If the flows do not commute, then their vector-fields do not commute either. This statement is defined by the commutator $[F,G] \neq 0$ called the *Lie bracket* of vector-fields F and G, which has the following three properties:

$$\begin{aligned} [F,G] &= -[G,F], \\ [F+G,H] &= [F,H] + [G,H] \ , \\ 0 &= [F,[G,H]] + [G,[H,F]] + [H,[F,G]], \end{aligned}$$

called anti-symmetry, bilinearity and Jacobi identity, respectively. The set of all tangent vector-fields $\mathfrak{X}(M)$ on the manifold M now (with the Lie bracket) becomes the *Lie algebra*.

Runge-Kutta-Munte-Kass family of Lie-group integrators

The *RCS* toolbox is developed around the main Lie-group integrator, called Runge-Kutta-Munte-Kass (RKMK) integrator [EMM99a, EMM99b], which combines standard Runge-Kutta family with Lie-group integration methods developed by A. Iserles (for a recent review, see [IMK05]) and H. Munthe-Kaas [MK95, MK98, MK99].

ODEs are solved in DiffMan usiing the following general 5-step procedure:[14]

1. Construct an initial domain object y in a homogeneous space;
2. Construct a vector-field object vf over the domain object y; DiffMan finds numerically the integral curve of this vector-field through the initial domain object;
3. Construct a time stepper object ts, ehich determines the numerical method used to advance the numerical solution along the integral line; it consists of two parts: coordinate and method;
4. Construct a flow object f, which is defined by the vector-field object; and
5. Solve the ODE by the flow object which is done by evaluating the flow object at the initial domain object, start time, end time, and step size.

[14] For technical details with worked examples, see [EMM99a, EMM99b].

10.3.2 Computational Newton-Euler Dynamics

First-order (velocities) equations of motion

Standard description of Newton-Euler dynamics starts with the first-order equations of motion in terms of translational and rotational velocities:[15]

$$\text{Newton}:\left\{\begin{array}{l} \dot{p}_1(t) \equiv m_1 v_1(t) = F_1(t) - m_3 v_3(t)\,\omega_2(t) + m_2 v_2(t)\,\omega_3(t), \\ \dot{p}_2(t) \equiv m_2 v_2(t) \;= F_2(t) - m_3 v_3(t)\,\omega_1(t) - m_1 v_1(t)\,\omega_3(t), \\ \dot{p}_3(t) \equiv m_3 v_3(t) = F_3(t) - m_2 v_2(t)\,\omega_1(t) + m_1 v_1(t)\,\omega_2(t). \end{array}\right\}$$

$$\text{Euler}:\left\{\begin{array}{l} \dot{\pi}_1(t) \equiv J_1\omega_1(t) \\ \qquad = T_1(t) + (J_2 - J_3)\,\omega_2(t)\,\omega_3(t) + (m_2 - m_3)\,v_2(t)\,v_3(t), \\ \dot{\pi}_2(t) \equiv J_2\omega_2(t) \\ \qquad = T_2(t) + (J_3 - J_1)\,\omega_1(t)\,\omega_3(t) + (m_3 - m_1)\,v_1(t)\,v_3(t), \\ \dot{\pi}_3(t) \equiv J_3\omega_3(t) \\ \qquad = T_3(t) + (J_1 - J_2)\,\omega_1(t)\,\omega_2(t) + (m_1 - m_2)\,v_1(t)\,v_2(t). \end{array}\right\}$$

Numerical solution of these equations (for some initial conditions) gives translational and rotational velocities ($v_i(t)$ and $\omega_i(t)$, $i = 1, 2, 3$). However, to be able to actually see the body motion in a virtual 3D environment, we need to evaluate these equations into the second-order equations in terms of translations (displacements x_i) and rotations (Euler angles θ_i).

Second-order (coordinates) equations of motion

The above standard first-order Newton-Euler velocity equations are expanded/evaluated into the following coordinate equations of motion:

$$\text{Newton}:\left\{\begin{array}{l} \dot{p}_1(t) \equiv m_1\ddot{x}_1(t) \\ \qquad = -m_3\dot{x}_3(t)\dot{\theta}_2(t) + m_2\dot{x}_2(t)\dot{\theta}_3(t) - b_1\dot{x}_1(t) - k_1x_1(t), \\ \dot{p}_2(t) \equiv m_2\ddot{x}_2(t) \\ \qquad = -m_3\dot{x}_3(t)\dot{\theta}_1(t) - m_1\dot{x}_1(t)\dot{\theta}_3(t) - b_2\dot{x}_2(t) - k_2x_2(t), \\ \dot{p}_3(t) \equiv m_3\ddot{x}_3(t) \\ \qquad = -m_2\dot{x}_2(t)\dot{\theta}_1(t) + m_1\dot{x}_1(t)\dot{\theta}_2(t) - b_3\dot{x}_3(t) - k_3x_3(t). \end{array}\right\}$$

$$\text{Euler}:\left\{\begin{array}{l} \dot{\pi}_1(t) \equiv J_1\ddot{\theta}_1(t) = (J_2 - J_3)\,\dot{\theta}_2(t)\dot{\theta}_3(t) \\ \qquad\qquad + (m_2 - m_3)\,\dot{x}_2(t)\dot{x}_3(t) - B_1\dot{\theta}_1(t) - K_1\theta_1(t), \\ \dot{\pi}_2(t) \equiv J_2\ddot{\theta}_2(t) = (J_3 - J_1)\,\dot{\theta}_1(t)\dot{\theta}_3(t) \\ \qquad\qquad + (m_3 - m_1)\,\dot{x}_1(t)\dot{x}_3(t) - B_2\dot{\theta}_2(t) - K_2\theta_2(t), \\ \dot{\pi}_3(t) \equiv J_3\ddot{\theta}_3(t) = (J_1 - J_2)\,\dot{\theta}_1(t)\dot{\theta}_2(t) \\ \qquad\qquad + (m_1 - m_2)\,\dot{x}_1(t)\dot{x}_2(t) - B_3\dot{\theta}_3(t) - K_3\theta_3(t). \end{array}\right\}$$

[15] "≡" means "ekvivalent", whatever is left from it is not part of the equations to be solved.

For simplicity, in these evaluated 2nd order equations of motion, forces $F_i(t)$ are replaced by springs $k_i x_i(t)$ and dampers $b_i \dot{x}_i(t)$; and similarly for rotations, instead of torques $T_i(t)$ we have angular springs $K_i \theta_i(t)$ and angular dampers $B_i \dot{\theta}_i(t)$.

Computational form of Newton-Euler equations of rigid motions

Newton-Euler acceleration ODEs.

Newton-Euler acceleration ODEs are defined as:

$$\text{Newton}: \begin{Bmatrix} \ddot{x}_1(t) = \Big[- m_3 \dot{x}_3(t) \dot{\theta}_2(t) + m_2 \dot{x}_2(t) \dot{\theta}_3(t) \\ \qquad -b_1 \dot{x}_1(t) - k_1 x_1(t) \Big] / m_1, \\ \ddot{x}_2(t) = \Big[- m_3 \dot{x}_3(t) \dot{\theta}_1(t) - m_1 \dot{x}_1(t) \dot{\theta}_3(t) \\ \qquad -b_2 \dot{x}_2(t) - k_2 x_2(t) \Big] / m_2, \\ \ddot{x}_3(t) = \Big[- m_2 \dot{x}_2(t) \dot{\theta}_1(t) + m_1 \dot{x}_1(t) \dot{\theta}_2(t) \\ \qquad -b_3 \dot{x}_3(t) - k_3 x_3(t) \Big] / m_3. \end{Bmatrix}$$

$$\text{Euler}: \begin{Bmatrix} \ddot{\theta}_1(t) = \Big[(J_2 - J_3) \, \dot{\theta}_2(t) \dot{\theta}_3(t) + (m_2 - m_3) \, \dot{x}_2(t) \dot{x}_3(t) \\ \qquad -B_1 \dot{\theta}_1(t) - K_1 \theta_1(t) \Big] / J_1, \\ \ddot{\theta}_2(t) = \Big[(J_3 - J_1) \, \dot{\theta}_1(t) \dot{\theta}_3(t) + (m_3 - m_1) \, \dot{x}_1(t) \dot{x}_3(t) \\ \qquad -B_2 \dot{\theta}_2(t) - K_2 \theta_2(t) \Big] / J_2, \\ \ddot{\theta}_3(t) = \Big[(J_1 - J_2) \, \dot{\theta}_1(t) \dot{\theta}_2(t) + (m_1 - m_2) \, \dot{x}_1(t) \dot{x}_2(t) \\ \qquad -B_3 \dot{\theta}_3(t) - K_3 \theta_3(t) \Big] / J_3. \end{Bmatrix}$$

Full set of first-order ODEs suitable for numerical integration.

The following set of 12 first-order ODEs has been implemented in the *RCS* toolbox for simulating a single intervertebral joint:

$$\begin{aligned} \dot{x}_1(t) &= v_1(t), \qquad (10.15) \\ \dot{v}_1(t) &= \Big[-m_3 \dot{x}_3(t) \dot{\theta}_2(t) + m_2 \dot{x}_2(t) \dot{\theta}_3(t) - b_1 \dot{x}_1(t) - k_1 x_1(t) \Big] / m_1, \\ \dot{x}_2(t) &= v_2(t), \\ \dot{v}_2(t) &= \Big[-m_3 \dot{x}_3(t) \dot{\theta}_1(t) - m_1 \dot{x}_1(t) \dot{\theta}_3(t) - b_2 \dot{x}_2(t) - k_2 x_2(t) \Big] / m_2, \\ \dot{x}_3(t) &= v_3(t), \\ \dot{v}_3(t) &= \Big[-m_2 \dot{x}_2(t) \dot{\theta}_1(t) + m_1 \dot{x}_1(t) \dot{\theta}_2(t) - b_3 \dot{x}_3(t) - k_3 x_3(t) \Big] / m_3, \\ \dot{\theta}_1(t) &= \omega_1(t), \end{aligned}$$

$$
\begin{aligned}
\dot{\omega}_1(t) &= \Big[(J_2 - J_3)\,\dot{\theta}_2(t)\dot{\theta}_3(t) + (m_2 - m_3)\,\dot{x}_2(t)\dot{x}_3(t) \\
&\quad - B_1\dot{\theta}_1(t) - K_1\theta_1(t) \Big] / J_1, \\
\dot{\theta}_2(t) &= \omega_2(t), \\
\dot{\omega}_2(t) &= \Big[(J_3 - J_1)\,\dot{\theta}_1(t)\dot{\theta}_3(t) + (m_3 - m_1)\,\dot{x}_1(t)\dot{x}_3(t) \\
&\quad - B_2\dot{\theta}_2(t) - K_2\theta_2(t) \Big] / J_2, \\
\dot{\theta}_3(t) &= \omega_3(t), \\
\dot{\omega}_3(t) &= \Big[(J_1 - J_2)\,\dot{\theta}_1(t)\dot{\theta}_2(t) + (m_1 - m_2)\,\dot{x}_1(t)\dot{x}_2(t) \\
&\quad - B_3\dot{\theta}_3(t) - K_3\theta_3(t) \Big] / J_3, \\
\text{init. conds} &: \; x_i(0) = a_i, \quad \dot{x}_i(0) = c_i, \quad \theta_i(0) = d_i, \quad \dot{\theta}_i(0) = e_i, \\
&\quad (\text{for } i = 1, 2, 3).
\end{aligned}
$$

The ODEs (10.15) have been solved using the RKMK integrator, as follows.

Matlab/DiffMan implementation

Testing the RKMK integrator.

For testing the RKMK integrator, we implemented three second-order Lorenz-like ODEs, rewritten as six coupled first-order ODEs. In *Mathematica*^{TM}, these equations are implemented as:

$$
\left\{
\begin{array}{c}
x'(t) = v_1(t), \qquad v_1'(t) = y(t) - x(t), \\
y'(t) = v_2(t), \qquad v_2'(t) = x(t)(-z(t)) + x(t) - y(t), \\
z'(t) = v_3(t), \qquad v_3'(t) = x(t)y(t) - z(t), \\
x(0) = z(0) = 0.001, \qquad y(0) = 1, \\
v_1(0) = v_2(0) = v_3(0) = 0.01.
\end{array}
\right\}
$$

and solved using the *NDSolve* integrator for 15 sec.

In DiffMan these ODEs are implemented in the following m-function:[16]

```
function [la] = vfexShady2Lorenz(t,y)
la = liealgebra(y);
ydat = getdata(y);
dat = [ 0 0 0 -1 1 0;
```

[16] More precisely, to implement any particular ODE-system in DiffMan, two m-functions are required. We are showing here only the first function (in which the ODEs are implemented), while we are skipping the second function (which calls the first one), because it is too long and out of scope of this paper.

```
          0 0 0 1 -1 -ydat(4);
          0 0 0 0 ydat(4) -1;
          1 0 0 0 0 0;
          0 1 0 0 0 0;
          0 0 1 0 0 0; ];
setdata(la,dat);
return;
```

The phase plots of this test problem for 15 sec are shown in Figure 10.4.

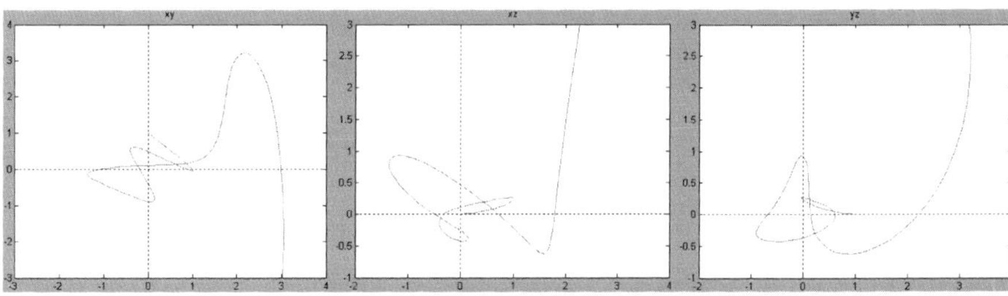

Fig. 10.4. *Test problem: solution of the 2nd-order Lorenz-like ODEs in Matlab, using the RKMK integrator. The phase plots are identical to those calculated by Mathematica's integrator NDSolve.*

DiffMan implementation of the system (10.15).

Matlab/DiffMan implementation of the system (10.15), using the RKMK integrator, is given by the following two m-function:

```
function [la] = vfexShady2(t,y)
la = liealgebra(y);
ydat = getdata(y);
global k1 k2 k3 b1 b2 b3 B1 B2 B3 K1 K2 K3 m1 m2 m3 J1 J2 J3
dat =     [0 1 0 0 0 0 0 0 0 0 0 0;
          -k1 -b1 0 0 0 0 0 0 0 -m3*ydat(6) 0 m2*ydat(4);
          0 0 0 1 0 0 0 0 0 0 0 0;
          0 0 -k2 -b2 0 0 0 -m3*ydat(6) 0 0 0 -m1*ydat(2);
          0 0 0 0 0 1 0 0 0 0 0 0;
          0 0 0 0 -k3 -b3 0 -m2*ydat(4) 0 m1*ydat(2) 0 0;
          0 0 0 0 0 0 0 1 0 0 0 0;
          0 0 0 0 0 (m2-m3)*ydat(4) -K1 -B1 0 0 0 (J2-J3)*ydat(10);
          0 0 0 0 0 0 0 0 0 1 0 0;
          0 0 0 0 0 (m3-m1)*ydat(2) 0 0 -K2 -B2 0 (J3-J1)*ydat(8);
          0 0 0 0 0 0 0 0 0 0 0 1;
          0 0 0 (m1-m2)*ydat(2) 0 0 0 0 0 (J1-J2)*ydat(8) -K3 -B3];
```

```
setdata(la,dat);
return;
```

Initial modeling of external impact forces for the crash simulation.

To simulate the action of an impact force on the single intervertebral joint, a 'soft form of' the *impulse Dirac delta function* term with amplitude A has been modeled by:

$$F(t) = A\,\mathrm{sech}(A\,t - A/2), \tag{10.16}$$

and added to translational Newtonian accelerations only. Due to translational/rotational coupling between Newton's and Euler's equations within the $SE(3)$-group dynamics, this translational impact force should cause both macroscopic angular change and microscopic displacement change within the intervertebral joint. This type of impact forces with amplitudes in the range 20g-100g are used to model road-vehicle crashes, while the amplitudes in the range 100g-400g are used to model land-mine related crashes and helicopter hard landings. In addition, for modeling effects of riding an operational watercraft with the speed of 20-30 knots on the high seas with waves of 2m-3m hight, the following sinus forces with frequency φ are used:

$$F(t) = A\sin(\varphi\,t). \tag{10.17}$$

To test RKMK integrator on the impact forces (10.16)-(10.17) the following two versions of the forced Van der Pol oscillator (with parameters a, b, w):

$$\begin{aligned} \mathrm{sech}: \quad & \ddot{x}(t) = A\,\mathrm{sech}(A\,t - A/2) - a\left[1 - 4b\,x(t)^2\right]\dot{x}(t) + w^2 x(t), \\ \sin: \quad & \ddot{x}(t) = A\sin(\varphi\,t) - a\left[1 - 4b\,x(t)^2\right]\dot{x}(t) + w^2 x(t), \end{aligned} \tag{10.18}$$

for the simulation with near-zero initial conditions, have been inplemented in the following two m-functions, respectively:

```
function [la] = vfexVdPolSech(t,y)
la = liealgebra(y);
ydat = [illegible]etdata(y);
a=1.5; [illegible]=5; w=2; A=20;
dat=[ [illegible];
   -[illegible] +A*sech(A*t-A/2)/ydat(1) a*(1-4*b*ydat(1)*ydat(1))];
setda[illegible] dat);
retur[illegible]

funct[illegible]] = vfexVdPolSin(t,y)
la = [illegible]bra(y);
ydat[illegible]ata(y);
a=1[illegible]w=2;A=20;fr=3;
dat[illegible]
   [illegible]A*sin(fr*t)/ydat(1) a*(1-4*b*ydat(1)*ydat(1))];
set[illegible]at);
ret[illegible]
```

Full set of the forced first-order ODEs for a single joint crash dynamics.

After successful tasting of the forced Van der Pol oscillators (10.18) against Mathematica's integrator NDSolve using the above m-functions, the following set of 12 first-order $SE(3)$-ODEs has been implemented in the RCS toolbox for simulating an impact force action on a single intervertebral joint:

$$
\begin{aligned}
\dot{x}_1(t) &= v_1(t), \\
\dot{v}_1(t) &= \Big[- m_3\dot{x}_3(t)\dot{\theta}_2(t) + m_2\dot{x}_2(t)\dot{\theta}_3(t) - b_1\dot{x}_1(t) - k_1 x_1(t) \\
&\quad + A_1 \operatorname{sech}(A_1 t - A_1/2) \Big] / m_1, \\
\dot{x}_2(t) &= v_2(t), \\
\dot{v}_2(t) &= \Big[- m_3\dot{x}_3(t)\dot{\theta}_1(t) - m_1\dot{x}_1(t)\dot{\theta}_3(t) - b_2\dot{x}_2(t) - k_2 x_2(t) \\
&\quad + A_2 \operatorname{sech}(A_2 t - A_2/2) \Big] / m_2, \\
\dot{x}_3(t) &= v_3(t), \\
\dot{v}_3(t) &= \Big[- m_2\dot{x}_2(t)\dot{\theta}_1(t) + m_1\dot{x}_1(t)\dot{\theta}_2(t) - b_3\dot{x}_3(t) - k_3 x_3(t) \\
&\quad + A_3 \operatorname{sech}(A_3 t - A_3/2) \Big] / m_3, \\
\dot{\theta}_1(t) &= \omega_1(t), \\
\dot{\omega}_1(t) &= \Big[(J_2 - J_3)\,\dot{\theta}_2(t)\dot{\theta}_3(t) + (m_2 - m_3)\,\dot{x}_2(t)\dot{x}_3(t) \\
&\quad - B_1\dot{\theta}_1(t) - K_1\theta_1(t) \Big] / J_1, \\
\dot{\theta}_2(t) &= \omega_2(t), \\
\dot{\omega}_2(t) &= \Big[(J_3 - J_1)\,\dot{\theta}_1(t)\dot{\theta}_3(t) + (m_3 - m_1)\,\dot{x}_1(t)\dot{x}_3(t) \\
&\quad - B_2\dot{\theta}_2(t) - K_2\theta_2(t) \Big] / J_2, \\
\dot{\theta}_3(t) &= \omega_3(t), \\
\dot{\omega}_3(t) &= \Big[(J_1 - J_2)\,\dot{\theta}_1(t)\dot{\theta}_2(t) + (m_1 - m_2)\,\dot{x}_1(t)\dot{x}_2(t) \\
&\quad - B_3\dot{\theta}_3(t) - K_3\theta_3(t) \Big] / J_3,
\end{aligned}
\tag{10.19}
$$

$$
\text{i.c.}:\; x_i(0) = a_i, \quad \dot{x}_i(0) = c_i, \quad \theta_i(0) = d_i, \quad \dot{\theta}_i(0) = e_i, \quad (\text{for } i = 1,2,3).
$$

DiffMan implementation of the full set of the forced first-order ODEs for a single joint crash dynamics.

The full set of forced $SE(3)$-ODEs (10.19) has been implemented in the following m-function:

```
function [la] = vfexShady4(t,y)
```

```
la = liealgebra(y);
ydat = getdata(y);
global k1 k2 k3 b1 b2 b3 B1 B2 B3 K1 K2 K3 m1 m2 m3 J1 J2 J3
global A1 A2 A3 A4 A5 A6
val1=A1*sech(A1*t-A1/2);
val2=A2*sech(A2*t-A2/2);
val3=A3*sech(A3*t-A3/2);
dat = [0 1 0 0 0 0 0 0 0 0 0 0;
    (-k1+val1/ydat(2)) -b1 0 0 0 0 0 0 0 -m3*ydat(6) 0 m2*ydat(4);
    0 0 0 1 0 0 0 0 0 0 0 0;
    val2/ydat(4) 0 -k2 -b2 0 0 0 -m3*ydat(6) 0 0 0 -m1*ydat(2);
    0 0 0 0 0 1 0 0 0 0 0 0;
    val3/ydat(6) 0 0 0 -k3 -b3 0 -m2*ydat(4) 0 m1*ydat(2) 0 0;
    0 0 0 0 0 0 0 1 0 0 0 0;
    0 0 0 0 0 (m2-m3)*ydat(4) -K1 -B1 0 0 0 (J2-J3)*ydat(10);
    0 0 0 0 0 0 0 0 0 1 0 0;
    0 0 0 0 0 (m3-m1)*ydat(2) 0 0 -K2 -B2 0 (J3-J1)*ydat(8);
    0 0 0 0 0 0 0 0 0 0 0 1;
    0 0 0 (m1-m2)*ydat(2) 0 0 0 0 0 (J1-J2)*ydat(8) -K3 -B3];
setdata(la,dat);
return;
```

10.3.3 Full Spine Crash Simulator

The full spine crash simulator, as implemented in the *RCS* toolbox, figures the forced $SE(3)$-equations of motion (10.19) at each spinal (intervertebral) joint independently. This *dynamical decoupling along the spine* is the only way to deal with the shear dimensionality of our problem: seven $SE(3)$-joints for the neck only (cervical spine) and 26 $SE(3)$-joints for the full spine. To compensate for this dynamical decoupling along the spine, at the same time we are inertially re-coupling all the joints along the spine: in the first joint (above the C1) the only mass is the head; in the second joint (above the C2) we have two masses: the head and C1; in the third joint (above C3) we have three masses: head + C1 + C2, etc. Regarding simulating various crashes, in the *RCS* toolbox, the so-called 'generic crash' is represented by a 3D force-vector which hits somewhere along the spine, at one only vertebral joint, so this force vector has 3 components:

$$F_{\text{crash}}(t) = \Big[A_1 \operatorname{sech}(A_1 t - A_1/2),\\ A_2 \operatorname{sech}(A_2 t - A_2/2),\ A_3 \operatorname{sech}(A_3 t - A_3/2)\Big].$$

The full spine crash simulator has been implemented in the form of the following vector of 26 $SE(3)$-equations of motion (with the joint-labeling superscript index j = 1, ..., 26):

$$
\begin{aligned}
\dot{x}_1^{\mathrm{j}}(t) &= v_1^{\mathrm{j}}(t), \qquad\qquad (10.20)\\
\dot{v}_1^{\mathrm{j}}(t) &= \Big[- m_3^{\mathrm{j}}\dot{x}_3^{\mathrm{j}}(t)\dot{\theta}_2^{\mathrm{j}}(t) + m_2^{\mathrm{j}}\dot{x}_2^{\mathrm{j}}(t)\dot{\theta}_3^{\mathrm{j}}(t) - b_1^{\mathrm{j}}\dot{x}_1^{\mathrm{j}}(t) - k_1^{\mathrm{j}}x_1^{\mathrm{j}}(t) \\
&\qquad + A_1^{\mathrm{j}}\,\mathrm{sech}(A_1^{\mathrm{j}}t - A_1^{\mathrm{j}}/2)\Big]/m_1^{\mathrm{j}},\\
\dot{x}_2^{\mathrm{j}}(t) &= v_2^{\mathrm{j}}(t),\\
\dot{v}_2^{\mathrm{j}}(t) &= \Big[- m_3^{\mathrm{j}}\dot{x}_3^{\mathrm{j}}(t)\dot{\theta}_1^{\mathrm{j}}(t) - m_1^{\mathrm{j}}\dot{x}_1^{\mathrm{j}}(t)\dot{\theta}_3^{\mathrm{j}}(t) - b_2^{\mathrm{j}}\dot{x}_2^{\mathrm{j}}(t) - k_2^{\mathrm{j}}x_2^{\mathrm{j}}(t) \\
&\qquad + A_2^{\mathrm{j}}\,\mathrm{sech}(A_2^{\mathrm{j}}t - A_2^{\mathrm{j}}/2)\Big]/m_2^{\mathrm{j}},\\
\dot{x}_3^{\mathrm{j}}(t) &= v_3^{\mathrm{j}}(t),\\
\dot{v}_3^{\mathrm{j}}(t) &= \Big[- m_2^{\mathrm{j}}\dot{x}_2^{\mathrm{j}}(t)\dot{\theta}_1^{\mathrm{j}}(t) + m_1^{\mathrm{j}}\dot{x}_1^{\mathrm{j}}(t)\dot{\theta}_2^{\mathrm{j}}(t) - b_3^{\mathrm{j}}\dot{x}_3^{\mathrm{j}}(t) - k_3^{\mathrm{j}}x_3^{\mathrm{j}}(t) \\
&\qquad + A_3^{\mathrm{j}}\,\mathrm{sech}(A_3^{\mathrm{j}}t - A_3^{\mathrm{j}}/2)\Big]/m_3^{\mathrm{j}},\\
\dot{\theta}_1^{\mathrm{j}}(t) &= \omega_1^{\mathrm{j}}(t),\\
\dot{\omega}_1^{\mathrm{j}}(t) &= \Big[\left(J_2^{\mathrm{j}} - J_3^{\mathrm{j}}\right)\dot{\theta}_2^{\mathrm{j}}(t)\dot{\theta}_3^{\mathrm{j}}(t) + \left(m_2^{\mathrm{j}} - m_3^{\mathrm{j}}\right)\dot{x}_2^{\mathrm{j}}(t)\dot{x}_3^{\mathrm{j}}(t) \\
&\qquad - B_1^{\mathrm{j}}\dot{\theta}_1^{\mathrm{j}}(t) - K_1^{\mathrm{j}}\theta_1^{\mathrm{j}}(t)\Big]/J_1^{\mathrm{j}},\\
\dot{\theta}_2^{\mathrm{j}}(t) &= \omega_2^{\mathrm{j}}(t),\\
\dot{\omega}_2^{\mathrm{j}}(t) &= \Big[\left(J_3^{\mathrm{j}} - J_1^{\mathrm{j}}\right)\dot{\theta}_1^{\mathrm{j}}(t)\dot{\theta}_3^{\mathrm{j}}(t) + \left(m_3^{\mathrm{j}} - m_1^{\mathrm{j}}\right)\dot{x}_1^{\mathrm{j}}(t)\dot{x}_3^{\mathrm{j}}(t) \\
&\qquad - B_2^{\mathrm{j}}\dot{\theta}_2^{\mathrm{j}}(t) - K_2^{\mathrm{j}}\theta_2^{\mathrm{j}}(t)\Big]/J_2^{\mathrm{j}},\\
\dot{\theta}_3^{\mathrm{j}}(t) &= \omega_3^{\mathrm{j}}(t),\\
\dot{\omega}_3^{\mathrm{j}}(t) &= \Big[\left(J_1^{\mathrm{j}} - J_2^{\mathrm{j}}\right)\dot{\theta}_1^{\mathrm{j}}(t)\dot{\theta}_2^{\mathrm{j}}(t) + \left(m_1^{\mathrm{j}} - m_2^{\mathrm{j}}\right)\dot{x}_1^{\mathrm{j}}(t)\dot{x}_2^{\mathrm{j}}(t) \\
&\qquad - B_3^{\mathrm{j}}\dot{\theta}_3^{\mathrm{j}}(t) - K_3^{\mathrm{j}}\theta_3^{\mathrm{j}}(t)\Big]/J_3^{\mathrm{j}},
\end{aligned}
$$

$$
\text{i.c.}:\; x_i^{\mathrm{j}}(0) = a_i^{\mathrm{j}}, \quad \dot{x}_i^{\mathrm{j}}(0) = c_i^{\mathrm{j}}, \quad \theta_i^{\mathrm{j}}(0) = d_i^{\mathrm{j}}, \quad \dot{\theta}_i^{\mathrm{j}}(0) = e_i^{\mathrm{j}}, \quad (\text{for } i = 1,2,3).
$$

In the full spine $SE(3)$-crash model (10.20), all crash amplitudes A_i^{j} (for $i = 1,..,3; j = 1,..,26$) are assumed zero, unless the impact joint has been selected with its 3 amplitudes only (e.g., A_i^7 - for the C7 cervical intervertebral joint).

In the Matlab implementation of (10.20) in the *RCS* toolbox (see Figure 10.5), the external forces are applied to each rigid body in the spinal system: all vertebras as well as the head and the pelvis. To propagate the force effect along the spine we make use of both graphical and numerical vertebral inter-dependency. In the graphical part, the whole spine is modeled as a tree structure. All vertebras are children to the pelvis and they are also connected to the head (which is the grandchild). So, the transformation of each spinal part affects all its children. In the numerical part, the transform propagation is modeled as a mass-propagation. Here, the mass of each spinal part is equal to the accumulated mass of all its children plus its own mass.

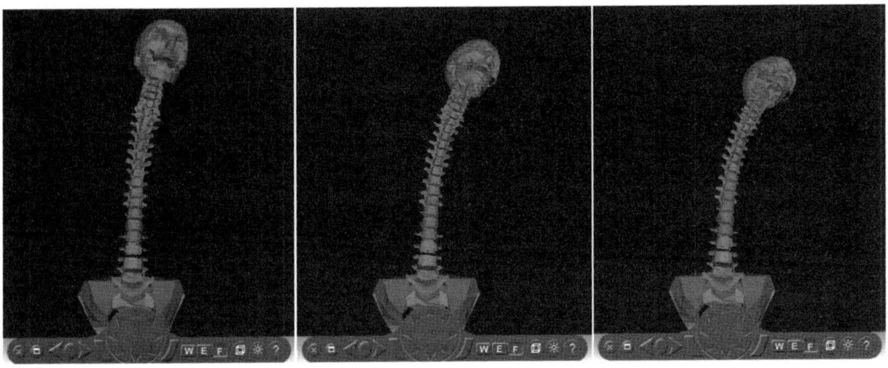

Fig. 10.5. *RCS toolbox at work: simulating a boxing punch (left hook) in the head.*

DiffMan implementation of the full spine crash simulator (10.20).

The full set of forced $SE(3)$-ODEs (10.20) has been implemented, for each spinal joint independently, in the following m-function:

```
function [la] = vfexShady5(t,y)
la = liealgebra(y);
ydat = getdata(y);
global k1 k2 k3 b1 b2 b3 B1 B2 B3 K1 K2 K3 m1 m2 m3 J1 J2 J3
global A1 A2 A3 A4 A5 A6
global prev_data
%x
if ~isWithin(ydat(1),-0.001,0.001)
ydat(1)=clip(ydat(1),-0.01,0.01);
ydat(2)=clip(ydat(2),10,10);
end
%y
if ~isWithin(ydat(3),-0.001,0.001)
ydat(3)=clip(ydat(3),-0.01,0.01);
ydat(4)=clip(ydat(4),10,10);
end
%z
if ~isWithin(ydat(5),-0.001,0.001)
ydat(5)=clip(ydat(5),-0.01,0.01);
ydat(6)=clip(ydat(6),10,10);
```

```
end
%th1
if ~isWithin(ydat(7),-0.01,0.01)
ydat(7)=clip(ydat(7),-0.01,0.01);
ydat(8)=clip(ydat(8),100,100);
end
%th2
if ~isWithin(ydat(9),-0.01,0.01)
ydat(9)=clip(ydat(9),-0.01,0.01);
ydat(10)=clip(ydat(10),100,100);
end
%th3
if ~isWithin(ydat(11),-0.01,0.01)
ydat(11)=clip(ydat(11),-0.01,0.01);
ydat(12)=clip(ydat(12),100,100);
end
val1=A1*sech(A1*t-A1/2);
val2=A2*sech(A2*t-A2/2);
val3=A3*sech(A3*t-A3/2);
dat = [0 1 0 0 0 0 0 0 0 0 0 0;
(-k1+val1/ydat(2)) -b1 0 0 0 0 0 0 0 -m3*ydat(6) 0 m2*ydat(4);
0 0 0 1 0 0 0 0 0 0 0 0;
val2/ydat(4) 0 -k2 -b2 0 0 0 -m3*ydat(6) 0 0 0 -m1*ydat(2);
0 0 0 0 0 1 0 0 0 0 0 0;
val3/ydat(6) 0 0 0 -k3 -b3 0 -m2*ydat(4) 0 m1*ydat(2) 0 0;
0 0 0 0 0 0 0 1 0 0 0 0;
0 0 0 0 0 (m2-m3)*ydat(4) -K1 -B1 0 0 0 (J2-J3)*ydat(10);
0 0 0 0 0 0 0 0 0 1 0 0;
0 0 0 0 0 (m3-m1)*ydat(2) 0 0 -K2 -B2 0 (J3-J1)*ydat(8);
0 0 0 0 0 0 0 0 0 0 0 1;
0 0 0 (m1-m2)*ydat(2) 0 0 0 0 0 (J1-J2)*ydat(8) -K3 -B3];
setdata(la,dat);
return;
```

10.3.4 Road-Vehicle Crash Simulation

Several crashes are implemented in the *RCS* toolbox, including road-vehicle crash, ejection seat and land-mine crash. In particular, Figure 10.6 shows the basic implementation of the road-vehicle crash. Note that, about 1 sec after the crash, the head moves back purely due to spinal elasticity. This particular simulated crash could cause severe TBI if the distance from the head and the steering wheel (or some other frontal part of cabin) is shorter that the movement amplitude. If the distance is safe, only a mild loss of consciousness for a few minutes could be expected, together with the strain in the cervical spine (neck).

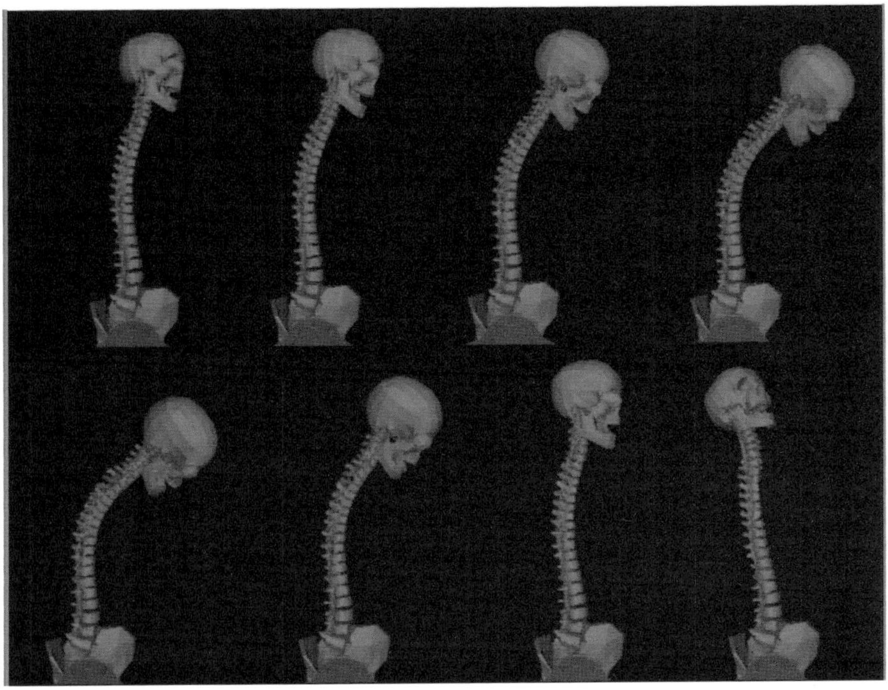

Fig. 10.6. *Road-vehicle crash simulation in the RCS toolbox: the sequence of 8 snapshots (starting with top-left and finishing with bottom-right) simulates the frontal (head-on) crash at the combined speed of two cars of about 80 km/h with the proper seat-belt on.*

Summary

We have presented the unique mechanics of traumatic brain-and-spine injury based on the new coupled loading-rate hypothesis, which states that the main cause of all mechanical human injuries is the Euclidean jolt, an impulsive loading that strikes head and spine (or any other part of the human body) - in several coupled degrees-of-freedom simultaneously, which causes two basic forms of brain and spinal injuries: (i) localized translational dislocations; and (ii) localized rotational disclinations. This model-theory of traumatic brain-and-spine injury is supported by the Rigorous Crash Simulator toolbox for *Matlab*, to be used for modeling high-impact crashes which can lead to both brain and spinal injuries. An example of a road-vehicle crash is included.

10.4 Appendix: Biodynamics and Control of Humanoid Robots

Here, starting from the basic biodynamics, we give a brief on a three-level control of humanoid robots.

10.4.1 Basics of Human Biodynamics

To describe general biodynamics of human–like movement, which is governed by the *covariant force law*:

$$F_i = m_{ij}a^j,$$

which states that:

the force covector-field F_i *equals the mass-distribution tensor* m_{ij} *times the acceleration vector-field* a^j (see Chapter 10 below and our references therein),

we usually start from the generalized Hamiltonian vector-field X_H, describing the basic behavior of the *human–like locomotor system*:

$$\dot{q}^i = \frac{\partial H}{\partial p_i} + \frac{\partial R}{\partial p_i}, \tag{10.21}$$

$$\dot{p}_i = F_i - \frac{\partial H}{\partial q^i} + \frac{\partial R}{\partial q^i}, \tag{10.22}$$

where the vector-field X_H is generating time evolution, or *phase–flow*, of $2n$ *system variables*: n generalized coordinates (joint angles q^i) and n generalized momenta (joint angular momenta p_i), $H = H(q,p)$ represents the system's conservative energy: kinetic energy + various mechano–chemical potentials, $R = R(q,p)$ denotes the nonlinear dissipation of energy, and $F_i = F_i(t,q,p,\sigma)$ are external control forces (biochemical energy inputs). The *system parameters* include inertia tensor with mass distribution of all body segments, stiffness and damping tensors for all joints (labeled by index i, which is, for geometric reasons, written as a subscript on angle variables, and as a superscript on momentum variables), as well as amplitudes, frequencies and time characteristics of all active muscular forces (supposed to be acting in all the joints; if some of the joints are inactive, we have the affine Hamiltonian control system, see chapter 6).

The equation (10.21) is called the *velocity equation*, representing the *flow* of the system (analogous to current in electrodynamics), while the equation (10.22) is a Newton–like *force equation*, representing the *effort* of the system (analogous to voltage). Together, these two functions represent Hamiltonian formulation of the *biomechanical force–velocity relation* of A.V. Hill [Hil38]. From engineering perspective, their (inner) product, $flow \cdot effort$, represents the total system's *power*, equal to the time–rate–of–change of the total system's energy (included in H, R and F_i functions). And energy itself is transformed into the *work* done by the system.

10.4.2 Spinal Control Level

The general form of Hamiltonian humanoid robotics on the symplectic cotangent bundle T^*M_{rob} of the configuration manifold M_{rob} is based on the *affine*

Hamiltonian function $H_a : T^*M \to \mathbb{R}$, in local canonical coordinates on T^*M given by (see [Iva10c], as well as references therein):

$$H_a(x,p,u) = H_0(x,p) - H^j(x,p)\, u_j, \tag{10.23}$$

where $H_0(x,p)$ is the physical Hamiltonian (kinetic + potential energy) dependent on joint coordinates x^i and canonical momenta p^i, $H^j = H^j(x,p)$, $(j = 1,\ldots, m \leq n$ are the coupling Hamiltonians corresponding to the system's active joints and $u_i = u_i(t,x,p)$ are (reflex) feedback–controls. Using (10.23) we come to the affine Hamiltonian control HBE–system, in deterministic form

$$\begin{aligned} \dot{x}^i &= \partial_{p_i} H_0 - \partial_{p_i} H^j\, u_j + \partial_{p_i} R, \\ \dot{p}_i &= \mathcal{F}_i(t,x,p) - \partial_{x^i} H_0 + \partial_{x^i} H^j\, u_j + \partial_{x^i} R, \\ o^i &= -\partial_{u_i} H_a = H^j, \\ x^i(0) &= x^i_0, \qquad p_i(0) = p^0_i, \\ &(i = 1,\ldots,n; \qquad j = 1,\ldots, M \leq n), \end{aligned} \tag{10.24}$$

$(\mathcal{F}_i = \mathcal{F}_i(t,x,p),\ H_0 = H_0(x,p),\ H^j = H^j(x,p),\ H_a = H_a(x,p,u),\ R = R(x,p))$, as well as in the fuzzy–stochastic form

$$\begin{aligned} dq^i &= \left(\partial_{p_i} H_0(\sigma_\mu) - \partial_{p_i} H^j(\sigma_\mu)\, u_j + \partial_{p_i} R\right) dt, \\ dp_i &= B_{ij}[x^i(t),t]\, dW^j(t) \qquad + \\ &\quad \left(\bar{\mathcal{F}}_i(t,x,p) - \partial_{x^i} H_0(\sigma_\mu) + \partial_{x^i} H^j(\sigma_\mu)\, u_j + \partial_{x^i} R\right) dt, \\ d\bar{o}^i &= -\partial_{u_i} H_a(\sigma_\mu)\, dt = H^j(\sigma_\mu)\, dt, \\ x^i(0) &= \bar{x}^i_0, \qquad p_i(0) = \bar{p}^0_i \end{aligned} \tag{10.25}$$

In (10.24)–(10.25), $R = R(x,p)$ denotes the joint (nonlinear) dissipation function, o^i are affine system outputs (which can be different from joint coordinates); $\{\sigma\}_\mu$ (with $\mu \geq 1$) denote fuzzy sets of conservative parameters (segment lengths, masses and moments of inertia), dissipative joint dampings and actuator parameters (amplitudes and frequencies), while the bar $(\bar{.})$ over a variable denotes the corresponding fuzzified variable; $B_{ij}[q^i(t),t]$ denote diffusion fluctuations and $W^j(t)$ are discontinuous jumps as the n–dimensional Wiener process.

Humanoid's *force servo–controller* is formulated as affine control Hamiltonian systems (10.24–10.25) (with possible extensions along the lines of the previous section), which resemble J. Houk's *autogenetic motor servo* [Hou79], acting on the spinal–reflex level of the human locomotion control. A voluntary contraction force F of human skeletal muscle is reflexly excited (positive feedback $+F^{-1}$) by the responses of its spindle receptors to stretch and is reflexly inhibited (negative feedback $-F^{-1}$) by the responses of its Golgi tendon organs to contraction. Stretch and unloading reflexes are mediated by combined actions of several autogenetic neural pathways, forming the so–called 'motor servo.' The term 'autogenetic' means that the stimulus excites receptors located in the same muscle that is the target of the reflex response. The most important of these muscle receptors are the primary and secondary endings in the muscle–spindles, which are sensitive to length change – positive length feedback $+F^{-1}$, and the Golgi tendon organs, which are sensitive to contractile force – negative force feedback $-F^{-1}$.

The gain G of the length feedback $+F^{-1}$ can be expressed as the positional stiffness (the ratio $G \approx S = dF/dx$ of the force–F change to the length–x change) of the muscle system. The greater the stiffness S, the less the muscle will be disturbed by a change in load. The autogenetic circuits $+F^{-1}$ and $-F^{-1}$ appear to function as servoregulatory loops that convey continuously graded amounts of excitation and inhibition to the large (alpha) skeletomotor neurons. Small (gamma) fusimotor neurons innervate the contractile poles of muscle spindles and function to modulate spindle–receptor discharge.

10.4.3 Cerebellum–Like Velocity and Jerk Control

Nonlinear velocity and jerk (time derivative of acceleration) servo–controllers (see [Iva10c] and references therein), developed using the Lie–derivative formalism, resemble self–stabilizing and adaptive tracking action of the cerebellum ([HBB96]). By introducing the vector-fields f and g, given respectively by

$$f = \left(\partial_{p_i} H_0, -\partial_{q^i} H_0\right), \qquad g = \left(-\partial_{p_i} H^j, \partial_{q^i} H^j\right),$$

we obtain the affine controller in the standard nonlinear MIMO–system form:

$$\dot{x}_i = f(x) + g(x)\, u_j. \tag{10.26}$$

Finally, using the standard *Lie derivative formalism*[17] (see, e.g. [Iva04] and references therein) and applying the constant relative degree r to all HB joints, the control law for asymptotic tracking of the reference outputs $o_R^j = o_R^j(t)$ could be formulated as:

$$u_j = \frac{\dot{o}_R^{(r)j} - L_f^{(r)} H^j + \sum_{s=1}^{r} c_{s-1}(o_R^{(s-1)j} - L_f^{(s-1)} H^j)}{L_g L_f^{(r-1)} H^j}, \tag{10.27}$$

[17] Let $F(M)$ denote the set of all smooth (i.e., C^∞) real valued functions $f : M \to \mathbb{R}$ on a smooth manifold M, $V(M)$ – the set of all smooth vector-fields on M, and $V^*(M)$ – the set of all differential one–forms on M. Also, let the vector-field $\zeta \in V(M)$ be given with its local flow $\phi_t : M \to M$ such that at a point $x \in M$, $\frac{d}{dt}|_{t=0}\, \phi_t x = \zeta(x)$, and ϕ_t^* representing the pull–back by ϕ_t. The Lie derivative differential operator $\mathcal{L}_\zeta$ is defined:

(i) on a function $f \in F(M)$ as

$$\mathcal{L}_\zeta : F(M) \to F(M), \qquad \mathcal{L}_\zeta f = \frac{d}{dt}(\phi_t^* f)|_{t=0},$$

where c_{s-1} are the coefficients of the linear differential equation of order r for the error function $e(t) = x^j(t) - o^j_R(t)$

$$e^{(r)} + c_{r-1}e^{(r-1)} + \cdots + c_1 e^{(1)} + c_0 e = 0.$$

The control law (10.27) can be implemented symbolically in *Mathematica*^{TM} in the following three steps:

1. Symbolic functions defining the gradient and Lie derivatives:[18]

$$\begin{aligned}
\mathrm{Grad}[s_, x_\mathrm{List}] &:= (D[s, \#1]\&)/@x;\\
\mathrm{LieDer}[v_\mathrm{List}, s_, x_\mathrm{List}] &:= \mathrm{Grad}[s, x]\cdot v;\\
\mathrm{KLieDer}[v_\mathrm{List}, s_, x_\mathrm{List}, k_] &:=\\
\mathrm{Block}[\{t\}, p := s;\ \ \mathrm{If}\ [k &== 0, p = s, \mathrm{Do}[p = \mathrm{LieDer}[v, p, x], \{k\}]]; p];
\end{aligned}$$

2. Control law defined (for simplicity, we show here only the first–order control law):

$$u[t\_] = (-\mathrm{LieDer}[F, y, X] + D[yR[t], t] + \alpha(yR[t] - y))/\mathrm{LieDer}[g, y, X];$$

3. Example for the reference output $yR[t]$, with the final time Tfin:

$$yR[t\_] = \mathrm{If}[t <= \mathrm{Tfin}/2, 5(1 - \mathrm{e}^{-5t}), (5(1 - \mathrm{e}^{-5t}))/\mathrm{e}^{(5(t-\mathrm{Tfin}/2))}];$$

The affine nonlinear MIMO control system (10.26) with the Lie–derivative control law (10.27) resembles the self–stabilizing and synergistic output tracking action of the human cerebellum. To make it adaptive (and thus more realistic), instead of the 'rigid' controller (10.27), we can use the adaptive Lie–derivative controller, as explained in the seminal paper on geometrical nonlinear control [SI89].

10.4.4 Cortical–Like Fuzzy–Topological Control

For the purpose of our cortical control, the dominant, rotational part of the human configuration manifold M^N, could be first, reduced to an N–torus,

---

(ii) on a vector-field $\eta \in V(M)$ as

$$\mathcal{L}_\varsigma : V(M) \to V(M), \qquad \mathcal{L}_\varsigma \eta = \frac{d}{dt}(\phi_t^* \eta)|_{t=0} \equiv [\zeta, \eta]$$

– the Lie bracket, and

(iii) on a one–form $\alpha \in V^*(M)$ as

$$\mathcal{L}_\varsigma : V^*(M) \to V^*(M), \qquad \mathcal{L}_\varsigma \alpha = \frac{d}{dt}(\phi_t^* \alpha)|_{t=0}.$$

In general, for any smooth tensor field $\mathbf{T}$ on M, the Lie derivative $\mathcal{L}_\varsigma \mathbf{T}$ geometrically represents a directional derivative of $\mathbf{T}$ along the flow ϕ_t.

[18] This is the code in *Mathematica*^{TM} version 7.

and second, transformed to an N–cube ('hyper–joystick'), using the following topological techniques[19] [Iva04].

Let S^1 denote the constrained unit circle in the complex plane, which is an Abelian Lie group. Firstly, we propose two reduction homeomorphisms, using the Cartesian product of the constrained $SO(2)$–groups:

$$SO(3) \approx SO(2) \times SO(2) \times SO(2) \qquad \text{and} \qquad SO(2) \approx S^1.$$

Next, let I^N be the unit cube $[0,1]^N$ in $\mathbb{R}^N$ and '$\sim$' an equivalence relation on $\mathbb{R}^N$ obtained by 'gluing' together the opposite sides of I^N, preserving their orientation. Therefore, M^N can be represented as the quotient space of $\mathbb{R}^N$ by the space of the integral lattice points in $\mathbb{R}^N$, that is an oriented and constrained N–dimensional torus T^N:

$$\mathbb{R}^N/Z^N \approx \prod_{i=1}^{N} S_i^1 \equiv \{(q^i,\, i=1,\dots,N) : \mathrm{mod}2\pi\} = T^N. \tag{10.28}$$

Its Euler–Poincaré characteristic is (by the de Rham theorem) both for the configuration manifold T^N and its momentum phase–space T^*T^N given by:

$$\chi(T^N, T^*T^N) = \sum_{p=1}^{N}(-1)^p b_p\,,$$

where b_p are the Betti numbers defined as

$$\begin{aligned} b^0 &= 1, \\ b^1 &= N, \dots b^p = \binom{N}{p}, \dots b^{N-1} = N, \\ b^N &= 1, \qquad\qquad (0 \le p \le N). \end{aligned}$$

Conversely by 'ungluing' the configuration space we obtain the primary unit cube. Let '$\sim^*$' denote an equivalent decomposition or 'ungluing' relation. According to Tychonoff's product–topology theorem, for every such quotient space there exists a 'selector' such that their quotient models are homeomorphic, that is, $T^N/\sim^* \approx A^N/\sim^*$. Therefore I_q^N represents a 'selector' for the configuration torus T^N and can be used as an N–directional '$\hat{q}$–command–space' for the feedback control (FC). Any subset of degrees of freedom on the configuration torus T^N representing the joints included in HB has its simple,

[19] This top control level has not yet been implemented. The main reason for this is its high dimensionality. For example, the Human Biodynamics Engine simulator has 270 degrees of freedom (both rotational and translational). Its rotational part includes 135 individual Lie-derivative controllers. The integration of so many individual controllers is a nontrivial problem that is currently beyond the capacity of pure fuzzy control.

rectangular image in the rectified $\hat{q}$–command space – selector I_q^N, and any joint angle q^i has its rectified image $\hat{q}^i$.

In the case of an end–effector, $\hat{q}^i$ reduces to the position vector in external–Cartesian coordinates z^r $(r = 1, \ldots, 3)$. If orientation of the end–effector can be neglected, this gives a topological solution to the standard inverse kinematics problem.

Analogously, all momenta $\hat{p}_i$ have their images as rectified momenta $\hat{p}_i$ in the $\hat{p}$–command space – selector I_p^N. Therefore, the total momentum phase–space manifold T^*T^N obtains its 'cortical image' as the $\widehat{(q,p)}$–command space, a trivial $2N$–dimensional bundle $I_q^N \times I_p^N$.

Now, the simplest way to perform the feedback FC on the cortical $\widehat{(q,p)}$–command space $I_q^N \times I_p^N$, and also to mimic the cortical–like behavior, is to use the $2N$– dimensional fuzzy–logic controller, in much the same way as in the popular 'inverted pendulum' examples.

We propose the fuzzy feedback–control map Ξ that maps all the rectified joint angles and momenta into the feedback–control one–forms

$$\Xi : (\hat{q}^i(t), \hat{p}_i(t)) \mapsto u_i(t, q, p), \tag{10.29}$$

so that their corresponding universes of discourse, $\hat{Q}^i = (\hat{q}^i_{max} - \hat{q}^i_{min})$, $\hat{P}_i = (\hat{p}_i^{max} - \hat{p}_i^{min})$ and $\hat{U}_i = (u_i^{max} - u_i^{min})$, respectively, are mapped as

$$\Xi : \prod_{i=1}^{N} \hat{Q}^i \times \prod_{i=1}^{N} \hat{P}_i \to \prod_{i=1}^{N} \hat{U}_i. \tag{10.30}$$

The $2N$–dimensional map Ξ (10.29,10.30) represents a fuzzy inference system, defined by:

1. Fuzzification of the crisp rectified and discretized angles, momenta and controls using Gaussian–bell membership functions

$$\mu_k(\chi) = \exp[-\frac{(\chi - m_k)^2}{2\sigma_k}], \qquad (k = 1, 2, \ldots, 9),$$

   where $\chi \in D$ is the common symbol for $\hat{q}^i$, $\hat{p}_i$ and $u_i(q,p)$ and D is the common symbol for $\hat{Q}^i$, $\hat{P}_i$ and $_i$; the mean values m_k of the nine partitions of each universe of discourse D are defined as $m_k = \lambda_k D + \chi_{min}$, with partition coefficients λ_k uniformly spanning the range of D, corresponding to the set of nine linguistic variables $L = \{NL, NB, NM, NS, ZE, PS, PM, PB, PL\}$; standard deviations are kept constant $\sigma_k = D/9$. Using the linguistic vector L, the 9×9 FAM (fuzzy associative memory) matrix (a 'linguistic phase–plane'), is heuristically defined for each human joint, in a symmetrical weighted form

$$\mu_{kl} = \varpi_{kl}\, exp\{-50[\lambda_k + u(q,p)]^2\}, \qquad (k, l = 1, ..., 9)$$

   with weights $\varpi_{kl} \in \{0.6, 0.6, 0.7, 0.7, 0.8, 0.8, 0.9, 0.9, 1.0\}$.

2. Mamdani inference is used on each FAM–matrix μ_{kl} for all human joints: (i) $\mu(\hat{q}^i)$ and $\mu(\hat{p}_i)$ are combined inside the fuzzy IF–THEN rules using AND (Intersection, or Minimum) operator,

$$\mu_k[\bar{u}_i(q,p)] = \min_l\{\mu_{kl}(\hat{q}^i),\, \mu_{kl}(\hat{p}_i)\}.$$

(ii) the output sets from different IF–THEN rules are then combined using OR (Union, or Maximum) operator, to get the final output, fuzzy–covariant torques,

$$\mu[u_i(q,p)] = \max_k\{\mu_k[\bar{u}_i(q,p)]\}.$$

3. Defuzzification of the fuzzy controls $\mu[u_i(q,p)]$ with the 'center of gravity' method

$$u_i(q,p) = \frac{\int \mu[u_i(q,p)]\, du_i}{\int du_i},$$

to update the crisp feedback–control one–forms $u_i = u_i(t,q,p)$.

Finally, it is easy to make this top–level controller adaptive, simply by weighting both the above fuzzy–rules and membership functions, by the use of any standard competitive neural–network (see [Kos92]).

11

Conclusion

It is often held that things should always be made simple, which presumes that either that they can always be made simple or that all the jetisoned logic doesn't matter anyway. Alledgedly, anything should be explainable so that anyone can understand it. Don't get bogged down in dreary details. It should be effortless for the reader: low-dimensional systems exhibit complex behaviour while high-dimensional systems exhibit simple behaviour (to return to our prolegomonal opening), competition is a universal solution, demand must increase as price falls, and everything under the sun neatly fits a power law. Or so the story goes.

There's just one problem with all of this: it's plain wrong. Yet neither do we give much credence to clattering jargon for jargon's sake, with simple concepts wrapped up in verbiage to make it appear impressive; think of Hegelian discourse. Equations can be used to escape reality just as readily as to investigate it, and writing a differential equation doesn't by itself amount to the conduct of science. What of the apparent contradiction? The impetus for simplicity appears to be two-fold. Firstly, the principle of Occam's Razor is sometimes held to dictate that everything be made simple, period, and secondly, it is just easier to market a simple message. Yet this this is a blatant abuse of Occam's Razor, for the principle actually holds that among all the possible explanations whose predictions do not conflict with the available evidence, we should prefer the simplest. In other words, we yearn to keep it as simple as we are able but strictly without oversimplifying and sacrificing explanatory and predictive power. As for the second: there is much more to science than merely touring the conference circuit.

What we have attempted to do in this book is specifically to avoid making simplifying assumptions that keep our theory development within the bounds of familiar tools yet limit the power of our theory, and to avoid conveniently forgetting about those assumptions on the way to declaring that our conclusions constitute immutable truths. We have proposed a basis for a rigorous approach to behavioural complexity suited to addressing problems of prediction and control pertaining to both arbitrarily large sets of agents and simultaneously to the individual agents making up such groupings. Despite our efforts, it is not without limitations and we do not claim that it constitutes the final word in complexity, behaviour and control. Our emphasis on mathematical rigour pertains to our impatience with mere narratives generally constructed with the full benefits of hindsight, and our conviction that the models of any theory worthy of the name must make definitive predictions that can concievably be refuted upon contact with the phenonema the theory and its models are supposed to address. We admit that the consequence is a rather more steep entry price than many interested in complex systems will want to pay; the result of our investigation is admittedly not the most gentle of reads nor the most amenable to reduction to intuitively appealing shibboleths and nice quotidian anecdotes.

We have implemented this approach throughout the book. The local geometrical and topological theory of crowds uses Käler manifolds to give us, simultaneously, Lagrangian dynamics, Hamiltonian dynamics, and a complex dynamics resembling quantum dynamics. This setting, we have argued, is the richest currently available for representing the behavioural dynamics of intelligent entities physically moving about in the complex plane; in other words, we submit that this approach makes the smallest concessions to the need to make simplifying assumptions we can currently muster.

The problem of burgeoning complexity in our models is bound to swamp us at some point in our search for rigorous models of ever greater explanatory and predictive power. Making simplifying assumptions is acceptable and, indeed, inevitable; what is neither acceptable nor inevitable is the practice of forgetting about the boundaries our assumptions set on the precision and generality of our conclusions. We have attempted wherever possible to adopt an approach that is distinctly computer science in flavour, whereby we abstract over details rather than assuming them away. Not suprisingly in light of the prominent role Category Theory plays in modern computer science precisely for its abstractive power, we use Categories to construct a global view that hides and generalises over the particulars of individual theories and models.

Our subsequent modelling framework for simulation and control of crowds and groups is built on this idea, using the language of commutative diagrams, and constructing a nonlinear attractor dynamics in the Langevin style. Rather than a nonlinear Schrödinger equation, this approach is built on a geometrical interpretation using the Kähler Manifolds and the consequent Kähler-Ricci flow introduced earlier. The simulation uses a Runge-Kutta-Cash-Carp numerical integrator, complexified, to yield a Langevin-type crowd dynamics and finite-type control of hundreds of agents moving in the complex plane.

We presented our socio-cognio-physical dynamics in the form of the open Liouville equation to describe dynamics simultaneously at the macroscopic interpersonal crowd level and the microscopic intrapersonal individual level. At the microscopic level, our approach amounts to a nonlinear generalisation of the linear correlation and factor dynamics, which gives us an arrow of time and shows a formal self-similarity property. This, in turn, gives us a unique control law that acts across different scales.

Next, we expored the problem of designing a topological evolution of adaptable complex systems, and proposed a novel generalisation of a stochastic gradient-descent algorithm. The model hinges on an Hamiltonian energy landscape cost funtion derived from a fleet configuration matrix. The resulting system adaption is topological in the sense that it satisfies a topological invariant and in that the evolution is graph-theoretic.

We have reviewed and analysed models of controlled complexity in nonlinear pulse conduction, starting with the Hodgkin-Huxley model of pulse conduction in neural fibres and advancing to rogue wave phenonema in optical fibres. We proposed a novel sine-Gordon alternative to the venerable Hodgkin-Huxley model, which explains pulse conduction more generally in terms of wave phenonema, including kinks/anti-kinks, solitons and breathers.

Our quantum-dynamical model based on Perceptual Control Theory uses large networks of Josephson Junctions in a multi-layered architecture, which returns us to the point about the need for abstraction rather than simplification wherever we can manage it: intuitive Category-Theoretic abstractions hide necessary but potentially overwhelming algebraic details from view while nonetheless enabling us to ensure rigorously correct function compositions. This represents a new take on Perceptual Control Theory, which provides a unifying framework for modelling psycho-physical systems, but is essentally model-free.

We built on this Perceptual Control Theory basis further in examining military Command and Control (and other organisational decision-making settings), where our modelling approach allows for the possibility that randomness plays a central role in selecting the path of realised events from all those that events are possible. This establishes entropy as a central order parameter by which we can, roughly speaking, quantify the degree of uncertainty about outcomes. We offer several views of this development: a physical view of partition functions and path integrals (which we can roughly describe as a sum over possible histories), a global view in which we employ our Category-Theoretic machinery to abstract away from details, a local geometrical view of the Kähler-Ricci flow, and a computational view using approximations of path integrals.

We have also presented a couple-loading rate hypothesis for brain and spinal injuries, holding that such injuries are caused by impulsive loading in several coupled degrees-of-freedom simultaneously – a Euclidean jolt. It is never a static force and never occurs in just one direction, but is impulsive transational and/or rotational and coupled to a mass eccentricity. The hypothesis allows prediction and thus prevention of injury, and pertains to both sudden motion injury and to mass distribution injury; it is motivated by the realisation that the traditional principal loading hypothesis that describes injury in terms of bending, shear, tension and compression does not suffice to predict and thus prevent the occurrence of injury.

The theme behind each of these components to our study of complexity, control and behaviour is that it doesn't suffice in the study of complexity to create anecdotes or narratives after the fact, and employing simplifying assumptions that keep us within the boundaries of our familiar tools will run out of steam surprisingly quickly. Perhaps it will not be so surprising to those who are naturally curious about the limitations and weaknesses that all assumptions necessarily impose on our conclusions and naturally suspicious of claims that overlook these caveats, whatever the intended ends. Applying more powerful – and necessarily more difficult – rigorous methods is necessary, but so too is using powerful mathematical means of preserving rigor while hiding as many details as possible under layers of abstraction. In fact, we submit that the future success of research in complexity topics in the broad hinges more, if anything, on the latter than on the former. We hope that our contribution will stimulate further investigation, and that in the application of our ideas, as attempts to explain phenonema in the real world, their weaknesses and limitations will eventually reveal themselves. In doing so, our readers will further the growth of knowledge and hopefully propose new theories behaviour, complexity and control of greater predictive power.

12

Code Samples Used for Complexity and Control

12.1 *Mathematica*® Code

12.1.1 Generic Chaotic Simulator

In this notebook we are experimenting with seven 3D nonlinear attractor dynamical systems, which are all chaotic for certain values of their parameters (a,b,c). Three forms of their presentation are given: (i) $x - y - z$ time-series, (ii) 3D phase space, and (iii) $x - y - z$ FFT-spectrum. Implemented chaotic attractor systems are defined by the following sets of three 1st-order attractor ODEs (see Figure 12.1 for the case of the Forced Van der Pol oscillator):

1. Lorenz attractor: $\dot{x} = a(y - x),\ \dot{y} = x(b - z) - y,\ \dot{z} = xy - cz.$
2. Rossler attractor: $\dot{x} = -y - z,\ \dot{y} = x + ay,\ \dot{z} = b + z(x - c).$
3. Forced Duffing oscillator: $\dot{x} = y,\ \dot{y} + a\dot{x} - bx + cx^3 = z,\ \dot{z} = \cos(5t).$
4. Forced Van der Pol oscillator: $\dot{x} = y,\ \dot{y} - a(1 - 4bx^2)y + c^2x = z,\ \dot{z} = \cos(5t).$
5. Another1 attractor: $\dot{x} = ax + yz,\ \dot{y} = by - xz,\ \dot{z} = cz + xy.$
6. Another2 attractor: $\dot{x} = -y - z,\ \dot{y} = x - 0.5ay,\ \dot{z} = 0.1 + |b| + xz + 10cz.$
7. Another3 attractor: $\dot{x} = ayz^2,\ \dot{y} = |b|xz,\ \dot{z} = cxy^2.$

```
Manipulate[
 Module[{Lorenz, Rossler, ForceDuffing, ForceVanDerPol,
 Another1,
   Another2, Another3, sol, x, y, z, t, pl1, pl2, pl3, data,
   fdata},

Lorenz = {{x'[t] == a (y[t] - x[t]), y'[t] == x[t] (b - z[t])
- y[t],
z'[t] == x[t] y[t] - c z[t], x[0] == y[0] == 20, z[0]
== 100}};
```

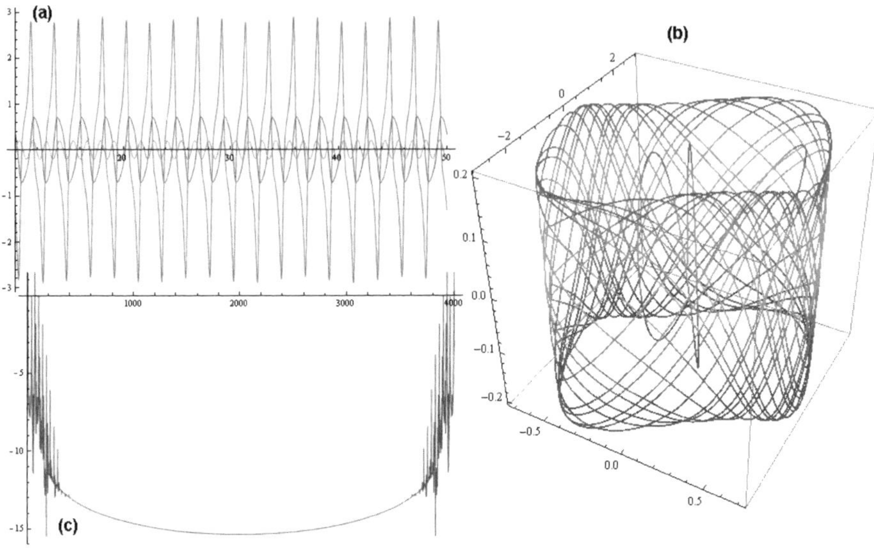

Fig. 12.1. *Numerical simulation of the Forced Van der Pol oscillator: (a) time plot, (b) phase space, and (iii) FFT spectrum.*

```
Rossler = {{x'[t] == -y[t] - z[t], y'[t] == x[t] + \[Alpha]1
y[t],
z'[t] == \[Beta]1 + z[t] (x[t] - \[Gamma]1), x[0] == 0.2,
y[0] == 0.3,
z[0] == 0.5}};

ForceDuffing = {{x'[t] == y[t], y'[t] + \[Alpha]2 x'[t] -
\[Beta]2 x[t]
+ \[Gamma]2 x[t] == z[t], z'[t] == Cos[5 t], x[0] == y[0] ==
z[0] == 0}};

ForceVanDerPol = {{x'[t] == y[t], y'[t] - \[Alpha]3
(1-4\[Beta]3 x[t])y[t]
+ \[Gamma]3 x[t] == z[t], z'[t] == Cos[5 t], x[0] == y[0] ==
z[0] == 0}};

Another1 = {{x'[t] == a1 x[t] + y[t] z[t], y'[t] == b1 y[t]
- x[t] z[t],
z'[t] == c1 z[t] + x[t] y[t], x[0] == -1, y[0] == 1, z[0]
== 2}};

Another2 = {{x'[t] == -y[t] - z[t], y'[t] == x[t] - 0.5 a2
y[t],
```

```
z'[t] == 0.1 + Abs[b2] + x[t] z[t] + 10 c2 z[t], x[0] == y[0]
== -1.2, z[0] == 1.2}};

Another3 = {{x'[t] == a3 y[t] z[t], y'[t] == Abs[b3] x[t]
z[t], z'[t] == c3 x[t] y[t], x[0] == 1, y[0] == 2,
z[0] == 1.8}};

sol = Quiet[NDSolve[ForceVanDerPol, {x[t], y[t], z[t]},
{t, 0, 50}, MaxSteps -> Infinity]];

data = Transpose[Table[Evaluate[x[t] y[t] z[t] /. sol],
{t,10,50,0.01}]];
fdata = Log[Abs[Fourier[data]] ];

pl1 = Plot[Evaluate[{x[t], y[t], z[t]} /. sol], {t,10,50},
PlotRange->All,
PlotStyle -> AbsoluteThickness[1.5], ImageSize -> {500,
350}];
pl2 = ParametricPlot3D[Evaluate[{x[t], y[t], z[t]} /. sol],
{t, 0, 50},
BoxRatios -> {1, 1, 1}, PlotRange -> All, ColorFunction ->
Rainbow,
PlotStyle -> AbsoluteThickness[1.5], ImageSize -> {500,
400}];
pl3 = ListLinePlot[fdata, PlotRange -> All, PlotStyle ->
AbsoluteThickness[1.1], ImageSize -> {500, 350}];

Which[Analysis == "Time series", Show[pl1], Analysis
== "Phase space",
    Show[pl2], Analysis == "FFT spectrum", Show[pl3]]],

 Delimiter, Style["Lorenz", 10],
 {{a, 16, "a"}, 16, 20, 0.1, ImageSize -> Small,
  Appearance -> "Labeled"}, {{b, 45.92, "b"}, 45, 56, 1,
  ImageSize -> Small, Appearance -> "Labeled"}, {{c, 4, "c"},
  1.0, 6, 0.1,
  ImageSize -> Small, Appearance -> "Labeled"},
 Delimiter, Style["Rossler", 10],
 {{\[Alpha]1, 0.2, "a"}, 0.15, 0.3, 0.01, ImageSize -> Small,
  Appearance -> "Labeled"}, {{\[Beta]1, 0.2, "b"}, 0.15, 0.4,
  0.01,
  ImageSize -> Small,
  Appearance -> "Labeled"}, {{\[Gamma]1, 3.0, "c"}, 1, 10,
  0.1,
  ImageSize -> Small, Appearance -> "Labeled"},
```

```
Delimiter, Style["ForceDuffing", 10],
{{\[Alpha]2, 1.5, "a"}, 0, 3, 0.01, ImageSize -> Small,
 Appearance -> "Labeled"}, {{\[Beta]2, 0.25, "b"}, 0, 1,
 0.01,
 ImageSize -> Small, Appearance -> "Labeled"}, {{\[Gamma]2,
 2, "c"}, 1, 3, 0.01,
 ImageSize -> Small, Appearance -> "Labeled"},
Delimiter, Style["ForceVanDerPol", 10],
{{\[Alpha]3 , 3, "a"}, 1, 9, 0.1, ImageSize -> Small,
 Appearance -> "Labeled"}, {{\[Beta]3, 2, "b"}, 2, 9, 0.1,
 ImageSize -> Small, Appearance -> "Labeled"}, {{\[Gamma]3,
 3, "c"}, 1, 9, 0.1,
 ImageSize -> Small, Appearance -> "Labeled"},
Delimiter, Style["Another1", 10],
{{a1, -0.4, "a"}, -1, -0.25, 0.001, ImageSize -> Small,
 Appearance -> "Labeled"}, {{b1, 0.3, "b"}, -1, 1, 0.001,
 ImageSize -> Small, Appearance -> "Labeled"}, {{c1, -0.1,
 "c"}, -1, 0, 0.001,
 ImageSize -> Small, Appearance -> "Labeled"},
Delimiter, Style["Another2", 10],
{{a2, -0.4, "a"}, -1, -0.25, 0.001, ImageSize -> Small,
 Appearance -> "Labeled"}, {{b2, 0.3, "b"}, -1, 1, 0.001,
 ImageSize -> Small, Appearance -> "Labeled"}, {{c2, -0.8,
 "c"}, -1,   0, 0.001, ImageSize -> Small, Appearance
 -> "Labeled"},
Delimiter, Style["Another3", 10],
{{a3, -0.4, "a"}, -1, -0.25, 0.001, ImageSize -> Small,
 Appearance -> "Labeled"}, {{b3, -0.7, "b"}, -1, 1, 0.001,
 ImageSize -> Small, Appearance -> "Labeled"}, {{c3, -0.8,
 "c"}, -1,   0, 0.001, ImageSize -> Small, Appearance ->
 "Labeled"},
Delimiter, {{Analysis, "Time series"}, {"Time series",
"Phase space", "FFT spectrum"}}, TrackedSymbols
-> Manipulate]
```

12.1.2 Vector Differential Operators

Here we give *Mathematica*® functions for definition/calculation of various vector differential operators. Note that, functional programming is used for this objective.

```
GM[x_List] := Times @@ x^(1/Length[x])      (* Geometric
mean *)

Nrm[x_List] := Sqrt[x^2 /. List -> Plus]      (* Norm *)
```

```
Grad[s_, x_List] := (D[s, #1] & ) /@ x      (* Gradient *)

GradSys[s_,x_List] := (D[#1,t] == -D[s,#1]&)/@x  (* Grad.
system *)

JacMat[v_List, x_List] := Outer[D, v, x]    (* Jacobian
matrix *)

Hess[s_, x_List] := Outer[D, Grad[s, x], x]   (* Hessian
matrix *)

Div[v_List, x_List] := Inner[D, v, x, Plus]   (* Divergence
*)

Laplacian[s_, x_List] := Div[Grad[s, x], x]   (* Laplacian *)

LieDer[v_List, s_, x_List] := Grad[s, x] . v   (* Lie
derivative *)

KLieDer[v_List, s_, x_List, k_] :=   (* k-th order Lie
derivative *)
  Block[{t}, p := s; If[k == 0, p = s, Do[p = LieDer[v, p,
  x],
      {k}]]; p]

LieBrc[u_List, v_List, x_List] := JacMat[v, x] . u
   - JacMat[u, x] . v                       (* Lie bracket
   *)

Adj[u_List, v_List, x_List, k_] := If[k == 0, v,
   LieBrc[u, Adj[u, v, x, k - 1], x]]   (* k-th order Lie
   bracket *)

PoisBrc[u_List, v_List, x_List, y_List] :=   (* Poisson
bracket *)
  JacMat[u, x] . JacMat[v, y] - JacMat[u, y] . JacMat[v, x]
```

12.1.3 NLS Explorer

The following *Mathematica*® code explores various forms of numerical solutions of the (1+1)D nonlinear Schrödinger equation (NLS).

```
V[t_, x_] := a*Sech[a*t*x^3];
IC[x_] := c*Exp[I*x]*Sech[c*x];
Tfin = 100; L = 60; a = b = c = 0.5;
```

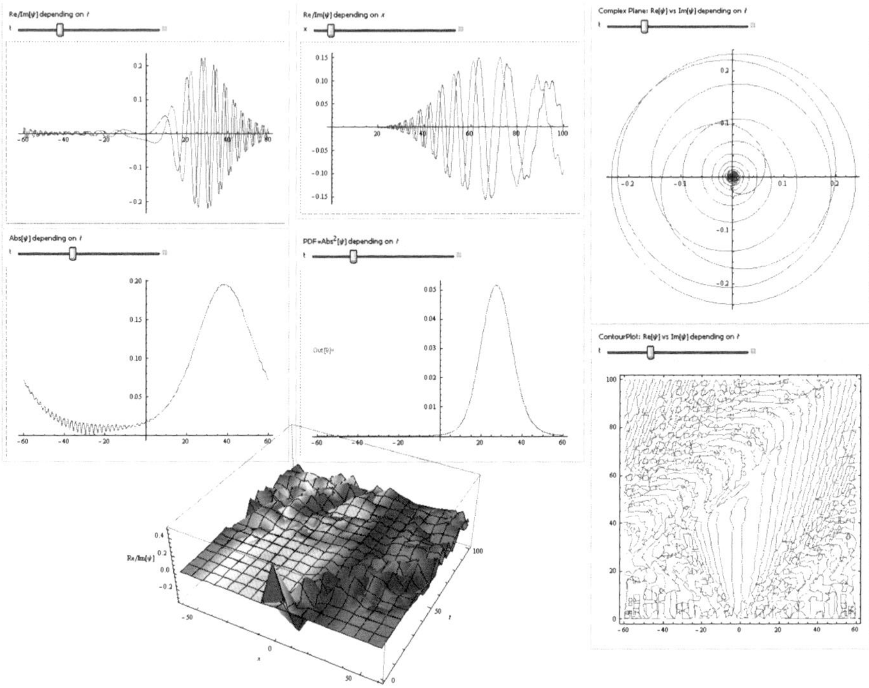

Fig. 12.2. NLS Explorer.

```
NLS = I*D[\[Psi][t, x], t] == (-(1/2))*D[\[Psi][t, x], x, x]
    + V[t, x]*Abs[\[Psi][t, x]]^2*\[Psi][t, x];
sol = NDSolve[{NLS, \[Psi][0, x] ==
     IC[x], \[Psi][t, -L] == \[Psi][t, L]}, \[Psi], {t, 0,
    Tfin}, {x, -L, L}, Method -> {"MethodOfLines",
     "SpatialDiscretization" -> {"TensorProductGrid",
       "DifferenceOrder" -> "Pseudospectral"}}];
Manipulate[pl1 = Plot[{Evaluate[Re[\[Psi][t, x] /.
First[sol]]],
    Evaluate[Im[\[Psi][t, x] /. First[sol]]]}, {x, -L, L},
       PlotRange -> All, PlotStyle -> Thickness[0.004]],
   Style["Re/Im[\[Psi]] depending on t"])], {t, 0, Tfin}]

Manipulate[pl2 = Plot[{Evaluate[Re[\[Psi][t, x] /.
First[sol]]],
    Evaluate[Im[\[Psi][t, x] /. First[sol]]]}, {t, 0, Tfin},
       PlotRange -> All, PlotStyle -> Thickness[0.004]],
   Style["Re/Im[\[Psi]] depending on x"])], {x, -L, L}]
```

```
Manipulate[pl3 = Plot[Evaluate[Abs[\[Psi][t, x]/.
First[sol]]], {x,-L,L},
   PlotRange -> All, PlotStyle -> Thickness[0.004]],
   Style["Abs[\[Psi]] depending on t"])], {t, 0, Tfin}]

Manipulate[pl4 = Plot[Evaluate[Abs[\[Psi][t,x]/.
First[sol]]^2],{x,-L,L},
   PlotRange -> All, PlotStyle -> Thickness[0.004]],
   Style["PDF=[\(Abs\), \(2\)]\)[\[Psi]]
   depending on t"])], {t, 0, Tfin}]

Manipulate[pl5 = ParametricPlot[{Evaluate[Re[\[Psi][t, x]/.
First[sol]]],
    Evaluate[Im[\[Psi][t, x] /. First[sol]]]}, {x, -L, L},
       PlotRange -> All, PlotStyle -> Thickness[0.004]],
   Style["Complex Plane: Re[\[Psi]] vs Im[\[Psi]]
   depending on t"])],
   {t, 0, Tfin}]

Manipulate[pl0 = Plot3D[{Evaluate[Re[\[Psi][t, x] /.
First[sol]]],
    Evaluate[Im[\[Psi][t, x] /. First[sol]]]}, {x, -L, L},
    {t, 0, Tfin}, PlotRange -> All, ColorFunction ->
    (Hue[#1] & ),
   AxesLabel -> x,t, "Re/Im[\[Psi]]"}, ImageSize -> 500],
   Style["3DPlot of Re[\[Psi]] and Im[\[Psi]] depending on t
   and x"])],
    {t, 0, Tfin}, {x, -L, L}]

Manipulate[pl6 = ContourPlot[{Evaluate[Re[\[Psi][t, x] /.
First[sol]]],
    Evaluate[Im[\[Psi][t, x] /. First[sol]]]}, {x, -L, L},
       {t, 0, Tfin}, PlotRange -> All],
   Style["ContourPlot: Re[\[Psi]] vs Im[\[Psi]] depending on
   t"])],
   {t, 0, Tfin}]

Manipulate[pl8 = StreamPlot[{Evaluate[Re[\[Psi][t, x] /.
First[sol]]],
    Evaluate[Im[\[Psi][t, x] /. First[sol]]]}, {x, -L, L},
       {t, 0, Tfin}, PlotRange -> All], Style["StreamPlot:
       Re[\[Psi]]
       vs Im[\[Psi]] depending on t"])], {t, 0, Tfin}]
```

The output of the code is given in Figure 12.2.

12.2 C++ Code

12.2.1 C++ Lambda Functions for Real Calculus

The following sample C++ code provides recursive numerical integration and differentiation of branching functions (see Figure 12.3) using auto-lambda functions (C++11 standard).

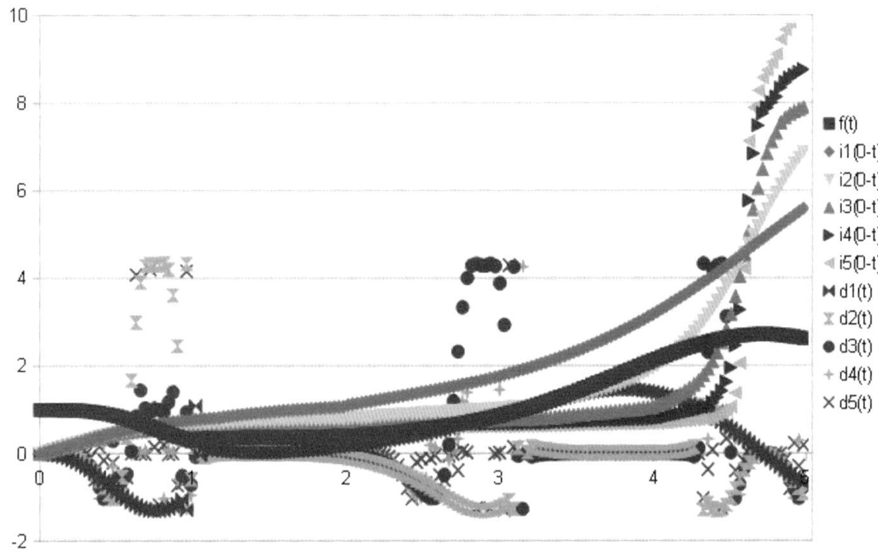

Fig. 12.3. *Recursive numerical integration and differentiation of branching functions using C++ auto-lambda functions.*

```
// Recursive integrator & differentiator of branching
functions
// using auto-lambda functions (C++11 standard)

#include <iostream>
#include <fstream>
#include <cmath>

using namespace std;

double t=0., h=0.03, tFin=5.;
// Define a branching function (using cond.op)
auto f = [] (double x)  { return (x > 1) ? exp(-sin(x))
*tanh(x*x) :
```

```
    exp(-sin(x*x*x))/cosh(x*x); };
// Weddle integration formula
auto ii = [] (double a,double b) { return
    3*(b-a)*(f(a)+5*f((5*a+b)/6)+f((4*a+2*b)/6)+6*f((3*a+3*b)
    /6)
    + f((2*a+4*b)/6)+5*f((a+5*b)/6)+f(b))/60; };
// 7-point central diff. formula
auto dd = [] (double x) { return
    (-f(x-3*h)+9*f(x-2*h)-45*f(x-h)
    + 45*f(x+h)-9*f(x+2*h)+f(x+3*h))/(60*h); };

int main() {
    ofstream file ("derInt.csv");
    file << "t" <<","<< "f(t)" <<","<< "i1(0-t)" <<","<<
    "i2(0-t)" <<","<< "i3(0-t)" <<","<< "i4(0-t)" <<","<<
    "i5(0-t)" <<","<< "d1(t)" <<","<< "d2(t)" <<","<< "d3(t)"
    <<","<< "d4(t)" <<","<< "d5(t)" << endl;
    while (t < tFin)
    {// Calling higher integrals and derivatives recursively
        file << t <<","<< f(t) <<","<< ii(0,t) <<","<<
    ii(0,ii(0,t)) <<","<< ii(0,ii(0,ii(0,t))) <<","<<
    ii(0,ii(0,ii(0,ii(0,t)))) <<","<<
    ii(0,ii(0,ii(0,ii(0,ii(0,t))))) <<","<< dd(t) <<","<<
    dd(dd(t)) <<","<< dd(dd(dd(t))) <<","<< dd(dd(dd(dd(t))))
    <<","<< dd(dd(dd(dd(dd(t))))) << endl;
        t+=h;
      }
    file.close(); return 0;
}
```

12.2.2 Accelerometer Data Processor

Here we give a C++ code that implements input reading of the N-column accelerometer data, and then performs its basic processing (calculating velocities, displacements and jerks) with an output to a csv-file.

```
// N-column accelerometer processor
// Simple Euler integration and derivative

#include <stdio.h>
#include <math.h>
#include <string.h>

#define n 9                 // number of data-columns
```

```
double Tfin = 19.5;      // final time
double dt = 0.1;         // time step
double t = 0.0;          // initial time
double Gdata[n],x[n],v[n],a[n],aPrev[n],j[n];

void header(FILE* pOut) {
    fputs("t", pOut);
    for(int i=0; i<n; i++)
       fprintf(pOut,",a%i,v%i,x%i,j%i",i,i,i,i);
       fputs("\n", pOut);
}

void print(FILE* pOut) {
    fprintf(pOut, "%.4f", t);
    for(int i=0; i<n; i++)
       fprintf(pOut,",%.7f,%.7f,%.7f,%.7f", a[i],v[i],x[i],
       j[i]);
       fprintf(pOut,"\n");
}

int main() {
    char sLine[512], *pch;
    FILE* pIn; FILE* pOut;
    for(int i=0; i<n; i++) {
        x[i] = 0.0;          // initial displacement
        v[i] = 0.0;          // initial velocity
        a[i] = 0.0;          // initial acceleration
        aPrev[i] = 0.0;      // initial accPrev
        j[i] =0.0;           // initial jerk
    }

    pIn = fopen("9cols.csv", "r");                  // open input
    file
    pOut = fopen("VelocJerk9.csv", "w");            // open
    output file
    header(pOut);                                   // print
    header

    while (t < Tfin)    {                              // main
    time loop

        fscanf(pIn,"%s",sLine);
        pch = strtok(sLine,",");

        for(int i=0; i<n; i++) {
```

```
        // G-data input:
            pch = strtok(NULL,",");
            Gdata[i] = atof(pch);

            // Euler integrations:
            a[i] = 9.81*Gdata[i];         // acceleration
            [m/s^2]
            v[i] += a[i]*dt;              // velocity [m/s]:
            v = v0 + a*dt
            x[i] += v[i]*dt;              // displacement [m]:
            x = x0 + v*dt

            // Derivative:
            j[i] = (a[i]-aPrev[i])/dt;  // jerk [m/s^3]:
            j = (a - a0)/dt
            aPrev[i] = a[i];
        }
        t += dt;                        // time increment
        print(pOut);                    // data printout
    }
    fclose(pIn);  fclose(pOut);
}
```

12.2.3 Simple Predictor-Corrector Integrator

Here we give the simplest C++ implementation of the *predictor-corrector ODE-integrator* (see Figure 12.5).

```
// Simple predictor-corector ODE-integrator
// Simulating a 1D particle under elastic force
#include <iostream>
#include <fstream>
#include <cmath>
using namespace std;

#define n 1001
#define dt 0.01
#define m 5

int main()
{
    ofstream fOut("fOut.csv");
    double t[n], x[n], v[n];
    t[0]=0;  x[0]=0;  v[0]=1;
    for (int i=0; i<n-1; ++i)
```

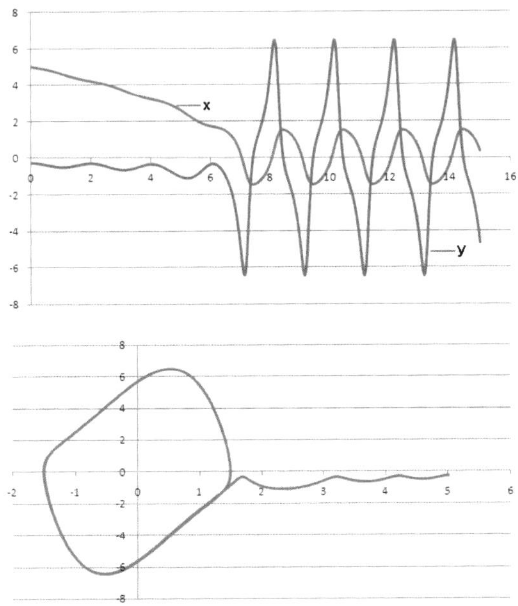

Fig. 12.4. *Example for the Shooting method: solving the BVP for the Forced Van der Pol oscillator (top: time series, bottom: phase-plot).*

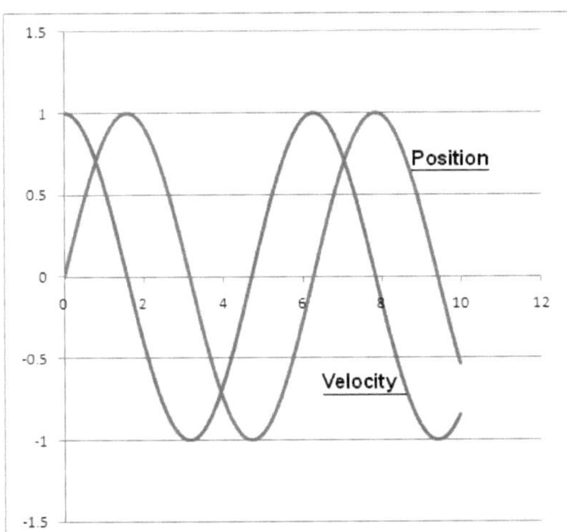

Fig. 12.5. *Simulation of a 1D particle under elastic force, using the simple predictor-corrector ODE-integrator.*

```
    {
        t[i+1] = (i+1)*dt;

        // Predictor for position and velocity
        x[i+1] = x[i]+v[i]*dt;
        v[i+1] = v[i]-x[i]*dt;

        // Corrector for position and velocity
        x[i+1] = x[i]+(v[i]+v[i+1])*dt/2;
        v[i+1] = v[i]-(x[i]+x[i+1])*dt/2;

        if (i == 0 || (i+1)%m == 0)  // printing every m-th
        step
            fOut << t[i] <<","<< x[i] <<","<< v[i] << endl;
    }
        fOut.close();
}
```

12.2.4 Solving the BVP with the Shooting Method

Here we give a C++ code that solves the boundary-value problem (BVP) for a 2nd-order nonlinear ODE, using the Shooting Method: combining the *RK4 integrator* with the *secant* root-finder (see Figure 12.4). The code also demonstrates the use of OpenMP for multi-threaded parallel processing.

```
// Solving BVP for the Forced Van der Pol Oscillator using
the Shooting Method:
// Combining the RK4 integrator with the Secant root-finder.

#include <stdio.h>
#include <math.h>
// using OpenMP for multi-threaded parallel processing
#include <omp.h>

#define n 10001        // total number of steps
#define ta 0.0         // initial and final times
#define tb 15.0
#define xa 5.0         // initial and final positions
#define xb 0.4
#define pi 3.141592653589793
#define a 12.0        // system parameters
#define b 9.0
#define c 0.6

// Using Lambda-functions
```

```
auto gX = [] (double x1,double x2,double t) { return x2; };
// x1=position, g=x2=velocity
auto gY = [] (double x1,double x2,double t)
{ return a*cos(pi*t) + c*(1-b*x1*x1)*x2 - pi*pi*x1; };

// Alternative - old C-style
//#define gX(x1,x2,t) (x2)    // System ODEs: x'=y=gX,
x''=y'=gY
//#define gY(x1,x2,t) (a*cos(pi*t) + c*(1-b*x1*x1)*x2 -
pi*pi*x1)

double t, h;  double x[2][n];  FILE* pOut;
void print(FILE*);          //function prototypes
void RK4(double [2][n]);

int main() {
  double d=0.1, dt=0.01, tol=1e-3;
  double t0, t1, t2, x1, x2, f0, f1;
  #pragma omp parallel    // using OpenMP
  pOut = fopen("ShootVDPforc.csv", "w");    // open output
  file
  h  = (tb-ta)/(n-1);       // calculated time-step
  t0 = (xb-xa)/(tb-ta);     // initial slope calculation
  t1 = t0 + dt;             // initial slope increment
  x[0][0] = xa;             // position = initial position
  while (fabs(d) > tol) {   // secant search for the root
    // RK4: first trial solution
    x[1][0] = t0;           // velocity = slope
    RK4(x);                 // call RK4
    f0 = x[0][n-1]-1;       // first function value
    calculated by RK4
    // RK4: second trial solution
    x[1][0] = t1;           // velocity = slope increment
    RK4(x);                 // call RK4
    f1 = x[0][n-1]-1;       // second function value
    calculated by RK4
    // perform secant search
    d = f1-f0;                // difference of the two
    function values
    t2 = t1-f1*(t1-t0)/d;     // second slope increment
    t0 = t1;                  // swap slopes and their
    increments
    t1 = t2;
  }
  print(pOut);        // data printout
```

```
  printf("tB,xB = %.2f,%.2f\n", tb,xb);
  printf("slope t0 = %.7f\n", t0);
}   // main

void print(FILE* pOut) {
  fputs("t,x,y\n", pOut);
  for (int i=0; i<n; i++) {
    t = h*i;
    fprintf(pOut, "%.4f,%.7f,%.7f\n", t, x[0][i], x[1][i]);
  }
}

void RK4(double x[2][n]) {     // RK4 algorithm for (gX,gY)
  double k11, k21, k12, k22, k13, k23, k14, k24;
  double x1, x2;
  for (int i=0; i<n-1; i++) {
    t = ta+h*(i+1);
    x1 = x[0][i];
    x2 = x[1][i];
    k11 = h*gX(x1,x2,t);
    k21 = h*gY(x1,x2,t);
    k12 = h*gX((x1+k11/2),(x2+k21/2),(t+h/2));
    k22 = h*gY((x1+k11/2),(x2+k21/2),(t+h/2));
    k13 = h*gX((x1+k12/2),(x2+k22/2),(t+h/2));
    k23 = h*gY((x1+k12/2),(x2+k22/2),(t+h/2));
    k14 = h*gX((x1+k13),(x2+k23),(t+h));
    k24 = h*gY((x1+k13),(x2+k23),(t+h));
    x[0][i+1] = x[0][i]+(k11+2*(k12+k13)+k14)/6;
    x[1][i+1] = x[1][i]+(k21+2*(k22+k23)+k24)/6;
  }
}
```

12.2.5 Linear Hyperbolic PDE Solver

Here, we show a finite-difference method based wave equation solver (see Figure 12.6).

```
// Linear Hyperbolic PDE Solver
// Algorithm: Finite-Difference Method

// Approximate solution of the Wave PDE: u_(tt)(x,t)
=c^2u_(xx)(x,t)
// over  R = {(x,t): 0 <= x <= a, 0 <= t <= b}  with
// u(0,t) = 0,  u(a,t) = 0   for  0 <= t <= b   and
// u(x,0) = f(x),  u_t(x,0) = g(x)   for  0 <= x <= a
```

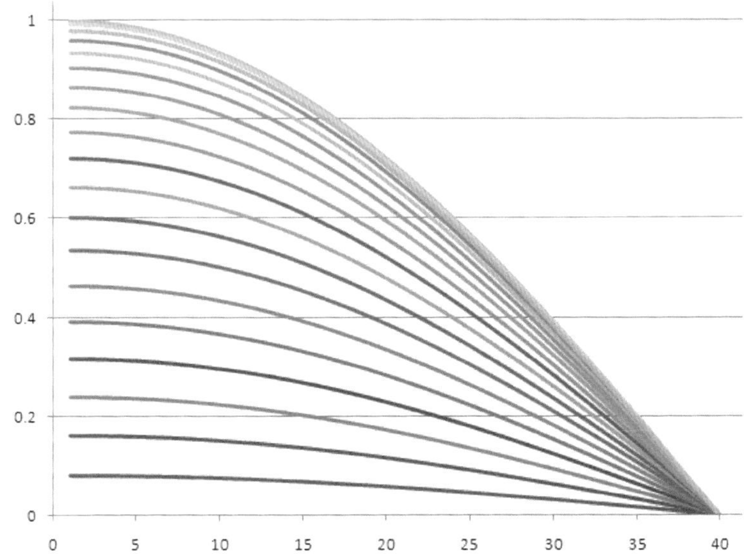

Fig. 12.6. *Solving the (1+1)-wave equation, using the finite-difference method.*

```
#include<stdio.h>
#include<stdlib.h>
#include<math.h>
#include<time.h>

#define Max 50

// Global Variables

double h;                    // step size

// Prototypes
double Fi(int i);
double Gi(int i);

// Grid function for amplitude
double Fi(int i)
{
extern double h;
double arg;
arg = h * (i - 1);
if( (arg >= 0) && (arg <= 1.0) )  return ( sin
```

```
(3.1415926*arg) );
else
{
printf(" Fi() :Argument not in range ! Exiting !\n");
printf(" arg  : %lf\n", arg);
exit(0);
}
}

// Grid function for velocity
double Gi(int i)
{
extern double h;
double arg;
arg = h * (i - 1);
if( (arg >= 0) && (arg <= 1) ) return ( 0.0 );
else
{
printf(" Gi() : Argument not in range ! Exiting !\n");
printf(" arg  : %lf\n", arg);
exit(0);
}
}

int main()
{
double A, B;                // INPUT : width and height of
Rectangle
double C;                   // INPUT : wave equation const
int N, M;                   // dimensions of the grid
double K, R, R2, R22, S1, S2;
double U[Max][Max];         // grid-amplitudes

double t  = 0.0;     // Initial time
int ticks;
double ms;
clock_t start, stop;
double prev_t = -1;  // Previous t

FILE* pOut;                          // File pointer to
output file
pOut = fopen("out.csv", "w");        // Write header
start = clock() * CLK_TCK;           // Clock setup

A = 1.0;          // interval boundary of x
```

```
B = 1.0;          // interval boundary of t
N = 40;           // dimension of grid in x-direction
M = 40;           // dimension of grid in y-direction
C = 0.5;          // wave equation constant C

// Compute step sizes
h   = A / ( N - 1 );
K   = B / ( M - 1 );
R   = C * K / h;
R2  = R * R;
R22 = R * R / 2.0;
S1  = 1.0 - R * R;
S2  = 2.0 - 2.0 * R * R;

// Boundary conditions
for ( int j = 1; j <= M; j++ )
{
U[1][j] = 0;
U[N][j] = 0;
}

// First and second rows
for ( int i = 2; i <= N - 1 ; i++ )
{
U[i][1] = Fi(i);
U[i][2] = S1 * Fi(i) + K * Gi(i) + R22 * ( Fi(i+1) +
Fi(i-1) );
}

// Generate new waves
for ( int j = 3; j <= M; j++ )
{
for ( int i = 2; i <= N - 1; i++ )
{
U[i][j]=S2*U[i][j-1] + R2*(U[i-1][j-1] +
U[i+1][j-1] ) - U[i][j-2];
}
}

// Output the solution
for ( int j = 1; j <= M; j++ )
{
fprintf(pOut,"%8.6lf ", K * (j - 1));
for ( int i = 1; i <= N; i++ ) fprintf(pOut,"
%8.6lf,", U[i][j]);
```

```
fprintf(pOut,"\n");
}
fclose(pOut);
stop  = clock() * CLK_TCK;                    // Timing
ticks = stop - start;
ms = (float)ticks / 1000.0f;
printf("finished... (ms = %f)\n", ms);
getchar();  return 0;
}
```

12.2.6 Linear Elliptic PDE Solver

Here, we show a *Dirichlet-grid algorithm* based Laplace PDE-solver (see Figure 12.7).

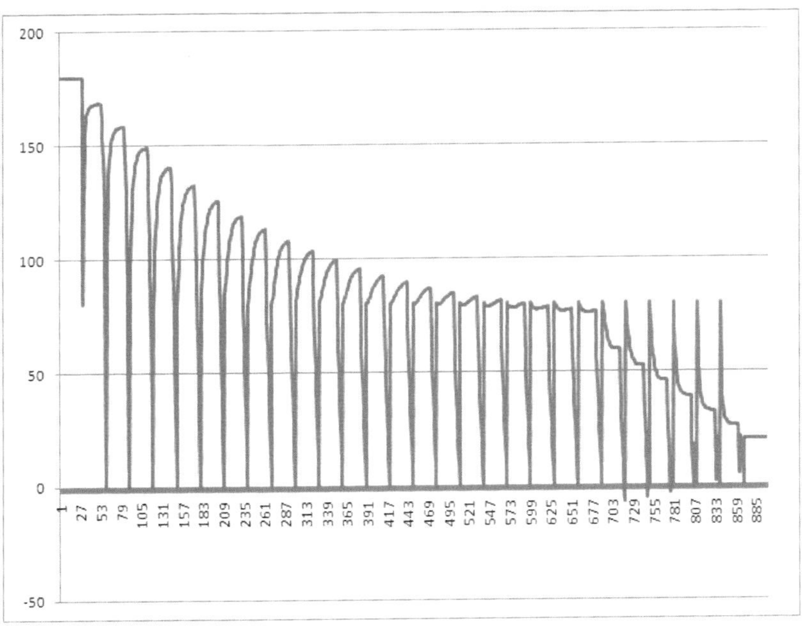

Fig. 12.7. *Solving the 2D Laplace equation, using the Dirichlet grid method.*

```
// Linear Elliptic PDE Solver
// Dirichlet Grid Method for the Laplace Equation
// Approximate solution of the PDE: u_(xx)(x,y) + u_(yy)(x,y)
= 0
// over R = {(x,y): 0 <= x <= a,  0 <= y <= b }   with
// u(x,0) = f_1(x),  u(x,b) = f_2(x)  for  0 <= x <= a   and
```

```
// u(0,y) = f_3(y),  u(a,y) = f_4(y)  for  0 <= y <= b

#include<stdio.h>
#include<stdlib.h>
#include<math.h>
#include<time.h>

#define Max 50
double h;                    // step size

// Prototypes
double F1i(int i);
double F2i(int i);
double F3i(int i);
double F4i(int i);

// Grid function for amplitude

double F1i(int i)
{
extern double h;
double arg;
arg = h * ( i - 1.0 );
if( (arg >= 0) && (arg <= 4.0) )  return ( 180.0 );
else
{
printf(" F1i() :Argument not in range !
Exiting !\n");
printf(" arg  : %lf\n", arg);
exit(0);
}
}

double F2i(int i)
{
extern double h;
double arg;
arg = h * ( i - 1.0 );
if( (arg >= 0) && (arg <= 4) ) return ( 20.0 );
else
{
printf(" F2i() :Argument not in range !
Exiting !\n");
printf(" arg  : %lf\n", arg);
exit(0);
```

```
}
}

double F3i(int i)
{
extern double h;
double arg;
arg = h * ( i - 1.0 );
if( (arg >= 0) && (arg <= 4) ) return ( 80.0 );
else
{
printf(" F3i() :Argument not in range !
Exiting !\n");
printf(" arg  : %lf\n", arg);
exit(0);
}
}

double F4i(int i)
{
extern double h;
double arg;
arg = h * ( i - 1.0 );
if( (arg >= 0) && (arg <= 4) ) return ( 0.0 );
else
{
printf(" F4i() :Argument not in range !
Exiting !\n");
printf(" arg  : %lf\n", arg);
exit(0);
}
}

int main()
{
double A, B;              // INPUT : Rectangle width and
height
double Ave;               // INPUT : Initial approximation
int N, M;                 // dimensions of the grid
double U[Max][Max];       // grid-amplitudes
int Count;                // counter for while loop
double Tol;               // stop criteria
double w, Pi;
double h;                 // grid spacing
double Relax, temp;
```

```
double t  = 0.0;     // Initial time
int ticks;
double ms;
clock_t start, stop;
double prev_t = -1;  // Previous t

FILE* pOut;                              // File pointer to
output file
pOut = fopen("out.csv", "w");            // Write header
start = clock() * CLK_TCK;               // Clock setup

A = 4;          // interval boundary of x
B = 4;          // interval boundary of y
h = 0.5;        // grid-spacing h
N = 30;         // dimension of grid in x-direction
M = 30;         // dimension of grid in y-direction
Ave = 70;       // initial approximation
Pi = 3.1415926535;

// Compute step sizes:

// Only to make the 0-indexes save we initialize them to
zero
for ( int i = 0; i < N; i++ ) U[i][0] = 0.0;
for ( int i = 1; i < M; i++ ) U[0][i] = 0.0;

// Initialize starting values at the interior points
for ( int i = 2; i <= N-1; i++ )
{
for ( int j = 2; j <= M-1; j++ )  U[i][j] = Ave;
}

// Store boundary values in the solution matrix
for ( int j = 1; j <= M ; j++ )
{
U[1][j] = F3i(j);
U[N][j] = F4i(j);
}

for ( int i = 1; i <= N ; i++ )
{
U[i][1] = F1i(i);
U[i][M] = F2i(i);
}
// The SQR parameter
```

```
temp =  cos( Pi/(N-1) ) + cos( Pi/(M-1) );
w = 4.0 / ( 2.0 + sqrt( 4.0 - temp * temp ) );

// Initialize the loop control parameters
Tol   = 1.0;
Count = 0.0;

while ( (Tol > 0.001) && (Count <= 140) )
{
Tol = 0.0;
for ( int j = 2; j <= M - 1; j++ )
{
for ( int i = 2; i <= N - 1; i++ )
{
Relax = w * ( U[i][j+1] + U[i][j-1] +
U[i+1][j] + U[i-1][j] - 4.0 * U[i][j] ) / 4.0;
U[i][j] += Relax;
if( fabs(Relax) > Tol ) Tol = fabs(Relax);
}
Count++;
}
}

// Output the solution
for ( int j = 1; j <= M; j++ )
{
for ( int i = 1; i <= N; i++ )
{
fprintf(pOut,"%8.4lf, ", U[i][j]);
fprintf(pOut,"\n");
}
//        printf("%8.4lf ", U[i][j]);
//        printf("\n");
}
fclose(pOut);
stop  = clock() * CLK_TCK;              // Timing
ticks = stop - start;
ms = (float)ticks / 1000.0f;
printf("finished... (ms = %f)\n", ms);
getchar();  return 0;
}
```

12.2.7 Method of Lines for a Set of the NLS Equations

Here we give a C++ code implementing the Method of Lines for solving a set of n nonlinear Schrödinger equations:

$$\mathrm{i}\partial_t\psi_j = -\partial_{xx}\psi_j + V(w)|\psi_j|^2\psi_j, \qquad (j = 1, \cdots, n)$$

weekly-coupled in a common adaptive potential field $V(w)$ (see Figure 12.8). The code implements the complex-valued Runge-Kutta-Fehlberg 4-5 integrator for a set of time-dependent ODEs, coupled with the following 2nd-order central finite difference scheme for the spatial discretization of each NLS:[1]

$$\partial_{xx}\psi \approx \frac{\psi_{k+1} - 2\psi_k + \psi_{k-1}}{\Delta x^2} \quad \text{(where } k \text{ is indexing the } x\text{-grid).}$$

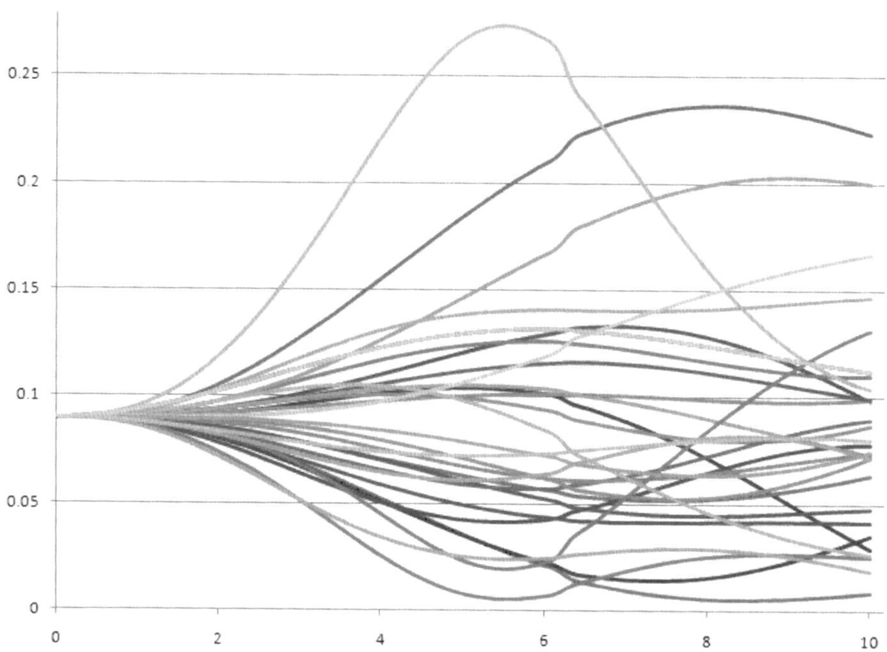

Fig. 12.8. *Numerical solution of a set of three weekly-coupled, adaptive, NLS equations with Hebbian learning: presenting PDFs for the first 10 lines for each PDE. Note that, besides PDFs, the code evaluates and outputs a number of other functions (wave-functions, Gaussian kernels, adaptive weights and potentials).*

[1] This code is the basis of the Crowd Dynamics Simulator, IP of Defence Science & Technology Organisation, Australia (see [IR12] for more technical details).

```
// Example: System of weakly-coupled, adaptive NLS equations
    with Hebbian learning
// Algorithm: Method of Lines (using complex-valued
    Runge-Kutta-Fehlberg 4-5 integrator for ODEs)

// Include statements
#include <stdio.h>
#include <math.h>
#include <time.h>
#include <stdlib.h>
#include <complex>

// Define statements
#define max(a, b) (((a) > (b)) ? (a) : (b)) // maximum macro
#define TRUE  1
#define FALSE 0
#define h0    0.01             // initial step-size
#define MAX_h h0*20
#define MIN_h h0/20
#define abstol 0.00001         // tolerance
#define reltol 0.001
#define TWO_PI 6.28318531      // 2pi
#define N 30                   // number of lines per PDE
#define M 3                    // number of PDEs

// Type declarations
typedef std::complex<double> _Complex;

// function pointer typedef for passing function as args.
typedef void (*Eqs)(double, _Complex[], _Complex*);

double h   = h0;            // Time step
double t   = 0.0;           // Initial time
double t1  = 10.0;          // Final time
_Complex x0 (0.0, 0.1);     // Initial x
_Complex x1 (50.0 + 0.1);  // Final x
_Complex I (0,1);

_Complex q[M+1][N+1];    _Complex w[N+1];
_Complex V[N+1];         _Complex gkf[N+1];

double g[N+1];  double beta = 0.1;  _Complex dx;

// Random generator defined
double RandomEqual(double Low, double High)
```

```
{
    return ((double) rand() / RAND_MAX) * (High-Low) + Low;
}

// Gaussian kernel functions
void gkf_func(int i, _Complex q, _Complex x)
{
    int j;
    double y, y2, pdf, vv, g_i;
    pdf = abs(q)*abs(q);
    y2 = -(1 / x.real() * pdf * dx.real());      // PDF
    y  = 2 * sin(TWO_PI * 10 * t);               // y = 2sin
    (2pi*10t)
    vv = y - y2;
    g_i = g[i] * vv;                        // Set-up means g[i]
    for gkf's
    gkf[i] =  exp(-(vv-g_i)*(vv-g_i));
}

// NLS PDEs: Method of Lines
void q_eqs(double t, _Complex q[], _Complex* dq)
{
    double coeff = -0.5;
    _Complex pdf, dx2 = pow(dx,2);

for(int i=2; i<=N; i++)
{
    pdf = abs(q[i])*abs(q[i]);

    if(i==N)  //u(t,x1)
    dq[i] = ((2.0*q[N-1] - q[i]) / dx2 * coeff)/ I +
                ((V[i] * q[i] * pdf / I));
else
    dq[i] = ((q[i+1] - 2.0*q[i] + q[i-1]) / dx2 * coeff) /
                 I + ((V[i] * q[i] * pdf / I));
}

dq[1] = dq[N];                  // BC: u(t,0)=u(t,x1);
}

// Weight ODE's: continuous Hebbian learning
void w_eqs(double t, _Complex w[], _Complex* dw)
{
    _Complex pdf, x, dx2 = pow(dx,2);
```

```
for(int i=1; i<=N; i++)
{
    pdf = abs(q[1][i])*abs(q[1][i]);
    x = (double)(i-1)/(N-1) * x1;
    gkf_func(i, q[1][i], x);
    dw[i] = -w[i] + beta * gkf[i] * pdf;          // Hebbian
    learning
}
}

// Initial conditions and defininitions
void Initialise()
{
    int i,j;  double x;

    for(i = 1; i <= N; i++)
{
    q[1][i] = 0.3;                  // ICs
    q[2][i] = 0.5;
    q[3][i] = 0.7;
}

    for(i = 0; i <= N; i++)
    w[i] = RandomEqual(-1.0, 1.0);           // Initialize
    weights

for(i = 0; i <= N; i++)
{
    x = (double)(i-1)/(N-1) * real(x1);          // Define x
}

for(i = 0; i <= N; i++)
    g[i] = RandomEqual(-1.0, 1.0);        // Initialize means
    for gkf's

dx  = (double)(real(x1)-real(x0))/(N-1);
// Define dx
}

// Complex RKF45 integrator
void RKF45(Eqs eqs, _Complex fx[], int increase_t)
{
_Complex k1[N+1],k2[N+1],k3[N+1],k4[N+1],k5[N+1],k6[N+1],
err[N+1],x[N+1];
double  delta, maxerr, tol[N+1];   int i=0,j=0;
```

```
_Complex dx  = (real(x1)-real(x0))/(N-1);

//k1
eqs(t,fx,k1);
for(i=1; i<=N; i++) k1[i]*=h;

//k2
for(i=1; i<=N; i++) x[i] = fx[i]+(k1[i]/4.0);
eqs(t+(h/4),x,k2);
for(i=1; i<=N; i++) k2[i]*=h;

//k3
for(i=1; i<=N; i++) x[i] = fx[i]+(3.0*k1[i]/32.0)+
(9.0*k2[i]/32.0);
eqs(t+(3*h/8),x,k3);
for(i=1; i<=N; i++) k3[i]*=h;

//k4
for(i=1; i<=N; i++) x[i] = fx[i]+(1932.0*k1[i]/2197.0)-
    (7200.0*k2[i]/2197.0)+(7296.0*k3[i]/2197.0);
eqs(t+(12*h/13),x,k4);
for(i=1; i<=N; i++) k4[i]*=h;

//k5
for(i=1; i<=N; i++) x[i] = fx[i]+(439.0*k1[i]/216.0)
    -(8.0*k2[i])+(3680.0*k3[i]/513.0)-(845.0*k4[i]/
    4104.0);
eqs(t+h,x,k5);
for(i=1; i<=N; i++) k5[i]*=h;

//k6
for(i=1; i<=N; i++) x[i] = fx[i]-(8.0*k1[i]/27.0)
    +(2.0*k2[i])-(3544.0*k3[i]/2565.0)+(1859.0*k4[i]
    /4104.0)-(11.0*k5[i]/40.0);
eqs(t+(h/2),x,k6);
for(i=1; i<=N; i++) k6[i]*=h;

maxerr = 0;
for(i=1; i<=N; i++)
{
x[i]   = fx[i]+(25.0*k1[i]/216.0)+(1408.0*k3[i]/
              2565.0)+(2197.0*k4[i]/4104.0)-(k5[i]/5.0);
err[i] = fx[i]+(16.0*k1[i]/135.0)+(6656.0*k3[i]/
              12825.0)+(28561.0*k4[i]/56430.0)-(9.0*k5[i]/
               50.0)+(2.0*k6[i]/55.0);
```

```
tol[i] = real(x[i])*reltol + abstol;
maxerr = max(maxerr, fabs(real(err[i]) - real(x[i]))/
tol[i]);
}

if(maxerr <= 1.0)
{
if(increase_t == TRUE) t += h;              // Only
update t if RKF45 goes
delta = 0.84*pow(maxerr, -0.25);

if(delta > 1.5 && h < MAX_h) h*= 1.5;     // Increase
step if required
else if(delta > 1.0 && h < MAX_h) h*=delta;
if(t+h > t1) h = t1-t;

for(i=1; i<=N; i++)
fx[i] = x[i];                            // Save
updated values
}
else
{
    delta = 0.84*pow(maxerr,-0.25);
    if(delta < 0.1) h *= 0.1;                // Reduce
    step-size
    else  h *= delta;
}
}

void header(FILE* pOut_r, FILE* pOut_i, FILE* pOut_pdf,
    FILE* pOut_w, FILE* pOut_gkf, FILE* pOut_V)  // changed
    by Vlad
{
    int i,j;
    fputs(",t\n u,", pOut_r);
    fputs(",t\n u,", pOut_i);
    fputs(",t\n u,", pOut_pdf);
    fputs(",t\n u,", pOut_w);
    fputs(",t\n u,", pOut_gkf);
    fputs(",t\n u,", pOut_V);
    for(i=1; i<=N; i++)
{
    fprintf(pOut_w, ",w%i", i);
    fprintf(pOut_gkf, ",gkf%i", i);
    fprintf(pOut_V, ",v%i", i);
```

```
}

for(j=1; j<=M;j++)
for(i=1; i<=N; i++)
{
    fprintf(pOut_r, ",x%i", i+(j-1)*N);
    fprintf(pOut_i, ",x%i", i+(j-1)*N);
    fprintf(pOut_pdf, ",x%i", i+(j-1)*N);
}

fputs("\n", pOut_r);
fputs("\n", pOut_i);
fputs("\n", pOut_pdf);
fputs("\n", pOut_w);
fputs("\n", pOut_gkf);
fputs("\n", pOut_V);
}

void print(FILE* pOut_r, FILE* pOut_i, FILE* pOut_pdf, FILE*
pOut_w,
    FILE* pOut_gkf, FILE* pOut_V)  // changed by Vlad
{
    int i,j;
    fprintf(pOut_r, ",%.4f", t);
    fprintf(pOut_i, ",%.4f", t);
    fprintf(pOut_pdf, ",%.4f", t);
    fprintf(pOut_w, ",%.4f", t);
    fprintf(pOut_gkf, ",%.4f", t);
    fprintf(pOut_V, ",%.4f", t);

for(i=1; i<=N; i++)
{
    fprintf(pOut_w, ",%.4f",  w[i]);
    fprintf(pOut_gkf, ",%.4f",  gkf[i]);
    fprintf(pOut_V, ",%.4f",  V[i]);
}

for(j=1; j<=M;j++)
for(i=1; i<=N; i++)
{
    fprintf(pOut_r, ",%.4f",  real(q[j][i]));
    fprintf(pOut_i, ",%.4f",  imag(q[j][i]));
    fprintf(pOut_pdf, ",%.4f", abs(q[j][i])*abs(q[j][i]));
}
```

```
    fprintf(pOut_r, "\n");
    fprintf(pOut_i, "\n");
    fprintf(pOut_pdf, "\n");
    fprintf(pOut_w, "\n");
    fprintf(pOut_gkf, "\n");
    fprintf(pOut_V, "\n");
}

int main()
{
int ticks,i,j;     _Complex x;
double ms, prev_t = -1;             // Previous t;
clock_t start, stop;

FILE* pOut_r  = fopen("out_r.csv",  "w");
FILE* pOut_i  = fopen("out_i.csv",  "w");
FILE* pOut_pdf = fopen("out_pdf.csv", "w");
FILE* pOut_w = fopen("out_w.csv", "w");
FILE* pOut_gkf = fopen("out_gkf.csv", "w");
FILE* pOut_V = fopen("out_v.csv", "w");

srand((unsigned int)time((time_t *)NULL));

header(pOut_r, pOut_i, pOut_pdf, pOut_w, pOut_gkf,
pOut_V);  // header
start = clock() * CLK_TCK;          // Clock setup
Initialise();                       // Initialize
Integrator

while (t < t1)                      // Main time
integration loop
{
    for(j=1; j<=M;j++)
    RKF45(&q_eqs, q[j], j-1);           // Call RKF
    RKF45(&w_eqs, w, FALSE);            // Call RKF
    for(i=1; i<=N; i++)  V[i] += w[i] * gkf[i];    // Update
    pot. V(x)
    if(prev_t != t)
    print(pOut_r, pOut_i, pOut_pdf, pOut_w, pOut_gkf,
    pOut_V);
    prev_t = t;
}
fclose(pOut_r);  fclose(pOut_i);  fclose(pOut_pdf);
fclose(pOut_w); fclose(pOut_gkf); fclose(pOut_V);
```

```
stop  = clock() * CLK_TCK;                // Timing
ticks = stop - start;
ms = (float)ticks / 1000.0f;
printf("finished... (ms = %f)\n", ms);
getchar();

return 0;
}
```

12.3 C# Code

12.3.1 Iterative Equation Solver

Here we give a C# class performing the iterative solution of various nonlinear algebraic equations with a given error (accuracy) and initial conditions. Note that the *iteration method* requires the equations to be written in iterative form.

```
using System;

class Iter  {
    static void Main()     {
        double[] x = new double[501];
        double err = 0.00000001;
        int i = 0;
        x[0] = 0.0;

        do {    // Equations to be solved:
            //x(i + 1) = 1 + 1 / Math.Cosh(x(i))
            //x(i + 1) = (1 + Math.Cos(x(i))) / 10.0
            //x(i + 1) = 1 / (4 + x(i) ^ 2)
            //x(i + 1) = (1 - x(i) ^ 3) / 3
            //x(i + 1) = 1 + 0.5 * Math.Atan(x(i))
            //x(i + 1) = Math.Exp(-x(i) ^ 2)
            //x(i + 1) = Math.Sin(Math.Exp(-x(i) ^ 3))
            //x[i + 1] = Math.Tanh(Math.Exp(-(Math.Pow(x[i],
            4))));
            x[i + 1] =
            Math.Tanh(Math.Sin(Math.Atan(Math.Exp(-(Math.Pow
            (x[i],4))))));
            i += 1;
        } while (Math.Abs(x[i] - x[i - 1]) >= err);

        Console.Write("Equation:  x = f(x)");   Console.Write
```

```
        ('\n');
        Console.Write("has a solution = ");     Console.Write
        (x[i]);
        Console.Write('\n');          Console.Write("with
        error = ");
        Console.Write(err);           Console.Write('\n');
        Console.Write("no. iterations = ");     Console.Write
        (i + 1);
        Console.Write('\n');                    Console.
        Read();
    }
}
```

12.3.2 Simulated Annealing: A Function Minimum

Here, we give a C# class that finds a local minimum of a highly-oscillating function (given analytically), using the basic *simulated annealing* algorithm.

```
using System;

namespace Anneal
{
    class Program
    {
        public static double f(ref double x)
        {
            return Math.Sin(x) * x * x * Math.Exp(-x / 15.0);
        }

        public static void randval(ref double s)
        {
            const double pi = 3.14159265;
            s = Math.IEEERemainder((s + pi) * (s + pi) *
            (s + pi)
                    * (s + pi) * (s + pi), 1.0);
        }

        public static int accept(ref double Ecurrent,
            ref double Enew, ref double T, ref double s)
        {
            double dE = Enew - Ecurrent;
            if (dE < 0.0)  { return 1; }
            if (s < Math.Exp(-dE / T))  { return 1; }
            else  { return 0; }
        }
```

```
        static void Main(string[] args)
        {
            Console.Write("Finding the minimum via
                simulated annealing:");
            Console.Write("\n");
            double xlow = 0.0, xhigh = 100.0;
            double Tmax = 500.0, Tmin = 1.0, Tstep = 0.1;
            double s = 0.118; // seed
            randval(ref s);
            double xcurrent = s * (xhigh - xlow);
            double Ecurrent = f(ref xcurrent);
            for (double T = Tmax; T > Tmin; T = T - Tstep)
            {  // main for loop
                randval(ref s);
                double xnew = s * (xhigh - xlow);
                double Enew = f(ref xnew);
                if (accept(ref Ecurrent, ref Enew,
                    ref T, ref s) != 0)
                {
                    xcurrent = xnew;
                    Ecurrent = Enew;
                }
            }
            Console.Write("The minimum found is ");
            Console.Write(Ecurrent);
            Console.Write(" at x = ");
            Console.Write(xcurrent);
            Console.Write(".\n");  Console.Read();
        }
    }
}
```

12.3.3 Simple Nonlinear Dynamics

Here we give a C# class performing (very) simple nonlinear dynamics, using Euler integrator (see Figure 12.9).

```
using System;
using System.Windows.Forms;
using System.Windows.Forms.DataVisualization.Charting;

namespace SimpleDynWhile  {
    public partial class Form1 : Form  {

        public Form1()  { InitializeComponent(); }
```

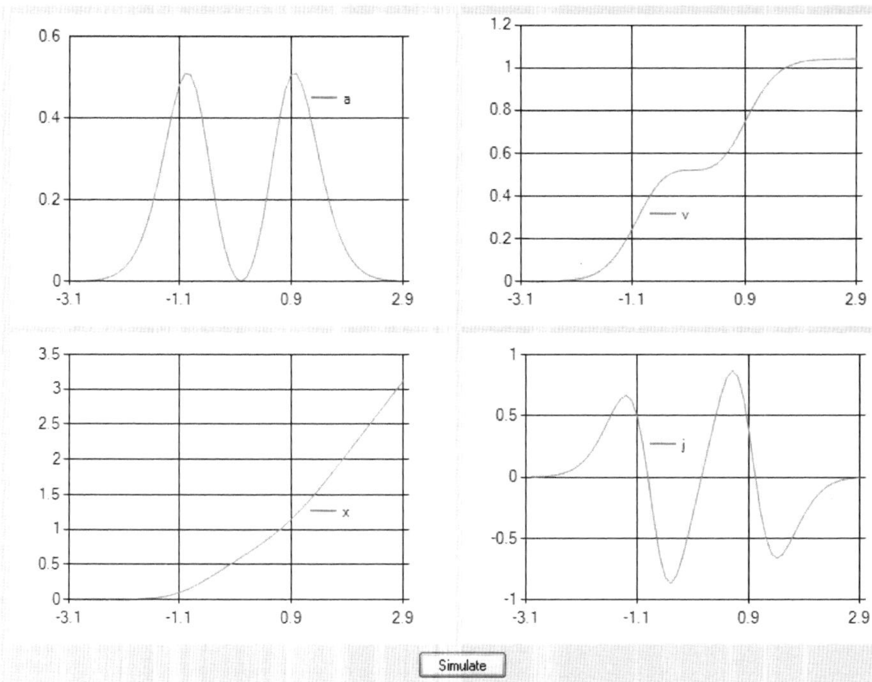

Fig. 12.9. *Simple nonlinear dynamics in C#.*

```
Series cs1 = new Series("a");  Series cs2 =
new Series("v");
Series cs3 = new Series("x");  Series cs4 =
new Series("j");

private void button1_Click(object sender, System.
EventArgs e)
{
    double t = -3.0, h = 0.1, a = 0.0, v = 0.0;
    double x = 0.0, j = 0.0, aPr = 0.0;
    while (t < 3.0)
    {
        a = Math.Atan(Math.Pow(t,2))/Math.Cosh(Math.
        Pow(t,2));
        v += a * h;
        x += v * h;
        j = (a - aPr) / h;
        aPr = a;
        cs1.Points.AddXY(t, a);  cs2.Points.AddXY(t,
```

```
                    v);
                    cs3.Points.AddXY(t, x);  cs4.Points.AddXY(t,
                    j);
                    t += h;
                }
                cs1.ChartType = SeriesChartType.FastLine;
                cs2.ChartType = SeriesChartType.FastLine;
                cs3.ChartType = SeriesChartType.FastLine;
                cs4.ChartType = SeriesChartType.FastLine;
                chart1.Series.Add(cs1);  chart2.Series.Add(cs2);
                chart3.Series.Add(cs3);  chart4.Series.Add(cs4);
            }
        }
    }
```

12.3.4 Nonlinear Pendulum Simulator

Here we give a C# class performing numerical integration of a 2nd order ODE system, solved by the standard Runge-Kutta 4 integrator (see Figure 12.10).

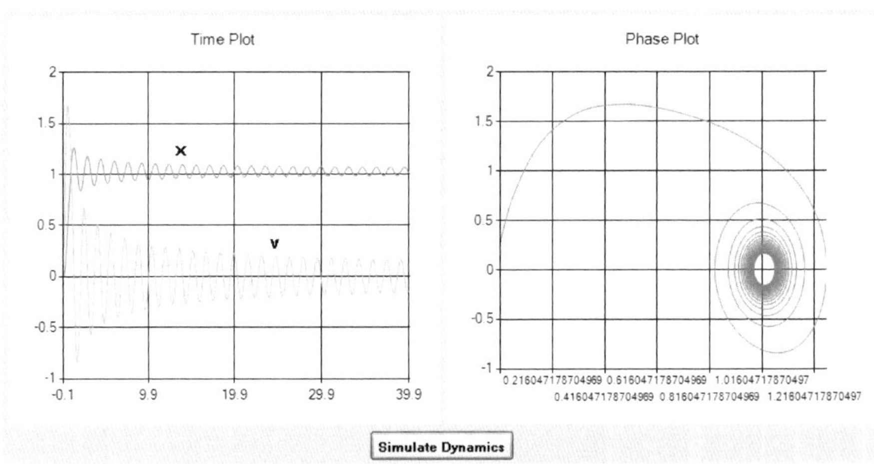

Fig. 12.10. *Simulating a hyperbolic pendulum in C# using the RK4 integrator.*

```
using System;
using System.Windows.Forms;
using System.Windows.Forms.DataVisualization.Charting;

namespace RK4hypen  {
```

```
public partial class Form1 : Form  {

    public Form1() { InitializeComponent(); }

    Series cs1 = new Series("x"); Series cs2 = new
    Series("v");
    Series cs3 = new Series("x-v");

    // 2nd order ODE system:
    Func<double, double, double, double> g = (x1, x2, t)
    => x2;
    Func<double, double, double, double> f = (x1, x2, t)
    => 12.0 * Math.Tanh(3.1416 * t) - 1.3 * x2*x2*x2
    - 9.8696 * Math.Sinh(x1);

    private void Button1_Click(object sender, System.
    EventArgs e)
    {
        double t = 0.0, x1 = 0.0, x2 = 0.1, h = 0.1;
        double k11 = 0.0, k21 = 0.0, k12 = 0.0,
        k22 = 0.0;
        double k13 = 0.0, k23 = 0.0, k14 = 0.0,
        k24 = 0.0;

        while (t < 40.0)
        {
            k11 = h * g(x1, x2, t);
            k21 = h * f(x1, x2, t);
            k12 = h * g(x1 + k11 / 2, x2 + k21 / 2, t +
            h / 2);
            k22 = h * f(x1 + k11 / 2, x2 + k21 / 2, t + h
            / 2);
            k13 = h * g(x1 + k12 / 2, x2 + k22 / 2, t + h
            / 2);
            k23 = h * f(x1 + k12 / 2, x2 + k22 / 2, t + h
            / 2);
            k14 = h * g(x1 + k13, x2 + k23, t + h);
            k24 = h * f(x1 + k13, x2 + k23, t + h);
            x1 += (k11 + 2 * (k12 + k13) + k14) / 6;
            x2 += (k21 + 2 * (k22 + k23) + k24) / 6;
            cs1.Points.AddXY(t, x1); cs2.Points.AddXY(t,
            x2);
            cs3.Points.AddXY(x1, x2);
            t += h;
        }
```

```
            cs1.ChartType = SeriesChartType.Spline;
            cs2.ChartType = SeriesChartType.Spline;
            cs3.ChartType = SeriesChartType.Spline;
            Chart1.Series.Add(cs1); Chart1.Series.Add(cs2);
            Chart2.Series.Add(cs3);
        }
    }
}
```

12.3.5 Lagrangian Dynamics Simulator

Here we give a C# class implementing a generic Lagrangian dynamics simulator, using the Runge-Kutta-Fehlberg 4-5 integrator (see Figure 12.11).

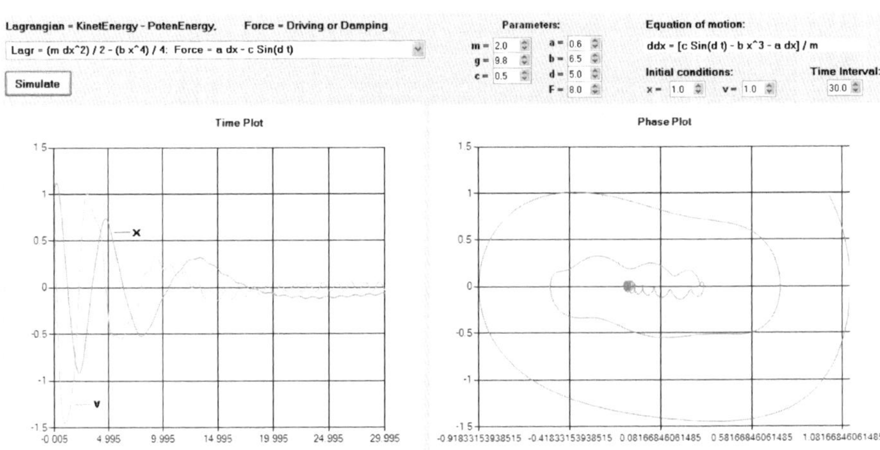

Fig. 12.11. A generic Lagrangian dynamics simulator in C#, based on the Runge-Kutta-Fehlberg integrator.

```
using System;
using System.Windows.Forms;
using System.Windows.Forms.DataVisualization.Charting;

namespace Lagrangian {
    public partial class Form1 : Form {

        public double h, t, prev_t;
        public const int N = 2;                   // System
        order
```

```
public Form1()
{
    InitializeComponent();

    this.comboBox1.Items.AddRange(items);
    this.comboBox1.SelectedIndex = 0;
    par1.Value = (decimal)(a); par2.Value = (decimal)
    (m);
    par3.Value = (decimal)(b);
    par4.Value = (decimal)(g); par5.Value = (decimal)
    (c);
    par6.Value = (decimal)(d);
    par7.Value = (decimal)(F); ic1.Value = (decimal)
    (xx);
    ic2.Value = (decimal)(vv);
    timeInt.Value = (decimal)(t1);
}

// default parameters
public double m = 2.0, g = 9.8, a = 0.8, b = 5.0;
public double c = 0.5, d = 5.0, F = 10.0;

// default ICs
public double xx = 1.0, vv = 1.0;
public double t1 = 20.0;                    // Final time

public String[] items = {
    "1. Lagr = (m a^2 dx^2) / 2 - m g a Cos(x):
    Force = 0",
    "2. Lagr = (m dx^2) / 2 - (b x^2) / 2:
    Force = 0",
    "3. Lagr = (m a^2 dx^2) / 2 - (b x^2) / 2 - m g a
    Cos(x):
    Force = 0",
    "4. Lagr = (m dx^2) / 2 - (b x^4) / 4:
    Force = 0",
    "5. Lagr = (m dx^2) / 2 - (b x^4) / 4 - (a b x^2)
    / 2:
    Force = 0",
    "6. Lagr = (m dx^2) / 2 - (b x^4) / 4 - (a b x^2)
    / 2
    - m g a Cos(x):  Force = 0",
    "7. Lagr = (m a^2 dx^2) / 2 - m g a Cos(x):
    Force = c dx",
    "8. Lagr = (m dx^2) / 2 - (b x^2) / 2:  Force = c
```

```
        dx",
        "9. Lagr = (m a^2 dx^2) / 2 - (b x^2) / 2 - m g a
        Cos(x):
        Force = c dx",
        "10. Lagr = (m dx^2) / 2 - (b x^4) / 4:
        Force = F [1 - exp(-d t)] + c dx",
        "11. Lagr = (m dx^2) / 2 - (b x^4) / 4 -
        (a b x^2) / 2:
        F [1 - exp(-d t)] + c dx",
        "12. Lagr = (m dx^2) / 2 - b x^4 / 4 - (a b x^2)
        / 2
        - m g a Cos(x):  Force = c dx",
        "VDP.  Lagr = dx^2 / 2 - (a^2 x^2) / 2:
        Force = - b (1 - x^2) dx",
        "VDP-forc.  Lagr = dx^2 / 2 - (a^2 x^2) / 2:
        Force = - F Cos(d t) - b (1 - x^2) dx",
        "Duffing.  Lagr = dx^2 / 2:  Force = - F Cos(d t)
        + c dx
        - a x (1 - x^2)",
        "Musc1.  Lagr = (m dx^2) / 2 - (b x^4) / 4:
        Force = a dx - c Sin(d t)",
        "Musc2.  Lagr = (m dx^2) / 2 + b Sech(x):
        Force = a dx - c Sin(d t)"
                            };

    public String[] eqns = {
        "1. ddx = g Sin(x) / a",
        "2. ddx = - b x / m",
        "3. ddx = (a g m Sin(x) - b x) / (a^2 m)",
        "4. ddx = - b x^3 / m",
        "5. ddx = - (a b x + b x^3) / m",
        "6. ddx = (a g m Sin(x) - a b x - b x^3) / m",
        "7. ddx = (a g m Sin(x) - c dx) / (a^2 m)",
        "8. ddx = - (b x + c dx) / m",
        "9. ddx = (a g m Sin(x) - b x - c dx) / (a^2 m)",
        "10. ddx = (F [exp(-d t) - 1] - b x^3 - c dx)
        / m",
        "11. ddx = (F [exp(-d t) - 1] - a b x - b x^3
        - c dx) / m",
        "12. ddx = (a g m Sin(x) - a b x - b x^3 - c dx)
        / m",
        "VDP.  ddx = b (1 - x^2) dx - a^2 x",
        "VDP-forc.  ddx = F Cos(d t) + b (1 - x^2) dx -
        a^2 x",
        "Duffing.  ddx = F Cos(d t) + a x (1 - x^2) - c
```

```
    dx",
    "Musc1.  ddx = [c Sin(d t) - b x^3 - a dx] / m",
    "Musc2.  ddx = [c Sin(d t) - b Sech(x) Tanh(x) -
    a dx] / m"
                        };

public double Fun(double t, double[] x)
{
    int n = comboBox1.SelectedIndex;
    switch (n)
    {
case 0: // 1.
    return g * Math.Sin(x[0]) / a;
case 1: // 2.
    return -b * x[0] / m;
case 2: // 3.
    return (a * g * m * Math.Sin(x[0]) - b * x[0]) /
    (a * a * m);
case 3: // 4.
    return -b * x[0] * x[0] * x[0] / m;
case 4: // 5.
    return -(a * b * x[0] + b * x[0] * x[0] * x[0])
    / m;
case 5: // 6.
    return (a * g * m * Math.Sin(x[0]) - a * b * x[0]
    - b * x[0] * x[0] * x[0]) / m;
case 6: // 7.
    return (a * g * m * Math.Sin(x[0]) - c * x[1]) /
    (a * a * m);
case 7: // 8.
    return -(b * x[0] + c * x[1]) / m;
case 8: // 9.
    return (a * g * m * Math.Sin(x[0]) - b * x[0]
    - c * x[1]) / (a * a * m);
case 9: // 10.
    return (F * (Math.Exp(-d * t) - 1) - b * x[0] *
    x[0] * x[0]
    - c * x[1]) / m;
case 10: // 11.
    return (F * (Math.Exp(-d * t) - 1) - a * b * x[0]
    - b * x[0] * x[0] * x[0] - c * x[1]) / m;
case 11: // 12.
    return (a * g * m * Math.Sin(x[0]) - a * b * x[0]
    - b * x[0] * x[0] * x[0] - c * x[1]) / m;
case 12: // 13.
```

```
            return b * (1 - x[0] * x[0]) * x[1] - a * a *
            x[0];
        case 13: // 14.
            return F * Math.Cos(d * t) + b * (1 - x[0] *
            x[0]) * x[1]
            - a * a * x[0];
        case 14: // 15.
            return F * Math.Cos(d * t) + a * (1 - x[0] *
            x[0]) *
            x[0]
            - c * x[1];
        case 15: // 16.
            return (c * Math.Sin(d * t) - b * x[0] * x[0] *
            x[0]
            - a * x[1]) / m;
        case 16: // 17.
            return (c * Math.Sin(d * t) - b * Math.Tanh(x[0])
            /
            Math.Cosh(x[0]) - a * x[1]) / m;
        default:
            return 0.0;
            }
        }

        // ODEs defined
        public void Eqs(double t, double[] y, double[] dy)
        {
            dy[0] = y[1];
            dy[1] = Fun(t, y);
        }

        // integrator parameters
        public const double h0 = 0.01, MAX_h = h0 * 10,
        MIN_h = h0 / 10;
        public const double abstol = 0.00001, reltol = 0.001;

        // Runge-Kutta-Fehlberg integrator
        public void RKF45(double[] y)
        {
            double[] k1 = new double[N];
            double[] k2 = new double[N];
            double[] k3 = new double[N];
            double[] k4 = new double[N];
            double[] k5 = new double[N];
            double[] k6 = new double[N];
```

```
double[] err = new double[N];
double[] x = new double[N];
double[] tol = new double[N];
double delta, maxerr;
int i = 0;

//k1
Eqs(t, y, k1);
for (i = 0; i < N; i++) k1[i] *= h;

//k2
for (i = 0; i < N; i++) x[i] = y[i] +
(k1[i] / 4);
Eqs(t + (h / 4), x, k2);
for (i = 0; i < N; i++) k2[i] *= h;

//k3
for (i = 0; i < N; i++) x[i] = y[i]
+ (3 * k1[i] / 32) + (9 * k2[i] / 32);
Eqs(t + (3 * h / 8), x, k3);
for (i = 0; i < N; i++) k3[i] *= h;

//k4
for (i = 0; i < N; i++) x[i] = y[i]
+ (1932 * k1[i] / 2197) - (7200 * k2[i] / 2197)
+ (7296 * k3[i] / 2197);
Eqs(t + (12 * h / 13), x, k4);
for (i = 0; i < N; i++) k4[i] *= h;

//k5
for (i = 0; i < N; i++) x[i] = y[i]
+ (439 * k1[i] / 216) - (8 * k2[i])
+ (3680 * k3[i] / 513) - (845 * k4[i] / 4104);
Eqs(t + h, x, k5);
for (i = 0; i < N; i++) k5[i] *= h;

//k6
for (i = 0; i < N; i++) x[i] = y[i]
- (8 * k1[i] / 27) + (2 * k2[i])
- (3544 * k3[i] / 2565) + (1859 * k4[i] / 4104)
- (11 * k5[i] / 40);
Eqs(t + (h / 2), x, k6);
for (i = 0; i < N; i++) k6[i] *= h;

maxerr = 0;
```

```
        for (i = 0; i < N; i++)
        {
            x[i] = y[i] + (25 * k1[i] / 216)
            + (1408 * k3[i] / 2565) + (2197 * k4[i]
            / 4104)
            - (k5[i] / 5);
            err[i] = y[i] + (16 * k1[i] / 135)
            + (6656 * k3[i] / 12825) + (28561 * k4[i]
            / 56430)
            - (9 * k5[i] / 50) + (2 * k6[i] / 55);
            tol[i] = x[i] * reltol + abstol;
            maxerr = Math.Max(maxerr, Math.Abs(err[i]
            - x[i]) / tol[i]);
        }

        if (maxerr <= 1.0)
        {
            t += h;
            delta = 0.84 * Math.Pow(maxerr, -0.25);

            // Increase step if required
            if (delta > 1.5 && h < MAX_h) h *= 1.5;
            else if (delta > 1.0 && h < MAX_h) h
            *= delta;
            if (t + h > t1) h = t1 - t;

            // Save updated values
            for (i = 0; i < N; i++) y[i] = x[i];
        }
        else
        {
            // Reduce step
            delta = 0.84 * Math.Pow(maxerr, -0.25);
            if (delta < 0.1) h *= 0.1;
            else h *= delta;
        }
    }

    private void butSimul_Click(object sender,
    EventArgs e)
    {
        this.chart1.Series.Clear();
        this.chart1.ChartAreas.Clear();
        ChartArea chartArea1 = new ChartArea();
        this.chart1.ChartAreas.Add(chartArea1);
```

```
this.chart2.Series.Clear();
this.chart2.ChartAreas.Clear();
ChartArea chartArea2 = new ChartArea();
this.chart2.ChartAreas.Add(chartArea2);

label1.Text = eqns[comboBox1.SelectedIndex];

a = (double)par1.Value;
m = (double)par2.Value;
b = (double)par3.Value;
g = (double)par4.Value;

c = (double)par5.Value;
d = (double)par6.Value;
F = (double)par7.Value;

xx = (double)ic1.Value;
vv = (double)ic2.Value;
t1 = (double)timeInt.Value;

h = h0;
t = 0.0;
prev_t = -1.0;

Series dSeries = new Series("x");
dSeries.ChartType = SeriesChartType.Spline;
Series vSeries = new Series("v");
vSeries.ChartType = SeriesChartType.Spline;
Series pSeries = new Series("Phase");
pSeries.ChartType = SeriesChartType.Spline;

double[] y = new double[2] { xx, vv };
// Initial conditions

while (t < t1)                                   // time
loop
{
    RKF45(y);                                    // Call
    RKF45
    if (prev_t != t)
    {
        dSeries.Points.AddXY(t, y[0]);
        vSeries.Points.AddXY(t, y[1]);
        pSeries.Points.AddXY(y[0], y[1]);
```

```
                }
                prev_t = t;
            }
            this.chart1.Series.Add(dSeries);
            this.chart1.Series.Add(vSeries);
            this.chart2.Series.Add(pSeries);
        }
    }
}
```

12.3.6 Complex-Valued Crowd Attractor Dynamics

Here we give a sample from the attractor-driven complex crowd dynamics simulation.

```
using System;
using System.Drawing;
using System.Drawing.Drawing2D;
using System.Windows.Forms;
using System.Windows.Forms.DataVisualization.Charting;

namespace IvCrowdLangevin
{
public partial class MultiAttrForm :
  Form
  {   // Initializations
    public int N = 200;
    // number of agents
    public double h = 0.05;                              // Time
    step
    public double t;
    // Initial time
    public double prev_t;
    // Previous t
    public static double tFin = 160.0;
    // final time
    public Complex I = new Complex(0, 1);             // define
    imagin unit
    public Random rand = new Random();
    // random generator
    public double FldStrn = 0.3;                 // attractor
    field strength
    public double alpha = 1.0;
    // linear parameter
    public double beta = 0.0001;
```

```
// cubic parameter

protected override void OnPaint(PaintEventArgs e)
{   // Drawing X-Y axes frame around the panel
  Graphics g = e.Graphics;
  Pen p = new Pen(Color.Black, 5);
  p.StartCap = LineCap.Round;
  p.EndCap = LineCap.ArrowAnchor;
  g.DrawLine(p, 60, 820, 910, 820);
  g.DrawLine(p, 95, 835, 95, 10);
  p.Dispose();
}

public MultiAttrForm()
{   // Initializing the form
  InitializeComponent();
  this.DoubleBuffered = true;
  numericUpDwnN.Value = (int)(N);
  numericUpDwnFld.Value = (decimal)(FldStrn);
}

// Attractors:
public Complex Attr1(double t) {
  return new Complex(Math.Sin(2 * t)) + I * (new Complex
  (Math.Cos(t)));
}
public Complex Attr2(double t) {
  return new Complex(Math.Sin(t)) + I * (new Complex
  (Math.Cos(2 * t)));
}
public Complex Attr3(double t) {
  return new Complex(-1) + I * (new Complex(Math.
  Cos(t)));
}
public Complex Attr4(double t) {
  return new Complex(1) + I * (new Complex(Math.Cos(t)));
}
public Complex Attr5(double t) {
  return new Complex(Math.Sin(t)) + I * (new
  Complex(-1));
}
public Complex Attr6(double t) {
  return new Complex(Math.Sin(t)) + I * (new Complex(1));
}
```

```
public void Eqs(double t, Complex[] x, Complex[] dx)
{
  for (int i = 0; i < N; i++)    // Langevin ODEs:
  {
    if (i < N / 6)
    {
      dx[i] = FldStrn * (3 * Attr1(t) - alpha * x[i] -
      beta * x[i] * x[i] * x[i] + beta * (rand.NextDouble
      () - 0.5) * x[i]);
    }
    else if (i < N / 3)
    {
      dx[i] = FldStrn * (3 * Attr2(t) - alpha * x[i] -
      beta * x[i] * x[i] * x[i] + beta * (rand.NextDouble
      () - 0.5) * x[i]);
    }
    else if (i < N / 2)
    {
      dx[i] = FldStrn * (3 * Attr3(t) - alpha * x[i] -
      beta * x[i] * x[i] * x[i] + beta * (rand.NextDouble
      () - 0.5) * x[i]);
    }
    else if (i < 2 * N / 3)
    {
      dx[i] = FldStrn * (3 * Attr4(t) - alpha * x[i] -
      beta * x[i] * x[i] * x[i] + beta * (rand.NextDouble
      () - 0.5) * x[i]);
    }
    else if (i < 5 * N / 6)
    {
      dx[i] = FldStrn * (3 * Attr5(t) - alpha * x[i] -
      beta * x[i] * x[i] * x[i] + beta * (rand.NextDouble
      () - 0.5) * x[i]);
    }
    else
    {
      dx[i] = FldStrn * (3 * Attr6(t) - alpha * x[i] -
      beta * x[i] * x[i] * x[i] + beta * (rand.NextDouble
      () - 0.5) * x[i]);
    }
  }
}
// ODE string printout on the form
public String eqStr = "  dx[i]  =  FldStrn . ( Attr(t) -
alpha . x[i] - beta . x[i]^3 + Rand(0) . x[i] )";
```

```
public const double h0 = 0.01;
public const double MAX_h = h0 * 10;
public const double MIN_h = h0 / 10;
public const double abstol = 0.00001;
public const double reltol = 0.001;
// Runge-Kutta-Fehlberg (Cash-Karp) adaptive integrator
public void RKF45(Complex[] y)
{
  Complex[] k1 = new Complex[N];
  Complex[] k2 = new Complex[N];
  Complex[] k3 = new Complex[N];
  Complex[] k4 = new Complex[N];
  Complex[] k5 = new Complex[N];
  Complex[] k6 = new Complex[N];
  Complex[] err = new Complex[N];
  Complex[] x = new Complex[N];
  double[] tol = new double[N];
  double delta, maxerr;
  int i = 0;

  //k1
  Eqs(t, y, k1);
  for (i = 0; i < N; i++) k1[i] = h * k1[i];

  //k2
  for (i = 0; i < N; i++) x[i] = y[i] + (k1[i] / 4);
  Eqs(t + (h / 4), x, k2);
  for (i = 0; i < N; i++) k2[i] = h * k2[i];

  //k3
  for (i = 0; i < N; i++) x[i] = y[i] + (3 * k1[i] / 32)
  + (9 * k2[i] / 32);
  Eqs(t + (3 * h / 8), x, k3);
  for (i = 0; i < N; i++) k3[i] = h * k3[i];

  //k4
  for (i = 0; i < N; i++) x[i] = y[i] + (1932 * k1[i]
  / 2197)
  - (7200 * k2[i] / 2197) + (7296 * k3[i] / 2197);
  Eqs(t + (12 * h / 13), x, k4);
  for (i = 0; i < N; i++) k4[i] = h * k4[i];

  //k5
  for (i = 0; i < N; i++) x[i] = y[i] + (439 * k1[i]
```

```
/ 216)
- (8 * k2[i]) + (3680 * k3[i] / 513) - (845 * k4[i]
/ 4104);
Eqs(t + h, x, k5);
for (i = 0; i < N; i++) k5[i] = h * k5[i];

//k6
for (i = 0; i < N; i++) x[i] = y[i] - (8 * k1[i] / 27)
+ (2 * k2[i]) - (3544 * k3[i] / 2565) + (1859 * k4[i]
/ 4104)
- (11 * k5[i] / 40);
Eqs(t + (h / 2), x, k6);
for (i = 0; i < N; i++) k6[i] = h * k6[i];

maxerr = 0.0;
for (i = 0; i < N; i++)
{
  x[i] = y[i] + (25 * k1[i] / 216) + (1408 * k3[i]
  / 2565)
  + (2197 * k4[i] / 4104) - (k5[i] / 5);
  err[i] = y[i] + (16 * k1[i] / 135) + (6656 * k3[i]
  / 12825)
  + (28561 * k4[i] / 56430) - (9 * k5[i] / 50) + (2 *
  k6[i] / 55);
  tol[i] = reltol * x[i].real + abstol;
  maxerr = Math.Max(maxerr, Math.Abs(err[i].real
  - x[i].real) / tol[i]);
}

if (maxerr <= 1.0)
{
  t += h;
  delta = 0.84 * Math.Pow(maxerr, -0.25);

  // Increase step if required
  if (delta > 1.5 && h < MAX_h) h *= 1.5;
  else if (delta > 1.0 && h < MAX_h) h *= delta;
  if (t + h > tFin) h = tFin - t;

  // Save updated values
  for (i = 0; i < N; i++) y[i] = x[i];
}
else
{
  // Reduce step
```

```
    delta = 0.84 * Math.Pow(maxerr, -0.25);
    if (delta < 0.1) h *= 0.1;
    else h *= delta;
  }
}

// Button butInit_Click event - Initialize
private void butInit_Click(object sender, EventArgs e)
{
  richTextBox3.AppendText(eqStr);
  //richTextBox1.AppendText(attrStr); // text boxes
  // Initializing dynamical variables
  Complex[] x = new Complex[N];
  Complex[] dx = new Complex[N];
  for (int j = 0; j < N; j++)
  {   // initial conditions for Langevin ODEs
    x[j] = new Complex(0.03 * (1 + j * (rand.NextDouble()
    - 0.5)), 0.03 * (1 + j * (rand.NextDouble() - 0.5)));
    dx[j] = new Complex(0.0, 0.0);
  }

  // Initializing 2D graphics on the panel
  Graphics graphics = panel1.CreateGraphics();
  graphics.Clear(Color.Navy);
  Image[] image = new Image[5];
  for (int j = 0; j < 5; j++)
  {   // read images for agents
    image[j] = Image.FromFile("image" + j + ".gif");
  }

  for (int j = 0; j < N - 1; j++)
  {   // initializing 2D animations of moving agents
    graphics.ResetTransform();
    // non-commutative translations and rotations
    graphics.TranslateTransform(400 +
    (int)(100 * x[j].real), 400 + (int)(100 * x[j].
    imagin));
    graphics.DrawImage(image[j % 5], 0, 0);
  }
}

// Button butSimul_Click event - Simulate (Run!)
private void butSimul_Click(object sender, EventArgs e)
{
  h = h0;
```

```
t = 0.0;
prev_t = -1.0;       // re-initializing
Complex[] x = new Complex[N];
Complex[] dx = new Complex[N];
double[] Angle = new double[N];

// Initializing Charts and Series
this.chart1.Series.Clear();
this.chart2.Series.Clear();
Series[] Xser = new Series[N];
Series[] Yser = new Series[N];
Series[] XYser = new Series[N];
Series Aser = new Series();
Aser.ChartType = SeriesChartType.FastLine;

for (int j = 0; j < N; j++)
{
  Xser[j] = new Series();         // initialize charts
  Yser[j] = new Series();
  XYser[j] = new Series();
  Xser[j].ChartType = SeriesChartType.FastLine;
  Yser[j].ChartType = SeriesChartType.FastLine;
  XYser[j].ChartType = SeriesChartType.FastLine;
}

// Initializing 2D graphics on the panel
Graphics graphics = panel1.CreateGraphics();
Image[] image = new Image[5];

for (int j = 0; j < 5; j++)
{  // read little images for agents
  image[j] = Image.FromFile("image" + j + ".gif");
}

while (t < tFin)
      // Main time loop: Langevin dynamics starts here!
{
  RKF45(x);                 // Call RKF45
  for (int j = 0; j < N; j++)
  {
    Xser[j].Points.AddXY(t, x[j].real);
    // real-time plots
    Yser[j].Points.AddXY(t, x[j].imagin);
    // imagin-time plots
    XYser[j].Points.AddXY(x[j].real, -x[j].imagin);
```

```
      // phase plots
    }
    prev_t = t;

    if (t % (tFin / 300) < h)
    {
      graphics.Clear(Color.Navy);   // clearing graphics
      for (int j = 0; j < N - 1; j++)
      {   // 2D animations of moving agents
        graphics.ResetTransform();
        // non-commutative translations and rotations
        if (x[j].real - x[j + 1].real < 0.1 && x[j].real
        > x[j + 1].real) {
          x[j].real += 0.1;
        }    // My simple ColDet
        if (x[j + 1].real - x[j].real < 0.1 && x[j].real
        < x[j + 1].real) {
          x[j].real -= 0.1;
        }
        if (x[j].imagin - x[j + 1].imagin < 0.1 && x[j].
        imagin
        > x[j + 1].imagin) {
          x[j].imagin += 0.1;
        }
        if (x[j + 1].imagin - x[j].imagin < 0.1 && x[j].
        imagin
        < x[j + 1].imagin) {
          x[j].imagin -= 0.1;
        }
        graphics.TranslateTransform(400 +
        (int)(400 * x[j].real), 400 + (int)(400 * x[j].
        imagin));
        graphics.RotateTransform((float)Angle[j]);
        graphics.DrawImage(image[j % 5], 0, 0);
      }
      graphics.Flush();        // flushing graphics
    }
  }
  for (int j = 0; j < N; j++)      // create charts
  {
    this.chart1.Series.Add(Xser[j]);
    this.chart1.Series.Add(Yser[j]);
    this.chart2.Series.Add(XYser[j]);
  }
}
```

```
    }
  }
```

12.4 Freeform Fortran Code

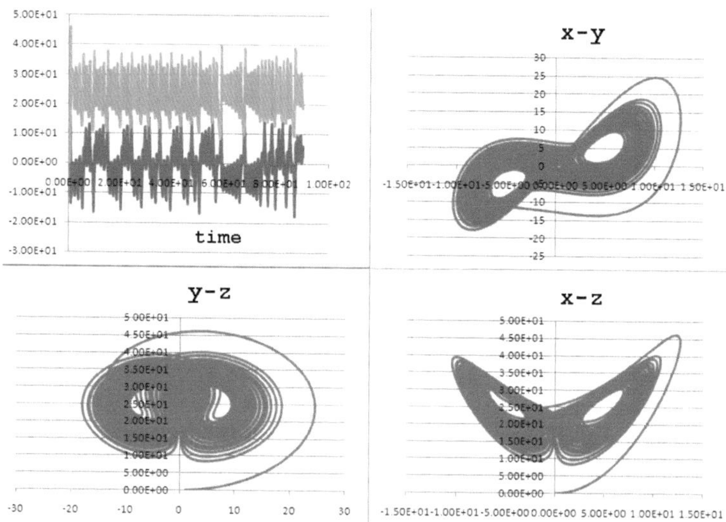

Fig. 12.12. Simulating the Lorenz butterfly-attractor in F90 using the RK4 integrator (plotting in Excel).

12.4.1 Lorenz Attractor Simulator

Here we give an F90 code simulating the Lorenz butterfly-attractor (see Figure 12.12), using the standard RK4 integrator.

```
program lorRK4        ! Lorenz attractor in RK4
    real :: ti=0.,xi=0.,yi=1.,zi=0.,dt=0.03,tFin=90.
    real tf,xf,yf,zf
    external dx,dy,dz                      ! declare functions
    open(1, file='lorRK4.csv')             ! file and header
    write(1,*) 't',',','x(t)',',','y(t)',',','z(t)'
    do while(ti<tFin)                      ! main time loop
       tf = ti + dt ! time increment;     ! call RK4
       call RK4(dx,dy,dz,ti,tf,xi,yi,zi,xf,yf,zf)
       write(1,*) tf,',',xf,',',yf,',',zf  ! printout
       ti = tf; xi = xf; yi = yf; zi = zf
```

```
    end do
    close(1)
end program

! Defining Lorenz equations:
function dx(t,x,y,z)
    real dx,t,x,y,z
    dx = -3*(x-y)
end function dx

function dy(t,x,y,z)
    real dx,t,x,y,z
    dy = -x*z + 26.5*x - y
end function dy

function dz(t,x,y,z)
    real dx,t,x,y,z
    dz = x*y - z
end function dz

! Runge-Kutta 4th-order ODE solver:
subroutine RK4(dx,dy,dz,ti,tf,xi,yi,zi,xf,yf,zf)
    real h,t,dx,dy,dz,ti,tf,xi,yi,zi,xf,yf,zf
    real k1x,k2x,k3x,k4x,k1y,k2y,k3y,k4y,k1z,k2z,k3z,k4z
    h = tf-ti;  t = ti
    k1x = h*dx(t,xi,yi,zi)
    k1y = h*dy(t,xi,yi,zi)
    k1z = h*dz(t,xi,yi,zi)
    k2x = h*dx(t+h/2.,xi+k1x/2.,yi+k1y/2.,zi+k1z/2.)
    k2y = h*dy(t+h/2.,xi+k1x/2.,yi+k1y/2.,zi+k1z/2.)
    k2z = h*dz(t+h/2.,xi+k1x/2.,yi+k1y/2.,zi+k1z/2.)
    k3x = h*dx(t+h/2.,xi+k1x/2.,yi+k1y/2.,zi+k1z/2.)
    k3y = h*dy(t+h/2.,xi+k1x/2.,yi+k1y/2.,zi+k1z/2.)
    k3z = h*dz(t+h/2.,xi+k1x/2.,yi+k1y/2.,zi+k1z/2.)
    k4x = h*dx(t+h,xi+k3x,yi+k3y,zi+k3z)
    k4y = h*dy(t+h,xi+k3x,yi+k3y,zi+k3z)
    k4z = h*dz(t+h,xi+k3x,yi+k3y,zi+k3z)
    xf = xi + (k1x + 2.*(k2x+k3x) + k4x)/6.
    yf = yi + (k1y + 2.*(k2y+k3y) + k4y)/6.
    zf = zi + (k1z + 2.*(k2z+k3z) + k4z)/6.
end subroutine RK4
```

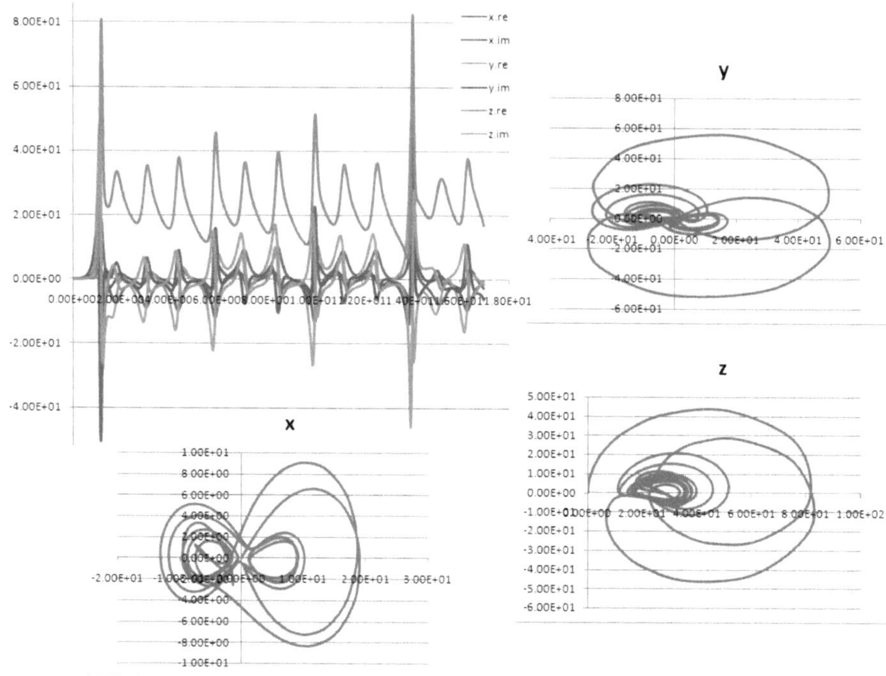

Fig. 12.13. *Complex-valued Lorenz attractor: time evolutions (top-left) and phase (Re vs. Im) plots.*

12.4.2 Complex Lorenz Attractor

Here we give an F90 code simulating the complex-valued Lorenz attractor (see Figure 12.13), using the complexified RK4 integrator.

```
program cmplxLor
! Complex Lorenz attractor via complex RK4-integrator
real :: ti=0., dt=0.03, tFin=17., tf
complex :: xi=(0.01,0.01), yi=(0.01,0.01),
    zi=(0.01,0.01), xf,yf,zf
external dx,dy,dz                   ! declare functions
open(1, file='cmplxLor.csv')          ! file and header
write(1,*) 't',',','x.re',',','x.im',',','y.re',',',
    'y.im',',','z.re',',','z.im'
    print *, 't',',','x.re',',','x.im',',','y.re',
    ',','y.im',',','z.re',',','z.im'
do while (ti<tFin)  ! main time loop
    tf = ti + dt        ! time increment
    call cmplxRK4(dx,dy,dz,ti,tf,xi,yi,zi,xf,yf,zf)  ! call
    RK4
```

```
    write(1,*) tf,',',real(xf),',',aimag(xf),',',real(yf)
    ,',',aimag(yf),
        ',',real(zf),',',aimag(zf) ! printout to file
    print *, tf,',',real(xf),',',aimag(xf),',',real(yf)
    ,',',aimag(yf),
        ',',real(zf),',',aimag(zf) ! printout to screen
    ti = tf; xi = xf; yi = yf; zi = zf
end do
close(1)
end program

function dx(t,x,y,z)
    real :: t; complex :: dx,x,y,z
    dx = -3*(x-y)
end function dx

function dy(t,x,y,z)
    real :: t; complex :: dy,x,y,z
    dy = -x*z + 26.5*x - y
end function dy

function dz(t,x,y,z)
    real :: t; complex :: dz,x,y,z
    dz = x*y - z
end function dz

subroutine cmplxRK4(dx,dy,dz,ti,tf,xi,yi,zi,xf,yf,zf)
    real :: h,t,ti,tf
    complex :: dx,dy,dz,xi,yi,zi,xf,yf,zf
    complex::k1x,k2x,k3x,k4x,k1y,k2y,k3y,k4y,k1z,k2z,k3z,k4z
    h = tf-ti;  t = ti
    k1x = h*dx(t,xi,yi,zi)
    k1y = h*dy(t,xi,yi,zi)
    k1z = h*dz(t,xi,yi,zi)
    k2x = h*dx(t+h/2.,xi+k1x/2.,yi+k1y/2.,zi+k1z/2.)
    k2y = h*dy(t+h/2.,xi+k1x/2.,yi+k1y/2.,zi+k1z/2.)
    k2z = h*dz(t+h/2.,xi+k1x/2.,yi+k1y/2.,zi+k1z/2.)
    k3x = h*dx(t+h/2.,xi+k1x/2.,yi+k1y/2.,zi+k1z/2.)
    k3y = h*dy(t+h/2.,xi+k1x/2.,yi+k1y/2.,zi+k1z/2.)
    k3z = h*dz(t+h/2.,xi+k1x/2.,yi+k1y/2.,zi+k1z/2.)
    k4x = h*dx(t+h,xi+k3x,yi+k3y,zi+k3z)
    k4y = h*dy(t+h,xi+k3x,yi+k3y,zi+k3z)
    k4z = h*dz(t+h,xi+k3x,yi+k3y,zi+k3z)
    xf = xi + (k1x + 2.*(k2x+k3x) + k4x)/6.
    yf = yi + (k1y + 2.*(k2y+k3y) + k4y)/6.
```

```
      zf = zi + (k1z + 2.*(k2z+k3z) + k4z)/6.
  end subroutine
```

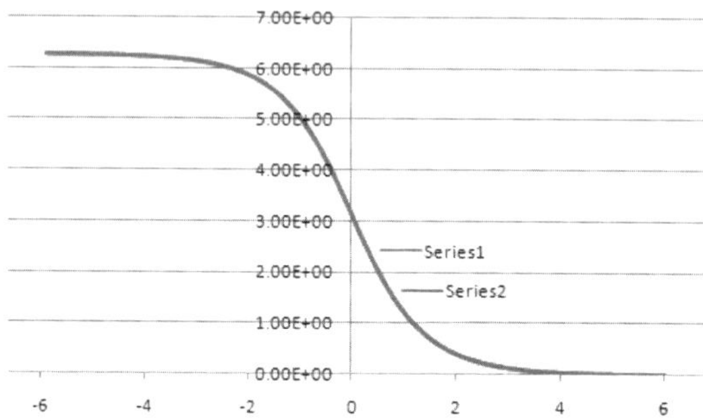

Fig. 12.14. *Simple Sine-Gordon soliton (antikink): comparing numerical solution with the exact one (two traces are on top of each other).*

12.4.3 Simple SGE Soliton

Here we give a simple F90 code that numerically solves the Sine-Gordon equation:

$$\phi_{tt} = \phi_{xx} - \sin\phi$$

and compares the numerical solution (see Figure 12.14) with the exact antikink solution:

$$\phi_{exact}(x) = 4\,\mathrm{atan}[\exp(-x/\sqrt{1-c^2})].$$

```
program SGsimple                ! Solitary SGE solutions
! f''(1-c**2)=sin(f),f(inf)=f'(inf)=0
! Based on f77-code by R.Baretti
    data c,xi,xf,nstep/0.5,6.,-6.,2000/
    yexact(x)=4.*atan(exp(-x/sqrt(1.-c**2)))
    dx=(xf-xi)/float(nstep)
    f0=4.*exp(-xi/sqrt(1.-c**2))
    f1=4.*exp(-(xi+dx)/sqrt(1.-c**2))
    kp=int(float(nstep)/60.)
    kount=kp
    print *,'   x       f          yex'; print *
    open(1, file='SG.csv')     ! file and header
    write(1,*) xi,',',f0,',',yexact(xi)
    do i=2,nstep
        x = xi + float(i)*dx
        f2 = 2.*f1 - f0 + sin(f1)/(1.-c**2)*(dx**2)
```

```
        if (i==kount) then
            write(1,*) x,',',f2,',',yexact(x)
            kount = kount + kp
        end if
        f0=f1
        f1=f2
    end do
    close(1)
end
```

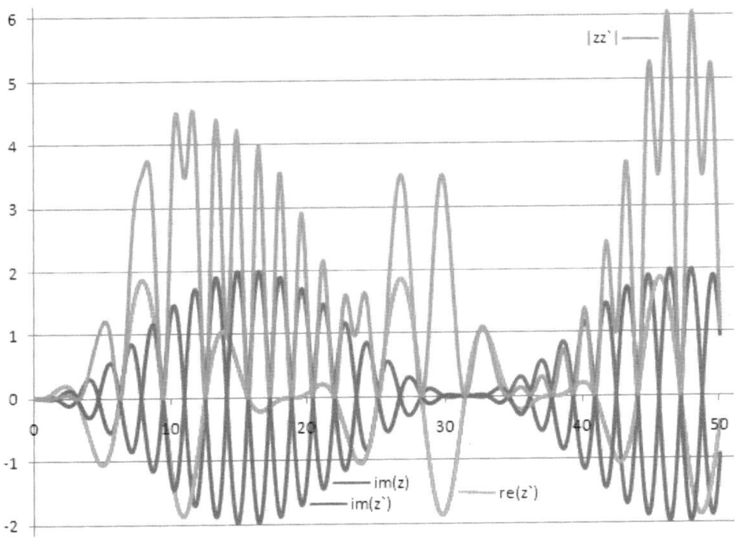

Fig. 12.15. *Evaluating and presenting a complex-valued signal out of two real-valued sampled signals.*

12.4.4 Complex Signal Presentation

Here we give a short F90 code that takes two real sampled signals and forms out of them a complex signal and its conjugate (see Figure 12.15).

```
program CmplxSignal  ! Complex Signal Presentation
    real :: t=0.,h=0.1,tFin=50.,x,y
    complex z,z1
    open(1,file='cmplxSig.csv')
    write(1,*) 't',',','re(z)',',','im(z)',',','re(z`)',',',
        'im(z`)',',','|zz`|'
    do while(t<tFin)
```

```
        x = sin(t)*(1-cos(t/3.0))           ! 2 real signals
        y = sin(2*t)*(1-cos(t/5.0))
        z = cmplx(x,y)                ! complex signal: z=x+iy
        z1= conjg(z)              ! conjugated z : z1=x-iy
        write(1,*) t,',',real(z),',',aimag(z),',',real(z1),','
         ,aimag(z1),',',abs(z*conjg(z))
        t = t+h
    end do
    close(1)
end program
```

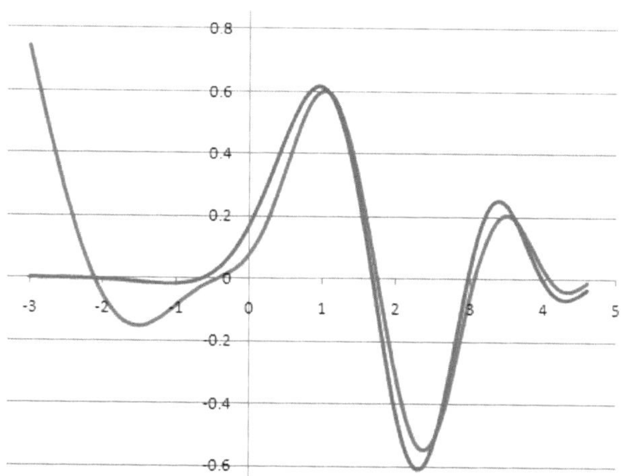

Fig. 12.16. *Evolution of a Gaussian wave packet by numerically solving the (1+1)D time-dependent Schrödinger equation.*

12.4.5 Gaussian Wave Packet

Here we give an F90 code that numerically solves the time-dependent Schrödinger equation and thus simulates the evolution of the Gaussian wave packet (see Figure 12.16). The algorithm uses the forward finite-difference scheme to solve the following discretized Schrödinger PDE:

$$\Psi(x, t+\Delta t) = \{1 - \mathrm{i}H\Delta t + (1/2)(-\mathrm{i}H\Delta t)^2\}\,\Psi(x,t),$$

where $H = -(1/2)\Delta^2 = -(1/2)d^2/dx^2$ is the Hamiltonian operator.

```
program GaussSchrod
! Gaussian wave packet
! Solution of time-dependent Schrodinger eq. by FD method
```

```
! Based on F77-code by Reinaldo Baretti Machn
   complex psiold,psinew,rooti,g,psi,psiG,d2psi,d4psi
   dimension psiold(0:1000),psinew(0:1000)
   data sigma,ak0,nstep/1.0,2.,150/
   g(x)=pi**(-.25)*sigma**(-.5)*exp(-.5*(x/sigma)**2)
       *exp(rooti*ak0*x)
! psiG from ref 2
   psiG(x,t)=pi**(-.25)*(sigma+rooti*beta*t)**(-.5)*  &
   exp(rooti*(ak0*x-w0*t))*exp(-.5*(x-vg*t)**2
       /(sigma+rooti*beta*t)**2)
   d2psi(n)=(psiold(n+1)-2.*psiold(n)+psiold(n-1))/dx**2
   d4psi(n)=(psiold(n+2)-4.*psiold(n+1)+       &
   6.*psiold(n)-4.*psiold(n-1)+psiold(n-2))/dx**4
   rooti=cmplx(0.0,1.0)
   pi=2.0*asin(1.0)
   vg=ak0
   beta=1./2.
   w0=ak0**2/2.
   aLength=sigma
   tscale= sigma**2
   tfinal=.8*tscale
   xi=-3.*alength
   xf=3.0*alength+vg*tfinal
   dx=(xf-xi)/float(nstep)
   dt=dx**2/100.
! dt=5.E-3*dx**2
   nt=int(tfinal/dt)
   ratio=dt/dx**2
   ! ratio dt/dx**2 is approx 5.0E-3
   print*,'tfinal,tscale,dt/dx**2=',tfinal,tscale,ratio
   print*,'nt,nstep,dt=',nt,nstep,dt
! initial psi(x,t=0) along x axis t=0
   do i=0,1000
       psiold(i)=0.
   end do
   do i=0,nstep
       x = xi + float(i)*dx
       psiold(i)=g(x)
   end do
! forward difference
   do it=1,nt
       t=dt*float(it)
       do i=0,nstep
           x = xi + float(i)*dx
 psinew(i)=psiold(i)+(rooti*dt/2.0)*d2psi(i)
```

```
      -1.*(dt**2/8.)*d4psi(i)
           end do
           do j=0,nstep
               psiold(j)=psinew(j)
           end do
       end do
   ! print group
       open(1,file='Schr.csv')
       print*,'    t,  x, Real psinum(x,t), Real psiG ';
       print *
       do i=0,nstep,2
           x = xi + float(i)*dx
           print'(2x,4(3x,e10.3))',t,x,real(psiold(i)),
               real(psiG(x,t))
           write(1,*) x,',',real(psiold(i)),',',real(psiG(x,t))
       end do
       close(1)
   end program
```

12.4.6 Hermitian Matrices

Here we give an F95 code that performs basic algebraic operations with Hermitian matrices.

```
   program Hermitian ! Operations with Hermitian Matrices
     implicit none; integer i
       complex A(2,2),At(2,2),Ac(2,2),An1(2,2),An2(2,2)
       open(1, file='Hermit.dat')

     A = reshape([ &       !matrix A
       (4., 0.), (1., 3.), &
       (1., -3.), (7., 0.)], [2, 2])
     write ( 1, '(a)' ) ' '
     write ( 1, '(a)' ) '  Matrix A:'
     do i=1,2
       write ( 1, '(2(2g8.2))' ) A(i,1:2)
     end do

     At=transpose(A)
     write ( 1, '(a)' ) ' '
     write ( 1, '(a)' ) '  At=transpose(A):'
     do i=1,2
       write ( 1, '(2(2g8.2))' ) At(i,1:2)
     end do
```

```
    Ac=conjg(At)
    write ( 1, '(a)' ) ' '
  write ( 1, '(a)' ) '  Ac=conjg(At):'
  do i=1,2
    write ( 1, '(2(2g8.2))' ) Ac(i,1:2)
  end do
  write(1,*) 'Hermitian criterion: Ac==A :', Ac==A

  An1=matmul(A,Ac)
  AN2=matmul(Ac,A)
  write ( 1, '(a)' ) ' '
  write ( 1, '(a)' ) '  A.Ac=matmul(A,Ac):'
  do i=1,2
    write ( 1, '(2(2g8.2))' ) An1(i,1:2)
  end do
    write ( 1, '(a)' ) ' '
  write ( 1, '(a)' ) '  Ac.A=matmul(Ac,A):'
  do i=1,2
    write ( 1, '(2(2g8.2))' ) AN2(i,1:2)
  end do
  write(1,*) 'Normality criterion: A.Ac==Ac.A :', An1==AN2
  close(1)
end program
```

The output of the program is:

```
  Matrix A:
4.0     0.0     1.0    -3.0
1.0     3.0     7.0     0.0

  At=transpose(A):
4.0     0.0     1.0     3.0
1.0    -3.0     7.0     0.0

  Ac=conjg(At):
4.0    -0.0     1.0    -3.0
1.0     3.0     7.0    -0.0
Hermitian criterion: Ac==A : T T T T

  A.Ac=matmul(A,Ac):
26.     0.0     11.    -33.
11.     33.     59.     0.0

  Ac.A=matmul(Ac,A):
26.     0.0     11.    -33.
11.     33.     59.     0.0
```

```
Normality criterion: A.Ac==Ac.A : T T T T
```

12.4.7 Euclidean L2-Norm

Here we give a short F95 code that calculates Euclidean L2-norm both directly and indirectly.

```
program L2Norm ! Euclidean L^2 Norm
    real*8 :: x(9) = [1.3, 2.1, 3.7, 4.2, 5.6, 6.3, 7.5, 8.4,
    9.8]
    open(1, file='norm.dat')
    write ( 1, '(a)' ) 'Given a sample 9D vector :'
    write ( 1, '(a)' ) 'x(9)=[1.3,2.1,3.7,4.2,5.6,6.3,7.5,
    8.4,9.8]'
    write ( 1, '(a)' ) 'Its Euclidean L2-norm is :';
    write ( 1, *) norm2(x)
    write ( 1, '(a)' ) 'L2-norm is computed as :';
    write ( 1, *) sqrt(sum(x(1:9)**2))
    close(1)
end program
```

The output of the program is:

```
Given a sample 9D vector :
x(9) = [1.3, 2.1, 3.7, 4.2, 5.6, 6.3, 7.5, 8.4, 9.8]
Its Euclidean L2-norm is :
   18.202472274145126
L2-norm is computed as :
   18.202472274145126
```

12.4.8 Vector/Matrix Operations

Here we give a short F95 code that performs basic vector and matrix operations.

```
program VecMat ! Vector and Matrix Operations
    implicit none
    integer :: i, j, n=5, m=5
    integer :: a(5)=[(i, i=1,5)], b(5)=[9,8,7,6,5]
    integer, dimension(5,5) :: c, d
    open(1, file='vecMat.dat')
    write(1,*) 'VECTOR AND MATRIX OPERATIONS IN F95/03';
    write(1,*)
    write(1,*) 'vec a(5)=[(i, i=1,5)]='; write(1,'(5i3)') a;
    write(1,*)
    write(1,*) 'size(a), sum(a), prod(a), min(a), max(a) = ';
```

```
    write(1,'(5i8)')size(a),sum(a),product(a),minval(a),
    maxval(a);
    write(1,*)
    write(1,*) 'vec b(5)=[9,8,7,6,5]='; write(1,'(5i3)') b;
    write(1,*)
    write(1,*) 'size(b), sum(b), prod(b), min(b), max(b) = ';
    write(1,'(5i8)')size(b),sum(b),product(b),minval(b),
    maxval(b);
    write(1,*)
    write(1,*) 'dot_prod(a,b)='; write(1,*) dot_product(a,b);
    write(1,*)
    forall (i=1:n, j=1:m) c(i,j)=i+j;  ! defining two
    matrices
    forall (i=1:n, j=1:m) d(i,j)=i*j;
    write(1,*) 'mat c(i,j)=i+j=';
    write(1,'(5i4)') c; write(1,*)
    write(1,*) 'mat d(i,j)=i*j=';
    write(1,'(5i4)') d; write(1,*)
    write(1,*) 'transp(c)='; write(1,'(5i4)') transpose(c);
    write(1,*)
    write(1,*) 'transp(d)='; write(1,'(5i4)') transpose(d);
    write(1,*)
    write(1,*) 'matmul(c,transp(dd))=';
    write(1,'(5i5)') matmul(c,transpose(d)); write(1,*)
    write(1,*) 'matmul(d,transp(cc))=';
    write(1,'(5i5)') matmul(d,transpose(c)); write(1,*)
    close(1)
end program
```

The output of the program is:

```
VECTOR AND MATRIX OPERATIONS IN F95/03

vec a(5)=[(i, i=1,5)]=
 1  2  3  4  5

size(a), sum(a), prod(a), min(a), max(a) =
       5      15     120        1       5

vec b(5)=[9,8,7,6,5]=
 9  8  7  6  5

size(b), sum(b), prod(b), min(b), max(b) =
       5      35   15120        5       9
dot_prod(a,b)=
         95
```

```
mat c(i,j)=i+j=
  2   3   4   5   6
  3   4   5   6   7
  4   5   6   7   8
  5   6   7   8   9
  6   7   8   9  10

mat d(i,j)=i*j=
  1   2   3   4   5
  2   4   6   8  10
  3   6   9  12  15
  4   8  12  16  20
  5  10  15  20  25

transp(c)=
  2   3   4   5   6
  3   4   5   6   7
  4   5   6   7   8
  5   6   7   8   9
  6   7   8   9  10

transp(d)=
  1   2   3   4   5
  2   4   6   8  10
  3   6   9  12  15
  4   8  12  16  20
  5  10  15  20  25

matmul(c,transp(dd))=
   70    85   100   115   130
  140   170   200   230   260
  210   255   300   345   390
  280   340   400   460   520
  350   425   500   575   650

matmul(d,transp(cc))=
   70   140   210   280   350
   85   170   255   340   425
  100   200   300   400   500
  115   230   345   460   575
  130   260   390   520   650
```

12.5 Plain C-Code: Levenberg-Marquardt Optimizer

Here we give a plain C-code implementing a fast Levenberg-Marquardt optimizer for M nonlinear functions in N variables.

```
// Levenberg-Marquardt Optimizer, an algorithm from MINPACK
// Minimizes the sum-of-squares of M nonlinear functions
(or a vector function FVEC) of N variables
// used values: M=7, N=7 and M=9, N=9
// This is a significantly shortened (almost 40%) and
cleaned-up modification of S. Moshier's C-translation of
the original MINPACK.F77 library by J. More, B. Garbow and
K. Hillstrom, Argonne National Laboratory, US, 1980.
// Modified by V. Ivancevic, July 2013

#define BUG 0
#define N 7
#define M 7
double ftol=1.0e-14, xtol=1.0e-14, gtol=1.0e-14;
double epsfcn=1.0e-15, factor=0.1, enorm();
double x[N]={0.0}, fvec[M]={0.0}, diag[N]={0.0};
double fjac[M*N]={0.0}, qtf[N]={0.0}, wa1[N]={0.0};
double wa2[N]={0.0}, wa3[N]={0.0}, wa4[M]={0.0};
int ipvt[N]={0}, maxfev = 200*(N+1), fcn();
double MACHEP = 1.2e-16; extern double MACHEP;
double DWARF = 1.0e-38; extern double DWARF;

int main()  // Main function
{
/* fcn is the user-supplied function which gives the
functions.
    the value of iflag should not be changed by fcn
    unless
    the user wants to terminate execution of lmdif1.
    in this case set iflag to a negative integer.
  m is a positive integer input variable set to the
    number of functions.
  n is a positive integer input variable set to the
    number
    of variables. n must not exceed m.
  x is an array of length n. on input x must contain
    an initial estimate of the solution vector. on
    output x contains the final estimate of the solution
    vector.
    fvec is an output array of length m which contains
```

```
      the functions evaluated at the output x. */
int m,n,info,mode,nfev,nprint,iflag,ldfjac;
double zero = 0.;  n = N; m = M;
fcn(m, n, x, fvec, &iflag);

printf("==============================================\n");
printf("       *** Levenberg-Marquardt Optimizer
(V.I., July 2013)  ***\n");
printf("Minimizes the sum-of-squares of M nonlinear
functions of N vars\n");
printf("Effectively minimizes the Euclidean L2-norm:
    norm2(f) = sqrt(sum(f(1:M)**2))\n");
printf("The norm2(f) induces the Euclidean metric:
    norm2(f1-f2)    ||\n");
printf("It is itself induced by the inner product:
    norm2(f) = dot_product(f,f)\n");
printf("If L2 Cauchy minimization sequence is convergent,
    a Banach space is constructed\n");
    printf("==========================================\n");
printf("\n");
printf("Sample initial solution x:\n");  pmat(1,n,x);
printf("\n");
printf("Sample initial vector function f:\n");
pmat(1,m,fvec);  printf("\n");
printf("----------------\n");
printf("Banach space construction  ->\n");
printf("Cauchy sequence for the L2-norm:\n");
printf("----------------\n");
ldfjac = m; mode = 1; nprint = 1; info = 0;
lmdif(m,n,x,fvec,ftol,xtol,gtol,maxfev,epsfcn,  // Call
lmdif
diag,mode,factor,nprint,&info,&nfev,fjac,
ldfjac,ipvt,qtf,wa1,wa2,wa3,wa4);
printf("\n"); printf("%d functional evalutations
performed\n",nfev);
printf("\n"); printf("Computed solution x:\n");
pmat(1,n,x); printf("\n");
printf("Computed minimum of the vector function f:\n");
pmat(1,m,fvec);
printf("\n");
printf("Final L2-norm = %.15e\n", enorm(m,fvec));
    printf("\n");
if (enorm(m,fvec) <= 1.0e-10)
    printf("1. Weak convergence : L2-norm <= 1.0e-10\n");
if (enorm(m,fvec) <= 1.0e-30)
```

```
        printf("2. Moderate convergence : L2-norm <=
        1.0e-30\n");
    if (enorm(m,fvec) <= 1.0e-60)
        printf("3. Strong convergence : L2-norm <=
        1.0e-60\n");
    if (enorm(m,fvec) <= 1.0e-100)
        printf("4. Very strong convergence : L2-norm <=
    1.0e-100\n");
    if (enorm(m,fvec) <= 1.0e-200)
    printf("5. Extremely strong convergence : L2-norm <=
    1.0e-150\n");
     printf("=========================================\n");
    printf("   :- Banach space is constructed (see attached
    PDF)\n");
printf("==================================================\n");
  getchar();
}

// Sample user-given function
int fcn(m,n,x,fvec,iflag)
int m,n,*iflag; double x[],fvec[];
/* m = number of functions
 n = number of variables
   x = vector of function arguments
   fvec = vector of function values
   iflag = error return variable */
{
double sin(), log(), sinh(), tanh(), atan();
// Sample test vector function, including 7 funs with
7 vars:
fvec[0] = tanh(x[0]*x[1]) - 1.7*x[2] + log(1.5 +
x[3]*x[4]);
fvec[1] = sin(-4.0*x[0]) - 3.*tanh(x[1]*x[5]) + 0.1*x[2]
*x[4] + x[3]*x[6];
fvec[2] = 0.5*x[1]*x[3] - tanh(x[2]*x[4]) + sinh(x[5]*
x[6]) + x[2]*x[1];
fvec[3] = x[5]*x[1] + x[2]*x[3] + x[0]*x[2] - x[6]*x[4];
fvec[4] = sinh(x[0]*x[1]) + x[2]*x[3] + x[6]*x[4] -
x[3]*x[4];
fvec[5] = 0.5*x[1]*x[4] - sinh(x[2]*x[3]) + tanh(x[5]*
x[6]) + x[2]*x[1];
fvec[6] = x[0]*x[1] + x[3]*x[4] + x[2]*x[5] -
atan(x[3]*x[6]);
}
```

```
#define BUG 0
extern double MACHEP;
// function lmdif (the CORE of the algorithm)
lmdif(m,n,x,fvec,ftol,xtol,gtol,maxfev,epsfcn,
diag,mode,factor,nprint,info,nfev,fjac,
ldfjac,ipvt,qtf,wa1,wa2,wa3,wa4)
int m,n,maxfev,mode,nprint,ldfjac;
int *info, *nfev, ipvt[];
double ftol, xtol, gtol, epsfcn, factor;
double x[],fvec[],diag[],fjac[],qtf[],wa1[],wa2[],wa3[],
wa4[];
{
/*    The function lmdif minimizes the sum of the squares of
*     m nonlinear functions in n variables by a
*     modification of
*     the levenberg-marquardt algorithm. The user
*     provides a
*     function which calculates the functions. the
*     jacobian is
*     then calculated by a forward-difference
*     approximation.
*     the function statement is
*   function lmdif(fcn,m,n,x,fvec,ftol,xtol,gtol,maxfev,
*   epsfcn,
*    diag,mode,factor,nprint,info,nfev,fjac,
*    ldfjac,ipvt,qtf,wa1,wa2,wa3,wa4)
*   where
*   fcn is the name of the user-supplied function which
*     calculates the functions.
*     the value of iflag should not be changed by fcn
*     unless
*     the user wants to terminate execution of lmdif.
*     in this case set iflag to a negative integer.
*   m is a positive integer input variable set to the
*   number of functions.
*   n is a positive integer input variable set to the
*   number
*     of variables. n must not exceed m.
*   x is an array of length n. on input x must contain
*     an initial estimate of the solution vector. on
*     output x
*     contains the final estimate of the solution vector.
*   fvec is an output array of length m which contains
*     the functions evaluated at the output x.
*   ftol is a nonnegative input variable. termination
```

```
*     occurs when both the actual and predicted relative
*     reductions in the sum of squares are at most ftol.
*     therefore, ftol measures the relative error desired
*     in the sum of squares.
*   xtol is a nonnegative input variable. termination
*     occurs when the relative error between two
*     consecutive
*     iterates is at most xtol. therefore, xtol measures the
*     relative error desired in the approximate solution.
*   gtol is a nonnegative input variable. termination
*     occurs when the cosine of the angle between fvec and
*     any column of the jacobian is at most gtol in
*     absolute value. therefore, gtol measures the
*     orthogonality desired between the function vector and
*     the columns of the jacobian.
*   maxfev is a positive integer input variable.
*   termination
*     occurs when the number of calls to fcn is at least
*     maxfev by the end of an iteration.
*   epsfcn is an input variable used in determining a
*   suitable
*     step length for the forward-difference
*     approximation. this
*     approximation assumes that the relative errors in the
*     functions are of the order of epsfcn. if epsfcn is
*     less
*     than the machine precision, it is assumed that the
*     relative
*     errors in the functions are of the order of the
*     machine precision.
*   diag is an array of length n. if mode = 1 (see
*     below), diag is internally set. if mode = 2, diag
*     must contain positive entries that serve as
*     multiplicative scale factors for the variables.
*   mode is an integer input variable. if mode = 1, the
*     variables will be scaled internally. if mode = 2,
*     the scaling is specified by the input diag. other
*     values of mode are equivalent to mode = 1.
*   factor is a positive input variable used in
*   determining the
*     initial step bound. this bound is set to the
*     product of
*     factor and the euclidean norm of diag*x if nonzero,
*     or else
*     to factor itself. in most cases factor should lie
```

```
*      in the
*      interval (.1,100.). 100. is a generally recommended
*      value.
*    nprint is an integer input variable that enables
*    controlled
*      printing of iterates if it is positive. in this case,
*      fcn is called with iflag = 0 at the beginning of
*      the first
*      iteration and every nprint iterations thereafter
*      and
*      immediately prior to return, with x and fvec
*      available
*      for printing. if nprint is not positive, no special
*      calls
*      of fcn with iflag = 0 are made.
*    info is an integer output variable. if the user has
*      terminated execution, info is set to the (negative)
*      value of iflag. see description of fcn. otherwise,
*      info is set as follows.
*      info = 0  improper input parameters.
*      info = 1  both actual and predicted relative
*      reductions
*            in the sum of squares are at most ftol.
*      info = 2  relative error between two consecutive
*      iterates
*            is at most xtol.
*      info = 3  conditions for info = 1 and info = 2 both
*      hold.
*      info = 4  the cosine of the angle between fvec and any
*            column of the jacobian is at most gtol in
*            absolute value.
*      info = 5  number of calls to fcn has reached or
*      exceeded maxfev.
*
*      info = 6  ftol is too small. no further reduction in
*            the sum of squares is possible.
*      info = 7  xtol is too small. no further improvement in
*            the approximate solution x is possible.
*      info = 8  gtol is too small. fvec is orthogonal to the
*            columns of the jacobian to machine precision.
*    nfev is an integer output variable set to the number
*    of calls to fcn.
*    fjac is an output m by n array. the upper n by n
*    submatrix of fjac contains an upper triangular matrix
*      r with diagonal elements of nonincreasing magnitude
```

```
*       such that
*          t     t       t
*         p *(jac *jac)*p = r *r,
*       where p is a permutation matrix and jac is the final
*       calculated jacobian. column j of p is column ipvt(j)
*       (see below) of the identity matrix. the lower
*       trapezoidal
*       part of fjac contains information generated during
*       the computation of r.
*     ldfjac is a positive integer input variable not less
*     than m
*       which specifies the leading dimension of the array
*       fjac.
*     ipvt is an integer output array of length n. ipvt
*       defines a permutation matrix p such that
*       jac*p = q*r,
*       where jac is the final calculated jacobian, q is
*       orthogonal (not stored), and r is upper triangular
*       with diagonal elements of nonincreasing magnitude.
*       column j of p is column ipvt(j) of the identity
*       matrix.
*     qtf is an output array of length n which contains
*       the first n elements of the vector (q transpose)*
        fvec. */
int i,iflag,ij,jj,iter,j,l;
double actred,delta,dirder,fnorm,fnorm1,gnorm;
double par,pnorm,prered,ratio,sum,temp,temp1,temp2,temp3,
xnorm;
double enorm(), fabs(), dmax1(), dmin1(), sqrt();
int fcn();  // user supplied function (given above)
double one=1., zero=0., p1=0.1, p5=0.5, p25=0.25;
double p75=0.75, p0001=1.0e-4, p05=0.05;
*info = 0;  iflag = 0;  *nfev = 0;
//  check the input parameters for errors.
if ( (n <= 0) || (m < n) || (ldfjac < m) || (ftol < zero)
|| (xtol < zero) || (gtol < zero) || (maxfev <= 0)
|| (factor <= zero) )  goto L300;
if (mode == 2) { // scaling by diag[]
for(j=0; j<n; j++) {
if (diag[j] <= 0.0)  goto L300;
}
}
#if BUG
printf("lmdif\n");
#endif
```

```
// evaluate the function at the starting point and calc.
its norm
iflag = 1;
fcn(m,n,x,fvec,&iflag);
*nfev = 1;
if(iflag < 0)  goto L300;
fnorm = enorm(m,fvec);
// initialize LM-parameter and iteration counter
par = zero;
iter = 1;
//  beginning of the outer loop
L30:
//  calculate the jacobian matrix
iflag = 2;
fdjac2(m,n,x,fvec,fjac,ldfjac,&iflag,epsfcn,wa4);
*nfev += n;
if (iflag < 0)  goto L300;
//   if requested, call fcn to enable printing of
iterates
if(nprint > 0) {
iflag = 0;
if (mod(iter-1,nprint) == 0) {
fcn(m,n,x,fvec,&iflag);
if (iflag < 0)  goto L300;
printf("L2-norm = %.15e\n", enorm(m,fvec));
}
}
//   compute the qr factorization of the jacobian
qrfac(m,n,fjac,ldfjac,1,ipvt,n,wa1,wa2,wa3);
//  on the first iteration and if mode is 1, scale
    according
// to the norms of the columns of the initial jacobian.
if (iter == 1) {
if (mode != 2) {
for (j=0; j<n; j++) {
diag[j] = wa2[j];
if (wa2[j] == zero)  diag[j] = one;
}
}
// on the first iteration, calculate the norm of the
   scaled x
    // and initialize the step bound delta
for(j=0; j<n; j++)  wa3[j] = diag[j] * x[j];
xnorm = enorm(n,wa3);
delta = factor*xnorm;
```

```
if (delta == zero)  delta = factor;
}
// form (q transpose)*fvec and store the first n
components in qtf
for (i=0; i<m; i++)  wa4[i] = fvec[i];
jj = 0;
for (j=0; j<n; j++) {
temp3 = fjac[jj];
if (temp3 != zero) {
sum = zero;
ij = jj;
for (i=j; i<m; i++) {
sum += fjac[ij] * wa4[i];
ij += 1;    // fjac[i+m*j]
}
temp = -sum / temp3;
ij = jj;
for (i=j; i<m; i++) {
wa4[i] += fjac[ij] * temp;
ij += 1;    // fjac[i+m*j]
}
}
fjac[jj] = wa1[j];
jj += m+1;  // fjac[j+m*j]
qtf[j] = wa4[j];
}
//  compute the norm of the scaled gradient
gnorm = zero;
if (fnorm != zero)
{
jj = 0;
for (j=0; j<n; j++) {
l = ipvt[j];
if (wa2[l] != zero) {
sum = zero;
ij = jj;
for (i=0; i<=j; i++) {
sum += fjac[ij]*(qtf[i]/fnorm);
ij += 1; // fjac[i+m*j]
}
gnorm = dmax1(gnorm,fabs(sum/wa2[l]));
}
jj += m;
}
}
```

```
//  test for convergence of the gradient norm
if (gnorm <= gtol)  *info = 4;
if (*info != 0)  goto L300;
//  rescale if necessary
if (mode != 2) {
for (j=0; j<n; j++)  diag[j] = dmax1(diag[j],wa2[j]);
}
//  beginning of the inner loop
L200:
//  determine the levenberg-marquardt parameter
lmpar(n,fjac,ldfjac,ipvt,diag,qtf,delta,&par,wa1,wa2,
wa3,wa4);
//  store the direction p and x + p. calculate the norm
    of p
for (j=0; j<n; j++) {
wa1[j] = -wa1[j];
wa2[j] = x[j] + wa1[j];
wa3[j] = diag[j]*wa1[j];
}
pnorm = enorm(n,wa3);
//  on the first iteration, adjust the initial step bound
if (iter == 1)  delta = dmin1(delta,pnorm);
//  evaluate the function at x + p and calculate its norm
iflag = 1;
fcn(m,n,wa2,wa4,&iflag);
*nfev += 1;
if (iflag < 0)  goto L300;
fnorm1 = enorm(m,wa4);
#if BUG
printf("pnorm %.10e  fnorm1 %.10e\n", pnorm, fnorm1);
#endif
//  compute the scaled actual reduction.
actred = -one;
if ((p1*fnorm1) < fnorm) {
temp = fnorm1/fnorm;
actred = one - temp * temp;
}
//  compute the scaled predicted reduction and the scaled
    directional derivative
jj = 0;
for (j=0; j<n; j++) {
wa3[j] = zero;
l = ipvt[j];
temp = wa1[l];
ij = jj;
```

```
for (i=0; i<=j; i++) {
wa3[i] += fjac[ij]*temp;
ij += 1; // fjac[i+m*j]
}
jj += m;
}
temp1 = enorm(n,wa3)/fnorm;
temp2 = (sqrt(par)*pnorm)/fnorm;
prered = temp1*temp1 + (temp2*temp2)/p5;
dirder = -(temp1*temp1 + temp2*temp2);
//  compute the ratio of the actual to the predicted
reduction
ratio = zero;
if (prered != zero)  ratio = actred/prered;
//   update the step bound
if (ratio <= p25) {
if (actred >= zero)  temp = p5;
else  temp = p5*dirder/(dirder + p5*actred);
if (((p1*fnorm1) >= fnorm) || (temp < p1))
temp = p1;
delta = temp*dmin1(delta,pnorm/p1);
par = par/temp;
}
else {
if ((par == zero) || (ratio >= p75)) {
delta = pnorm/p5;
par = p5*par;
}
}
//   test for successful iteration
if (ratio >= p0001)
{
//  successful iteration. update x, fvec, and their
    norms
for (j=0; j<n; j++) {
x[j] = wa2[j];
wa2[j] = diag[j]*x[j];
}
for (i=0; i<m; i++)  fvec[i] = wa4[i];
xnorm = enorm(n,wa2);
fnorm = fnorm1;
iter += 1;
}
//  tests for convergence
if ( (fabs(actred) <= ftol)
```

```
&& (prered <= ftol)
&& (p5*ratio <= one) )  *info = 1;
if (delta <= xtol*xnorm)    *info = 2;
if ( (fabs(actred) <= ftol)
&& (prered <= ftol)
&& (p5*ratio <= one)
&& ( *info == 2) )  *info = 3;
if (*info != 0)  goto L300;
//  tests for termination and stringent tolerances
if (*nfev >= maxfev)  *info = 5;
if ( (fabs(actred) <= MACHEP)
&& (prered <= MACHEP)
&& (p5*ratio <= one) )  *info = 6;
if (delta <= MACHEP*xnorm)  *info = 7;
if (gnorm <= MACHEP)  *info = 8;
if (*info != 0)  goto L300;
//   end of the inner loop. repeat if iteration
unsuccessful
if (ratio < p0001)  goto L200;
//  end of the outer loop
goto L30;
L300:
//  termination, either normal or user imposed
if (iflag < 0)  *info = iflag;
iflag = 0;
if (nprint > 0)  fcn(m,n,x,fvec,&iflag);
}

#define BUG 0
// function lmpar
lmpar(n,r,ldr,ipvt,diag,qtb,delta,par,x,sdiag,wa1,wa2)
int n,ldr,ipvt[];
double r[],diag[],qtb[],x[],sdiag[],wa1[],wa2[],delta,*par;
{
/*    Given an m by n matrix a, an n by n nonsingular
*     diagonal matrix d, an m-vector b, and a positive number
*     delta, the problem is to determine a value for the
*     parameter par such that if x solves the system
*     a*x = b ,     sqrt(par)*d*x = 0 ,
*     in the least squares sense, and dxnorm is the
*     euclidean norm of d*x, then either par is zero and
*     (dxnorm-delta) .le. 0.1*delta ,
*     or par is positive and abs(dxnorm-delta) .le. 0.1*delta
*     this function completes the solution of the problem
*     if it is provided with the necessary information
```

```
*      from the qr factorization, with column pivoting, of a.
*      that is, if a*p = q*r, where p is a permutation matrix,
*      q has orthogonal columns, and r is an upper triangular
*      matrix with diagonal elements of nonincreasing
*      magnitude, then lmpar expects
*      the full upper triangle of r, the permutation matrix p,
*      and the first n components of (q transpose)*b. on
*      output lmpar also provides an upper triangular matrix s
*      such that
*         t   t           t
*      p *(a *a + par*d*d)*p = s *s .
*      s is employed within lmpar and may be of separate
*      interest. only a few iterations are generally needed
*      for convergence of the algorithm. if, however, the
*      limit of 10 iterations is reached, then the output
*      par will contain the best value obtained so far.
*      the function statement is
* function lmpar(n,r,ldr,ipvt,diag,qtb,delta,par,x, sdiag,
* wa1,wa2)
*      where
* n is a positive integer input variable set to the
* order of r.
* r is an n by n array. on input the full upper
* triangle
*   must contain the full upper triangle of the matrix r.
*   on output the full upper triangle is unaltered, and the
*   strict lower triangle contains the strict upper
*   triangle (transposed) of the upper triangular matrix s.
* ldr is a positive integer input variable not less than n
*   which specifies the leading dimension of the array r.
* ipvt is an integer input array of length n which
*   defines the permutation matrix p such that a*p = q*r.
*   column j of p is column ipvt(j) of the identity matrix.
* diag is an input array of length n which must contain the
* diagonal elements of the matrix d.
* qtb is an input array of length n which must contain
*   the first n elements of the vector (q transpose)*b.
* delta is a positive input variable which specifies an
*   upper bound on the euclidean norm of d*x.
* par is a nonnegative variable. on input par contains
*   an initial estimate of the levenberg-marquardt
*   parameter. on output par contains the final estimate.
* x is an output array of length n which contains the
*   least squares solution of the system a*x = b,
*   sqrt(par)*d*x = 0, for the output par.
```

```
* sdiag is an output array of length n which contains
*   the diagonal elements of the upper triangular matrix s.
* wa1 and wa2 are work arrays of length n. */
int i,iter,ij,jj,j,jm1,jp1,k,l,nsing;
double dxnorm,fp,gnorm,parc,parl,paru,sum,temp;
double enorm(), fabs(), dmax1(), dmin1(), sqrt();
static double zero=0.,one=1.,p1=0.1, p001=0.001;
extern double MACHEP, DWARF;
#if BUG
printf("lmpar\n");
#endif
//    Compute and store in x the gauss-newton direction.
If the
//    jacobian is rank-deficient, obtain a least-squares
solution
nsing = n;
jj = 0;
for (j=0; j<n; j++) {
wa1[j] = qtb[j];
if ((r[jj] == zero) && (nsing == n))  nsing = j;
if (nsing < n)  wa1[j] = zero;
jj += ldr+1; // [j+ldr*j]
}
#if BUG
printf("nsing %d ", nsing);
#endif
if (nsing >= 1) {
for (k=0; k<nsing; k++) {
j = nsing - k - 1;
wa1[j] = wa1[j]/r[j+ldr*j];
temp = wa1[j];
jm1 = j - 1;
if (jm1 >= 0) {
ij = ldr * j;
for (i=0; i<=jm1; i++) {
wa1[i] -= r[ij]*temp;
ij += 1;
}
}
}
}

for (j=0; j<n; j++) {
l = ipvt[j];
x[l] = wa1[j];
```

```
}
/* initialize the iteration counter.
evaluate the function at the origin, and test
for acceptance of the gauss-newton direction */
iter = 0;
for (j=0; j<n; j++)  wa2[j] = diag[j]*x[j];
dxnorm = enorm(n,wa2);
fp = dxnorm - delta;
if (fp <= p1*delta) {
#if BUG
printf("going to L220\n");
#endif
goto L220;
}
/*  if the jacobian is not rank deficient, the newton
 step provides a lower bound, parl, for the zero of
 the function. otherwise set this bound to zero. */
parl = zero;
if (nsing >= n) {
for (j=0; j<n; j++) {
l = ipvt[j];
wa1[j] = diag[l]*(wa2[l]/dxnorm);
}
jj = 0;
for (j=0; j<n; j++) {
sum = zero;
jm1 = j - 1;
if (jm1 >= 0) {
ij = jj;
for (i=0; i<=jm1; i++) {
sum += r[ij]*wa1[i];
ij += 1;
}
}
wa1[j] = (wa1[j] - sum)/r[j+ldr*j];
jj += ldr; // [i+ldr*j]
}
temp = enorm(n,wa1);
parl = ((fp/delta)/temp)/temp;
}
// calculate an upper bound, paru, for the zero of the
function
jj = 0;
for (j=0; j<n; j++) {
sum = zero;
```

```
ij = jj;
for (i=0; i<=j; i++) {
sum += r[ij]*qtb[i];
ij += 1;
}
l = ipvt[j];
wa1[j] = sum/diag[l];
jj += ldr; // [i+ldr*j]
}
gnorm = enorm(n,wa1);
paru = gnorm/delta;
if (paru == zero) paru = DWARF/dmin1(delta,p1);
// if the input par lies outside of the interval
(parl,paru),
// set par to the closer endpoint
*par = dmax1( *par,parl);
*par = dmin1( *par,paru);
if (*par == zero)  *par = gnorm/dxnorm;
#if BUG
printf("parl %.4e  par %.4e  paru %.4e\n", parl, *par,
paru);
#endif
// beginning of an iteration
L150:
iter += 1;
// evaluate the function at the current value of par
if (*par == zero)  *par = dmax1(DWARF,p001*paru);
temp = sqrt( *par );
for (j=0; j<n; j++)  wa1[j] = temp*diag[j];
qrsolv(n,r,ldr,ipvt,wa1,qtb,x,sdiag,wa2);
for (j=0; j<n; j++)  wa2[j] = diag[j]*x[j];
dxnorm = enorm(n,wa2);
temp = fp;
fp = dxnorm - delta;
/* If the function is small enough, accept the current
 value
 of par. also test for the exceptional cases where
 parl
 is zero or the number of iterations has reached 10. */
if ( (fabs(fp) <= p1*delta)
|| ((parl == zero) && (fp <= temp) && (temp < zero))
|| (iter == 10) )  goto L220;
//   compute the Newton correction
for (j=0; j<n; j++) {
l = ipvt[j];
```

```
wa1[j] = diag[l]*(wa2[l]/dxnorm);
}
jj = 0;
for (j=0; j<n; j++) {
wa1[j] = wa1[j]/sdiag[j];
temp = wa1[j];
jp1 = j + 1;
if (jp1 < n) {
ij = jp1 + jj;
for (i=jp1; i<n; i++) {
wa1[i] -= r[ij]*temp;
ij += 1; // [i+ldr*j]
}
}
jj += ldr; // ldr*j
}
temp = enorm(n,wa1);
parc = ((fp/delta)/temp)/temp;
// depending on the sign of the function, update parl or
paru
if (fp > zero)  parl = dmax1(parl, *par);
if (fp < zero)  paru = dmin1(paru, *par);
// compute an improved estimate for par
*par = dmax1(parl, *par + parc);
// end of an iteration
goto L150;
L220:
// termination
if (iter == 0)  *par = zero;
}

#define BUG 0
// function qrfac
qrfac(m,n,a,lda,pivot,ipvt,lipvt,rdiag,acnorm,wa)
int m,n,lda,lipvt,ipvt[],pivot;
double a[],rdiag[],acnorm[],wa[];
{
/*    This function uses Householder transformations with
*     column pivoting (optional) to compute a qr
*     factorization of the m by n matrix a. that is, qrfac
*     determines an orthogonal matrix q, a permutation matrix
*     p, and an upper trapezoidal matrix r with diagonal
*     elements of nonincreasing magnitude, such that
*     a*p = q*r. the householder transformation for column k,
*     k = 1,2,...,min(m,n), is of the form
```

```
*     t
*     i - (1/u(k))*u*u
*
*     where u has zeros in the first k-1 positions. the
*     form of this transformation and the method of pivoting
*     first appeared in the corresponding linpack function.
*     the function statement is
* function qrfac(m,n,a,lda,pivot,ipvt,lipvt,rdiag, acnorm,wa)
*     where
* m is a positive integer input variable set to the number
*   of rows of a.
* n is a positive integer input variable set to the number
*   of columns of a.
* a is an m by n array. on input a contains the matrix for
*   which the qr factorization is to be computed. on output
*   the strict upper trapezoidal part of a contains the
*   strict upper trapezoidal part of r, and the lower
*   trapezoidal part of a contains a factored form of q (the
*   non-trivial elements of the u vectors described above).
* lda is a positive integer input variable not less than m
*   which specifies the leading dimension of the array a.
* pivot is a logical input variable. if pivot is set true,
*   then column pivoting is enforced. if pivot is set false,
*   then no column pivoting is done.
* ipvt is an integer output array of length lipvt. ipvt
*   defines the permutation matrix p such that a*p = q*r.
*   column j of p is column ipvt(j) of the identity matrix.
*   if pivot is false, ipvt is not referenced.
* lipvt is a positive integer input variable. if pivot
*   is false, then lipvt may be as small as 1. if pivot is
*   true, then
*   lipvt must be at least n.
* rdiag is an output array of length n which contains the
*   diagonal elements of r.
* acnorm is an output array of length n which contains the
*   norms of the corresponding columns of the input matrix a.
*   if this information is not needed, then acnorm can
*   coincide with rdiag.
* wa is a work array of length n. if pivot is false, then wa
*   can coincide with rdiag. */
int i,ij,jj,j,jp1,k,kmax,minmn;
double ajnorm,sum,temp,zero=0.,one=1.,p05=0.05;
extern double MACHEP;
double enorm(), dmax1(), sqrt();
// compute the initial column norms and initialize
```

```
several arrays
ij = 0;
for (j=0; j<n; j++) {
acnorm[j] = enorm(m,&a[ij]);
rdiag[j] = acnorm[j];
wa[j] = rdiag[j];
if (pivot != 0)  ipvt[j] = j;
ij += m; // m*j
}
#if BUG
printf("qrfac\n");
#endif
// reduce a to r with Householder transformations
minmn = min0(m,n);
for (j=0; j<minmn; j++)
{
if (pivot == 0)  goto L40;
//   bring the column of largest norm into the pivot
position
kmax = j;
for (k=j; k<n; k++) {
if (rdiag[k] > rdiag[kmax])  kmax = k;
}
if (kmax == j)  goto L40;
ij = m * j;
jj = m * kmax;
for (i=0; i<m; i++) {
temp = a[ij]; // [i+m*j]
a[ij] = a[jj]; // [i+m*kmax]
a[jj] = temp;
ij += 1;
jj += 1;
}
rdiag[kmax] = rdiag[j];
wa[kmax] = wa[j];
k = ipvt[j];
ipvt[j] = ipvt[kmax];
ipvt[kmax] = k;
L40:
/* Compute the Householder transformation to reduce
the
 j-th column of a to a multiple of the j-th unit
 vector */
jj = j + m*j;
ajnorm = enorm(m-j,&a[jj]);
```

```
if (ajnorm == zero)  goto L100;
if (a[jj] < zero)  ajnorm = -ajnorm;
ij = jj;
for (i=j; i<m; i++) {
a[ij] /= ajnorm;
ij += 1; // [i+m*j]
}
a[jj] += one;
// apply the transform. to the remaining columns and
update the norms
jp1 = j + 1;
if (jp1 < n) {
for (k=jp1; k<n; k++) {
sum = zero;
ij = j + m*k;
jj = j + m*j;
for (i=j; i<m; i++) {
sum += a[jj]*a[ij];
ij += 1; // [i+m*k]
jj += 1; // [i+m*j]
}
temp = sum/a[j+m*j];
ij = j + m*k;
jj = j + m*j;
for (i=j; i<m; i++) {
a[ij] -= temp*a[jj];
ij += 1; // [i+m*k]
jj += 1; // [i+m*j]
}
if ((pivot != 0) && (rdiag[k] != zero)) {
temp = a[j+m*k]/rdiag[k];
temp = dmax1(zero, one-temp*temp);
rdiag[k] *= sqrt(temp);
temp = rdiag[k]/wa[k];
if ((p05*temp*temp) <= MACHEP) {
rdiag[k] = enorm(m-j-1,&a[jp1+m*k]);
wa[k] = rdiag[k];
}
}
}
}
L100:
rdiag[j] = -ajnorm;
}
}
```

```
#define BUG 0
// function qrsolv
qrsolv(n,r,ldr,ipvt,diag,qtb,x,sdiag,wa)
int n,ldr,ipvt[];
double r[],diag[],qtb[],x[],sdiag[],wa[];
{
/*   Given an m by n matrix a, an n by n diagonal matrix d,
*    and an m-vector b, the problem is to determine an x
*    which solves the system:
*    a*x = b ,     d*x = 0 ,
*    in the least squares sense.
*    This function completes the solution of the problem
*    if it is provided with the necessary information
*    from the qr factorization, with column pivoting, of a.
*    that is, if a*p = q*r, where p is a permutation matrix,
*    q has orthogonal columns, and r is an upper triangular
*    matrix with diagonal elements of nonincreasing
*    magnitude, then qrsolv expects the full upper triangle
*    of r, the permutation matrix p, and the first n
*    components of (q transpose)*b. the system
*    a*x = b, d*x = 0, is then equivalent to
*   t       t
*    r*z = q *b ,  p *d*p*z = 0 ,
*    where x = p*z. if this system does not have full rank,
*    then a least squares solution is obtained. on
*    output qrsolv also provides an upper triangular matrix
*    s such that
*       t   t       t
*    p *(a *a + d*d)*p = s *s .
*    s is computed within qrsolv and may be of separate
*    interest. the function statement is
* function qrsolv(n,r,ldr,ipvt,diag,qtb,x,sdiag,wa), where
* n is a positive integer input variable set to the
*   order of r.
* r is an n by n array. on input the full upper
*   triangle must contain the full upper triangle of the
*   matrix r. on output the full upper triangle is unaltered,
*   and the strict lower triangle contains the strict upper
*   triangle (transposed) of the upper triangular matrix s.
* ldr is a positive integer input variable not less
*   than n which specifies the leading dimension of the
*   array r.
* ipvt is an integer input array of length n which defines
*   the permutation matrix p such that a*p = q*r. column j
```

```
*   of p is column ipvt(j) of the identity matrix.
* diag is an input array of length n which must contain
*   the diagonal elements of the matrix d.
* qtb is an input array of length n which must contain
*   the first n elements of the vector (q transpose)*b.
* x is an output array of length n which contains the
*   least squares solution of the system a*x = b, d*x = 0.
* sdiag is an output array of length n which contains
*   the diagonal elements of the upper triangular matrix s.
* wa is a work array of length n. */
int i,ij,ik,kk,j,jp1,k,kp1,l,nsing;
double cos,cotan,qtbpj,sin,sum,tan,temp;
static double zero=0., p25=0.25, p5=0.5;
double fabs(), sqrt();
/* copy r and (q transpose)*b to preserve input and
initialize s.
   in particular, save the diagonal elements of r in x.
*/
kk = 0;
for (j=0; j<n; j++) {
ij = kk;
ik = kk;
for (i=j; i<n; i++) {
r[ij] = r[ik];
ij += 1;   // [i+ldr*j]
ik += ldr; // [j+ldr*i]
}
x[j] = r[kk];
wa[j] = qtb[j];
kk += ldr+1; // j+ldr*j
}
#if BUG
printf("qrsolv\n");
#endif
// eliminate the diagonal matrix d using a givens
rotation
for (j=0; j<n; j++)
{
/* prepare the row of d to be eliminated, locating
   the
*  diagonal element using p from the qr
   factorization. */
l = ipvt[j];
if (diag[l] == zero)  goto L90;
for (k=j; k<n; k++)  sdiag[k] = zero;
```

```
sdiag[j] = diag[l];
/*   the transformations to eliminate the row of d
   modify only a single element of (q transpose)*b
   beyond the first n, which is initially zero. */
qtbpj = zero;
for (k=j; k<n; k++) {
/*  determine a givens rotation which eliminates
    the appropriate element in the current row of d. */
if (sdiag[k] == zero)  continue;
kk = k + ldr * k;
if (fabs(r[kk]) < fabs(sdiag[k]))   {
cotan = r[kk]/sdiag[k];
sin = p5/sqrt(p25+p25*cotan*cotan);
cos = sin*cotan;
}
else {
tan = sdiag[k]/r[kk];
cos = p5/sqrt(p25+p25*tan*tan);
sin = cos*tan;
}
/*  compute the modified diagonal element of r
    and the modified element of ((q transpose) *b,0). */
r[kk] = cos*r[kk] + sin*sdiag[k];
temp = cos*wa[k] + sin*qtbpj;
qtbpj = -sin*wa[k] + cos*qtbpj;
wa[k] = temp;
//    accumulate the tranformation in the row
of s.
kp1 = k + 1;
if (n > kp1) {
ik = kk + 1;
for (i=kp1; i<n; i++)   {
temp = cos*r[ik] + sin*sdiag[i];
sdiag[i] = -sin*r[ik] + cos*sdiag[i];
r[ik] = temp;
ik += 1; // [i+ldr*k]
}
}
}
L90:
/*  store the diagonal element of s and restore
    the corresponding diagonal element of r.   */
kk = j + ldr*j;
sdiag[j] = r[kk];
r[kk] = x[j];
```

```
}
/*   solve the triangular system for z. if the system is
     singular, then obtain a least squares solution. */
nsing = n;
for (j=0; j<n; j++) {
if ((sdiag[j] == zero) && (nsing == n))  nsing = j;
if (nsing < n)  wa[j] = zero;
}
if (nsing < 1)  goto L150;
for (k=0; k<nsing; k++) {
j = nsing - k - 1;
sum = zero;
jp1 = j + 1;
if (nsing > jp1) {
ij = jp1 + ldr * j;
for (i=jp1; i<nsing; i++) {
sum += r[ij]*wa[i];
ij += 1; // [i+ldr*j]
}
}
wa[j] = (wa[j] - sum)/sdiag[j];
}
L150:
// permute the components of z back to components of x
for (j=0; j<n; j++) {
l = ipvt[j];
x[l] = wa[j];
}
}

// function enorm
double enorm(n,x)
int n;  double x[];
{
/*    Given an n-vector x, this function calculates the
*     Euclidean norm of x.
*     The Euclidean norm is computed by accumulating the
*     sum of squares in three different sums. the sums of
*     squares for the small and large components are scaled
*     so that no overflows occur. non-destructive underflows
*     are permitted. underflows and overflows do not occur
*     in the computation of the unscaled sum of squares for
*     the intermediate components. the definitions of small,
*     intermediate and large components depend on two
*     constants, rdwarf and rgiant. the main restrictions
```

```
*     on these constants are that rdwarf**2 not underflow and
*     rgiant**2 not overflow. the constants given here are
*     suitable for every known computer. the function
*     statement is
* double precision function enorm(n,x),      where
* n is a positive integer input variable.
* x is an input array of length n. */
int i;
double agiant,floatn,s1,s2,s3,xabs,x1max,x3max,ans,temp;
double rdwarf=3.834e-20, rgiant=1.304e19,zero=0.,one=1.;
double fabs(), sqrt();
s1 = zero; s2 = zero; s3 = zero;
x1max = zero; x3max = zero;
floatn = n; agiant = rgiant/floatn;
for(i=0; i<n; i++)
{
xabs = fabs(x[i]);
if ((xabs > rdwarf) && (xabs < agiant)) {
//   sum for intermediate components
s2 += xabs*xabs;
continue;
}
if (xabs > rdwarf)
{
//   sum for large components
if (xabs > x1max)   {
temp = x1max/xabs;
s1 = one + s1*temp*temp;
x1max = xabs;
}
else {
temp = xabs/x1max;
s1 += temp*temp;
}
continue;
}
//   sum for small components
if (xabs > x3max) {
temp = x3max/xabs;
s3 = one + s3*temp*temp;
x3max = xabs;
}
else {
if (xabs != zero) {
temp = xabs/x3max;
```

```
s3 += temp*temp;
}
}
}
// calculation of norm
if (s1 != zero) {
temp = s1 + (s2/x1max)/x1max;
ans = x1max*sqrt(temp);
return(ans);
}
if (s2 != zero) {
if (s2 >= x3max)  temp = s2*(one+(x3max/s2)*
(x3max*s3));
else  temp = x3max*((s2/x3max)+(x3max*s3));
ans = sqrt(temp);
}
else  ans = x3max*sqrt(s3);
return(ans);
}

#define BUG 0
// function fdjac2
fdjac2(m,n,x,fvec,fjac,ldfjac,iflag,epsfcn,wa)
int m,n,ldfjac,*iflag;
double epsfcn,x[],fvec[],fjac[],wa[];
{
/*  Computes a forward-difference approximation to the
*      m by n jacobian matrix associated with a specified
*      problem of m functions in n variables.
*      the function statement is
* function fdjac2(fcn,m,n,x,fvec,fjac,ldfjac,iflag,epsfcn,wa)
*      where
* fcn is the name of the user-supplied function which
*   calculates the functions.
*   the value of iflag should not be changed by fcn unless
*   the user wants to terminate execution of fdjac2.
*   in this case set iflag to a negative integer.
* m is a positive integer input variable set to the number
*   of functions.
* n is a positive integer input variable set to the number
*   of variables. n must not exceed m.
* x is an input array of length n.
* fvec is an input array of length m which must contain the
*   functions evaluated at x.
* fjac is an output m by n array which contains the
```

```
*   approximation to the jacobian matrix evaluated at x.
* ldfjac is a positive integer input variable not less than m
*   which specifies the leading dimension of the array fjac.
* iflag is an integer variable which can be used to terminate
*   the execution of fdjac2. see description of fcn.
* epsfcn is an input variable used in determining a suitable
*   step length for the forward-difference approximation.
*   this approximation assumes that the relative errors in
*   the functions are of the order of epsfcn. if epsfcn is
*   less than the machine precision, it is assumed that the
*   relative errors in the functions are of the order of the
*   machine precision.
* wa is a work array of length m. */
int i,j,ij;
double eps,h,temp;
double fabs(), dmax1(), sqrt(), zero = 0.0;
extern double MACHEP;
temp = dmax1(epsfcn,MACHEP);
eps = sqrt(temp);
#if BUG
printf("fdjac2\n");
#endif
ij = 0;
for (j=0; j<n; j++) {
temp = x[j];
h = eps * fabs(temp);
if (h == zero)  h = eps;
x[j] = temp + h;
fcn(m,n,x,wa,iflag);
if (*iflag < 0)  return;
x[j] = temp;
for (i=0; i<m; i++) {
fjac[ij] = (wa[i] - fvec[i])/h;
ij += 1;    // fjac[i+m*j]
}
}
#if BUG
pmat(m, n, fjac);
#endif
}

// other functions
double dmax1(a,b)
double a,b;
{
```

```
if (a >= b) return(a);
else    return(b);
}

double dmin1(a,b)
double a,b;
{
if (a <= b) return(a);
else    return(b);
}

int min0(a,b)
int a,b;
{
if (a <= b) return(a);
else    return(b);
}

int mod(k, m)
int k, m;
{ return(k % m); }

pmat(m, n, y)
int m,n;
double y[];
{
int i,j,k;
k = 0;
for (i=0; i<m; i++) {
for (j=0; j<n; j++) {
printf("%.5e ", y[k]);
k += 1;
}
printf("\n");
}
}
```

12.6 Free Basic Code: 2D Crowd Dynamics with 3000 Agents

Here we give a fast FB code using Chipmunk 2D Physics Engine and simulating 2D crowd dynamics ('an obstacle course through force-fields') with 3000 particle-type agents (see Figure 12.17).

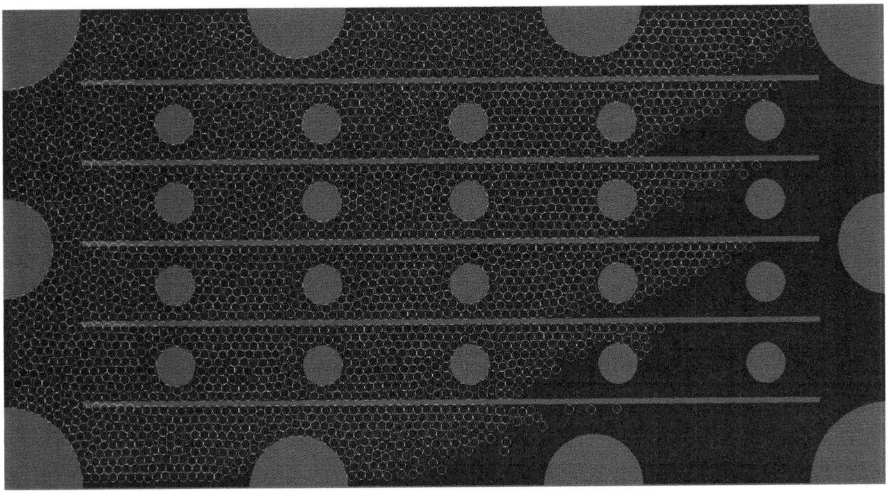

Fig. 12.17. Molecular 2D crowd dynamics with 3000 agents moving through 'an obstacle course through force-fields' using Chipmunk 2D Physics Engine.

```
''
'' Molecular Crowd Dynamics using Chipmunk 2D Physics Engine
''

#include "chipmunk/chipmunk.bi"
#inclib "chipmunk"
#include "fbgfx.bi"
using fb

#define GRABABLE_MASK_BIT (1 shl 31)
#define NOT_GRABABLE_MASK (not GRABABLE_MASK_BIT)

' define balls-agents' parameters
const nball as integer = 3000
const radius as double = 9.0
const mass as double = 50.0

' define screen's parameters
const as integer xx = 1800
const as integer yy = 950
const as integer aa = 1400
const as integer bb = 30

' Define type to make the program more readible
type PhysEngine
```

```
' This allocates/initializes all the main/global chipmunk
variables
declare constructor( byval Hz as integer )
declare destructor( ) ' De-allocate/initialize all the
chipmunk stuff

declare sub simulate( ) ' Starts the simulation ticks
declare sub stop( ) ' Stop simulation ticks
declare function iskilled( ) as ubyte  ' checks if the
simulation stopped

declare sub getinput() ' Input handling
declare sub update() ' Update the screen

mousePoint  as cpVect ' Mouse position
mouseBody   as cpBody ptr ' A static body attached to the
mouse position

' This connects a non-static body to the mouse temporarily
mouseJoint  as cpConstraint ptr

space as cpSpace ptr ' Simulation space
staticbody as cpBody ptr ' Main static body (for attaching
any shape)

kill as unsigned integer ' Flag to stop the physics engine
mutex as any ptr ' Mutex for thread safety
physthread  as any ptr ' Thread to make physics simulation
consistent

deltaTime as cpFloat ' dt used in cpSpaceStep
currentTime as double ' Used to help smooth out the
simulation ' Helps to know how many more cpSpaceStep calls
are needed accumulator as double
end type

' Physics engine tick thread
sub physicstick( byval e as any ptr )
var physics = cptr( PhysEngine ptr, e )
do
mutexlock( physics->mutex )
var newTime = timer()
var ftime = newTime - physics->currentTime
physics->accumulator += ftime
while physics->accumulator >= physics->deltaTime
```

```
cpSpaceStep( physics->space, physics->deltaTime )
physics->accumulator -= physics->deltaTime
wend
physics->currentTime = newTime
mutexunlock( physics->mutex )

' Sleep a consistent amount of time to make a smooth
simulation sleep 1, 1                     ' without affecting
the frame-rate loop until physics->iskilled()
end sub


' Main function
function main( byval argc as integer,
    byval argv as zstring ptr ptr ) as integer
dim i as integer
    const ff = 160
    const fg = 100

    dim cc as Integer = xx-10
    dim c3c as Integer = xx-50
    dim dd as Integer = yy-10
    dim d3d as Integer = yy-50
    dim ee as Integer = xx-300

    dim gg as Integer = yy-ff
    dim hh as Integer = yy\3
    dim kk as Integer = 2*yy\3
    dim ll as Integer = yy\2
    dim mm as Integer = xx\2
    dim nn as Integer = xx\3

' Create a new instance of the physics engine (60 hertz)
dim instance as PhysEngine ptr = new PhysEngine( 60 )
' Define ground
dim as cpShape ptr ground(0 to 3)
ground(0) = cpSegmentShapeNew( NULL, cpv( 10, 10 ),
cpv( cc, 10 ), 1 )
ground(1) = cpSegmentShapeNew( NULL, cpv( cc, 10 ),
cpv( cc, dd ), 1 )
ground(2) = cpSegmentShapeNew( NULL, cpv( cc, dd ),
cpv( 10, dd ), 1 )
ground(3) = cpSegmentShapeNew( NULL, cpv( 10, dd ),
cpv( 10, 10 ), 1 )
for i = 0 to 3
```

```
    ground(i)->e = 1.0
    ground(i)->u = 10.0
    ground(i)->layers = NOT_GRABABLE_MASK
    cpSpaceAddStaticShape( instance->space, ground( i ) )
next i

dim as cpShape ptr Hline1          ' Defining Horizontal
Obstacle 1
dim verts1(0 to 3) as cpVect =
        { cpv(0,10), cpv(ee,10), cpv(ee,0), cpv(ff,0) }
Hline1 = cpPolyShapeNew(instance->staticBody,
        4, @verts1(0), cpv( ff, ff ))
Hline1->e = 1.0
Hline1->u = 5.0
Hline1->layers = NOT_GRABABLE_MASK
cpSpaceAddShape( instance->space, Hline1 )

dim as cpShape ptr Hline2          ' Defining Horizontal
Obstacle 2
dim verts2(0 to 3) as cpVect =
        { cpv(0,10), cpv(ee,10), cpv(ee,0), cpv(ff,0) }
Hline2 = cpPolyShapeNew(instance->staticBody,
        4, @verts2(0), cpv( ff, 2*ff ))
Hline2->e = 1.0
Hline2->u = 5.0
Hline2->layers = NOT_GRABABLE_MASK
cpSpaceAddShape( instance->space, Hline2 )

dim as cpShape ptr Hline3          ' Defining Horizontal
Obstacle 3
dim verts3(0 to 3) as cpVect =
        { cpv(0,10), cpv(ee,10), cpv(ee,0), cpv(ff,0) }
Hline3 = cpPolyShapeNew(instance->staticBody,
        4, @verts3(0), cpv( ff, 3*ff ))
Hline3->e = 1.0
Hline3->u = 5.0
Hline3->layers = NOT_GRABABLE_MASK
cpSpaceAddShape( instance->space, Hline3 )

dim as cpShape ptr Hline4          ' Defining Horizontal
Obstacle 4
dim verts4(0 to 4) as cpVect =
        { cpv(0,10), cpv(ee,10), cpv(ee,0), cpv(ff,0) }
Hline4 = cpPolyShapeNew(instance->staticBody,
        4, @verts4(0), cpv( ff, 4*ff ))
```

```
Hline4->e = 1.0
Hline4->u = 5.0
Hline4->layers = NOT_GRABABLE_MASK
cpSpaceAddShape( instance->space, Hline4 )

dim as cpShape ptr Hline5         ' Defining Horizontal
Obstacle 5
dim verts5(0 to 5) as cpVect =
        { cpv(0,10), cpv(ee,10), cpv(ee,0), cpv(ff,0) }
Hline5 = cpPolyShapeNew(instance->staticBody,
        4, @verts5(0), cpv( ff, 5*ff ))
Hline5->e = 1.0
Hline5->u = 5.0
Hline5->layers = NOT_GRABABLE_MASK
cpSpaceAddShape( instance->space, Hline5 )

    ' Defining Corner Circular Obstacles
    dim as cpShape ptr shape1 =
        cpCircleShapeNew( NULL, ff, cpv(0,0) )
    cpSpaceAddStaticShape( instance->space, shape1 )

    dim as cpShape ptr shape2 =
        cpCircleShapeNew( NULL, ff, cpv(0,yy) )
    cpSpaceAddStaticShape( instance->space, shape2 )

    dim as cpShape ptr shape3 =
        cpCircleShapeNew( NULL, ff, cpv(xx,0) )
    cpSpaceAddStaticShape( instance->space, shape3 )

    dim as cpShape ptr shape4 =
        cpCircleShapeNew( NULL, ff, cpv(xx,yy) )
    cpSpaceAddStaticShape( instance->space, shape4 )

    ' Defining Side Circular Obstacles
    dim as cpShape ptr shape5 =
        cpCircleShapeNew( NULL, fg, cpv(0,ll) )
    cpSpaceAddStaticShape( instance->space, shape5 )

    dim as cpShape ptr shape6 =
        cpCircleShapeNew( NULL, fg, cpv(xx,ll) )
    cpSpaceAddStaticShape( instance->space, shape6 )

    ' Defining Top Circular Obstacles
    dim as cpShape ptr shape7 =
        cpCircleShapeNew( NULL, fg, cpv(nn,0) )
```

```
cpSpaceAddStaticShape( instance->space, shape7 )
dim as cpShape ptr shape71 =
    cpCircleShapeNew( NULL, fg, cpv(2*nn,0) )
cpSpaceAddStaticShape( instance->space, shape71 )

' Defining Bottom Circular Obstacles
dim as cpShape ptr shape8 =
    cpCircleShapeNew( NULL, fg, cpv(nn,yy) )
cpSpaceAddStaticShape( instance->space, shape8 )
dim as cpShape ptr shape81 =
    cpCircleShapeNew( NULL, fg, cpv(2*nn,yy) )
cpSpaceAddStaticShape( instance->space, shape81 )

' Defining middle small Circular Obstacles 9
dim as cpShape ptr shape90 =
    cpCircleShapeNew( NULL, 40, cpv(350,400))
cpSpaceAddStaticShape( instance->space, shape90 )
dim as cpShape ptr shape91 =
    cpCircleShapeNew( NULL, 40, cpv(350+300,400))
cpSpaceAddStaticShape( instance->space, shape91 )
dim as cpShape ptr shape92 =
    cpCircleShapeNew( NULL, 40, cpv(350+600,400))
cpSpaceAddStaticShape( instance->space, shape92 )
dim as cpShape ptr shape93 =
    cpCircleShapeNew( NULL, 40, cpv(350+900,400))
cpSpaceAddStaticShape( instance->space, shape93 )
dim as cpShape ptr shape94 =
    cpCircleShapeNew( NULL, 40, cpv(350+1200,400))
cpSpaceAddStaticShape( instance->space, shape94 )

' Defining middle small Circular Obstacles 10
dim as cpShape ptr shape100 =
    cpCircleShapeNew( NULL, 40, cpv(350,565))
cpSpaceAddStaticShape( instance->space, shape100 )
dim as cpShape ptr shape101 =
    cpCircleShapeNew( NULL, 40, cpv(350+300,565))
cpSpaceAddStaticShape( instance->space, shape101 )
dim as cpShape ptr shape102 =
    cpCircleShapeNew( NULL, 40, cpv(350+600,565))
cpSpaceAddStaticShape( instance->space, shape102 )
dim as cpShape ptr shape103 =
    cpCircleShapeNew( NULL, 40, cpv(350+900,565))
cpSpaceAddStaticShape( instance->space, shape103 )
dim as cpShape ptr shape104 =
    cpCircleShapeNew( NULL, 40, cpv(350+1200,565))
```

```
    cpSpaceAddStaticShape( instance->space, shape104 )

    ' Defining middle small Circular Obstacles 11
    dim as cpShape ptr shape110 =
        cpCircleShapeNew( NULL, 40, cpv(350,240))
    cpSpaceAddStaticShape( instance->space, shape110 )
    dim as cpShape ptr shape111 =
        cpCircleShapeNew( NULL, 40, cpv(350+300,240))
    cpSpaceAddStaticShape( instance->space, shape111 )
    dim as cpShape ptr shape112 =
        cpCircleShapeNew( NULL, 40, cpv(350+600,240))
    cpSpaceAddStaticShape( instance->space, shape112 )
    dim as cpShape ptr shape113 =
        cpCircleShapeNew( NULL, 40, cpv(350+900,240))
    cpSpaceAddStaticShape( instance->space, shape113 )
    dim as cpShape ptr shape114 =
        cpCircleShapeNew( NULL, 40, cpv(350+1200,240))
    cpSpaceAddStaticShape( instance->space, shape114 )

    ' Defining middle small Circular Obstacles 12
    dim as cpShape ptr shape120 =
        cpCircleShapeNew( NULL, 40, cpv(350,720))
    cpSpaceAddStaticShape( instance->space, shape120 )
    dim as cpShape ptr shape121 =
        cpCircleShapeNew( NULL, 40, cpv(350+300,720))
    cpSpaceAddStaticShape( instance->space, shape121 )
    dim as cpShape ptr shape122 =
        cpCircleShapeNew( NULL, 40, cpv(350+600,720))
    cpSpaceAddStaticShape( instance->space, shape122 )
    dim as cpShape ptr shape123 =
        cpCircleShapeNew( NULL, 40, cpv(350+900,720))
    cpSpaceAddStaticShape( instance->space, shape123 )
    dim as cpShape ptr shape124 =
        cpCircleShapeNew( NULL, 40, cpv(350+1200,720))
    cpSpaceAddStaticShape( instance->space, shape124 )

    ' Defining balls-agents
    dim ballb(0 to nball - 1) as cpBody ptr,
        balls(0 to nball - 1) as cpShape ptr
for i = 0 to nball - 1
    ballb(i) = cpBodyNew(mass, cpMomentForCircle(mass,
            0.0, radius, cpvzero))
    ballb(i)->p = cpv( int(rnd * c3c) + 30, int(rnd * d3d) +
    30)
    cpSpaceAddBody( instance->space, ballb(i) )
```

```
        balls(i) = cpCircleShapeNew( ballb(i), radius, cpvzero )
        balls(i)->e = 1.0
        balls(i)->u = 1.0
        cpSpaceAddShape( instance->space, balls(i) )
    next i

    instance->simulate( )           ' Begin the simulation

    do
    instance->getinput()
    mutexlock( instance->mutex )
    screenlock()
    cls

                ' Draw horizontal lines
            for i = 0 to 4
                line ( ff, ff+i*ff )-step( ee, 10 ), rgb( 0, 0,
                255 ), bf
            next i

                ' Draw corner circles
                circle (0, 0), ff, rgb( 0, 0, 255 )
                ,,,,f
                circle (xx, 0), ff, rgb( 0, 0, 255 )
                ,,,,f
                circle (0, yy), ff, rgb( 0, 0, 255 )
                ,,,,f
                circle (xx, yy), ff, rgb( 0, 0, 255 )
                ,,,,f

                ' Draw side circles
                circle (0, ll), fg, rgb( 0, 0, 255 ),,,,f
                circle (xx, ll), fg, rgb( 0,0, 255 ),,,,f

                ' Draw top and bottom circles
            for i = 1 to 2
                circle (nn*i, 0), fg, rgb( 0, 0, 255 ),,,,f
            next i
            for i = 1 to 2
                circle (nn*i, yy), fg, rgb( 0, 0, 255 ),,,,f
            next i

                ' Draw middle small circles
            for i = 0 to 4
```

```
            circle (350+i*300, 400), 40, rgb( 0, 0, 255 )
            ,,,,f
        next i
        for i = 0 to 4
            circle (350+i*300, 565), 40, rgb( 0, 0, 255 )
            ,,,,f
        next i
        for i = 0 to 4
            circle (350+i*300, 240), 40, rgb( 0, 0, 255 )
            ,,,,f
        next i
        for i = 0 to 4
            circle (350+i*300, 720), 40, rgb( 0, 0, 255 )
            ,,,,f
        next i

        ' Draw balls-agents
        for i = 0 to nball - 1
circle (ballb(i)->p.x, ballb(i)->p.y), radius,
rgb( 255, 255, 0 )
next i

screenunlock()
mutexunlock( instance->mutex )
instance->update()
sleep 1, 1
loop until instance->iskilled()

delete instance              ' Cleanup
return 0                     ' Nothing went wrong :)
end function                  ' End main function

end main( __FB_ARGC__, __FB_ARGV__ )    ' Main Entry Point

'' ------------------------------ PhysEngine - Wrapper class

' Constructor PhysEngine
constructor PhysEngine( byval Hz as integer )
screenres  xx, yy, 32
screenset 0, 1
Window ( 0, yy ) - ( xx, 0 )        ' Chipmunk flips the y
axis

    '2D Force Field defined as a function of "timer"
    dim ForceField as cpVect = cpv(100*(sin(cos(timer))),
```

```
        100*(cos(exp(-timer))))

cpInitChipmunk()                ' Initialize the Chipmunk
Engine

this.kill = 0
this.mutex = mutexcreate()

this.staticBody = cpBodyNew(INFINITY, INFINITY)
this.mouseBody = cpBodyNew(INFINITY, INFINITY)

this.space = cpSpaceNew()
this.space->iterations = 10
cpSpaceResizeStaticHash( this.space, bb, aa )
    cpSpaceResizeActiveHash( this.space, bb, aa )
    this.space->gravity = ForceField      'Force Field

WindowTitle "Molecular Crowd Dynamics: An Obstacle Course
    through 2D Force Fields, with "& nball &" Agents,
    using Chipmunk Phys.Eng.v" & cpVersion()

this.deltaTime = 1.0 / csng( Hz )       ' 1/60 Hz
this.currentTime = timer
this.accumulator = 0.0
end constructor

' Destructor PhysEngine
destructor PhysEngine( )
if this.iskilled() = 0 then
    this.stop()
    mutexdestroy this.mutex
end if
cpBodyFree( this.staticBody )
cpBodyFree( this.mouseBody )
cpSpaceFree( this.space )
end destructor

' Other subs and funs
sub PhysEngine.simulate( )
    this.physthread = threadcreate( @physicstick, @this )
end sub

sub PhysEngine.stop()
    mutexlock( this.mutex )
    this.kill = 1
```

```
    mutexunlock( this.mutex )
    threadwait this.physthread
    cpSpaceFreeChildren( this.space )
end sub

function PhysEngine.iskilled() as ubyte
    mutexlock( this.mutex )
    var status = this.kill
    mutexunlock( this.mutex )
    return status
end function

sub PhysEngine.getinput()
dim e as Event
dim killit as integer = 0
    const yy = 950
if ScreenEvent( @e ) then
mutexlock( this.mutex )
select case e.type
case EVENT_MOUSE_EXIT
' Remove any joint between the mouse and a shape if the mouse
if this.mouseJoint <> 0 then                ' leaves the window
    cpSpaceRemoveConstraint( this.space, this.mouseJoint )
    cpConstraintFree( this.mouseJoint )
    this.mouseJoint = NULL
end if
case EVENT_MOUSE_MOVE
' Update the mouse location
var newPoint = cpv( e.x, yy - e.y )
'this.mouseBody->v = cpvmult(cpvsub(newPoint, this.
mousePoint), 0.01)
this.mousePoint = newPoint
this.mouseBody->p = this.mousePoint
case EVENT_MOUSE_BUTTON_PRESS
if e.button = BUTTON_LEFT then
    var shape = cpSpacePointQueryFirst(this.space,
    mousePoint, _GRABABLE_MASK_BIT, 0 )
    if shape <> 0 then
        ' If you click on a shape, then attach it to the
        mouse var body = shape->body
        this.mouseJoint = cpPivotJointNew2( this.mouseBody,
        body,_ cpvzero, cpBodyWorld2Local( body, this.
        mousePoint ) )
        this.mouseJoint->maxForce = INFINITY
        this.mouseJoint->biasCoef = 1.0
```

```
            cpSpaceAddConstraint( this.space, this.mouseJoint )
        end if
    end if
    case EVENT_MOUSE_BUTTON_RELEASE
        if e.button = BUTTON_LEFT then
            if this.mouseJoint <> 0 then
            ' Release any attached shape from the mouse
                cpSpaceRemoveConstraint( this.space, this.
                mouseJoint )
                cpConstraintFree( this.mouseJoint )
                this.mouseJoint = NULL
            end if
        end if
    case EVENT_KEY_RELEASE
    ' Exit if the user presses Escape
    if e.scancode = SC_ESCAPE then killit = 1
        end select
        mutexunlock( this.mutex )
    end if
    if killit = 1 then this.stop()
    end sub

    sub PhysEngine.update()
        screencopy
        flip
    end sub
```

References

AS98. Abbott, L., Sejnowski, T.J.: Neural Codes and Distributed Representations, MIT Press, Cambridge, (1998)

AM78. Abraham, R., Marsden, J.E.: Foundations of Mechanics (2nd ed.) Addison-Wesley, Reading, MA, (1978)

AMR88. Abraham, R., Marsden, J., Ratiu, T.: Manifolds, Tensor Analysis and Applications. Springer, New York, (1988)

AC04. Abramsky, S., Coecke, B.: A categorical semantics of quantum protocols. In LICS Proceedings, pp. 415-425. IEEE Computer Society, (2004)

AHH58. Abbott, B.C., Hill, A.V., Howarth, J.V.: The positive and negative heat production associated with a nerve impulse, Proc. R. Soc. London B **148**(931), 149-87, (1958)

AKN73. Ablowitz, M.J., Kaup, D.J., Newell, A.C., Segur, H.: Method for solving the sine-Gordon equation, Phys. Rev. Let. **30**, 1262-1264, (1973)

AC91. Ablowitz, M.J., Clarkson, P.A.: Solitons, nonlinear evolution equations and inverse scattering, Cambridge Univ. Press, (1991)

Adr26. Adrian, E.D.: The mechanism of nervous action, J. Physiol. (London) **61**, 49-72, (1926)

Agr01. Agrawal, G.P.: Nonlinear Fiber Optics, Academic Press, San Diego, (2001)

AE75. Allen, L., Eberly, J.H.: Optical Resonance and Two-Level Atoms, Wiley Press, New York, (1975)

Ama77. Amari, S.: Dynamics of pattern formation in lateral-inhibition type neural fields. Biological Cybernetics, **27**, 77-87, (1977)

Agr01. Agrawal, G.: Nonlinear fiber optics (3rd ed.) Academic Press, San Diego, (2001)

Ale928. Alexander, J.W.: Topological invariants of knots and links, Trans. Amer. Math. Soc. **30**, 275-306, (1928)

AGI98. Amaral, L.A.N., Goldberger, A.L., Ivanov, P.Ch. & Stanley H.E.: Scale-independent measures and pathologic cardiac dynamics. Phys. Rev. Lett. **81**, 2388-2391, (1998)

And04. Anderson, M.T.: Geometrization of 3-manifolds via the Ricci flow, Not. Am. Math. Soc. **512**, 184-193, (2004)

And84. Anderson, T.W.: An Introduction to Multivariate Statistical Analysis. Wiley, New York, (1984)

AG05. Apps, R., Garwicz, M.: Anatomical and physiological foundations of cerebellar information processing. Nature Rev. Neurosci. **6**, 297-311, (2005)

Arb98. Arbib, M. (ed.): Handbook of Brain Theory and Neural Networks (2nd ed.) MIT Press, Cambridge, (1998)

Art47. Artin, E.: Theory of Braids. Ann. Math. **48**, 101-126, (1947)

Arn78. Arnold, V.I.: Ordinary Differential Equations. MIT Press, Cambridge, MA, (1978)

Arn88. Arnold, V.I.: Geometrical Methods in the Theory of Ordinary differential equations. Springer, New York, (1988)

Arn89. Arnold, V.I.: Mathematical Methods of Classical Mechanics (2nd ed). Springer, New York, (1989)

Arn92. Arnold, V.I.: Catastrophe Theory. Springer, Berlin, (1992)

Arn93. Arnold, V.I.: Dynamical systems. Encyclopaedia of Mathematical Sciences, Springer, Berlin, (1993)

AK98. Arnold, V.I., Khezin, B.: Topological Methods in Hydrodynamics, Springer, (1998)

Abr98. Abreu, M.: Kähler geometry of toric varieties and extremal metrics, Int. J. Math. **9**, 641-651, (1998)

AB82. Atiyah, M.F., Bott, R.: The Yang-Mills equations over Riemann surfaces, Phil. Trans. Roy. Soc. London, Ser. A **308**, 523-615, (1982)

AS72. Abramowitz, M., Stegun, I.A.: Handbook of Mathematical Functions. Dover, New York, (1972)

AI95. de Azcarraga, J.A., Izquierdo, J.M.: Lie groups, Lie algebras, cohomology, and some applications in physics, Cambridge Univ. Press, (1995)

ALW08. Ambjorn, J., Loll, R., Watabiki, Y., Westra, W., Zohren, S.: A matrix model for 2D quantum gravity defined by Causal Dynamical Triangulations, Phys.Lett. B **665**, 252-256, (2008)

AKK98. Amemiya, T., Kettunen, P., Kadar, S.: Waves in Photosensitive Excitable Media, Chaos **8**, 872-878, (1998)

AM74. Arbib, M., Manes, E.: Machines in a Category: an Expository Introduction, SIAM Rev. **16**(2), 163-192, (1974)

AM80. Arbib, M., Manes, E.: Machines in a Category, J. Pur. Appl. Alg. **19**, 9-20, (1980)

AGH88. Arneodo, A., Grasseau, G., Holschneider, M.: Wavelet Transform of Multifractals. Phys. Rev. Lett. **61**, 2281-2284, (1988)

BLW94. Bassingthwaighte, J.B., Liebovitch, L.S., West, B.J.: Fractal Physiology. Oxford Univ. Press, New York, (1994)

Bla76. Black, F.: Studies of Stock Price Volatility Changes Proc. 1976 Meet. Ame. Stat. Assoc. Bus. Econ. Stat. 177–181, (1976)

Bus94. Bushev, M.: Synergetics: Chaos, Order, Self-organization, World Scientific, Singapore, (1994)

BBW94. Bohr, T., Bosch, E, van de Water, W.: Spatiotemporal Chaos, Nature **372**, 48, (1994)

Buc88. Buck, J.: Synchronous rhythmic flashing of fireflies, Q. Rev. Biol. **63**, 265, (1988)

Bur08. Burdujan, I.: Clifford Kähler Manifolds, Balkan J. Geom. Appl. **13**(2), 12-23, (2008)

BDL87. Blaquiere, A., Diner, S., Lochak, G. (Eds.): Information Complexity and Control in Quantum Physics, Springer, New York, (1987)

Bac96. Back, T.: Evolutionary Algorithms in Theory and Practice: Evolution Strategies, Evolutionary Programming, Genetic Algorithms, Oxford Univ. Press, (1996)

Ber01. Berezowski, M..: Effect of delay time on the generation of chaos in continuous systems. One-dimensional model. Two-dimensional modeltubular chemical reactor with recycle, Cha. Sol. Frac. **12**, 83-89, (2001)

BK98. Braun, O.M., Kivshar, Yu.S.: Nonlinear dynamics of the Frenkel-Kontorova model, Phys. Rep. **306**, 1-108, (1998)

BK04. Braun, O.M., Kivshar, Yu.S.: The Frenkel-Kontorova model, Springer, New York, (2004)

BM94. Baez, J.C., Muniain, J.P.: Gauge Fields, Knots and Gravity, Series on Knots and Everything, Vol. 4, World Scientific, Singapore, (1994)

Bel64. Bell, J.S: On the Einstein Podolsky Rosen Paradox. Physics **1**, 195, (1964)

Bel66. Bell, J.S: On the problem of hidden variables in quantum mechanics. Rev. Mod. Phys. **38**, 447, (1966)

Bel87. Bell, J.S: Speakable and Unspeakable in Quantum Mechanics. Cambridge Univ. Press, Cambridge, (1987)

BC80. Bullough, R.K., Caudrey, P.J. (Eds.): Solitons, Springer, Berlin, (1980)

BG79. Barrow-Green, J.: Poincaré and the Three Body Problem. American Mathematical Society, Providence, RI, (1997)

BG96. Baker, G.L., Gollub, J.P.: Chaotic Dynamics: An Introduction (2nd ed.) Cambridge Univ. Press, Cambridge, (1996)

Bir912. Birkhoff, G.D.: A determinant formula for the number of ways of coloring a map, Ann. of Math. **2**(14), 42-46, (1912)

Bir75. Birman, J.: Braids, Links, and Mapping Class Groups. Ann. Math. Studies 82, Princeton Univ. Press, (1975)

BS73. Black, F., Scholes, M.: The Pricing of Options and Corporate Liabilities, J. Pol. Econ. **81**, 637-659, (1973)

BMZ05. Bloch, A.M., Marsden, J.E., Zenkov, D.V.: Nonholonomic Dynamics, Notices of the AMS, **52**(3), 320-329, (2005)

BGL00. Boccaletti, S., Grebogi, C., Lai, Y.-C., Mancini, H., Maza, D.: The Control of Chaos: Theory and Applications. Physics Reports **329**, 103-197, (2000)

BC020. Boccaletti, S., Kurths, J., Osipov, G., Valladares, D.L., Zhou, C.S., The synchronization of chaotic systems, Phys. Rep. **366**(1-2), 1-101, (2002)

Bil65. Billingsley, P.: Ergodic theory and information, Wiley, New York, (1965)

Bla86. Blackman, S.S.: Multiple-Target Tracking with Radar Applications, Artech House, Norwood, MA, (1986)

BDS12. Barri, A., Dooms, A., Schelkens, P.: The near shift-invariance of the dual-tree complex wavelet transform revisited, J. Math. Anal. Appl. **389**(2), 1303-1314, (2012)

BLV01. Boffetta, G., Lacorata, G., Vulpiani, A.: Introduction to chaos and diffusion. Chaos in geophysical flows, ISSAOS, (2001)

Boa88. Boashash, B.: Note on the Use of the Wigner Distribution for Time Frequency Signal Analysis, IEEE Trans. Acoustics, Speech, Signal Proc. **36**(9), 1518-1521, (1988)

BEW03. Bracken, A.J., Ellinas, D., Wood, J.G.: Group theory and quasiprobability integrals of Wigner functions. J. Phys. A **36**(20), L297-L305, (2003)

BPT97. Broeck, C. van, Parrondo, J.M.R., Toral, R., Kawai, R., Nonequilibrium phase transitions induced by multiplicative noise, Phys. Rev. E **55**(4), 4084-4094, (1997)

Bae97. Baez, J.: An introduction to $n-$categories. *7th* Conference on Category Theory and Computer Science, E. Moggi and G. Rosolini (eds), Lecture Notes in Computer Science, Springer, Berlin, (1997)
BD98. Baez, J., Dolan, J.: Higher-Dimensional Algebra III: $n-$categories and the Algebra of Opetopes. Adv. Math. **135**(2), 145-206, (1998)
BN01. Bastian J., Nguyenkim J.: Dendritic Modulation of Burst-Like Firing in Sensory Neurons. J. Neurophysiol. **85**, 10-22, (2001)
BP82. Barone, A., Paterno, G.: Physics and Applications of the Josephson Effect. Wiley, New York, (1982)
BO95. Basar, T., Olsder, G.J.: Dynamic Noncooperative Game Theory (2nd ed.), Academic Press, New York, (1995)
B-KMR05. Balint-Kurti, G., Manby, F., Ren, Q., Artamonov, M., Ho, T.-S., Rabitz, H.: Quantum control of molecular motion including electronic polarization effects with a two-stage toolkit, J. Chem. Phys. **122**, 084110, (2005)
BBS97. Baumert, T., Brixner, T., Seyfried, V., Strehle, M., Gerber, G.: Femtosecond pulse shaping by an evolutionary algorithm with feedback, Appl. Phys. B **65**, 779-782, (1997)
BHD13. Beale, M.H., Hagan, M.T., Demuth, H.B.: Neural Network Toolbox^{TM}, Matlab® R2013a User's Guide, (2013)
BK64. Bellman, R., Kalaba, R.: Selected papers on mathematical trends in control theory. Dover, New York, (1964)
Ben67. Bénabou, J.: Introduction to bicategories. In: Lecture Notes in Mathematics. Springer, New York, (1967)
BDR07. Beltrani, V., Dominy, J., Rabitz, H.A.: Laboratory observation of quantum control level sets, Phys. Rev. A **74**, 043414, (2007)
Bal87. Ballentine, L.E.: Foundations of Quantum Mechanics Since the Bell Inequalities. Amer. J. Phys. **55**, 785, (1987)
BBC93. Bennett, C.H., Brassard, G., Crepeau, C., Jozsa, R., Peres, A., Wootters, W.K.: Teleporting an Unknown Quantum State via Dual Classical and Einstein-Podolsky-Rosen Channels. Phys. Rev. Lett. **70**, 1895-1899, (1993)
BV93. Bernstein, E., Vazirani, U.: Quantum complexity theory. SIAM J. Comput. **26**(5), 1411-1473, (1997)
Ben82. Benioff, P.A.: Quantum mechanical Hamiltonian models of Turing machines. J. Stat. Phys. **29**(3), 515-546, (1982)
BP92. Benvenuto, N., Piazza, F.: On the complex backpropagation algorithm. IEEE Trans. Sig. Proc., **40**(4), 967-969, (1992)
BS90. Butkovskiy, A., Samoilenko, Y.L.: Control of Quantum-Mechanical Processes and Systems, Kluwer Academic, Dordrecht, (1990)
BP97. Badii, R. Politi, A: Complexity: Hierarchical Structures and Scaling in Physics, Cambridge Univ. Press, Cambridge, (1997)
BDK03. Brixner, T., Damrauer, M.H., Kiefer, B., Gerber, G.: Liquid-phase adaptive femtosecond quantum control: Removing intrinsic intensity dependencies, J. Chem. Phys **118**, 3692, (2003)
BCG02. Boscain, U., Charlot, G., Gauthier, J.P., Guerin, S., Jauslin, H.R.: Optimal Control in laser-induced population transfer for two and three level quantum systems, J. Math. Phys. **43**, 2017, (2002)
Ble81. Bleecker, D.: Gauge Theory and Variational Principles, Addison-Wesley, Reading, MA, (1981)
BT82. Bott, R., Tu, L.W.: Differential Forms in Algebraic Topology. Graduate Texts in Mathematics, Springer, New York, (1982)

Bra01. Branke, J.: Evolutionary Optimization in Dynamic Environments. Kluwer, Dordrecht, (2001)

Bro77. Broadbent, D.E.: Levels, hierarchies and the locus of control. Quarterly J. Exper. Psychology, **29**, 181-201, (1977)

BCG03. Berlekamp, E.R., Conway, J.H., Guy, R.K.: Winning Ways for Your Mathematical Plays, Vol. 3 (2nd ed.), A.K. Peters, MA, (2003)

BPM97. Bouwmeester, D., Pan, J.-W., Mattle, K., Eibl, M., Weinfurter, H., Zeilinger, A.: Experimental quantum teleportation, Nature **390**, 575-579, (1997)

Bro72. Brockett, R.W.: System theory on group manifolds and coset spaces, SIAM J. Con. **10**(2), 265-284, (1972)

Bro82. Brockett, R.W.: Control theory and singular Riemannian geometry. In New Directions in Applied Mathematics, Springer-Verlag, (1982)

Bro01. Brockett, R.: New Issues in the Mathematics of Control. In Mathematics Unlimited - 2001 and Beyond, Springer, New York, (2001)

Bro86. Brooks, R.A.: A robust layered control system for a mobile robot. IEEE Trans. Rob. Aut. **2**(1), 14-23, (1986)

BN09. Buluta, I., Nori, F.: Quantum Simulators, Science, 326(5949), 108-111, (2009)

BS93. Beck, C., Schlogl, F.: Thermodynamics of chaotic systems. Cambridge Univ. Press, Cambridge, (1993)

Bou02. Bourbaki, N.: Elements of Mathematics, Lie Groups and Lie Algebras, Springer, (2002)

BT93. Busemeyer, J.R., Townsend, J.T.: Decision field theory: A dynamic-cognitive approach to decision making in an uncertain environment. Psych. Rev., **100**, 432-459, (1993)

BD02. Busemeyer, J.R., Diederich, A.: Survey of decision field theory. Math. Soc. Sci., **43**, 345-370, (2002)

BY97. Bar-Yam, Y.: Dynamics of Complex Systems. Perseus Books, Reading, Mas, (1997)

Cha98. deCharms, R.C.: Information coding in the cortex by independent or coordinated populations, Proc. Natl. Acad. Sci USA **95**, 15166-15168, (1998)

CZ00. deCharms, R.C., Zador, A.: Neural representation and the cortical code, Ann. Rev. Neurosci. **23**, 613-647, (2000)

Che46. Chevalley, C.: Theory of Lie groups, Princeton Univ. Press, Princeton, (1946)

CD82. Choquet-Bruhat, Y., DeWitt-Morete, C.: Analysis, Manifolds and Physics (2nd ed). North-Holland, Amsterdam, (1982)

CD00. Choquet-Bruhat, Y., DeWitt-Morete, C.: Analysis, Manifolds and Physics, Part II, 92 Applications (rev. ed). North-Holland, Amsterdam, (2000)

Chu98. Chua, L.O.: CNN: A Paradigm for Complexity, World Scientific, Singapore, (1998)

Chu95. Chua, L.O. (Ed): Special issue on 'Nonlinear waves, patterns and spatio-temporal chaos in dynamic arrays,' IEEE. Trans. Circ. Syst. **42**, 557, (1995)

CK08. Chitra, R.N., Kuriakose, V.C.: Phase synchronization in an array of driven JJs, Chaos **18**, 013125, (2008)

Che46. Chern, S.S.: Characteristic classes of Hermitian Manifolds, Ann. Math. **47**(1), 85-121, (1946)

CDH98. Chow, E., Delsing, P., Haviland, D.B.: Length-Scale Dependence of the Superconductor-to-Insulator Quantum Phase Transition in One Dimension, Phys. Rev. Lett. **81**, 204-207, (1998)

CT00b. Chen, X.X., Tian, G.: Ricci flow on Kähler-Einstein surfaces, arXiv:math/0010008 [math.DG], (2000)

CMQ91. Coifman, R., Meyer, Y., Quake, S., Wickerhauser, M.V.: Wavelet Analysis and Signal Processing, Proceedings, conference on Wavelets, Lowell, MA, (1991)

Cox92. Cox, E.: Fuzzy Fundamentals, IEEE Spectrum, 58–61, (1992)

CH93. Cross, M.C., Hohenberg, P.C.: Pattern-Formation Outside of Equilibrium, Rev. Mod. Phys. **65**, 851, (1993)

CFP91. Crisanti, A., Falcioni, M., Paladin, G., Vulpiani, A.: Lagrangian Chaos: Transport, Mixing, Diffusion in Fluids. Riv. Nuovo Cim. **14**, 1, (1991)

CFP94. Crisanti, A., Falcioni, M., Paladin, G., Vulpiani, A.: Stochastic Resonance in Deterministic Chaotic Systems. J. Phys. A **27**, L597, (1994)

CG03. Carpenter, G.A., Grossberg, S.: Adaptive Resonance Theory. In M.A. Arbib (ed.) The Handbook of Brain Theory, Neural Networks, Second Edition, MIT Press, Cambridge, MA, 87-90, (2003)

CT-HR05. Cardoza, D., Trallero-Herrero, C., Langhojer, F., Rabitz, H.A., Weinacht, T.: Transformations to diagonal bases in closed loop quantum learning control problems, J. Chem. Phys. **122**, 124306, (2005)

CK90. Cash, J.R., Karp, A.H.: A variable order Runge-Kutta method for initial value problems with rapidly varying right-hand sides, ACM Trans. Math. Software **16**, 201-222, (1990)

CCC08. Chen, Z., Cao, J., Cao, Y., Zhang, Y., Gu, F., Zhu, G., Hong, Z., Wang, B., Cichocki, A.: An empirical EEG analysis in brain death diagnosis for adults. Cogn. Neurodyn. **2**, 257-271, (2008)

CDK97. Chen, Y., Ding, M., Kelso, S.J.A.: Long memory processes ($1/f\alpha$-type) in human coordination, Phys. Rev. Lett. **79**, 4501-504, (1997)

Cal83. Calabi, E.: Extremal Kähler metrics In: Seminar in Differential geometry (ed. Yau) Princeton Univ. Press (1983)

CG83. Cohen, M.A., Grossberg, S.: Absolute stability of global pattern formation, parallel memory storage by competitive neural networks. IEEE Trans. Syst., Man, Cybern., **13**(5), 815-826, (1983)

CCN85. Conway, J.H., Curtis, R.T., Norton, S.P., Parker, R.A., Wilson, R.A.: Atlas of Finite Groups: Maximal Subgroups and Ordinary Characters for Simple Groups. Clarendon Press, Oxford, (1985)

Chi79. Chirikov, B.V.: A universal instability of many-dimensional oscillator systems. Phys. Rep. **52**, 264-379, (1979)

CL84. Cheng, T.-P., Li, L.-F.: Gauge Theory of Elementary Particle Physics. Clarendon Press, Oxford, (1984)

CT07. Craddock, T.J.A., Tuszynski, J.A.: On the Role of the Microtubules in Cognitive Brain Functions, NeuroQuant. **5**(1), 32-57, (2007)

Col06. Collins, G.P.: Computing with Quantum Knots. Scientific American, April, (2006)

CMN98. Carinena, J.F., Marmo, G., Nasarre, J.: The nonlinear superposition principle and the Wei-Norman method, Int. J. Mod. Phys. A **13**, 3601-3627, (1998)

Cao85. Cao, H.D.: Deformation of Kähler metrics to Kähler-Einstein metrics on compact Kähler manifolds. Invent. Math. **81**, 359-372, (1985)

CC99. Cao, H.D., Chow, B.: Recent developments on the Ricci flow, Bull. Amer. Math. Soc. **36**, 59-74, (1999)

Cho75. Choi, M.-D., Completely Positive Linear Maps on Complex Matrices, Lin. Alg. Appl. **10**, 285, (1975)

CL08. Chen, X.X., Li, H.: The Kähler-Ricci flow on Kähler manifolds with 2 traceless bisectional curvature operator, Chin. Ann. Math. **29B**(5), 543-556, (2008); preprint arXiv:math/0503645 [math.DG], (2005)

CT00. Chen, X.X., Tian, G.: Ricci flow on Kähler manifolds, arXiv:math/0010007 [math.DG], (2000)

CW95. Ciufolini, I., Wheeler, J.A.: Gravitation and Inertia, Princeton Series in Physics, Princeton University Press, Princeton, New Jersey, (1995)

Coe03. Coecke, B.: The logic of entanglement. Research Report PRG-RR-03-12, (2003); http://web.comlab.ox.ac.uk/oucl/publications/tr/rr-03-12.html

Coe05. Coecke, B.: Kindergarten quantum mechanics - lecture notes. In A. Khrennikov, ed. Quantum Theory: Reconsiderations of the Foundations III, pp 81-98. AIP Press, (2005)

Coe06. Coecke, B.: Introducing categories to the practicing physicist. in: What is Category Theory? Advanced Studies in Mathematics and Logic 30, Polimetrica Publishing, (2006)

CP09. Coecke, B., Paquette, E.O.: Categories for the practising physicist. in: New Structures for Physics, B. Coecke (ed.), Springer Lecture Notes in Physics, (2009)

Coe09. Coecke, B.: Quantum picturalism. Contem. Phys. **51**, 59-83, (2009)

CU10. Chaudhury, K.N., Unser, M.: On the Shiftability of Dual-Tree Complex Wavelet Transforms, IEEE Trans. Sig. Proc. **58**, 221–232, (2010)

Cor84. Cornwell, J.F.: Group theory in Physics, Vol. II, Academic Press, London, (1984)

CT65. Cooley, J.W., Tukey, J.W.: An algorithm for the machine calculation of complex Fourier series. Math. Comput. **19**, 297–301, (1965)

CL71. Cooley, W.W., Lohnes, P.R.: Multivariate Data Analysis. Wiley, New York, (1971)

Cus77. Cushing, J.: Integrodifferential Equations and Delay Models in Population Dynamics, Springer, Berlin, (1977)

Con. Conway, J.H.: An enumeration of knots and links, Computational problems in abstract algebra (ed. J.Leech), Pergamon Press, NY, (1969)

CRV92. Celeghini, E., Rasetti, M., Vitiello, G.: Quantum Dissipation. Annals Phys., **215**, 156, (1992)

CS00. Cristianini, N., Shawe-Taylor, J.: Support Vector Machines. Cambridge Univ. Press, Cambridge, (2000)

CGP88. Cvitanovic, P., Gunaratne, G., Procaccia, I.: Topological, metric properties of Hénon-type strange attractors. Phys. Rev. A **38**, 1503-1520, (1988)

CE91. Cvitanovic, P., Eckhardt, B.: Periodic orbit expansions for classical smooth flows. J. Phys. A **24**, L237, (1991)

CR07. Chakrabarti, R., Rabitz, H.: Quantum control landscapes, Int. Rev. Phys. Chem. **26**(4), 671-735, (2007)

CDS12a. Chen, X.X., Donaldson, S., Sun, S.: Kähler-Einstein metrics on Fano manifolds, I: approximation of metrics with cone singularities, arXiv:1211.4566, (2012)

CDS12b. Chen, X.X., Donaldson, S., Sun, S.: Kähler-Einstein metrics on Fano manifolds, II: limits with cone angle less than 2π, arXiv:1212.4714, (2012)

CDS12c. Chen, X.X., Donaldson, S., Sun, S.: Kähler-Einstein metrics on Fano manifolds, III: limits as cone angle approaches 2π and completion of the main proof, arXiv:1302.0282, (2012)

CS08. Childs, L.M., Strogatz, S.H.: Stability diagram for the forced Kuramoto model, Chaos **18**, 043128, (2008)

Cvi00. Cvitanovic, P.: Chaotic field theory: a sketch. Physica A **288**, 61-80, (2000)

Cvi91. Cvitanovic, P.: Periodic orbits as the skeleton of classical and quantum chaos. Physica, D **51**, 138, (1991)

CAM05. Cvitanovic, P., Artuso, R., Mainieri, R., Tanner, G., Vattay, G.: Chaos: Classical and Quantum. ChaosBook.org, Niels Bohr Institute, Copenhagen, (2005)

D'AD01. D'Alessandro, D., Dahleh, M.: Optimal control of two-level quantum systems, IEEE Trans. Autom. Control **46**, 866, (2001)

D'Al00. D'Alessandro, D.: Topological properties of reachable sets and the control of quantum bits, Syst. Control Lett. **41**, 213, (2000)

Dau92. Daubechies, I.: Ten Lectures on Wavelets. SIAM, Philadelphia, PA, (1992)

Dau01. Daubechies, I.: Ten Lectures on Wavelets. R.Kh.D., Moscow, (2001)

DR93. Demiralp, M., Rabitz, H.A.: Optimally controlled quantum molecular dynamics: The effect of nonlinearities on the magnitude and multiplicity of control-field solutions, Phys. Rev. A **47**, 809, (1993)

DS95. Das, P., Schwarz, W.H.: Solitons in cell membranes, Phys. Rev. E **51**, 3588-3612, (1995)

DCH94. Delsing, P., Chen, C.D., Haviland, D.B., Harada, Y., Claeson, T.: Charge solitons and quantum fluctuations in two-dimensional arrays of small JJs, Phys. Rev. B **50**, 3959-3971, (1994)

DB95. Deissler, R.J., Brand, H.R.: Interaction of Breathing Localized Solutions for Subcritical. Bifurcations, Phys. Rev. Lett. **74**, 4847, (1995)

Deu85,Fey85. Deutsch, D.: Quantum theory, the Church-Turing principle and the universal quantum computer. Proc. Roy. Soc. (London), A **400**, 97-117, (1985); also, Feynman R.P., Quantum mechanical Computers, Found. of Phys. **16**(6), 507-31, (1985)

Deu92. Deutsch, D., Jozsa, R.: Rapid solution of problems by quantum computation. Proc. Roy. Soc. (London), A **439**, 553-8, (1992)

Die69. Dieudonne, J.A.: Foundations of Modern Analysis (in four volumes). Academic Press, New York, (1969)

Die88. Dieudonne, J.A.: A History of Algebraic and Differential Topology 1900-1960. Birkhauser, Basel, (1988)

DHS91. Domany, E., van Hemmen, J.L., Schulten, K. (eds.): Models of Neural Networks. Springer, Berlin, (1991)

Don99. Donaldson, S.K.: Symmetric spaces, Kähler geometry and Hamiltonian dynamics. A chapter in Northern California Symplectic Geometry Seminar (Ed. Y. Eliashberg *et al.*), Adv. Math. Sci., AMS Press, (1999)

Dai94. Daido, H.: Generic scaling at the onset of macroscopic mutual entrainment in limit-cycle oscillators with uniform all-to-all coupling, Phys. Rev. Lett. 73, 760, (1994)

Das85. Dash, P.C.: Correct dynamics of sine-Gordon 2π-kink in presence of periodic perturbing force, J. Phys. C: Solid State Phys. **18**, L125, (1985)

Day67. Day, C.: Quipus and Witches' Knots, With a Translation and Analysis of "Oribasius De Laqueis", The Univ. Kansas Press, Lawrence, (1967)

DDG07. Derks, G., Doelman, A., van Gils, S.A., Susanto, H.: Stability analysis of π-kinks in a 0-π JJ, SIAM J. Appl. Dyn. Syst. **6**, 99-141, (2007)

Dun55. Dunnington, G.W.: Carl Friedrich Gauss, Titan of Science, Hafner Publishing Co., New York, (1955)

Dir49. Dirac, P.A.M.: The Principles of Quantum Mechanics. Oxford Univ Press, Oxford, (1949)

Dir58. Dirac, P.A.M.: Generalized Hamiltonian Dynamics. Proc. Roy. Soc. London A, **246**(1246), 326-332, (1958)

DM01. Digalakis, J.G., Margaritis, K.G.: On benchmarking functions for genetic algorithms, Int. J. Comp. Math. **77**, 481, (2001)

DJ99. Doherty, A.C., Jacobs, K.: Feedback-control of quantum systems using continuous state-estimation. Phys. Rev. A **60**, 2700, (1999)

DHJ00. Doherty, A.C., Habib, S., Jacobs, K., Mabuchi, H., Tan, S.M.: Quantum feedback control and classical control theory, Phys. Rev. A **62**, 012105, (2000)

Don87. Donaldson, S.K.: Infinite determinants, stable bundles and curvature, Duke Math. J. **54**, 231-47, (1987)

Don97. Donaldson, S.K.: Remarks on gauge theory, complex geometry and four-manifold topology Fields Medallists' Lectures (Atiyah, Iagolnitzer eds.), World Scientific, (1997)

Don90. Donaldson, S.K., Kronheimer, P.: The geometry of four-manifolds, Oxford Univ. Press, (1990)

Don02. Donaldson, S.K.: Scalar curvature and stability of toric varieties, J. Diff. Geom. **62**, 289-349, (2002)

Don08a. Donaldson, S.K.: Kähler geometry on toric manifolds, and some other manifolds with large symmetry, arXiv:0803.0985v2 [math.DG], (2008)

Don08b. Donaldson, S.K.: Extremal metrics on toric surfaces: A continuity method, J. Diff. Geom. **79**, 389-432, (2008)

DP01. Dosen, K., Petric, Z.: The Maximality of Cartesian Categories. Math. Logic Quart. **47**, 137-144, (2001)

DP02. Dosen, K., Petric, Z.: Bicartesiaan Coherence. Studia Logica **71**, 331-353, (2002)

DP04. Dosen, K., Petric, Z.: Proof-Theoretical Coherence. Studies in Logic 1, King's College Publ. London, (2004)

DH81. Dutta, P. Horn, P.M.: Low-frequency fluctuations in solids: 1/f noise, Rev. Mod. Phys. **53**, 497-516, (1981)

Don95. Donoho, D.L.: De-Noising by soft-thresholding. IEEE Trans. Inf. Theory, **41**(3), 613-627, (1995)

DGT77. Dattoli G., Gallardo J.C., Torre A.: An algebraic view to Operatorial Ordering and Its Applications to Optics, Riv. Nuovo Cim. **11**, 1-79, (1988)

DW07. Dynnikov, I., Wiest, B.: On the Complexity of Braids. J. Eur. Math. Soc. **9**, 801-840, (2007)

Dus84. Dustin, P.: Microtubules. Springer, Berlin, (1984)

Ecc64. Eccles, J.C.: The Physiology of Synapses. Springer, Berlin, (1964)

EIS67. Eccles, J.C., Ito M., Szentagothai J.: The Cerebellum as a Neuronal Machine. Springer, Berlin, (1967)

Ede80. Edelen, D.G.B.: A four-dimensional formulation of defect dynamics and some of its consequences, Int. J. Eng. Sci. **18**, 1095, (1980)

Ego94. Egolf, D.A., Greenside, H.S.: Relation between fractal dimension and spatial correlation length for exten- sive chaos, Nature **369**, 124, (1994)

EMN92. Ellis, J., Mavromatos, N., Nanopoulos, D.V.: String theory modifies quantum mechanics. CERN-TH/6595, (1992)

EMN99. Ellis, J., Mavromatos, N., Nanopoulos, D.V.: A microscopic Liouville arrow of time. Chaos, Solit. Fract., **10**(2-3), 345-363, (1999)

EEI07. Elgin, J., Enolski, V., Its, A.: Effective integration of the nonlinear vector Schrödinger equation. Physica D **225**(22), 127-152, (2007)

ES11. Eliasson, B., Shukla, P.K.: Nonlinear propagation of light in Dirac matter, Phys. Rev. E 84, 036401 [8 pages] (2011)

ES12. Eliasson, B., Shukla, P.K.: The formation of electrostatic shocks in quantum plasmas with relativistically degenerate electrons, EPL 97, 15001, (2012)

EMM99a. Engø, K., Marthinsen, A., Munthe-Kaas, H.Z.: DiffMan - an object oriented MATLAB toolbox for solving differential equations on manifolds, Tech. Rep. **164**, Dep. Informatics, Univ. Bergen, Norway, (1999)

EMM99b. Engø, K., Marthinsen, A., Munthe-Kaas, H.Z.: DiffMan User's Guide, Version 1.6, Tech. Rep. **166**, Dep. Informatics, Univ. Bergen, Norway, (1999)

EK91. Ermentrout, G.B., Kopell, N.: Multiple pulse interactions and averaging in coupled neural oscillators, J. Math. Biol. **29**, 195-217, (1991)

EPG95. Ernst, U., Pawelzik, K., Geisel, T.: Synchronization induced by temporal delays in pulse-coupled oscillators. Phys. Rev. Lett. **74**, 1570, (1995)

EBW87. Ernst, R.R., Bodenhausen, G., Wokaun, A.: Principles of Nuclear Magnetic Resonance in One and Two Dimensions, Clarendon Press, Oxford, (1987)

Egg98. Eggermont, J.J.: Is there a Neural Code?, Neurosci. Biobehav. Rev, **22**, 355-700, (1998)

EPR35a. Einstein, A., Podolsky, B., Rosen, N.: Quantum theory and Measurement. Zurek, W.H., Wheeler, J.A. (ed.), (1935)

EPR35b. Einstein, A., Podolsky, B., Rosen, N.: Can Quantum Mechanical Description of Physical Reality Be Considered Complete? Phys. Rev. **47**, 777, (1935)

EPR99. Eckmann, J.P., Pillet, C.A., Rey-Bellet, L.: Non-equilibrium statistical mechanics of anharmonic chains coupled to two heat baths at different temperatures. Commun. Math. Phys., **201**, 657-697, (1999)

EB05. Edwards, S.C., Belavkin, V.P.: Optimal quantum filtering and quantum feedback control. arxiv.quant-ph/0506018, (2005)

ER85. Eckmann, J.P., Ruelle, D.: Ergodic theory of chaos, strange attractors, Rev. Mod. Phys., **57**, 617-630, (1985)

EM45. Eilenberg, S., MacLane, S.: General theory of natural equivalences. Trans. Am. Math. Soc. **58**, 231-294, (1945)

Eve94. Evensen, G.: Sequential data assimilation with a nonlinear quasi-geostrophic model using Monte Carlo methods to forecast error statistics, J. Geophys. Res. **99 (C5)**, 10,143–10 162, (1994)

EL96. Evensen, G., van Leeuwen, P.: Assimilation of Geosat altimeter data for the Agulhas Current using the ensemble Kalman filter with a quasigeostrophic model, Mon. Wea. Rev. **124**, 85–96, (1996)

Eve03. Evensen, G.: The ensemble Kalman filter: Theoretical formulation and practical implementation, Ocean dynamics, **53**, 343367, (2003)

Eul736. Euler, L.: Solutio problematis ad geometriam situs pertinentis, Commentarii Academiae Scientiarum Imperialis Petropolitanae **8**, 128-140, (1736)

Fei78. Feigenbaum, M.J.: Quantitative universality for a class of nonlinear transformations. J. Stat. Phys. **19**, 25-52, (1978)

Fei79. Feigenbaum, M.J.: The universal metric properties of nonlinear transformations. J. Stat. Phys. **21**, 669-706, (1979)

FS80. Faddeev, L.D., Slavnov, A.A.: Gauge Fields: Introduction to Quantum Theory, Benjamin-Cumming, (1980)

FVB08. Fistul, M.V., Vinokur, V.M. Baturina, T.I.: Macroscopic Coulomb blockade in large JJ arrays, arXiv:0806.4311 (cond-mat.supr-con), (2008)

Fin96. Fineberg, J.: Physics in a jumping sandbox, Nature **382**, 763, (1996)
Fri95. Frisch, U.: Turbulence, Cambridge University Press, Cambridge, (1995)
FK38. Frenkel, Y.I., Kontorova, T.: Phys. Z. Sowjetunion **13**, 1, (1938); J. Phys. Acad. Sci. USSR **1**, 137, (1939)
Fey39. Feynman, R.P.: Forces in Molecules, Phys. Rev. **56**, 340, (1939)
Fei78. Feigenbaum, M.J.: Quanitative universality for a class of nonlinear transformations, J. Stat. Phys. **19**(1), 2552, (1978)
Fey48. Feynman, R.P.: Space-time Approach to Nonrelativistic Quantum Mechanics. Rev. Mod. Phys. **20**, 367-387, (1948)
Fey55. Feynman, R.P.: Application of quantum mechanics to liquid helium. Progress in Low Temperature Physics Vol.1 (Gorter, C.J. ed.) Amsterdam. North-Holland, 17-53, (1955)
FH65. Feynman, R.P., Hibbs, A.R.: Quantum Mechanics and Path Integrals, McGraw-Hill, New York, (1965)
Fey98. Feynman, R.P.: Quantum Electrodynamics. Advanced Book Classics, Perseus Publishing, (1998)
Fey82. Feynman, R.P.: Simulating physics with computers. Int. J. Th. Phys. **21**(6/7), 467-488, (1982)
Fey72. Feynman, R.P.: Statistical Mechanics, a Set of Lectures. WA Benjamin, Inc., Reading, Massachusetts, (1972)
Fey98. Feynman, R.P.: Quantum Electrodynamics. Advanced Book Classics, Perseus Publishing, (1998)
FYH85. Freyd, P., Yetter, D., Hoste, J., Lickorish, W.B.R., Millett, K., Ocneanu, A.: A new polynomial invariant of knots and links, Bull. Amer. Math. Soc. **12**, 239-249, (1985)
Fit55. FitzHugh, R.: Mathematical models of threshold phenomena in the nerve membrane, Bull. Math. Biophys. **17**, 257-269, (1955)
Fit61. FitzHugh, R.A.: Impulses, physiological states in theoretical models of nerve membrane. Biophys J., **1**, 445-466, (1961)
Fla63. Flanders, H.: Differential Forms: with Applications to the Physical Sciences. Acad. Press, (1963)
FW98. Flach, S., Willis, C.R.: Discrete breathers, Phys. Rep. **295**, 182-264, (1998)
FMF03. Flach, S., Miroshnichenko, A.E., Fistul, M.V.: Wave scattering by discrete breathers, Chaos **13**, 596-609, (2003)
FG05. Flach, S., Gorbach, A.V.: Discrete breathers in Fermi-Pasta-Ulam lattices, Chaos **15**, 015112, (2005)
FG08. Flach, S., Gorbach, A.V.: Discrete breathers - Advances in theory and applications, Phys. Rep. **467**, 1-116, (2008)
Fra86. Frampton, P.H.: Gauge Field Theories, Frontiers in Physics. Addison-Wesley, (1986)
FB12. di Franco, C., Ballester, D.: Optimal path for a quantum teleportation protocol in entangled networks. Phys. Rev. A **85**, 010303(R), (2012)
FK83. Frolich, H., Kremer, F.: Coherent Excitations in Biological Systems. Springer, New York, (1983)
Fre00. Freeman, W.J.: Neurodynamics: An exploration of mesoscopic brain dynamics. Springer, Berlin, (2000)
Fre90. Freeman, W.J.: On the the problem of anomalous dispersion in chaotic phase transitions of neural masses, its significance for the management of perceptual information in brains. In H.Haken, M.Stadler (ed.) Synergetics of cognition 45, 126-143. Springer Verlag, Berlin, (1990)

Fre91. Freeman, W.J.: The physiology of perception. Sci. Am., **264**(2), 78-85, (1991)

Fre92. Freeman, W.J.: Tutorial on neurobiology: from single neurons to brain chaos. Int. J. Bif. Chaos. **2**(3), 451-82, (1992)

Fre96. Freeman, W.J.: Random activity at the microscopic neural level in cortex sustains is regulated by low dimensional dynamics of macroscopic cortical activity. Int. J. Neural Sys. **7**, 473, (1996)

FS92. Freeman, J.A., Skapura, D.M.: Neural Networks: Algorithms, Applications and Programming Techniques. Addison-Wesley, Reading, MA, (1992)

Fra98. Frasca, M.: Duality in Perturbation Theory and the Quantum Adiabatic Approximation, Phys. Rev. A **58**(5), 3439, (1998)

FP09. Frost, A.J., Prechter, R.R. Jr.: Elliott Wave Principle: Key to Market Behavior. Wiley, New York, (1978); (10th Edition) Elliott Wave International, (2009)

Fra62. Frankel, T.: In Differential and Combinatorial Topology: a symposium in honor of Marston Morse, 37-53, (Ed. S.S. Cairns), Princeton Univ. Press, Princeton, NJ, (1962)

Fuk75. Fukushima, K.: Cognitron: a self-organizing multilayered neural network. Biol. Cyb., **20**, 121-136, (1975)

Fut83. Futaki, A.: An obstruction to the existence of Einstein Kähler metrics, Inv. Math. Fasc. **73**(3), 437-443, (1983)

FH91. Fulton, W., Harris, J.: Representation theory. A first course, Graduate Texts in Mathematics, Springer, New York, (1991)

FSB98. Furusawa, A., Srensen, J., Braunstein, S., Fuchs, C., Kimble, J., Polzik, E.: Unconditional quantum teleportation. Science **282**, 706-709, (1998)

Gar85. Gardiner, C.W.: Handbook of Stochastic Methods for Physics, Chemistry, Natural Sciences, (2nd ed.) Springer-Verlag, New York, (1985)

GR96. García-Ojalvo, J., Roy, R.: Noise Amplification in a Stochastic Ikeda Model, Phys. Lett. A **224**, 51-56, (1996)

GPG89. Geerligs, L.J., Peters, M., de Groot, L.E.M., Verbruggen, A., Mooij, J.E.: Charging effects and quantum coherence in regular JJ arrays, Phys. Rev. Lett. **63**, 326-329, (1989)

GMS97. Giachetta, G., Mangiarotti, L., Sardanashvily, G.: New Lagrangian and Hamiltonian Methods in Field Theory, World Scientific, Singapore, (1997)

Gom94. Gómez, J.C.: Using symbolic computation for the computer aided design of nonlinear (adaptive) control systems, Tech. Rep. EE9454, Dept. Electr. and Comput. Eng., Uni. Newcastle, Callaghan, NSW, AUS, (1994)

GRF12. Godinez, H.C., Reisner, J.M., Fierro, A.O., Guimond, S.R., Kao, J.: Determining Key Model Parameters of Rapidly Intensifying Hurricane Guillermo (1997) Using the Ensemble Kalman Filter, J. Atmos. Sci., **69**, 3147–3171, (2012)

GC98. Gershenfeld, N., Chuang, I.L.: Quantum Computing with Molecules. Scientific American, June, (1998)

GBM94. Gross, P., Bairagi, D., Mishra, M., Rabitz, H.: Optimal control of IBr curve-crossing reactions, Chem. Phys. 223, 263, (1994)

Gro58. Grothendieck, A.: La théorie des classes de Chern, Bul. Soc. Math. de France **86**, 137-154, (1958)

GM49. Gorter, C.J., Mellink, J.H.: On the irreversible processes in liquid helium II. Physica, **15**, 285-304, (1949)

GM88. Glass, L., Mackey, M.C.: From Clocks to Chaos: The Rhythms of Life, Princeton Univ. Press, Princeton, (1988)

Gol99. Gold, M.: A Kurt Lewin Reader, the Complete Social Scientist, Am. Psych. Assoc., Washington, (1999)
GP05. Goldobin, D., Pikovsky, A.: Synchronization and desynchronization of self-sustained oscillators by common noise, Phys. Rev. E **71**, 045201, (2005)
GH78. Griffiths, P., Harris, J.: Principles of Algebriac Geometry, Wiley, New York, (1978)
Gui94. Guillemin, V.: Kähler structures on toric varieties, J. Diff. Geom. **40**, 285-309, (1994)
GTW12. Guo, R., Tian, B., Wang, L.: Soliton solutions for the reduced Maxwell-Bloch system in nonlinear optics via N-fold Darboux transformation, Nonlinear Dyn. **69**, 2009-2020, (2012)
GHS92. Golomb, D., Hansel, D., Shraiman, B., Sompolinsky, H.: Clustering in globally coupled phase oscillators, Phys. Rev. A **45**, 3516-3530, (1992)
GL88. Guenther, R.B., Lee, J.W.: Partial Differential Equations of Mathematical Physics and Integral Equations, Dover, Englewood Cliffs, (1988)
GGS81. Guevara, M.R., Glass, L., Shrier, A.: Phase locking, period-doubling bifurcations and irregular dynamics in periodically stimulated cardiac cells. Science **214**, 1350-53, (1981)
GK06. Gulevich, D.R., Kusmartsev, F.V.: Perturbation theory for localized solutions of sine-Gordon equation: decay of a breather and pinning by microresistor, Phys. Rev. B **74**, 214303, (2006)
GH13. Guo, R., Hao, H.-Q.: Breathers and localized solitons in erbium doped fibers on constance backgrounds, Nonlin. Dyn. (submitted, 2013)
GH00. Gade, P.M., Hu, C.K.: Synchronous chaos in coupled map lattices with small-world interactions. Phys. Rev. E **62**, 6409-6413, (2000)
Gil02. Gilmore, R.: Lie Groups, Lie Algebras and Some of their Applications (2nd ed.), Dover, (2002)
GH83. Guckenheimer, J., Holmes, P.: Nonlinear Oscillations, Dynamical Systems and Bifurcations of vector fields. Springer-Verlag, Berlin, (1983)
GK94. Gegenberg, J., Kunstatter, G.: The Partition Function for Topological Field Theories, Ann. Phys. **231**, 270-289, (1994)
Ghe96. McGhee, R.B.: Research Notes: A Quaternion Attitude Filter Using Angular Rate Sensors, Accelerometers, and a 3-Axis Magnetometer, Computer Science Department, Naval Postgraduate School, Monterey, CA, (1996)
Gla63. Glauber, R.J.: The Quantum Theory of Optical Coherence. Phys. Rev. **130**, 2529-2539, (1963)
Gre96. Greene, B.R.: String Theory on Calabi-Yau Manifolds. Lectures given at the TASI-96 summer school on Strings, Fields and Duality, (1996)
GSL98. Girardeau, M.D., Schirmer, S.G., Leahy, J.V., Koch, R.M.: Kinematical Bounds on Optimization of Observables for Quantum Systems, Phys. Rev. A **58**, 2684, (1998)
GS-HS98. Glaser, S.J., Schulte-Herbruggen, T., Sieveking, M., Scheletzky, O., Nielsen, N.C., Sorensen, O.W., Griesinger, C.: Unitary Control in Quantum Ensembles, Maximizing Signal Intensity in Coherent Spectroscopy, Science **280**, 421-424, (1998)
GBR07. Grace, M., Brif, C., Rabitz, H.A., Walmsley, I., Kosut, R.L., Lidar, D.A.: Encoding a qubit into multilevel subspaces, New J. Phys. **8**, 35, (2006)
Ghr01. Ghrist, R.: Configuration spaces and braid groups on graphs in robotics, in Knots, braids and mapping class groups, AMS/IP Studies in Advanced Mathematics, Amer. Math. Soc. and Int. Press, 29-40, (2001)

Gol89. Goldberg, D.E.: Genetic Algorithms in Search, Optimization and Machine Learning. Kluwer Acad. Pub., Boston, MA, (1989)

GOY87. Grebogi, C., Ott, E., Yorke, J.A.: Chaos, strange attractors, and fractal basin boundaries in nonlinear dynamics. Science, **238**, 632-637, (1987)

GP83a. Grassberger, P., Procaccia, I.: Measuring the Strangeness of Strange Attractors. Phys. D **9**, 189-208, (1983)

GP83b. Grassberger, P., Procaccia, I.: Characterization of Strange Attractors. Phys. Rev. Lett. **50**, 346-349, (1983)

Gra90. Granato, E.: Phase transitions in Josephson-junction ladders in a magnetic field. Phys. Rev. B **42**, 4797-4799, (1990)

GWA01. Grewal, M.S., Weill, L.R., Andrews, A.P.: Global Positioning Systems, Inertial Navigation, and Integration. Wiley, New York, (2001)

GZ94. Glass, L., Zeng, W.: Bifurcations in flat-topped maps and the control of cardiac chaos. Int. J. Bif. Chaos **4**, 1061-1067, (1994)

GM95. Granata, K.P., Marras, W.S. An EMG-assisted model of trunk loading during free-dynamic lifting. J. Biomech. **28**(11), 1309-17, (1995)

GWP12. Guo, D., Wang, Q., Perc, M.: Complex synchronous behavior in interneuronal networks with delayed inhibitory and fast electrical synapses, Phys. Rev. E **85**, 061905, (2012)

GZ00. Gardiner, C., Zoller, P.: Quantum Noise: A Handbook of Markovian and Non-Markovian Quantum Stochastic Methods with Applications to Quantum Optics (2nd Ed). Springer Series in Synergetics, Berlin, (2000)

Gri05. Griffiths, D.J.: Introduction to Quantum Mechanics (2nd ed.), Pearson Educ. Int., (2005)

HK95. Hasegawa, A., Kodama, Y.: Solitons in Optical Communications. Clarendon, Oxford, (1995)

Ham82. Hamilton, R.S.: Three-manifolds with positive Ricci curvature, J. Diff. Geom. **17**, 255-306, (1982)

Ham86. Hamilton, R.S.: Four-manifolds with positive curvature operator, J. Dif. Geom. **24**, 153-179, (1986)

Ham88. Hamilton, R.S.: The Ricci flow on surfaces, Cont. Math. **71**, 237-261, (1988)

Ham93a. Hamilton, R.S.: The Harnack estimate for the Ricci flow, J. Dif. Geom. 37, 225-243, (1993)

Ham93b. Hamilton, R.: The formation of singularities in the Ricci flow, Vol. II, Inter. Press, (1993)

Ham99. Hamilton, R.S.: Non-singular solutions of the Ricci flow on three-manifolds, Comm. Anal. Geom. **7**(4), 695-729, (1999)

H96P. Hawking, S., Penrose, R.: The Nature of Space and Time, Princeton Univ. Press, (1996)

HMM93. Hansel, D., Mato, G., Meunier, C.: Clustering and slow switching in globally coupled phase oscillators, Phys. Rev. E **48**, 3470-3477, (1993)

Has02. Hasegawa, H.: Stochastic Resonance of Ensemble Neurons for Transient Spike Trains: A Wavelet Analysis. Phys. Rev. E **66**, 21902, (2002)

Has03a. Hasegawa, H.: Dynamical mean-field theory of spiking neuron ensembles: response to a single spike with independent noises, Phys. Rev. E **67**, 041903, (2003)

Has03b. Hasegawa, H.: Dynamical mean-field theory of noisy spiking neuron ensembles: Application to the Hodgkin-Huxley model, Phys. Rev. E **68**, 041909, (2003)

Has04a. Hasegawa, H.: An augmented moment method for stochastic ensembles with delayed couplings: I. Langevin model, Phys. Rev. E **70**, 021911, (2004)

Has04b. Hasegawa, H.: An augmented moment method for stochastic ensembles with delayed couplings: II. FitzHugh-Nagumo model, Phys. Rev. E **70**, 021912, (2004)

HZW12. He, J.S., Zhang, H.R., Wang, L.H., Porsezian, K., Fokas, A.S.: A generating mechanism for higher order rogue waves, arXiv:1209.3742, (2012)

HM95. Helbing, D., Molnar, P.: Social force model for pedestrian dynamics. Phys. Rev. E **51**(5), 4282-4286, (1995)

HJ05. Heimburg, T., Jackson, A.D.: On soliton propagation in biomembranes and nerves, Proc. Natl. Acad. Sci. US, **102**(28), 9790-9795, (2005)

HJ07. Heimburg, T., Jackson, A.D.: On the action potential as a propagating density pulse and the role of anesthetics, Biophys. Rev. Let. **2**, 57-78, (2007)

HP84. Hudson, R.L., Parthasarathy, K.R.L Quantum Ito's formula and stochastic evolution. Commun. Math. Phys. **93**, 301-323, (1984)

Hay01. Haykin, S. (Ed.): Kalman Filtering and Neural Networks. Wiley, New York, (2001)

Hay91. Haykin, S.: Adaptive Filter Theory. Prentice-Hall, Englewood Cliffs, (1991)

Hay94. Haykin, S.: Neural Networks: A Comprehensive Foundation. Macmillan, (1994)

Hel37. Hellmann, H.: Einführung in die Quantenchemie. Franz Deuticke, Leipzig, (1937)

Hil38. Hill, A.V.: The heat of shortening and the dynamic constants of muscle. Proc. Roy. Soc. B **76**, 136-195, (1938)

HWC02. Herek, J.L., Wohlleben, W., Cogdell, R.J., Zeidler, D., Motzkus, M.: Quantum control of energy flow in light harvesting, Nature 417, 533, (2002)

Hee90. Heermann, D.W.: Computer Simulation Methods in Theoretical Physics (2nd ed). Springer, Berlin, (1990)

HWR08. Hsieh, M., Wu, R., Rosenthal, C., Rabitz, H.: The topology and statistical properties of quantum control transition landscapes, J. Phys. B **41**, 074020, (2008)

HV56a. Hall, H.E., Vinen, W.F.: The rotation of liquid helium II I. Experiments on the propagation of second sound in uniformly rotating helium II. Proc. Roy. Soc. London, A **238**, 204-214, (1956)

HV56b. Hall, H.E., Vinen, W.F.: The rotation of liquid helium II II. The theory of mutual friction in uniformly rotating helium II. Proc. Roy. Soc. London, A **238**, 215-234, (1956)

HBB96. Houk, J.C., Buckingham, J.T., Barto, A.G.: Models of the cerebellum, motor learning. Behavioral, Brain Sciences, **19**(3), 368-383, (1996)

HR06. Ho, T.-S., Rabitz, H.: Why do effective quantum controls appear easy to find? J. Photochem. Photobiol. A **180**, 226, (2006)

HM98. Houtekamer, P., Mitchell, H.: Data assimilation using an ensemble Kalman filter technique, Mon. Wea. Rev. **126**, 796–811, (1998)

HSK92. Hauser, J., Sastry, S., Kokotovic, P.: Nonlinear control via approximate input-output linearization: The ball and beam example, IEEE Trans. Aut. Con. AC-37, 392-398, (1992)

HL98. Horn, D., Levanda, S.: Fast temporal encoding and decoding with spiking neurons, Neural Comput. **10**, 1705-1720, (1998)

Hor89. Horowitz, G.T.: Exactly soluble difieomorphism invariant theories. Comm. Math. Phys. **125**, 417, (1989)

HS90. Horowitz, G.T., Srednicki, M.: A quantum field theoretic description of linking numbers and their generalization, Comm. Math. Phys. **130**, 83-94, (1990)
HCK02a. Hong, H., Choi, M.Y., Kim, B.J.: Synchronization on small-world networks. Phys. Rev. E **65**, 26139, (2002)
HCK02b. Hong, H., Choi, M.Y., Kim, B.J.: Phase ordering on small-world networks with nearest-neighbor edges. Phys. Rev. E **65**, 047104, (2002)
Hak83. Haken, H.: Synergetics: An Introduction (3rd ed). Springer, Berlin, (1983)
Hak91. Haken, H.: Synergetic Computers and Cognition. Springer-Verlag, Berlin, (1991)
Hak93. Haken, H.: Advanced Synergetics: Instability Hierarchies of Self-Organizing Systems and Devices (3nd ed.) Springer, Berlin, (1993)
Hak96. Haken, H.: Principles of Brain Functioning: A Synergetic Approach to Brain Activity, Behavior and Cognition. Springer, Berlin, (1996)
Hak00. Haken, H.: Information, Self-Organization: A Macroscopic Approach to Complex Systems. Springer, Berlin, (2000)
Hak02. Haken, H.: Brain Dynamics, Synchronization and Activity Patterns in Pulse-Codupled Neural Nets with Delays and Noise. Springer, New York, (2002)
HK99. Hanm, S.-H., Koh, I.G.: Stability of neural networks and solitons of field theory. Phys. Rev. E **60**, 7608-7611, (1999)
HS94. Haelterman, M., Sheppard, A.P.: Bifurcation phenomena and multiple soliton bound states in isotropic Kerr media. Phys. Rev. E **49**, 3376-3381, (1994)
Ham87. Hameroff, S.R.: Ultimate Computing: Biomolecular Consciousness and Nanotechnology. North-Holland, Amsterdam, (1987)
Ham98. Hameroff, S.: Quantum computation in brain microtubules? The Penrose-Hameroff Orch OR model of consciousness. Philos. Trans. R. Soc. London Ser. A **356**, 1869-1896, (1998)
HP93. Hameroff, S.R., Penrose, R.: Conscious events as orchestrated spacetime selections. Journal of Consciousness Studies, **3**(1), 36-53, (1996)
HP96. Hameroff, S.R., Penrose, R.: Orchestrated reduction of quantum coherence in brain microtubules: A model for consciousness. In: Hameroff, S. R., Kaszniak, A.W., Scott, A.C. Eds: Toward a Science of Consciousness: the First Tucson Discussion, Debates, 507-539. MIT Press, Cambridge, MA, (1996)
Har75. Harris, R.J.: A Primer of Multivariate Statistics. Academic Press, New York, (1975)
HS98. Haskins, M., Speight, J.M.: Breathers in the weakly coupled topological discrete sine-Gordon system, Nonlinearity **11**, 1651-1671, (1998)
HT05. He, B., Teixeira, F.L.: On the degrees of freedom of lattice electrodynamics. Phys. Let. A **336**, 1-7, (2005)
Hec87. Hecht-Nielsen, R.: Counterpropagation networks. Applied Optics, **26**(23), 4979-4984, (1987)
Hec90. Hecht-Nielsen, R.: NeuroComputing. Addison-Wesley, Reading, (1990)
Hel01. Helgason, S.: Differential Geometry, Lie Groups and Symmetric Spaces. (2nd ed.) American Mathematical Society, Providence, RI, (2001)
Hen66. Hénon, M.: Sur la topologie des lignes de courant dans un cas particulier. C. R. Acad. Sci. Paris A, **262**, 312-314, (1966)
Hen69. Hénon, M.: Numerical study of quadratic area preserving mappings. Q. Appl. Math. **27**, (1969)
Hen76. Hénon, M.: A twoo-dimensional mapping with a strange attractor. Com. Math. Phys. **50**, 69-77, (1976)

HHC97. Hirota, O., Holevo, A.S., Caves, C.M. (Eds.): Quantum Communication, Computing, and Measurement, Plenum Press, New York, (1997)

HDR09. Ho, T.S., Dominy, J., Rabitz, H.: Landscape of unitary transformations in controlled quantum dynamics, Phys. Rev. A **79**, 013422, (2009)

HS74. Hirsch, M.W., Smale, S.: Differential Equations, Dynamical Systems and Linear Algebra. Academic Press, New York, (1974)

HI01. Hoppensteadt, F.C., Izhikevich, E.M.: Canonical Neural Models. In Arbib MA (ed.) Brain Theory, Neural Networks (2nd ed.) MIT press, Cambridge, MA, (2001)

HI97. Hoppensteadt, F.C., Izhikevich, E.M.: Weakly Connected Neural Networks. Springer, New York, (1997)

HI99. Hoppensteadt, F.C., Izhikevich, E.M.: Oscillatory Neurocomputers With Dynamic Connectivity, Phys. Rev. Let., **82**(14), 2983-86, (1999)

HL84. Horsthemke, W., Lefever, R.: Noise-Induced Transitions. Springer, Berlin, (1984)

HH52. Hodgkin, A.L., Huxley, A.F.: A quantitative description of membrane current and application to conduction and excitation in nerve. J. Physiol., **117**, 500-544, (1952)

Hod64. Hodgkin, A.L.: The Conduction of the Nervous Impulse. Liverpool Univ. Press, Liverpool, (1964)

Hug00. Hughes, J.: Generalising monads to arrows. Sci. Comp. Prog. **37**, 67–111, (2000)

Hug04. Hughes, J.: Programming with arrows, In Advanced Functional Programming, Springer Ser. LNCS, (2004)

Hux57. Huxley, A.F.: Muscle structure and theories of contraction. Progr. Biophys. Chem., **7**, 255-328, (1957)

HN54. Huxley, A.F., Niedergerke, R.: Changes in the cross-striations of muscle during contraction and stretch and their structural interpretation. Nature, **173**, 973-976, (1954)

HT93. Hunt, L, Turi, J.: A new algorithm for constructing approximate transformations for nonlinear systems. IEEE Trans. Aut. Con., AC-38,1553-1556, (1993)

HOG88. Hsu, G.H., Ott, E., Grebogi, C.: Strange Saddles and Dimensions of their Invariant Manifolds. Phys. Lett. A **127**, 199-204, (1988)

Hol92. Holland, J.H.: Adaptation in Natural, Artificial Systems (2nd ed.) MIT Press, Cambridge, MA, (1992)

Hou79. Houk, J.C.: Regulation of stiffness by skeletomotor reflexes. Ann. Rev. Physiol., **41**, 99-123, (1979)

Hop82). Hopfield, J.J.: Neural networks, physical systems with emergent collective computational activity. Proc. Natl. Acad. Sci. USA., **79**, 2554-2558, (1982)

Hop84. Hopfield, J.J.: Neurons with graded response have collective computational properties like those of two-state neurons. Proc. Natl. Acad. Sci. USA, **81**, 3088-3092, (1984)

HT85. Hopfield, J.J., Tank, D.W.: Neural computation of decisions in optimisation problems. Biol. Cybern., **52**, 114-152, (1985)

Hop95. Hopfield, J.J.: Pattern recognition computing using action potential timing for stimulus representation, Nature **376**, 33-36, (1995)

Ing97. Ingber, L.: Statistical mechanics of neocortical interactions: Applications of canonical momenta indicators to electroencephalography. Phys. Rev. E, **55**(4), 4578-4593, (1997)

Ing98. Ingber, L.: Statistical mechanics of neocortical interactions: Training, testing canonical momenta indicators of EEG. Mathl. Computer Modelling **27**(3), 33-64, (1998)

IDA80. Ikeda, K., Daido, H., Akimoto, O.: Optical Turbulence: Chaotic Behavior of Transmitted Light from a Ring Cavity, Phys. Rev. Lett. **45**, 709-712, (1980)

IM87. Ikeda, K., Matsumoto, K.: High-dimensional chaotic behavior in systems with time-delayed feedback, Physica D **29**, 223-235, (1987)

IGF99. Ioffe, L.B., Geshkenbein, V.B., Feigel'man, M.V., Fauchère, A.L., Blatter, G.: Environmentally decoupled sds-wave JJs for quantum computing. Nature **398**, 679-681, (1999)

Isi89. Isidori, A.: Nonlinear Control Systems, An Introduction (2nd ed.) Springer, Berlin, (1989)

Ito51. Itô, K.: On Stochastic Differential Equations, Mem. Am. Math. Soc. **4**, 1-51, (1951)

Ito60. Ito, K.: Wiener Integral, Feynman Integral. Proc. Fourth Berkeley Symp. Math., Stat., Prob., **2**, 227-238, (1960)

IP01a. Ivancevic, V., Pearce, C.E.M.: Topological duality in humanoid robot dynamics. ANZIAM J. **43**, 43183-194, (2001)

IP01b. Ivancevic, V., Pearce, C.E.M.: Poisson manifolds in generalized Hamiltonian biomechanics. Bull. Austral. Math. Soc. **64**, 515-526, (2001)

IS01. Ivancevic, V., Snoswell, M.: Fuzzy-stochastic functor machine for general humanoid robot dynamics. IEEE Trans. Syst., Man, Cybern. B **31**(3), 319-330, (2001)

Iva02. Ivancevic, V.: Generalized Hamiltonian biodynamics, topology invariants of humanoid robots. Int. J. Math., Math. Sci. **31**(9), 555-565, (2002)

Iva04. Ivancevic, V.: Symplectic rotational geometry in human biomechanics, SIAM Rev., **46**(3), 455-474, (2004)

II05. Ivancevic, V., Ivancevic, T.: Human-Like Biomechanics. Springer, Dordrecht, (2005)

IB05. Ivancevic, V., Beagley, N.: Brain-like functor control-machine for general humanoid biodynamics. Int. J. Math., Math. Sci. **11**, 1759-1779, (2005)

II06a. Ivancevic, V., Ivancevic, T.: Natural Biodynamics. World Scientific, Singapore, (2006)

II06b. Ivancevic, V., Ivancevic, T.: Geometrical Dynamics of Complex Systems. Springer, Dordrecht, (2006)

Iva06a. Ivancevic, V.: Lie-Lagrangian model for realistic human bio-dynamics. Int. J. Hum. Rob., **3**(2), 205-218, (2006)

Iva06b. Ivancevic, V.: Dynamics of Humanoid Robots: Geometrical, Topological Duality. In Biomathematics: Modelling, Simulation, ed. J.C. Misra, World Scientific, Singapore, (2006)

II07. Ivancevic, V., Ivancevic, T., Neuro-Fuzzy Associative Machinery for Comprehensive Brain and Cognition Modelling. Springer, Berlin, (2007)

II07. Ivancevic, V., Ivancevic, T.: Applied Differential Geometry: A Modern Introduction. World Scientific, Singapore, (2007)

II07d. Ivancevic, V., Ivancevic, T.: Computational Mind: A Complex Dynamics Perspective. Springer, Berlin, (2007)

II08a. Ivancevic, V., Ivancevic, T.: Complex Nonlinearity: Chaos, Phase Transitions, Topology Change and Path Integrals, Springer, (2008)

II08b. Ivancevic, V., Ivancevic, T.: Quantum Leap: From Dirac and Feynman, Across the Universe, to Human Body and Mind. World Scientific, Singapore, (2008)

IA07. Ivancevic, V., Aidman, E.: Life-space foam: A medium for motivational and cognitive dynamics. Physica A 382, 616-630, (2007)

IAY09. Ivancevic, V., Aidman, E., Yen, L.: Extending Feynman's Formalisms for Modelling Human Joint Action Coordination. Int. J. Biomath. 2(1), 1-7, (2009)

IS08. Ivancevic, V., Sharma, S.: Complexity in Human Bio-Mechanics. IJHR, 5(4), 679-698 (2008)

II08. Ivancevic, V., Ivancevic, T.: Human versus Humanoid Biodynamics, IJHR, 5(4), 699-713, (2008)

II06d. Ivancevic V., Ivancevic T.: High-Dimensional Chaotic and Attractor Systems. Springer, (2006)

II09. Ivancevic, V., Ivancevic, T.: Quantum Neural Computation, Springer, (2009)

Iva09a. Ivancevic, V.G.: New mechanics of traumatic brain injury, Cogn. Neurodyn. **3**, 281-293, (2009)

Iva09b. Ivancevic, V.G.: New mechanics of spinal injury, IJAM, **1**(2), 387-401, (2009)

Iva09c. Ivancevic, V.G.: New mechanics of generic musculo-skeletal injury, BRL, 4(3), 273-287, (2009)

Iva10a. Ivancevic, V.G.: A unique cause of traumatic brain injury and all neuro-musculo-skeletal injuries, Brain Research J. **3**(2), 1935-2875, (2010)

Iva10b. Ivancevic, V.G.: Adaptive-Wave Alternative for the Black-Scholes Option Pricing Model, Cogn. Comput., 2(1), 17-30, (2010)

Iva10c. Ivancevic, V.G.: Nonlinear complexity of human biodynamics engine, Nonlin. Dynamics, **61**(1-2), 123-139, (2010)

Iva11. Ivancevic, V.G.: Adaptive Wave Models for Sophisticated Option Pricing, J. Math. Fin. (in press), (2011)

II11a. Ivancevic, V., Ivancevic, T., Dynamics and control of humanoid robots: A geometrical approach, Paladyn J. Behav. Rob. **1**(4), 204-218, (2011)

II11b. Ivancevic, V., Ivancevic, T.: Ricci flow and nonlinear reaction-diffusion systems in biology, chemistry, and physics, Nonl. Dyn. **65**(1-2), 35-54, (2011)

IR09. Ivancevic, V., Reid, D.: Dynamics of Confined Crowds Modelled Using Entropic Stochastic Resonance and Quantum Neural Networks. Int. J. Intel. Defence Sup. Sys. **2**(4), 269-289, (2009)

IR10a. Ivancevic, V., Reid, D., Aidman, E.: Crowd behavior dynamics: entropic path-integral model. Nonl. Dyn. **59**, 351-373, (2010)

IR10b. Ivancevic, V., Reid, D.: Entropic geometry of crowd dynamics. A Chapter in Nonlear Dynamics (T. Evancs, Ed.), Intech, Vienna, (2010)

IR10c. Ivancevic, V., Reid, D.: Geometrical and Topological Duality in Crowd Dynamics. Int. J. Biomath. **3**(4), (2010), 493-507.

IR14. Ivancevic, V., Reid, D., Scholz, J.: Action-Amplitude Approach to Controlled Entropic Self-Organization. Entropy. 16, 2699-2712 (2014)

Iva10. Ivancevic, V.: Adaptive-wave alternative for the Black-Scholes option pricing model. Cogn. Comput. **2**, 17-30, (2010)

Iva11a. Ivancevic, V.: Adaptive Wave Models for Sophisticated Option Pricing, J. Math. Fin. **1**(3), 41-49, (2011)

Iva11b. Ivancevic, T. *et al:* Gateway to Future Science with Nikola Tesla, Nova Sci. Pub. (2011)

II12. Ivancevic, V., Ivancevic, T.: New Trends in Control Theory, World Scientific, Singapore, (2012)

IR12. Ivancevic, V., Reid, D.: Turbulence and shock-waves in crowd dynamics, Nonl. Dyn. **68**, 285-304, (2012)

IJ12. Ivancevic, V.G., Johnson, W.T.: Macro-Microscopic Self-Similarity in Socio-Cognio-Physical Dynamics, MESA 3(3), 325-339, (2012)

II13. Ivancevic, V., Ivancevic, T.: Sine-Gordon Solitons, Kinks and Breathers as Physical Models of Nonlinear Excitations in Living Cellular Structures, J. Geo. Sym. Phys. **31**, 1-56, (2013)

IIG02. Ivanov, D.A., Ioffe, L.B., Geshkenbein, V.B., Blatter, G.: Interference effects in isolated JJ arrays with geometric symmetries, Phys. Rev. B **65**, 024509, (2002)

Izh00. Izhikevich, E.M.: Neural Excitability, Spiking and Bursting. International Journal of Bifurcation and Chaos, **10**, 1171-1266, (2000)

Izh01. Izhikevich, E.M.: Synchronization of Elliptic Bursters, SIAM Rev., **43**(2), 315-344, (2001)

Izh01. Izhikevich, E.M.: Resonate-and-fire neurons. Neu. Net. **14**, 883-894, (2001)

IDW03. Izhikevich, E.M., Desai, N.S., Walcott, E.C., Hoppensteadt, F.C.: Bursts as a unit of neural information: selective communication via resonance. Trends in Neurosci. **26**, 161-167, (2003)

Izh04. Izhikevich, E.M.: Which model to use for cortical spiking neurons? IEEE Trans. Neu. Net. **15**, 1063-1070, (2004)

Izh07. Izhikevich, E.M.: Dynamical Systems in Neuroscience: The Geometry of Excitability and Bursting. The MIT Press, Cambridge, MA, (2007)

Izh99a. Izhikevich, E.M.: Class 1 neural excitability, conventional synapses, weakly connected networks and mathematical foundations of pulse-coupled models. IEEE Trans. Neu. Net., **10**, 499-507, (1999)

Izh99b. Izhikevich, E.M.: Weakly Connected Quasiperiodic Oscillators, FM Interactions and Multiplexing in the Brain. SIAM J. Appl. Math., **59**(6), 2193-2223, (1999)

IMK05. Iserles, A., Munthe-Kaas, H.Z., Nørsett, S.P., Zanna, A.: Lie-group methods, Acta Numer. **9**, 1-148, (2005)

Jos. Joshi, S.S.: Inverted Pendulum with Damping, Am. J. Phys. **34**(6), 533-533, (1966)

JNP06. James, M.R., Nurdin, H.I., Petersen, I.R.: H^∞ control of linear quantum stochastic systems. IEEE Trans. Automat. Contr. **53**(8), 1787-1803, (2008)

JS96. Jang, J.-S.R., Sun, C.-T.: Neuro-fuzzy and soft computing: a computational approach to learning and machine intelligence, Prentice-Hall, Inc., Upper Saddle River, NJ, (1996)

Jen07. Jensen, F.: Introduction to Computational Chemistry, Wiley, Sussex, (2007)

Joa00. Joao, M.L.: An Extended Kalman Filter for Quaternion-Based Attitude Estimation. Master thesis, Computer Science Department, Naval Postgraduate School, Monterey, CA, (2000)

Jon85. Jones, V.F.R.: A polynomial invariant for knots via Von Neuman algebras, Bull. Amer. Math. Soc **12**, 103-111, (1985)

Jon87. Jones, V.F.R.: Hecke algebra representations of Braid groups and link polynomials. Ann. Math. **126**, 335-388, (1987)

Jon90. Jones, V.F.R.: Knot theory and statistical mechanics, Scientific American, 263, **5**(4), 98-103, (1990)

JFH03. Jezek, M., Fiurasek, J., Hradil, Z.: Quantum inference of states and processes, Phys. Rev. A **68**, 012305, (2003)

Jos74. Josephson, B.D.: The discovery of tunnelling supercurrents. Rev. Mod. Phys. **46**(2), 251-254, (1974)

JS72. Jurdjevic, V., Sussmann, J.: Control systems on Lie groups. J. Diff. Equat. **12**, 313-329, (1972)

Jur97. Jurdjevic, V.: Geometric control theory, Cambridge Univ. Press, (1997)

JQ78. Jurdjevic, V., Quinn, J.: Controllability and stability. J. Diff. Eq. **28**(3), 381-389, (1978)

JK04. Justh, E., Krishnaprasad, P.: Equilibria and steering laws for planar formations. Systems and Control Letters **52**, 25-38, (2004)

JR92. Judson, R.S., Rabitz, H.: Teaching lasers to control molecules, Phys. Rev. Lett. **68**, 1500, (1992)

Jos74. Josephson, B.D.: The discovery of tunnelling supercurrents. Rev. Mod. Phys. **46**(2), 251-254, (1974)

KFA69. Kalman, R.E., Falb, P., Arbib, M.A.: Topics in Mathematical System Theory. McGraw Hill, New York, (1969)

Kal60. Kalman, R.E.: A new approach to linear filtering and prediction problems. Transactions of the ASME, Ser. D, J. Bas. Eng., **82**, 34-45, (1960)

KH66. Kao, K.C., Hockham, G.A.: Dielectric-fibre surface waveguides for optical frequencies, Proc. IEEE **113**(7), 1151-1158, (1966)

KBG01. Khaneja, N., Brockett, R., Glaser, S.J.: Time optimal control in spin systems, Phys. Rev. A **63**, 032308, (2001)

KGB02. Khaneja, N., Brockett, R., Glaser, S.J.: Time optimal control in spin systems, Phys. Rev. A **65**, 032301, (2002)

Kas02. Kasabov, N.: Evolving connectionist systems: Methods and applications in bioinformatics, brain study and intelligent machines. Springer, London, (2002)

Kaw99. Kawato, M.: Internal models for motor control and trajectory planning. Current Opinion in Neurobiology, **9**, 718-727, (1999)

Kir847. Kirchhoff, G.R.: Über die Auflosung der Gleichungen, auf welche man bei der Untersuchung der linearen Vertheilung galvanischer strome geführt wird, Annalen d. Physik und Chemie **72**, 497-508, (1847)

KE83. Kadic, A., Edelen, D.G.B.: A Gauge theory of Dislocations and Disclinations. Springer, New York, (1983)

Kla00. Klauder, J.R.: Beyond Conventional Quantization. Cambridge Univ. Press, Cambridge, (2000)

Kla97. Klauder, J.R.: Understanding Quantization. Found. Phys. **27**, 1467-1483, (1997)

Kno911. Knott, C.G.: Life and Scientific work of Peter Guthrie Tait, Supplementing the two volumes of scientific papers published in 1898 and 1900, Cambridge Univ. Press, Cambridge, (1911)

KES96. König, P., Engel, A.K., Singer, W.: Integrator or coincidence detector? The role of the cortical neuron revisited, Trends Neurosci. **19**, 130-137, (1996)

Kre84. Krener, A.: Approximate linearization by state feedback and coordinate change, Systems Control Lett. 5, 181-185, (1984)

KSW05. Kiesel, N., Schmid, C., Weber, U., Ursin, R., Weinfurter, H.: Linear optics controlled-phase gate made simple. Phys. Rev. Lett.

KMM94. Konen, W., Maurer, T., von der Malsburg, C.: A fast dynamic link matching algorithm for invariant pattern recognition. Neural Networks, **7**, 1019-1030, (1994)

KMY84. Kaplan, J.L., Mallet-Paret, J., Yorke, J.A.: The Lyapunov dimension of a nowhere dierentiable attracting torus. Ergod. Th. Dynam. Sys. **4**, 261 (1984)

Koh82. Kohonen, T.: Self-Organized Formation of Toplogically Correct Feature Maps. Biol. Cybern. **43**, 59-69, (1982)

Koh88. Kohonen, T.: Self Organization, Associative Memory. Springer, (1988)

Koh91. Kohonen, T.: Self-Organizing Maps: Optimization Approaches. In: Artificial Neural Networks, ed. T. Kohonen etal. North-Holland, Amsterdam, (1991)

Kog96. Kogan, S.: Electronic Noise and Fluctuations in Solids (Cambridge Univ. Press, Cambridge, (1996)

KM75. Kohler, G., Milstein, C.: Continuous cultures of fused cells secreting antibody of predefined specificity. Nature **256**, 495, (1975)

Kos86. Kosko, B.: Fuzzy Cognitive Maps. Int. J. Man-Mach. Stud. **24**, 65-75, (1986)

Kos88. Kosko, B.: Bidirectional Associative Memory. IEEE Trans. Sys. Man Cyb. **18**, 49-60, (1988)

Kos92. Kosko, B.: Neural Networks, Fuzzy Systems, A Dynamical Systems Approach to Machine Intelligence. Prentice-Hall, New York, (1992)

KI93. Kosko, B., Isaka, S.: Fuzzy Logic, Sci. Amer. **269**, 76-81, July, (1993)

Kos93. Kosko, B.: Fuzzy Thinking. Disney Books, Hyperion, (1993)

Kos96. Kosko, B.: Fuzzy Engineering. Prentice Hall, New York, (1996)

Kos98. Kosko, B.: Global Stability of Generalized Additive Fuzzy Systems, IEEE Trans. SMC, C, **28**(3), 441-452, (1996)

Kos99. Kosko, B.: The Fuzzy Future: From Society, Science to Heaven in a Chip. Random House, Harmony, (1999)

KMG94. Kaluza, M., Muckerman, J., Gross, P., Rabitz, H.J.: Optimally Controlled Five-Laser Infrared Multiphoton Dissociation of HF, Chem. Phys. **100**, 4211, (1994)

KTS05. Kral, P., Thanopulos, I., Shapiro, M.: Quantum-field coherent control: Preparation of broken-symmetry entangled states, Phys. Rev. A **72**, 020303, (2005)

KP95. Kocarev, L., Parlitz, U.: General Approach for Chaotic Synchronization with Applications to Communication. Phys. Rev. Lett. **74**, 5028-5031, (1995)

KP96. Kocarev, L., Parlitz, U.: Generalized Synchronization, Predictability and Equivalence of Unidirectionally Coupled Dynamical Systems. Phys. Rev. Lett. **76**, 1816-1819, (1996)

KTU03. Kasamatsu, K., Tsubota, M., Ueda, M.: Nonlinear dynamics of vortex lattice formation in a rotating Bose-Einstein condensate, Phys. Rev. A **67**, 033610, (2003)

KMS05. Kasamatsu, K., Machida, M., Sasa, N., Tsubota, M.: Three-dimensional dynamics of vortex-lattice formation in Bose-Einstein condensate Phys. Rev. A **71**, 063616, (2005)

KS02. Kasabov, N., Song, Q.: Denfis: Dynamic evolving neural fuzzy inference systems and its application for time series prediction. IEEE Trans. Fuz. Sys. **10**(2), 144-154, (2002)

Kra83. Kraus, K.: States, Effects and Operations: Fundamental Notions of Quantum Theory, Springer, Berlin, (1983)

KLM03. Kosut, R.L., Levis, R.J., Mabuchi, H., Rabitz, H., Walmsley, I.A., Yablonovich, E.: Managing uncertainties in the control of quantum systems, Princeton-Oxford Workshop on Control of Quantum Systems, (2003)

KK01. Kawahara, G., Kida, S.: Periodic motion embedded in plane Couette turbulence: regeneration cycle and burst. J. Fluid Mech. **449**, 291-300, (2001)

KMS93. Kolar, I., Michor, P.W., Slovak, J.: Natural Operations in Differential Geometry. Springer, Berlin, (1993)

KT01. Kye, W.-H., Topaj, D.: Attractor bifurcation and on-off intermittency. Phys. Rev. E **63**, 045202(R), (2001)

KN06. Kumar, S., Narayan, Y.: Torque and EMG in rotation extension of the torso from pre-rotated and flexed postures. Clin. Biomech. **21**(9), 920-931, (2006)

Kur84. Kuramoto, Y.: Chemical Oscillations. Waves, Turbulence. Springer, New York, (1984)

Kuz95. Kuznetsov, Y.A.: Elements of Applied Bifurcation Theory. Applied Mathematical Sciences **112**, Springer-Verlag, Berlin, (1995)

Kap38. Kapitza, P.: Viscosity of liquid helium below the λ point. Nature **141**, 74, (1938)

KY75. Kaplan, J.L., Yorke, J.A.: On the stability of a periodic solution of a differential delay equation. SIAM J. Math. Ana. **6**, 268-282, (1975)

KY79. Kaplan, J.L., Yorke, J.A.: Numerical Solution of a Generalized Eigenvalue Problem for Even Mapping. Peitgen, H.O., Walther, H.O. (Eds): Functional Differential Equations, Approximations of Fixed Points, Lecture Notes in Mathematics, **730**, 228-256, Springer, Berlin, (1979)

KY79. Kaplan, J.L., Yorke, J.A.: Preturbulence: a regime observed in a fluid flow of Lorenz. Commun. Math. Phys. **67**, 93-108, (1979)

KY91. Kennedy, J., Yorke, J.A.: Basins of Wada. Physica D **51**, 213-225, (1991)

KZH02. Kiss, I.Z., Zhai, Y., Hudson, J.L.: Emerging coherence in a population of chemical oscillators. Science **296**, 1676-1678, (2002)

KNA10. Kawamura, Y., Nakao, H., Arai, K., Kori, H., Kuramoto, Y.: Phase synchronization between collective rhythms of globally coupled oscillator groups: noiseless nonidentical case, Chaos **20**, 043110, (2010)

KZH05. Kiss, I.Z., Zhai, Y., Hudson, J.L.: Predicting mutual entrainment of oscillators with experiment-based phase models. Phys Rev Lett. **94**, 248301, (2005)

Kir07. Kirillov, A.: Image Processing Lab in C#, Code Project, (2007)

KGS92. Kivshar, Yu.S., Gronbech-Jensen, N., Samuelsen, M.R.: Pi kinks in a parametrically driven sine-Gordon chain, Phys. Rev. B **45**(14), 7789-7794, (1992)

Kle02. Kleinert, H.: Path Integrals in Quantum Mechanics, Statistics, Polymer Physics, and Financial Markets (3rd ed), World Scientific, Singapore, (2002)

KN63/9. Kobayashi, S., Nomizu, K.: Foundations of Differential Geometry, Vols. 1,2., Interscience Publ., New York, (1963/1969)

KM94. Koch, A.J., Meinhardt, H.: Biological pattern formation: from basic mechanisms to complex structures, Rev. Mod. Phys. **66**, 1481-1507, (1994)

KB93. Kruglyak, L., Bialek, W.: Statistical mechanics for a network of spiking neurons, Neural. Comp. **5**, 21-31, (1993)

Kee87. Keener, J.P.: Propagation and its failure in coupled systems of discrete excitable cells, SIAM J. Appl. Math. **47**, 556, (1987)

LPM96. Lafarge, P., Matters, M., Mooij, J.E.: Charge representation of a small two-dimensional Josephson-junction array in the quantum regime, Phys. Rev. B **54**, 7380, (1996)

Lai94. Lai, Y.-C.: Controlling chaos. Comput. Phys., **8**, 62-67, (1994)

LM96. Lakshmanan, M., Murali, K.: Chaos in Nonlinear Oscillators: Controlling and Synchronization, World Scientific, Singapore, (1996)

Lak97. Lakshmanan, M.: Bifurcations, Chaos, Controlling and Synchronization of Certain Nonlinear Oscillators. In Lecture Notes in Physics, **495**, 206, Y. Kosmann-Schwarzbach, B. Grammaticos, K.M. Tamizhmani (ed), Springer-Verlag, Berlin, (1997)

LM00. Lakshmanan, M., Muruganandam, P.: Spatiotemporal patterns in array of coupled nonlinear oscillators, Proc. INSA A: Phys. Sci. 66, 393-413, (2000)

Lak03. Lakshmanan, M., Rajasekar, S: Nonlinear Dynamics: Integrability, Chaos and Patterns, Springer-Verlag, New York, (2003)

Lan99. Lang, S.: Fundamentals of Differential Geometry. Graduate Texts in Mathematics, Springer, New York, (1999)

Lan02. Lang, S.: Introduction to Differentiable Manifolds (2nd ed.). Graduate Texts in Mathematics, Springer, New York, (2002)

LCD87. Leggett, A.J., Chakravarty, S., Dorsey, A.T., Fisher, M.P.A., Chang, A., Zwerger, W.: Dynamics of the dissipative two-state system. Rev. Mod. Phys. **59**, 1, (1987)

Lew86. Lewis, F.L.: Optimal Control, Wiley, New York, (1986)

Lei02. Leinster, T.: A survey of definitions of $n-$category. Theor. Appl. Categ. **10**, 1-70, (2002)

Lei03. Leinster, T.: Higher Operads, Higher Categories, London Mathematical Society Lecture Notes Series, Cambridge Univ. Press, (2003)

Lei04. Leinster, T.: Operads in higher-dimensional category theory. Theor. Appl. Categ. **12**, 73-194, (2004)

Lep78. Lepage, G.P.: A New Algorithm for Adaptive Multidimensional Integration, J. Comp. Phys. **27**, 192-203, (1978)

Lep98. Lepage, G.P.: Lattice QCD for Novices, Proc. HUGS 98, (J.L. Goity ed.), World Sci. (2000)

Lep14. Lepage, G.P.: Vegas-2.1.4. Adaptive multidimensional Monte Carlo integration, https://github.com/gplepage/vegas, (2014)

Lei04. Chair for Computer Aided Medical Procedures & Augmented Reality: Sensor Fusion using the Kalman Filter, (2013); `http://campar.in.tum.de/Chair/KalmanFilter#Literature`

Lev44. Levenberg, K.: A Method for the Solution of Certain Non-Linear Problems in Least Squares, Quart. Appl. Math. **2**, 164-168, (1944)

Lew95. Lewis, A.D.: Aspects of Geometric Mechanics and Control of Mechanical Systems. Technical Report CIT-CDS 95-017 for the Control and Dynamical Systems Option, California Institute of Technology, Pasadena, CA, (1995)

LM97. Lewis, A.D., Murray, R.M.: Controllability of simple mechanical control systems, SIAM J. Con. Opt., **35**(3), 766-790, (1997)

Lew98. Lewis, A.D.: Affine connections and distributions with applications to nonholonomic mechanics, Reports on Mathematical Physics, **42**(1/2), 135-164, (1998)

Lew99. Lewis, A.D.: When is a mechanical control system kinematic?, in Proceedings of the 38th IEEE Conf. Decis. Con., 1162-1167, IEEE, Phoenix, AZ, (1999)

LM99. Lewis, A.D., Murray, R.M.: Configuration controllability of simple mechanical control systems, SIAM Review, **41**(3), 555-574, (1999)

Lew00a. Lewis, A.D.: Simple mechanical control systems with constraints, IEEE Trans. Aut. Con., **45**(8), 1420-1436, (2000)

Lew00b. Lewis, A.D.: Affine connection control systems. Proceedings of the IFAC Workshop on Lagrangian and Hamiltonian Methods for Nonlinear Control 128-133, Princeton, (2000)

LHP13. Li, C., He, J., Porsezian, K.: Rogue waves of the Hirota and the Maxwell-Bloch equations, Phys. Rev. E **87**, 012913, (2013)

Lis847. Listing, J.B.: Vorstudien zur Topologie, Göttinger Studien (Abtheilung 1) **1**, 811-875, (1847)

LP10. Z. Levnajic and A. Pikovsky, Phase resetting of collective rhythm in ensembles of oscillators, Phys. Rev. E **82**, 056202, (2010)

LOA09. Lee, W.S., Ott, E., Antonsen, T.M.: Large Coupled Oscillator Systems with Heterogeneous Interaction Delays, Phys. Rev. Lett. **103**, 044101, (2009)

LRO11. Lee, W.S., Restrepo, J.G., Ott, E., Antonsen, T.M.: Dynamics and pattern formation in large systems of spatially-coupled oscillators with finite response times, Chaos **21**, 023122, (2011)

LC10. Leaw, J.N., Cheong, S.A.: Strategic Insights From Playing the Quantum Tic-Tac-Toe, J. Phys. A **43**(45), 455304, (2010)

LWS01. Leadbeater, M., Winiecki, T., Samuels, D.C., Barenghi, C.F., Adams, C.S.: Sound emission due to superfluid vortex reconnections, Phys. Rev. Lett. **86**, 1410-1413, (2001)

LFL01. Liu, S., Fu, Z., Liu, S., Zhao, Q.: Jacobi elliptic function expansion method and periodic wave solutions of nonlinear wave equations. Phys. Let. A **289**, 69-74, (2001)

LF05. Liu, G-T., Fan, T-Y.: New applications of developed Jacobi elliptic function expansion methods. Phys. Let. A **345**, 161-166, (2005)

Lo04. Lo, A.W.: The Adaptive Markets Hypothesis: Market Efficiency from an Evolutionary Perspective, J. Portf. Manag. **30**, 15-29, (2004)

Lo05. Lo, A.W.: Reconciling Efficient Markets with Behavioral Finance: The Adaptive Markets Hypothesis, J. Inves. Consult. **7**, 21-44, (2005)

Lol08. Loll, R.: The emergence of spacetime or quantum gravity on your desktop, Class. Quantum Grav. **25**, 114006, (2008)

LEA00. Lehnertz, K., Elger, C.E., Arnhold, J., Grassberger, P. (ed): Chaos in Brain. World Scientific, Singapore, (2000)

Lee90. Lee, C.C.: Fuzzy Logic in Control Systems. IEEE Trans. Sys., Man, Cybern., **20**(2), 404-435, (1990)

LAJ11. Lautrup, B., Appali, R., Jackson, A.D., Heimburg, T.: The stability of solitons in biomembranes and nerves, Eur. Phys. J. E **34**(6), 1-9, (2011)

LR89. De Leon, M., Rodrigues, P.R.: Methods of Differential Geometry in Analytical Mechanics, North-Holland Mathematics Studies, Vol.152, Elsevier, Amsterdam, (1989)

Llo00. Lloyd, S.: Coherent quantum feedback, Phys. Rev. A **62**, 022108 (2000)

Llo96. Lloyd, S.: Universal Quantum Simulators, Science **273**, 1073-1078, (1996)

LV00. Lloyd, S., Viola, L.: Control of open quantum system dynamics, arXiv:quant-ph/0008101, (2000)

LV01. Lloyd, S., Viola, L.: Engineering quantum dynamics, Phys. Rev. A **65**, 010101 (2001)

LYA01. Loth, F., Yardimci, M.A., Alperin, N., Hydrodynamic modelling of cerebrospinal fluid motion within the spinal cavity. J. Biomech. Eng. **123**(1), 71-79, (2001)

LBW01. Lidar, D.A., Bihary, Z., Whaley, K.B.: From completely positive maps to the quantum Markovian semigroup master equation, Chem. Phys. **268**, 35, (2001)

Lam68. Lambek, J.: Deductive systems and categories I: Syntactic calculus and residuated categories, Math. Sys. Th. **2**, 287-318, (1968)

Lam69. Lambek, J.: Deductive systems and categories II: Standard constructions and closed categories, in: Category Theory, Homology Theory and their Applications I, Lecture Notes in Mathematics 86, Springer, Berlin, 76-122, (1969)

Lam72. Lambek, J.: Deductive systems and categories III: Cartesian closed categories, intuitionist propositional calculus, and combinatory logic, in: F. W. Lawvere ed., Toposes, Algebraic Geometry and Logic, Lecture Notes in Mathematics 274, Springer, Berlin, 57-82, (1972)

LS86. Lambek, J., Scott, P.J.: Introduction to Higher-Order Categorical Logic, Cambridge Univ. Press, Cambridge, (1986)

Lan41. Landau, L.: The theory of superfluidity of helium II. J. Phys. U.S.S.R. **5**, 71-90, (1941)

LL77. Landau, L., Lifshitz, E.M.: Quantum Mechanics, Non-Relativistic Theory. Pergamon Press, Oxford, (1977)

Lan02. Lang, S.: Algebra, Springer, Berlin, (3rd ed), (2002)

Lon38. London, F. On the Bose-Einstein condensation. Phys. Rev. **54**, 947-954, (1938)

Leo95a. Leonard, N.E.: Control synthesis and adaptation for an underactuated autonomous underwater vehicle. IEEE J. Oceanic Eng. (1995)

LL97. Labastida, J.M.F., Lozano, C.: Lectures on Topological Quantum Field Theory. CERN-TH/97-250; US-FT-30/97; hep-th/9709192, (1997)

LML06. Lee, T., McClamroch, N.H., Leok, M.: Optimal Control of a Rigid Body using Geometrically Exact Computations on SE(3), Proc. IEEE Conf. Decis. Con. 2710-2715, (2006)

LLM05. Lee, T., Leok, M., McClamroch, N.H.: A Lie group variational integrator for the attitude dynamics of a rigid body with applications to the 3D pendulum, Proc. IEEE Con. Con. Appl. 962-967, (2005)

Len04. Lenz, F.: Topological concepts in gauge theories, FAU-TP3-04/3; arXiv:hep-th/0403286, (2004)

LW98. Leonard, N.E., Woolsey, C.A.: Internal Actuation for Intelligent Underwater Vehicle Control, 10th Yale Works. Adapt. Learn. Sys. (1998)

LD04. de Leon, M., de Diego, M.D., Santamaria Merino, A.: Geometric numerical integration of nonholonomic systems and optimal control problems, E.J. Con. **10**, 520-526, (2004)

LP95. Lebovitz, N.R., Pesci, A.I.: Dynamics bifurcation in Hamiltonian systems with one degree of freedom. SIAM J. Appl. Math. **55**, 1117-1133, (1995)

LB99. Lloyd, S., Braunstein, S.L.: Quantum Computation over Continuous Variables, Phys. Rev. Lett. **82**, 1784, (1999)

Lut96. Lütkenpohl, H.: Handbook of Matrices, Wiley, New York, (1996)

LMR01. Levis, R., Menkir, G., Rabitz, H.A.: Selective Covalent Bond Dissociation and Rearrangement by Closed-Loop, Optimal Control of Tailored, Strong Field Laser Pulses, Science **292**, 709-713, (2001)

Li08. Li, M., Vitańyi., P.: An Introduction to Kolmogorov Complexity and Its Applications. 3rd Ed. Springer, New York,(2008)

Lin98. Linblad, G.: Brownian motion of harmonic oscillators: Existence of a subdynamics. J. Math. Phys. **39**(5), 2763-2780, (1998)

Lax68. Lax, P.: Integrals of nonlinear equations of evolution and solitary waves. Comm. Pure Applied Math. **21**, 467-490, (1968)

LMB90. Longtin, A., Milton, J.G., Bos, J.E., Mackey, M.C.: Noise and critical behavior of the pupil light reflex at oscillation onset, Phys. Rev A **41**, 6992-7005, (1990)

Mac63. MacLane, S.: Natural associativity and commutativity. Rice University Studies, Papers in Mathematics, **49**, 28-46, (1963)

Mac71. MacLane, S.: Categories for the Working Mathematician. Springer, New York, (1971)

May79. Maybeck, P.S.: Stochastic Models, Estimation, and Control, Vol. 1, Academic Press, New York, (1979)

Man74. Manakov, S.V.: On the theory of two-dimensional stationary self-focusing of electromagnetic waves. (in Russian) Zh. Eksp. Teor. Fiz. **65**, (1973), 505-516; (transleted into English) Sov. Phys. JETP **38**, 248-253, (1974)

MS98. Mangiarotti, L., Sardanashvily, G.: Gauge Mechanics. World Scientific, Singapore, (1998)

Mer98. Merzbacher, E.: Quantum Mechanics (3rd ed.) Wiley, New York, (1998)

Mal798. Malthus, T.R.: An essay on the Principle of Population. Originally published in 1798. Penguin, (1970)

MM78. Marmarelis, P.Z., Marmarelis, V.Z.: Analysis of Physiological Systems: The White Noise Approach, Plenum Press, (1978)

Mar63. Marquardt, D.: An Algorithm for Least-Squares Estimation of Nonlinear Parameters, SIAM J. Appl. Math. **11**, 431-441, (1963)

May76. May, R.M.: Simple mathematical models with very complicated dynamics, Nature **261**(5560), 459-467, (1976)

Mat13. MathWorks, Optimization Toolbox^{TM}, Matlab$^{®}$ R2013a User's Guide, (2013)

MHR08. Moore, K., Hsieh, M., Rabitz, H.J.: On the relationship between quantum control landscape structure and optimization complexity, Chem. Phys. **128**, 154117, (2008)

Mor07. Moroianu, A.: Lectures on Kähler Geometry. London Mathematical Society Student Texts **69**, Cambridge Univ Press, (2007)

Mos97. Mostafazadeh, A.: Quantum adiabatic approximation and the geometric phase, Phys. Rev. A **55**(3), 1653, (1997)

MHJ94. Maier, S.E., Hardy, C.J., Jolesz, F.A.: Brain and cerebrospinal fluid motion: real-time quantification with M_mode MR imaging, Radiology, **193**(2), 477-483, (1994)

Moy49. Moyal, J.E.: Quantum mechanics as a statistical theory, Proc. of Cambridge Phil. Soc. **45**, 99-124, (1949)

Mal85. Von der Malsburg, C.: Nervous structures with dynamical links. Ber. Bunsenges. Phys. Chem. **89**, 703-710, (1985)

Mal88. Von der Malsburg, C.: Pattern recognition by labelled graph matching. Neural Networks, **7**, 1019-1030, (1988)

Man80a. Mandelbrot, B.: Fractal aspects of the iteration of $z \mapsto \lambda z(1-z)$ for complex λ, z, Annals NY Acad. Sci. **357**, 249-259, (1980)

Man80b. Mandelbrot, B.: The Fractal Geometry of Nature. WH Freeman, Co., New York, (1980)

Mar99. Marsden, J.E.: Elementary Theory of Dynamical Systems. Lecture notes. CDS, Caltech, (1999)

MR99. Marsden, J.E., Ratiu, T.S.: Introduction to Mechanics and Symmetry: A Basic Exposition of Classical Mechanical Systems. (2nd ed), Springer, New York, (1999)
MT03. Marsden, J.E., Tromba, A.: Vector Calculus (5th ed.), W. Freeman and Company, New York, (2003)
MW01. Marsden, J.E., West, M.: Discrete mechanics and variational integrators, Acta Numerica **10**, 357-514, (2001)
May73. May, R.M. (ed.): Stability and Complexity in Model Ecosystems. Princeton Univ. Press, Princeton, NJ, (1973)
May76. May, R.: Simple Mathematical Models with Very Complicated Dynamics. Nature, **261**(5560), 459-467, (1976)
MDC85. Martinis, J.M., Devoret, M.H., Clarke, J.: Energy-Level Quantization in the Zero-Voltage State of a Current-Biased Josephson Junction. Phys. Rev. Lett. **55**, 1543-1546, (1985)
MCB09. Martinez, S., Cortes, J., Bullo, F.: A Catalog of Inverse-Kinematics Planners for Underactuated Systems on Matrix Groups. Journal of Geometric Mechanics, **1**(4), 445-460, (2009)
Mal09. Mallat, S.: A Wavelet Tour of Signal Processing: The Sparse Way (3rd ed.), Academic Press, New York, (2009)
Mos13. Mosi_62: Quick Fourier Transformation, Code Project, (2013)
MTW73. Misner, C.W., Thorne, K.S., Wheeler, J.A.: Gravitation. Freeman, San Francisco, (1973)
MS74. Milnor, J.W., Stasheff, J.D.: Characteristic classes, Annals of Mathematics Studies 76, Princeton Univ. Press, (1974)
Mil99. Milnor, J.W: Periodic Orbits, Externals Rays and the Mandelbrot Set. Stony Brook IMS Preprint # 1999/3, (1999)
Mer73. Merton, R.C.: Theory of Rational Option Pricing, Bell J. Econ. and Management Sci. **4**, 141-183, (1973)
MH92. Meyer, K.R., Hall, G.R.: Introduction to Hamiltonian Dynamical Systems and the N-body Problem. Springer, New York, (1992)
MS00. Meyer, K.R., Schmidt, D.S.: From the restricted to the full three-body problem. Trans. Amer. Math. Soc. **352**, 2283-2299, (2000)
MS92. Matveev, V.B., Salle, M.A.: Darboux transformations and solitons, Springer, Berlin, (1991)
Mat92a. Matveev, V.B.: Generalized Wronskian formula for solutions of the KdV system: first applications, Phys. Let. A **166**, 205-208, (1992)
Mat92b. Matveev, V.B.: Positon-positon and soliton-positon collisions: KdV case, Phys. Let. A **166**, 209-212, (1992)
Mat02. Matveev, V.B.: Positons: Slowly decreasing analogues of solitons, The. Math. Phys. **131**(1), 483-497, (2002)
MSF05. Miroshnichenko, A.E., Schuster, M., Flach, S., Fistul, M.V., Ustinov, A.V.: Resonant plasmon scattering by discrete breathers in JJ ladders, Phys. Rev. B **71**, 174306, (2005)
Mil03. Milnor, J.: Towards the Poincaré Conjecture and the Classification of 3-Manifolds, Not. Am. Math. Soc. **50**(10), 1226-1233, (2003)
MLS04. Murray, R.M., Li, Z.X., Sastry, S.S.: A Mathematical Introduction to Robotic Manipulation. CRC Press, (1994)
Mit96. Mitchell, M.: An Introduction to Genetic Algorithms. MIT Press, Cambridge, MA, (1996)

MAH99. Matthews, M.R., Anderson, B.P., Haljan, P.C., Hall, D.S., Wieman, C.E., Cornell, E.A.: Vortices in a Bose-Einstein condensate, Phys. Rev. Lett. **83**, 2498-2501, (1999)

Max73. Maxwell, J.C.: A treatise on electricity and magnetism, Oxford, (1873)

MCW00. Madison, K.W., Chevy, F., Wlhlleben, W., Dalibard, J.: Vortex formation in a stirred Bose-Einstein condensate, Phys. Rev. Lett. **84**, 806-809, (2000)

MCW01. Madison, K.W., Chevy, F., Wlhlleben, W., Dalibard, J.: Statonary states of a rotating Bose-Einstein condensate: Routes to vortex nucleation, Phys. Rev. Lett. **86**, 4443-4446, (2001)

MM93. McElhaney, J.H., Myers, B.S. Biomechanical Aspects of Cervical Trauma, in: A.M. Nahum and J.W. Melvin (Eds.), Accidental injury: Biomechanics and Prevention, Springer, New York, (1993)

MC98. Melin, P., Castillo, O.: A New Method for Adaptive Model-Based Neuro-Fuzzy-Fractal Control of Non-Linear Dynamic Plants: The Case of Biochemical Reactors. Proceedings of IPMU '98, **1**, 475-482, EDK Publ. Paris, (1998)

Mer93. Mermin, D.: Hidden Variables and the Two Theorems of John Bell. Rev. Mod. Phys. **65**, 803, (1993)

MC03. Melin, P., Castillo, O.: A new method for adaptive model-based control of dynamic industrial plants using neural networks, fuzzy logic and fractal theory, Systems Analysis Modelling Simulation, **43**(1), 1-15, (2003)

MN95a. Mavromatos, N.E., Nanopoulos, D.V.: A Non-critical String (Liouville) Approach to Brain Microtubules: State Vector reduction and Memory coding, Capacity. ACT-19/95, CTP-TAMU-55/95, OUTP-95-52P, (1995)

MN95b. Mavromatos, N.E., Nanopoulos, D.V.: Non-Critical String Theory Formulation of Microtubule Dynamics and Quantum Aspects of Brain Function. ENSLAPP-A-524/95, (1995)

MN97. Mavromatos, N.E., Nanopoulos, D.V.: A Non-Critical String (Liouville) Approach to Brain Microtubules: State Vector Reduction, Memory Coding and Capacity, Int. J. Mod. Phys. B **11**, 851, (1997)

MB99. Maimistov, A.I., Basharov, A.M.: Nonlinear Optical Waves, Springer, Berlin, (1999)

MS09. Marvel, S.A., Strogatz, S.H.: Invariant submanifold for series arrays of Josephson junctions, Chaos **19**, 013132, (2009)

Mas01. Masoller, C.: Anticipation in synchronization of chaotic semiconductor lasers with optical feedback, Phys. Rev. Lett. **86**, 2782-2785, (2001)

MSO09. Martens, E.A., Strogatz, S.H., Ott, E., So, P., Antonsen, T.M.: Exact results for the Kuramoto model with a bimodal frequency distribution, Phys. Rev. E **79**, 026204 (2009)

MP43. McCulloch W., Pitts W.: A logical calculus of the ideas imminent in the nervous activity. Bull. Math. Biophys. **5**, 115-133, 1943

MRR53. Metropolis, N., Rosenbluth, A., Rosenbluth, M., Teller, A., Teller, E.: Equation of State Calculations by Fast Computing Machines, J. Chem. Phys. **21**, 1087, (1953)

MP69. Minsky, M., Papert, S.: Perceptrons. MIT Press, Cambridge, MA, (1969)

MWG90. Mooij, J.E., van Wees, B.J., Geerligs, L.J., Peters, M., Fazio, R., Schön, G.: Unbinding of charge-anticharge pairs in two-dimensional arrays of small tunnel junctions, Phys. Rev. Lett. **65**, 645-648, (1990)

Mou74. Moussouris, J.: Gibbs and Markov random systems with constraints, J. Stat. Phys. **10**(1), 11-33, (1974)

Mor69. Morrison, N.: Introduction to Sequential Smoothing and Prediction, McGraw-Hill, New York, (1969)
Mur03. Murray, J.D.: Mathematical Biology I: An Introduction. Springer, New York, (2003)
MK95. Munthe-Kaas, H.Z.: Lie-Butcher theory for Runge-Kutta methods. BIT, **35**(4), 572-587, (1995)
MK98. Munthe-Kaas, H.Z.: Runge-Kutta methods on Lie groups. BIT, **38**(1), 92-111, (1998)
MK99. Munthe-Kaas, H.Z.: High order Runge-Kutta methods on manifolds. Appl. Numer. Math. **29**, 115-127, (1999)
Mur02. Murray, J.D.: Mathematical Biology, Vol. I: An Introduction (3rd ed.), Springer, New York, (2002)
MWH01. Michel, A.N., Wang, K., Hu, B.: Qualitative Theory of Dynamical Systems (2nd ed.) Dekker, New York, (2001)
MT03. Maday, Y., Turinici, G.: New formulations of monotonically convergent quantum control algorithms, J. Chem. Phys. **118**, 8191, (2003)
MG77. Mackey, M.C., Glass, L., Science **197**, 287, (1977)
Mor03. Morawetz, K.: Bifurcation in kinetic equation for interacting Fermi systems, CHAOS **13**, 572, (2003)
MH93. Malley, J., Hornstein, J.: Quantum Statistical Inference, Stat. Sci. **8**, 433-457, (1993)
Man09. Mandel, J.: A Brief Tutorial on the Ensemble Kalman Filter, arXiv[physics.ao-ph]:0901.3725, (2009)
MS08. Mauroy, A., Sepulchre, R.: Clustering behaviors in networks of integrate-and-fire oscillators, Chaos 18, 037122, (2008)
MHR08. Moore, K., Hsieh, M., Rabitz, H.: On the relationship between quantum control landscape structure and optimization complexity, J. Chem. Phys. **128**, 154117, (2008)
MT07. Morgan, J.W., Tian, G.: Ricci Flow and the Poincaré Conjecture, Clay Mathematics Monographs, Vol. 3, AMS, Boston, (2007)
Mur96. Murasugi, K.: Knot theory and its applications. Birkhäuser, Boston, (1996)
NAY62. Nagumo, J., Arimoto, S., Yoshizawa, S.: An active pulse transmission line simulating nerve axon, Proc. IRE **50**, 2061-2070, (1962)
NOF02. Nakagami, K., Ohtsuki, Y., Fujimura, Y.: Quantum optimal control of unbounded molecular dynamics: Application to NaI predissociation, J. Chem. Phys. **117**, 6439-6445, (2002)
Nak03. Nakahara, M.: Geometry, Topology and Physics (2nd ed.), IOP, Bristol and Philadelphia, (2003)
Nan95. Nanopoulos, D.V.: Theory of Brain Function, Quantum Mechanics and Superstrings. CERN-TH/95128, (1995)
Nay73. Nayfeh, A.H.: Perturbation Methods. Wiley, New York, (1973)
NSS08. Nayak, C., Simon, S.H., Stern, A., Freedman, M., Sarma, S.D.: Non-Abelian Anyons and Topological Quantum Computation. Rev. Mod. Phys. **80**, 1083 (77 pages), (2008)
New00. Newman, M.E.J.: Models of the small world. J. Stat. Phys. **101**, 819, (2000)
NLG00. Newrock, R.S., Lobb, C.J., Geigenmüller, U., Octavio, M.: The Two-Dimensional Physics of JJ Arrays, Solid State Phys. **54**, 263, (2000)
NF91. Nitta, T., Furuya, T.: A complex back-propagation learning. Trans. Inf. Proc. Soc. Jpn., **32**(10), 1319-1329, (1991)

Nik95. Nikitin, I.N.: Quantum string theory in the space of states in an indefinite metric. Theor. Math. Phys. **107**(2), 589-601, (1995)

Nil02. Nilsson, H., Courtney, A., Peterson, J.: Functional reactive programming, continued, In Proc. 2002 ACM SIGPLAN Haskell Workshop (Haskell'02), 51–64, Pittsburgh, Pennsylvania, USA, ACM Press, (2002)

Nit00. Nitta, T.: An analysis on fundamental structure of complex-valued neuron. Neu. Proc. Let., **12**(3), 239-246, (2000)

Nit04. Nitta, T.: Reducibility of the Complex-valued Neural Network. Neu. Inf. Proc., **2**(3), 53-56, (2004)

Nit97. Nitta, T.: An extension of the back-propagation algorithm to complex numbers. Neu. Net., **10**(8), 1392-1415, (1997)

NKL98. Nielsen, M.A., Knill, E., LaFlamme, R.: Complete quantum teleportation using nuclear magnetic resonance. Nature **396**, 52-55, (1998)

NC00. Nielsen, M., Chuang, I.: Quantum Computation and Quantum Communication, Cambridge Univ. Press, Cambridge, (2000)

NML03. Nishikawa, T., Motter, A.E., Lai, Y.C., Hoppensteadt, F.C.: Heterogeneity in Oscillator Networks: Are Smaller Worlds Easier to Synchronize? Phys. Rev. Lett. **91**, 014101, (2003)

NNM00. Nagao, N., Nishimura, H., Matsui, N.: A Neural Chaos Model of Multistable Perception. Neural Processing Letters **12**(3): 267-276, (2000)

NK10. Nagai, K.H., Kori, H.: Noise-induced synchronization of a large population of globally coupled nonidentical oscillators, Phys. Rev. E **81**, 065202, (2010)

NPT99. Nakamura, Y., Pashkin, Yu.A., Tsai, J.S.: Coherent control of macroscopic quantum states in a single-Cooper-pair box, Nature, **398**, 786-788, (1999)

Nit97. Nitta, T.: An extension of the back-propagation algorithm to complex numbers. Neu. Net. **10**(8), 1392-1415, (1997)

NE09. Niyogi, R.K., English, L.Q.: Learning-rate-dependent clustering and self- development in a network of coupled phase oscillators, Phys. Rev. E **80**, 066213, (2009)

Nob62. Noble, D.: A modification of the Hodgkin-Huxley equations applicable to Purkinie fibre action and peace-maker potentials. J. Physiol, **160**, 317-330, (1962)

NN94. Nesterov, Y., Nemirovskii, A.L Interior-point polynomial algorithms in convex programming. SIAM, Philadelphia, (1994)

NIH02. NIH: Traumatic Brain Injury: Hope Through Research. NIH Publ. No. 02-2478. Nat. Inst. Health. (Feb. 2002)

NJP09. Nurdin, H.I., James, M.R., Petersen, I.R.: Coherent Quantum LQG Control, Automatica **45**, 1837-1846, (2009)

Nor99. Normile, D.: Complex Systems: Building Working Cells 'in Silico'. Science, **284**, 80, (1999)

Neu27. Von Neumann, J.: Wahrscheinlichkeitstheoretischer Aufbau der Quantenmechanik, Göttinger Nachrichten **1**, 245-272, (1927)

NAB97a. Nore, C., Abid, M., Brachet, M.E.: Kolmogorov turbulence in low-temperature superflows, Phys. Rev. Lett. **78**, 3296-3299, (1997)

NAB97b. Nore, C., Abid, M., Brachet, M.E.: Decaying Kolmogorov turbulence in a model of superflow, Phys. Fluids **9**, 2644-2669, (1997)

NOY95. Nusse, H.E., Ott, E., Yorke, J.A.: Saddle-Node Bifurcations on Fractal Basin Boundaries. Phys. Rev. Lett. **75**(13), 2482, (1995)

NPT99. Nakamura, Y., Pashkin, Yu.A., Tsai, J.S.: Coherent control of macroscopic quantum states in a single-Cooper-pair box, Nature, **398**, 786-788, (1999)

NR02. Neiman, A.B., Russell, D.F.: Synchronization of Noise-Induced Bursts in Noncoupled Sensory Neurons. Phys. Rev. Lett. **88**, 138103, (2002)

NS90. Nijmeijer, H., van der Schaft, A.J.: Nonlinear Dynamical Control Systems, Springer, New York, (1990)

NP77. Nicolis, G., Prigogine, I.: Self-Organization in Nonequilibrium Systems: From Dissipative Structures to Order through Fluctuations. Wiley: Europe, (1977)

Nic86. Nicolis, J.S.: Dynamics of hierarchical systems: An evolutionary approach. Springer: Berlin, (1986)

NHP89. Noakes, L., Heinzinger, G., Paden, B.: Cubic splines on curved spaces, IMA J. Math. Con. Inf. , **6**(4), 465-473, (1989)

Nun00. Nunez, P.L.: Toward a quantitative description of largescale neocortical dynamic function, EEG. Beh. Brain Sci. **23**, 371-437, (2000)

Nun81. Nunez, P.L.: Electric fields of the brain: the neurophysics of EEG. Oxford Univ. Press, New York, (1981)

Nus87. Nusse, H.E.: Asymptotically Periodic Behaviour in the Dynamics of Chaotic Mappings. SIAM J. Appl. Math. **47**, 498, (1987)

NY89. Nusse, H.E., Yorke, J.A.: A procedure for finding numerical trajectories on chaotic saddles. Physica D **36**, 137, (1989)

Oku8. Okubo, A.: Diffusion and Ecological Problems: Mathematical Models, Springer-Verlag, New York, (1980)

Oku93. Okuda, K.: Variety and generality of clustering in globally coupled oscillators, Physica D **63**, 424-436, (1993)

OA08. Ott, E., Antonsen, T.M.: Low dimensional behavior of large systems of globally coupled oscillators, Chaos **18**, 037113, (2008)

OA09. Ott, E., Antonsen, T.M.: Long time evolution of phase oscillator systems, Chaos **19**, 023117, (2009)

ONF01. Ohtsuki, Y., Nakagami, K., Fujimura, Y., Zhu, W., Rabitz, H.J.: Quantum optimal control of multiple targets: Development of a monotonically convergent algorithm and application to intramolecular energy redistribution control, Chem. Phys. **114**, 8867, (2001)

OTR04. Ohtsuki, Y., Turinici, G., Rabitz, H.J.: Generalized monotonically convergent algorithms for solving quantum optimal control equations in product spaces, Chem. Phys. **120**, 5509, (2004)

OFM07. Olfati-Saber, R., Fax, J., Murray, R.: Consensus and cooperation in networked multi-agent systems. Proc. IEEE, **95**(1), 215-233, (2007)

Ohl99. Ohl, T.: Vegas revisited: Adaptive Monte Carlo integration beyond factorization, Comp. Phys. Comm. **120**(1), 13-19, (1999)

OS83. Olsen, O.H., Samuelsen, M.R.: Sine-Gordon 2π-kink dynamics in the presence of small perturbations, Phys. Rev. B **28**(1), 210-217, (1983)

Ons49. Onsager, L.: Nuovo Cimento Suppl. **6**, 249-250, (1949)

Ora88. Oran, B.R.: The fast Fourier transform and its applications. Englewood Cliffs, Prentice Hall, NJ, (1988)

OPK02. Osipov, G., Pikovsky, A., Kurths, J.: Phase synchronization of chaotic rotators, Phys. Rev. Lett, **88**, 054102, (2002)

OGS08. O'Sullivan, B., Goerzen, J., Stewart, D.: Real World Haskell, O'Reilly Media, Inc, (2008)

OGY90. Ott, E., Grebogi, C., Yorke, J.A.: Controlling chaos. Phys. Rev. Lett. **64**, 1196-1199, (1990)

OSB02. Ott, E., So, P., Barreto, E., Antonsen, T.: The onset of synchronization in systems of globally coupled chaotic and periodic oscillators. Physica D **173**(1-2), 29-51, (2002)

Ose68. Oseledets, V.I.: A Multiplicative Ergodic Theorem: Characteristic Lyapunov Exponents of Dynamical Systems. Trans. Moscow Math. Soc., **19**, 197-231, (1968)

Ott89. Ottino, J.M.: The kinematics of mixing: stretching, chaos and transport. Cambridge Univ. Press, Cambridge, (1989)

Ott93. Ott, E.: Chaos in dynamical systems. Cambridge Univ. Press, Cambridge, (1993)

Osb59. Osborne, M.F.M.: Brownian Motion in the Stock Market, Operations Research **7**, 145-173, (1959)

OB07. Oxtoby, O., Barashenkov, I.: Moving solitons in the discrete nonlinear Schrödinger equation, Phys. Rev. **E76**, 036603, (2007)

Per83. Peregrine, D.H.: Water Waves, Nonlinear Schrödinger Equations and Their Solutions, J. Austral. Math. Soc. Ser. B **25**, 16-43, (1983)

Par65. Pars, L.A.: A Treatise on Analytical Dynamics, Wiley, New York, (1965)

PC05. Park, J., Chung, W-K.: Geometric Integration on IEEE Trans. Robotics, **21**(5), 850-863, (2005)

PC90. Pecora, L.M., Carroll, T.L.: Synchronization in chaotic systems. Phys. Rev. Lett. **64**, 821-824, (1990)

PC91. Pecora, L.M., Carroll, T.L.: Driving systems with chaotic signals. Phys. Rev. A **44**, 2374–2383, (1991)

PC98. Pecora, L.M., Carroll, T.L.: Master stability functions for synchronized coupled systems. Phys. Rev. Lett. **80**, 2109–2112, (1998)

PR10. Pelayo, A., and Ratiu, T., Applying Hodge theory to detect Hamiltonian flows, arXiv:1005.2163 [math.SG], (2010)

Per02. Perelman, G.: The entropy formula for the Ricci flow and its geometric applications, arXiv:math.DG/0211159, (2002)

PPM00. Perello, J., Porra, J.M., Montero, M., Masoliver, J.: Black-Scholes option pricing within Ito and Stratonovich conventions. Physica A **278**, **1-2**, 260-274, (2000)

PSM08. Perello, J., Sircar, R., Masoliver, J.: Option pricing under stochastic volatility: the exponential Ornstein-Uhlenbeck model. J. Stat. Mech. P06010, (2008)

PLS00. Pappas, G.J., Lafferriere, G., Sastry, S.: Hierarchically consistent control systems. IEEE Trans. Aut. Con., **45**(6), 1144-1160, (2000)

PS02. Pappas, G.J., Simic, S.: Consistent hierarchies of affine nonlinear systems. IEEE Trans. Aut. Con., **47**(5), 745-756, (2002)

PNW06. Prokhorenko, V.I., Nagy, A.M., Waschuk, S.A., Brown, L.S., Birge, R.R., Miller, R.: Coherent Control of Retinal Isomerization in Bacteriorhodopsin, Science **313**, 1257-1261, (2006)

PR99. Phan, M.Q., Rabitz, H.J.: A self-guided algorithm for learning control of quantum-mechanical systems, Chem. Phys. **110**, 34-41, (1999)

Poi895. Poincaré, H.: Analysis Situs (&12), Journal d'Ecole Polytechnique Normale, **1**, 1-121, (1895)

Pos86. Postnikov, M.M.: Lectures in Geometry V, Lie Groups and Lie Algebras, Mir Publ., Moscow, (1986)

PD95. Pritchard, W.S., Duke, D.W.: Measuring 'Chaos' in the brain: a tutorial review of EEG dimension estimation. Brain Cog., **27**, 353-97, (1995)

PEL00. Principe, J., Euliano, N., Lefebvre, C.: Neural and Adaptive Systems: Fundamentals Through Simulations. Wiley, New York, (2000)

Pel85. Pelikan, S.: A dynamical meaning of fractal dimension. Trans. Am. Math. Soc. **292**, 695-703, (1985)

Pes76. Pesin, Ya.B.: Invariant manifold families which correspond to non-vanishing characteristic exponents. Izv. Akad. Nauk SSSR Ser. Mat. **40**(6), 1332-1379, (1976)

Pes77. Pesin, Ya.B.: Lyapunov Characteristic Exponents, Smooth Ergodic Theory. Russ. Math. Surveys, **32**(4), 55-114, (1977)

Pet02. Petras, I.: Control of Fractional-Order Chua's System. J. El. Eng. **53**(7-8), 219-222, (2002)

Pet93. Peterson, I.: Newton's Clock: Chaos in the Solar System. W.H. Freeman, San Francisco, (1993)

PS95. Peskin, M.E., Schroeder, D.V.: An Introduction to Quantum Field Theory, Addison Wesley, (1995)

Pet96. Petrov, V., Showalter, K.: Nonlinear Control from Time-Series. Phys. Rev. Lett. **76**, 3312, (1996)

Pop00. Pope, S.B.: Turbulent Flows. Cambridge Univ. Press, Cambridge, (2000)

PTV07. Press, W., Teukolsky, S.A., Vetterling, W.T., Flannery, B.P.: Numerical recipes: the art of scientific computing, Cambridge Univ. Press, (2007)

PK02. Palao, J., Kosloff, R.: Quantum computing by an optimal control algorithm for unitary transformations, Phys. Rev. Lett. **89**, 188301, (2002)

Par92. Parthasarathy, K.R.: An Introduction to Quantum Stochastic Calculus. Birkhauser, Berlin, (1992)

Per02. Perelman, G.: The entropy formula for the Ricci flow and its geometric applications, arXiv:math.DG/0211159, (2002)

Per03. Perelman, G.: Ricci flow with surgery on three-manifolds, arXiv:math.DG/0303109, (2003)

POR97. Pikovsky, A., Osipov, G., Rosenblum, M., Zaks, M., Kurths, J.: Attractor-repeller collision and eyelet intermittency at the transition to phase synchronization. Phys. Rev. Lett. **79**, 47-50, (1997)

PN95. Porsezian, K., Nakkeeran, K.: Optical Soliton Propagation in an Erbium Doped Nonlinear Light Guide with Higher Order Dispersion, Phys. Rev. Lett. **74**, 2941, (1995)

PN96. Porsezian, K., Nakkeeran, K.: Singularity structure analysis and the complete integrability of the higher order nonlinear Schrödinger-Maxwell-Bloch equations, Chaos Sol. Frac. **7**(3), 377-382, (1996)

PDZ00. Pouget, A., Dayan, P., Zemel, R.: Information processing with population codes, Nature Neurosci. **1**, 125-132, (2000)

PZR97. Pikovsky, A., Zaks, M., Rosenblum, M., Osipov, G., Kurths, J.: Phase synchronization of chaotic oscillations in terms of periodic orbits. Chaos **7**, 680, (1997)

PBK62. Pontryagin, L., Boltyanskii, V., Gamkrelidze, R., Mishchenko, E.: The mathematical theory of optimal processes. Interscience, (1962)

PS62. Perring, J.K., Skyrme, T.H.: A Model Unified Field Equation, Nucl. Phys. **31**, 550-555, (1962)

PZR97. Pikovsky, A., Zaks, M., Rosenblum, M., Osipov, G., Kurths, J.: Phase synchronization of chaotic oscillations in terms of periodic orbits, Chaos **7**(4), 680-687, (1997)

PRK00. Pikovsky, A., Rosenblum, M., Kurths, J.: Phase synchronization in regular and chaotic systems, Int. J. Bif. Chaos **10**(10), 2291-2305, (2000)

PRK01. Pikovsky, A., Rosenblum, M., Kurths, J.: Synchronization: A Universal concept in Nonlinear Science, Cambridge University press, Cambridge, (2001)

PZ04. Polyanin, A.D., Zaitsev, V.F.: Handbook of Nonlinear Partial Differential Equations, Chapman & Hall/CRC Press, New York, (2004)

PZ04. Polyanin, A.D., Zaitsev, V.F.: Handbook of Integral Equations (2nd ed.), Chapman & Hall/CRC Press, New York, (2008)

Pow73a. Powers, W.T.: Feedback: beyond behaviorism, Science **179**, 351-356, (1973)

Pow73b. Powers, W.T.: Behaviorism and feedback control, Science **181**, 1118-1120, (1973)

Pow05. Powers, W.T.: Behavior: the control of perception, Benchmark Publications Inc., New Canaan, (2005)

Pow11. Powers, W.T.: The neglected phenomenon of negative feedback control, in: D. Forssell (Ed.), Perceptual Control Theory, Science and Applications, (2011)

Pri71. Pribram, K.H.: Languages of the brain. Prentice-Hall, Englewood Cliffs, N.J., (1971)

Pri91. Pribram, K.H.: Brain and perception. Lawrence Erlbaum, Hillsdale, N.J., (1991)

Prz10. Przytycki, J.: Knots: From combinatorics of knot diagrams to combinatorial topology based on knots, arXiv:math/0703096, 2007; see also Knots in Washington XXX: Categorification, Quantum knots and Quantum computing, (2010)

PRK01. Pikovsky, A., Rosenblum, M., Kurths, J.: Synchronization: A Universal Concept in Nonlinear Sciences. Cambridge Univ. Press, Cambridge, UK, (2001)

PS90. Preparata, F.P., Shamos, M.I.: Computational geometry. Springer, New York, 1990.

PT93. Palis, J., Takens, F.: Hyperbolicity, sensitive-chaotic dynamics at homoclinic bifurcations. Cambridge Univ. Press, (1993)

Pen89. Penrose, R.: The Emperor's New Mind, Oxford Univ. Press, Oxford, (1989)

Put93. Puta, M.: Hamiltonian Mechanical Systems and Geometric Quantization, Kluwer, Dordrecht, (1993)

Per93. Peres, A.: Quantum Theory: Concepts and Methods, Kluwer Dordrecht, (1993)

Pet02. Peterka, R.J.: Sensorimotor Integration in Human Postural Control. J. Neurophysiol. **88**, 1097-1118, (2002)

PS03. Pitaevskii, L., Stringari, S.:Bose-Einstein Condensation. Oxford University Press, Oxford, (2003)

PV03. Pessa, E., Vitiello, G.: Quantum noise, entanglement and chaos in the quantum field theory of mind/brain states, Mind, Matter, **1**, 59-79, (2003)

PV04. Pessa, E., Vitiello, G.: Quantum noise induced entanglement and chaos in the dissipative quantum model of brain, Int. J. Mod. Phys. B, **18**, 841-858, (2004)

PV99. Pessa, E., Vitiello, G.: Quantum dissipation and neural net dynamics. Biolectrochem. Bioener., **48**, 339-342, (1999)

PBL05. Purwins, H.-G., Bodeker, H.U., Liehr, A.W.: Dissipative Solitons in Reaction-Diffusion Systems, in Dissipative Solitons (ed. N. Akhmediev, A. Ankiewicz), Lecture Notes in Physics, Springer, (2005)

Pyr92. Pyragas, K.: Continuous control of chaos, by self-controlling feedback. Phys. Lett. A, **170**, 421-428, (1992)

Pyr95. Pyragas, K.: Control of chaos via extended delay feedback. Phys. Lett. A, **206**, 323-330, (1995)

Rha84. de Rham, G.: Differentiable Manifolds. Springer, Berlin, (1984)

RSD95. Ramakrishna, V., Rabitz, H., Salapaka, M.V., Dahleh, M., Peirce, A.: Controllability of Molecular Systems, Phys. Rev. A **51**, 960-966, (1995)

RB-KM06. Ren, Q., Balint-Kurti, G.G., Manby, F.R., Artamonov, M., Ho, T.-S., Rabitz, H.J.: Quantum control of molecular vibrational and rotational excitation in a homonuclear diatomic molecule: a full three-dimensional treatment, Chem. Phys. **124**, 014111, (2006)

RHH06. Rabitz, H., Ho, T.-S., Hsieh, M., Kosut, R., Demiralp, M.: Topology of optimally controlled quantum mechanical transition probability landscapes, Phys. Rev. A **74**, 012721, (2006)

RHR04. Rabitz, H.A., Hsieh, M.M., Rosenthal, C.M.: Quantum Optimally Controlled Transition Landscapes, Science **303**, 1998, (2004)

RHR06. Rabitz, H., Hsieh, M., Rosenthal, C.: Optimal control landscape for quantum observables, J. Chem. Phys. **124**, 204107, (2006)

RHR05. Rabitz, H.A., Hsieh, M.M., Rosenthal, C.M.: Landscape for Optimal Control of Quantum-Mechanical Unitary Transformations, Phys. Rev. A **72**, 052337, (2005)

RHK07. Rehacek, J., Hradil, Z., Knill, E., Lvovsky, A.I.: Diluted maximum- likelihood algorithm for quantum tomography, Phys. Rev. A **75**, 042108, (2007)

ROH05. Restrepo, J.G., Ott, E., Hunt, B.R.: Onset of synchronization in large networks of coupled oscillators, Phys. Rev. E **71**, 036151, (2005)

RPD08. Roman, H.E., Porto, M., Dose, C.: Skewness, long-time memory, and non-stationarity: Application to leverage effect in financial time series. EPL **84**, 28001, (5pp), (2008)

RHR05. Rothman, A., Ho, T.S., Rabitz, H.A.: Observable-Preserving control of quantum dynamics over a family of related systems, Phys. Rev. A **72**, 023416, (2005)

RHR06a. Rothman, A., Ho, T.S., Rabitz, H.A.: Quantum observable homotopy tracking control, J. Chem. Phys. **123**, 134104, (2006)

RHR06b. Rothman, A., Ho, T.S., Rabitz, H.A.: Exploring the level sets of quantum control landscapes, Phys. Rev. A **73**, 053401, (2006)

RT71. Ruelle, D., Takens, F., On the nature of turbulence, Comm. Math. Phys. **20**(2), (1971), 167-192; Comm. Math. Phys. **23**(3), 343-344, (1971)

RRR06. Roslund, J., Roth, M., Rabitz, H.A.: Laboratory observation of quantum control level sets, Phys. Rev. A **74**, 043414, (2006)

RR09. Roslund, J., Rabitz, H.A.: Gradient algorithm applied to laboratory quantum control, Phys. Rev. A **79**, 053417, (2009)

Roy68. Royden, H.: Real Analysis (2nd ed). Macmillan, New York, (1968)

RCH04. Robertson, D.G.E., Caldwell, G.H., Hamill, J., Kamen,G., Whittlesey, S.N.: Research Methods in Biomechanics. Human Kinetics, Champaign, IL, (2004)

RAO00. Robert, C., Alligood, K.T., Ott, E., Yorke, J.A.: Explosions of chaotic sets. Physica D **144**, 44, (2000)

RN03. Russell, Norvig: Artificial Intelligence. A modern Approach. Prentice Hall, (2003)

RBT01. Roe, R.M., Busemeyer, J.R., Townsend, J.T.: Multi-alternative decision field theory: A dynamic connectionist model of decision making. Psych. Rev., **108**, 370-392, (2001)

RG98. Rao, A.S., Georgeff, M.P.: Decision Procedures for BDI Logics. Journal of Logic and Computation, **8**(3), 292–343, (1998)

RVM00. Rabitz, H., de Vivie-Riedle, R., Motzkus, M., Kompa, K.: Whither the Future of Controlling Quantum Phenomena? Science **288**, 824, (2000)

RCR01. Rannou, F., Corvol, M., Revel, M., Poiraudeau, S. Disk degeneration and disk herniation: the contribution of mechanical stress. Joint, Bone, Spine: Revue du Rheumatisme, **68**(6), 543-546, (2001)

RU06. Rau, A.R.P., Uskov, D.: Effective Hamiltonians in quantum physics: resonances and geometric phase, Phys. Scr. **74**, C31, (2006)

RWM08. Reiser, R.F., Wickel, E.E., Menzer, H.H., Lumbar mechanics of floor to knuckle height lifting on sloped surfaces. Int. J. Ind. Erg. **38**(1), 47-55, (2008)

RH89. Rose, R.M., Hindmarsh, J.L.: The assembly of ionic currents in a thalamic neuron. I The three-dimensional model. Proc. R. Soc. Lond. B, **237**, 267-288, (1989)

RPK97a. Rosenblum, M., Pikovsky, A., Kurths, J.: From Phase to Lag Synchronization in Coupled Chaotic Oscillators, Phys. Rev. Lett. **78**, 4193, (1997)

RPK97b. Rosenblum, M., Pikovsky, A., Kurths, J.: Phase synchronization in driven and coupled chaotic oscillators, IEEE Trans. Circ. Sys. I **44**(10), 874-881, (1997)

RN03. Russel, S., Norvig, P.: Artificial Intelligence: A Modern Approach. Prentice Hall, New Jersey, (2003)

Rob95. Robinson, C.: Dynamical Systems. CRC Press. Boca Raton, FL, (1995)

ROH98. Rosa, E., Ott, E., Hess, M.H.: Transition to Phase Synchronization of Chaos. Phys. Rev. Lett. **80**, 1642-1645, (1998)

Ros58b. Rosenblatt F.: The perceptron: a probabilistic model for information storage and organization in the brain. Physiol. Rev., **65**, 386-408, (1958)

RPK96. Rosenblum, M., Pikovsky, A., Kurths, J.: Phase synchronization of chaotic oscillators. Phys. Rev. Lett. **76**, 1804, (1996)

Rot81. Rothaus, O.: Logarithmic Sobolev inequality and the spectrum of Schrödinger operators, J. Funct. Anal. **42**, 110-120, (1981)

RWS96. Rieke, F., Warland, D., Steveninck, R., Bialek, W.: Exploring the Neural Code, MIT Press, Cambridge, (1996)

RPK97. Rosenblum, M., Pikovsky, A., Kurths, J.: From Phase to Lag Synchronization in Coupled Chaotic Oscillators. Phys. Rev. Lett. **78**, 4193-4196, (1997)

RS75. Reed, M., Simon, B.: Methods of modern mathematical physics, Vol. 2: Fourier analysis, self-adjointness. Academic Press, San Diego, (1975)

RU67. Ricciardi, L.M., Umezawa, H.: Brain physics and many-body problems, Kibernetik, **4**, 44, (1967)

RT96. Rodriguez, R., Tuckwell, H.C.: Statistical properties of stochastic nonlinear dynamical models of single spiking neurons and neural networks, Phys. Rev. E **54**, 5585-5590, (1996)

Rud83. Rudolph, L.: Some knot theory of complex plane curves, Enseign. Math. **29**, 185-208, (1983)

Rud05. Rudolph, L.: Knot theory of complex plane curves, in Handbook of Knot Theory (ed. W. Menasco and M. Thistlethwaite), Elsevier (2005)

Rt01. Van Rullen, R., Thorpe, S.J.: Rate coding vs. temporal order coding: what the retinal ganglion cells tell the visual cortex, Neural Comput. **13**, 1255-1283, (2001)

Ryd96. Ryder, L.: Quantum Field Theory. Cambridge Univ. Press, (1996)

RYD96. Ryu, S., Yu, W., Stroud, D.: Dynamics of an underdamped Josephson-junction ladder. Phys. Rev. E **53**, 2190-2195, (1996)

RLH97. Radhakrishnan, R., Lakshmanan, M., Hietarinta, J.: Inelastic collision and switching of coupled bright solitons in optical fibers. Phys. Rev. E **56**, 2213, (1997)

Ray894. Rayleigh, J.W.S.: Theory of Sound I (2nd ed.) Macmillan, London, (1894)

Sch26. Schrödinger, E.: Quantisierung als Eigenwertproblem [Quantification of the eigen-value problem]. Annalen der Physik **80**(13), 437-490, (1926)

SRK07. Solli, D.R., Ropers, C., Koonath, P., Jalali, B.: Optical Rogue Waves, Nature **450**, 1054-1057, (2007)

SL03. Stevens, B.L., Lewis, F.L.: Aircraft Control and Simulation (2nd ed.). Wiley, Hoboken, NJ, (2003)

SB98. Sutton, R.S., Barto, A.G.: Reinforcement Learning: An Introduction. MIT Press, Cambridge, MA, (1998)

SGC97. Scherer, C., Gahinet, P., Chilali, M.: Multiobjective output-feedback control via LMI optimization. IEEE Trans. Automat. Contr. **42**(7), 896-911, (1997)

SPJ07. Shaiju, A.J., Petersen, I.R., James, M.R.: Guaranteed cost LQG control of uncertain linear stochastic quantum systems. In Proc. 2007 American Control Conf. New York, (2007)

Sch06. Scholarpedia, http://www.scholarpedia.org, (2011)

SchFit13. Scholarpedia: FitzHugh-Nagumo model, 2013; `http://www.scholarpedia.org/article/FitzHugh-Nagumo_model`

SchMor13. Scholarpedia: Morris-Lecar model, 2013; `http://www.scholarpedia.org/article/Morris-Lecar_model`

SI10. Sharma, S., Ivancevic, V.G.: Nonlinear dynamical characteristics of situation awareness, Th. Iss. Erg. Sci. **11**(5), 448-460, (2010)

SKW09. Schmid, C., Kiesel, N., Weber, U., Ursin, R., Zeilinger, A., Weinfurter, H.: Quantum teleportation and entanglement swapping with linear optics logic gates. New J. Phys. **11**, 033008, (2009)

Sch81. Schulman, L.S.: Techniques and Applications of Path Integration. New York: Wiley, (1981)

Sch88. Schuster, H.G.(ed): Handbook of Chaos Control. Wiley-VCH, (1999)

Sch07. Schöner, G.: Dynamical Systems Approaches to Cognition. In: Cambridge Handbook of Computational Cognitive Modeling. Cambridge Univ. Press: Cambridge, (2007)

Sch94. Schiff S.J. *et al.*: Controlling chaos in the brain. Nature **370**, 615-620, (1994)

Sch96. Schafer, R.D.: An Introduction to Nonassociative Algebras. Dover, New York, (1996)

SI89. Sastry, S.S., Isidory, A.: Adaptive control of linearizable systems, IEEE Trans. Aut. Con. **34**, 1123-1131, (1989)

Siv77. Sivashinsky, G.I.: Nonlinear analysis of hydrodynamical instability in laminar flames - I. Derivation of basic equations. Acta Astr. **4**, 1177, (1977)

STZ93. M. V. Satarić, J. A., Tuszyński, and R. B. Zakula, "Kinklike excitations as an energy-transfer mechanism in microtubules," *Phys. Rev. E*, vol. 48, pp. 589-597, 1993.

SI89. Sastri, S.S., Isidori, A.: Adaptive control of linearizable systems, IEEE Trans. Aut. Con. **34**(11), 1123-1131, (1989)

SJD94. Schiff, S.J., Jerger, K., Duong, D.H., Chang, T., Spano, M.L., Ditto, W.L.: Controlling chaos in the brain. Nature, **370**, 615-620, (1994)

SHR06. Shen, Z., Hsieh, M., Rabitz, H.: Quantum Optimal Control: Hessian analysis of the control landscape, J. Chem. Phys. **124**, 204106, (2006)

SHF99. Siegelmann, H.T., Hur, A.B., Fishman, S.: Computational complexity for continuous time dynamics, Phys. Rev. Lett. **83**, 1463, (1999)

SGL00. Schirmer, S.G., Girardeau, M.D., Leahy, J.V.: Efficient algorithm for optimal control of mixed-state quantum systems, Phys. Rev. A **61**, 012010, (2000)

SMT99. Sola, I., Malinovsky, V.S., Tannor, D.J.: Optimal pulse sequences for population transfer in multilevel systems, Phys. Rev. A **60**, 3081, (1999)

Sok08. Sokoloff, L.: The physiological and biochemical bases of functional brain imaging. Cogn. Neurodyn. **2**, 1-5, (2008)

Son12. Song, J., Weinkove, B.: Lecture notes on the Kähler-Ricci flow, preprint, arXiv:1212.3653 [math.DG], (2012)

SR90. Shi, S., Rabitz, H.: Quantum mechanical optimal control of physical observables in microsystems, J. Chem. Phys. **92**, 364, (1990)

Spi65. Spivak, M.: Calculus on Manifolds, A Modern Approach to Classical Theorems of Advanced Calculus. HarperCollins Publishers, (1965)

Spi75. Spivak, M.: A comprehensive introduction to differential geometry, Vol.I-V, Publish or Perish Inc., Berkeley, (1970-75)

SG01. Seidel, H., Griffin, M.J. Modelling the response of the spinal system to whole-body vibration and repeated shock. Clin. Biomech. **16**(1), S3-7, (2001)

SS09. Sarlette, A., Sepulchre, R.: Consensus optimization on manifolds, SIAM J. Con. Opt. **48**(1), 56-76, (2009)

Ste93. Stengel, R.: Optimal control and estimation. Dover, New York, (1993)

SBS10. Sarlette, A., Bonnabel, S., Sepulchre, R.: Coordinated motion design on Lie groups, IEEE Trans. Aut. Con. 55(5), 1047-1058, (2010)

SJK. Sepulchre, R., JankoviC, M., Kokotovic, P.: Constructive Nonlinear Control. Springer, (1997)

SPL05. Sepulchre, R., Paley, D., Leonard, N.: Stabilization of planar collective motion with all-to-all communication. IEEE Trans. Aut. Con. **52**(5), 811-824, (2007)

Shu06. Shukla, P.K., Eliasson, B.: Formation and Dynamics of Dark Solitons and Vortices in Quantum Electron Plasmas, PRL **96**, 245001 [4 pages] (2006)

Shu09. Shukla, P.K.: Plasma physics: A new spin on quantum plasmas, Nature Physics 5, 92-93, (2009)

SE10. Shukla, P.K., Eliasson, B.: Nonlinear aspects of quantum plasma physics, UFN, 180(1), 55-82, (2010)

SE11. Shukla, P.K., Eliasson, B.: Colloquium: Nonlinear collective interactions in quantum plasmas with degenerate electron fluids, Rev. Mod. Phys. 83, 885-906, (2011)

SE12. Shukla, P.K., Eliasson, B.: Novel attractive force between ions in quantum plasmas, PRL 108, 165002 [5 pages] (2012)

SL00. Sprott, J.C., Linz, S.J.: Algebraically Simple Chaotic Flows. Int. J. Chaos Theory, Appl., **5**(2), 3-22, (2000)

Sm88. Strogatz, S.H., Mirollo, R.E.: Phase-locking and critical phenomena in lattices of coupled nonlinear oscillators with random intrinsic frequencies. Physica D **31**, 143, (1988)

Ste03. Steven, P.: Applying Elliott Wave Theory Profitably. Wiley, New York, (2003)

SO00. Sweet D., Ott E.: Fractal dimension of higher-dimensional chaotic repellors. Physica D **139**(1), 1-27, (2000)

Spa82. Sparrow, C.: The Lorenz Equations: Bifurcations, Chaos and Strange Attractors. Springer, New York, (1982)

Spr93b. Sprott, J.C.: Strange Attractors: Creating Patterns in Chaos. M&T Books, New York, (1993)

Str01. Strogatz, S.H.: Exploring complex networks. Nature, **410**, 268, (2001)

Str00. Strogatz, S.H.: From Kuramoto to Crawford: exploring the onset of synchronization in populations of coupled oscillators. Physica D, **143**, 1-20, (2000)

Str94. Strogatz, S.: Nonlinear Dynamics and Chaos. Addison-Wesley, Reading, MA, (1994)

SJ72. Sussmann, H., Jurdjevic, V.: Controllability of nonlinear systems J. Diff. **12**, 95-116, (1972)

SB98. Sutton, R.S., Barto, A.G.: Reinforcement Learning: An Introduction. MIT Press, Cambridge, MA, (1998)

SMO02. Spooner, J.T., Maggiore, M., Ordonez, R., Passino, K.M.: Stable Adaptive Control and Estimation for Nonlinear Systems: Neural and Fuzzy Approximator Techniques. Wiley, New York, (2002)

SYT02. Seliger, P., Young, S.C., Tsimring, L.S.: Plasticity and learning in a network of coupled phase oscillators, Phys. Rev. E **65**, 041906, (2002)

SOR11. Skardal, P.S., Ott, E., Restrepo, J.G.: Cluster synchrony in systems of coupled phase oscillators with higher-order coupling, Phys. Rev. E **84**, 036208, (2011)

SCB08. So, P., Cotton, B.C., Barreto, E.: Synchronization in interacting populations of heterogeneous oscillators with time-varying coupling, Chaos **18**, 037114, (2008)

Sod94. Soderkvist, J.: Micromachined gyroscopes. Sensors and Actuators A, **43**, 65-71, (1994)

Sol81. Soliverez, C.E.: General Theory of Effective Hamiltonians, Phys. Rev. A **24**, 4-9, (1981)

SK11. Shirdhonkar, M.S., Kokare, M.: Off-Line Handwritten Signature Identification Using Rotated Complex Wavelet Filters, Int. J. Comp. Sci. Issu. **8**(1), 1694-0814, (2011)

SB10. Solomon, C.J., Breckon, T.P.: Fundamentals of Digital Image Processing: A Practical Approach with Examples in Matlab. Wiley-Blackwell, (2010)

SAM05. Strogatz, S.H., Abrams, D.M., McRobie, A., Eckhardt, B., Ott, E.: Theoretical mechanics: Crowd synchrony on the Millennium Bridge, Nature (London) **438**, 43, (2005)

Sar98. Sardanashvily, G.: Hamiltonian time-dependent mechanics. J. Math. Phys. **39**, 2714, (1998)

Str66. Stratonovich, R.L.: A new representation for stochastic integrals and equations. SIAM J. Control **4**, 362-371, (1966)

Sun82. Sundermeyer, K.: Constrained Dynamics (with Applications to Yang-Mills Theory, General Relativity, Classical Spin, Dual String Model), Springer-Verlag, Berlin, (1982)

Sus83. Sussmann, H.: Lie brackets and local controllability: A sufficient condition for scalar-input systems, SIAM J. Cntr. Opt. **21**, 686-713, (1983)

Sus87. Sussmann, H.: General theorem on local controllability. SIAM J. Cntr. Opt. 25, 158-194, (1987)

Sch85. Schwarz, K.W.: Three-dimensional vortex dynamics in superfluid ^{4}He: Line-line and line-boundary interactions. Phys. Rev. B **31**, 5782-5803, (1985)

Sch88. Schwarz, K.W.: Three-dimensional vortex dynamics in superfluid ^{4}He: Homogeneous superfluid turbulence. Phys. Rev. B **38**, 2398-2417, (1988)

Swi75. Switzer, R.K.: Algebraic Topology - Homology and Homotopy. Series Classics in Mathematics, Springer, New York, (1975)

Sco75. Scott, A.C.: The electrophysics of a nerve fiber, Rev. Mod. Phys. **47**, 487, (1975)

SZ98. Stevens, C.F., Zador, A.M.: Input synchrony and the irregular firing of cortical neurons, Nature Neurosci. **1**, 210-217, (1998)

SK93. Softky, W.R., Koch, C.: The highly irregular firing of cortical cells is inconsistent with temporal integration of random EPSPs, J. Neurosci. **13**, 334-350, (1993)

SK88. Sugeno, M., Kang, G.T.: Structure identification of fuzzy model. Fuzzy Sets and Systems, **28**, 15-33, (1988)

TS85. Takagi, T., Sugeno, M., Fuzzy identification of systems and its applications to modelling and control. IEEE Transactions on Systems, Man and Cybernetics, **20**(2), 116-132, (1985)

TP01. Tabuada, P., Pappas, G.J.: Abstractions of Hamiltonian Control Systems. Proceedings of the 40th IEEE Conf. Decis. Con., Orlando, FL, (2001)

TP01. Tanabe, S., Pakdaman, K.: Dynamics of moments of FitzHugh-Nagumo neuronal models and stochastic bifurcations, Phys. Rev. E **63**, 31911, (2001)

TR98. Tuckwell, H.C., Rodriguez, R.: Analytical and simulation results for stochastic Fitzhugh-Nagumo neurons and neural networks, J. Comput. Neurosci. **5**, 91-113, (1998)

Tek05. Tekkoyun, M.: On Para-Euler-Lagrange and Para-Hamiltonian Equations, Phys. Lett. A **340**(1-4), 7-11, (2005)

Tek09a. Tekkoyun, M.: Hamiltonian Mechanics on Quaternion Kähler Manifolds, arXiv[math-ph]:0902.3727, (2009)

Tek09b. Tekkoyun, M.: Lagrangian Mechanics on Quaternion Kähler Manifolds, arXiv[math-ph]:0902.4079, (2009)

Tur52. Turing, A.M.: The Chemical Basis of Morphogenesis, Phil. Trans. Roy. Soc. London B **237**, 37-72, (1952)

Tao06. Tao, T.: Nonlinear dispersive equations: local and global analysis, CBMS regional series in mathematics, (2006)

Tha00. Thaller, B.: Visual Quantum Mechanics, Springer, New York, (2000)

Thu82. Thurston, W.: Three-dimensional manifolds, Kleinian groups and hyperbolic geometry, Bull. Amer. Math. Soc. **6**, 357-381, (1982)

TKU02. Tsubota, M., Kasamatsu, K., Ueda, M.: Vortex lattice formation in a rotating Bose-Einstein condensate, Phys. Rev. A **65**, 023603, (2002)

Tou82. Tough, J.T.: Superfluid turbulence. Progress in Low Temperature Physics Vol. 8 (Gorter, C. J. ed.). Amsterdam. North-Holland, 133-220, (1982)

Tho75. Thom, R.: Structural Stability and Morphogenesis. Addison-Wesley, Reading, (1975)

TR85. Tannor, D.J., Rice, S.A.: Control of selectivity of chemical reaction via control of wavepacket evolution, J. Chem. Phys. **83**, 5013, (1985)

TTH93. Tighe, T.S., Tuominen, M.T., Hergenrother, J.M., Tinknam, M.: Measurements of charge soliton motion in two-dimensional arrays of ultrasmall JJs, Phys. Rev. B **47**, 1145-1148, (1993)

TC99. Teixeira, F.L., Chew, W.C.: Lattice electromagnetic theory from a topological viewpoint, J. Math. Phys. **40**, 169-187, (1999)

TW01. Tegmark, M., Wheeler, J.A.: 100 Years of the Quantum. Scientific American, 68-75, February, (2001)

TSS05. Trees, B.R., Saranathan, V., Stroud, D.: Synchronization in disordered Josephson junction arrays: Small-world connections and the Kuramoto model. Phys. Rev. E **71**, 016215, (2005)

TS12. Transtruma, M.K., Sethna, J.P.: Improvements to the Levenberg-Marquardt algorithm for nonlinear least-squares minimization, arXiv[physics.data-an]1201.5885, (2012)

TVP99. Tabony, J., Vuillard, L., Papaseit, C.: Biological self-organisation, pattern formation by way of microtubule reaction-diffusion processes. Adv. Complex Syst. **2**(3), 221-276, (1999)

Tab89. Tabor, M.: Chaos and integrability in nonlinear dynamics. John Wiley & Sons, New York, (1989)

Thi05. Thiffeault, J-L.: Measuring topological chaos. Phys. Rev. Let. **94**, 084502, (2005)

Thi10. Thiffeault, J-L.: Braids of entangled particles trajectories. CHAOS **20**, 017516, (2010)

Tia97. Tian, G.: Kähler-Einstein metrics with positive scalar curvature, Invent. Math., 130 (1997), 1-39.

Tia12. Tian, G.: K-stability and Kähler-Einstein metrics, arXiv:1211.4669

Ti13. Tian, G.: Partial C^0-estimates for Kähler-Einstein metrics, Commun. Math. Stat., 1 (2013), 105-113.

TZ13. Tian, G., Zhang, Z.: Regularity of Kähler-Ricci flows on Fano manifolds, arXiv:1310.5897 [math.DG], (2013)

TK93. Turner, R.H., Killian, L.M. Collective Behavior (4th ed.) Englewood Cliffs, NJ, (1993)

TU00. Terng, C.L., Uhlenbeck, K.: Geometry of Solitons, Notices of AMS **47**, 17-25, (2000)

TFT98. Thurner, S., Feurstein, M.C., Teich, M.C.: Multiresolution wavelet analysis of heartbeat intervals discriminates healthy patients from those with cardiac pathology, Phys. Rev. Lett. **80**, 1544-1547, (1998)

TP98. Tomlin, C., Pappas, G.J., Sastry, S.: Conflict resolution for air traffic management: A case study in multi-agent hybrid systems, IEEE Trans. Aut. Con., **43**, 509-521, (1998)

Tow79. Townsend, P.K.: Covariant quantization of antisymmetric tensor gauge fields, Phys. Let. B, **88**(1-2), 97-101, (1979)

Tho867. Thomson, W.H. (Lord Kelvin): On vortex motion, Trans. Roy. Soc. Edin. **25**, 217-260, (1867)

TT93. Turcott, R.G., Teich, M.C.: Fractal Character of the Electrocardiogram: Distinguishing Heart-Failure and Normal Patients. Proc. SPIE 2036 (Chaos in Biology and Medicine), 22-39, (1993)

TT96. Turcott, R.G., Teich, M.C.: Fractal Character of the Electrocardiogram: Distinguishing Heart-Failure and Normal Patients. Ann. Biomed. Eng. **24**, 269-293, (1996)

Ume93. Umezawa, H.: Advanced field theory: micro, macro and thermal concepts. Am. Inst. Phys. New York, (1993)

UMS96. Umbanhowar, P.B., Melo, F., Swinney, H.L.: Localized excitations in a vertically vibrated granular layer, Nature **382**, 793-796, (1996)

UR99. Ursey, W.M., Reid, R.C.: Synchronous activity in the visual system, Annu. Rev. Physiol. **61**, 435-456, (1999)

Vap95. Vapnik, V.: The Nature of Statistical Learning Theory. Springer, New York, (1995)

GVW99. Goldenberg, L., Vaidman, L., Wiesner, S.: Quantum Gambling, Phys. Rev. Lett. **82**, 3356-3359, (1999)

Vap98. Vapnik, V.: Statistical Learning Theory. Wiley, New York, (1998)

Van771. Vandermonde, A.T.: Remarques sur les problèmes de situation, Mémoires de l'Académie Royale des Sciences, Paris, 566-574, (1771)

VKL99. Viola, L., Knill, E., Lloyd, S.: Dynamical Decoupling of Open Quantum Systems, Phys. Rev. Lett. **82**, 2417-2421, (1999)

Vil48. Ville, J.: Théorie et Applications de la Notion de Signal Analytique, Câbles et Transmission **2**, 61-74, (1948)

VL98. Viola, L., Lloyd, S.: Dynamical suppression of decoherence in two-state quantum systems, Phys. Rev. A **58**, 2733-2744, (1998)

Vir00. Virgin, L.N.: Introduction to Experimental Nonlinear Dynamics. Cambridge Univ. Press, Cambridge, (2000)

Vit01. Vitiello, G.: My Double Unveiled. John Benjamins, Amsterdam, (2001)

Vit95. Vitiello, G.: Dissipation and memory capacity in the quantum brain model, Int. J. Mod. Phys. B **9**, 973-989, (1995)

Vin57a. Vinen, W.F.: Mutual friction in a heat current in liquid helium II I. Experiments on steady heat currents, Proc. Roy. Soc. London, A **240**, 114-127, (1957)

Vin57b. Vinen, W.F.: Mutual friction in a heat current in liquid helium II. II. Experiments on transient effects. Proc. Roy. Soc. London, A **240**, 128-143, (1957)

Vin57c. Vinen, W.F.: Mutual friction in a heat current in liquid helium II III. Theory of mutual friction. Proc. Roy. Soc. London, A **242**, 493-515, (1957)

Vin57d. Vinen, W.F.: Mutual friction in a heat current in liquid helium II IV. Critical heat currents in wide channels. Proc. Roy. Soc. London, A **243**, 400-413, (1957)

Vin61. Vinen, W.F.: The detection of single quanta circulation in liquid helium II. Proc. Roy. Soc. London, A **260**, 218-236, (1961)

Vic87. Victor, J.D.: The fractal dimension of a test signal: implications for system identification procedures, Biol. Cyber. **57**, 421-426, (1987)

Voi02. Voisin, C.: Hodge Theory and Complex Algebraic Geometry, Cambridge Univ. Press, Cambridge, (2002)

Voi05. Voit, J.: The Statistical Mechanics of Financial Markets. Springer, (2005)

Wad98. Waddell, G. The Back Pain Revolution. Churchill Livingstone, Edinburgh, (1998)

WTW12. Wang, Y., Tan, G., Wang, Y., Yin, Y.: Perceptual control architecture for cyber-physical systems in traffic incident management, J. Sys. Arch. **58**, 398-411, (2012)

WZ98. Whiting, W.C., Zernicke, R.F. Biomechanics of Musculoskeletal Injury. Human Kinetics, Champaign, IL, (1998)

Wat99. Watts, D.J.: Small Worlds. Princeton Univ. Press, Princeton, (1999)
WC02. Wang, X.F., Chen, G.: Synchronization in small-world dynamical networks. Int. J. Bifur. Chaos **12**, 187, (2002)
WMM06. Wang, L., Meyer, H.-D., May, V.: Computational studies on the pyrazine molecule, J. Chem. Phys. **125**, 014102, (2006)
WRB03. Wen, H., Rangan, C., Bucksbaum, P.: Control of angular momentum evolution in Stark wave packets, Phys. Rev. A **68**, 053405, (2003)
Wer89. Werbos, P.J.: Backpropagation, neurocontrol: A review and prospectus. In IEEE/INNS Int. Joint Conf. Neu. Net., Washington, D.C. **1**, 209-216, (1989)
Wer90. Werbos, P.: Backpropagation through time: what it does, how to do it. Proc. IEEE **78** (10), (1990)
WF49. Wheeler, J.A., Feynman, R.P.: Classical Electrodynamics in Terms of Direct Interparticle Action. Rev. Mod. Phys. **21**, 425-433, (1949)
Whe89. Wheeler, J.A.: Information, Physics and Quantum: the Search for the Links. Proc. 3rd Int. symp. Foundations of Quantum Mechanics, Tokyo, 354-368, (1989)
WP74. Williams, G.A., Packard, R.E.: Photographs of quantized vortex lines in rotating He II, Phys. Rev. Lett. **33**, 280-283, (1974)
Wit82. Witten, E.: Supersymmetry and Morse theory. J. Diff. Geom. **17**, 661-692, (1982)
WAN11. Wang, H., Ashhab, S., Nori, F.: Quantum algorithm for simulating the dynamics of an open quantum system, Phys. Rev. A **83**, 062317 (2011)
Wei08. Weiss, U.: Quantum Dissipative Systems (3rd Ed.) World Scientific, Singapore, (2008)
Wig90. Wiggins, S.: Introduction to Applied Dynamical Systems, Chaos. Springer, New York, (1990)
Win98. Wineland, D.J. et al.: Experimental Primer on the trapped ion quantum computer, Proc. Roy. Soc. London Ser. A 454, 411-429, (1998)
Wal96. Walgraef, D.: Spatio-temporal Pattern Formation Springer-Verlag, New York, (1996)
Wik13a. Wikipedia, the free encyclopedia; http://wikipedia.org, (2013)
Wik13b. Wikipedia: Fourier transform, (2013)
Wik13c. Wikipedia: Euler class, (2013)
Wik13d. Wikipedia: Chern class, (2013)
Wik13e. Wikipedia: Wavelet transform, (2013)
Per13. Wikipedia: Perturbation theory (quantum mechanics), (2013)
Qua13. Wikipedia: Quaternion group, (2013)
LV13. Wikipedia, Lotka-Volterra equation, (2013)
FC13. Wikipedia, Fractional Calculus, (2013)
Win67. Winfree, A.T.: Biological rhythms and the behavior of populations of coupled oscillators. J. Theor. Biol. **16**, 15, (1967)
Win80. Winfree, A.T.: The Geometry of Biological Time. Springer, New York, (1980)
Wol02. Wolfram, S.: A New Kind of Science. Wolfram Media, (2002)
Wol84. Wolfram, S.: Cellular Automata as Models of Complexity. Nature **311**, 419-424, (1984)
WS98. Watts, D.J., Strogatz, S.H.: Collective dynamics of small-world networks. Nature **393**, 440, (1998)
WN63. Wei, J., Norman, E.: Lie algebraic solution of linear differential equations, J. Math. Phys. **4**, 575-81, (1963)

WN64. Wei, J., Norman, E.: On the global representations of the solutions of linear differential equations as a product of exponentials. Proc. Amer. Math. Soc. **15**, 327-334, (1964)

Wil56. Wilkie, D.R.: The mechanical properties of muscle. Brit. Med. Bull. **12**, 177-182, (1956)

Wie61. Wiener, N.: Cybernetics: Or Control and Communication in the Animal and the Machine (2nd ed.), Wiley, New York, (1961)

Wey31. Weyl, H.: The Theory of Groups and Quantum Mechanics. Dover, New York, (1931)

WB95. Welch, G., Bishop, G.: An Introduction to the Kalman Filter. Univ. North Carolina, Dep. Comp. Sci. TR 95-041, (1995)

Wig32. Wigner, E.P.: On the quantum correlation for thermodynamic equilibrium, Phys. Rev. **40**, 749-759, (1932)

Wil00. Wilson, D.: Nonlinear Control, Advanced Control Course (Student Version), Karlstad Univ., (2000)

Wis06. Wise, D.K.: $p-$form electrodynamics on discrete spacetimes. Class. Quantum Grav. **23**, 5129-5176, (2006)

Wis94. Wiseman, H.M.: Quantum theory of continuous feedback, Phys. Rev. A **49**, 2133, (1994)

Wit89. Witten, E.: Quantum field theory and the Jones polynomial, Comm. Math. Phys. **121**, 351-399, (1989)

WM10. Wiseman, H.M., Milburn, G.J.: Quantum Measurement and Control, Cambridge University Press, Cambridge, UK, (2010)

Wol09. Wolfram MathWorld: Burau Representation, (2009) http://mathworld.wolfram.com/BurauRepresentation.html

WPB07. Wu, R., Pechen, A., Brif, C., H. Rabitz, Controllability of open quantum systems with Kraus-map dynamics, J. Phys. A **40**, 5681, (2007)

WDH09. Wu, R., Dominy, J., Ho, T.-S. Rabitz, H.: Singularities of quantum control landscapes, arXiv:0907.2354, (2009)

WR07. Wu, R., Rabitz, H., Hsieh, M.: Characterization of the critical submanifolds in quantum ensemble control landscapes, J. Phys. A: Math. Theor. **41**, 015006, (2007)

WHR07a. Wu, R., Rabitz, H., Hsieh, M.: Characterization of the critical submanifolds in quantum ensemble control landscapes, J. Phys. A: Math. Theor. 41, 015006, (2007)

WCR07. Wu, R., Chakrabarti, R., Rabitz, H.: Optimal Control Theory for Continuous Variable Quantum Gates, arXiv:0708.2118 [quant-ph], (2007)

WTL06. Wu, R., Tarn, T.J., Li, C.W.: Smooth controllability of infinite dimensional quantum mechanical systems, Phys. Rev. A **73**, 012719, (2006)

WPR08. Wu, R., Pechen, A., Rabitz, H., Hsieh, M., Tsou, B.: Control landscapes for observable preparation with open quantum systems, J. Math. Phys. **49**, 022108, (2008)

XH94. Xu, Z., Hauser, J.: Higher order approximate feedback linearization about a manifold. J. Math. Sys. Est. Con., **4**, 451-465, (1994)

XH95. Xu, Z., Hauser, J.: Higher order approximate feedback linearization about a manifold for multi-input systems. IEEE Trans. Aut. Con, AC **40**, 833-840, (1995)

XtZ09. Xue, Y.S., Tian, B., Zhang, H.Q., Liu, W.J., Guo, R., Qi, F.H.: Soliton-like solutions of the coupled Hirota-Maxwell-Bloch system in optical fibers with symbolic computation, Phys. Scr. **79**, 065016, (2009)

XTA12. Xue, Y.S., Tian, B., Ai, W.B., Qi, F.H., Guo, R., Qin, B.: Soliton interactions in a generalized inhomogeneous coupled Hirota-Maxwell-Bloch system, Nonlinear Dyn. **67**, 2799-2806, (2012)

Yan97. Yang, J.: Classification of the solitary wave in coupled nonlinear Schrödinger equations. Physica D **108**, 92-112, (1997)

Yan01. Yang, J.: Interactions of vector solitons. Phys. Rev. E **64**, 026607, (2001)

Yao93. Yao, A.: Quantum circuit complexity. In Proc 34th IEEE Symposium on Found. Comp. Sci. 352-361, (1993)

YAS96. Yorke, J.A., Alligood, K., Sauer, T.: Chaos: An Introduction to Dynamical Systems. Springer, New York, (1996)

Yag87. Yager, R.R.: Fuzzy Sets and Applications: Selected Papers by L.A. Zadeh, Wiley, New York, (1987)

YP82. Yarmchuck, E.J., Packard, R.E.: Photographic studies of quantized vortex lines, J. Low Temp. Phys. **46**, 479-515, (1982)

YZ99. Yong, J., Zhou, X.: Stochastic controls. Hamiltonian Systems and HJB Equations. Springer, New York, (1999)

YK84. Yano, K., Kon, M.: Structures on Manifolds, Series: Pure Mathematics, Vol 3, World Scientific, Singore, (1984)

Yau78. Yau, S.T.: On the Ricci curvature of a compact Kähler manifold and the complex Monge-Ampere equation, I^*. Comm. Pure Appl. Math. **31**, 339-441, (1978)

Yau06. Yau, S.T.: Structure of Three-Manifolds - Poincaré and geometrization conjectures. Morningside Center of Mathematics, June 20, (2006)

Ye07. Ye, R.: The log entropy functional along the Ricci flow. arXiv:math. DG/0708.2008v3, (2007)

Yam03. Yamaguchi, S. et al.: Synchronization of Cellular Clocks in the Suprachiasmatic Nucleus, Science **302**, 1408, (2003)

Zad65. Zadeh, L.A.: Fuzzy sets. Inform. Contr. 8, 338-353, (1965)

Zad78. Zadeh, L.A.: Fuzzy sets as a basis for a theory of possibility. Fuzzy Sets Sys. **1**(1), 3-28, (1978)

ZMN80. Zakharov, V.E., Manakov, S.V., Novikov, S.P., Pitaevskii, L.P.: Soliton theory: inverse scattering method. Nauka, Moscow, (1980)

ZFE92. van der Zant, H.S.J., Fritschy, F.C., Elion, W.J., Geerligs, L.J., Mooij, J.E.: Field-induced superconductor-to-insulator transitions in Josephson-junction arrays, Phys. Rev. Lett. **69**, 2971-2974, (1992)

ZEG96. van der Zant, H.S.J., F.C., Elion, W.J., Geerligs, L.J., Mooij, J.E.: Quantum phase transitions in two dimensions: Experiments in Josephson-junction arrays, Phys. Rev. B **54**, 10081-10093, (1996)

ZML98. Zharnitsky, V., Mitkov, I., Levi, M.: Parametrically forced sine-Gordon equation and domain walls dynamics in ferromagnets, Phys. Rev. B **57**(9), 5033, (1998)

ZYZ09. Zhang, J., Yuan, Z., Zhou, T.: Synchronization and clustering of synthetic genetic networks: A role for cis-regulatory modules, Phys. Rev. E **79**, 041903, (2009)

Zee03. Zee, A.: Quantum Field Theory in a Nutshell, Princeton Univ. Press, (2003)

ZBR98. Zhu, W., Botina, J., Rabitz, H.: Rapidly convergent iteration methods for quantum optimal control of population, J. Chem. Phys. **108**, 1953, (1998)

Index